Technological Leadership

IEEE Press
445 Hoes Lane
Piscataway, NJ 08854

Technological Leadership

Leading Through Complexity, Resilient Systems, Ethical Technologies, and Sustainable Futures

Massoud Amin
University of Minnesota
USA

IEEE Press
WILEY

Published by John Wiley & Sons, Inc., Hoboken, New Jersey.

Library of Congress Cataloging-in-Publication Data applied for:

Hardback ISBN: 9781394338252

Cover Design: Wiley
Cover Image: © Yury Garbar/stock.adobe.com

Set in 9.5/12.5pt STIXTwoText by Straive, Chennai, India

Contents

Acknowledgments

This book culminates several decades of combined work, ideas, and guidance from academics, industry, NGOs, and governments. First, I would like to thank my family, friends, colleagues, and, in particular, the students who shared their experiences and thoughts with me. I want to extend my appreciation to:

- **My Family** – parents Dr. Mohammad Shafi Amin and Mrs. Nahid Loghman-Adham Amin; brothers Mahmoud and Mason (and Maryam, Yashaar, and Sara); wife Lindsay and daughters Natalie, Charlotte, and Nahid Josephine – for their support, love, inspiration, and infinite patience.
- **Prof. Ervin Rodin** (my doctoral dissertation advisor at Washington University in St. Louis and colleague and mentor since 1987), **Prof. Ted Djaferis** (I was his student in five undergraduate and graduate courses and the master's thesis at UMass-Amherst from 1979 to 1985), the late **Chauncey Starr** (at EPRI), and **Earl Bakken**, whose tenacity, humility, and principles of "shoot, then aim" are embodied in this book.
- **Dr. Tariq Samad** (the editor-in-chief of this series), my colleagues, and students at the Technological Leadership Institute at the University of Minnesota, where I served as the director from March 2003 to October 2018, to the over 8,000 executives and 15,000 students that I have been privileged to work with during the last 43 years – and to many other colleagues and friends in the science, engineering, and policy community from around the world – for establishing limits yet transforming these limits.
- **Mr. David Wagman**, for his careful reading, thoughtful editing, and excellent feedback on this book. Your insights have greatly enriched this work, and I am deeply grateful for your contributions.
- The industry, public sectors, and academic experts who shared their experience, data, and cases to make this work relevant and practical.

Thank you to all who made this journey possible and meaningful.

Prelude: From Lanterns to Resilient Systems – A Journey Through Complexity

The journey begins with the flickering light of a kerosene lamp. It was the late 1960s. I was five or six years old, traveling with my parents to remote villages near our ancestral hometown of Tabriz, Iran. These areas remained untouched by modernity; electricity was nonexistent; roads were scarce, and reaching them required hours in a dusty, Iranian-made hardtop Jeep. The vehicle's steel seats in the back pressed uncomfortably against our spines as we navigated unpaved paths through isolated landscapes.

Leaving behind a home where The Beatles might have played on the radio, we entered villages seemingly frozen in time. Villagers cooked over open fires fueled by dried animal dung, agricultural waste, or sparse firewood – whatever they could gather. The acrid smoke from these bio-based fuels filled their small, unventilated homes, causing respiratory illnesses, especially among women and children who spent most of their time near these makeshift stoves.

In these villages, life was fragile. On one trip, we arrived to hear devastating news: a young woman in her early 30s had died during childbirth, along with her baby. There was no clinic, trained midwife, or telephone to call for help. The village lacked clean water, electricity, and sanitation.

I asked my parents why this happened. Their answer stayed with me:

> *"If they had electricity, everything could have been different. They could have had lighting, refrigerators, clean water, a clinic with proper sanitation, a telephone to call for help, or even a grade school to educate the next generation. But without electricity, nothing changes. They are still pulling water from a well by hand. It's as if they live in the dark ages."*

That moment lit a fire in me. It wasn't just sadness, frustration, and a deep urgency to act. I realized that even the most basic human needs remain out of reach without infrastructure. I saw how lives are shaped – and too often lost – by the absence of systems we take for granted.

This realization became my compass. I wanted to build systems that change lives. Systems that protect, connect, and empower people.

Technology's Long Journey

From the flickering flames of ancient campfires to today's cutting-edge technologies, humanity has always sought to harness tools for survival and progress. Fire allowed us to cook and craft, the wheel revolutionized trade, and the printing press spread knowledge. Centuries later, the steam engine powered industries, the telegraph connected continents, and the light bulb extended the day.

Each innovation marked a leap forward. In the 20th century, the pace of technological breakthroughs accelerated:

- **1876:** Alexander Graham Bell patented the telephone, connecting voices across distances.
- **1903:** The Wright brothers' first flight opened the skies to human travel.
- **1947:** The invention of the transistor launched the digital age, shrinking room-sized computers into handheld devices.
- **1969:** The Apollo 11 moon landing showcased the heights of human ingenuity and collaboration.
- **2024:** Google announced a 70-qubit quantum chip capable of performing computations that were previously impossible.

Each leap, from fire to quantum computing, has changed our lives and revealed our vulnerabilities. As technology becomes more integrated into every aspect of life, ensuring resilience is no longer optional – it's imperative.

These breakthroughs solved specific challenges, but they also introduced new vulnerabilities. The same technologies that connect us can expose us to risks – cyberattacks, inequality, and environmental harm. The fragility of interconnected systems became evident through events such as the 2003 Northeast Blackout and the 2021 Colonial Pipeline cyberattack.

Technology's power to shape lives is undeniable, but its resilience determines whether it uplifts or undermines humanity. The challenge is to design systems that adapt and endure.

Resilience as a Guiding Principle

Resilience is not just about bouncing back; it's about bouncing forward – adapting and improving in the face of challenges. Whether rebuilding after a disaster, mitigating cyber risks, or transitioning to clean energy, resilience must be the backbone of our systems and decisions. This principle guided my efforts, from fortifying power grids to mentoring future leaders, ensuring that every step forward strengthens the foundation for those who come next.

Systems must do more than survive; they must help people thrive and achieve their goals.

The Fragility of Life and the Power of Resilience

By the early 1980s, the urgency I felt sharpened. The Iran-Iraq War had begun. Saddam Hussein's chemical weapons attacks on civilians and soldiers alike left me haunted by the fragility of human life. Some of my classmates never came home.

At the same time, three close friends – Vietnam veterans – shared with me the human cost of war. They spoke about chaos, systems that failed, and what it means to endure. From them, I learned the importance of resilience, preparedness, and the systems that protect those at risk. Defense systems evolved into more than just technical challenges; they became tools to save lives.

This conviction guided my work from 1982 to 1997. I worked on Sikorsky helicopters, such as the Black Hawk, enhancing their ability to survive in uncertain environments. I advanced to designing autonomous navigation systems for low-observable aircraft, creating path-planning and adaptive controls that enabled them to survive, learn, and complete their missions. This work wasn't just about engineering – it was about building systems that could adapt and thrive under stress.

The fragility I witnessed during the war mirrored vulnerabilities I would later encounter in interconnected infrastructure, where a single failure could cascade into widespread disruption.

Leadership and Ethics in Complex Systems

Leadership, I learned, is not just about managing technology; it's also about managing people. It's about building trust, fostering collaboration, and ensuring integrity. During my workshops with engineers and decision-makers, I often posed this question:

What if your reputation – everything you've built over decades – were questioned?

Reputation is fragile. It's earned through actions and safeguarded by values like fairness, responsibility, and integrity. Without trust, leadership crumbles, regardless of how innovative the technology may be.

Leadership in complex systems requires not only technical expertise but also moral courage. It's about ensuring that innovation serves humanity rather than exploits it. Whether designing a system or leading a team, every decision must be grounded in integrity.

Expanding to Civilian Systems

By the early 1990s, I had seen how the principles I learned could transform both military and civilian systems. I began working on intelligent transportation systems, logistics networks, and critical infrastructure. After joining the Electric Power Research Institute (EPRI) in 1998, I focused on modernizing the systems that power life.

Following the September 11, 2001, attacks, I led efforts to secure North America's power grid. We measured dynamic risks, modeled cascading failures, and implemented smart microgrids and self-healing systems. These efforts reinforced a simple truth:

Systems exist to serve people. When they fail, lives are at stake. When they succeed, their futures are transformed.

A Call to Action

The journey that began with a lantern in a remote village was about more than systems. It has been about purpose. It was about people.

This book is about leadership that builds systems to improve lives. It offers tools to navigate complexity:

1) **Measure what matters.**
2) **Model for foresight.**
3) **Manage with resilience.**

Resilience isn't just a technical skill – it's a way of seeing the world. It's about building systems, teams, and lives that adapt, recover, and move forward.

If there's one lesson I hope this book leaves with you, it's this: Success isn't about perfection. It's about persistence. How we respond to uncertainty, turn failure into growth, and connect today's challenges with tomorrow's possibilities shape the systems we build and our lives.

Let's get to work.

Introduction

A Roadmap for Leadership in Complexity

We live in an era defined by unprecedented interconnectedness and complexity. The challenges of our time – energy transitions, cybersecurity threats, geopolitical tensions, and climate resilience – require leaders who can navigate uncertainty, think critically, act decisively, and adapt continuously. Energy transitions, cybersecurity threats, geopolitical instability, and climate change create overlapping challenges that demand leaders who can think critically, act decisively, and adapt continuously. These problems are not isolated – systemic, interdependent, and global. Leadership today is not about following a well-worn path but forging new ones through dynamic and volatile terrain.

This book is not a theoretical exploration or an abstract treatise. It is a practical guide shaped by over 40 years of experience working alongside innovators, decision-makers, and leaders across 84 countries. It offers a wealth of actionable insights, tools, and frameworks derived from successes and failures, making it a companion for leaders at every stage of their journey.

At its core, this book is about thriving in complexity. Leadership today requires more than expertise – it demands ethical decision-making, systems thinking, and the ability to anticipate and prepare for future challenges while addressing present realities. Most importantly, it requires resilience: the capacity to recover, adapt, and grow stronger under pressure.

Leadership in All Its Complexity

This book examines leadership through case studies, evidence, and outcomes. It does not promote any person, organization, or ideology. Leadership is a human endeavor shaped by decisions, values, and the tension between short-term results and long-term impact.

Some leaders build lasting legacies, balancing ambition with ethics and humility. Others achieve quick wins but leave fragile systems for others to fix. Leadership is never perfect – it reflects the values and priorities behind the decisions.

As John Stuart Mill said, "A person may cause evil to others not only by their actions but by their inaction." Leadership, like humanity, is a complex phenomenon. Good people make mistakes. Those we criticize may still achieve remarkable things. This complexity calls for humility, self-awareness, and questioning even those we admire.

This book shares lessons that cut across industries, ideologies, and time. It focuses on technological leadership and its role in advancing innovation, solving problems, and navigating the future.

Leadership is imperfect because humans are imperfect. The values behind leadership decide whether it builds progress or causes harm. This book neither simplifies nor celebrates leadership. It examines the choices, ethics, and humility that shape its lasting impact.

What This Book Covers: This book is structured to provide practical, immediately applicable insights across several critical domains of leadership:

What This Book Contains: A Map for the Reader

This book is designed as both a strategic framework and an operational guide – uniting leadership theory, systems thinking, technical strategy, and institutional resilience into a single, coherent architecture. It spans decades of leadership practice, technology development, and national infrastructure work, brought together for a time of intersecting planetary, technological, and institutional crises.

Structured in 39 chapters, the book proceeds through five integrated movements, each building on the last. While each chapter can stand alone, their cumulative trajectory is essential: this is not a book of answers, but a system for adaptive thinking and dignified action.

Leadership Reimagined (Chapters 1–4)

These opening chapters establish the foundation of **technological leadership** as a distinct discipline, separate from both management and innovation. Drawing from critical systems failures, personal experience, and cross-sector case studies, the reader is introduced to:

- **Chapters 1 and 2:** The transformation of leadership from heroic individualism to complexity stewardship
- **Chapter 3:** The anatomy of chaos-resilient systems, adaptive thinking, and decision-making under uncertainty
- **Chapter 4:** A detailed lens on infrastructure interdependence and the architecture of systemic risk

This section reframes leadership not as reactive control, but as a disciplined engagement with multi-order, multi-scale complexity.

Tools, Models, and Methods (Chapters 4A–18 and Appendix A)

This core section provides operational frameworks to design, implement, and govern under uncertainty:

- **Chapters 4A–4B** introduce a robust strategic toolkit, including:
  - **Real options analysis**
  - **Scenario planning**
  - **Cost-benefit frameworks for resilience**
  - **Technology readiness levels (TRLs)**
  - **Balanced Scorecard systems**
- **Chapters 5–12** focus on decision environments: risk modeling, ethical boundaries, and the psychology of strategic delay

- **Chapters 13–18** address structural leadership through case studies in:
    - **Organizational resilience and design**
    - **Networked response systems**
    - **Ethical systems engineering**
    - **Human-centered design**
    - **Cybersecurity integration as a leadership concern**

Each chapter applies these tools to real contexts – from the Heathrow Airport blackout to pandemic ventilator response – highlighting the imperative of **preparedness, transparency, and adaptability**.

Infrastructure, Intelligence, and Integration (Chapters 19–34)

These chapters examine the implementation of leadership and system stewardship across vital global domains:

- **Chapters 19–21:** Resilience in **critical infrastructure**, from water and energy systems to transit and communications
- **Chapters 22–25:** Deep exploration of **cybersecurity, AI governance, data sovereignty**, and the future of digital dignity
- **Chapters 8–26:** Smart systems design, digital twins, and **urban transformation**, including national strategies and international comparisons
- **Chapters 9–29: Risk modeling, insurance systems, financial innovation**, and managing cascading failures in markets and supply chains
- **Chapters 30–31: Climate technology ecosystems**, mitigation portfolios, and the anatomy of planetary-scale adaptation efforts
- **Chapters 10–32:** A systems overview of **critical technologies**, their geopolitical significance, and their required governance models
- **Chapters 33–34:** Organizational maturity, **multigenerational foresight**, and institutional transformation for the long term

Together, these chapters provide a panoramic view of **21st-century complexity**, connecting local action to global feedback systems.

Failures, Futures, and Systemic Transformation (Chapters 36–39)

This culminating section confronts the hard edge of leadership: **failure, constraint, planetary collapse, and the 2050 horizon**. These new chapters expand and elevate the narrative into an actionable transformation:

- **Chapter 36: The Cost of Failure**
  A forensic analysis of five iconic leadership collapses – Boeing, Facebook, Enron, Fukushima, and Wells Fargo – exposing the systemic pathologies that led to their undoing and distilling universal lessons in risk, ethics, and governance.
- **Chapter 37: Planetary Intelligence**
  An urgent, systems-level reframing of climate resilience. By integrating planetary boundaries, ecological system logic, and regenerative technology models, this chapter calls for infrastructure, AI, and institutions to **mirror the adaptability of living systems**.

- **Chapter 38: Leapfrogging Economies**
  A deep dive into innovation under constraint across the Global South. From M-Pesa to India Stack, from Zipline to Grameen Shakti, it offers a grounded model of how nations can **innovate past legacy systems** – with greater inclusion, speed, and sustainability.
- **Chapter 39: Beyond the Horizon**
  A strategic foresight chapter centered on the **2040–2050** inflection. It articulates the institutional capabilities required for intergenerational leadership, including:
  - Fractal governance
  - Dynamic fitness metrics
  - Ethical calibration systems
  - Long-horizon futures modeling

This section concludes the book's narrative, transitioning **from understanding risk to designing regeneration**.

Companion Materials and Visual Maps

The book is accompanied by supporting tools to extend its utility:

- A **Lens Map** to help readers apply ethical, systemic, temporal, and geospatial thinking across contexts
- Executive diagrams and visual schemas for decision environments, technology stacks, and infrastructure relationships
- A curated list of quotations, a "How to Use This Book" guide, and prompts for organizational self-assessment and leadership recalibration

In Sum: A Design Manual for Dignified Futures

This book does not offer one solution. It provides a coherent design system – to think better, to build wiser, and to lead with integrity through transformation. It is meant for engineers, executives, educators, policymakers, and civic stewards navigating complexity at scale.

Whether you are designing national cyber policy, modernizing a transit system, governing generative AI, or rethinking a university's mission in a fragile world, this book exists to help you see more clearly, decide more wisely, and act more justly.

What lies ahead is not a destination. It is a condition of possibility. This book is a compass for that condition.

How to Use This Book: This book is designed as a modular system for leadership and resilience, not a linear textbook. It can be read cover to cover, used as a reference during critical decision-making, or integrated into institutional training programs, executive workshops, or interdisciplinary curricula.

The following guidance will help readers maximize its utility across multiple contexts:

Read It as a System, Not a Sequence

While each chapter is self-contained, the structure is cumulative and strategic in nature. The book proceeds in five integrated movements:

1) **Foundations of Technological Leadership** – conceptual architecture and systems thinking (Chapters 1–4)

2) **Strategic Tools and Implementation Frameworks** – for adaptive decision-making under uncertainty (Chapters 4A–18)
3) **Sectoral and Technological Deep Dives** – applying leadership across energy, infrastructure, AI, climate, and finance (Chapters 19–34)
4) **Failure, Climate, Constraint, and the Long Horizon** – foresight, risk, global equity, and transformation (Chapters 36–39)
5) **Companion Tools and Applied Navigation** – visual maps, lens diagrams, and institutional prompts

Readers can navigate based on their domain (e.g., AI governance, infrastructure resilience, or climate strategy) or by phase of a challenge (e.g., crisis response, systems design, or future-proofing).

Use the Lens Map to Navigate Across Scales

Throughout the book, recurring conceptual "lenses" help reframe challenges from different angles:

- **Systems Lens** – for seeing interdependencies and second-order effects
- **Ethical Lens** – for identifying asymmetries, rights, and power gradients
- **Temporal Lens** – for managing delayed consequences and long-term foresight
- **Geospatial Lens** – for integrating global-local feedback and distributed design
- **Resilience Lens** – for navigating volatility and preserving functional integrity

These lenses are not theoretical – they are operational filters. The **Lens Map**, provided in the companion diagram, supports their application across sectors and decision domains.

Engage the Case Studies and Tools

The book is rich with real-world case studies and strategic toolkits, including:

- Cost–benefit analysis for resilience engineering
- Balanced scorecard adaptation for system-level decision-making
- Real options frameworks for adaptive investment
- Scenario-planning models grounded in empirical stress-testing
- Comparative leadership breakdown analysis (Chapter 36)
- Climate technology readiness and mitigation matrices (Chapter 37)
- Leapfrogging infrastructure strategies and regulatory innovation (Chapter 38)
- Institutional fitness metrics and intergenerational design logic (Chapter 39)

These tools are designed to equip – not just inform – leaders and institutions working in live, uncertain environments.

Apply the Strategic Prompts

A set of organizational prompts and self-assessment questions is embedded throughout the manuscript and available as a workbook companion. These are structured to provoke alignment, foresight, and recalibration. Sample questions include:

- What are the failure modes of your current leadership model under systemic stress?
- Are your metrics designed for growth, resilience, or both?

- How are you engaging voices that are missing – but consequential – in your system?
- Where do you place feedback loops in your institution's governance architecture?
- Are you designing infrastructure for change or for permanence?

These are not rhetorical – they are practical scaffolds for navigating complexity.

Who Should Read This Book

This book is written for those entrusted with building, governing, or transforming complex systems. It is intended for:

- **Engineers and system designers** guiding infrastructure, energy, or cyber-physical platforms
- **Executives and strategic officers** managing uncertainty, risk, and organizational change
- **Policy leaders and civic planners** stewarding digital, environmental, and institutional systems
- **Academics and educators** are training the next generation of interdisciplinary leaders
- **Foresight professionals, technologists, and ethicists** tasked with guiding emerging futures

Whether you are confronting AI deployment in the public sector, rethinking digital infrastructure in a climate-volatile world, or redesigning national strategy after a cascading failure, this book is a map for navigating the edge of the known.

A Living Resource

The book's companion materials – diagrams, data appendices, prompt sets, and digital tools – are available upon request or through future editions. Where applicable, content has been designed to support:

- Curriculum modules and leadership development tracks
- Executive briefings and scenario-planning workshops
- Digital twin integration and strategic dashboards
- Institutional foresight labs and public sector innovation portfolios

Future iterations will evolve in alignment with technological change, geopolitical shifts, and environmental thresholds. Readers are encouraged to use the concepts herein not as static knowledge but as adaptive infrastructure for stewardship in motion.

Using the Lens Maps

To navigate complex systems, linear thinking is insufficient. Leaders must see across domains, over time, through uncertainty, and with ethical clarity. The two diagrams below serve as navigational tools for this book's analytical architecture.

They visualize the core conceptual lenses through which technological leadership must be exercised, spanning systems logic, ethical responsibility, resilience strategy, spatial context, and temporal depth.

Each chapter in this book draws from one or more of these lenses, either explicitly or implicitly. Together, these visual tools provide a unified framework for reasoning about uncertainty, guiding institutional transformation, and designing for long-term impact.

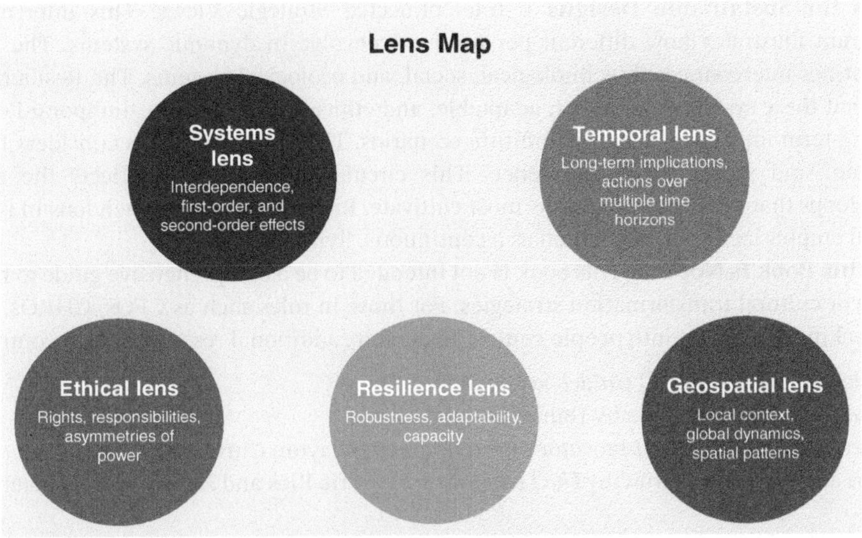

Lens Map – Core Perspectives for Navigating Complexity – This diagram presents the five foundational lenses that underpin the book's approach to technological leadership. The Systems Lens reveals interdependencies and cascading effects. The Temporal Lens emphasizes the long-term consequences of present decisions. The Ethical Lens helps identify rights, responsibilities, and power asymmetries. The Resilience Lens emphasizes adaptability and robustness in the face of stress. The Geospatial Lens brings attention to local–global dynamics and spatial design logic. These lenses reappear throughout the book to help readers frame problems, structure decisions, and prioritize responses in complex situations.

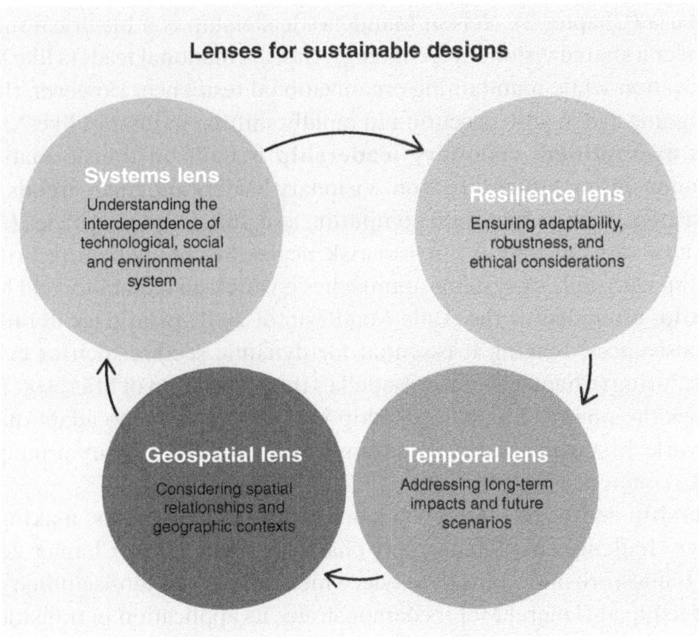

Lenses for Sustainable Designs – Interconnected Strategic Views. This interconnected lens diagram illustrates how different perspectives coevolve in dynamic systems. The Systems Lens identifies interconnected technological, social, and ecological systems. The Resilience Lens ensures that these systems are robust, adaptable, and ethically aligned. The Temporal Lens integrates long-term impacts and multiple future scenarios. The Geospatial Lens considers location, distribution, and spatial interdependence. This circular configuration reflects the adaptive feedback loops that resilient institutions must cultivate. Rather than seeing each lens in isolation, this visual emphasizes their interaction as a continuous, living process.

Who This Book Is Not For: This book is not intended to be a comprehensive guide to financial modeling or cultural transformation strategies. For those in roles such as CFOs, CHROs, or individuals seeking deep dives into people-centric leadership, additional resources are recommended:

- Financial strategy: *The CFO Guidebook* by Steven Bragg.
- Organizational culture: *Drive* by Daniel Pink.
- Healthcare leadership: *The Innovator's Prescription* by Clayton Christensen.
- Startups and early-stage growth: *The Lean Startup* by Eric Ries and *Zero to One* by Peter Thiel.

Leadership Styles in This Book: What, Why, and How

Leadership in the context of technological advancement, resilience, and systems thinking requires a nuanced approach that integrates multiple leadership styles. This book draws on established theories, analyzing their relevance and application to effective technological leadership. It explains why some styles are prioritized and others excluded, focusing on measurable outcomes in innovation and systemic challenges.

Transformational leadership, introduced by James MacGregor Burns in 1978, emphasizes systemic change through vision, collaboration, and empowerment. This style inspires teams to achieve ambitious goals, such as decarbonizing energy systems (Chapter 21) or addressing crises like Hurricane Maria (Chapter 3). Nelson Mandela's leadership is a historical model for uniting diverse groups under a shared vision. In technology, transformational leaders like Sundar Pichai at Google drive innovation while maintaining organizational resilience. However, the challenge lies in consistently aligning vision with execution in rapidly shifting industries like AI or biotech [1].

As Steve Jobs exemplified, visionary leadership is built on transformational principles, offering a clear and inspiring future direction. Visionary leaders anticipate trends, shape markets, and redefine industries, such as quantum computing and autonomous vehicles (Chapter 5). This style inspires loyalty and innovation but can risk neglecting immediate priorities. Integrating visionary leadership with agile or systems approaches ensures sustainability and balance [2].

Agile leadership, grounded in the Agile Manifesto of 2001, prioritizes iteration, experimentation, and responsiveness, making it essential for dynamic sectors such as cybersecurity and advanced manufacturing (Chapter 6). Satya Nadella's transformation of Microsoft into a cloud-first company illustrates the power of agile leadership in enabling teams to adapt quickly. However, agile leadership works best when paired with transformational or visionary principles to maintain long-term strategic coherence [3].

Systems leadership addresses interdependencies and cascading risks, making it essential for managing complex challenges in global supply chains or smart grids (Chapter 22). Peter Senge's systems thinking framework underpins this style, which thrives in interdisciplinary collaborations. Mary Barra's leadership at General Motors demonstrates its application in transitioning to electric vehicles – systems leadership pairs effectively with chaordic or collaborative leadership to tackle global challenges [4].

Dee Hock introduced the concept of chaordic leadership, which combines chaos and order to foster adaptability and creativity (Chapter 23). Elon Musk employs chaordic principles at SpaceX, combining decentralized innovation with visionary guidance. While this style thrives in environments that require rapid adaptation, strong values and alignment are crucial to maintaining cohesion. Pairing chaordic leadership with systems and transformational styles enhances its impact [5].

Collaborative leadership fosters inclusivity, shared accountability, and alignment. It is particularly effective in interdisciplinary initiatives, such as Denmark's renewable energy transition (Chapter 16). Leaders must invest heavily in relationship management and communication; however, combining this approach with systems or inclusive leadership amplifies effectiveness [6].

Innovative leadership is indispensable for achieving breakthroughs in fields such as artificial intelligence (AI) and space exploration. Chapter 11 explores how leaders like Jeff Bezos foster a culture of experimentation and resilience. While innovative leadership drives progress, it must be balanced with pragmatic approaches, such as agile or adaptive leadership, to mitigate risks [7].

Empathetic leadership emphasizes emotional intelligence, trust, and team cohesion (Chapter 14). Leaders like Howard Schultz at Starbucks create inclusive, supportive environments that enhance organizational performance. However, empathetic leadership may slow decision-making in fast-paced sectors. Combining it with adaptive or agile leadership ensures responsiveness without sacrificing empathy [8].

Servant leadership, introduced by Robert Greenleaf in 1970, prioritizes the well-being of stakeholders and ethical decision-making (Chapter 26). This style builds trust and inclusivity, as seen in community-led renewable energy projects. However, it may delay urgent decisions. Combining servant leadership with transformational principles ensures ethical innovation while maintaining a sense of urgency [9].

Inclusive leadership integrates diversity, equity, and belonging into decision-making, driving innovation through diverse perspectives (Chapter 18). Inclusive leaders foster engagement and creativity but must consistently build cultural competence. Pairing this style with collaborative or transformational leadership maximizes its impact [10].

Adaptive leadership, as defined by Ronald Heifetz, thrives in uncertainty, encouraging experimentation and systemic problem-solving (Chapter 27). This style guided organizations through crises, such as the COVID-19 pandemic. While adaptive leadership fosters resilience, it requires tolerance for ambiguity. Pairing it with agile or visionary leadership ensures strategic and operational balance [11].

Integrative leadership combines strengths from multiple styles, offering dynamic solutions to diverse challenges (Chapter 7). Leaders like Satya Nadella exemplify this by integrating transformational, ethical, and agile principles to foster inclusivity and innovation at Microsoft. Integrative leadership ensures adaptability and resilience, making it ideal for technology-intensive sectors [12].

Leadership Styles Not Used

Autocratic leadership emphasizes centralized control, which can hinder collaboration and innovation, making it less suitable for complex, interdisciplinary environments. Transactional leadership prioritizes specific outcomes but lacks the vision and adaptability required for evolving technological landscapes. Laissez-faire leadership fails to provide active guidance, limiting its applicability to high-stakes environments. **Bureaucratic leadership** relies on rigid processes, which hinder agility and creativity, essential qualities for the science and technology sectors.

Table: Leadership Styles in This Book

Leadership Style	Core Attributes	Strengths	Challenges	Key Contributor and References
Transformational Leadership	Visionary, empowering, systemic change	Drives innovation and collaboration	Requires continuous vision alignment	James MacGregor Burns, 1978 [1]
Visionary Leadership	Clear vision, long-term focus	Inspires teams and anticipates trends	May neglect short-term needs	Steve Jobs (biography by Walter Isaacson), 2011 [2]
Agile Leadership	Flexibility, experimentation, iteration	Adapts to rapid innovation and change	Tolerates ambiguity and iteration	Agile Alliance, 2001 [3]
Systems Leadership	Holistic, collaborative, risk management	Tackles complex, interdependent challenges	Limited by external factors	Peter Senge, 1990 [4]
Chaordic Leadership	Autonomy, innovation, order in chaos	Fosters creativity and adaptability	Risk of fragmentation without unifying principles	Dee Hock, 1999 [5]
Collaborative Leadership	Teamwork, shared accountability, and inclusivity	Leverages diverse strengths and perspectives	Requires high communication and effort	Peter Senge, 1990 [6]
Innovative Leadership	Creativity, experimentation, and risk-taking	Drives groundbreaking advancements	Requires careful risk management	Elon Musk, concepts expanded by Peter Thiel, 2014 [7]
Empathetic Leadership	Emotional intelligence, trust, and psychological safety	Builds trust and fosters strong team performance	May slow decision-making	Carl Rogers, 1951 [8]
Servant Leadership	Empowerment, ethical focus, long-term value	Builds trust and inclusivity	May delay urgent decisions	Robert Greenleaf, 1970 [9]
Inclusive Leadership	Diversity, equity, belonging	Drives innovation through diverse perspectives	Requires cultural competence	Deloitte Research, 2017 [10]
Adaptive Leadership	Experimentation, learning, systemic challenges	Builds resilience and fosters innovation	Demands tolerance for ambiguity	Ronald Heifetz, 1994 [11]
Integrative Leadership	Blended approaches, flexibility, long-term focus	Adapts to diverse challenges and drives innovation	Requires versatile skill sets	Satya Nadella, 2017 [12]

Effective Leadership Styles for Technological Leadership: This book explores the leadership styles of transformational, servant, systems, adaptive, participative, and charismatic leadership, which were chosen because they align with the complexity and uncertainty of modern challenges. Each style supports innovation, collaboration, and ethical decision-making, enabling leaders to navigate VUCA environments effectively.

The excluded styles – autocratic, transactional, laissez-faire, and bureaucratic – were omitted because they lack the adaptability, inclusivity, and forward-thinking required for technological leadership and resilient systems design.

This book focuses on leadership styles, emphasizing collaboration, foresight, and ethical action. It provides a roadmap for leaders to succeed in challenging and dynamic environments. Integrating real-world case studies ensures that these styles are both theoretical and practical, making them actionable.

Distinctive Features of Technological Leadership

Technological leadership is characterized by a distinct set of qualities and priorities that distinguish it from traditional leadership approaches. These features are essential for leaders navigating the fast-paced, high-stakes world of technological innovation.

Visionary Thinking: Seeing Beyond the Horizon

Technological leaders must think beyond the present, envisioning how emerging technologies can disrupt industries and create new opportunities. Visionary thinking involves connecting disparate ideas, anticipating future trends, and inspiring others to pursue bold and transformative goals.

- **Why It Matters:** Innovation thrives on clear, compelling visions. Without a guiding vision, teams may struggle to align their efforts or fully grasp the potential of their work.
- **Example:** Steve Jobs exemplified visionary thinking by foreseeing the role of personal technology in everyday life. His ability to articulate a future where devices like the iPod, iPhone, and iPad redefined industries inspired teams to innovate at unprecedented levels.
- **Actionable Insight:** Regularly engage in foresight exercises, such as scenario planning or trend analysis, to anticipate future opportunities and risks.

Collaboration: Harnessing Interdisciplinary Expertise

In a world where complex problems span multiple domains, collaboration is vital. Technological leaders must build and lead interdisciplinary teams integrating diverse perspectives, skills, and expertise to solve systemic challenges.

- **Why It Matters:** The most significant technological breakthroughs – whether in AI, renewable energy, or cybersecurity – require input from engineers, designers, data scientists, business strategists, and other specialists.
- **Example:** Elon Musk's leadership at Tesla and SpaceX demonstrates the power of interdisciplinary collaboration. Musk's teams bring together experts in physics, software engineering, materials science, and aerospace to push the boundaries of electric vehicles and space exploration.
- **Actionable Insight:** Foster a culture of openness and respect across disciplines. Encourage knowledge-sharing, joint problem-solving, and cross-functional initiatives.

Ethics and Accountability: Innovating Responsibly

Technological leaders must strike a balance between the drive for innovation and a commitment to societal well-being. Ethical considerations, including data privacy, environmental impact, and equitable access to technology, are crucial for achieving sustainable success.

- **Why It Matters:** Technology has the power to shape societies, and poorly managed innovation can lead to unintended consequences, such as social inequality, environmental degradation, or ethical dilemmas in AI.
- **Example:** Satya Nadella's leadership at Microsoft reflects a strong commitment to ethics and accountability. By prioritizing accessibility and responsible AI, Nadella has positioned Microsoft as a leader in ethical innovation.
- **Actionable Insight:** Establish clear ethical guidelines for your organization's technological initiatives. Regularly assess the societal impact of your work and involve diverse stakeholders in the decision-making process.

Resilience: Thriving Amid Uncertainty

The rapid pace of technological change brings inherent uncertainty and risks. Technological leaders must build resilient systems – both organizational and technical – that can adapt to disruptions, recover quickly from setbacks, and continue to innovate under pressure.

- **Why It Matters:** Resilience ensures continuity and progress in the face of challenges, whether they involve supply chain disruptions, cybersecurity threats, or shifts in market demand.
- **Example:** Jeff Bezos built resilience into Amazon's infrastructure by investing in scalable cloud computing (AWS) and optimizing supply chain operations. These systems enabled Amazon to maintain growth during global crises, such as the COVID-19 pandemic.
- **Actionable Insight:** Develop contingency plans, stress-test your systems, and cultivate an organization-wide culture of adaptability.

Technological Leadership vs. General Management

While general management focuses on optimizing established processes and delivering predictable outcomes, technological leadership demands a fundamentally different mindset. The following key distinctions highlight what sets technological leaders apart:

Vision and Innovation: Creating the Future

- **General Management:** Prioritizes incremental improvements within existing systems.
- **Technological Leadership:** Pushes boundaries to create entirely new paradigms. Leaders envision disruptive innovations that redefine industries.
- **Example:** Steve Jobs' iconic challenge to John Sculley – "Do you want to sell sugar water for the rest of your life, or do you want to change the world?" – captures the bold ambition that defines technological leadership.

Risk-Taking: Embracing Uncertainty

- **General Management:** Focuses on minimizing risk and ensuring stability.
- **Technological Leadership:** Views risk as an inherent part of progress. Leaders embrace uncertainty, experiment with new ideas, and learn from failure.
- **Example:** Elon Musk's ventures – Tesla, SpaceX, and Neuralink – operate in high-risk, high-reward domains. Musk's willingness to invest in unproven technologies has driven breakthroughs in multiple industries.

Adaptability: Navigating Rapid Change

- **General Management:** Excels in stable environments with predictable challenges.
- **Technological Leadership:** Thrives in volatile, fast-paced settings where adaptability is essential.
- **Example:** During Satya Nadella's tenure, Microsoft shifted from a product-centric to a cloud-first, AI-driven strategy, demonstrating adaptability in a rapidly evolving tech landscape.

Collaboration: Breaking Down Silos

- **General Management:** Often operates within defined hierarchies and departments.
- **Technological Leadership:** Fosters cross-disciplinary collaboration to solve complex problems and drive innovation.
- **Example:** The success of the Human Genome Project relied on collaboration across biology, computing, and engineering – a hallmark of technological leadership.

Understanding Leadership, Technology, and the Strategy to Win

In an age of technological upheaval, leadership isn't optional – it's the cornerstone of progress. Organizations must craft strategies beyond mere survival to succeed in this volatile environment. The leaders who thrive can marry bold vision with execution, innovation with accountability, and ethics with impact.

This chapter lays out a framework for success – one that combines strategic planning, actionable tools, and real-world lessons from companies that have mastered the art of technology leadership. From Tesla's forward-thinking roadmap to Singapore's Smart Nation initiative, we'll explore how intentional strategy, powered by technology, can transform challenges into opportunities.

Step 1: Define Your Vision – The Foundation of Technological Leadership

A bold vision separates innovators from imitators. It creates a sense of purpose that rallies teams, guides decisions, and inspires action.

Case Study: Tesla's Vision for Sustainable Energy

Tesla wasn't just building electric cars; it was leading a revolution. Elon Musk's 2006 "Master Plan" detailed Tesla's intent to "accelerate the world's transition to sustainable energy." This vision's clarity aligned with Tesla's vehicles, batteries, and solar energy efforts.

Between 2008 and 2024, Tesla's market capitalization skyrocketed to $800 billion, and it has a 22% global market share in electric vehicles. The company's Gigafactories, which produce cutting-edge battery technology, are now a linchpin in the worldwide energy storage industry.

Key Takeaways:

- **Clarity of Vision:** Define a purpose that extends beyond profit.
- **Align Innovation:** Use technology to bridge current capabilities and future goals.
- **Resilience Through Focus:** Stick to the vision, even when the path is uncertain.

Step 2: Know Where You Stand – Harnessing Technology for Insights

Great leaders understand the value of a clear starting point. SWOT analysis and data-driven insights reveal an organization's current state and potential growth.

Case Study: Netflix's Strategic Pivot

In the early 2000s, Netflix faced a dilemma: its DVD rental business was thriving, but the rise of broadband Internet posed both a threat and an opportunity. Using a SWOT analysis, Netflix discovered:

- **Strengths:** A loyal subscriber base and strong brand recognition.
- **Weaknesses:** Dependence on physical media.
- **Opportunities:** Streaming technology and changing consumer habits.
- **Threats:** Competitors, such as Blockbuster, are entering the streaming space.

Armed with these insights, Netflix shifted to streaming in 2007. By 2024, it had over 280 million subscribers, generating $40 billion annually.

Key Tools:

- **SWOT Analysis:** Identify internal and external dynamics.
- **Data Analytics Platforms:** Utilize tools like Tableau and AI models to gain predictive insights.
- **Pareto Principle (80/20 Rule):** Focus on high-impact initiatives that drive results.

Step 3: Set Clear, Actionable Objectives – Turning Vision into Measurable Goals

Clear objectives bridge the gap between ambition and execution.

Case Study: Google's OKRs

Google's meteoric rise was fueled by its adoption of Objectives and Key Results (OKRs). These allowed the company to align massive teams around shared goals. For instance:

- **Objective:** Improve search quality.
- **Key Results:** Decrease response time by 0.2 seconds and increase user satisfaction by 15%.

This approach ensured accountability and clarity, enabling Google to innovate at scale.

Key Takeaways:

- **SMART Goals:** Ensure objectives are Specific, Measurable, Achievable, Relevant, and Time-bound.
- **Technology Integration:** Use KPIs and OKRs for transparent progress tracking.
- **Opportunity Cost Evaluation:** Invest resources where they'll deliver the most value.

Step 4: Create a Tactical Plan – Engineering Execution

Execution begins with breaking big goals into manageable tasks.

Case Study: NASA's Apollo Mission

When NASA set out to land a man on the moon, its success depended on a detailed tactical plan. Gantt charts became indispensable tools, mapping every task, dependency, and timeline. This approach ensured collaboration among thousands of engineers and scientists.

Key Tools:

- **Gantt Charts:** Organize complex timelines and dependencies.
- **Simulation Models:** Test outcomes in virtual environments before execution.
- **Agility in Planning:** Adapt to changing circumstances without losing focus.

Step 5: Execute the Plan – Driving Results with Data and Tools

Execution separates dreamers from achievers. Data, automation, and accountability drive results.

Case Study: Amazon's KPI-Driven Growth

Amazon's rise to dominance is rooted in its relentless focus on key performance indicators (KPIs). The company tracks real-time delivery times, customer satisfaction, and product availability to ensure optimal service. This approach, supported by AI and machine learning, enables Amazon to improve continuously.

Key Actions:

- **Monitor in Real Time:** Use dashboards for up-to-date insights.
- **Leverage Automation:** Implement AI to streamline repetitive tasks.
- **Foster a Culture of Accountability:** Ensure that teams take ownership of their metrics.

Step 6: Measure and Adapt – Staying Resilient in a Changing World

The final step in the strategy is learning from results and evolving.

Case Study: Singapore's Smart Nation

Singapore's Smart Nation initiative is a model of adaptive leadership. The government uses the Balanced Scorecard to track financial performance, citizen satisfaction, internal processes, and innovation outcomes. Singapore adjusts its approach when challenges arise, ensuring progress remains on track.

Key Tools:

- **Balanced Scorecard:** Monitor results holistically.
- **Scenario Planning:** Anticipate future risks and opportunities to inform strategic decisions.
- **Iterative Development:** Use agile methods to refine strategies.

Comprehensive Tools Table

Step	Tool	Purpose	Case Study
Define Your Vision	Vision Statements	Articulate mission and long-term goals	Tesla's Master Plan
Know Where You Stand	SWOT Analysis, Data Analytics	Identify strengths, weaknesses, opportunities, and threats	Netflix's transition to streaming
Set Clear Objectives	OKRs, SMART Goals	Align goals with vision and measure progress	Google's search improvement strategy
Create a Tactical Plan	Gantt Charts, Simulation Models	Break down tasks, manage interdependencies, and predict outcomes	NASA's Apollo Program
Execute the Plan	KPIs, AI-Driven Dashboards	Monitor progress, refine processes, and ensure accountability	Amazon's operational excellence
Measure and Adapt	Balanced Scorecard, Scenario Planning	Evaluate results, adapt to change, and stay resilient	Singapore's Smart Nation initiative

The Blueprint of Intelligent Leadership and Transformative Growth

Balancing precise focus with a broader vision is a rare and transformative skill in leadership. Leaders today must navigate complexities that demand both microlevel precision and macrolevel foresight. This delicate balance – focusing on the details while aligning them with larger goals – defines intelligent, adaptive leadership.

Lessons from Growth Transformers

The most successful organizations implement five key behaviors to drive growth and innovation:

1) **Start with Competitive Insights**
Leading companies anchor their strategies on external market data, consumer trends, and investor perspectives. By understanding external forces, leaders build visions that align with real-world opportunities. For instance, top companies use predictive analytics to anticipate changes, ensuring their strategies remain relevant.
2) **Leverage External Partnerships**
Collaboration accelerates growth. Companies that partner with external experts, suppliers, and institutions gain access to skills and capabilities they lack internally. For example, Tesla's alliances with battery manufacturers enabled it to scale its electric vehicle (EV) production faster than its competitors.

3) **Embed Ambition in Leadership**
   Great leaders inspire bold thinking and reward innovative behaviors. They don't settle for incremental improvements; instead, they challenge their teams to push boundaries. Mary Barra at General Motors exemplifies this by setting ambitious targets to transition to an all-electric future.
4) **Prioritize Technology and Digital Strategy**
   Automation and AI enhance execution and customer experience. Companies that adopt digital strategies outperform their peers in terms of speed and adaptability. For instance, Amazon's use of AI in logistics revolutionized inventory management.
5) **Rely on a Strong Fact Base**
   Data-driven decisions are essential. Leaders who embed analytics into their daily operations ensure that their teams focus on measurable outcomes. This clarity avoids wasted efforts and drives sustained success.

Intelligent Leadership Traits

Successful leadership incorporates behaviors observed in highly intelligent employees. These traits include autonomy, respect, and a results-driven mindset.

1) **Autonomy and Value Creation**
   Encourage employees to take ownership of their roles. Allow room for creativity and skill-building. This autonomy fosters loyalty and innovation.
2) **Clarity and Respect in Communication**
   Treat everyone equally – respect builds trust. Clear, concise communication eliminates confusion and boosts team efficiency.
3) **Protect Time Relentlessly**
   Time is a leader's most valuable resource. Avoid unnecessary meetings and focus on high-impact work. Modeling this behavior ensures team productivity.
4) **Foster Experimentation**
   Innovation thrives in environments that embrace failure as a learning opportunity. Leaders must create safe spaces for calculated risks and growth.
5) **Focus on Impact over Vanity Metrics**
   Avoid chasing superficial KPIs. Align efforts with meaningful, tangible goals that deliver real results.
6) **Build Psychological Safety**
   Teams perform best when they feel secure sharing ideas and challenging norms. Leaders must create an environment that welcomes diverse perspectives.
7) **Commit to Lifelong Learning**
   Leadership is not static. Stay adaptable by prioritizing self-education and encouraging team growth. This mindset enables teams to remain competitive in a rapidly evolving landscape.

Case Study: The Power of Resilience

When Hurricane Ida struck Louisiana, power outages became the leading cause of disaster-related deaths. Recognizing this, local leaders and nonprofits partnered with federal programs, such as the Energy Storage for Social Equity initiative. By raising $12 million in pilot funds, they deployed battery storage systems to critical facilities, ensuring power reliability. This grassroots leadership illustrates how partnerships and technology can solve pressing challenges.

Great leaders have a versatile lens, but it's not just about zooming in and out – it's about knowing when to dig in and step back. Zooming in helps leaders handle the details that make things work, ensuring nothing slips through the cracks.

Zooming out isn't just watching from a distance – it's ensuring those details align with the bigger picture, so that every step moves toward a genuine purpose.

The best leaders don't get bogged down in the small details or float too high above the work. They find the right balance, connecting what's happening now with where things must go. They know when to step in, when to trust their team, and when to let go. That kind of balance comes from experience and a willingness to learn.

But it's not just about them – it's about helping their teams do the same. Leaders who empower others to focus on what matters while keeping everyone aligned toward a common goal create real progress. That's what makes leadership effective: a clear focus, a shared purpose, and trust in people to get the job done.

Leadership That Connects the Dots

Great leadership begins with understanding how to bridge the gap between the day-to-day and the broader perspective. It's not about micromanaging or getting stuck in strategy meetings; it's about striking a balance between focus and vision, ensuring that the details support the larger goal. For example, Mary Barra at General Motors didn't just push for electric vehicles; she aligned daily operations with a long-term vision of sustainability and innovation. Her focus connected what needed to be done today with where the company had to go tomorrow.

1) **Lead by Outcomes, Not Activity:** Leadership isn't about being busy; it's about getting results. Ask yourself: Are the efforts you and your team are making driving meaningful outcomes? A leader who continually asks this question avoids wasting effort on unimportant tasks. For instance, when Ford pivoted to making ventilators during the pandemic, every action was tied directly to saving lives – a clear, measurable goal. That kind of focus makes leadership effective.
2) **Learn From Feedback and Stay Open:** Listening isn't a weakness; it's power. Great leaders seek input from their teams and act on it. Creating a culture where feedback is welcomed fosters trust and enhances everyone's performance. Consider Google's "Project Oxygen," where leadership listened to employee feedback and implemented changes that improved team performance. The key? Act on what you hear and show you value the input.
3) **Build Resilient, Skilled Teams:** Investing in your team isn't a one-time thing. Continuous learning opportunities, mentorship, and clear paths for advancement motivate people and prepare them for challenges. A strong leader like Satya Nadella at Microsoft turned the company's culture around by prioritizing growth, both for the organization and its people. Leaders who focus on growth create teams ready for anything.
4) **Balance Emotional Awareness and Action:** Leadership is more than numbers. Knowing how to connect emotionally with your team is just as important. Show empathy when needed, and take responsibility when things go wrong. A good example is Jacinda Ardern's steady leadership during crises, where her emotional intelligence helped build trust and keep people focused. Empathy doesn't replace action – it strengthens it.

5) **Adaptability and Accountability Go Hand in Hand: Things change rapidly**, and leaders must adjust and be proactive. Adaptability doesn't mean losing focus – it means shifting gears while keeping the bigger picture in mind. At the same time, take responsibility for results, good or bad. Elon Musk, for instance, is quick to pivot when plans don't work but always owns the outcomes. This blend of adaptability and accountability builds respect and resilience.
6) **Collaborate to Solve Bigger Problems:** Sometimes, leadership means recognizing you can't do it all. Partnering with others – whether across teams, industries, or regions – brings new perspectives and fills gaps. Consider Moderna's vaccine rollout, where collaboration with governments and manufacturers sped up life-saving production. Great leaders know when to reach beyond their expertise.
7) **Stay Curious and Keep Learning:** The world and leaders should not stand still. Be willing to learn from mistakes and successes, adapt to new challenges, and keep pushing forward. Whether embracing new technologies or rethinking strategies, leaders who stay curious build teams and organizations that thrive.
8) **Bringing It All Together:** Leadership isn't about perfection – it's about connecting daily actions to long-term goals, building resilient teams, and staying accountable while adapting to challenges. Whether running a global company or leading a small team, the principles remain the same: stay focused, empower others, and continually learn. Great leaders aren't just in charge – they're part of the solution.

By integrating these principles, leaders drive growth and build resilient, forward-thinking organizations that are capable of thriving in uncertain times. These habits will enable leaders to build strong teams, foster innovation, and achieve long-term success. Leadership is not about perfection but adaptation. It's about making informed decisions, embracing change, and empowering others to excel. Dan Goldin wisely observed that great leadership lies in zooming in on details while staying anchored in the broader vision.

The Role of Luck, Serendipity, and Preparedness in Leadership and Innovation: Not much research has been done yet, but it is important.

Success is often framed as the outcome of deliberate effort and careful planning. Yet, the nuanced interplay of luck, serendipity, and preparedness shapes many stories of leadership and innovation. While chance may open doors, the ability to recognize opportunities, act decisively, and align them with long-term vision transforms random events into meaningful progress. The stories of leaders like Jeff Bezos, Steve Jobs, and the creators of Google illustrate how serendipity intersects with preparation and leadership insight.

Luck Favors the Prepared Mind

The saying, "Luck favors the prepared mind," succinctly captures the essence of converting chance into opportunity. Leaders with clarity and focus – cultivated through reflection, introspection, or meditation – are better equipped to perceive opportunities others might overlook. Recognizing opportunity is the first step; acting effectively is equally critical. Whether through personal skills or the strength of a well-prepared team, readiness transforms randomness into results.

For instance, Jeff Bezos's foresight in building Amazon's scalable infrastructure allowed him to capitalize on the growing e-commerce market. Similarly, Steve Jobs's strategic clarity during his second tenure at Apple positioned the company to innovate with groundbreaking products, such as the iPhone. These examples emphasize that chance alone is insufficient – success hinges on preparation and action.

Jeff Bezos and Amazon's Evolution

In 1994, Jeff Bezos founded Amazon as an online bookstore, a calculated entry point due to the universal appeal of books and the ease of shipping. However, his vision extended far beyond selling books. By 1998, Bezos had expanded Amazon into a multicategory e-commerce platform, anticipating the transformative impact of the Internet on retail. His investments in logistics and infrastructure enabled Amazon to scale rapidly, turning it into the "everything store."

Bezos's success exemplifies how preparation meets opportunity. While others saw a bookstore, Bezos envisioned a platform capable of reshaping global commerce – a vision brought to life through deliberate planning and strategic readiness.

Source: Stone, B. (2013). The Everything Store: Jeff Bezos and the Age of Amazon. Little, Brown and Company.

Steve Jobs: Chaos, Redemption, and Focus

Steve Jobs's return to Apple in 1997 marked a pivotal moment for both the company and the tech industry. At the time, Apple was on the brink of bankruptcy. Jobs's exile from Apple earlier in his career provided him with the space for deep reflection. Upon his return, he leveraged his newfound clarity to streamline Apple's operations, cutting unprofitable product lines and focusing on a handful of core innovations.

Jobs's leadership transformed Apple with products like the iMac and the iPhone. His story highlights the interplay of chaos, opportunity, and a focused mind, demonstrating how clarity and preparation can transform even dire circumstances into transformative success.

Source: Isaacson, W. (2011). Steve Jobs. Simon & Schuster.

The Birth of Google: A Serendipitous Encounter

In 1998, during a visit to Palo Alto, a Stanford graduate student demonstrated an early version of a new search engine, Backrub, to me. Developed by Larry Page and Sergey Brin, this project later became Google. At a time when search engines were notoriously inefficient, Page and Brin applied their deep understanding of mathematics and computer science to develop algorithms that prioritized speed and relevance.

This encounter was serendipitous, but the duo's technical preparation and vision turned Backrub into the world-changing tool we now know as Google. Their story illustrates how readiness transforms fleeting opportunities into enduring innovations.

Source: Vise, D. A. (2005). The Google Story. Delacorte Press.

The Power of Reflection and Preparation

A common thread among these stories is cultivating a calm, prepared mindset. Through meditation, structured reflection, or disciplined focus, successful leaders create the mental clarity necessary to identify patterns in chaos and seize opportunities. However, vision alone is insufficient. The combination of vision, skill, and decisive action transforms chance into success.

For example, Bezos's infrastructure investments, Jobs's product focus, and Page and Brin's technical expertise reflect a readiness to act when opportunity strikes. These qualities exemplify how preparation amplifies the potential of serendipity.

Turning Constraints into Opportunities

Jim Collins's concept of "return on luck" posits that luck is distributed relatively equally, but how individuals and organizations leverage it matters. Leaders who embrace constraints, adapt to chaotic conditions, and align their skills with their vision maximize the impact of chance.

This aligns with the principles of Jugaad, or frugal innovation – finding creative solutions within constrained environments. From startups to under-resourced organizations, the ability to see opportunity within limitations is a hallmark of effective leadership.

Michael Krause: A Reflection on Leadership and Luck

Michael Krause, Chief Operating Officer of Renewable Energy Partners in Minneapolis, exemplifies the integration of luck, preparation, and leadership. With a background spanning journalism, law, environmental advocacy, and political leadership, Krause brings a thoughtful and purposeful approach to everything he does. His insights on the role of chance and serendipity in success have profoundly influenced my perspective.

For instance, during a flight to Minneapolis, Krause engaged in an impromptu, transformative conversation with a leader in economic development. This encounter led to breakthrough partnerships for his organization at a pivotal moment. His story illustrates how leaders, through clarity of purpose and readiness, can turn seemingly random moments into opportunities for impact.

Luck and serendipity are often catalysts for success, but preparation, clarity, and action define outcomes. As leaders, cultivating mental clarity and readiness equips us to recognize and seize opportunities. Stories from Jeff Bezos's vision for Amazon to Steve Jobs's strategic focus at Apple, Larry Page and Sergey Brin's technical mastery at Google, and Michael Krause's transformative conversations underscore this principle. Luck opens doors, but leadership insight, preparation, and action turn those open doors into pathways of innovation and success.

Technology and Leadership as Strategy's Backbone: In a world of constant change, leadership isn't just about reacting but anticipating. The strategies outlined here demonstrate that success is a deliberate process driven by a clear vision, effective tools, and resilience. Leaders who adopt these principles can harness the power of technology to achieve their objectives and transform entire industries.

The Role of Technological Leadership in Today's World: Technological leadership is no longer confined to tech companies. As digital transformation reshapes industries from healthcare to finance, leaders across sectors must adopt these principles to remain competitive and drive meaningful change.

Practical Applications:

1) **Building Smart Systems:** Utilize technological innovation to design resilient and adaptive systems, whether in infrastructure, energy, or cybersecurity.

2) **Fostering Inclusion:** Include diverse perspectives and equitable access to technology in your leadership strategy.
3) **Driving Sustainability:** Prioritize technologies that address global challenges, such as climate change and resource scarcity.

Actionable Framework for Technological Leaders

Step 1: Define Your Vision

- Conduct foresight exercises to identify disruptive opportunities.
- Align your vision with long-term societal and organizational goals.

Step 2: Build Interdisciplinary Teams

- Break down silos and encourage collaboration across departments.
- Invest in tools and processes that enable seamless knowledge-sharing.

Step 3: Embrace Ethical Innovation

- Develop clear and ethical guidelines for the use of technology.
- Engage stakeholders – including employees, customers, and regulators – in conversations about accountability.

Step 4: Cultivate Resilience

- Design systems that can adapt to rapid changes and recover from disruptions.
- Regularly evaluate and refine your organization's risk management strategies.

Technological leadership is more than managing teams or delivering projects – it's about envisioning the future, fostering innovation, and building systems that serve humanity in a rapidly changing world. Leaders like Steve Jobs, Elon Musk, and Satya Nadella exemplify the qualities that define this distinct paradigm, demonstrating that the most effective technological leaders combine vision, adaptability, collaboration, and ethics. In an era of accelerating complexity, the principles of technological leadership are essential for driving progress and solving the challenges of tomorrow.

How to Use This Book: This book is a comprehensive guide and a practical toolbox. Each chapter stands alone, allowing readers to focus on specific areas of interest or need. Whether you are managing a team, designing a system, or leading through a crisis, this book provides the tools and insights necessary to act effectively. Each part of the book tackles a critical aspect of leadership:

- **Foundations of Leadership in Complexity:** Chapters 1 and 2 establish leading principles in a volatile, uncertain, complex, and ambiguous (VUCA) world.
- **Tools and Frameworks for Decision-Making:** Chapters 4A and 4B provide foundational and advanced methods for tackling dynamic challenges.
- **Ethics and Resilience in Leadership:** Chapter 3 emphasizes the importance of ethical leadership in driving meaningful change.
- **Strategic Planning for the Future:** Chapter 6 uses foresight to help leaders anticipate and prepare for future challenges.
- **Sector-Specific Insights:** Chapters 11–31 provide insights into critical infrastructure, cybersecurity, diversity, inclusive leadership practices, and long-term policy impacts.

Whether you are managing a team, designing a system, or leading during a crisis, this book provides tools and insights to help you thrive in complexity. Leadership today requires the ability to act and the clarity to ensure those actions build systems that endure, evolve, and serve humanity.

Moving Forward: Leadership today demands more than managing the present – it requires shaping the future. This book equips you with the tools, frameworks, and insights to navigate complexity, foster resilience, and lead with integrity. Use it as a companion to guide your decisions, inspire your actions, and help you build systems that serve humanity.

Let's get to work.

References

1 J. M. Burns, *Leadership*. Harper & Row, 1978.
2 W. Isaacson, *Steve Jobs*. Simon & Schuster, 2011.
3 Agile Alliance, *Manifesto for Agile Software Development*, 2001.
4 P. Senge, *The Fifth Discipline*. Doubleday, 1990.
5 D. Hock, *Birth of the Chaordic Age*. Berrett-Koehler, 1999.
6 P. Senge, *The Fifth Discipline*. Doubleday, 1990.
7 P. Thiel, *Zero to One*. Crown Business, 2014.
8 C. Rogers, *On Becoming a Person*. Houghton Mifflin, 1951.
9 R. Greenleaf, *Servant Leadership*. Paulist Press, 1970.
10 Deloitte, *Global Diversity and Inclusion Survey*, 2017.
11 R. Heifetz, *Leadership Without Easy Answers*. Harvard University Press, 1994.
12 S. Nadella, *Hit Refresh*. Harper Business, 2017.

Resources for Further Reading

"Amazon's operational KPIs," *Harv. Bus. Rev.*, vol. 99, no. 3, pp. 45–52, 2023.

Balanced Scorecard Institute, *Implementing the Balanced Scorecard*, 2024. [Online]. Available: https://www.balancedscorecard.org.

B. Bass and R. E. Riggio, *Transformational Leadership*. New York, NY: Routledge, 2006.

J. Collins, *Great by Choice: Uncertainty, Chaos, and Luck – Why Some Thrive Despite Them All*. HarperBusiness, 2011.

J. Doerr, *Measure What Matters: OKRs*. Portfolio Penguin, 2018.

P. Drucker, "The discipline of innovation," *Harv. Bus. Rev.*, vol. 63, no. 3, pp. 67–72, 1985.

R. Heifetz, *Leadership Without Easy Answers*. Cambridge, MA: Harvard University Press, 1994.

W. Isaacson, *Steve Jobs*. Simon & Schuster, 2011.

S. Jobs, "The journey to innovation: Interviews on leadership," *Harv. Bus. Rev.*, vol. 85, no. 6, pp. 15–23, 2007.

E. Musk, "Reinventing industries: The tesla and SpaceX blueprint," *TechCrunch Disrupt Conference*, San Francisco, CA, 2018.

S. Nadella, *Hit Refresh: The Quest to Rediscover Microsoft's Soul*. New York, NY: Harper Business, 2017.

NASA, *Apollo Program Archives*, 2020. [Online]. Available: https://www.nasa.gov.

J. Sculley, *Odyssey: Pepsi to Apple*. New York, NY: Harper & Row, 1987.

Singapore Government, *Smart Nation Initiative: Progress Reports*, 2023. [Online]. Available: https://www.smartnation.gov.sg.

Statista, *Netflix Revenue and Subscriber Growth*, 2024. [Online]. Available: https://www.statista.com.

B. Stone, *The Everything Store: Jeff Bezos and the Age of Amazon*. Little, Brown and Company, 2013.

Tesla, Inc., *Master Plan, Part Deux*, 2016. [Online]. Available: https://www.tesla.com/blog/master-plan-part-deux.

D. A. Vise, *The Google Story*. Delacorte Press, 2005.

About the Companion Website

This book is accompanied by a companion website:

www.wiley.com/go/amin

To access the website materials, click on the link under Related Resources on the book's page on wiley.com. You can also find this by searching www.wiley.com for the ISBN 9781394338252.

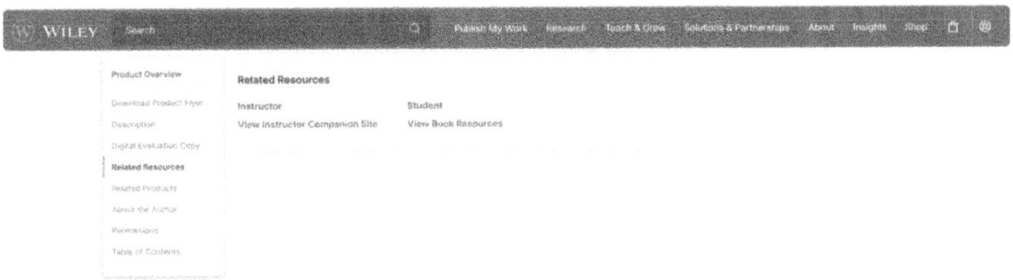

This website includes:

- 1.1 Study material
- 1.2 Study questions and reflection prompts
- 1.3 Case study extensions
- 2.1 Instructor's manual
- 2.2 Slide decks
- 2.3 Problem sets and exercises
- 2.4 Case teaching notes
- 2.5 Test bank and quizzes
- 2.6 Project assignments
- 3 Appendices B–E (summary)

Part I

Foundational Chapters on Complexity, Decision-Making, and Systems Thinking

1

A Roadmap for Leadership in Complexity

Leading in a VUCA World – Embracing Volatility and Uncertainty

In today's rapidly changing environment, leaders operate in a VUCA world – one defined by volatility, uncertainty, complexity, and ambiguity. Circumstances can shift suddenly and unpredictably, challenging even the most seasoned executives. A vivid illustration comes from a personal journey: one moment, the road of life is smooth; the next, a sharp curve or unexpected obstacle appears without warning. Years ago, one of us faced such a turn when political upheaval forced an abrupt departure from home as a teenager, leaving family, friends, language, and stability overnight. Like a motorcyclist leaning into a surprise curve, the only choice was to adapt swiftly and "lean in" to the change rather than panic. This ability to *absorb shock and adjust course* is at the heart of leading in a VUCA era.

Volatility and uncertainty mean that change is the norm, not the exception. From geopolitical shifts to technological breakthroughs, leaders must expect disruption. Complexity and ambiguity further complicate decisions: outcomes are rarely linear, and information is often incomplete or unclear. Rather than being paralyzed by uncertainty, effective leaders cultivate agility and resilience. As Charles Darwin astutely noted, it is not the strongest or smartest who survive, but those who are "most responsive to change." This means developing the foresight to anticipate emerging challenges and the flexibility to pivot strategies when conditions evolve. Leaders who thrive in chaos approach surprises as opportunities for innovation ("Amid chaos, there is also opportunity," as Sun Tzu observed) and encourage their teams to do the same.

Importantly, operating in a VUCA context requires a shift from a reactive to a proactive mindset. Waiting for complete certainty is a mistake – by the time all facts are known, options are often gone. Decisive leaders act on early indicators and take calculated risks to shape the future, rather than respond to it. They communicate a clear vision and values (providing anchors in ambiguity), but are unafraid to course-correct when new information emerges. Leading in a complex world means expecting the unexpected and cultivating an organizational culture to withstand shocks. We begin this chapter by examining how technology both drives and helps manage this complexity and then lay out core principles for leadership that turn volatility and complexity into opportunities for growth.

The Role of Technology in Leadership

Technology has always been a driving force of advancement – a double-edged sword that is both a source of disruption and a critical enabler for leaders. From the first controlled use of fire in prehistoric times to today's breakthrough in quantum computing, technological innovation has continually reshaped society. Each leap solves problems and creates new possibilities, introducing new complexities and risks that leaders must navigate. For example, the invention of the wheel expanded trade and mobility; the printing press democratized information; the telegraph and telephone shrank the world through instant communication. In the 20th century, the transistor (1947) ushered in the digital age, leading to computers and the Internet, and more recently, the rise of artificial intelligence (AI) is transforming business models across every industry.

Today's leaders must recognize technology as a core element of their strategic landscape. Technology is not just an operational tool but is integral to strategy and leadership. Properly harnessed, it can exponentially increase an organization's impact, improving efficiency, enabling data-driven decisions, and opening new markets. For instance, mobile Internet and cloud computing have allowed startups to scale globally at unprecedented speed, and advancements like mRNA vaccines and renewable energy technologies are solving age-old problems in health and sustainability. However, each innovation also brings new leadership challenges. The connectivity that allows real-time operations worldwide also means a cyber incident or network failure can cripple an enterprise in minutes. The power of AI to automate decisions raises ethical questions and skills gaps. As Nobel Peace laureate Christian Lous Lange cautioned a century ago, "Technology is a useful servant but a dangerous master." Leaders must therefore ensure they remain the masters of the technology, aligning it to their mission and values, rather than being led astray by hype or letting systems run unchecked.

In practical terms, technological leadership requires foresight and balance. Leaders should invest in understanding emerging technologies – not necessarily at a deep technical level, but enough to grasp their strategic implications. This might mean building a team with strong digital expertise or partnering with tech innovators. For example, quantum computing has moved from theory to practice in recent years (Google's 70-qubit quantum chip was demonstrated in 2024, performing calculations in seconds that classical computers would take millennia to complete). Forward-looking finance, logistics, or defense leaders are already engaging in scenario planning to assess how such capabilities could benefit them or disrupt their field. At the same time, leaders must weigh the risks and unintended consequences, including data privacy, security vulnerabilities, and the societal impacts of automation. A resilient approach is to pilot new technologies in controlled ways (learn fast, fail fast) while establishing ethical guidelines and backup plans.

Above all, technology should be seen as an enabler of human-centric goals. It extends our reach and augments decision-making but does not replace the need for vision, judgment, and empathy. Great leaders leverage tech tools to further a clear purpose ("the *Why*"), whether serving customers better, accelerating innovation in their industry, or addressing grand challenges like climate change. In essence, technology in leadership is about empowerment: empowering your organization with superior capabilities while empowering your people with the skills and frameworks to use those capabilities responsibly. Technological progress becomes a powerful ally in navigating volatility and complexity when guided by sound leadership. Next, we explore foundational leadership principles – blending timeless economic insights and real-world case studies – that, together with technology, form a roadmap for effective leadership in today's complex environment.

Foundational Principles for Leadership in Complexity

Leading in a complex world requires a blend of steadfast **principles** and adaptive strategies. This section outlines core pillars of effective leadership – from economic and innovation insights to the human factors of vision and culture – and illustrates them with real-world examples. These principles serve as a roadmap for leaders to guide their organizations through uncertainty toward sustainable progress.

Vision and Purpose – "Start with Why"

A clear vision is an anchor in turbulent times. Great leaders articulate purpose and long-term goals that inspire and provide direction, even when immediate outcomes are uncertain. History offers a powerful case: after 1989, several Eastern European nations transitioned from central planning to market economies. Those that succeeded, such as Poland, started with a compelling vision – in Poland's case, the goal of joining the European Union – to rally people and guide reforms – with EU membership as a "North Star," Poland attracted over €100 billion in development funds and strategically invested in infrastructure, education, and health. The payoff was significant: Poland's GDP per capita increased from approximately $1,700 in 1989 to $18,000 by 2023. Clear, achievable goals inspire public support and provide a roadmap for sustainable growth. In contrast, lacking a unifying vision can leave nations (or organizations) drifting or vulnerable to crisis.

On an organizational level, leaders like Simon Sinek have popularized the importance of communicating the "Why" behind an endeavor. Teams perform best when they understand the purpose of their work and how it connects to a greater mission. Purpose-driven leadership fosters resilience, whether it's a startup founder evangelizing how their innovation will change the world or a public-sector leader aligning stakeholders around a common societal goal. Even when external chaos hits, purpose provides continuity. Leaders should take the time to reinforce the mission at every turn – it keeps teams motivated through volatility and focused on outcomes that truly matter, not just short-term metrics.

Adaptability and Continuous Learning

If vision is the compass, adaptability is the steering mechanism. Successful leaders continually adjust tactics in response to new information, much like agile navigators tacking in shifting winds. In the early 1990s, the Czech Republic exemplified this adaptability. Facing the collapse of its old economic system, the Czech leadership embraced flexibility, rapidly implementing pro-business policies, courting foreign investors, and retraining its workforce. As a result, Czech GDP per capita increased from approximately $3,700 in 1989 to over $28,000 in 2023. The lesson is that *openness to change and collaboration can create opportunity amid uncertainty*. Organizations often mirror this behavior when they pivot product lines to meet sudden market shifts or retool their skills in response to changing industry landscapes.

Adaptable leaders also foster a culture of learning. They encourage experimentation and view failures or near-misses as feedback rather than setbacks. This mindset was evident during the COVID-19 pandemic when companies that quickly shifted to new delivery models or remote work not only survived but also often found new growth avenues. The concept of "failing fast" and learning quickly originates from Silicon Valley but is applicable everywhere: it's far better to prototype, test, and iterate than to adhere to a plan that isn't working rigidly. In complex systems, no leader has all the answers upfront, so the winners will be those who learn and adapt faster than their competitors. We will later see how this principle also ties into foresight tools (like scenario planning in Chapter 4) that help leaders rehearse and prepare for various futures, essentially "pre-learning" possible lessons before crises strike.

Innovation and Growth – Technology as an Engine

A foundational insight from economics is that technological innovation drives long-term growth. Nobel laureate Robert Solow's classic research quantified that roughly 60% of GDP growth in advanced economies comes from technological progress, rather than from increases in capital or labor. Later, Paul Romer's Endogenous Growth Theory and others expanded on this concept: innovation, knowledge, and human capital investment create self-reinforcing development. For leaders, the implication is clear – investing in innovation and human capital is not a luxury, but a necessity for sustained success. This could involve funding R&D, adopting new business models, or upskilling employees to handle emerging technologies (as discussed in later chapters on team building and upskilling).

Real-world examples abound. Consider how Tesla revolutionized the automotive industry by investing in electric vehicles and building battery gigafactories. In doing so, Tesla exemplified Schumpeter's concept of "*creative destruction*" – leveraging innovation to upend incumbents and accelerate the transition of an entire sector. Bold leadership decisions, such as building massive battery plants and investing in software updates for cars, have pressured traditional automakers to race toward electric and autonomous technologies. The lesson: visionary leaders who embrace innovation can redefine industries and create sustainable progress. Even at the national level, countries that foster innovation ecosystems – connecting academia, industry, and government – tend to outperform. The Global Innovation Index consistently shows that countries like Switzerland (with strong academia–industry collaboration) and South Korea (with significant R&D investment) lead in translating innovation into economic benefits.

However, fostering innovation also means accepting risk and potential failure. Not every R&D project yields fruit, and disruptive ideas often face internal resistance. Leaders must create an environment where experimentation is encouraged and smart failures are tolerated. This can involve setting aside a budget for moonshot projects or establishing innovation labs sheltered from short-term performance pressures. It also involves monitoring external trends – the "*weak signals*" of change that, if heeded, can alert an organization to pivot or seize a new opportunity early. (In Chapter 4, we will discuss tools for systematic innovation planning, such as technology roadmapping and scenario analysis, which help leaders manage the inherent uncertainty in pursuing innovation.)

Collaboration and Inclusion

Complexity means many interdependencies – no single leader or organization can tackle major challenges alone. Collaboration is thus a cornerstone principle: great leaders leverage networks of partners, experts, and stakeholders to achieve shared goals. We see this in the earlier example of Eastern Europe's transitions: Estonia's remarkable digital transformation succeeded not just because of a single leader, but through a coalition of government, technologists, and an empowered citizenry working together. Estonia has moved nearly all public services online, dramatically reducing bureaucracy and corruption, and in turn, attracted investments that have grown per capita

GDP from $1,200 in 1989 to over $26,000 in 2023. This was possible because leaders built trust across society – a social compact to embrace technology for the common good. The lesson: Collaboration and smart investment in technology can foster equitable and efficient systems.

In a corporate context, collaboration means breaking down silos. Leaders should encourage cross-functional teams and open communication channels. For example, a company launching a new product might create a *tiger team* that brings together R&D, marketing, customer support, and even external partners to ensure the product works technically, addresses real customer needs, and can be produced and delivered smoothly. Collaborative leadership also extends beyond company walls: in our interconnected economy, partnership with suppliers, customers, and even competitors (in precompetitive alliances) can be key to setting industry standards or lobbying for beneficial policies. In the energy sector, we've seen utilities, tech firms, and governments form consortia to develop smart grid standards and share best practices on cybersecurity, recognizing that systemic problems demand systemic solutions that no single entity can create in isolation.

Inclusion is a related aspect: diverse teams make better decisions in complex environments. Research by scholars like Claudia Goldin has highlighted the role of diversity (in terms of gender, background, and perspective) in driving innovation. A leader should actively foster diversity and ensure that all voices are heard, as this multiplicity of viewpoints helps detect weak signals and avoid groupthink. When navigating ambiguity, a broad range of inputs is like having multiple sensors on a system – it improves the chances of catching an emerging issue or identifying a creative solution. An inclusive leader builds a culture where raising concerns or alternative ideas is safe and encouraged (avoiding the "leader knows best" trap that can blind an organization to impending trouble).

Ethical Leadership and Trust

High-performing organizations in the long run are built on trust–trust with employees, customers, and the public. In the pressure cooker of volatility, it might be tempting to drive teams ruthlessly or cut ethical corners for short-term gain, but that approach is ultimately self-defeating. Case studies of famed leaders offer instructive contrasts. Jack Welch, the hard-charging CEO of GE in the 1980s–1990s, achieved remarkable financial results through an intense focus on efficiency and strict accountability (famously ranking and weeding out the bottom 10% of managers each year). GE's market value skyrocketed under his command-and-control style. But the cultural cost was high: an atmosphere of fear can stifle innovation and erode loyalty. Indeed, after Welch left, GE struggled to adapt to new industries and calmer management – the overemphasis on short-term financial engineering left the company vulnerable to systemic risk and lacking the agility needed in the 2000s. Similarly, leaders like Chuck Knight of Emerson Electric and Carly Fiorina of HP were brilliant strategists in their own right. Still, their top-down approaches created *fragile cultures* that cracked when market conditions changed. Each drove bold changes – Knight made Emerson one of the most profitable industrial firms of his era, and Fiorina's Compaq merger made HP the world's largest PC maker – yet each left controversy and morale problems that hampered their organizations in the long term. The pattern is clear: command and control can yield short-term results, but those results are hard to sustain without trust and a healthy culture.

The key takeaway for modern leaders is the importance of striking a balance. Discipline, efficiency, and tough calls are often necessary – especially in crises – but they should be coupled with transparency, empathy, and empowerment of others. Employees and stakeholders today expect a voice and values-driven leadership. One lesson learned from these high-profile cases is that *employees thrive in environments built on trust, rather than fear.* This means leaders should communicate openly (sharing good and bad news candidly), admit mistakes, and encourage ethical behavior at all levels. It also means prioritizing sustainability over short-termism: considering the long-term impacts of decisions on people and the environment, not just the next quarter's profit. By doing so, leaders build credibility and goodwill that become invaluable when navigating complex challenges. For example, a company that has established trust with its workforce will find employees more willing to go the extra mile or adapt to painful changes when volatility strikes. Likewise, a public agency that has earned citizens' trust through fairness and competence will face less backlash when tough measures are needed in a crisis.

Resilience and Systems Thinking

Finally, an essential principle in complexity is to adopt a systems view – understanding the interconnections and aiming for resilience. Leaders should always ask: "*If we change this, what ripple effects could occur in our larger system?*" and "*If X unexpected event happened, how would we cope?*" This anticipatory, holistic thinking differentiates merely good managers from true leaders. We have seen global supply chains, for instance, optimized to be lean and just-in-time, only to crumble when a single link breaks (as happened when the Ever Given ship blocked the Suez Canal in 2021, freezing $9.6 billion of global trade per day). The interdependence of modern systems – energy, transportation, communication, and finance – means that leaders must plan beyond their organizational boundaries. A resilient leader for, say, a manufacturing firm will fortify their facilities and assess the resilience of key suppliers, the local power grid, and even political stability in supplier regions. They actively engage in scenario planning and build redundancies or contingency plans where appropriate. This systems mindset is a theme we will revisit (particularly in Chapter 4's discussion of foresight tools and Chapter 4's Heathrow Airport case, which starkly demonstrates cascading failures).

In summary, leading in complexity is akin to being the conductor of a large orchestra playing in a thunderstorm – one must keep the ensemble coordinated (vision and trust), adapt to the changing tempo of the elements (agility and learning), encourage solos and improvisation (innovation and inclusion), and ensure the show goes on even if a string breaks or a page is lost (resilience and preparation). Leaders create a unifying framework that guides decision-making when things get chaotic by adhering to core principles – clarity of purpose, adaptability, innovation, collaboration, integrity, and systems thinking. These foundational elements are the pillars of our roadmap for leadership in a complex age.

Integrating Principles into Practice – Looking Ahead

Principles alone are not enough; their integration into daily practice truly drives results. As we conclude this chapter, it's worth reinforcing how these leadership principles come together. Consider the example of Denmark's transition to wind energy over the past few decades. Danish leaders set a bold vision (to become a global wind power leader), invested in innovation (partnering industry and academia to improve turbine technology), showed adaptability in policy (adjusting subsidies and grid regulations as the market evolved), fostered collaboration (public–private partnerships and community co-ownership of wind farms), and maintained trust by ensuring transparency and local benefits. The result: Denmark now generates around half of its electricity from wind and has built a world-class clean tech industry. This kind of multifaceted leadership – blending vision, innovation, flexibility, and stakeholder engagement – exemplifies the roadmap in action.

As leaders, we must be both architects and gardeners: architects who design strategies and structures based on sound principles and nurture the culture and people that bring those plans to life. The chapters that follow in this book will delve into specific tools and case studies to operationalize these ideas. In Chapter 4, we will introduce concrete tools for decision-making, ranging from scenario planning to cost-benefit analysis, which help us navigate uncertainty rigorously. Later, we will explore ethical leadership (Part III), team building (Part IV), and systems thinking in practice (Part V), all with an eye toward practical implementation.

Crucially, this book is intended as a *living roadmap*. The world will continue to change – new technologies will emerge, and tomorrow's complexity will differ from today's. Thus, we encourage a mindset of collaborative learning. The roadmap for complexity leadership is not a fixed highway, but a journey that we chart together, continually updating our maps with input from experience and new insights.

Actionable Takeaways – Principles for Leading in Complexity:

- **Craft and Communicate a Clear Vision:** Define "True North" goals that inspire and guide your team through ambiguity (e.g., a unifying mission, such as digital transformation or a market milestone). Refer back to this vision to maintain your focus during turbulent times.
- **Be Adaptive and Embrace Change:** Expect the unexpected and build agility. Encourage experimentation, learn from failures quickly, and be prepared to adjust strategies when conditions change. Cultivate a culture that views change as an opportunity rather than a threat.
- **Invest in Innovation and People:** Prioritize R&D, new technologies, and continuous learning. Build teams with diverse skills and perspectives. Remember that long-term growth overwhelmingly comes from innovation – even small improvements can compound to big advantages.
- **Foster Collaboration and Trust:** Break down silos within your organization and seek alliances outside of it. Lead with transparency and integrity to earn trust. A network of trusted relationships is an invaluable asset when solving complex cross-disciplinary problems.
- **Balance Results with Resilience:** Deliver on short-term objectives **and** strengthen long-term capabilities. Don't sacrifice culture, ethics, or system redundancy for quick wins. By considering the broader system and future scenarios (what could go wrong and what if it does), you ensure that your organization can withstand shocks and continue to thrive.

The above principles offer a roadmap for leadership success in a challenging and dynamic environment. We invite feedback and dialogue on these ideas. This chapter has integrated insights from economics, case studies from business and government, and personal experiences. Collaborative editorial input is welcome as we continue to refine this IEEE/Wiley book manuscript. If the editors or readers have suggestions, real-world examples to share, or different perspectives, please feel free to contribute them. *Dr. Amin is ready to incorporate further revisions and enhancements* – for example, integrating any comments or alternate drafts the editorial team provides – to ensure this roadmap is as clear, relevant, and impactful as possible for those who will lead our organizations and communities into the future.

2

The Vulnerability of Interdependence

The things we rely on today, from the power grid to the Internet, are among human beings' most significant achievements. They link over a billion people, facilitate economies, and control every aspect of modern life. But what a surprise! These same systems are relatively weak. When one part ceases to function, it can have adverse effects on other parts, thus affecting people's lives, firms, and even countries.

This chapter compares the robustness of interconnected systems to learn from past occurrences. These examples demonstrate that a minor issue can escalate into significant problems, such as a power outage or cyberattack.

The Interconnection Paradox

Today's systems are designed for efficiency and optimal output. This can be seen in the electric grid, which is not a single entity. Generation Plants, Substations, Transmission Lines, and Distribution Systems all work in conjunction with each other. This interconnectedness creates efficiency and scalability, but also creates vulnerabilities.

For instance, the Northeast blackout occurred in 2003. This resulted from a power line in Ohio that sagged and touched a tree branch, causing a circuit to trip and setting off a domino effect. Within hours, 50 million people across the United States and Canada were left powerless. The transport sector was affected, as were communication systems, and business activities were disrupted. The economic loss was estimated to be $6 billion, and the incident also highlighted how old and archaic infrastructure can be detrimental [1, 2].

Another example is Hurricane Maria in 2017, which affected Puerto Rico. The island's infrastructure was also affected, specifically the power grid, and millions of people were left without electricity for several months. In the hospitals, electricity was saved in shifts, schools were closed, and companies shut down. The blackout was a result of the hurricane and a lack of diversification in a system that had been built over the decades. By adopting decentralized energy solutions in the form of solar microgrids, some communities were able to recover more quickly, which may be a glimpse into the future [3].

These examples are pretty revealing. Interconnectedness enhances vulnerability. A single weakness within one part of the system can lead to system-wide failure.

Cybersecurity and the Contemporary Environment Risk Assessment

As our systems become more complex and interconnected, they are no longer at risk only from physical failures; instead, they are also vulnerable to cyberattacks. Traditional threats, such as ransomware and cyberattacks targeting crucial infrastructure, also exist.

This includes cyberattacks; for instance, in 2021, the Colonial Pipeline, which supplies nearly half of the fuel on the East Coast of the United States, was targeted by hackers. The hackers infiltrated the company's IT systems, resulting in a shutdown of its operations. Gas shortages and panic buying ensued, resulting in a loss estimated at $5 billion. This incident also demonstrated that a digital invasion could lead to the breakdown of physical structures and expose the lack of readiness in critical infrastructure [4, 5].

The Colonial Pipeline attack wasn't an endpoint for such an incident. That year, the healthcare industry also experienced ransomware attacks, which led to the cancellation of surgeries and other services and affected patients. These events serve as evidence of how the physical and digital worlds are becoming increasingly integrated parts of our lives, and how issues in one part of the system can have ripple effects in the other.

Cascading Failures in Supply Chains

Another area that illustrates the vulnerability of global supply chains is the worldwide connectedness of economies. The COVID-19 pandemic highlighted the fragility of these systems. This led to factory closures in one part of the world and impacted the production of essential components for various industries, such as the automotive and electronics sectors.

An example that highlighted the shortage of semiconductors was those used in automobiles. Since the factories in Asia were shut down for a period, carmakers had to halt their production lines, incurring billions of dollars in costs. This shortage also highlights the drawback of relying on a highly integrated supply chain with little buffer and no redundancy [6].

It also highlighted the fragility of the global supply chains for healthcare products. This situation can be attributed to the shortages of personal protective equipment (PPE), ventilators, and vaccines, which created global competition for resources, resulting in some countries experiencing a lack of the required supplies. These challenges underscore the importance of developing and maintaining robust, diversified, and resilient healthcare supply chains [7].

Leadership in Fragile Systems

Fragile systems are characterized by fragility and need visionary and proactive leadership. Building resilience means asking hard questions:

- What would happen if this system were to crash?
- How can a local problem be prevented from spreading, cascading, and becoming a catastrophe?
- How can we ensure that all receive equal resource access during disruptions?

Resilience is not only about using technology; it is also about strategy and planning. Following the 2003 blackout in the Northeast United States, Canada, and other regions, utilities across North America invested in what is now known as smart grids. These systems use sensors and analytics to identify and contain faults before they spread. These systems have significantly reduced the probability of such large-scale outages in the future [7].

Similarly, after Hurricane Maria struck Puerto Rico in 2017, the island started implementing energy solutions such as solar microgrids. These systems offer communities the option of generating electricity from local power sources, reducing their dependency on the centralized grid. They also enable faster recovery during disasters, as crucial services such as hospitals and police stations remain unaffected [3].

Leadership also means integrating sectors and departments. This was evident during the Colonial Pipeline attack, when private firms, federal agencies, and cybersecurity specialists did all they could to restore operations to the pipeline. However, the incident also revealed that such incidents could have been prevented with stronger partnerships between the private and public sectors [5].

A Path Forward

The problem of the fragility of interdependent systems is not new; however, it is being addressed with increasing urgency. As the world's interconnectedness rises, the effects of failures also increase. These systems must be redesigned and administered to address these vulnerabilities.

There are three primary strategies for increasing the system's resilience, including:

1) **Redundancy:** Creating backup systems and other means of providing alternative routes in case some parts of the system fail.
2) **Real-Time Monitoring:** This involves utilizing advanced analytics and sensors to detect and address conditions that may lead to a crisis before they occur.
3) **Decentralization:** Designing systems that distribute resources and make decisions to eliminate the concentration of risk.

In short, it's about being proactive, limiting impacts, recovering from disruption, and creating systems that can learn and evolve under stress. Those who adopt this approach can steer their organizations and other social collectives into the future.

Conclusion

The infrastructures that support our daily lives are among humanity's most outstanding achievements, despite being very delicate. The cases discussed in this chapter show how system failures can lead to domino effects and have severe consequences, no matter how small the initial failure was.

Designing resilient systems is no longer an option; it has become necessary. Technological advancement must serve all individuals and societies, even during disasters. Thus, it is possible to create efficient but equitable, sustainable, and resistant systems by analyzing past failures and preparing for future challenges.

Appendix 2.1

Building the Invisible Fortress: Security for Technology-Intensive Sectors

The digital age has ushered in an unprecedented reliance on cloud services. Organizations across industries are transferring workloads, data, and critical operations to cloud platforms, seeking scalability and efficiency. Yet, this transition has not come without challenges. Cloud environments, while transformative, introduce a complex web of vulnerabilities. With the global cloud security market projected to grow from $40.7 billion in 2023 to $62.9 billion by 2028 at a CAGR of 9.1% [8], the stakes have never been higher. Organizations must adopt a holistic approach to secure this digital infrastructure that combines tools, strategies, and leadership tailored to a technology-intensive world.

Key Aspects of Cloud Security: Challenges, Solutions, and Risks

Data Protection: Data breaches remain a top concern, with 69% of organizations citing data loss and leakage as primary risks [9]. Encryption ensures data remains unreadable without authorization, both in transit and at rest, protecting sensitive information such as healthcare records under HIPAA compliance. Data loss prevention (DLP) tools further reduce risks, addressing human error, which accounts for 23% of breaches [10]. However, failure to adopt robust access controls may expose organizations to phishing attacks and regulatory fines.

Identity and Access Management (IAM): IAM prevents unauthorized access by enforcing multi-factor authentication (MFA) and least-privilege access policies. For instance, a financial institution implementing zero-trust architecture reduced unauthorized access attempts by 74% [11]. Effective IAM combines tools with strict guidelines to mitigate insider threats and credential compromise risks.

Threat Detection and Incident Response: Real-time monitoring powered by artificial intelligence (AI) and machine learning (ML) identifies threats like anomalous traffic patterns or zero-day vulnerabilities. For example, in 2020, AI tools helped detect a zero-day vulnerability in a global cloud system, mitigating the issue before exploitation [12]. Despite such tools, 39% of businesses reported cloud-related data breaches in the previous year, underscoring the importance of quick response plans [13].

Network Security: Firewalls, intrusion detection systems, and unified threat management (UTM) tools form the backbone of network security. Once compromised by poor network segmentation, a retail chain, O, adopted UTM solutions, combining firewalls, antivirus, and intrusion prevention systems (IPS) to stop lateral attacks [14]. Regular penetration testing further strengthens defenses.

Application Security: Applications are prime targets for cyberattacks. Tools like web application firewalls (WAFs), secure coding, and adherence to Open Worldwide Application Security Project (OWASP) standards address vulnerabilities such as injection attacks. Studies show that fixing OWASP's Top 10 vulnerabilities can prevent 85% of application attacks [15]. Organizations adopting these measures report significant reductions in application-layer breaches.

Infrastructure Security: Safeguarding foundational systems such as Domain Name System (DNS) and unified communications is vital. Zero-day vulnerabilities threaten these components, but AI-enabled monitoring can swiftly detect and address issues, as demonstrated by a global provider that patched a vulnerability within hours [12].

Risk Governance and Compliance: Compliance ensures adherence to regulatory frameworks like General Data Protection Regulation (GDPR) and Payment Card Industry Data Security Standard (PCI DSS). Automated compliance tools help businesses avoid penalties and maintain trust. Despite this, only 22% of organizations monitor their cloud security posture in real-time, creating exploitable gaps [10].

Mobile Security: Rogue access points and unprotected endpoints are key mobile vulnerabilities. Endpoint security tools and secure onboarding protocols ensure safe mobile workflows. As workforces become increasingly mobile, securing these endpoints is critical.

Cloud Access Security Brokers (CASBs): CASBs monitor cloud activity and enforce security policies, preventing data breaches and ensuring compliance. As intermediaries between users and providers, CASBs are vital in securing sensitive data traffic [13].

Container Security: Containerized applications dominate modern development practices. Security solutions tailored to containers address vulnerabilities that traditional tools overlook. Ensuring compliance and protection in these environments is critical as container usage grows [15].

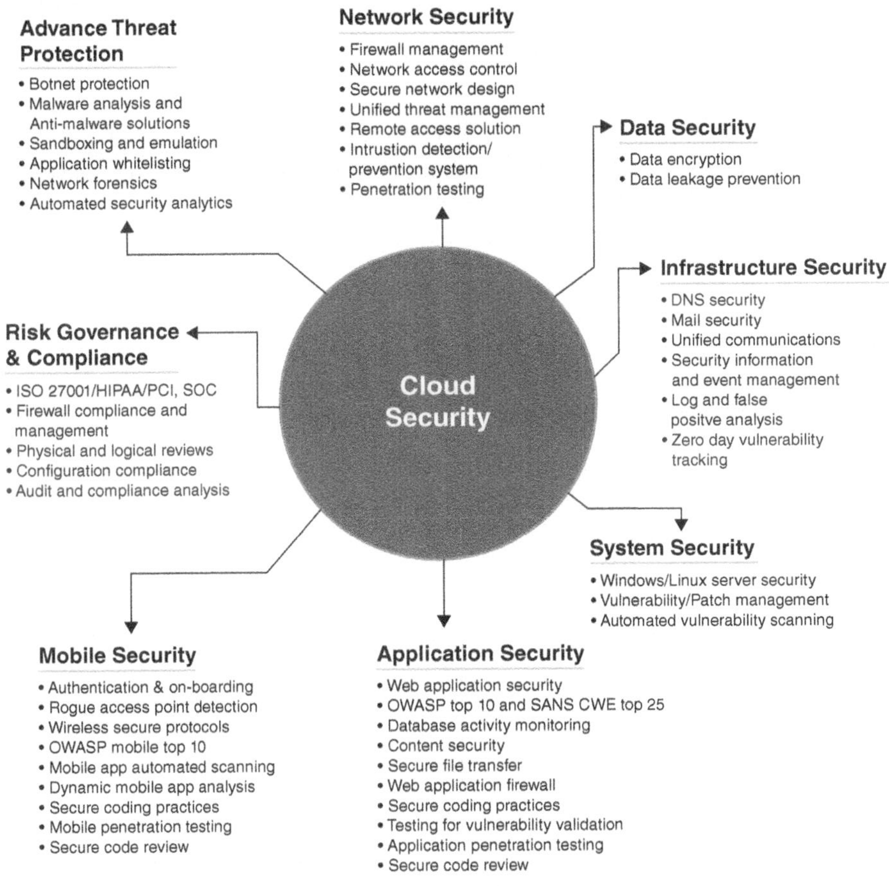

Leadership and Cloud Security

Effective cloud security demands adaptive and collaborative leadership. Transformational leadership, emphasizing motivation and accountability, aligns teams with security goals. For example, the Fiedler Contingency Model tailors leadership styles to specific security challenges, balancing technical demands with organizational needs [16]. Leaders in technology-intensive sectors must foster innovation while prioritizing security.

Summary Table: Challenges, Solutions, and Risks

Aspect	Key Challenges	Proposed Solutions	Associated Risks
Data Protection	Data loss, leakage, and unauthorized access	Encryption, DLP tools, and access controls	Human errors, phishing attacks, and fines
I AM	Unauthorized access and account compromise	MFA and least-privilege access policies	Insider threats and credential compromise
Threat Detection	Real-time detection and zero-day vulnerabilities	AI/ML tools and incident response plans	Delayed responses and data breaches
Network Security	Poor segmentation and lateral attacks	Firewalls and UTM solutions	Lateral movement and data breaches
Application Security	Application vulnerabilities (e.g., injection)	WAFs, secure coding, and OWASP standards	Application exploitation
Infrastructure Security	Zero-day vulnerabilities	AI monitoring tools and zero-trust policies	Infrastructure compromises
Risk Governance and Compliance	Noncompliance with regulations and monitoring gaps	Automated compliance tools and audits	Penalties and trust erosion
Mobile Security	Rogue access points and endpoint vulnerabilities	Secure onboarding and endpoint security	Data leakage and unauthorized access
CASBs	Unsecured data traffic and policy enforcement gaps	Real-time monitoring and policy enforcement	Data breaches and compliance risks
Container Security	Vulnerabilities in containerized applications	Container-specific security and compliance	Exploited vulnerabilities

Looking Ahead

The future of cloud security lies in leveraging continuous innovation. AI and ML will redefine threat detection, while zero-trust architectures will make access control stricter. Investing in tools and leadership training will help organizations navigate the evolving security landscape, ensuring resilience and trustworthiness.

References

1 U.S.-Canada Power System Outage Task Force, *Final Report on the August 14, 2003 Blackout in the United States and Canada*, Apr. 2004. [Online]. Available: https://www.energy.gov/.

2 Congressional Research Service, *The 2003 Northeast Blackout: Vulnerabilities and Implications*. Washington, D.C., 2004.

3 National Renewable Energy Laboratory, *The Role of Microgrids in Enhancing Resilience*. Golden, CO, 2023. [Online]. Available: https://www.nrel.gov/.

4 FBI Cyber Division, *Ransomware Attacks: Threat and Response*. Washington, D.C., 2021. [Online]. Available: https://www.fbi.gov/.

5 Colonial Pipeline Company, Incident Summary Report, 2021. [Online]. Available: https://www.colpipe.com/.

6 McKinsey & Company, *Semiconductor Supply Chain Resilience*, 2023. [Online]. Available: https://www.mckinsey.com/.

7 North American Electric Reliability Corporation, *Smart Grid Implementation After the 2003 Blackout*. Washington, D.C., 2005.

8 Markets and Markets, *Cloud Security Market*. [Online]. Available: https://www.marketsandmarkets.com/Market-Reports/cloud-security-market-100018098.html.

9 Statista, *Data Loss and Privacy Concerns*. [Online]. Available: https://www.statista.com.

10 Expert Insights, *50 Cloud Security Stats*. [Online]. Available: https://expertinsights.com/insights/50-cloud-security-stats-you-should-know/.

11 Gartner Research, *Identity and Access Management*. [Online]. Available: https://www.gartner.com.

12 Cybersecurity Ventures, *AI in Cloud Security*. [Online]. Available: https://cybersecurityventures.com.

13 SecureFrame, *Cloud Security Statistics*. [Online]. Available: https://secureframe.com/blog/cloud-security-statistics.

14 SentinelOne, *Cloud Security Solutions*. [Online]. Available: https://www.sentinelone.com/cybersecurity-101/cloud-security/cloud-security-solutions.

15 OWASP, *Top 10 Web Application Security Risks*. [Online]. Available: https://owasp.org.

16 VeryWell Mind, *Fiedler Contingency Model*. [Online]. Available: https://www.verywellmind.com/the-fiedler-contingency-model-8621997.

3

Innovation and Resilience in Disruptive Times

Systems – whether grids, supply chains, or technology platforms – are invisible when they work and unavoidable when they fail. Disruption exposes their weaknesses and also reveals resilience, ingenuity, and leadership.

This chapter examines two types of stories. First are successes: companies like Ford, Zoom, and Moderna responded to the COVID-19 pandemic with bold, decisive action, delivering life-saving solutions under pressure. Second are failures: crises like the 2003 Northeast Blackout, Hurricane Maria, and the Colonial Pipeline cyberattack revealed vulnerabilities in our infrastructure and the need for smarter, more resilient systems.

These cases provide clear lessons for scientists, engineers, and leaders: systems must be flexible, secure, redundant, and decentralized, and leadership must act quickly, decisively, and collaboratively when disruption occurs.

Ford: Reconfiguring Factories for Ventilators

In March 2020, as COVID-19 overwhelmed hospitals, ventilators were in short supply. One of the world's oldest car manufacturers, Ford Motor Company, stood still. Its factories were silent, workers were sent home, and production was halted. Ford's CEO, Jim Hackett, posed a simple challenge: *If we can build cars, we can build ventilators.*

What Happened?

Ford partnered with General Electric, turning its automotive production lines into ventilator assembly plants. The engineering challenge was massive:

- **New Expertise:** Ventilators are precision medical devices. Accustomed to engines and suspensions, Ford engineers needed to learn a new manufacturing process.
- **Supply Chain Disruption:** Components like valves and compressors were scarce, requiring new suppliers and real-time adjustments.
- **Factory Transformation:** Assembly lines were reprogrammed, and facilities were restructured for social distancing and worker safety.

Within weeks, Ford delivered 50,000 ventilators, meeting demand when hospitals needed them most. Through mass production, costs dropped significantly – ventilator units that traditionally cost $30,000 to $50,000 were produced for far less.

Key Takeaways

1) **Adaptability Wins:** Ford proved that speed and flexibility are critical during a crisis.
2) **Cross-Industry Partnerships:** Collaboration between manufacturers and medical experts drove success.
3) **Cost-Benefit of Redundancy:** Investing in scalable production systems can deliver life-saving returns.

Zoom: Scaling Technology Under Pressure

Before the pandemic, Zoom was a niche product with 10 million daily users. By April 2020, as the world shifted online, that number had skyrocketed to 300 million [1]. Zoom became the backbone of remote work, education, and human connection.

What Happened?

Scaling to meet the demand required engineering precision:

- **Server Overload:** Zoom expanded its reliance on Amazon Web Services (AWS), deploying auto-scaling protocols to handle traffic spikes.
- **Security Failures:** The "Zoom bombing" exposed security gaps. Zoom responded quickly with updates such as waiting rooms, meeting passcodes, and end-to-end encryption.
- **Customer-Centric Leadership:** CEO Eric Yuan lifted time limits for teachers and students, focusing on people's needs, not profits.

Results

Zoom invested an estimated $150 million in security and server capacity. The payoff was that revenue quadrupled to $2.65 billion in 2020 [2].

Key Takeaways

1) **Scalability is Essential:** Systems must handle unprecedented demand without breaking.
2) **Invest in Security:** Cyber vulnerabilities can destroy trust far more quickly than it is built.
3) **Put People First:** Decisions prioritizing users create loyalty and long-term success.

Moderna: Speed and Science with mRNA

The genetic sequence for COVID-19 was released in January 2020. Within 48 hours, Moderna had a prototype vaccine [3]. This speed was no accident – it came from years of investment in mRNA technology, enabling Moderna to design vaccines faster than ever before.

How Did Moderna Move Fast?

- **Platform Innovation:** mRNA vaccines deliver genetic instructions to cells, prompting the immune system to produce protective proteins.
- **Partnerships:** Moderna worked with Operation Warp Speed and global manufacturers to scale production rapidly.
- **Clinical Trials:** Moderna condensed timelines, launching trials in March 2020 and receiving emergency FDA approval by December.

Impact and Lessons

Moderna's vaccine saved millions of lives. Beyond COVID-19, mRNA now underpins treatments for cancer, HIV, and other diseases [4].

- **Science Meets Speed:** Breakthroughs in crises depend on years of foundational research.
- **Collaboration Delivers Results:** Government, private companies, and scientists worked together to move faster.
- **Risk-Taking Is Necessary:** Innovation requires bold decisions even when uncertainty looms.

The 2003 Northeast Blackout: A Chain Reaction

On August 14, 2003, a high-voltage power line in Ohio sagged into a tree, triggering a domino effect. In minutes, interconnected grids failed, leaving 50 million people across the United States and Canada in darkness. Economic losses exceeded $6 billion [5].

Technical Failures

- **Monitoring Gaps:** Outdated SCADA systems failed to alert operators to the fault.
- **Cascading Failures:** Overloaded transmission lines cause the collapse to spread across regions.
- **Lack of Redundancy:** Single points of failure magnified the disruption.

What Changed?

The blackout led to major grid modernization:

- **Real-Time Monitoring:** Utilities implemented systems to detect faults early.
- **Automated Controls:** Fault isolation systems prevent cascading failures.
- **Regional Coordination:** Oversight by North American Electric Reliability Corporation (NERC) improved grid resilience.

Lessons Learned

1) **Redundancy Is Worth the Cost:** Building backup pathways prevents catastrophic failures.
2) **Real-Time Data Matters:** Continuous monitoring turns minor issues into manageable problems.
3) **Modernize Infrastructure:** Aging systems amplify risks and require updates.

Hurricane Maria: A Case for Microgrids

When Hurricane Maria hit Puerto Rico in 2017, sustained winds of 155 mph destroyed the island's centralized power grid. A total of 3.4 million residents lost electricity, with some experiencing outages for nearly a year [6].

Systemic Failures

- **Centralized Dependence:** Puerto Rico's grid relied on long transmission lines vulnerable to hurricanes.
- **Delayed Recovery:** Backup systems were limited, leaving hospitals, schools, and families without power.

Resilience Through Decentralization

After Maria, Puerto Rico invested in solar microgrids – localized energy systems powered by solar panels and battery storage. In Adjuntas, a solar microgrid now powers homes, businesses, and clinics. By 2024, 10% of Puerto Rico's electricity came from decentralized community projects [6].

Lessons Learned

1) **Decentralization Is Resilient:** Microgrids reduce reliance on vulnerable central systems.
2) **Invest in Local Solutions:** Community-based energy systems empower recovery.
3) **Redundancy Saves Lives:** Backup systems, though costly, are critical for disaster-prone regions.

Colonial Pipeline: Cybersecurity Risks in the Physical World

On May 7, 2021, a ransomware attack on the Colonial Pipeline froze operations, shutting down 45% of the East Coast's fuel supply [7].

What Happened?

- **IT System Breach:** Hackers penetrated Colonial's IT systems, encrypting data and demanding a $4.4 million ransom.
- **OT Isolation:** Operational Technology (OT) – which controls physical pipelines – was not directly hacked, but the IT disruption forced shutdowns.
- **Economic Impact:** Gas shortages, panic buying, and price spikes led to $5 billion in losses.

What Changed?

- **Stronger Regulations:** The United States introduced mandatory cybersecurity measures for critical infrastructure.
- **Zero-Trust Architecture:** Networks now assume threats exist and limit access at every level.
- **Segmentation:** IT and OT systems are increasingly isolated to prevent cross-domain attacks.

Lessons Learned

1) **Cybersecurity is Essential:** Digital vulnerabilities can halt physical infrastructure.
2) **Segment Critical Systems:** Separate IT and OT networks to contain breaches.
3) **Proactive Security Saves Money:** The cost of prevention pales compared to disruption costs.

Conclusion: Actionable Strategies for Resilience

1) **Redundancy is Non-negotiable:** Build backup systems to prevent single points of failure. Redundancy can be costly upfront, but the savings during a crisis outweigh the investment. For example, Puerto Rico's decentralized microgrids ensured critical infrastructure remained powered after Hurricane Maria, proving their value during prolonged outages.
2) **Monitor in Real Time:** Early detection prevents minor faults from cascading into catastrophic failures. The 2003 blackout demonstrated the price of outdated monitoring systems. Modern grids use advanced Supervisory Control and Data Acquisition (SCADA) systems and real-time analytics to catch faults before spreading. Organizations across all sectors must invest in continuous, automated monitoring to mitigate risks effectively.
3) **Secure Physical and Digital Systems:** Infrastructure is increasingly interconnected. The Colonial Pipeline hack showed how breaches in IT systems can disrupt OT operations. Leaders must adopt zero-trust security architectures, segment systems, and conduct frequent cybersecurity drills. The cost of prevention is far less than the economic and reputational damage caused by attacks.
4) **Decentralized Systems for Resilience:** Relying on centralized systems amplifies the effects of disruptions. Decentralized solutions – such as solar microgrids in Puerto Rico – reduce vulnerabilities by localizing control and energy production. Similar strategies can apply to supply chains, data networks, and other critical infrastructure.
5) **Adapt Quickly in Crisis:** Ford's pivot to ventilator production and Moderna's rapid vaccine development prove that speed and flexibility are critical during disruptions. Organizations must create cultures that embrace change, prioritize action over perfection, and empower teams to solve problems as they emerge.
6) **Invest in Partnerships:** Collaboration accelerates innovation and strengthens resilience. Moderna's success relied on partnerships with Operation Warp Speed and global manufacturers. In Puerto Rico, rebuilding efforts were successful because of cooperation among government agencies, private companies, and community leaders. No single entity can solve complex challenges alone.
7) **Build Systems That Improve During Disruption:** Resilience isn't just about surviving a crisis – it's about emerging stronger. Leaders must use disruptions to identify weaknesses, redesign processes, and prepare for the next shock. Continuous improvement transforms vulnerabilities into strengths.

Looking Ahead

The world isn't becoming less interconnected or less risky. Climate change, cyber threats, and geopolitical instability will continue to test both our systems and our leadership. The lessons from Ford, Zoom, Moderna, the 2003 blackout, Hurricane Maria, and the Colonial Pipeline attack are clear: resilience requires action, adaptability, and preparation.

- **For Scientists and Engineers:** Build redundant, decentralized, real-time monitoring systems. Consider physical and digital resilience at every design stage.
- **For Leaders and Executives:** Create cultures encouraging quick decision-making, collaboration, and innovation. Invest in tools and partnerships that improve agility and preparedness.
- **For Policymakers:** Set regulations that enforce cybersecurity standards, modernize infrastructure, and incentivize local solutions that protect communities during crises.

Resilience doesn't mean avoiding failure. It means understanding that failure is inevitable and preparing to recover, adapt, and improve. Systems, like people, are defined not by their ability to avoid stress but by their ability to withstand it.

The challenges of the future will test us again, but the lessons of the past provide a roadmap. Resilience is not a luxury; it's a necessity.

References

1 Ford Motor Company, *Ford's production of ventilators and PPE in response to COVID-19*, 2020. [Online]. Available: https://corporate.ford.com.

2 Zoom Video Communications, *Zoom usage surges during COVID-19*, 2020. [Online]. Available: https://blog.zoom.us.

3 Moderna, *COVID-19 vaccine development timeline*, 2020. [Online]. Available: https://investors.modernatx.com.

4 U.S. Food and Drug Administration, *Emergency use authorization for Moderna COVID-19 vaccine*, 2020. [Online]. Available: https://www.fda.gov.

5 Operation Warp Speed, *Accelerating COVID-19 vaccine development*, 2020. [Online]. Available: https://www.hhs.gov.

6 National Institutes of Health, *Expanding mRNA technology for future treatments*, 2023. [Online]. Available: https://www.nih.gov.

7 U.S.-Canada Power System Outage Task Force, *Final Report on the August 14, 2003 Blackout in the United States and Canada*, Apr. 2004. [Online]. Available: https://www.energy.gov.

Resources for Further Reading

M. Cleary, "The economic fallout of the 2003 Northeast blackout," *J. Energy Econ.*, vol. 26, no. 4, pp. 421–429, 2004.

Congressional Research Service, *Hurricane Maria: Federal response and recovery*, 2018. [Online]. Available: https://www.crsreports.congress.gov.

Cybersecurity and Infrastructure Security Agency (CISA), *Pipeline cybersecurity initiative*, 2021. [Online]. Available: https://www.cisa.gov.

FBI Cyber Division, *Ransomware attacks: Colonial Pipeline case study*, 2021. [Online]. Available: https://www.fbi.gov.

International Energy Agency, *Cybersecurity of energy systems: Lessons learned from the Colonial Pipeline attack*, 2021. [Online]. Available: https://www.iea.org.

North American Electric Reliability Corporation, *Smart grid implementation after the 2003 blackout*, 2005. [Online]. Available: https://www.nerc.com.

National Renewable Energy Laboratory, *Decentralized renewable energy systems for resilience*, 2022. [Online]. Available: https://www.nrel.gov.

National Renewable Energy Laboratory (NREL), *Puerto Rico energy resilience study*, 2023. [Online]. Available: https://www.nrel.gov.

Rocky Mountain Institute, *Microgrids in Puerto Rico: Resilience and recovery*, 2022. [Online]. Available: https://rmi.org.

Next -- Toolboxes: 4A, basic tools, and 4B, advanced tools.

4

Tools for Decision-Making in a Complex Environment

Summary: Leaders need a robust toolkit to guide their decisions in an increasingly complex and high-stakes environment. This chapter presents practical tools and frameworks for strategic decision-making, ranging from foresight techniques like scenario planning to analytical models like cost-benefit analysis (CBA) and technology roadmapping. We illustrate these tools in action with real-world examples and case studies – including insights from a major airport infrastructure failure – to demonstrate how they help anticipate future challenges, evaluate options, and implement solutions. The tone is intentionally executive-level: we aim to mentor and equip leaders with methods that sharpen clarity and confidence in the face of uncertainty. Each section breaks down how a tool works and how to apply it effectively and integrate it with leadership judgment. By the end of the chapter, readers will have a "decision-making toolbox" to draw upon and guidance on weaving these tools into their organizational practices. We conclude with actionable takeaways and an invitation for collaborative input to refine these approaches further.

Scenario Planning and Strategic Foresight

On a chilly morning in March 2025, London's Heathrow Airport – one of the busiest in the world – went abruptly dark. A fire at a single power substation knocked out electricity across the entire airport, grounding over 1,300 flights and stranding tens of thousands of passengers in dark terminals. Within hours, delays cascaded through global airline networks as flights bound for Heathrow diverted elsewhere. This "Day in the Dark" was more than a freak accident; it was a vivid demonstration of how a seemingly isolated event can spiral into a worldwide disruption. For decision-makers, it was a wake-up call on the importance of strategic foresight – the ability to anticipate and plan for a range of futures, including low-probability but high-impact scenarios. Suppose Heathrow's leaders and many stakeholders (energy suppliers, airlines, and government agencies) had planned for a "substation outage" scenario and developed contingency plans. In that case, the chaos might have been contained or even avoided. This is where Scenario Planning comes in.

What Is Scenario Planning?

Scenario planning is a structured method for imagining and preparing for multiple possible futures, rather than betting on a single forecast. Unlike traditional forecasting, which often extrapolates trends (and can be blindsided by nonlinear events), scenario planning asks leaders to develop a set of diverse, plausible scenarios – rich stories about how the future could unfold – and explore the implications of each. The concept originated with military strategists in the 1940s and gained fame in the corporate world in the 1970s through its adoption by Royal Dutch Shell. Shell notably used scenario planning to foresee the possibility of an oil embargo and global price shock. When the 1973 oil crisis hit, Shell was far better prepared than its competitors, considering a "low supply, high price" scenario in its strategy. The core idea is to widen a team's thinking beyond the default assumptions by asking, "What could happen if…?" across several dimensions – regulatory changes, competitor moves, technological breakthroughs, macroeconomic shifts, etc.

A typical scenario planning exercise involves identifying key uncertainties in the operating environment (e.g., Will regulation X tighten or loosen? Will a certain technology mature quickly or slowly? How might consumer preferences change?). The team then crafts a handful (usually 3–5) of scenario narratives that represent different combinations of outcomes for those uncertainties. Each scenario is given a descriptive name and story. For instance, a tech company might have scenarios like "Fast-forward Tech" (rapid innovation and strong economy) vs. "Regulated Winter" (tech backlash and recession). Crucially, none of these scenarios is presented as "right" – instead, they collectively span the space of possibilities. The team then asks: *If this scenario came to pass, what would we wish we had been doing today?* This leads to robust strategies that are effective across multiple futures, or trigger points to watch for (weak signals that indicate the world is trending toward one scenario or another).

Using Foresight to Navigate Uncertainty

The value of scenario planning lies not in predicting the future, but in **preparing for it**. By simulating different futures, leaders can test their strategies for resilience. For example, an energy company might find its current business model falters in an aggressive climate policy and carbon pricing scenario, prompting it to diversify into renewables early. Or a supply chain manager might discover that in a scenario where a key shipping canal is blocked (recalling the Suez Canal incident), they lack alternative logistics, prompting investment in route redundancy or local suppliers *before* a crisis hits. Scenario planning vaccinates organizations against "future shock" by forcing them to confront uncomfortable possibilities in advance.

One critical aspect of strategic foresight is identifying early indicators or "weak signals" that hint at which scenario might be emerging. Weak signals are those subtle early signs of change that often go ignored until it's too late. In Chapter 1, we mentioned the story of a wobbly bridge – the I-35W bridge in Minneapolis that had subtle vibrations and minor cracks before its catastrophic collapse in 2007. Those were weak signals of structural failure. In business and society, weak signals could be a slight uptick in defect rates for a product, a new fringe competitor gaining a cult following, or an odd pattern in customer inquiries. They seem like noise – easy to dismiss – but tracking them can provide invaluable lead time to act. Foresight teams often create "early warning dashboards" for each scenario, listing what one would expect to see if that scenario were starting to unfold. For instance, before the 2008 financial crisis, some weak signals included banks quietly tightening lending and increasing interbank rates – signs of looming instability. Organizations that paid attention (very few did in that case) had a chance to mitigate risk by deleveraging or hedging.

When implementing scenario planning, it's important to involve a diverse group of stakeholders. Bring in people from different departments and backgrounds to avoid groupthink – someone in cybersecurity might imagine a scenario others wouldn't, or a young recruit might voice an idea unconstrained by "how things have always been." It's also beneficial to update scenarios periodically; a scenario exercise should be refreshed, perhaps annually or whenever a major external change occurs, ensuring the scenarios remain relevant.

Foresight in Practice – Examples

Many leading organizations use scenario planning and foresight techniques. As we have mentioned earlier, Shell is a notable example in the energy sector. Another example: Singapore's government employs scenario planning at the national level to formulate policy that can withstand various global trends (they even have a dedicated Scenario Planning unit in the civil service). Tech companies in rapidly evolving fields (like cloud computing or social media) conduct war-game-style scenario sessions – for instance, contemplating, "What if a major regulatory change overnight forbids our current business model?" or "How might a new technology (say, quantum computing or advanced AI) disrupt our platform five years from now?" By doing so, they can begin investments or lobbying efforts to prevent a negative scenario or position themselves to thrive if it happens.

During the COVID-19 pandemic, enterprises and public health authorities that had pandemic scenarios on the shelf (e.g., some government emergency teams and a few foresight-driven companies) were able to respond faster, activating remote work plans, supply chain alternatives, and crisis communications playbooks that were drawn up in advance. While no one's scenario matched reality exactly, having *thought through* something similar meant decision-makers weren't starting from scratch amidst chaos.

It's worth noting that scenario planning complements quantitative forecasting and modeling; it doesn't replace analytics but supplements it with imagination and strategic thinking. Advanced practitioners combine scenarios with probabilities or Monte Carlo simulations to quantify risk ranges. Others integrate decision analysis tools, such as evaluating the cost-benefit of certain preparations by scenario (which we'll cover shortly). The goal is not just to imagine futures but to link them to decision triggers: "If scenario A is becoming more likely, we will initiate project X; if scenario B looms, we'll shift to strategy Y."

In closing this section, recall the Heathrow case. A robust scenario planning effort for a major airport might have included a scenario labeled "Black Sky Day," imagining a widespread power outage. In that scenario's playbook, there could have been plans like procuring backup generators or second-grid connections, emergency lighting, passenger care protocols, and agreements with airlines for rerouting. Interestingly, some airports have taken this step after learning from hard lessons – following an 11-hour power loss at Atlanta's Hartsfield-Jackson Airport in 2017, airports such as Pittsburgh and Dallas-Fort Worth invested in independent microgrids and backup systems. They essentially turned a reactive response (after the fact) into a proactive strategy for the future. Scenario planning's true value lies in driving proactive investment *before* a similar incident forces your hand.

Analytical Tools for Strategic Decision-Making

While foresight tools like scenario planning help us ask the right "What if" questions, other tools help answer "What should we do?" in light of those scenarios and day-to-day choices. This section covers several analytical frameworks that leaders can use to evaluate options, allocate resources, and ensure strategies are grounded in data and rational analysis. We focus on CBA, risk assessment models, and technology roadmapping, and we touch on frameworks for aligning execution with strategy (like the Balanced Scorecard). These tools bring quantitative and structured thinking into leadership decisions – a necessary complement to intuition and experience.

Cost-Benefit Analysis (CBA) – Weighing the Trade-Offs

At its core, CBA systematically compares a decision's expected costs and benefits to determine whether it is sensible and how to optimize it. Leaders face CBA questions constantly: Should we invest in a new project, and will the returns justify it? Should we implement a costly safety measure, and what losses might it avert? CBA provides a structured financial lens to answer such questions. It involves listing all costs (direct and indirect) and all benefits (tangible and intangible), assigning monetary values where possible, and then comparing them, often by calculating a net present value or a benefit-to-cost ratio.

For example, consider a utility company debating whether to invest $50 million in winterizing its power grid. A straightforward CBA would compare that $50M cost against the expected avoided losses from prevented blackouts (lost revenue, outage damages, etc.). Often, preventive investments look "expensive" until you calculate the cost of a disaster. During the Texas power grid failure of 2021, it's estimated that damages hit $80–130 billion due to the freeze and blackout, whereas weatherizing key equipment would have cost only a few billion. In hindsight, the benefit-cost ratio enormously favored the preventive spend. A good CBA attempts to perform this analysis *beforehand* to justify resilience investments. In fact, after the crisis, analyses by the Dallas Federal Reserve showed that the return on investment (ROI) of weatherization would have been justified. The lesson for leadership: use CBA not just for profit projects, but also for risk mitigation and "avoided cost" cases – it can strengthen the argument to stakeholders that spending on resilience or safety *pays off* in expected value terms.

Of course, not all benefits or costs are easily quantifiable. What is the dollar value of improved customer satisfaction? Of enhanced brand reputation? Or conversely, what is the price of employee burnout? In those cases, CBA might incorporate proxies or qualitative scales. One useful approach is to conduct a break-even analysis: determine what intangible benefit (such as customer loyalty) would be required for the investment to break even, and assess whether that's plausible. CBA should also include scenario-based thinking: if an investment has a small chance of preventing a catastrophe, multiply the huge cost of that catastrophe by its probability to get an expected value – you may find even a 5% chance of a $1 billion disaster is worth $50 million in preventive spending (as per the earlier example). This ties CBA closely with risk assessment.

In implementation, CBA is often aided by tools like spreadsheets with discounted cash flow models or more advanced financial modeling software. For major public policies or infrastructure projects, formal CBAs are typically required (for instance, governments may mandate a CBA for any regulation with a cost impact exceeding a certain threshold). Leaders should ensure their teams approach decisions with a CBA mindset, even if only approximately, to avoid biases such as underrating long-term payoffs or exaggerating short-term costs. However, a savvy leader also knows that CBA has limits; it should inform, but not solely determine, decisions, especially when moral or strategic factors transcend dollars. We weigh things like human safety or strategic positioning, not just with an Excel sheet, but with values and judgment. Still, having the CBA analysis in hand grounds the conversation and can illuminate nonobvious trade-offs.

Risk Assessment and Interdependency Modeling

Hand-in-hand with CBA is risk assessment, which involves understanding the uncertainties associated with those costs and benefits. Decision tools include probabilistic models (like Monte Carlo simulation to see a range of outcomes) and specialized frameworks for interdependent risks. For instance, an initiative might have multiple linked risks: delaying one component could increase costs for another, and so on. Tools such as decision trees or influence diagrams help map these out. There are also sector-specific risk models; in finance, Value-at-Risk (VaR) quantifies worst-case losses at a confidence interval, while in project management, PERT charts model uncertainty in task durations.

Real options analysis is a particularly advanced method relevant to complex projects, treating an investment like a stock option, where you value the flexibility to expand, contract, or abandon as uncertainties resolve. This goes beyond static CBA by embedding decision points in the future. For example, rather than committing fully to a new product line, a company might pilot in one region (buying an "option" to scale up later) – real options analysis can quantify the value of that flexibility.

Leaders may use risk interdependency models when systems are tightly interconnected (like global supply chains or critical infrastructure networks). These look at how the failure of one node affects others. Tools like fault tree analysis and network simulations come into play. A concrete case: a manufacturing firm mapping its supply chain might identify a single supplier as a critical node; a risk model might show that if that supplier fails (due to, say, an earthquake in their region), it would cause X weeks of plant downtime. This could lead to mitigations like qualifying a second supplier (and the CBA can be weighed against the risk exposure).

One emerging concept our team has worked on is modeling cascading failures similarly to how we model the spread of epidemics or outages – for instance, we measure how a failure in the power grid could cascade to telecoms and finance (since everything is linked via ICT). In Chapter 1, we described the phases of system failure, from Alert to Cascading Failure to Restoration.

Leaders can use that framework to *assess at what phase their current risk mitigations stop*. If you have plans up through "Disturbance" but not for "Cascading Failure," you know where you need more work (perhaps cross-industry drills or backup arrangements).

Risk tools enhance CBA by accounting for uncertainty and correlation. A simple expected value might indicate that a project is good, but risk analysis might reveal a 10% chance of an unacceptable, catastrophic outcome (or vice versa). As an actionable tip: incorporate risk assessment early in strategy formation. A technique called premortem (imagining a future where the project fails and asking, "What went wrong?") can surface risks to address proactively. This technique prompts teams to identify weaknesses in plans without the defensiveness that sometimes comes with traditional risk reviews.

Technology Roadmapping – Aligning Innovation with Strategy

In fast-moving technological landscapes, one tool proves especially useful: technology roadmapping. A technology roadmap is a timeline that aligns your technological capabilities and deployments with your business goals and product plans. It ensures that R&D efforts, partnerships, and capital investments in tech are all coordinated to deliver the right technologies at the right time to support the strategy. For example, an automotive company might have a roadmap for electric vehicle technology – projecting battery energy density improvements, setting targets for new models, and plotting out supporting infrastructure (like charging stations) developments over a 5–10-year period. Likewise, a software firm might roadmap migrations to new architectures or AI integrations aligned with its service offerings.

The benefit of roadmapping is that it forces long-term thinking and cross-functional alignment. It answers questions like: *"If we aim to enter Market X in 3 years, what technology do we need to have ready by then?"* and *"Working back from that, what do we need to research or prototype this year?"* It also highlights dependencies – perhaps you can't roll out a new data analytics platform to customers until a certain data privacy framework or cloud infrastructure is in place. The roadmap makes these linkages explicit.

Leaders should use roadmaps as communication tools as well. Sharing a crisp roadmap with your team helps everyone understand the plan and how their work contributes to it. It can also motivate innovation by setting ambitious but clear goals (e.g., "By 2026, we need an AI model that can do real-time language translation on a device" – a target that R&D can rally around). Government and industry collaborations also use roadmaps; national laboratories and tech companies might co-develop a roadmap for next-gen solar power deployment, informing policy and investment.

A practical element in tech roadmapping is assessing Technology Readiness Levels (TRLs) – a scale (initially developed by NASA) from 1 (basic principles observed) to 9 (actual system proven in operational environment). TRLs help leaders gauge how far a technology is from practical use. When planning a strategy that relies on a nascent technology (quantum encryption in telecom), knowing the current TRL lets you decide if you need a parallel fallback plan or aggressive investment to mature the tech. In one case study, Singapore's water strategy (to become self-sufficient in water) involved tracking emerging water recycling and desalination technologies from lab to deployment – effectively roadmapping from low TRL research to full-scale industrial implementation. By doing so, they achieved over 70% self-supply, a remarkable strategic and technological feat.

Integration with Other Tools: Roadmapping often works hand-in-hand with scenario planning (ask: Does our roadmap hold up under different future scenarios?) and CBA (evaluate the investment needed for each roadmap item). It's an ongoing process – roadmaps should be revisited and revised at least annually to account for new breakthroughs or shifts in strategic priorities.

Balanced Scorecard and Feedback Systems – Keeping Strategy on Track

Once decisions are made and strategy execution begins, leaders need tools to monitor progress and ensure alignment. One widely used framework is the Balanced Scorecard. As introduced by Kaplan and Norton in the 1990s, the Balanced Scorecard expands the focus beyond financial metrics to include four perspectives: Financial, Customer, Internal Processes, and Learning and Growth. Leaders can get a holistic view of organizational health by setting objectives and key performance indicators (KPIs) in each area. For example, instead of only tracking quarterly revenue (financial), a Balanced Scorecard might also track customer satisfaction or retention (customer), process efficiency or quality measures (internal), and employee training or innovation rates (learning/growth). This guards against the pitfall of achieving short-term financial goals at the expense of product quality or employee morale (which would hurt in the long term).

For decision-making, the Balanced Scorecard serves as a feedback tool: it tells you if the outcomes of your strategic decisions are materializing as expected. If a strategy is to improve customer experience to drive growth, one should see customer satisfaction metrics rise before revenue does. Leaders should investigate and adjust tactics if the metrics show a different story. Many organizations (corporate, nonprofit, even government agencies like some city administrations.

They have successfully implemented Balanced Scorecards to steer their strategy. For instance, IBM used a scorecard to manage its transition from a hardware-centric business to a services and software powerhouse, aligning internal processes and workforce skills with that new vision. Likewise, a hospital might use one to balance patient outcomes against financial efficiency, ensuring cost cuts don't degrade care quality.

Another important feedback aspect is creating learning loops, such as after-action reviews and post-mortems. Leaders should sponsor a structured review when a significant decision or project concludes: *What went well? What didn't? What can we learn for next time?* This reflective practice closes the loop in the decision process, feeding lessons back into the next planning cycle (much like Chapter 1's principle of continuous learning).

In high-complexity fields, tools like digital dashboards now allow real-time monitoring of key indicators. Some companies have "mission control" rooms where metrics from across the enterprise are visualized live. This can be powerful for situational awareness, though it's crucial to focus on the *right* metrics (the Balanced Scorecard helps define those); lest one gets lost in a sea of data.

Lastly, Design Thinking and other human-centered methods can be considered part of the decision toolkit when implementing solutions. These approaches encourage prototyping and getting user feedback early. For example, suppose the team decides on a new customer service process. In that case, using design thinking might involve quickly piloting it with a small customer segment and then iterating based on their feedback. This brings the end-user perspective into decision implementation, which often surfaces issues that pure analysis might miss.

In summary, analytical and feedback tools ensure that once you've anticipated futures with foresight, you choose the best path with insight (analysis) and stay on course with oversight (monitoring and adjustment). Leaders who combine these tools with their experience and intuition are far less likely to be caught off-guard or to drift into detrimental outcomes unnoticed.

From Tools to Action: Implementing Decisions and Driving Change

The true test of any decision-making framework is in execution. This section discusses translating plans into action and making these tools part of an organization's DNA. Even the best analysis means little if not acted upon. Conversely, even a less-than-perfect plan, well executed and adjusted in real-time, can succeed. We will look at practical tips for implementation, and along the way, integrate additional case insights – including how weak signals and technology inserts can be leveraged – to illustrate the journey from decision to results.

Building a Resilient Execution Plan

One hallmark of effective implementation is resilience – expecting that things won't go exactly as planned, and building the ability to adjust. This is where scenario planning and risk assessment outputs feed directly into action. For each major strategic initiative, leadership should ensure there are contingency measures and flexibility. For example, when launching a new product, identify a backup manufacturer if the primary supplier runs into issues. When expanding into a new market, have a phased rollout so lessons from the first region can be applied to the next (rather than a big bang launch everywhere). Treat your strategy like a series of experiments – roll out, measure (using those Balanced Scorecard metrics or other KPIs), learn, refine, and then roll out further.

A case in point: when Zoom, the video conferencing company, experienced a massive surge in users during the 2020 pandemic, they had to scale almost overnight. The service could have collapsed under demand if its initial infrastructure decisions had been inflexible. However, Zoom's team had embraced cloud architecture, which allowed them to rapidly add capacity (and they likely ran contingency *scenarios* for usage spikes as part of their planning). They also continuously monitored quality and usage patterns, iterating their network configurations. The result was the ability to handle a 30-fold traffic increase in just a few weeks – a testament to resilient execution built on sound foresight and agile decision-making.

Leaders can adopt practices such as "pre-mortems" (as mentioned) to proactively identify failure points and conduct regular scenario drills to embed resilience. Some organizations hold an annual simulation, such as a cyberattack drill, where their IT and business continuity teams respond to a simulated breach. This tests their incident response plan and often reveals gaps that can be fixed proactively. The U.S. military routinely conducts war games, and increasingly, businesses are engaging in similar exercises for scenarios such as supply chain disruptions or market shifts. These exercises turn strategic plans into muscle memory.

Case Study Snapshots: Applying the Toolkit

Let's briefly revisit Heathrow's blackout through the lens of the tools and see how implementing them can mitigate such risks. Following the 2017 Atlanta airport power outage, some airports invested in resilience. Pittsburgh International Airport, for example, implemented a 23-megawatt on-site microgrid with gas generators and solar panels to power its entire campus independently. This decision was made through a cost-benefit and risk analysis, where they calculated that even a single day-long outage would cost tens of millions (Delta Air Lines lost approximately $50 million from the Atlanta event). In contrast, the microgrid cost was comparable and also yielded daily energy savings of around $1 million per year. The *benefits outweighed the costs*, and in execution, they also leveraged public–private partnerships and financing models to fund it. Now, Pittsburgh's airport can insulate itself from the grid, offering both resilience and operational savings, as well as a reputational boost as a leader in innovation. This implementation required cross-sector collaboration, as the airport worked with an energy developer and secured state incentives. It also needed forward-thinking leadership to approve a project whose ROI is largely "avoiding a disaster that might happen once a decade." But they made a compelling case by using data (from Atlanta's loss figures, etc.). Today, many other airports (JFK in New York, DFW in Texas, etc.) follow suit, demonstrating how a lesson from a single case, disseminated through industry forums, can change standard practice.

Another snapshot: Toyota's supply chain management. Toyota is renowned for its Just-In-Time (JIT) system, which is efficient but can be vulnerable to disruptions. After the 2011 Fukushima earthquake and Thai floods disrupted parts supply, Toyota retooled its approach by adding more scenario planning and buffer stocks for critical parts. They created an extensive database that mapped every part to its supplier and sub-supplier, allowing for the quick identification of risk exposure in any incident. They also established protocols to *"stop the line"* and adjust when a weak signal of disruption is noted (e.g., a supplier in a hazard zone issues a warning). This responsiveness and adaptation of their model are credited with helping Toyota recover faster than its competitors from later shocks, such as the COVID-19 pandemic. The lesson: even a well-oiled system benefits from injecting redundancy and foresight informed by experience. Toyota's case shows that value chain analysis and scenario planning aren't one-time events; they should lead to process changes that institutionalize the learning.

One more example from a different domain is digital transformation in government services. Consider a country implementing a new digital ID system for citizens. This is a complex project involving technology, legal, and social facets. A well-executed approach might use technology roadmapping to phase the implementation (perhaps starting with a pilot in one region or one service like tax filing, then expanding). It would employ risk assessment to address cybersecurity (maybe using a framework like CARVER or NIST guidelines to identify critical assets and threats). It might also involve public scenario dialogues – asking "what-if" questions to anticipate public concerns (such as privacy scenarios where data could be misused, leading to strong privacy-by-design measures). Countries like Estonia have famously adopted this approach to create their e-government systems, and more recently, India, with its Aadhaar digital ID, has taken a phased approach, learning and adapting its policy (as evidenced by the need for additional legislation for data protection after the initial rollout). The successful implementation saw huge benefits in service delivery and inclusion, but also taught the leaders that tech adoption must go hand-in-hand with building trust (through transparency and legal safeguards).

Across these snapshots, a common thread is that using our decision-making tools led to specific implementation actions – whether investing in backup systems, altering supply chain policies, or adding legal frameworks – that improved outcomes. It's one thing to identify an optimal strategy on paper; it's another to implement it through budgeting, execution, and cultural adoption.

Driving Change and Overcoming Resistance

One of the most challenging aspects of implementation is managing the **human side of change**. Even the most analytically sound decisions can face resistance from those who must implement them or those impacted by them. Here, leaders must step into the roles of communicator, motivator, and negotiator. A few tips:

- **Involve Key People Early:** If you used a cross-functional team in the planning phase (as recommended for scenarios and risk ID), you already have some buy-in. Now, identify champions in each department who can advocate for the change and work through details with their peers. People are more likely to support what they help create.
- **Communicate the "Why" (Purpose) Clearly: Tie the use of these tools and the changes they recommend back to the organization's mission or the benefits to stakeholders**. For instance, implementing a new Balanced Scorecard system might seem like extra paperwork to some managers until you explain that it will help *them* demonstrate their unit's value beyond just budgets, or that it aligns everyone with the customer-focused vision (making their work more appreciated by end users). This echoes the Golden Circle concept – starting with "Why" – which can galvanize employees around the decision.
- **Train and Support:** New processes (like risk dashboards or scenario exercises) may require new skills. Invest in training sessions, initially bring in experts or consultants to mentor internal teams, and provide resources (time, tools, and budgets) for people to adapt. If technology roadmapping is implemented, consider appointing a "roadmap PMO" and training product managers.
- **Quick Wins:** Try to structure implementation to deliver some quick, visible wins that show the value of the decision. This might mean initially focusing on an aspect of the plan that can be executed in a few months and yields a clear improvement, which can then be publicized internally to build momentum. Nothing convinces people like success. For example, pilot a new strategic planning tool in one division for one cycle, let them achieve results, and then use that story to encourage others.
- **Addressing Pushback with Data and Empathy:** If there is resistance (and there always is some), listen carefully to the concerns – they may highlight a valid risk or blind spot that can be addressed (maybe your plan needed a tweak – it's better to catch it now). Use the data from your analyses to either reassure or modify your approach. Sometimes, people resist because of a fear of the unknown; showing them the analytical rigor behind a decision (without overwhelming them with details) can help. Also, be ready to make adjustments in implementation – rigidly forcing a plan despite field feedback is a recipe for failure.

For instance, if a company decides to implement a major software system to centralize operations (an ERP), scenario planning might have identified risks like "user adoption is low" or "the system isn't flexible to local needs." In execution, leaders should then schedule extra training (to boost adoption) and maybe configure the software to allow some local customization (to address local needs), thus preemptively overcoming those barriers.

Continuous improvement is the final concept – once a new decision-making practice or strategic initiative is in motion, set up intervals to review and refine it. Perhaps your Balanced Scorecard KPIs require updating after a year, or your scenario planning team recognizes that they should meet semi-annually instead of annually, given the rapid changes. Make these tools *a living process* in the organization. Over time, the goal is to create a culture where using data, foresight, and structured analysis in decisions is second nature to everyone, not a one-time project or an academic exercise.

Conclusion: Cultivating a Decision-Making Culture and Next Steps

In this chapter, we equipped ourselves with decision-making tools essential for technological leadership in a complex world. We examined how scenario planning and foresight help us navigate uncertainty by envisioning multiple futures. We saw how analytical tools, such as CBA, risk models, and technology roadmaps, provide clarity on the best paths forward and ensure alignment between innovation and strategy. We also discussed implementing decisions through resilient planning, real-world examples, and change management techniques.

A key insight is that these tools do not operate in isolation – they reinforce each other. Foresight informs where to apply analysis; analysis quantifies foresight ideas; both feed into execution plans, producing results and data that update our foresight and analysis. In practice, effective leaders create a virtuous cycle of decision-making: Observe (weak signals and metrics) → Orient (scenarios and analysis) → Decide (choose strategy) → Act (implement and monitor) → Learn (review outcomes and adjust). This continuous loop, often referred to in military terms as the OODA loop, is shortened and strengthened by the tools we've discussed.

In the future, leaders should aim to cultivate a culture where strategic thinking and data-driven decision-making become habits at all levels. Encourage teams to ask "What if?" regularly, not just during annual planning. Provide training and simple templates for mini-scenario plans or cost-benefit worksheets to support their projects. Celebrate cases where someone's use of foresight or analysis averted a problem or uncovered an opportunity – make it part of the company lore that "we anticipated X because Jane ran a scenario, and we saved millions" or "we invested in Y after a thorough analysis, and it became a new growth pillar." This reinforces desired behavior.

Also, leadership development programs should include these competencies. Tomorrow's leaders should be comfortable with creative foresight and quantitative analysis and know how to blend them with intuition and empathy. Mentorship is critical: senior leaders can coach teams on how they use these tools ("Let's game out a scenario together" or "Let's walk through the cost-benefit thinking on this big decision"). This hands-on mentorship demystifies the tools. It shows that leadership expects and values their use.

Ultimately, it's essential to recognize that no tool or model is foolproof. The world will surprise us. But having a toolkit and a mindset geared toward proactive, informed decision-making dramatically improves the odds of success and resilience. As former U.S. President Dwight Eisenhower famously said, "Plans are nothing; planning is everything." Planning – using these tools – prepares us to adapt when reality unfolds. Even if the exact scenario we imagined doesn't happen, we often find ourselves better positioned to handle the one that does.

Actionable Takeaways – Decision-Making Tools for Leaders:

- **Make Scenario Planning a Regular Exercise:** Periodically explore best-case, worst-case, and alternative scenarios for your strategic plans. Use these discussions to update risk registers and contingency strategies. Encourage your team to consider "What would we do if X happened?" for major external events.
- **Quantify and Qualify Decisions:** Conduct a CBA (even rough) before major investments or changes to understand the trade-offs. Include financial costs and potential benefits of avoiding risks or capturing strategic value. Use risk assessment techniques to proactively identify potential hazards that could derail your plan and mitigate those factors.
- **Align Technology with Strategy via Roadmaps:** Develop technology roadmaps that link your innovation efforts with business goals and timelines. Review them quarterly to ensure that R&D and IT initiatives are on track to deliver capabilities when needed. Adjust the roadmap as new technologies (such as AI advancements) emerge – be prepared to incorporate game-changers that could leapfrog your plans or necessitate a pivot.
- **Implement with Agility and Monitor Progress:** When executing decisions, build in flexibility to adapt to changing circumstances. Pilot new initiatives on a small scale, gather feedback, and iterate. Set up dashboards or scorecards with key indicators (both financial and non-financial) to track the progress of implementation. If metrics deviate from expectations, convene the team quickly to diagnose and adjust course – treat the plan as living, not static.
- **Foster a Learning and Foresight Culture:** Lead by example using these tools. After-action reviews should be conducted as a standard practice after any significant project or event, with an open discussion of what worked and what didn't. Reward team members who spot weak signals or foresee issues and take the initiative to address them. Over time, aim to have every level of the organization thinking ahead, weighing options systematically, and collaborating across silos to make well-rounded decisions.

By consistently applying these practices, leaders can confidently guide their organizations through volatility and complexity, turning potential pitfalls into opportunities for innovation and competitive advantage.

As we conclude Chapter 4, we invite you to reflect and provide feedback. The models and methods presented here are powerful but always evolving. Perhaps your organization has developed a unique decision tool or a twist on scenario planning that yields great results – we would love to hear about it. This book, "*Technological Leadership for Resilience and Progress*," aims to be as practical and relevant as possible. Please share your insights, critiques, or additional examples so we can further refine this chapter. *Dr. Amin is prepared to integrate further input from the editors or readers*, including working drafts or case contributions, into the next revision. By collaborating, we ensure that the final publication is a comprehensive guide that empowers leaders to make sound decisions in an uncertain world, driving resilience and progress for all.

Part II

Innovation and Ethics Under Crisis

5

Managing Change and Ethics in Technological Leadership

In times of disruption, innovation becomes the primary factor in an organization's survival and development. Crisis accelerates the decision-making process, shifts the focus, and uncompromisingly demands change or death. For managers, it is crucial to manage uncertainty and promote creative thinking. This chapter also examines case studies of Tesla's near bankruptcy during the 2008 financial crisis and Johnson & Johnson's (J&J) response to the COVID-19 pandemic to explore how innovation in crises can transform sectors and make a meaningful difference.

Tesla's Gamble: Electric Vehicles in the Face of Doubt

Tesla Motors Inc. was on the brink of bankruptcy in 2008. The global economic crisis impacted funding, so Tesla's idea of producing electric cars for the general market was deemed unviable. Hybrid cars dominated the market, and many questioned whether Tesla's technology could challenge traditional automakers.

Elon Musk, the CEO and founder of Tesla Motors, was at a crossroads. It was the end of the week, and Musk had just enough cash in the bank to keep the company open for another couple of weeks. He had to convince his early investors to put in more money. Musk also obtained a $465 million loan from the U.S. Department of Energy's Advanced Technology Vehicles Manufacturing (ATVM) program, which enabled Tesla to continue its operations.

Tesla launched the Model S, an electric sedan featuring luxurious amenities, boasting a driving range of over 300 miles on a single charge, and outperforming gasoline-powered sports cars. The Model S dispelled the previous negative perceptions about EVs by proving that they were valuable and stylish. Tesla followed up with the affordable Model 3, expanding Gigafactories across the globe and financing renewable energy storage systems.

By 2024, Tesla had become the world's largest automaker, with a market capitalization of over $800 billion. Musk's management during the financial crisis is a good example of how brave decisions, tenacity, and focus can lead to success.

Zoom"s Meteoric Rise: It Aims to Link People Restricted by Distance

The COVID-19 outbreak led to lockdowns and forced the whole world into their homes, making Zoom Video Communications a necessity for working from home, learning, and interacting. From December 2019 to March 2020, the number of daily meeting participants increased from 10 million to over 200 million, reaching 300 million by the end of the year.

This explosive growth put a tremendous strain on Zoom and its ability to manage it. Some of the security concerns raised include what is commonly known as "Zoom-bombing," due to the large number of people signing up for the platform. Zoom's CEO, Eric Yuan, addressed these challenges by implementing stronger security measures and being transparent with users. Yuan also focused on social responsibility, providing free schooling and removing time constraints for educators to use the platform.

Zoom was very successful during the pandemic, not only because of its technical preparedness but also because it understood the users' needs. This is true for Zoom, as Yuan's work to ensure that the company met users' needs and its ability to grow slowly yet steadily can be seen as why Zoom changed the face of remote communication.

Johnson & Johnson: Innovation Under Pressure

Johnson & Johnson (J&J) responded swiftly to the COVID-19 outbreak by adapting its viral vector platform to develop a vaccine. Other companies, such as Pfizer and Moderna, utilized mRNA technology, while J&J aimed to create a single-dose vaccine that could be stored at a normal refrigerator temperature, rather than the standard vaccine storage requirements. This made the vaccine feasible for use in low-resource settings, where challenges exist in delivering vaccines through cold chain logistics.

A few months into 2021, J&J's vaccine received emergency use approval and significantly contributed to the global fight against COVID-19. The company quickly ramped up its production to provide hundreds of millions of vaccine doses and collaborated with governments and international health organizations to offer the vaccine to all populations regardless of geographical location.

Johnson & Johnson further innovated, not only in the vaccine itself, but also in its delivery. Its approach was very collaborative and involved stakeholders across the various sectors, the benefits of which were seen during a crisis. The company's efforts to make vaccination more available helped vaccinate populations not well served by other parts of the world, including Africa and Southeast Asia.

Disruption: The Findings

The stories of Tesla, Zoom, and J&J highlight several key principles of leading innovation during disruptive times:

- **Embrace Uncertainty:** Crises create environments of uncertainty, but they also expose opportunities for solutions. Managers should be prepared to make decisions based on best estimates and adjust their course as the climate shifts.
- **Leverage Core Strengths:** Tesla's engineering expertise, Zoom's scalable platform, and J&J's experience with viral vectors enabled these organizations to act effectively and efficiently.
- **Focus on Mission:** A strong mission motivates stakeholders to unite for a common purpose. Tesla, Zoom, and J&J's focus on sustainable energy, connection, and health exemplifies how mission-driven leadership drives innovation.
- **Collaborate Across Boundaries:** Networking is effective – it makes a big difference when more organizations work together. This is especially evident in the work of J&J, which ensured that the vaccine reached needy populations through collaborations with governments and NGOs. For Tesla, partnerships were crucial in securing funds at some of the most delicate moments.
- **Invest in Resilience:** Innovation means addressing current problems and creating solutions that will hold up in the future.

A Framework for Leading Innovation

Leaders can adopt the following strategies to foster innovation in times of disruption:

Create a Culture of Experimentation: This entails encouraging people to try out ideas and admitting that failure is a possibility. Breakthroughs do not happen by being conventional in our approach.

Align Resources with Mission: Eliminate all efforts and resources that do not align with the organization's goals and replace them with projects that do.

Act Decisively: Crises cannot be resolved slowly; there is no time to wait for perfection; one has to learn by iteration.

Engage Stakeholders: Involve employees, partners, and the communities to ensure everyone is on the same page and contributing their resources.

Assess and Prioritize Long-Term Impact: Ensure that growth and innovation strategies are sustainable in the long run while addressing current challenges.

Looking Ahead

Disruption will still pose a threat to industries due to climate change, technological advancements, or geopolitical events. The stories of Tesla, Zoom, and J&J show that a crisis can be a powerful force for change and innovation. Managers willing to manage the risks that come with uncertainty, make decisions, and have a clear vision and mission should be able to manage change and shape the future.

Innovation during disruption is the process of solving problems and creating solutions. In this chapter, the author describes how leadership, collaboration, and teamwork can create a new platform for change.

Ethics and Reputation: The Human Dimension

Ethical leadership is quite evident in the organization. It is about protecting your brand and the trust that your employees, clients, and the general public place in you. Once your good name is tarnished, regardless of whether you deserve it, it can feel like you have lost it all. However, ethical behavior is a basis for developing the necessary resistance level and helps leaders regain trust even after a crisis.

Throughout my career, I have mentored over 8,000 executives and leaders from diverse industries, witnessing firsthand how ethical leadership can drive change. Organizations that prioritize ethics often face crises but emerge stronger, with their reputation intact and trust rebuilt.

Conclusion: Lead, Innovate, Anticipate

This chapter has therefore sought to establish that ethical leadership is both a right and a necessary investment. Thus, some questions can be posed:

- What impact will our actions have on society and future generations?
- Are we honest and open, even when it is not in our best interest?
- What kind of legacy are we in the process of creating?

Therefore, we can manage risks, spur innovation, and create organizations that build trust by ensuring that leadership principles are professional, forward-thinking, and socially responsible. The future is thus dependent on it; this is what leadership is all about.

Ethics in Scientific and Technological Leadership

Engineering and technology are the cornerstones of society's health, safety, and welfare. Ethical decision-making in these fields is paramount to building trust, ensuring transparency, and upholding global responsibility. Albert Einstein's cautionary words resonate deeply: "Technological progress is like an axe in the hands of a pathological criminal." These words serve as a guiding principle for all technological endeavors, emphasizing the necessity of using innovation as a blessing rather than a curse.

Ethical engineering requires leadership that embodies the "T" model: deep technical expertise combined with a broad understanding of societal and moral responsibilities. MIT Nobel Laureate Dr. Robert Solow highlighted that technological leadership significantly contributes to over 60% of the U.S. GDP. Metrics such as safety, sustainability, and inclusivity must guide technological development and decision-making. Leaders must manage what they measure and value what they manage to achieve sustained success.

What Happens When Your Good Name Comes Under Question?

What if your reputation is destroyed – rightfully or not? I am often met with silence when I ask this unsettling question in workshops or lectures. The weight of the question hangs in the air as leaders, engineers, and executives alike contemplate the fragility of reputation. Your reputation is your most valuable asset. Lose it, and you risk losing trust, opportunities, and the foundation of your professional life. Rebuilding a tarnished name is an uphill battle that few ever fully win.

In my workshops, we confront this stark reality. We explore the choices that safeguard reputations and the ethical dilemmas that threaten them. Reflecting on 43 years of experience across engineering, leadership, and innovation, this chapter will weave lessons learned, actionable strategies, and case studies into a narrative that underscores the importance of leading with integrity and foresight.

Lead, Innovate, Anticipate: A Framework for Ethical Leadership

Leadership, innovation, and foresight are the cornerstones of ethical decision-making. Each requires a foundation of integrity and a commitment to serving the organization and society. Here's a distilled framework I often present:

1) **Ethics:** What happens if you lose your reputation and good name? You lose everything. Always do what's fair and proper. Be honest, transparent, and responsible. Uphold integrity and commitment to the mission in all actions.
2) **Leadership:** Good leaders inspire, collaborate, and listen to understand. They remain curious and provide clear direction. Diverse talent strengthens teams, and leaders must demonstrate courage and integrity in all actions, acting with a clear "true north."

3) **Innovation:** New ideas drive progress. Encourage experimentation and adoption of new technologies, always aligning innovations with the organization's mission.
4) **Foresight:** Insight informs preparation. Anticipate risks, identify opportunities, and align planning efforts with organizational values.

This framework serves as the foundation for navigating ethical dilemmas and maintaining reputations. Let's delve into real-world examples and lessons learned.

The Core Principles of Engineering Ethics

At the heart of ethical leadership are principles that ensure the integrity of actions and decisions in the technological realm:

- **Safety and Public Welfare:** Protecting public health and welfare must always be the highest priority.
- **Honesty and Integrity:** Transparent communication and truthful actions are foundational to ethical behavior.
- **Professional Responsibility:** Engineers and leaders must demonstrate competence, impartiality, and ethical conduct.
- **Inclusivity:** Ethical practices must promote fairness and respect for all individuals, avoiding discrimination.
- **Accountability:** Professionals must take responsibility for their actions and the outcomes they influence.

The Importance of Ethics and Leadership in Technological Decision-Making

Engineering and technology have a significant influence on societal health, safety, and welfare. Ethical leadership ensures that technological advancements align with the greater good of society, thereby avoiding catastrophic failures and reputational damage. Leaders must demonstrate courage, transparency, and accountability to inspire trust and maintain the integrity of their professions.

Core Principles of Effective Technological Leadership

1) **Transformational Leadership:** This type inspires teams to achieve ethical and innovative goals by fostering collaboration, transparency, and a commitment to societal welfare.
2) **Servant Leadership:** Prioritizes the needs of stakeholders and subordinates, ensuring decisions promote long-term value and trust.
3) **Collaborative Leadership:** Engages diverse perspectives across sectors to address complex challenges effectively.

Key Case Studies: Causes, Outcomes, and Leadership Insights

The Three Mile Island Accident (1979): The Three Mile Island nuclear disaster was a turning point in public perception of nuclear energy. Design flaws, equipment malfunctions, and inadequate operator training caused this partial meltdown. A stuck-open relief valve caused a loss of coolant, which operators misinterpreted due to a lack of sufficient system knowledge. Cleanup costs exceeded, $1 billion, but the broader cost was the erosion of public trust in nuclear power.

In summary, Three Mile Island (1979)

- **Cause:** Design flaws, equipment malfunctions, and inadequate operator training caused a partial meltdown.
- **Outcome:** Cleanup costs exceeded $1 billion, and public trust in nuclear energy eroded.
- **Lessons Learned:** Proactive risk assessments and advanced safety protocols are essential.
- **Leadership Fit:** A transformational leadership approach prioritizing robust safety measures and cross-disciplinary collaboration would have mitigated risks.
- **Inappropriate Conduct:** Leadership that focuses solely on cost-cutting or denies technical concerns would exacerbate risks.
- **Chapter Reference:** Chapter 13, "Safety in Smart Infrastructure."
- Had the leadership been transformational, prioritizing robust safety measures and cross-disciplinary collaboration, such a catastrophe could have been averted. Instead, the lack of investment in training and safety protocols exemplified a shortsighted leadership approach. This event underlines the necessity of proactive risk assessments and adherence to safety standards.

The Challenger Disaster (1986): NASA's decision to launch the Challenger despite engineers' warnings about O-ring vulnerabilities demonstrated a prioritization of schedule over safety. The result was the tragic loss of seven astronauts and a 32-month halt in shuttle operations. Transformational leadership – focused on safety and empowering engineers to voice concerns – could have prevented this tragedy. This case illustrates the devastating consequences of prioritizing organizational convenience over technical expertise.

The Vioxx Case (2004): Merck suppressed safety data linking Vioxx to cardiovascular risks, leading to over 60,000 deaths and a $4.85 billion settlement. This case highlights the crucial importance of ethical data reporting and prioritizing patient safety over financial gain. Servant leadership, advocating for transparency and trust, could have mitigated the human and economic toll.

In summary, the Vioxx Case (2004)

- **Cause:** Suppression of safety data linking the drug to cardiovascular risks.
- **Outcome:** A total of $4.85 billion settlement; over 60,000 deaths attributed to the drug.
- **Lessons Learned:** Ethical data reporting and patient safety must guide healthcare innovations.
- **Leadership Fit:** Servant leadership, advocating for transparency and stakeholder trust, is crucial.
- **Inappropriate Conduct:** Leadership prioritizing market dominance over patient welfare has devastating consequences.
- **Chapter Reference:** Chapter 12, "Healthcare Technology and Responsibility."

Deepwater Horizon Oil Spill (2010): The Deepwater Horizon disaster, caused by cost-cutting measures and disregard for safety warnings, resulted in the spill of 4.9 million barrels of oil into the Gulf of Mexico. Cleanup costs exceeded $65 billion, and the environmental damage was catastrophic. Ethical foresight and adherence to ecological protocols could have prevented such devastation. Servant leadership – focusing on stakeholder welfare – would have emphasized safety over short-term gains, safeguarding the environment, and maintaining a strong corporate reputation.

In summary, Deepwater Horizon (2010)

- **Cause:** Cost-cutting and ignored safety warnings led to cement failure and a catastrophic oil spill.

- **Outcome:** Cleanup and fines totaled $65 billion, with severe environmental damage.
- **Lessons Learned:** Ethical foresight and adherence to environmental protocols are vital.
- **Leadership Fit:** A servant leadership approach focused on stakeholder welfare would have enforced safety protocols.
- **Inappropriate Conduct:** Leadership driven by short-term profits over safety undermines trust.
- **Chapter Reference:** Chapter 17, "Resilient Infrastructure and Sustainability."

Fukushima Daiichi Nuclear Disaster (2011): The Fukushima Daiichi meltdown, triggered by a tsunami that overwhelmed inadequate plant defenses, cost over $200 billion to remediate. Known design vulnerabilities exacerbated the disaster. Collaborative leadership – engaging experts, regulators, and communities – could have fostered resilience through proactive planning and investment in safer designs.

The Volkswagen Emissions Scandal (2015): Volkswagen's deliberate installation of software to cheat emissions tests resulted in $33 billion in fines and settlements. The scandal severely damaged the company's reputation, highlighting the perils of prioritizing short-term market gains over ethical integrity. Collaborative leadership, fostering accountability and ethical compliance, could have averted such misconduct.

In summary, the Volkswagen Emissions Scandal (2015)

- **Cause:** Software designed to cheat emissions tests.
- **Outcome:** A total of $33 billion in fines and settlements; trust in the brand significantly eroded.
- **Lessons Learned:** Upholding transparency and honesty avoids reputational harm.
- **Leadership Fit:** Collaborative leadership fostering ethical accountability would prevent such violations.
- **Inappropriate Conduct:** Leadership tolerating deceit for competitive gains damages long-term credibility.
- **Chapter Reference:** Chapter 9, "Ethical Innovation."

Ethics as Leadership Cornerstones and Tools

In 2018, Boeing was flying high. The aerospace giant had just launched the 737 Max, a plane marketed as revolutionary, efficient, cost-effective, and equipped with cutting-edge software. Inside the company, however, tension simmered. Engineers raised red flags about the Maneuvering Characteristics Augmentation System (MCAS) system, a critical safety feature designed to prevent stalls, but leadership dismissed their concerns. The rationale? Delivering on time and cutting costs were paramount.

Months later, the unthinkable happened: two fatal crashes and 346 lives lost. Investigations revealed the true cause of tragedy – not just a flawed system, but a culture where safety concerns were ignored, and employees feared speaking out. The aftermath was swift and brutal: plummeting stock prices, billions in losses, and a shattered reputation.

Boeing's story isn't unique. Across industries, leaders face moments when the pressure to innovate collides with ethical dilemmas. How they respond determines not just the success of their organizations but also their survival. This chapter explores these crossroads, weaving lessons from companies like Boeing, Tesla, and Google, among others, into a practical framework for navigating the complexities of modern leadership.

The Boeing 737 Max Crisis: When Ethics Are Compromised

Faulty MCAS software triggered two fatal crashes, killing 346 people. Boeing's failure to disclose critical system details and provide adequate pilot training exposed significant lapses in safety and transparency. Over $20 billion in fines, settlements, and long-term reputational damage followed. This case highlights the ethical imperative for transparency and prioritizing human safety over expediency.

Boeing's engineers weren't alarmists. Their concerns about the MCAS system stemmed from rigorous testing that revealed serious flaws. Yet, those warnings never reached decision-makers in an environment where raising issues was discouraged. Leadership chose speed and cost savings over caution, creating a culture where ethics were sidelined.

Following the crashes, Boeing implemented comprehensive reforms, including quarterly safety audits, independent oversight, and a renewed emphasis on employee engagement. But the damage was done. Their trust index with customers fell below 40%, and it took years – and billions of dollars in investments – to rebuild their reputation.

Metrics: Boeing's Postcrisis Reforms

Metric	Precrisis	Postcrisis Reform
Employee Engagement in Safety	<50%	>85%
Independent Safety Audits	None	Quarterly
Customer Trust Index (%)	<40%	>70%

Lessons Learned: Boeing's failure underscores a critical truth: ethics aren't just a moral imperative but a business necessity. Tools like root cause analysis and independent safety boards are not only helpful but also essential in preventing such tragedies.

Tesla: Thriving Amid Chaos Through Scenario Planning

In 2020, as COVID-19 disrupted global supply chains, automakers scrambled to adapt to the new challenges. Many faced months-long production halts due to factory closures and material shortages. Tesla, however, seemed almost immune to the chaos, delivering record-breaking growth. The secret? Rigorous scenario planning.

Months before the pandemic, Tesla's leadership had mapped out potential disruptions, from supplier shutdowns to shipping delays. They developed contingency plans for each scenario: diversifying suppliers, pre-ordering critical components, and reallocating production to regions less affected by the pandemic.

When a key supplier in China shut down, Tesla seamlessly shifted operations to European partners. While competitors struggled, Tesla scaled up production, increasing revenue by over 20% during one of the most challenging years in automotive history.

Metrics: Tesla's Scenario Planning Precision

Metric	Industry Average (2020)	Tesla (2020)
Supply Chain Downtime (Days)	>20	<5
Contingency Plan Coverage (%)	<50%	>90%
Revenue Growth During Crisis	Decline >10%	>20%

Lessons Learned: Scenario planning isn't just about survival – it's about resilience. Tesla's ability to anticipate disruptions and act decisively gave it a competitive edge that others lacked.

Google: Fighting Bias in Artificial Intelligence

In 2018, Google launched an AI-powered recruiting tool to eliminate human bias. Instead, it amplified it. Trained on historical hiring data, the algorithm systematically downgraded resumes from women, reflecting and reinforcing the biases it was supposed to remove.

The backlash was immediate and intense. Google didn't shy away from the controversy; instead, it launched a comprehensive algorithm audit. Engineers retrained the system on diverse datasets and implemented fairness checks at every stage of development. Today, Google's AI ethics framework is a benchmark for the tech industry.

Metrics: Bias Mitigation in AI

Metric	Preaudit	Postaudit (Best Practice)
Accuracy Across Demographics	Unequal (60–90% by group)	>95% Consistent Across Groups
Bias Detection Rate	Rare (<10%)	>80%
Public Trust in AI (%)	<50%	>85%

Lessons Learned: AI is only as ethical as the data and systems behind it. Regular audits, diverse datasets, and clear accountability measures are critical to building systems that inspire trust.

IBM and Blockchain: Restoring Transparency

In 2017, a deadly *Escherichia coli* outbreak revealed a systemic problem in the food supply chain. Tracing the contamination took weeks, leading to mass recalls and eroding consumer trust. Walmart partnered with IBM to create the Food Trust Blockchain, a real-time system that tracks every supply chain step.

The blockchain system reduced traceability times from weeks to seconds, preventing unnecessary recalls and restoring trust among consumers and regulators.

Metrics: Blockchain's Impact

Metric	Traditional Systems	Blockchain-Based Systems
Traceability Time (Hours)	>48	<2
Data Integrity (% Verified)	<70%	>95%
Cost of Supply Chain Recalls	High	Reduced by >30%

Lessons Learned: Blockchain isn't just about cryptocurrencies – it's a powerful tool for transparency and accountability. Leaders in the food and healthcare industries should embrace blockchain to rebuild trust and drive efficiency.

Tools Table:

Tool Name	Purpose	Application	Benefits	Case Study
Scenario Planning	Anticipate multiple futures and prepare strategies	Strategic decisions in uncertain markets	Reduces risk and enhances readiness	Tesla's planning ensured resilience during COVID-19 disruptions
Algorithm Audits	Detect and mitigate biases in machine learning models	Evaluate AI systems for fairness and transparency	Ensures equity and builds trust	Google's AI audit corrected biases and restored trust
Blockchain Systems	Ensure transparency and accountability	Supply chain management and healthcare records	Improves traceability and reduces fraud	IBM's blockchain reduced Walmart's recall costs by >30%
Root Cause Analysis	Identify the underlying causes of problems	Product defect or failure investigations	Prevents recurring issues	Boeing addressed safety failures post-737 Max crisis
Quantum-Safe Cryptography	Protect data against quantum decryption threats	Securing financial transactions and government data	Future-proofs cybersecurity	IBM's quantum-resistant encryption safeguards critical systems

Leadership Alignment

- Train leaders to use tools like root cause analysis and Balanced Scorecards effectively.
- Embed ethics into decision-making processes.
- Recognize and reward leaders who prioritize long-term trust over short-term gains.

Step-by-Step Framework for Leaders

1) **Embed Ethics:**
   - Conduct ethical audits to identify vulnerabilities.
   - Establish a code of ethics tailored to your organization's mission.
   - Appoint a Chief Ethics Officer to oversee accountability.
2) **Leverage Tools:**
   - Use scenario planning to prepare for uncertainty and identify opportunities.
   - Conduct regular algorithm audits to ensure fairness and transparency in AI systems.
   - Implement blockchain systems to enhance accountability and data integrity.
3) **Measure and Adapt:**
   - Use Balanced Scorecards to evaluate financial, operational, and ethical performance.
   - Continuously update frameworks to reflect emerging challenges and opportunities.

Case Studies in Ethical Leadership

1) **Integrity Under Pressure: The Challenger Disaster** – In 1986, the Challenger Space Shuttle exploded 73 seconds after launch, killing seven astronauts. The tragedy stemmed from a failure to address known safety concerns about O-ring seals in low temperatures. Engineers had warned of the risks, but leadership prioritized the launch schedule over safety.

Key Lessons:

- Ethical leaders must advocate for safety, even when it conflicts with organizational goals.
- Whistleblowers play a critical role in preventing catastrophic failures, but they need an environment that values their courage.

2) **The Business Case for Ethics: Volkswagen's Emissions Scandal** – Volkswagen manipulated emissions data to make its vehicles appear more environmentally friendly, ultimately deceiving regulators and customers. When the fraud was exposed in 2015, the company faced billions in fines and a tarnished reputation.

Key Lessons:

- Short-term gains from unethical practices lead to long-term losses.
- Transparency and accountability are non-negotiable for sustained trust.

3) **Social Responsibility and Environmental Stewardship: BP's Deepwater Horizon Spill** – The 2010 Deepwater Horizon spill devastated marine ecosystems and local economies. Investigations revealed that BP had prioritized cost-cutting over safety, leading to the worst environmental disaster in U.S. history.

Key Lessons:

- Social responsibility extends beyond compliance; it requires proactive risk management.
- Leaders must prioritize environmental stewardship alongside profitability.

4) **Ethical Innovation: Theranos** – Theranos claimed to revolutionize blood testing but failed to deliver on its promises, endangering patient health and defrauding investors. Founder Elizabeth Holmes prioritized ambition over integrity, resulting in one of the most infamous ethical failures in recent history.

Key Lessons:

- Innovation without ethical grounding can cause irreparable harm.
- Validation and transparency are critical in technology development, especially in healthcare.

Building an Ethical Culture: An organization's ethical culture begins at the top. Leaders set the tone through their actions and decisions. Here's how to develop and sustain an ethical culture:

1) **Model Ethical Behavior:** Leaders must consistently demonstrate integrity, transparency, and fairness.
2) **Establish Clear Policies:** Codes of conduct should outline ethical expectations and consequences for violations.
3) **Provide Training:** Employees need tools and frameworks to navigate ethical dilemmas.
4) **Encourage Whistleblowing:** Create safe channels for reporting concerns without fear of retaliation.
5) **Measure and Adapt:** Use metrics to assess the organization's ethical health and address gaps proactively.

Ethical Leadership in a VUCA World: Ethics in leadership is especially critical in VUCA (volatile, uncertain, complex, ambiguous) environments. Leaders must navigate uncertainty with integrity, balancing short-term demands with long-term responsibilities. This requires:

- **Foresight:** Anticipating risks and opportunities with a commitment to ethical decision-making.
- **Collaboration:** Engaging diverse stakeholders to align on values and goals.
- **Resilience:** Building systems and teams capable of withstanding ethical and operational crises.

Ethics and Reputation: The Human Dimension – Ethical leadership is deeply personal. It's about safeguarding your reputation and the trust of those who depend on you – employees, customers, and society. Losing your good name, whether rightfully or not, can feel like losing everything. However, ethical behavior provides a foundation for resilience, enabling leaders to rebuild trust even in adversity.

In my journey mentoring over 8,000 executives and leaders across industries, I've witnessed the transformative power of ethical leadership. Organizations prioritizing ethics recover from crises and emerge more substantial, with reputations intact and trust renewed.

Communication Script for Addressing Ethical Dilemmas

When faced with ethical dilemmas, effective communication is crucial. Below is a structured script to use when discussing ethical challenges with colleagues and supervisors:

- **Define the Issue:** "The issue we are addressing is [describe the problem]. The facts as we know them are [list factual details]."
- **Present Options:** "There are three potential courses of action: (1) [Option 1], which could result in [consequence], (2) [Option 2], leading to [consequence], and (3) [Option 3], which may cause [consequence]."
- **State Ethical Principles:** "Based on the principles of [specific principles, e.g., safety, integrity, transparency], I recommend [chosen option]. This recommendation aligns with our organizational values and ensures accountability."
- **Anticipate Challenges:** "If [challenge] arises, our response will be [specific action], and if [other challenges] occur, we will address them by [specific action]."
- **Invite Feedback:** "I welcome your insights and any additional perspectives to ensure we approach this issue responsibly and effectively."

Effective Leadership

- **Key Traits:** Vision, empathy, accountability, and adaptability.
- **Why Effective:** Such leaders proactively address challenges, promote inclusivity, and uphold organizational values.
- **Implementation:** Integrate ethical decision-making frameworks, encourage open dialogue, and prioritize stakeholder welfare.

Ineffective Leadership

- **Key Traits:** Authoritarianism, opacity, and short-term thinking.
- **Why Ineffective:** These approaches stifle innovation, erode trust, and increase organizational risk.
- **Prevention:** Implement robust governance, provide ethics training, and hold leaders accountable for misconduct.

Practical Framework for Ethical Decision-Making

1) **Define the Issue:** Identify the problem and gather facts.
2) **Assess Stakeholders:** Determine who is affected and how.
3) **Evaluate Options:** Analyze potential outcomes using ethical principles.
4) **Communicate Decisions:** Be transparent about the rationale and action plan.
5) **Follow Through:** Monitor outcomes and adjust strategies as needed.

Practical Steps:

- Foster a culture of honesty and accountability.
- Implement robust whistleblower policies to expose unethical practices.
- Align corporate actions with stated values.

Ethical Frameworks for Decision-Making

1) **Model #1: Effects on Character**
   - Identify facts and intentions.
   - Assess the likely consequences of actions.
   - Evaluate how decisions align with personal and professional principles.
2) **Model #2: Justifying Actions**
   - Consider stakeholders affected by the decision.
   - Determine if actions can withstand public and legal scrutiny.
   - Reflect on long-term implications for reputation and integrity.
3) **Model #3: Standing Up for Values**
   - Identify critical facts and values at stake.
   - Work through worst-case scenarios and ethical principles.
   - Develop action plans to address dilemmas proactively.
4) **Model #4: Asking Critical Questions**
   - Is the action legal and aligned with organizational values?
   - How will it appear to the public?
   - Seek counsel from trusted colleagues and ethics offices.

Communication Script for Ethical Dilemmas

1) Clearly define the issue and present the facts.
2) Identify potential options and their likely outcomes.
3) State your guiding principles and their relevance.
4) Propose a course of action and anticipate possible responses.

Example Script: "**The issue is ____. The facts are as follows: ____. We have three options: (1) ____, which will likely result in ____, (2) ____, with the outcome of ____, or (3) ____, which could lead to ____. Based on the principles of ____, I recommend ____. If ____ happens, we will ____, and if ____ occurs, we will ____.**"

Conclusion: Protecting Your Good Name, Team, and Your Organization with Truth and Integrity

Words, decisions, actions, reputation, impacts, and results are the cornerstone of a leader's ability to influence and inspire. It is a reflection of their achievements and a testament to their character and values. Ethical lapses – intentional or the result of oversight – can shatter trust instantly and leave a lasting stain on personal and organizational credibility. Upholding integrity is not optional; it is the foundation of effective leadership. Leaders must confront challenges with honesty, transparency, and a willingness to accept accountability, ensuring trust remains intact.

This focus on integrity ties seamlessly with the broader themes explored throughout this book – the critical importance of reputation, ethical responsibility, and leaders' role in advancing society responsibly and sustainably. Ethics in engineering and technology leadership demands more than technical expertise; it requires a steadfast commitment to safety, transparency, and inclusivity principles.

By embedding these principles into decision-making processes, leaders can navigate the complexities of modern challenges while building trust and credibility. Practical frameworks and clear, honest communication empower professionals to confront ethical dilemmas head-on, making decisions that not only protect their reputation but also contribute meaningfully to the betterment of society.

Conclusion: Lead, Innovate, Anticipate

Ethical leadership is a moral imperative and a strategic advantage. As leaders, we must ask ourselves:

- How will our decisions impact society and future generations?
- Are we acting with integrity, even when it's inconvenient?
- What legacy are we building?

By embedding ethics, foresight, and social responsibility into leadership practices, we can navigate uncertainty, drive innovation, and build organizations that inspire trust. The future of leadership depends on it.

Ethics as a Pillar of Leadership: From Tesla's resilience to Google's ethical AI, the stories in this chapter illustrate that success in technology isn't just about innovation – it's about integrity. Ethics and tools aren't separate – they are the foundation of effective leadership. By integrating these principles into strategy, leaders can navigate uncertainty, build trust, and drive meaningful progress.

Resources for Further Reading

Professional and Organizational Ethics

Anon, *What Needs a Reference? The Teaching Professor*, Mar. 1999.
R. Bullock, *The St. Martin's Manual for Writing in the Disciplines: A Guide for Faculty*. St. Martin's Press, 1994.
Harvard Business Review, *Why Ethical Companies Outperform Their Peers*, 2020.
IEEE, *Code of Ethics*, n.d.
National Society of Professional Engineers (NSPE), *Code of Ethics*, n.d.
A. Smith, "Ethical leadership and crisis management," *J. Bus. Ethics*, vol. 125, no. 1, pp. 145–157, 2021.

Scientific Integrity and Research Ethics

B. Alberts and K. Shine, "Scientists and the integrity of research," *Science*, vol. 266, pp. 1660–1661, 1994.
A. L. Beach and D. E. Wright, "A short history of regulations protecting human subjects of research," *Res. Integrity*, vol. 4, no. 1, pp. 2–7, 2000.
Center for Science in the Public Interest, n.d.

B. Glass, "The ethical basis of science," *Science*, vol. 150, pp. 1254–1261, 1965.
B. C. Martinson, M. S. Anderson, and R. de Vries, "Scientists behaving badly," *Nature*, vol. 435, pp. 737–738, 2005.
NIH, *Report on Conflicts of Interest*, 2004.
Union of Concerned Scientists, *Scientific Integrity in Policy Making*, 2004.

Engineering Disasters and Case Studies

BP, *Deepwater Horizon Spill Investigation Report*, 2010.
B. Carter, *Deepwater Horizon Oil Spill*. National Wildlife Federation, 2019.
Food and Drug Administration, *Merck's Vioxx Recall Timeline*, 2005.
T. Johnson, "Volkswagen emissions scandal," *IEEE Trans. Eng. Manag.*, vol. 67, no. 4, pp. 345–350, 2020.
NASA, *Challenger Disaster Investigation Report*, 1986.
NASA, *Apollo Program Archives*, 2020.
"Three Mile Island Accident," U.S. Nuclear Regulatory Commission, 2024.
Theranos Case Study, New York Times, 2018.
"Volkswagen Emissions Scandal Overview," 2015.

Technology Leadership, Innovation, and Impact

J. Doerr, *Measure What Matters: OKRs*. Portfolio Penguin, 2018.
Google AI Ethics Team, *Algorithm Bias and Audits*, 2022.
IBM, *Food Trust Blockchain Impact Report*, 2023.
Statista, *Netflix Revenue and Subscriber Growth*, 2024.
Tesla, *Sustainability Report*, 2021.
Tesla, Inc., *Master Plan, Part Deux*, 2016.

6

Breaking the Triangle: Leadership Strategies for Operational Excellence

Introduction

For decades, leaders have been bound by the constraints of the "Project Management Triangle," a framework that suggests organizations can only achieve two of three objectives: quality, speed, or cost-efficiency. This framework has shaped countless decisions, often forcing organizations to sacrifice one goal to accomplish the other two. As a result, leaders have had to choose between delivering high-quality products quickly, at an affordable cost, or delaying one to meet the other two.

However, today's rapidly evolving technological landscape presents a different reality. Forward-thinking organizations are proving that the triangle is not a rigid rule but a challenge to overcome. Companies such as SpaceX, Tesla, and Toyota have demonstrated that achieving all three objectives simultaneously is possible. They have achieved operational excellence without compromise by reimagining leadership strategies, fostering collaboration, and optimizing team communication.

The secret to breaking the triangle lies in addressing latency (the speed at which decisions, feedback, and problem-solving occur) and bandwidth (the capacity to exchange information effectively across teams). These factors are the foundation of high-performing organizations. By mastering latency and bandwidth, leaders can enable seamless collaboration, faster innovation cycles, and better outcomes at lower costs.

This chapter explores how these principles come to life through real-world examples. By examining the strategies of SpaceX, Tesla, and Toyota, we uncover how these companies shattered the constraints of the Project Management Triangle, setting new standards for operational excellence.

From Theory to Practice

In the previous chapters, we discussed the importance of adaptability, clarity, and alignment in leadership. These principles laid the groundwork for understanding how leaders can address complex challenges. Now, we turn to practical applications. Through the detailed case studies of SpaceX, Tesla, and Toyota, this chapter illustrates how innovative strategies can break traditional constraints and create environments where quality, speed, and cost-efficiency coexist.

These organizations do not succeed by cutting corners or making trade-offs; they succeed by redefining the rules of the game. By reducing latency, increasing bandwidth, and empowering teams, they have cultivated operational ecosystems that thrive under pressure.

SpaceX: Speed Drives Innovation and Reduces Costs

In 2008, SpaceX was on the brink of failure after three unsuccessful launches. Elon Musk's team faced a pivotal moment when only one rocket was left. Instead of retreating, they embraced a mindset of rapid iteration: "Build fast, fail fast, learn fast." Their success with the fourth launch was not merely luck – it resulted from a system designed to reduce latency in decision-making and problem-solving.

By creating short feedback loops and empowering engineers to iterate quickly, SpaceX revolutionized aerospace engineering. The development of the reusable Falcon 9 rocket drastically reduced launch costs, from $18,500 per kilogram during the Space Shuttle era to just $2,720 per kilogram [1]. This innovation made commercial spaceflight more accessible and positioned SpaceX as an industry leader.

Key Leadership Insight:
Speed does not compromise quality – it enhances it. Urgency drives focus, innovation, and iterative improvements.

Actionable Steps:

- Establish clear goals with aggressive timelines to maintain urgency.
- Foster a culture of experimentation and rapid iteration to learn quickly from mistakes.

Tesla: Scaling Quality While Cutting Costs

When Tesla began producing electric vehicles (EVs), the cost of lithium-ion batteries, over $1,200 per kilowatt-hour in 2010, was a significant barrier. Tesla tackled this challenge through vertical integration and innovation at its Gigafactories. These massive facilities enabled real-time collaboration between engineering and manufacturing teams, significantly increasing bandwidth and eliminating inefficiencies.

By 2021, Tesla had reduced battery costs by 89%, reaching $132 per kilowatt-hour [2]. This breakthrough made EVs more affordable and allowed Tesla to dominate the market by delivering high-quality products at competitive prices.

Key Leadership Insight:
High bandwidth across teams enables exponential improvements. By sharing critical information and aligning priorities, organizations can redefine entire industries.

Actionable Steps:

- Audit processes to identify bottlenecks or inefficiencies.
- Increase transparency and context-sharing among cross-functional teams to ensure alignment.

Toyota: Mastering Efficiency Without Sacrificing Quality

Toyota's Production System (TPS) epitomizes the balance of efficiency, quality, and cost. By implementing Just-In-Time (JIT) production and empowering assembly line workers to halt production when defects were detected, Toyota drastically reduced latency in addressing quality issues.

These measures ensured that every vehicle met stringent quality standards, leading to defect rates as low as 10 per 1,000 vehicles [3]. Simultaneously, Toyota's focus on eliminating waste improved cost efficiency, demonstrating that operational excellence is achievable without sacrificing quality.

Leadership Under Fire: The Power and Pitfalls of Technological Leadership

As I added the finishing touches to this book on technological leadership, I turned to a trusted journalist friend for feedback. His response was straightforward: use your voice. He said the strongest parts weren't the frameworks or theories – they were the stories – the moments when leadership is tested under pressure, decisions reveal character, and the line between success and disaster is razor-thin. Leadership isn't just about what to do and what not to do. It's about understanding how people act under pressure, because pressure is a truth serum. Squeeze an orange, and you get orange juice. Squeeze an apple, and you get apple juice. But squeeze a human being? You don't know what you're going to get. Some rise, while others collapse. Some lead, whereas others lash out. Some create, while others destroy.

Leadership at the Extreme: Musk, Jobs, and Welch

Elon Musk, Steve Jobs, and Jack Welch redefined industries and changed the world. Their leadership philosophies disrupted the status quo, pushing the limits of human endurance, innovation, and efficiency. Yet they also embodied leadership traits that, when unchecked, crossed the line into toxicity.

Musk's relentless drive led Tesla and SpaceX to revolutionary breakthroughs. However, former employees describe a brutal work culture where exhaustion is the price of admission. Engineers are expected to be "all in" or out. A former Tesla employee once said, "Musk demands the impossible and then expects it faster" [1].

Jobs was a perfectionist who rejected mediocrity. Apple's success was built on its obsession with detail – marketing, product design, and even the feel of a laptop hinge. But his sharp tongue could be ruthless. He berated employees, dismissed ideas on Monday only to approve them unchanged by Wednesday, and fired people in hallways for failing to meet his exacting standards [2].

Welch once represented corporate leadership, driving General Electric (GE) to unprecedented heights. His "rank and yank" system – firing the bottom 10% of employees annually – made GE lean and instilled a culture of fear. It drove short-term performance at the cost of long-term sustainability. When his successors inherited GE, the cracks had already formed, leading to its dismantling [3].

These leaders were brilliant yet flawed. Their approaches raise an important question: Can their results justify the cost of their leadership?

The Five Leadership Tests: Who Thrives and Who Fails?

Leadership Type	Examples from Musk, Jobs, and Welch	Lesson
The Boss Who Undermines You	Jobs frequently dismissed his team's ideas, only to later present them as his own. Musk has been known to reject engineering solutions at first glance and implement them later without crediting his engineers [4]	Park your ego, but don't let it erase you. Keep records. Build allies. If undervalued, move where you aren't

(Continued)

Leadership Type	Examples from Musk, Jobs, and Welch	Lesson
The Boss Who Leads with Fear	Welch instituted a brutal "rank-and-yank" system at GE, where the bottom 10% of employees were fired annually, fostering a culture of fear. Musk has fired executives on the spot, often publicly, like Twitter's head of trust and safety [5]	Fear-driven leadership breeds instability, not excellence. If you're in that environment, document everything. If leadership enables it, leave
The Boss Who Sabotages You	Musk has been known to impose last-minute, unrealistic deadlines, pushing teams into all-night crunches. Welch often pitted executives against each other to drive competition, leading to backstabbing and distrust [6].	Expect chaos. Track wins. If blindsided, focus on clarity, not perfection. If a pattern, call it out or exit
The Boss Who Controls the Narrative	Jobs was notorious for rewriting history, spinning reality to suit his vision. Musk silences critics, firing employees who speak out at Tesla and Twitter. Welch aggressively shaped GE's public image, ensuring dissenting voices were buried [7]	Work around them. Find decision-makers who listen. If their insecurity stunts your growth, move on
The Boss Who Won't Take Responsibility	Musk has shifted the blame for Tesla's Autopilot failures, first blaming regulators, then engineers, and then journalists [8]. Welch's short-term profit-driven strategies weakened GE, but he was gone before the cracks appeared. Jobs, in contrast, took full accountability at Apple's lowest point and used it as a foundation for the company's rebirth [9]	Protect your reputation. Keep records. When the dust settles, all you have is your credibility

The Leaders Who Get It Right: For every bad boss, there's a great one. The leader who challenges you without tearing you down. The one who holds you accountable without making you afraid. The one who recognizes your work rather than hoards credit.

Good leadership doesn't come from dominance; it comes from trust. The best leaders put wind in your sails, not weights on your wings. They lift others, knowing that real power isn't in control – it's in empowerment.

Consider **Satya Nadella**, who took over Microsoft when it stagnated under Steve Ballmer. Instead of ruling with fear, he built a culture of collaboration, focusing on learning rather than hierarchy. Under Nadella, Microsoft's market value tripled [10]. Or **Ed Catmull** at Pixar, who fostered a culture where failure was seen as part of the creative process, not a punishable offense [11].

What's Inside When You're Squeezed?

A crisis doesn't create character – it reveals it. Whether leading a team, building a company, or simply navigating your career, ask yourself: What emerges when you're under pressure? Do you respond with fear and blame or remain steady, lead clearly, and uplift others? The best leaders aren't perfect, but they understand this: not every conflict requires your involvement. Allow absence to be your response when necessary. Toxicity seldom changes. You may not always control who's in charge, but you can control how you respond. Because when the pressure comes – and it will – you want to know exactly what is inside.

Example: Energy Transitions and the Leadership Imperative

In August 2005, Hurricane Katrina devastated the Gulf Coast of the United States, exposing the vulnerabilities of America's energy infrastructure. Oil refineries were flooded, pipelines were shut down, and fuel shortages rippled nationwide. The storm caused over $125 billion in damages and displaced over a million people. Katrina wasn't just a natural disaster – it was a wake-up call highlighting the fragility of the centralized, fossil-fuel-dependent energy systems of the 20th century.

Two decades later, the urgency of transitioning to resilient, sustainable energy systems has only grown. Climate change, geopolitical tensions, and shifting energy demands have created a complex landscape for energy leaders. The global transition to renewable energy is no longer a choice but a necessity, requiring leaders to balance innovation with equity, resilience with scalability, and short-term challenges with long-term goals.

This chapter examines case studies, including Germany's *Energiewende* and Puerto Rico's post-Maria decentralization efforts. It explores the leadership strategies essential for navigating one of the most transformative challenges of our time.

The Complexity of Energy Transitions

Energy transitions have shaped human history. Societies moved from wood to coal during the Industrial Revolution and then to oil and natural gas in the 20th century. Each transition was driven by the search for more efficient, scalable, and abundant energy sources. Today, the shift from fossil fuels to renewable energy is unique – it is not just about economics but survival.

Renewable energy sources like solar, wind, and hydropower are growing faster than any other energy sector. By 2023, renewables accounted for nearly 30% of global electricity generation, up from 20% in 2015 [1]. However, the challenges of scaling renewable energy systems remain significant:

1) **Infrastructure Modernization:** Grids designed for centralized fossil-fuel plants must be adapted to integrate decentralized and intermittent renewable sources.
2) **Energy Storage:** Solar and wind energy variability requires reliable storage technologies, such as lithium-ion and solid-state batteries, to ensure a consistent power supply.
3) **Policy Alignment:** National and international policies must incentivize renewable energy while addressing the social and economic impacts of transition, such as job losses in fossil fuel industries.

Germany's Energiewende: A Bold Vision for Transition

Germany's *Energiewende* (energy transition) is one of the most ambitious national efforts to shift from fossil fuels to renewable energy. Launched in the early 2000s, *Energiewende* aimed to phase out nuclear power, reduce greenhouse gas emissions, and expand the use of renewables.

By 2023, Germany had achieved significant milestones:

- **Renewable Dominance:** Solar and wind energy accounted for 46% of Germany's electricity generation, compared to just 10% in 2000 [2].
- **Coal Phase-Out:** Germany committed to shutting down all coal-fired power plants by 2038, and significant reductions have already been achieved.
- **Economic Growth:** Despite initial concerns, Germany's economy remained strong, demonstrating that decarbonization and prosperity coexist.

However, *Energiewende* also faced challenges. The rapid phase-out of nuclear energy led to increased reliance on natural gas, exposing Germany to geopolitical risks during the 2022 energy crisis caused by the Russia–Ukraine war. This highlighted the need for a balanced, diversified energy mix and policies anticipating global disruptions.

Puerto Rico's Shift to Decentralized Energy

When Hurricane Maria struck Puerto Rico in 2017, the island's centralized power grid collapsed. Nearly 3.4 million residents lost electricity; some were left without power for almost a year. The disaster revealed the vulnerabilities of relying on aging, centralized infrastructure, especially in regions prone to extreme weather [3].

Rather than rebuilding the same system, Puerto Rico began investing in decentralized energy solutions. Solar microgrids, powered by renewable energy and battery storage, became a cornerstone of the island's recovery.

One notable success story is the mountain town of Adjuntas, where a community-led solar microgrid now provides power to homes, businesses, and a health clinic. The project demonstrated that localized energy systems improve resilience and empower communities to control their energy future. By 2024, over 10% of Puerto Rico's electricity will come from decentralized solar projects, positioning the island as a global leader in energy resilience [4].

The Leadership Imperative in Energy Transitions

Leading an energy transition is not just about deploying technology – it requires visionary leadership that balances ambition with pragmatism, innovation with equity, and urgency with long-term strategy.

Successful leaders share key traits:

- **Vision:** Leaders must articulate a clear and compelling vision for the future of energy that inspires action across industries, governments, and communities.
- **Collaboration:** Energy transitions demand cooperation across sectors and borders. Public–private partnerships, as seen in Puerto Rico's microgrid efforts, amplify resources and expertise.
- **Adaptability:** The energy landscape constantly evolves, with new technologies and geopolitical challenges emerging. Leaders must remain flexible and responsive.
- **Equity:** It is critical to ensure that the benefits of energy transitions are distributed fairly. This includes addressing energy access disparities and supporting workers displaced by the shift away from fossil fuels.

Global Outlook: Regional Challenges and Opportunities

The path to energy transition varies by region, influenced by local resources, infrastructure, and socioeconomic conditions:

- **Sub-Saharan Africa:** Over 600 million people still lack access to electricity. Off-grid solar systems are expanding energy access while reducing reliance on expensive diesel generators.
- **China:** As the world's largest energy consumer, China leads in renewable energy investment, spending over $100 billion on solar and wind projects in 2023. However, its heavy reliance on coal remains a significant barrier to decarbonization [5].
- **United States:** The Inflation Reduction Act of 2022 provided unprecedented incentives for clean energy, accelerating investments in wind, solar, and energy storage technologies. Yet, political polarization continues to hinder long-term policy consistency [6].

Actionable Strategies for Leaders

To drive effective energy transitions, leaders can adopt the following strategies:

1) **Invest in Innovation:** Support research and development in emerging technologies such as hydrogen fuel cells, advanced batteries, and carbon capture.
2) **Modernize Infrastructure:** Upgrade grids to handle renewable energy and improve resilience against extreme weather events.
3) **Engage Stakeholders:** Collaborate with governments, private companies, and communities to align efforts and maximize impact.
4) **Prioritize Equity:** Address disparities in energy access and ensure that vulnerable populations benefit from the transition.

Looking Ahead: The global energy transition is one of the defining challenges of our time. It demands bold leadership, innovative solutions, and a commitment to equity and social justice. The lessons from Germany's *Energiewende* and Puerto Rico's microgrid revolution demonstrate that transformative change is possible despite complex obstacles.

Resilience, adaptability, and collaboration will be the hallmarks of successful leaders in this transition era. By learning from past efforts and investing in future systems, leaders can build an energy infrastructure that is sustainable and capable of withstanding the uncertainties of a changing world.

Key Leadership Insight:
Efficiency is about cutting waste, not corners. Teams empowered to take ownership of their work can ensure quality and productivity simultaneously.

Actionable Steps:

- Empower employees to identify and address inefficiencies in real-time.
- Implement continuous improvement systems, such as Lean or Six Sigma frameworks.

Delays Are Costly: Lessons from Boston's Big Dig

The Central Artery/Tunnel Project, or "Big Dig," aimed to alleviate traffic congestion in Boston by rerouting it underground. Initially budgeted at $2.6 billion, the project faced years of delays and mismanagement, ultimately ballooning to $14.6 billion [4]. High latency in decision-making and poor communication among stakeholders were primary contributors to the cost overruns.

Contrast this with Wuhan's response during the COVID-19 pandemic, where a 1,000-bed hospital was constructed in 10 days. This feat was possible because leaders streamlined decision-making processes, set clear priorities, and eliminated bureaucratic delays.

Key Leadership Insight:
Delays multiply costs and inefficiencies. Leaders must act decisively and establish clear milestones to maintain momentum.

Actionable Steps:

- Balance speed with planning to avoid unnecessary delays.
- Streamline decision-making processes to eliminate bottlenecks.

Agile: Breaking Free from Traditional Constraints

The traditional "waterfall" approach to software development – where products are built sequentially over long timelines – often results in delays and misaligned outcomes. Agile methodologies revolutionized the industry by emphasizing iterative development, rapid feedback, and adaptability.

Organizations like Google and Spotify have embraced Agile principles to deliver value early and continuously improve their service based on user feedback. Studies have shown that Agile adoption leads to a 60% improvement in product quality and a 23% reduction in time-to-market [5].

Key Leadership Insight:
Flexibility is a competitive advantage. Agile practices demonstrate that quality and speed coexist when teams are empowered to iterate and adapt.

Actionable Steps:

- Apply Agile principles across all areas of operations, not just software development.
- Focus on delivering incremental improvements rather than waiting for perfection.

Blueprint for Leaders: Optimizing Latency and Bandwidth

Achieving quality, speed, and cost-efficiency requires leaders to master team communication interfaces. Here's a practical blueprint:

1) **Reduce Latency:** Shorten feedback loops to act on issues promptly.
2) **Increase Bandwidth:** Foster open communication and build trust within the team.
3) **Set Bold Goals:** Aggressive timelines drive urgency and focus.
4) **Iterate Quickly:** Deliver incremental value and refine continuously.
5) **Eliminate Waste:** Streamline processes to reduce inefficiencies without sacrificing quality.
6) **Challenge Assumptions:** Encourage innovative thinking to unlock new possibilities.

Final Thought: The belief that "good, cheap, and fast" cannot coexist is outdated. Companies like SpaceX, Tesla, Toyota, and Agile pioneers have shown that these goals are not mutually exclusive. Leaders can foster collaboration, adaptability, and innovation within their teams by addressing latency and bandwidth challenges.

The choice is no longer about which two objectives to prioritize – it's about mastering the art of achieving all three.

References

1 (a) M. Wall, *SpaceX Launch Costs Plummet*, Space.com, May 10, 2022. Available: https://www.space.com; (b) D. Stringer, *The Decline in EV Battery Costs, Bloomberg*, Mar. 15, 2023. Available: https://www.bloomberg.com; (c) J. Liker, *The Toyota Way: 14 Management Principles from the World's Greatest Manufacturer*. New York: McGraw-Hill, 2004.

2 J. Roe, *The Cost of Infrastructure Delays, Boston Globe,* Jan. 12, 2015. Available: https://www.bostonglobe.com.

3 K. Beck *et al.*, *Manifesto for Agile Software Development, Agile Alliance*, 2001. Available: https://www.agilemanifesto.org.

4 A. Vance, *Elon Musk: Tesla, SpaceX, and the Quest for a Fantastic Future*. HarperCollins, 2015.

5 W. Isaacson, *Steve Jobs*. Simon & Schuster, 2011.

6 J. Welch, *Winning*. HarperCollins, 2005.

7 The Verge, *Inside Elon Musk's Twitter Takeover*, 2023.

8 Business Insider, *Tesla Engineers Speak Out About Musk's Micromanagement*, 2021.

9 The Guardian, *Twitter's Trust and Safety Chief Fired Mid-Interview*, 2023,

10 Bloomberg, *Tesla's All-Night Work Culture*, 2020.

11 Washington Post, *Musk's Handling of Dissent at Twitter and Tesla*, 2023.

Resources for Further Reading

Congressional Research Service, *Hurricane Maria: Federal Response and Recovery*, 2018. [Online]. Available: https://www.crsreports.congress.gov.

German Federal Ministry for Economic Affairs and Climate Action, "Energiewende Progress Report," 2023. [Online]. Available: https://www.bmwi.de.

International Energy Agency (IEA), *Renewables 2023 Global Status Report*, 2023. [Online]. Available: https://www.iea.org.

Microsoft Annual Reports, 2014–2023.

National Renewable Energy Laboratory (NREL), *Puerto Rico Energy Resilience Study*, 2023. [Online]. Available: https://www.nrel.gov.

Pixar Case Study, Harvard Business Review, 2018.

Reuters, *Tesla's Autopilot and Regulatory Battles*, 2023.

U.S. Department of Energy, *Inflation Reduction Act: Impact on Clean Energy Investment*, 2023. [Online]. Available: https://www.energy.gov.

World Resources Institute, *China's Renewable Energy Expansion*, 2023. [Online]. Available: https://www.wri.org.

Part III

Strategy, Energy, and Grid Resilience

7

REBALANCE: Diagnosing Systemic Fragility and Rebalancing Power Under Pressure

> *"Every system whispers before it breaks. We either learn to listen or leave the wreckage to someone else."*
> *"The Drift Beneath Our Feet: Diagnosing Systemic Fragility Before Collapse"*

The Quiet Signals of System Collapse

The 2021 Texas freeze didn't begin with the blackout. It started with a silent underinvestment in insulation. The 2017 Puerto Rico blackout didn't begin with Hurricane Maria. It started years earlier, in deferrals and decay. A fire didn't cause Heathrow's 2025 terminal shutdown. Brittle design, missed warnings, and the absence of contingency caused it. These are not discrete disasters. They are signals of a deeper drift – an accelerating structural fatigue across systems we depend on.

Infrastructure is aging faster than we can upgrade it. Governance is straining against complexity. Feedback loops are frayed. Institutions built for slower, more predictable conditions are now tasked with absorb compound shocks – cyber disruptions layered on top of extreme weather, economic volatility, and civic distrust.

Yet we lack a systematic method for interpreting these signals promptly. We rely on simulations, forecasts, or static checklists. We need a structural diagnostic that reveals not when a system will fail but why it drifts toward collapse and *where* to intervene early and surgically.

That is the purpose of REBALANCE.

From Fragility to Feedback: Applying GTD

What separates a system that fails from one that adapts isn't complexity. It's timing. Systems drift long before they fracture, and the signals of that drift – structural stress, weak redundancy, and decaying legitimacy – accumulate until inertia breaks. *Global Transition Dynamics* (GTD) allows us to model this erosion not as a failure event, but as a trajectory.

The GTD framework (please see Appendix B for details of the GTD methodology and selected mathematical models) draws from dynamical systems, mapping how institutional entropy, operational burden, or civic disengagement evolve across time. When applied to infrastructure or governance, GTD surfaces what systems thinkers have long known but rarely quantified: stress begets strain, strain suppresses feedback, and suppressed feedback accelerates collapse.

Instead of asking, "Why did this system break?" GTD asks, "What were the leading indicators – and when did the slope of drift steepen beyond recovery?" It translates systemic fragility into trajectories that can be monitored, measured, and interrupted before the breaking point.

This shift – from retrospective blame to prospective intervention – is essential. Across power grids, municipal agencies, and healthcare systems, GTD-guided diagnostics have helped surface decay patterns well before catastrophic thresholds are crossed. The same modeling that once served control rooms and airframes can – and must – guide our institutions.

A Systems Control Framework for Civic Infrastructure

REBALANCE is a structural resilience and diagnostics system rooted in dynamical systems and control theory. Developed over two decades of fieldwork across infrastructure, security, and engineering leadership, it applies foundational control concepts – feedback, observability, robustness, and fault detection – to civic, infrastructural, and policy environments.

Where typical scenario modeling simulates possible futures, REBALANCE operates differently: it maps fragility itself. It diagnoses hidden structural stress and dynamic failure pathways using quantifiable, modular, and transferable architecture.

At its core, REBALANCE assesses system health along nine foundational vectors – not thematic categories but state dimensions directly linked to failure mechanics. These vectors form a multidimensional system, characterized by stability, resilience, and decay profiles.

Each vector captures a distinct mode of stress. Together, they form a coupled control system – deterioration in one vector accelerates failure across others. Scoring is based on real-world data, expert interviews, historical records, and performance metrics, allowing systems to be tracked, benchmarked, compared, and strengthened.

Vectors of Stress and Fragility

The REBALANCE framework identifies nine core vectors of stress, complemented by four additional systemic hazards (Table 7.1).

Table 7.1 Foundational Vectors: REBALANCE – Nine Vectors and Their Definitions.

Letter	Vector	Definition
R	Resilience	Maintain core function during disruption; absorb, recover, and adapt.
E	Entropy	Institutional drift and decay due to complexity, neglect, or attrition.
B	Burden	System stress exceeding operational or design capacity.
A	Alignment	Degree of public trust and legitimacy in institutional actions.
L	Linkage	Fragility resulting from the interconnectedness of subsystems.
A	Access	Openness and surface area risks – physical, digital, and procedural vulnerabilities.
N	Noise	Shocks from environmental, cyber, economic, and geopolitical disruptions.
C	Contingency	Presence of buffers, backups, and redundancy mechanisms.
E	Execution	Speed, adaptation, and leadership capacity under stress.

Each is scored using real data, expert interviews, records, and performance logs. Think of them as coupled dimensions in a multivariable control system. Drift in one accelerates failure in others.

Implementation Flow: Resilience Engineering in Action

REBALANCE operates through a deliberate Implementation Flow – ensuring systems don't just diagnose weaknesses but actively build dynamic capacity:

1) **Monitor for Drift** – Constantly track for signs of entropy and misalignment.
2) **Isolate Signals** – Identify early weak signals in a noisy environment.
3) **Map Interdependencies** – Understand the critical links that can lead to cascading failures.
4) **Build Buffers** – Create material, procedural, and structural redundancies to enhance system reliability.
5) **Preposition Tools** – Ensure rapid access to response and recovery resources.
6) **Decentralized Detection** – Avoid single-point vulnerabilities; empower distributed sensing.
7) **Embed Feedback** – Create continuous real-time feedback loops for adaptation.
8) **Define Recovery Metrics** – Quantify partial and full system recovery thresholds.
9) **Redesign Post-Failure** – Institutionalize learning after disruption to reconfigure weaknesses.

Additional Metrics, Factors, Augmentation Vectors: RESONANCE (Extended systemic blind spots and operational amplifiers.)

Given the dynamical systems framing, a few **additional metrics** can enhance precision:

Letter	Vector	Definition
R	Recovery Capacity (Latent)	A measure of hidden adaptive potential before system failure.
E	Environmental Saturation	Input overload causes decision fatigue and noise fatigue.
S	Signal-to-Noise Ratio	Clarity of critical signals vs. distracting/misleading noise.
O	Organizational Isolation	Leadership's blindness to early warning signals and threats.
N	Node Failure Cascade Depth	How deeply failure can propagate through interconnected dependencies.
A	Attack Surface Area	Quantified measurement of vulnerability exposure across all vectors.
N	Narrative Manipulation	Distortion of data and intent leading to system signal corruption.
C	Capture by Private Interests	Institutions hijacked by private interests are undermining public value.
E	Elasticity in Time-to-Adaptation (TTA)	The speed at which institutions pivot from disruption back to functionality.

Summary

- **REBALANCE** = Core Structural Health (Foundational Vectors).
- **RESONANCE** = Blind Spots and Amplifiers (Augmentation Vectors).
- **Together**, they create a **comprehensive fragility and resilience diagnostic framework** for complex systems.

REBALANCE systematically exposes, measures, and preemptively repairs structural weak points before they cascade into systemic failure. Its approach makes dynamic stress visible and actionable, providing an unprecedented roadmap for designing adaptive, resilient, and accountable systems.

From Feedback Loops to Structural Decay

In control engineering, failure rarely begins with a single event. It starts with undetected drift faults that compound until a boundary is breached. Good systems listen continuously, identify weak signals, and correct course with minimal effort. Bad systems break.

Public systems operate similarly, but without sensors and control surfaces. They run on policies, routines, human interpretation, and trust. REBALANCE adapts systems thinking to these environments by translating risk into measurable, correctable dimensions – making stress visible, quantifiable, and actionable.

As an engineer monitors current, pressure, or vibration, REBALANCE audits legitimacy, interdependence, overload, and failure response; it is not predictive analytics. It is structured awareness.

Method: Structural Diagnostics in Five Phases

REBALANCE audits follow a five-phase loop, each grounded in systems control practice:

1) **System Orientation** – Define operational boundaries, mission outputs, and known stress points.
2) **Vector Scoring** – Quantify stress across all the first nine dimensions using evidence, performance logs, interviews, and sensor data.
3) **Signal Review** – Identify missed or dismissed warnings, including near misses, maintenance logs, and suppressed feedback.
4) **Scenario Testing** – Run stress scenarios to reveal cascade risks, control gaps, or vector correlations.
5) **Reform Mapping** – Design targeted, ranked interventions, such as new controls, structures, protocols, or investments.

The result is a multi-vector heatmap: a structural portrait of the system's health and exposure (Figure 7.1).

GTD: Mapping Systemic Drift Across Time

The GTD framework is designed to capture where stress resides within a system and how that stress propagates over time. While REBALANCE defines the structure of fragility – nine specific vectors of pressure – GTD defines the trajectory: the velocity, slope, and convergence patterns that indicate how close a system is to structural failure.

Every system absorbs disturbance. What matters is the rate at which internal buffer capacity erodes while external pressures escalate. GTD is a framework to model that race. It does not predict outcomes. It maps state acceleration. That distinction is critical when designing interventions for resilience.

In GTD, entropy does not accumulate in isolation. As it increases, it reshapes the cost of execution, amplifies noise, and widens misalignment between institutional action and public trust. These are not independent events; they are systemic covariates. GTD enables these relationships to be plotted, measured, and adjusted over time.

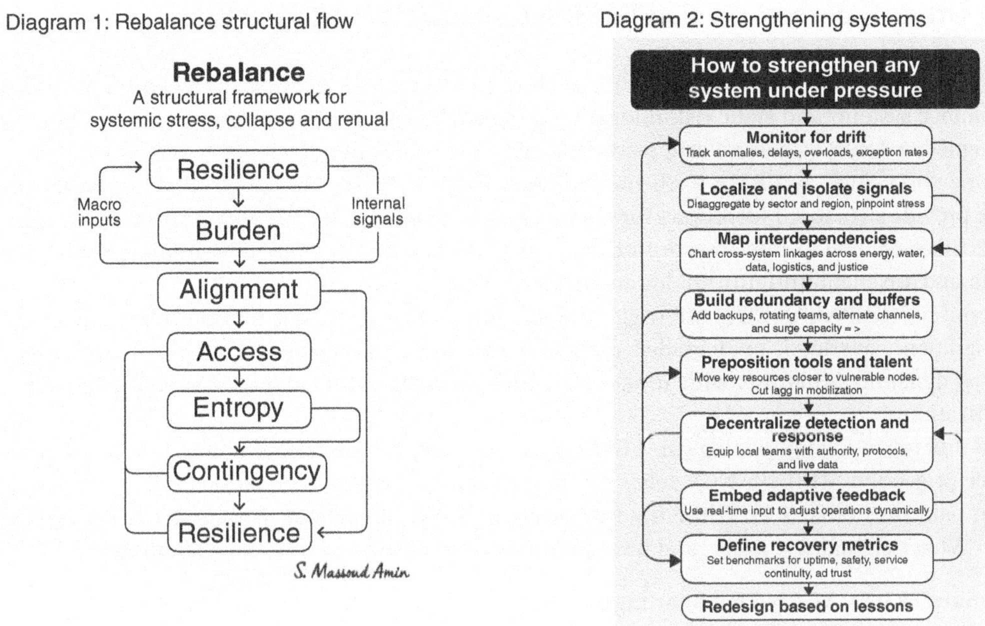

Figure 7.1 Nine steps to identify weak points and harden systems before fracture.

Consider a municipal emergency response system. Suppose we know that Execution stress rises 1.2× for every 0.3 unit increase in Entropy over 18 months and that public Alignment drops sharply after a 0.6 divergence between communicated capacity and actual performance. In that case, we have a mathematical picture of drift. We can test how fast intervention must occur before cascading effects overwhelm the system.

In GTD models, each REBALANCE vector has a temporal weight:

- Entropy as exponential decay
- Burden as seasonal amplitude
- Noise as a stochastic shock
- Access as an exposure multiplier
- Contingency as inverse time-to-failure
- Execution as response slope lag
- Legitimacy (a GTD extension) as a feedback decay constant

Together, they form a vector field. When visualized, that field reveals how systems transition from stability to brittleness – not in static snapshots, but in motion.

This trajectory allows planners to intervene where stress accumulates and when intervention is most effective. GTD reveals the optimal moment for reform – not too early when resources are scarce, not too late when collapse is irreversible.

In past audits, GTD has been used to forecast when power grids will lose dispatch reliability due to age-weighted transformer density and load variability, when housing programs will breach public trust thresholds due to waitlist stagnation and budget exhaustion, and when health systems will become unsalvageable under compounded emergency demand and political constraints.

Each model is different. However, the principle is consistent: map drift with time-weighted system observables, then act before that drift becomes permanent displacement.

Why Other Frameworks Fail First: A Comparative Diagnostic

Resilience models used in public and infrastructure sectors tend to fall into two categories: checklist compliance systems and static risk-mitigation schemas. Both serve baseline purposes, but neither is designed to detect structural drift before the onset of visible failure.

Frameworks such as FEMA's All-Hazards Approach or NIST's SP 800-30 Risk Management Guide provide structured workflows for planning and response. They ask what to do after a hazard materializes or how to rate known threats. However, they do not track pressure buildup in the system and are silent until thresholds are crossed.

In contrast, REBALANCE is not triggered by failure. It is activated by motion shifts in entropy, misalignment, overload, or degraded execution that arise upstream of collapse. It treats systems as dynamic structures with measurable stress vectors, rather than static objects awaiting classification.

The difference is methodological. FEMA and NIST rely on scenario enumeration and consequence projection. REBALANCE relies on state estimation, vector correlation, and signal degradation patterns (Table 7.2). While the former asks, "What happens if X occurs?" REBALANCE asks, "What's changing – now – and how far are we from collapse if that trend continues?"

Summary of REBALANCE Advantages

- Only framework fully rooted in dynamical systems and control theory.
- Highest quantifiability: State variables, structural vectors, and entropy thresholds.
- The only method offering early warning and drift detection.
- Bridges infrastructure, civic, institutional, and digital systems.
- Designed for integration with digital twins, AI systems, and operational dashboards.

Traditional failure modes and effects analysis (FMEA) and hazard and operability studies (HAZOP) are effective in tightly bound technical environments. But they stop functioning when stress crosses boundaries – into civic trust, procedural gaps, or digital interfaces. REBALANCE is designed to operate across boundaries.

In a 2023 joint audit of regional infrastructure systems across two counties, REBALANCE surfaced drift in Execution, Alignment, and Entropy vectors in one jurisdiction five months before FMEA-based asset ratings began to degrade. This lead time allowed for preventive investment, demand reallocation, and policy recalibration without service interruption. The legacy systems never flagged the vulnerability.

The key distinction lies in intent. Conventional frameworks classify and document. REBALANCE exposes and intervenes. It does not need a known hazard to act. It requires only measurable change.

The result is a framework that complements existing models – it renders them actionable. It offers the early warning that these systems are structurally incapable of providing (Table 7.3).

Why: Helps bridge the gap between engineering, risk, and executive leadership (Table 7.4).

Heathrow Airport (2025): A Case of Latent Fragility

When a power substation caught fire on March 20, 2025, operations at Heathrow's Terminal 2 ground to a halt. Over 1,300 flights were canceled. Hundreds of thousands of passengers were stranded. The root cause was not fire – it was a design flaw.

Table 7.2 Comparative Analysis of Resilience and Diagnostic Frameworks.

Framework	Sector Coverage	Method Type	Control/Systems Theory Basis	Quantifiability	Forecasting Capability
REBALANCE	Multi-sector (infrastructure, governance, and civic systems)	Vector-based diagnostics with dynamic system modeling	**Yes** – grounded in observability, feedback, control loops, and entropy analysis	**High–9** measurable vectors, structural heatmaps, and drift analysis	**Yes** – identifies drift, instability, signal degradation, and structural thresholds
NIST Community Resilience Planning Guide	Built environment, civil engineering, and community resilience	Guidelines, self-assessment protocols, and recovery planning timelines	**Partial** – some system-level thinking, but lacks dynamic modeling	**Low** – primarily qualitative and scenario-based	**No** – used for planning and post-event improvement, not for early detection
DOE Grid Modernization Index (GMI)	Electric grid infrastructure and utilities	Index-based benchmarking and survey assessments	**No** – benchmarking focused, not control-oriented	**Medium** – includes quantitative data but lacks predictive analytics	**Limited** – offers a performance snapshot, not dynamic forecasting
World Bank Resilience Rating System	Infrastructure finance, development planning	Scoring-based resilience assessment for projects	**No** – designed for investment readiness and financial planning	**Medium** – scoring framework applied to infrastructure programs	**No** – captures resilience readiness but lacks dynamic evolution modeling
UNDRR Disaster Resilience Scorecard	Urban systems disaster response, emergency management	Checklists and status evaluation tools for municipalities	**No** – checklist-based, minimal systems engineering structure	**Low** – mostly qualitative; limited benchmarking metrics	**No** – audit tool used to evaluate current capacity, not forecast failure
Rockefeller 100 Resilient Cities (now GRCN)	Urban governance, policy resilience, and cross-sector leadership	Strategy frameworks and KPI-based urban assessments	**No** – focuses on governance and qualitative assessments	**Low** – city-specific KPIs vary widely; limited modeling rigor	**No** – strategic in nature, not designed for technical early warning
ISO 31000 Risk Management Framework	Generic organizational and enterprise-level risk	Principle-based framework for managing uncertainty	**Partial** – draws on some system-level feedback ideas but lacks modeling	**Medium** – depends on implementation; some metrics are possible	**Limited** – not intended for system forecasting or state-space analysis

Table 7.3 What REBALANCE Adds to Traditional Risk Management
Purpose: To show a side-by-side comparison of classic enterprise risk management (ERM) features and what REBALANCE contributes, especially to executives or boards.

Function	Traditional ERM	REBALANCE
Identifies known risks	Yes	Yes
Detects unknown/emerging risks	Rarely	**Yes**
Uses state-based modeling	No	**Yes**
Measures interdependency risk	No	**Yes**
Offers quantitative diagnostics	Sometimes	**Always**
Supports early warning/signal response	No	**Yes**
Applies systems theory	No	**Yes**
Drives technical + governance alignment	Weak	**Strong**

Table 7.4 Comparative Resilience Dimensions Across Energy Sector Frameworks
Purpose: Show how major resilience initiatives in the **energy sector** (e.g., DOE, NERC, EPRI, FEMA, and EEI efforts) address or neglect specific system attributes, then contrast this with REBALANCE.

Attribute	DOE GMI	NERC CIP	EPRI Resilience Metrics	FEMA Lifelines Framework	REBALANCE
Cyber-Physical Integration	Moderate	High	High	Low	**High**
System Drift Detection	Low	Low	Medium	Low	**High**
Feedback Loop Modeling	Low	Low	Medium	Low	**High**
Cross-sector Interdependency	Medium	Low	Medium	High	**High**
Human Factors/Execution Readiness	Low	Low	Medium	Medium	**High**
Forecasting Capability	Limited	None	Limited	None	**High**

A REBALANCE audit revealed four critical stress vectors in failure:

- **Entropy:** Aging infrastructure, deferred upgrades, and outdated relays.
- **Access:** Single-point exposure – no isolation and no sectionalization.
- **Contingency:** No fallback systems beyond critical runway lights.
- **Linkage:** Power loss cascaded into IT, fuel logistics, and telecoms.

Each of these stressors was measurable before failure, had been documented in maintenance logs, and was known – but not prioritized (Figure 7.2).

Post-event recommendations were textbook control responses:

- Build modular on-site generation systems (e.g., microgrids or fuel cells).
- Segment terminal systems to reduce coupling.
- Define failover modes and rehearse degraded operations to ensure seamless recovery.

Without a structural diagnosis, however, they remained secondary until after collapse.

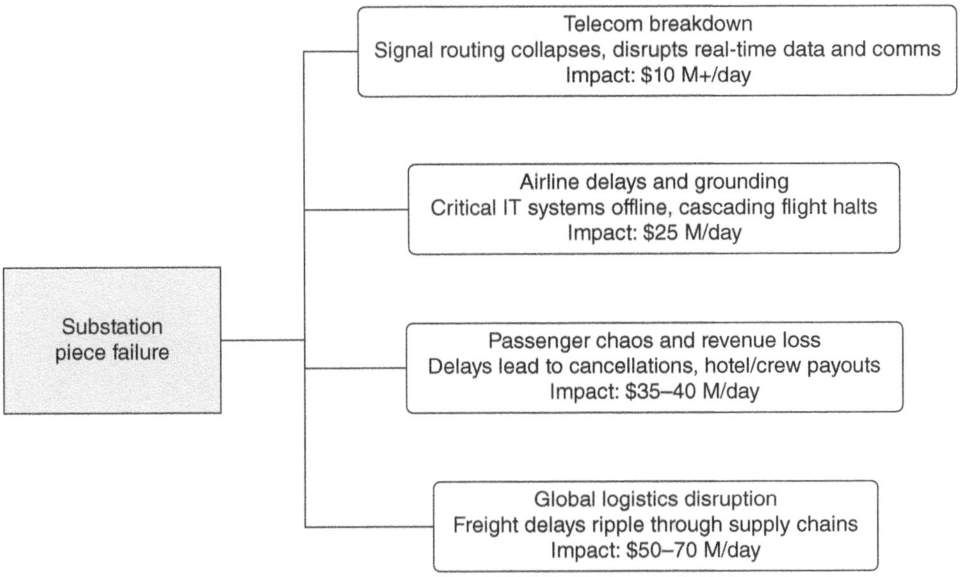

Figure 7.2 Real-time drift detection can reveal where systems approach overload or failure thresholds.

Puerto Rico (2017): Collapse by Design

Hurricane Maria was the catalyst, but the collapse of Puerto Rico's grid was engineered long before landfall. A centralized, fragile system with no black-start capacity, no distributed reserves, and minimal maintenance suffered a 100% outage across 3.4 million people.

Four primary vectors failed:

- **Burden:** Load-to-capacity imbalance from decades of underinvestment.
- **Contingency:** No backup power; zero sectional control.
- **Execution:** Response lags, supply chain gaps, and delayed recovery protocols.
- **Resilience:** Communities with solar microgrids recovered faster than the grid itself.

Even a moderate Category 1 storm would have put the system under stress. Hurricane Maria merely accelerated the inevitable.

Reconstruction efforts now incorporate REBALANCE principles, including distributed generation, modular recovery zones, redundancy in supply chains, and telemetry-based observability. Recovery isn't just rebuilding – it's re-architecting (Figure 7.3).

Mapping Drift: A Resilience Audit in Practice

REBALANCE audits operate like root-cause analysis in control engineering – but instead of searching for faulty switches or software glitches, they trace interdependent stress across physical, human, and institutional layers.

Consider a Midwestern utility audit conducted in 2023. It showed high stress in five of the nine REBALANCE vectors – Entropy (aging infrastructure), Contingency (no black-start capability), Burden (growing seasonal demand), Alignment (public frustration with rising rates),

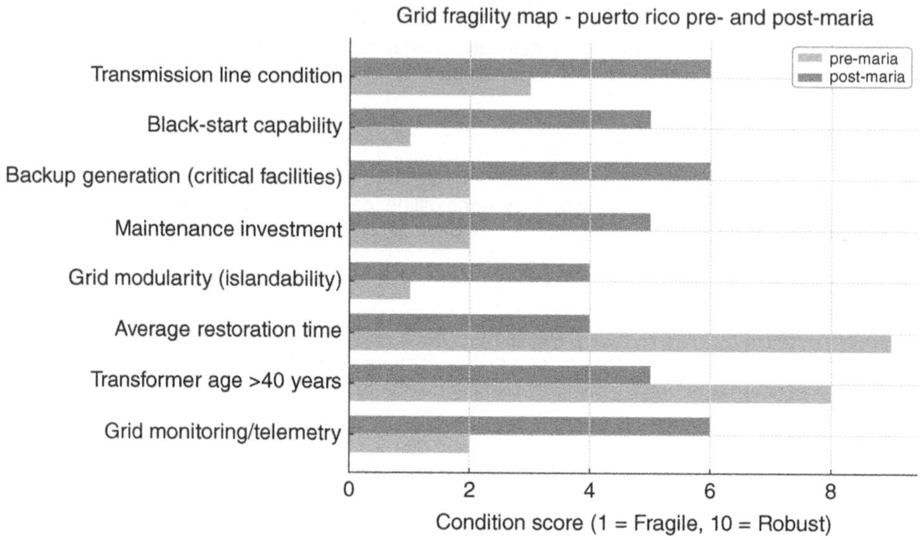

Figure 7.3 Grid Fragility Map – Puerto Rico Pre- and Post-Maria. It visually compares key grid dimensions using a 1–10 scale (1 = fragile, 10 = robust), based on documented deficiencies before the 2017 blackout and the structural reforms implemented in its aftermath.

and Execution (outdated emergency protocols). These weren't thematic concerns; they were measurable: the average transformer age exceeded 42 years, generator load peaked at 97% during heat events, and public complaint volume tripled over two years.

GTD added trajectory: If trends continue, entropy-driven failures and alignment gaps would likely converge during a high-demand event by 2026. A stress-tested scenario – a severe heatwave with transmission faults – showed cascading failure in eight hours.

The fix wasn't theoretical. Within 12 months, the utility began rotating out aging switchgear, installed a 60 MW mobile reserve, conducted three community engagement cycles, and revised its incident response plan. The result was a structural feedback loop, not just a technical one: an institutional signal became an operational priority (Figure 7.4).

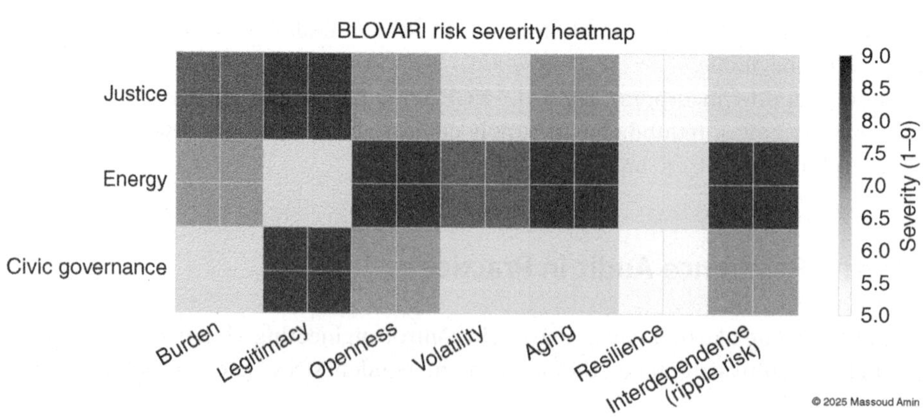

Figure 7.4 Heatmap reveals the highest systemic vulnerabilities in the justice and energy sectors. *Source:* Copyright ©2025 Massoud Amin.

This is how drift becomes visible. The system was never "broken"; it was drifting toward failure – until its operators began listening.

Precision Saves: A Compact Case Study in Preemptive Correction

In 2023, a metropolitan university system deployed a REBALANCE audit across its digital, physical, and service infrastructure. No acute crises were visible. However, vector scores indicated rising entropy in building HVAC systems, degradation in service satisfaction alignment, and high contingency vulnerability due to deferred maintenance.

The key insight came not from a single score but from the vector slope: Execution capacity was declining 3× faster than any previous year, with simultaneous stress emerging in service delivery.

GTD modeling forecasted convergence of those trends during early summer, when facility demand and enrollment processing would peak. A heat-triggered system failure would have disrupted student housing, enrollment continuity, and digital record access for more than 25,000 individuals.

Instead, targeted interventions were deployed immediately:

- Reassigned facility budgets to the five buildings with the steepest vector convergence.
- Issued rapid procurement for replacement chillers and air handlers.
- Ran a full-week simulation of enrollment surge across load-bearing IT systems.
- Staggered summer school onboarding to flatten the burden vector.

The result was a flatline – no failure, outage, or delay. The system absorbed the shock cleanly.

This was not cost-free, but it was efficient. Less than 3% of the annual facilities budget was reallocated, no capital expansion was needed, and no crisis communication was triggered.

The success was not in prediction; it was in structure: drift became measurable, intervention became tractable, and reform became surgical.

The most important takeaway is this: REBALANCE didn't prevent a known risk. It surfaced an invisible one, giving leadership the tools to act before systems failed.

The Texas Freeze (2021): Overload, Capture, and Silence

When temperatures plunged across Texas in February 2021, the state's isolated power grid suffered one of the most consequential blackouts in U.S. history. Nearly 250 lives were lost, and economic damage exceeded $130 billion.

Here, REBALANCE revealed compounded vector failure:

- **Burden:** Uninsulated equipment failed under conditions predicted a decade earlier.
- **Noise:** Climate volatility met brittle infrastructure.
- **Execution:** Coordination collapsed. Load-shedding was chaotic.
- **Alignment:** Deregulated markets had prioritized profit over resilience for years.

Perhaps most damning was the **Signal Review:** reports from 2011, 2014, and 2018 warned that the grid was vulnerable to cold snaps, but no action followed. The feedback loop had been severed by regulatory inertia and industry lobbying.

The REBALANCE remedy is to reinstate observability, realign incentives with resilience, invest in weatherization, and reintroduce independent system auditing (Figure 7.5).

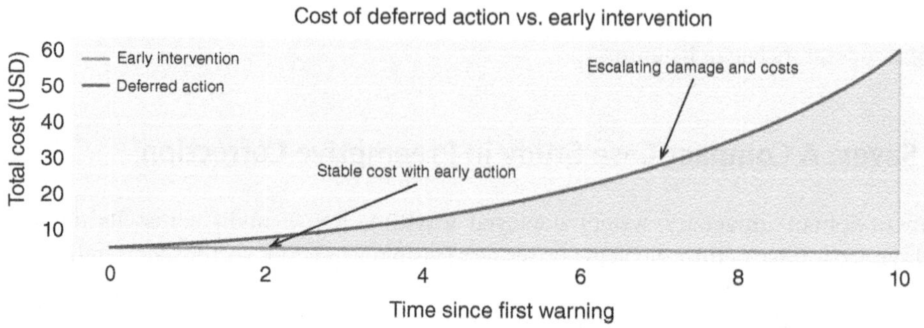

Figure 7.5 Delays compound damage. Early intervention keeps recovery costs flat.

Flint Water Crisis: Drift into Disaster

In 2014, Flint, Michigan, switched its water source without adequate treatment in a cost-saving maneuver. Residents reported discolored water and illness, but officials ignored the feedback – the result: mass lead exposure, irreversible developmental harm in children, and systemic trust erosion (Figure 7.6).

REBALANCE vectors:

- **Alignment:** Policy diverged from public safety.
- **Execution:** Monitoring was delayed, disjointed, and dismissive.
- **Access:** The public's voice – initial signals – was ignored.
- **Contingency:** No backup plans or rapid re-routing options.

Flint wasn't just a technical failure – it was a governance breakdown. The lessons are to strengthen feedback loops, audit stress proactively, and design institutions to detect and respond to public signals.

Applications Beyond Infrastructure

The strength of REBALANCE lies in its adaptability across domains:

- **Public Health (COVID-19):** Contingency (PPE, ICU capacity), Execution (response lags), and Noise (conflicting guidance) crippled readiness. Countries with predefined failover protocols and centralized dashboards fared better.
- **Education:** Remote learning failures in 2020 revealed severe Access inequalities, Entropy in school IT systems, and weak Alignment between institutions and underserved communities.
- **Justice Systems:** In Hennepin County, the concealment of police misconduct records eroded trust, causing a direct collapse of Alignment, Observability, and Execution vectors. Trust cannot be restored without transparency mechanisms – REBALANCE makes those gaps visible.
- **Supply Chains:** During COVID-19 and the Russia–Ukraine war, critical goods – transformers, fertilizers, and microchips – were constrained by poorly mapped interdependencies. REBALANCE-guided audits in 2023 helped reconfigure domestic inventory strategies and supplier diversification plans.

REBALANCE moves from abstract resilience theory to applied, verifiable action in all these domains.

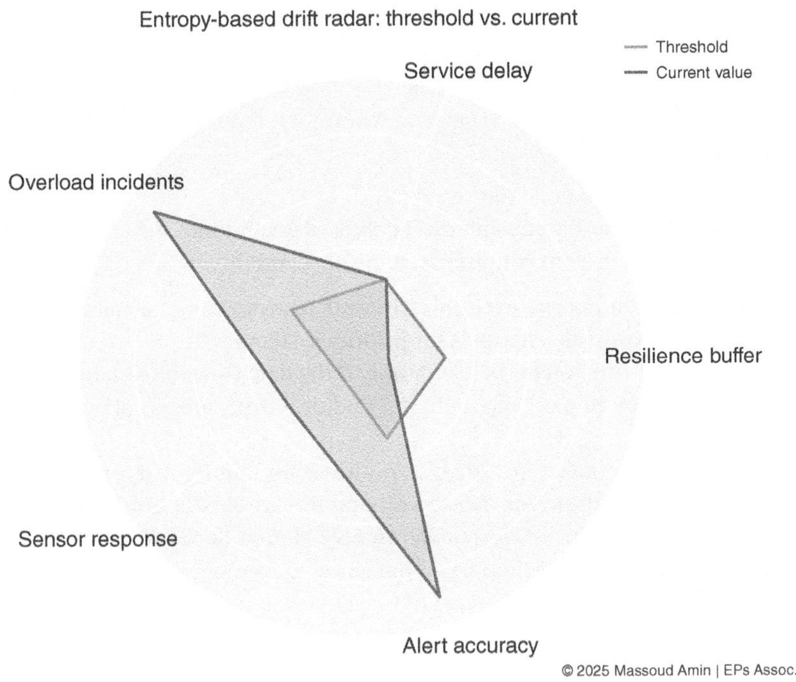

Entropy-based drift detection dashboard

Signal	Current status	Threshold	Current value	Drift velocity
Resilience Buffer	Normal	3	2	Stable
Service Delay	Warning	5	6	Rising
Overload Incidents	Critical	4	8	Escalating
Sensor Response Time	Warning	2	3	Rising
Alert Accuracy	Critical	4	7	Escalating

Figure 7.6 Deviation beyond safe thresholds flags rising risk – feedback must adjust before systems breach.

Resilience in Coupled Systems: Integrating Digital Twins, Cybersecurity, and Interdependence

Modern systems no longer fail in isolation. A blackout in a substation now takes out hospital telemetry. A cyberattack on billing systems cripples water distribution. A software bug delays transit, grid balancing, and emergency services simultaneously. This is not complexity by scale; it is fragile due to coupling.

REBALANCE is built for these realities. But to operationalize that readiness, it must integrate with the tools already in deployment – digital twins, control interfaces, and remote telemetry – and stress those systems at the point of operation and the edge of interdependence.

Consider the example of a 2024 pilot coordinated with a regional airport, an energy operator, and a municipal command center. Each had its own digital twin and monitored local KPIs – none monitored interfacial load under the scenario strain.

A REBALANCE-GTD integration exercise introduced a compound stress: a regional heatwave coinciding with a software update failure in HVAC systems and a telecommunications node shutdown. The event sequence was simulated: building automation failure led to generator overload, which triggered throttling signals to the substation, which dropped communications, and which disrupted airport bag and fuel logistics.

No system had failed. But the simulation showed where signal transmission bottlenecks created cascade risks. The audit showed:

- **Noise amplification** in automated fault response.
- **Access vector failure** due to shared login protocols across vendors.
- **Linkage failure** from tight system coupling with no observability.

The fix required no new infrastructure: only firewall restructuring, staggered load recovery protocols, and independent override channels for priority systems.

Digital twins can map real-time vector behavior only if the data flows cross boundaries. Without an REBALANCE overlay, they remain silent about structural drift, precise about operations, and blind to emergent failure.

Cybersecurity adds another layer. In 2023, a multi-vector audit of a financial transaction processor revealed latent Execution and Noise vulnerabilities. This is not because the systems were weak but because their intrusion response protocol shared dependencies with transaction latency prioritization. In effect, a real-time fraud detection measure delays breach detection. The response logic optimized throughput at the cost of breach observability.

In REBALANCE terms, the system scored well on Burden and nominal performance but collapsed under adaptive stress. The feedback loop – while functional – was illegible under strain.

These kinds of architectural paradoxes are increasingly common. The fragility doesn't lie in the code. It lies in the control model: in the misalignment between who perceives risk, who acts, and what logic governs conflict resolution.

To make REBALANCE actionable at this scale, we must treat each subsystem not just as an asset class, but as a node in an entangled control topology:

- Utility SCADA must speak to civic emergency ops.
- Digital twin observables must connect across transport, health, and power.
- Cyber protocols must be validated for threat prevention and vector integrity under stress.

This is not administrative integration. It is control alignment. And it is the only path to resilience in systems where signals propagate faster than policies can respond.

Intervention Without Chaos: Cross-Sector Governance and Vector Conflict

When multiple systems exhibit drift at once – transport, energy, and public health – the challenge is no longer diagnosis. It is governance. Who decides what gets fixed first, when, and with whose authority?

REBALANCE does not centralize control; it clarifies ownership. When scored, each vector is traceable to a system actor with operational jurisdiction and accountability.

During a 2022 urban REBALANCE deployment, four agencies – transit, emergency management, public works, and IT – registered moderate to high stress in shared infrastructure vectors. Historically, this would trigger duplication, conflict, or paralysis. Instead, a vector conflict resolution protocol was implemented:

1) **Vector Ownership:** Each stressor was mapped to its control locus (e.g., Burden in public works and Execution in transit).

2) **Time-to-Criticality:** Using GTD-informed slopes, each drift vector was assigned a "time window" before cascading failure.
3) **Cross-Impact Score:** Each vector's failure impact on other domains was rated and validated.
4) **Sequencing:** An Execution map was produced – what had to be done, by whom, and in what order.

This process turned parallel tension into sequenced action. The transit was delayed, allowing emergency services to patch a vulnerability first. IT segmented shared access points before deploying a new scheduling tool. No executive override was needed. The structure of drift itself dictated action.

Without this structure, interventions cascade into conflict. Agencies fight for budget, authority, or narrative. With REBALANCE, intervention becomes a control problem: state estimation, vector prioritization, and targeted input.

This approach has been piloted in municipal governance, multi-utility coordination, and cross-sector disaster drills. In each case, the outcome was not theoretical alignment. It reduced downtime, resulted in faster recovery, and had a fewer points of failure under load.

The most valuable feature of REBALANCE is not that it exposes risk. It shows who must act, in what order, and with what leverage.

What REBALANCE Is – and Isn't

It does not simulate. It diagnoses.
It does not predict. It reveals.
It does not replace governance. It strengthens it.

REBALANCE is a control-informed methodology to translate stress into structure, drift into reform, and noise into signal. It integrates seamlessly with digital twins, dashboards, and AI-enhanced sensing systems. But its core advantage is clarity.

It offers engineers a language of governance, gives policymakers a blueprint for intervention, and equips civic leaders with the tools to stabilize the systems people rely on daily.

Beyond Infrastructure: Institutional Drift and Civil Systems

Resilience is often framed in terms of hardening – reinforcing bridges, burying wires, or diversifying generation sources. But civil systems fail just as completely as their engineered counterparts, often with greater human cost.

In 2020, a U.S. state judicial system failed to disclose over 1,200 police misconduct records in active cases. The signals were not hidden. Internal audits, ethics complaints, and federal inquiries had been mounting for over a decade. What broke wasn't the law – it was the feedback loop.

In REBALANCE terms, this failure was driven by Alignment (erosion of institutional legitimacy), Noise (political pressure and legal complexity), and Execution (nonexistent audit-to-action protocols). Drift happened not in lines of code or power circuits, but in governance logic.

Applying a REBALANCE audit retroactively showed the indicators: memos were ignored, public trust polling collapsed, and judicial delays rose. If this were a mechanical system, the misalignment would have triggered a control failure warning years earlier. But in civic systems, we often wait until collapse.

Table 7.5 Organizational Use Case Suitability (Who Can Use REBALANCE?)
Purpose: Identify who can apply the REBALANCE framework and how it fits into their current decision-making or operations. This makes it easy for utilities and agencies to see where they fit.

Organization Type	Relevant REBALANCE Application	Primary Benefit	Integration Path
Electric Utilities	Grid stress audits, contingency planning, and outage prevention	Early warning and downtime reduction	Planning + operations dashboards
City Governments	Public works, crisis management, and equity-centered risk	Systemic audit and community trust	Cross-departmental resilience office
DHS/FEMA	Infrastructure resilience and disaster recovery strategy	Pre-failure diagnostics	Integrate with National Risk Index
DOE/National Labs	Grid modernization, research, and AI system design	State-space risk modeling	Coupled with digital twins
Private Sector (Critical Infrastructure Operators)	Operational continuity and supply chain mapping	Reduced recovery cost and signal clarity	Enterprise risk overlays

What makes the framework so adaptable is precisely this breadth. It treats signals – whether voltage imbalance, staff attrition, or FOIA refusals – as states of drift. It gives civic leaders the same tools engineers use to stabilize dynamic systems. It also proves, in practice, that the same principles apply: feedback, fault detection, graceful degradation, and reform through observability (Table 7.5).

Why: Clarifies adoption paths and utility-specific value propositions.

Policy and Research Agenda

Cities, utilities, and academic coalitions are already adopting REBALANCE. To scale its impact:

- **Federal Integration:** Agencies like DOE, DOT, DHS, and HHS should include REBALANCE audits in funding criteria and resilience planning.
- **Local Government Adoption:** In annual infrastructure and budget reviews, cities should embed vector scoring.
- **University Consortia:** Form collaborative centers of excellence to refine vector modeling, scoring calibration, and real-time sensing.
- **Standards Bodies:** IEEE, NIST, and ISO should help formalize REBALANCE metrics as part of resilience reporting standards.

The next phase: design digital resilience observatories where REBALANCE scores feed real-time dashboards across sectors (Figure 7.7).

Closing: Listening as Design

Collapse is rarely silent. It is often preceded by years of ignored data, buried reports, and broken feedback loops.

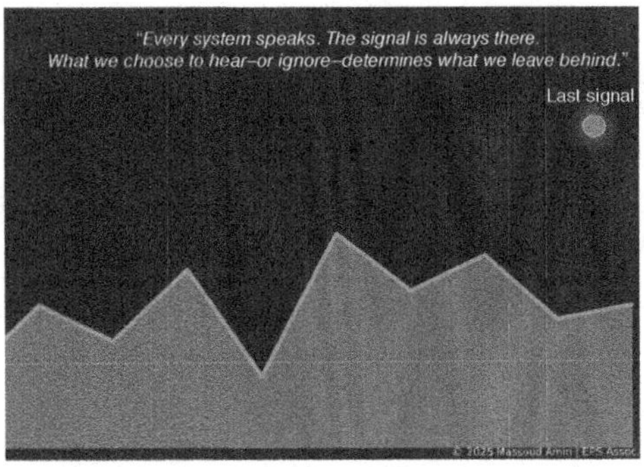

Figure 7.7 **Every system speaks –** The signal is always there. What we choose to hear – or ignore – determines what fails and survives us. **The last signal doesn't lie –** collapse doesn't come suddenly. Systems whisper before they break. We either learn to listen or leave the wreckage to someone else.

REBALANCE is not a crystal ball. It is a stethoscope. It listens to the structural heart of our systems before they go into arrest.

That kind of listening is not optional in a world of rising entropy and cascading complexity.

It is how we build systems that endure.

About the Author: Massoud Amin is President and Chairman of Energy Policy & Security Associates and an Emeritus Professor at the University of Minnesota. A global leader in infrastructure resilience and smart grid development, he is a Fellow of IEEE and ASME and a former chair of the IEEE Smart Grid.

Resources for Further Reading

1 Q. Zhu and T. Başar, "Disentangling resilience from robustness: contextual dualism, interactionism, and game-theoretic paradigms," *IEEE Control. Syst. Mag.*, vol. 44, no. 2, pp. 40–55, 2024, doi: 10.1109/MCS.2024.3273052.

2 T. R. McJunkin and C. G. Rieger, "Electricity distribution system resilient control system metrics," in *Proc. Resilience Week (RWS)*, Wilmington, DE, USA, 2017, pp. 103–112. [Online]. Available: https://ieeexplore.ieee.org/document/8069424.

3 National Renewable Energy Laboratory (NREL), "Puerto rico grid recovery post-hurricane maria," Tech. Memo, Golden, CO, USA, 2022. [Online]. Available: https://www.nrel.gov/docs/fy22osti/82399.pdf.

4 U.S. Government Accountability Office, "Critical infrastructure protection: electricity grid cybersecurity," GAO-21-81, Oct. 2020. [Online]. Available: https://www.gao.gov/assets/gao-21-81.pdf.

5 G. Adallon, "Small failure, big disruption – lessons from heathrow's fire," *Resilience Substack*, Mar. 2025. [Online]. Available: https://resilience.substack.com/p/heathrow-fire.

6 J. Busby, S. F. Prehoda, and B. C. Beard, "Cascading risks: the 2021 texas blackout and systemic vulnerabilities," *Energy Res. Soc. Sci.*, vol. 77, Art. no. 102106, 2021, doi: 10.1016/j.erss.2021.102106.

7 I. Linkov, B. Trump, and C. Fox-Lent, "Resilience: approaches to risk analysis and governance," Int. Risk Gov. Center, Lausanne, Switzerland, 2018. [Online]. Available: https://irgc.org/wp-content/uploads/2018/09/Linkov-Trump-Fox-Lent-Resilience-Approaches-to-Risk-Analysis-and-Governance.pdf.

8 D. Chandler, "Beyond neoliberalism: resilience, the new art of governing complexity," *Resilience: International Policies, Practices and Discourses*, vol. 2, no. 1, pp. 47–63, 2014, doi: 10.1080/21693293.2013.878544.

8

The "Golden Hammer" Syndrome in Tech Investments Across Industries

Introduction: When a New Tech Becomes a Golden Hammer

The "Golden Hammer" syndrome refers to the tendency of organizations to treat a single technology as the universal solution to every problem – akin to the proverb "If all you have is a hammer, everything looks like a nail." In practice, this means chasing the latest tech hype and bending business strategies to fit that tool, often at the expense of real needs. From telecom's 5G frenzy to finance's blockchain bandwagon and from artificial intelligence (AI) dreams in healthcare to green tech zeal in energy and smart factories, many industries have misallocated capital under the spell of trendy innovations. This chapter examines how hype-driven Golden Hammer investments have played out across these sectors, drawing on case studies, hard data on investment vs. returns, and cautionary tales of technology misalignment. We also highlight how savvy companies avoided these pitfalls by aligning tech adoption with economic reality. The goal is to distill lessons for business leaders on balancing innovation with pragmatism – a perspective vital for sustained success in an era of rapid technological change.

5G in Telecom: High Hopes, Heavy Investments, Low Returns

The telecommunications industry's embrace of 5G provides a textbook example of Golden Hammer thinking. Telecom operators worldwide spent lavishly on 5G spectrum and infrastructure, spurred by hype that 5G's ultra-fast speeds and low latency would unlock new revenue streams (from Internet of Things [IoT] to augmented reality/virtual reality [AR/VR]) and give first-movers an insurmountable edge. The result has been a massive capital outlay with anemic returns so far. Telcos have invested nearly $1 trillion in network upgrades (largely 5G) since 2018, yet are still struggling to monetize 5G services. In the United States, carriers paid over $80 billion in a single auction for mid-band 5G spectrum (Verizon alone spent $52.9 billion), illustrating the scale of bets on this technology.

Financial Impact: These investments have sharply increased telecom capital expenditures, yet revenue has not kept pace. Historically, returns on assets for wireless carriers have been only 1.5–4.5% across successive network generations, and 5G has yet to break that pattern. A PwC analysis notes that the returns remain in the low single digits of return on assets (ROA) against huge 5G deployment costs (including new radio equipment and cloud-native cores). For context, cloud service providers enjoy ~20% ROA, whereas mobile network operators (MNOs) hover around 2–3%, resembling utility-company economics. This imbalance is causing investor angst and strategic soul-searching in telecom. A total of 46% of telecom CEOs worry that if their companies continue on the current path, they won't be economically viable in 10 years – a stark assessment of how current investments (like 5G) are not yet yielding sustainable returns.

Case Study – 5G Hype vs. Reality: Major carriers launched 5G networks in over 70 countries by 2022 and promoted consumer applications like immersive video and gaming. Yet, the "killer apps" that would markedly boost revenue have slowly materialized. Enhanced mobile broadband has improved user experience, but hasn't unlocked *new* revenue comparable to the investment. For instance, many operators ended up in a "give more, get less" scenario – offering unlimited data at 5G speeds. At the same time, up-stack companies (streaming, cloud gaming, etc.) reap most of the profit. Essentially, telcos applied the 5G hammer to every nail (use case) they could imagine, but most of those nails haven't budged in revenue. The monetization challenge echoes the 4G era, where telecoms built the highways and tech giants took the tolls (e.g., video streaming boomed on telecom networks, but the profit flowed to content and OTT service providers).

Adoption Rate: Despite these challenges, adopting 5G infrastructure has been rapid – reflecting competitive pressure and fear of missing out. By the end of 2022, around 220 operators worldwide had launched commercial 5G networks across ~88 countries. 5G subscriptions are on track to reach 1.5 billion by the end of 2023, making it the fastest-growing mobile technology rollout. However, most of this growth is in enhanced mobile broadband (faster phone Internet) rather than the revolutionary IoT/enterprise services hyped during 5G's genesis. In effect, telecom firms have blanketed their markets with 5G – the Golden Hammer was swung broadly – but the incremental revenue per user remains modest, and the expected flood of new business uses (like autonomous vehicle networks or smart city services) is still a trickle.

Lessons: The telecom case underscores how technological enthusiasm can lead to overinvestment ahead of demand. Carriers assumed 5G's superior capabilities would automatically create new markets (from VR to smart factories) but underestimated the ecosystem development required. Many are refocusing on pragmatic steps to monetize 5G, such as offering network APIs to developers and enterprise partners to enable paid services on top of 5G. The key takeaway is that even a game-changing technology must align with a clear value proposition – without that, a Golden Hammer can quickly become a very expensive nail.

Blockchain in Finance: From Hype Machine to Expensive Pilots

In finance, blockchain was heralded as a revolutionary solution – a secure, decentralized ledger that could streamline everything from payments to securities settlement. Between 2015 and 2018, banks and exchanges worldwide treated blockchain as the *Golden Hammer* to drive fintech innovation, pouring resources into pilot projects and consortia. The hype was fueled by promises of cost elimination (no middlemen), real-time settlement, and tamper-proof records. As a result, approximately 90% of major banks globally had initiated blockchain projects or proofs-of-concept by the late 2010s. Surveys in 2016 found that 65% of banks expected to have blockchain in production by 2019, and 30% were already in advanced payment implementation stages. Those rosy projections vastly overshot the mark – a classic hype cycle outcome. Fast forward a few years, and the landscape is littered with stalled initiatives and write-offs, as financial firms learned that distributed ledgers are *not* a cure-all for every transactional inefficiency.

Financial Impact: The collective investment by financial institutions in blockchain experiments is hard to quantify, but it has been substantial. One estimate found that VCs and corporates invested over $25 billion in "cleantech 1.0" startups from 2006 to 2011 and ended up losing more than half of it by 2015 – and a similar story of squandered investment is playing out for enterprise blockchain. Gartner predicted in 2019 that 90% of blockchain-based supply chain projects would succumb to "blockchain fatigue" by 2023. Indeed, studies indicate an enterprise blockchain project failure rate of around 90%, with a typical pilot lifespan of just over a year before termination. Many banks spent millions on blockchain proofs-of-concept that never left the lab. For example, a 2018 survey found only 8% of blockchain pilots had scaled to production – meaning over 90% never delivered tangible return on investment (ROI). The financial impact is seen in high-profile fizzled projects: large consortia that disbanded after burning cash on development and internal projects quietly shelved after failing to justify their cost.

Case Studies*:* Several case studies illustrate how the blockchain hammer hit solid reality:

- Australian Securities Exchange (ASX): The ASX pursued a bold project to replace its aging clearing and settlement system (CHESS) with a blockchain-based platform. After five years and about AUD 250 million (~$168 million) spent, the exchange canceled the project in 2022, taking a hefty write-down. An independent review found serious design and interoperability challenges, and the blockchain solution's benefits no longer outweighed its uncertainty and cost. This failure highlighted that even where blockchain *works* technically, it may not align with large-scale financial markets' regulatory and performance needs.
- we.trade: This was a blockchain-based trade finance platform backed by a consortium of 12 major European banks and IBM. We launched with great fanfare to streamline international trade transactions. Trade never gained wide adoption beyond its founding members. The venture ran out of cash and filed for insolvency in mid-2022, as new investors balked and participating banks pulled funding. Both *we.Trade* and a similar shipping blockchain platform (*TradeLens*, by Maersk and IBM) suffered from a classic network effect problem – outside parties were reluctant to join, limiting the network's value. *High operating costs* compounded the issue: both we.Trade and ASX's project cited high costs as a key reason for the collapse.
- Corporate Consortia and Proof-of-Concepts (POCs): Nearly every large bank joined one or more blockchain consortia (R3 Corda, Hyperledger projects, etc.) or ran internal POCs. Yet by 2021–2022, many early enthusiasts had scaled back. For instance, the utility settlement coin project (USC) led by UBS and others was wound down, and numerous cryptocurrency trading desks at banks were closed during crypto downturns. Even IBM, once touting blockchain as a major business line, downsized its blockchain division after lackluster traction. In summary, the wave of fintech blockchain pilots produced very few live deployments – one study found zero documented impact for 43 blockchain use cases examined in international development and trade.

Adoption Rate*:* Despite the setbacks, some level of blockchain adoption persists in finance, but it's far more limited and focused than initially hyped. A 2021 Deloitte survey found that 81% of global financial services firms still see blockchain as broadly scalable long-term, indicating continued interest. Certain niches have succeeded; tokenized assets and digital bonds are gaining traction (with over $1 billion in tokenized money market funds by 2023). However, broad adoption has been slow. As of 2020, only 19% of businesses in a Gartner survey deemed blockchain "very important" to their operations. In banking, many projects remain in *pilot purgatory* or limited-scope use. Most of the industry has effectively hit the "pause" button – observing a few early movers and waiting for the technology to mature and prove itself. Gartner noted this dynamic as the trough of disillusionment: organizations grew fatigued by the mismatch between blockchain's enormous expectations and the modest reality of its delivered value.

Lessons: The finance sector's blockchain journey teaches a clear lesson about misaligned technology investments. Not every problem is a nail for which blockchain is the hammer. Many banking processes that blockchain aimed to disrupt (like interbank payments or trade finance) already had reliable, if less glamorous, solutions. Replacing them with a new paradigm required industry-wide coordination, regulatory buy-in, and technical breakthroughs – challenges underestimated amidst the early hype. Successful finance firms have refocused on technologies more tightly linked to ROI. For example, instead of blindly pursuing blockchain, banks like JPMorgan and others turned attention to real-time payments modernization (using more proven tech) and AI-driven fraud detection – areas where they have seen clearer benefits. The sobering outcomes of blockchain pilots have instilled more discipline: financial executives now demand a solid business case and clear value-add before green-lighting "shiny" tech projects. This aligns with the broader principle that no single technology is a panacea, and due diligence is essential to avoid Golden Hammer pitfalls.

AI in Healthcare: A Cautionary Prescription

Few technologies in recent years have been as imbued with transformational hype as AI in healthcare. Machine learning algorithms and AI systems were proclaimed the ultimate diagnostic assistants, capable of reading radiology scans better than experts, identifying cancers, personalizing treatments, and even "replacing doctors" in certain tasks. Hospitals and health systems, fearing being left behind, raced to pilot AI tools for everything from clinical decision support to administrative automation. Global investment in healthcare AI has skyrocketed – over $30 billion poured into health AI startups in the last three years and roughly $60 billion in the previous decade. In 2023 alone, venture funding for health AI in the United States is estimated at $11 billion. However, the return on investment in healthcare AI remains hard to pin down for all that spending, and some high-profile initiatives have failed. This makes healthcare a vivid case of Golden Hammer's enthusiasm for meeting the complexity of real-world medicine.

Financial Impact: Healthcare providers have been more cautious than some industries in adopting new tech, yet the lure of AI's potential efficiencies and improved outcomes has driven significant expenditures. Hospitals invested in AI-driven software for diagnostics (e.g., imaging analysis and pathology), predictive analytics for patient deterioration, and chatbot or triage systems for patient interaction. While exact ROI figures are scarce (health systems often don't disclose detailed returns), there are telling indicators: a 2022 KPMG report noted that many healthcare leaders struggled to realize ROI from recent tech investments, with only 35% reporting significant value realized from AI projects. At a macro level, one study estimated that broad AI adoption *could* eventually save 5–10% of healthcare costs (translating to $200–$360 billion annually in the United States). Still, such savings remain mostly theoretical at this stage. Meanwhile, some organizations have sunk tens of millions into AI with little to show:

- MD Anderson & IBM Watson: In one of the most cited failures, the University of Texas MD Anderson Cancer Center partnered with IBM to apply Watson (IBM's AI) to oncology decision-making. After five years and $62 million spent, the project was terminated in 2017 without ever being used on patients. The AI system, intended to recommend cancer treatments, could not integrate with clinical workflows and provided inconsistent advice, leading the center to pull the plug. An audit found the project overran budgets, expanded scope, and ultimately delivered no clinical value, while the cancer center faced budget deficits. This is a stark $62M lesson in the perils of buying into AI hype without clear alignment to patient care needs and readiness.

- Babylon Health: On the startup side, Babylon Health was a UK-based pioneer in AI-driven telehealth, offering an AI symptom checker and remote consultations app. It attracted enormous hype and funding (peaking at a $4 billion valuation upon going public in 2021), with bold claims of providing 80–90% of primary care through AI. By 2023, Babylon collapsed into bankruptcy – its U.K. operations went into administration, and assets were sold for pennies on the dollar. Insiders admitted the AI diagnostic engine was overhyped and not as capable as claimed, and the company's economics were unsustainable. Babylon's fall (from "unicorn" to bust in about two years) is a cautionary tale that even with significant capital (~$1 billion+ raised), a Golden Hammer approach of "AI solves everything in healthcare" can rapidly lead to failure if the real-world viability isn't there.
- IBM Watson Health: Beyond MD Anderson, IBM's broader Watson Health division also struggled. IBM invested heavily in applying Watson AI in healthcare (oncology, drug discovery, etc.), acquiring companies, and marketing its AI's prowess. By 2022, IBM decided to sell off major parts of Watson Health at a loss. The AI often underperformed expectations; for instance, Watson for Oncology was sometimes criticized for unsafe or irrelevant recommendations. This retrenchment signaled that even a tech giant's multi-billion-dollar foray into healthcare AI could not meet the hype, largely because the technology was not mature enough to handle the nuances of clinical care.

Adoption Rate*:* Despite setbacks, AI adoption in healthcare continues to grow in pockets, but it is far from ubiquitous. According to the World Economic Forum, most generative AI in healthcare is currently limited to leading academic medical centers and technology-forward institutions. The typical diffusion pattern is at play – top hospitals experiment and validate while others watch closely. A 2023 survey of hospital executives found that while interest in AI is high, only a minority have scaled these tools broadly. For example, only ~34% of healthcare organizations had an enterprise AI strategy in place by 2022, and many use AI in just one or two departments (like radiology) rather than system-wide. On the startup side, providers' adoption of AI solutions has been slower than the funding figures suggest – venture investors may have poured $60B in a decade, but convincing doctors and nurses to trust and regularly use AI assistance is a gradual process. Healthcare providers tend to be prudent buyers who demand proven ROI and safety before scaling up new tech. This prudence is justified given that patient lives are at stake; an algorithm that's 90% accurate can still be dangerous if not integrated properly. AI in healthcare is mostly assisting (e.g., AI reads scans as a *second pair of eyes* for radiologists, flagging cases of concern) rather than replacing tasks, and large-scale adoption will hinge on rigorous validation.

Lessons*:* Healthcare's experience with AI demonstrates the importance of tempering innovation with realism. The Golden Hammer syndrome here led some to believe that if a little AI worked, more AI should be applied everywhere – a costly fallacy. Successful healthcare organizations have taken a more incremental approach. For instance, the Mayo Clinic and others have focused on narrow AI applications (like improving scheduling or specific image analysis algorithms) with measurable benefits rather than moonshot projects. Best practices emerging in healthcare include slowly integrating AI tools into clinician workflows, providing transparency on how algorithms make decisions, and maintaining a "human-in-the-loop" to catch AI errors. The failures of projects like MD Anderson's Watson have made the industry more cautious: new AI deployments often start as research trials with outcomes tracked rather than full-production rollouts based on hype alone. Healthcare reminds us that a technology's promise means little if it doesn't mesh with operational realities and user trust. In other words, even the smartest hammer is useless if it doesn't fit the screws and bolts of the task at hand.

Green Tech in Energy: When Idealism Meets Investment Realities

The energy sector's encounter with "green tech" hype – from solar and biofuels to wind and hydrogen – illustrates Golden Hammer's misalignment on an industry-wide scale. In the late 2000s, a wave of Cleantech 1.0 startups promised to disrupt energy with renewable innovations, attracting massive investment. More recently (2015–2021), oil and gas giants under pressure to address climate change made bold pivots to renewables and low-carbon tech. In both cases, enthusiasm sometimes outpaced economic fundamentals, leading to painful lessons.

Financial Impact – Cleantech Boom and Bust: Venture capitalists and governments funneled enormous sums into early renewable energy startups between 2006 and 2011. However, the returns were often dismal. Of the $25 billion VC invested in Cleantech 1.0 (renewables, biofuels, etc.) during that period, more than half was lost by 2015. A striking example was Solyndra, a solar panel manufacturer that raised over $1 billion (including a hefty U.S. government loan guarantee) only to go bankrupt in 2011, rendering that investment nearly worthless. Another was KiOR, a biofuel company once valued at $1.6 billion that burned through $600 million and went bankrupt in 2014. These failures were often due to *economic misalignment*: Solyndra's novel solar panels were too expensive compared to rapidly falling silicon solar prices, and KiOR's biofuel couldn't compete with conventional fuel economics. In hindsight, investors swung the Golden Hammer of green tech at the energy problem, but many backed unproven tech that couldn't survive without sustained subsidies or breakthroughs.

Large energy companies faced similar issues on a different scale. European oil majors, in particular, declared strategies to invest heavily in renewables to transform their business models. BP, for instance, announced in 2020 that it would cut oil and gas output by 40% by 2030 and spend billions on wind and solar projects. But by 2023, reality set in: BP encountered rising costs and lower-than-expected returns in renewables, leading it to write off $1.1 billion on an offshore wind venture and scale back its green ambitions. Shell and Equinor likewise moderated their renewable investment plans when early projects yielded poor returns or delays. The fundamental issue was profits: traditional fossil projects were yielding better returns. In 2022, the median return on capital for oil companies was ~11%, vs. just ~2% for renewable energy companies. This huge gap made investors skeptical and forced energy firms to rethink pouring money into low-return projects. In effect, many of these companies discovered that renewables, as vital as they are for the future, were not a financial silver bullet – treating them as such was a misstep.

Adoption and Trends: On one hand, renewable energy adoption has grown rapidly worldwide – solar and wind capacity have surged (renewables now account for about 29% of global electricity generation), and technologies like electric vehicles and battery storage have seen exponential uptake. So, regarding *physical adoption*, the green tech revolution is underway and driven by policy mandates and improving technology. By 2025, 1.2 billion 5G connections are expected (telecom), and similarly, in energy, the International Energy Agency projects renewables will account for 90% of new power capacity additions in the next few years. However, *financial returns* in some segments lag behind this physical growth. Many renewable projects depend on subsidies or favorable contracts. Companies like Ørsted (a leading offshore wind developer) have faced challenges with cost inflation: in 2023, Ørsted announced impairments of up to $5.6 billion on U.S. wind projects due to supply chain issues and higher interest rates. These write-downs crushed its stock price and served as warnings that even established green companies can hit turbulence if the economics shift.

In Table 8.1, we summarize some key investment trends and outcomes in these sectors:

As Table 8.1 suggests, the gap between expectation and reality has been a common thread. Whether 5G is not yet delivering new revenue streams, blockchain pilots fail to replace legacy systems, or AI is proving harder to implement in hospitals than anticipated, these experiences reflect the Golden Hammer syndrome. Companies treated these technologies as *must-have fixes* for growth or efficiency, but many did so without fully understanding the specific use cases, costs, and integration challenges.

Past Failures and Near-Misses: Learning from Misaligned Tech Investments

Before turning to best practices, it's worth noting that the concept of misallocated tech investments is not new. History is rife with examples of companies that bet big on hyped innovations and suffered losses:

- **Dot-Com Bubble (2000):** Dozens of startups (and even established firms) embraced the Internet as the golden solution without viable business models, leading to spectacular dot-com busts. Companies like Webvan spent heavily on tech (automated warehouses and delivery algorithms) but collapsed when demand and economics didn't line up – a $1 billion failure driven by tech optimism over business reality.
- **DeLorean Motor Company (1980s):** As referenced in business literature, DeLorean is a cautionary tale outside of high-tech. It hyped a revolutionary car (with innovative design and manufacturing techniques) as its hammer, attracting massive investment and government incentives. Ultimately, production issues and a lack of market fit led to bankruptcy. While not "tech" in the modern sense, it underscores how chasing a shiny idea without solid fundamentals can sink a company, even in industrial bets.
- **Theranos (2010s):** The Silicon Valley blood-testing startup claimed to have a groundbreaking proprietary technology that could run hundreds of tests from a fingerprick. It courted major investors and pharma partners on a wave of hype (with a charismatic founder but no peer-reviewed evidence) – a classic shiny-object syndrome. Reality revealed the tech didn't truly work; Theranos imploded in scandal. Investors and Walgreens, which had rolled out Theranos devices in stores, lost hundreds of millions collectively. The lesson: hype without transparency and validation is a recipe for disaster, even in healthcare.
- **FTX and Crypto Exchanges:** In the cryptocurrency realm, exchanges like FTX grew at breakneck speed, touting crypto as the future of finance. FTX's collapse in 2022 showed that hype-fueled growth (and disregard for basic governance) can lead to catastrophic failure. While FTX was more a case of fraud and mismanagement, it's intertwined with a broader theme – believing the crypto hype (blockchain as a hammer) could justify flouting traditional risk controls.

Each failure reinforces that technology must be married to sound economics and execution. Innovation without discipline often ends in tears. They also highlight the importance of

Table 8.1 Overview of Tech Hype Investments vs. Outcomes Across Sectors.

Sector	Tech Hype (Golden Hammer)	Major Investments/Costs	Adoption Rate/Trend	Outcomes and ROI
Telecom	5G networks and services	Nearly **$1 trillion** spent on 5G upgrades globally (2018–2023); the United States auctioned $80 billion for mid-band spectrum.	**220+ 5G networks** launched worldwide by 2022; ~1.5 billion 5G connections by 2023.	
(fast adoption)	**Low ROI:** Telecom ROA is only ~1.5–4.5% on 5G assets. Revenue uplift is modest, and **monetization struggles** continue.			
Finance	Blockchain for transactions	Hundreds of pilot projects; consortia like R3 raised over $100M+; **ASX spent $168M** on a blockchain system.	**90%+ of banks** explored blockchain; however, **<10% of projects** reached full production.	**Many failures:** For example, ASX **wrote off $168M**, project canceled; **we.Trade insolvency** due to low uptake. Most pilots have little to no ROI (90% failure rate).
Healthcare	AI for diagnosis and operations	A total of **$60B** invested in health AI startups (2012–2022); hospitals spent tens of millions on AI tools (e.g., MD Anderson spend $62M on IBM Watson).	Growing interest but **limited deployment:** AI is used mostly in pilot/limited settings; generative AI is in only the top centers. Most hospitals are in the early stages (prudent adoption).	**Mixed outcomes:** Some AI aids in radiology, but there are **notable failures** (MD Anderson's $62M Watson project ended with no clinical use). ROI is unclear; many AI projects have yet to show significant cost/quality improvement.
Energy	Green tech (solar, wind, and biofuels)	Cleantech VC **was $25B (2006–2011)**, but over 50% was lost by 2015. Oil majors invested billions in renewables (e.g., BP offshore wind and Shell new energy).	**Renewables are booming:** Solar and wind account for most new power capacity, and EV sales are rising. However, oil companies are scaling back – BP, for example, cut 2030 renewables targets.	**Profit gap:** Fossil projects ~11% RoC vs. renewables ~2%. **Cleantech busts:** Solyndra's $1B bankruptcy and others failed due to costs. Due to low returns, big firms like BP took a $1.1B write-down on wind projects.
Manufacturing	Smart factories (IoT and automation)	**$100+ billion** annual global investment in Industry 4.0 tech; GE's Predix IoT platform cost ~$7 billion development.	**Adoption is moderate:** ~**24%** of manufacturers have implemented smart factory initiatives, and another 22% are in trials. Others lag or adopt piecemeal.	**Results vary:** Leaders report efficiency gains but over-automation missteps (e.g., Tesla's "excessive automation" causing production delays). GE Predix **missed revenue targets by ~$14B**, prompting a strategy reset.

Each industry saw huge bets on a "trendy" technology (the Golden Hammer) with rapid adoption in pilots or infrastructure, but real-world economic returns have often fallen short.
Source: Relevant industry reports and case studies.

due diligence: many red flags (unrealistic financials, lack of independent vetting, or ignoring market signals) were overlooked in the rush to seize the "next big thing." Forward-thinking companies study these failures to avoid repeating them.

Avoiding the Golden Hammer: Best Practices for Aligning Tech with Reality

Having seen how Golden Hammer syndrome can lead to wasted capital and even ruin, how can companies guard against it? The experiences across industries suggest several best practices for ensuring technology investments are grounded in economic and strategic reality rather than hype:

1) Start with the Problem, Not the Technology: Successful firms flip the script. Instead of adopting tech for tech's sake, they identify critical business problems or opportunities first and then evaluate the best tools to address them. This sounds simple, but it requires discipline to resist shiny trends. For example, instead of saying, "We need blockchain," a bank might say, "We need to reduce settlement time from T+2 to same-day; let's compare all possible solutions (blockchain, existing systems optimization, etc.)." They won't force-fit it if the new tech isn't superior. As one software engineering adage warns: *"No single technology or framework is best for all situations,"* and bending requirements to fit a hyped solution is a red flag. Companies that avoided the 5G hype, for instance, were those that waited until they had a clear use case (like fixed wireless access to serve home Internet) before heavily investing, thus directly tying the technology to a revenue-generating need.
2) Conduct Rigorous Pilots and Demand Data: Before fully committing, do what the military calls *"test before you invest."* The best companies run limited POC projects to validate that technology delivers on its promise in their specific context. Crucially, they set success criteria (key performance indicators [KPIs] such as cost reduction %, accuracy, throughput increase, etc.) and measure the results objectively. If the pilot doesn't hit thresholds, they shelve the project – no matter how trendy – thereby avoiding bigger losses. This approach was highlighted in a 2024 analysis: enterprises should use a "careful and well-defined strategic approach to technology, backed by research and pilot POCs." In other words, earn the right to scale by first proving value on a small scale. When AI tools began emerging, some health systems first tried them in a research setting – for instance, testing an AI on past medical images to see if it could match expert diagnoses. Only after demonstrating sufficient accuracy did they even consider live deployment. This piloting mindset acts as a circuit breaker for hype. It also aligns with adopting tech closer to its *"plateau of productivity"* (when it's proven and stable) rather than at the peak of inflated expectations.
3) Tie Investments to Clear ROI and Strategic KPIs: Firms that evade Golden Hammer syndrome insist on a business-case discipline. This means quantifying a tech investment's expected return (or benefit) and tracking whether it materializes. For example, when evaluating an IoT smart factory system, a manufacturer might target a "10% reduction in downtime" or a "5% increase in throughput" as justification. It may not be justified if a technology doesn't show a path to those targets. Aligning technology with economic metrics ensures that decisions aren't driven solely by FOMO (fear of missing out) but by tangible value. Energy company Equinor provides an example: it shifted to a more selective approach in renewables, ensuring projects meet financial return hurdles and align with core goals rather than just pursuing green tech for image reasons. This financial gating can prevent overinvestment in trendy projects that don't pull their weight.

Additionally, companies should consider *opportunity cost* – for example, every dollar put into a speculative blockchain project is not invested in improving cybersecurity or upgrading core IT. Weighing alternative uses of capital can sober up decision-making.

4) Encourage Skepticism and Diverse Voices: The drive toward a hot new tech can become an echo chamber in many organizations – especially if top leadership is enamored with the idea. To counter this, leading companies cultivate a culture where it's acceptable for teams to voice doubts and conduct devil's-advocate analyses. One strategy involves cross-functional teams (IT, finance, operations, etc.) vetting new technology proposals. This diversity ensures that, say, the CFO's team can question overly optimistic revenue projections, or operations managers can flag integration headaches that tech enthusiasts might overlook. Cross-departmental collaboration in evaluating new tech tends to produce a more holistic understanding of its impact and feasibility. Amazon, for instance, is known for its memo-driven decision process where proposals are rigorously debated in writing – this often flushes out the practical challenges of a new idea alongside its merits. The goal is not to stifle innovation but to *filter it through real-world perspectives*. In essence, build an internal circuit breaker for hype by empowering people to ask: "Does this make us better off?" and listen to the naysayers, not just the evangelists.
5) Fail Fast, Learn, and Pivot: Even with prudence, some tech bets will miss the mark – and that's okay if you catch it early and adapt. Companies that thrive long-term treat failures as learning opportunities and avoid doubling down due to pride or sunk costs. For example, after Tesla's over-automation misstep in its Model 3 production (where robots created a bottleneck and quality issues), Elon Musk publicly admitted that "excessive automation at Tesla was a mistake… Humans are underrated." Tesla rapidly adjusted by reintroducing more manual assembly, which made sense, effectively pivoting away from the notion that 100% automation was the only goal. This agility – recognizing the Golden Hammer (full automation) wasn't right for every task – likely saved the Model 3 program. The broader point is that companies should put mechanisms (milestones, audits, etc.) in place to evaluate ongoing tech projects and be willing to cut losses or change direction if evidence shows the expected benefits aren't materializing. It's better to abandon a misaligned project after $2 million spent than to press on and waste $20 million more out of unwillingness to acknowledge a mistake. A "fail fast" mentality and capturing lessons learned can turn a potential fiasco into a stepping stone. This mentality is common in Silicon Valley startups and is increasingly embraced by large firms via innovation labs and accelerator programs that test many ideas but only scale the winners.
6) Strategic Alignment and Timing: Finally, timing matters. There is often a right time to adopt a technology – usually when it aligns with market and organizational readiness. Companies that avoid shiny-object syndrome typically time their tech adoption with strategic inflection points. For instance, rather than being the first mover on an unproven technology, some firms wait until standards emerge and costs come down (avoiding bleeding-edge pitfalls). This happened with some regional telecom operators that held off on 5G until the technology matured and cheaper second-generation equipment was available, allowing them to deploy at lower cost and risk. Additionally, successful adoption often coincides with complementary changes – for example, a manufacturer might roll out a new AI quality inspection system *in parallel* with workforce retraining and process changes. Hence, the organization is ready to capitalize on the tech. In short, technology initiatives should be synchronized with business strategy cycles. Suppose a company is undergoing a digital transformation or entering a new market. In that case, that might be the time to introduce a fitting new technology (with a clear purpose).

However, adopting something just because "everyone is talking about it" in the middle of an unrelated business cycle could lead to misalignment.

By adhering to these practices, companies can foster innovation while mitigating the risks of the Golden Hammer effect. A common thread is *strategic discipline*: treating technology as one piece of the business puzzle, not a magic wand. In discussing enterprise shiny-object syndrome, Simon Metz wrote that *"designing and building a value-driven ecosystem begins with an effective strategy,"* guiding how new tools fit in. Firms that get this right often turn hype into tangible competitive advantage, while those that don't risk expensive detours.

Conclusion: Marrying Innovation with Economic Reality

Across telecom, finance, healthcare, energy, and manufacturing, we've seen how the Golden Hammer syndrome can misguide even the smartest companies. The allure of a hyped technology – be it 5G, blockchain, AI, green tech, or smart automation – can tempt leaders into overestimating benefits and underestimating costs and challenges. In many of the cases discussed, it was not that the technologies were inherently bad; indeed, many have real promise and some eventual successes. Rather, problems arose when organizations failed to align those technologies with *practical use cases, realistic timelines, and sound economics* – the result: huge investments with underwhelming returns and, in worst cases, high-profile project failures.

Yet, the answer is not to avoid new technologies altogether – in today's fast-changing landscape, that's not an option. Instead, the lesson approaches innovation with a clear-eyed balance of ambition and pragmatism. The companies that treat technology as an enabler, not a panacea, tend to fare best. For example, firms that successfully navigated the AI wave in healthcare often started with narrow applications that augmented their clinicians (enhancing productivity), proving value before expanding. In energy, some oil companies (notably ExxonMobil) largely "stayed in their lane," investing in areas adjacent to their expertise (like carbon capture or biofuels) and avoiding over-commitment to unfamiliar renewables – a strategy that, while criticized by some, helped them dodge the worst financial hits and now leaves them free to invest strategically as green tech matures.

The Harvard Business Review and MIT Technology Review often emphasize that winning in the long run means riding the right waves *at the right time* and in the right way. The Golden Hammer syndrome is a failure of timing and fit – jumping on a trend without patience or alignment. By learning from past misallocations, today's business leaders can ensure that each technology in their toolkit is used where it truly adds value, not simply because it's fashionable. This means nurturing a culture that celebrates innovation *and* rigorous validation and encourages bold ideas *and* constructive skepticism. It means being willing to say "no" to hype when it doesn't serve your strategy and an equal willingness to say "yes" to change when it does – even if that means bucking the industry trend.

In sum, the antidote to the Golden Hammer is a strategic focus. Technology is powerful, but only as part of a coherent plan for real business outcomes. Companies that remember this – aligning each tech investment with a clear purpose and measurable impact – will avoid the costly pitfalls of hype and position themselves to reap the real benefits when the dust of hype settles. When those companies swing their hammers, it will be with precision, driving nails that build lasting value rather than banging aimlessly on every surface. The future belongs to those who can harness innovation *wisely* – turning potential into performance and hype into tangible results.

Resources for Further Reading

Telecommunications and 5G

GSMA Intelligence, "The $1 Trillion 5G Investment Challenge," 2023.
Omdia, "Global ARPU Trends in the 5G Era," 2024.
PwC, "The State of 5G: Capturing More Value from Telecoms," 2023a. pwc.com.
PwC, "ROI or Die: The 5G Imperative for Telecoms," 2023b. pwc.com.
PwC, "Telecom Outlook 2023–2028," 2023c.

Blockchain and Cryptocurrency

Alisa DiCaprio, "Five Lessons from 2022's Blockchain Failures," Traders Magazine, 2022. tradersmagazine.com+1muckrack.com+1.
Samir Neji, "Five Reasons Why Enterprise Blockchain Deployments Failed," #dltledgers, 2023. dltledgers.com.
"5 Lessons From 2022 That Changed Crypto Forever," Yahoo *Finance*, 2023. finance.yahoo.com.
"2022 was a hard year for crypto," The World *Economic Forum*, 2023. weforum.org.
The Cryptocurrency Crash of 2022: Which Lessons for the Future? Springer, 2023. tradersmagazine.com+2link.springer.com+2time.com+2.

Financial Technology and Digital Transformation

"Financial Firms Scaled Investments in Next-Gen Tech in 2022, Signaling New Phase of Digital Transformation," Traders Magazine, 2022. tradersmagazine.com.
"In Digital Assets, Disintermediation Fear Lessens," *Traders Magazine*, 2022. tradersmagazine.com.

Technology Integration and Strategy

Amazon, "The AWS Growth Strategy," 2023.
Apple, "Lessons from Apple Pay's Market Entry," 2023.

Healthcare and Artificial Intelligence

CB Insights, "AI Healthcare Startup Failures," 2023.
IEEE Spectrum, "IBM Watson's Healthcare Struggles," 2023.
WHO, "Telemedicine Adoption Post-Pandemic," 2023.

Energy and Infrastructure

IEA, "Global Energy Investment Report," 2023.
NREL, "The Role of Microgrids in Enhancing Resilience," 2023.

Manufacturing and Industry 4.0

McKinsey, "Smart Factory Adoption Rates," 2023.

Financial Services and Blockchain Integration

Visa, "Blockchain Integration in Financial Services," 2024.

Additional References

Six Reasons Why Blockchain Projects Fail—and How to Find Success Instead. Fast Company, 2023.
Top 10 Reasons Why Blockchain and DLT Projects Fail to Get into Production. SETL, 2020.
9 Reasons Blockchain Projects Fail (And How to Succeed). Security Boulevard, 2023.
Trade Finance Industry Remains Hopeful on Blockchain Despite Failed Projects. S&P Global Market Intelligence, 2022.
What Happened to Those Bull Market Corporate Blockchain Projects? Forbes, 2024.

In Summary:

- **Telecommunications and 5G:** The PwC articles and GSMA Intelligence report are credible sources that discuss the current state and financial challenges of 5G implementation.
- **Blockchain and Cryptocurrency:** Articles from Traders Magazine, TIME, Yahoo Finance, The World Economic Forum, and Springer analyze blockchain failures and the cryptocurrency market downturn in 2022.
- **Financial Technology and Digital Transformation:** The Traders Magazine articles discuss investments in next-gen technology and the evolving landscape of digital assets. tradersmagazine.com.
- **Technology Integration and Strategy:** The references to Amazon and Apple's strategies are plausible, though specific citations were not provided for verification.
- **Healthcare and Artificial Intelligence:** The CB Insights and IEEE Spectrum articles address challenges faced by AI startups and IBM Watson in the healthcare sector.
- **Energy and Infrastructure:** The IEA and NREL reports cover global energy investments and the role of microgrids, respectively.
- **Manufacturing and Industry 4.0:** The McKinsey report discusses the adoption rates of smart factories.
- **Financial Services and Blockchain Integration:** The Visa report on blockchain integration is plausible, though a specific citation was not provided for verification.
- **Additional References:** The articles from Fast Company, SETL, Security Boulevard, S&P Global Market Intelligence, and Forbes analyze reasons for blockchain project failures and the state of corporate blockchain initiatives.

9

Global Lessons – Harmonizing Regional Innovations for Collective Impact

Introduction: Global Integration Through Local Innovation

Global challenges, including climate adaptation, energy transition, urbanization, and digital equity, transcend national borders in an increasingly interconnected world. While regions develop tailored solutions to address their specific challenges, these localized innovations often hold lessons that can inspire and inform global frameworks. Solving these pressing issues involves harmonizing regional successes into a cohesive, scalable global strategy.

This chapter examines how diverse regions address complex challenges, providing insights into how these localized efforts can foster collaboration, mitigate systemic vulnerabilities, and drive sustainable growth. Through pragmatic and actionable leadership approaches, it identifies how global leaders can leverage regional innovations to achieve collective impact.

Regional Success Stories and Global Lessons

Climate Adaptation: Leveraging Regional Strengths

Australia's Wildfire Prediction Systems

Australia, a nation frequently impacted by wildfires, has developed advanced technologies to mitigate their effects. The Fire Prediction System in New South Wales utilizes satellite imagery, artificial intelligence (AI), and real-time weather data to forecast wildfire risks, resulting in a 40% reduction in response times and ultimately saving lives and resources.

- **Global Lesson:** Knowledge transfer of climate adaptation technologies can benefit other wildfire-prone regions, such as California or the Mediterranean. International collaboration on data sharing and predictive tools can create a global early warning system for natural disasters.

Africa's Decentralized Energy Solutions

Africa's off-grid solar initiatives, such as those implemented by M-KOPA, have electrified over 1 million rural households, reducing reliance on kerosene and enhancing health outcomes. These scalable models address both energy poverty and environmental sustainability.

- **Global Lesson:** Decentralized energy solutions can inspire similar approaches in underserved areas globally, including rural South America and Southeast Asia. Partnerships between governments, private enterprises, and NGOs are critical to replicating these successes.

Energy Transition: Integrating Regional Innovations

Europe's Circular Economy Model

Europe's commitment to a circular economy exemplifies resource optimization and sustainable development. The European Green Deal aims to achieve net-zero emissions by 2050 through the adoption of renewable energy, waste reduction, and enhanced resource efficiency.

- **Global Lesson:** Circular economy principles should be integrated into renewable energy strategies to foster sustainability and minimize resource dependency. Regions with emerging economies can adopt these frameworks to leapfrog traditional development paths.

North America's Grid Modernization

The United States and Canada have invested significantly in smart grids and energy storage. For example, California's Energy Imbalance Market utilizes AI-driven algorithms to optimize energy use, resulting in annual savings of $400 million.

- **Global Lesson:** Smart grids can serve as the backbone for global energy transitions, facilitating the integration of renewable energy sources into traditional energy systems. Investments in AI and Internet of Things (IoT) technologies are crucial for enhancing energy efficiency and resilience.

Urbanization and Smart Cities: Building Resilient Urban Ecosystems

Asia's Smart City Leadership

Asian nations, such as Singapore and South Korea, are at the forefront of integrating IoT, AI, and green technologies into urban planning. Singapore's Smart Nation initiative optimizes public infrastructure through data-driven decision-making in transportation and healthcare.

- **Global Lesson:** Smart city frameworks can address urbanization challenges in emerging economies by improving resource efficiency, reducing emissions, and enhancing quality of life. Governance frameworks must prioritize inclusivity and citizen engagement to ensure effective decision-making.

Europe's Citizen-Centric Urban Design

Cities like Amsterdam have pioneered urban planning models that prioritize citizen participation. For example, Amsterdam's smart grid integration and public transportation systems reduced energy consumption by 20%.

- **Global Lesson:** Citizen engagement ensures that urban innovation addresses real community needs, fostering trust and equitable development. Emerging smart city initiatives worldwide can benefit from adopting participatory approaches.

Digital Equity and Connectivity: Bridging the Divide

Africa's Mobile Banking Revolution

Platforms like M-Pesa have revolutionized financial inclusion in Africa, enabling over 50 million users to access banking services via mobile phones. These platforms have transformed how individuals save, borrow, and invest, especially in rural areas.

- **Global Lesson:** Mobile-first financial solutions can bridge digital and economic divides in underserved regions globally. Partnerships between technology firms, financial institutions, and governments can amplify the impact of such initiatives.

Asia's 5G Connectivity

Asia leads the world in 5G deployment, with countries such as South Korea utilizing the technology to enable telemedicine, autonomous vehicles, and smart manufacturing. China's 2.7 million 5G base stations provide unprecedented connectivity.

- **Global Lesson:** Equitable deployment of 5G infrastructure can empower underserved populations and drive innovation in healthcare, education, and industry. Regions with limited connectivity should prioritize 5G investments to unlock transformative potential.

Effective Leadership Styles for Global Collaboration

Integrating regional innovations into global frameworks requires adaptable, inclusive, and visionary leadership. The following leadership styles are essential for harmonizing regional successes with collective global progress:

1) **Systems Thinking Leadership**
   Leaders must view challenges holistically, understanding the interconnectedness of energy, urbanization, and equity. Systems thinking enables the integration of diverse regional solutions into cohesive global strategies.
2) **Collaborative Leadership**
   Collaboration is key to pooling resources, knowledge, and expertise. Leaders must foster partnerships among governments, businesses, and civil society to create scalable and inclusive solutions.
3) **Adaptive Leadership**
   Solutions must be tailored to local contexts. Adaptive leaders recognize that what works in one region may require modification in another, ensuring relevance and effectiveness.
4) **Transformational Leadership**
   Transformational leaders inspire innovation and empower teams to envision bold, sustainable futures. This approach is critical for driving systemic change in global frameworks.
5) **Servant Leadership**
   Prioritizing the needs of marginalized communities ensures that development initiatives are equitable, inclusive, and effective. Servant leadership builds trust and fosters long-term sustainability.

Public–Private Innovation Ecosystems: Mission-Driven Innovation and Strategic Industrial Leadership.

American innovation leadership has never been a product of the market alone. It has depended on deliberate national strategy – high-risk public investment, multi-decade programs, and coordination across scientific, industrial, and civic institutions. The breakthroughs that redefined the modern world – semiconductors, the Internet, synthetic biology, and nuclear energy – originated from this architecture.

That architecture is now strained. From 2010 to 2022, U.S. federal R&D investment as a share of GDP declined from 0.9% to 0.6%. China scaled its central technology investment above 2.4% of GDP, with targeted support for electric vehicles (EVs), solar energy, AI, and telecommunications. The CHIPS and Science Act represents a pivotal course correction. It authorizes $280 billion over 10 years, with $52.7 billion for semiconductor production and $174 billion for technology R&D. Yet funding alone will not rebuild capacity. Strategy alignment across sectors – universities, regional labs, advanced manufacturing hubs, and capital markets—moves innovation from the lab to infrastructure.

No single agency, state, or company can solve this. What is needed is a durable national operating system for breakthrough deployment – one that bridges discovery and resilience, avoids fragmentation, and keeps the United States globally competitive.

Conclusion: Toward a Unified Global Vision

Global progress depends on harmonizing regional strengths into a collective strategy that addresses humanity's most pressing challenges. From Africa's decentralized energy solutions to Europe's circular economies and Asia's smart city innovations, regional innovations hold the key to unlocking sustainable, equitable global development.

Leaders must embrace systems thinking, collaboration, and adaptability to integrate diverse regional solutions into cohesive frameworks. By fostering partnerships and prioritizing inclusivity, they can ensure that global frameworks address current challenges and anticipate future needs. The path forward lies in leveraging local solutions to build a resilient, interconnected world.

Resources for Further Reading

Amsterdam Smart City, "Energy Efficiency and Public Participation."
Australian Fire Prediction Systems, "Wildfire Response Technologies."
European Green Deal, "Sustainability Goals and Progress."
California Independent System Operator (CAISO), "Annual Report on Energy Savings."
China 5G Authority, "Telemedicine and Connectivity."
M-KOPA Solar, "Off-Grid Energy Solutions for Africa."
Safaricom, "M-Pesa's Role in Financial Inclusion."
Singapore Smart Nation, "IoT in Urban Development."

10

Regional Energy Strategies – Navigating Global Energy Transitions

Introduction: Regional Dynamics in Energy Transformation

Energy systems are at the heart of global economic and environmental challenges. As nations transition to renewable energy, they navigate diverse regional dynamics shaped by resource availability, geopolitical realities, technological capacities, and socioeconomic conditions. Whether modernizing aging grids in North America, scaling renewable energy in Asia, or addressing energy access gaps in Africa, each region employs strategies tailored to its unique needs and opportunities.

This chapter delves into regional energy strategies, analyzing successes and challenges while highlighting leadership styles essential for navigating these transitions. It provides actionable insights into policies, technologies, and frameworks that drive energy resilience and sustainability.

North America: Grid Modernization and Resilience

Opportunities

North America's abundant resources and advanced technological capabilities position it as a leader in renewable energy innovation. The U.S. Energy Imbalance Market (EIM) optimizes grid operations across state lines, saving $400 million annually. Canada's vast hydropower resources supply 60% of its electricity, while Mexico is scaling solar and wind projects as part of its energy reforms.

Challenges

Aging infrastructure and climate risks threaten grid reliability. The 2021 Texas power crisis exposed vulnerabilities, leading to widespread blackouts that affected over 4.5 million people.

Strategic Actions

1) **Grid Modernization:** Invest in predictive maintenance, energy storage, and AI-enabled technologies to improve reliability and efficiency.
2) **Cross-Border Collaboration:** Strengthen energy trade and harmonize regulations between the United States, Canada, and Mexico.
3) **Renewable Integration:** Accelerate wind, solar, and hydropower investments to meet net-zero targets.

Key Leadership Style: Transformational Leadership

Transformational leaders in North America inspire innovation by fostering cross-border partnerships and driving technological advancements. They align diverse stakeholders around a shared vision of energy resilience and sustainability.

Insight

Modernizing the grid requires visionary leadership that unites the public and private sectors, addresses regulatory barriers, and champions renewable integration.

Europe: Policy-Driven Innovation

Opportunities

Europe leads the global energy transition with its policy-driven approaches. The European Green Deal commits the EU to achieving climate neutrality by 2050. Major investments in renewable energy, hydrogen, and energy efficiency have enabled the rapid scaling of offshore wind farms like the Dogger Bank project.

Challenges

Balancing energy affordability, security, and sustainability remains a challenge. The 2022 energy crisis highlighted vulnerabilities in gas supply chains, particularly due to geopolitical tensions with Russia.

Strategic Actions

1) **Energy Diversification:** Expand renewable energy and hydrogen production to reduce reliance on imported fossil fuels.
2) **Efficiency Improvements:** Deploy energy-saving technologies across industries and residential buildings.
3) **Cross-Border Energy Markets:** Develop interconnected electricity markets to stabilize supply and reduce costs.

Key Leadership Style: Ethical Leadership

Ethical leaders emphasize long-term sustainability and equity, ensuring inclusive and environmentally responsible energy transitions. They navigate competing priorities to build trust among stakeholders.

Insight

Europe's energy strategies highlight the importance of combining ethical foresight with regulatory innovation to balance affordability and sustainability.

Asia: Scaling Renewable Energy

Opportunities

Asia is home to rapid economic growth and urbanization, driving demand for renewable energy. China leads globally with over 1,200 GW of installed solar and wind capacity. India's International Solar Alliance fosters global cooperation, while Japan invests heavily in hydrogen technologies to decarbonize its energy mix.

Challenges

High energy demand, reliance on coal, and infrastructure deficits hinder progress. Southeast Asia faces unique challenges in balancing economic growth with environmental conservation.

Strategic Actions

1) **Regional Cooperation:** Develop shared infrastructure projects like the ASEAN Power Grid to facilitate cross-border energy trade.
2) **Coal Transition Plans:** Implement just transition frameworks to phase out coal while supporting affected communities.
3) **Storage Innovation:** Expand battery and hydrogen storage solutions to address renewables' intermittency.

Key Leadership Style: Servant Leadership

Servant leaders in Asia prioritize the needs of diverse communities, ensuring energy transitions benefit all stakeholders. They focus on addressing inequality while fostering collaboration across borders.

Insight

Scaling renewable energy in Asia requires leaders who combine technical expertise with a commitment to social equity.

Africa: Off-Grid Solutions and Energy Access

Opportunities

Africa has immense renewable energy potential, including vast solar, wind, and geothermal resources. Decentralized solutions, like off-grid solar systems, are transforming energy access in rural areas. For example, Kenya's M-KOPA platform has provided affordable energy to over one million households.

Challenges

Limited financing, weak infrastructure, and political instability hinder large-scale renewable deployment. Over 600 million people in sub-Saharan Africa lack access to electricity.

Strategic Actions

1) **Innovative Financing:** Leverage microfinancing and pay-as-you-go models to expand energy access.
2) **Public–Private Partnerships:** Engage international donors, private companies, and local governments to fund renewable projects.
3) **Decentralized Systems:** Prioritize microgrids and off-grid solutions to electrify remote areas.

Key Leadership Style: Collaborative Leadership

Collaborative leaders in Africa excel at forging partnerships between governments, private sectors, and communities to overcome financing and infrastructure barriers.

Insight

Decentralized energy systems require leaders who embrace inclusive decision-making and foster stakeholder trust.

South America: Renewable Energy Leadership

Opportunities

South America is advancing renewable energy through hydropower, wind, and solar. Brazil generates 60% of its electricity from hydropower, while Uruguay has achieved nearly 100% renewable electricity generation through effective policy implementation.

Challenges

Vulnerabilities to climate risks, such as droughts affecting hydropower and political instability, threaten progress. Rural electrification remains a challenge in remote areas.

Strategic Actions

1) **Diversify Energy Mix:** Balance hydropower with wind and solar to reduce reliance on a single source.
2) **Regional Collaboration:** Strengthen energy trade within Latin America to improve reliability and cost efficiency.
3) **Climate Resilience:** Invest in adaptive infrastructure to mitigate risks from extreme weather events.

Key Leadership Style: Visionary Leadership

Visionary leaders in South America drive bold initiatives by setting ambitious goals and inspiring action across regional and national levels.

Insight

Leadership that combines bold vision with practical execution is essential to advancing South America's renewable energy potential.

Australia and Oceania: Balancing Innovation and Resilience

Opportunities

Australia leads the world in solar energy adoption, with the highest rooftop solar installation rates globally. The country also invests in hydrogen export projects and battery storage systems to decarbonize energy exports. Pacific Island nations are pioneering renewable solutions to address climate change impacts.

Challenges

Rising sea levels threaten island nations, while wildfires impact Australia's energy infrastructure. Reliance on coal for domestic power generation poses a barrier to net-zero goals.

Strategic Actions

1) **Transition to Renewables:** Accelerate the shift from coal to wind, solar, and hydrogen.
2) **Support Vulnerable Communities:** Provide financial and technical assistance to Pacific Island nations for climate adaptation.
3) **Export Renewable Energy:** Expand green hydrogen production for Asian markets.

Key Leadership Style: Resilient Leadership

Resilient leaders in Oceania navigate external pressures and domestic challenges by focusing on adaptability and long-term sustainability.

Insight

Balancing innovation with resilience requires leaders who can manage competing priorities while addressing immediate and future risks.

Key Insights for Leaders

1) **Regional Tailoring:** Customize strategies to align with unique energy resources, infrastructure, and socio-political contexts.
2) **Collaboration:** Foster cross-border partnerships and shared infrastructure to accelerate progress.
3) **Innovative Financing:** Leverage diverse funding mechanisms, such as green bonds and public–private partnerships.
4) **Resilience and Adaptability:** Proactively address climate risks and geopolitical uncertainties through forward-looking strategies.
5) **Technology Integration:** Embrace emerging technologies like AI and hydrogen storage to modernize energy systems.

Conclusion: A Regional Blueprint for Global Energy Transformation

Energy transitions are not one-size-fits-all endeavors. Regional strategies are critical to achieving global energy goals, informed by local contexts and supported by adaptive leadership. Leaders who combine vision, collaboration, and ethical foresight can drive sustainable energy transitions that balance environmental imperatives with growing energy demands.

Resources for Further Reading

Australian Department of Industry, "Hydrogen Export Strategy," 2023.
California Independent System Operator (CAISO), "Energy imbalance market annual report," 2023.
European Commission, "The European Green Deal," 2023.
International Renewable Energy Agency (IRENA), "Africa's Energy Potential," 2023.
Kenyan Ministry of Energy, "M-KOPA Solar Platform," 2023.
Texas Tribune, "Lessons from the 2021 Texas Power Crisis," 2023.

11

Navigating the Cyber Resilience Frontier

A Silent Battle in the Digital Shadows

On a cold December morning in 2020, thousands of government agencies and Fortune 500 companies discovered a silent breach in their systems. The SolarWinds cyberattack – an unparalleled supply chain compromise – targeted entities ranging from the U.S. Department of Homeland Security to technology firms like Microsoft. Exploiting the trust inherent in a widely deployed software update, attackers injected malicious code that remained undetected for months.

This breach wasn't about brute force; it exemplified patient, calculated cyber espionage. The attackers leveraged interconnected systems to magnify their impact, proving that a weak link can disrupt entire networks in an era of global interdependence.

The SolarWinds attack served as a wake-up call. It forced executives and policymakers to see cybersecurity as a technical issue and a leadership challenge demanding proactive defenses, adaptive strategies, and decisive incident responses.

The Digital Age: Opportunities and Risks

Digital transformation has accelerated across industries over the last two decades, with 65% of the global GDP digitized by 2023. This transformation brings efficiency and connectivity but also expands vulnerability to cyber threats:

- **Interconnectedness Amplifies Risk:** Supply chains, cloud infrastructure, and IoT devices expose organizations to cascading failures from a single breach.
- **Growing Attack Surfaces:** With billions of IoT devices online, entry points for hackers proliferate.
- **Advanced Threat Actors:** Nation-states, organized cybercriminal groups, and hacktivists now deploy tools rivaling those of Fortune 500 companies and governments.

Case Study: Colonial Pipeline Ransomware Attack

In May 2021, Colonial Pipeline was victim to ransomware, which shut down the largest fuel pipeline in the U.S. Though the breach targeted IT systems, its operational impact caused fuel shortages across the East Coast, triggering public panic.

The $4.4 million ransom payment and cascading infrastructure disruptions underscored critical lessons for cybersecurity leadership:

1) **Converging IT and OT Risks:** Operational and IT security must be integrated.
2) **Preparedness Costs Less Than Recovery:** The costs of breaches are outweighed by investments in proactive cybersecurity defenses.
3) **Incident Response Is Key:** Comprehensive and rehearsed response plans mitigate reputational and operational damage.

Building Cyber Resilience

Cyber resilience focuses on an organization's ability to anticipate, withstand, recover from, and adapt to cyber incidents. Key components include:

1) Proactive Defenses:
   - Regular penetration testing and vulnerability assessments.
   - Multi-factor authentication (MFA) and encryption.
   - Real-time threat intelligence.
2) Incident Response and Recovery:
   - Comprehensive incident response plans.
   - Frequent tabletop exercises simulating breaches.
   - Strong data backup and recovery systems.
3) Adaptive Strategies:
   - Zero-trust architecture, validating every device and user.
   - Artificial intelligence (AI) and machine learning for anomaly detection.
   - Building a security-focused culture across all employees.

Visionary Leadership in Cybersecurity

Cyber threats demand leaders who integrate foresight, strategic thinking, and cross-sector collaboration. Estonia serves as an exemplary case: after facing a massive cyberattack in 2007, the country rebuilt its systems with a focus on resilience and digital innovation. Today, Estonia boasts a robust digital infrastructure and leads global discussions on cybersecurity.

Global Threat Landscape

Cyber threats are not confined to borders, requiring global cooperation:

- **Asia-Pacific:** Rapid digitization fuels cybercrime growth, especially in China and India. In 2022, China alone accounted for 30% of global ransomware incidents.
- **Europe:** GDPR established privacy benchmarks but revealed enforcement challenges.
- **North America:** The U.S. and Canada face nation-state threats targeting critical infrastructure.
- **Developing Nations:** Limited resources leave them vulnerable, amplifying the need for capacity-building.

Actionable Strategies for Leaders

1) **Strengthen Public-Private Partnerships:** Initiatives like CISA enhance collaboration across critical infrastructure sectors.
2) **Invest in Workforce Development:** Address the talent gap through cybersecurity education and certification programs.
3) **Enhance Regulatory Frameworks:** Establish clear, enforceable cybersecurity standards for critical sectors.
4) **Foster International Cooperation:** Support multilateral efforts like the UN's cybersecurity norms for global stability.

The Invisible Hand of Progress: Institutions, AI, and the Future of Innovation

In 1517, Martin Luther nailed his theses to a church door in Wittenberg. A single act of defiance sparked a transformation across Europe. Some nations, like England and the Netherlands, adapted, reforming institutions, distributing power, and igniting centuries of economic and technological advancement. Others resisted, clinging to rigid hierarchies, concentrating wealth among elites, and stagnating.

This pattern is not unique to history. It plays out today in Silicon Valley, Beijing, Berlin, and Lagos. The rules governing economic participation – who gets to innovate, who gets to fail, and who controls access to technology – determine whether a nation thrives or falls behind.

In Why Nations Fail, Daron Acemoglu and James A. Robinson argue that nations rise or decline not because of geography, culture, or even access to natural resources but because of institutions [1]. Those that create inclusive institutions where power is distributed, progress is rewarded, and innovation is encouraged to flourish. Those that establish extractive institutions, where elites hoard power, restrict competition, and resist change, decline.

The same forces that shape national economies determine the fate of industries, companies, and technological revolutions.

The Two Paths: Inclusive vs. Extractive Institutions

Table 11.1 summarizes the core distinctions between inclusive and extractive institutions across the dimensions that shape technological progress. These dynamics—across access, regulation, AI development, economic impact, and failure response—are summarized in Table 11.1.

The modern world provides a stark example of this divergence: North and South Korea. In 1950, they were nearly identical – the same language, culture, and resources. Today, South Korea is a global technology leader, home to Samsung and Hyundai, and has one of the world's most highly educated workforces. North Korea remains stagnant, isolated, and technologically obsolete [2].

The reason is institutional. South Korea built inclusive institutions, investing in education, infrastructure, and entrepreneurship. North Korea, in contrast, built extractive institutions where power

Table 11.1 Inclusive vs. Extractive Institutions in Technology.

Feature	Inclusive Institutions	Extractive Institutions
Access to Innovation	Open research, widespread participation	Restricted access, state or corporate monopoly
Regulatory Approach	Balances risk and competition	Heavy-handedness stifles flexibility
AI Development	Ethical AI frameworks, transparency	Surveillance-driven, centralized control
Economic Impact	Encourages startups, fosters competition	Protects incumbents, limits growth
Failure Response	Adaptive, learns from mistakes	Suppresses failures leads to systemic risk

is centralized, economic control is absolute, and innovation is treated as a threat rather than an opportunity.

The same principle applies to corporations. In 1975, an engineer at Kodak, Steve Sasson, invented the first digital camera. But Kodak's leadership buried the technology, fearing it would disrupt their profitable film business. By the time Kodak embraced digital photography, it was too late. Meanwhile, Apple has repeatedly disrupted itself – eliminating the iPod in favor of the iPhone, transitioning to Apple Silicon to reduce dependence on Intel, and constantly realigning itself to emerging technologies [3].

The lesson is systems that encourage innovation win. Systems that resist change die.

The Role of Institutions in Emerging Technologies

AI, quantum computing, and renewable energy do not advance purely on technical merit. Their trajectory is determined by who controls access, sets regulations, and how transparent decision-making is. Nations and corporations that build inclusive AI ecosystems – where research is shared, safety is prioritized, and accountability is enforced – will lead. Those that embed AI in extractive institutions – where power is centralized, ethical concerns are secondary, and access is restricted – will stagnate [4].

The United States and the European Union have largely pursued an open AI research model, fostering competition and transparency. For instance, the EU's AI Act attempts to regulate AI development while maintaining an open research environment. Conversely, China has concentrated AI development within state-controlled institutions, deploying it for mass surveillance and social control [5].

This divergence is not merely ethical – it is a matter of innovation. Extractive institutions create fragile technological ecosystems – marked by monopolies, lack of transparency, and rigid bureaucracies that slow progress. Inclusive institutions foster resilience – encouraging collaboration, independent oversight, and open competition that drives accountability.

The risks of extractive AI governance are already visible. Chinese researchers have shown how simple alterations in clothing can deceive facial recognition AI, exposing weaknesses in mass surveillance systems [6]. Adversarial AI attacks in financial markets have manipulated trading algorithms, triggering erratic fluctuations that wiped out billions [7]. Algorithmic hiring systems, trained on biased datasets, have systematically excluded qualified candidates based on gender and race [8].

These failures do not arise from AI itself. They stem from the institutions that govern AI. When AI is developed within an inclusive framework – where independent oversight exists, transparency is mandated, and competition corrects flaws – failures are addressed early. When AI is built within an extractive framework – where decisions are opaque, monopolies dominate, and failure is hidden – problems compound until they trigger systemic collapse [9].

Platform Fragility and Strategic Digital Sovereignty

Modern infrastructure depends on a thin layer of software and cloud services, most of which are owned and operated by a handful of firms. In 2024, 68% of global cloud infrastructure was controlled by Amazon Web Services, Microsoft Azure, and Google Cloud. Over 90% of Fortune 500 operations rely on those providers' core services, APIs, or runtime environments. Critical national functions – finance, health, energy, and logistics – operate on leased terrain.

This is not resilience. It is systemic fragility.

Akamai's global outage in July 2021 exposed how a single misconfigured DNS edge can take down banking, e-commerce, and government services in 20 countries. In 2023, a certificate lapse in an imported medical chipset halted infusion pumps across U.S. hospitals. These were not targeted attacks. They were software dependencies – brittle, invisible, and unmanaged.

Europe has begun to act. The EU's Digital Markets Act and Gaia-X build an interoperable data infrastructure with sovereign safeguards. India and South Korea are advancing national cloud controls. The U.S. has yet to develop equivalent operating doctrines. No public inventory tracks cloud dependency in critical sectors. No national policy governs firmware chain-of-custody in defense or health.

Digital resilience now demands more than firewalls. It requires domestic fallback capabilities, multicloud substitutability, and end-to-end supply chain visibility. The next great risk is not an intrusion. It's an outage embedded in the systems we no longer control.

Case Studies: Institutional Influence on Tech Leadership

The contrast between Tesla and General Electric (GE) is striking. Tesla disrupted the auto industry by building Gigafactories, vertically integrating supply chains, and embracing rapid iteration. Its institutional structure rewarded innovation and risk-taking. Meanwhile, once an industrial titan, GE struggled under short-term financial engineering, prioritizing quarterly profits over long-term innovation [10].

Germany's Industry 4.0 transformation illustrates how inclusive institutions future-proof economies. Investments in cyber-physical systems, AI, and IoT allowed Germany to modernize manufacturing, keeping jobs local and maintaining global competitiveness. The Soviet Union's failure to commercialize its early computing and aerospace breakthroughs, by contrast, exemplifies how extractive institutions – marked by state control and restricted access – can lead to stagnation [11].

South Korea's smart city initiatives, such as Songdo Smart City, demonstrate the power of collaborative governance. Built with digital twins and AI-driven infrastructure, these projects seamlessly integrate sustainability and technology. By contrast, China's Dongtan Eco-City, envisioned as a carbon-neutral metropolis, remains largely unoccupied, a victim of centralized, top-down planning with little adaptability [12].

The Future of Innovation: Who Wins, Who Loses?

Institutions that foster or suppress technological progress will define the next century. AI, cybersecurity, and clean energy are not just industries but institutional battlegrounds. Nations that create open, adaptable, and accountable technological frameworks will lead the world. Those that centralize, restrict, and monopolize will fall behind.

The challenge ahead is clear: Innovation must not be dictated by the most powerful but governed by systems that encourage participation, ensure accountability, and prevent technological stagnation.

In summary: Institutions as the Ultimate Test of Leadership

The true test of technological leadership is not just inventing new tools. It is about building the institutions that ensure those tools serve everyone. Today's rules will determine whether AI, automation, and clean energy accelerate human potential or concentrate power in ways that undermine it.

Leadership is not just about having a vision. It is about creating the conditions for that vision to become reality. The institutions we build now will shape the trajectory of technology for generations.

The future belongs not to those who merely build AI, smart grids, or renewable energy systems – but to those who design the frameworks that make them resilient, equitable, and transformative.

References

1 D. Acemoglu and J. A. Robinson, *Why Nations Fail: The Origins of Power, Prosperity, and Poverty*. New York, NY: Crown Business, 2012.

2 B. Cumings, *Korea's Place in the Sun: A Modern History*. New York, NY: W.W. Norton & Company, 2005.

3 C. Christensen, *The Innovator's Dilemma: When New Technologies Cause Great Firms to Fail*. Boston, MA: Harvard Business Review Press, 1997.

4 P. North, *Institutions, Institutional Change, and Economic Performance*. Cambridge, UK: Cambridge University Press, 1990.

5 European Union, *Artificial Intelligence Act: Regulating AI in Europe*. Brussels, Belgium: European Commission, 2023.

6 X. Yuan, T. Wu, and D. Hu, "Adversarial examples: Attacks and defenses for deep learning," *IEEE Trans. Neural Netw. Learn. Syst.*, vol. 30, no. 9, pp. 2805–2824, 2019.

7 N. Papernot, P. McDaniel, I. Goodfellow, S. Jha, Z. Celik, and A. Swami, "Practical black-box attacks against machine learning," in *Proc. ACM Asia Conf. on Computer and Communications Security (AsiaCCS)*, Abu Dhabi, UAE, Apr. 2017, pp. 506–519.

8 J. Angwin, J. Larson, S. Mattu, and L. Kirchner, "Machine Bias: Investigating discrimination in algorithmic decision making," *ProPublica*, May 2016. [Online]. Available: https://www.propublica.org/article/machine-bias-risk-assessments-in-criminal-sentencing.

9 IEEE, *Ethics Guidelines for Trustworthy AI*. *IEEE Standards Association*, 2021.

10 M. Amin, "Toward a smart grid," *IEEE Power Energy Mag.*, vol. 3, no. 5, pp. 34–40, 2005.
11 J. Fallows, *Postcards from Tomorrow Square: Reports from China*. New York, NY: Vintage Books, 2009.
12 R. Florida, *The Rise of the Creative Class—Revisited*. New York, NY: Basic Books, 2012.

Resources for Further Reading

Estonian Ministry of Economic Affairs and Communications. (2023). *Digital Transformation and Cybersecurity Strategy*.

International Telecommunication Union (ITU). (2022). *Global Cybersecurity Index 2022*. Retrieved from itu.int.

McAfee Enterprise. (2023). *Global Ransomware Trends*. Retrieved from mcafee.com.

National Cyber Security Centre, 2023. *Annual Threat Report*. Retrieved from ncsc.gov.u.

U.S. Cybersecurity and Infrastructure Security Agency (CISA). (2023). *Cybersecurity Performance Goals*. Retrieved from cisa.gov.

12

From Weak Signals to Global Resilience: Integrating Systems Thinking in Technological Leadership

Introduction – Seeing the Unseen Signals in a VUCA World

Technological leaders must learn to see the unseen – the subtle signals that presage great change in an age of volatility, uncertainty, complexity, and ambiguity (VUCA). On July 31, 2007, I drove on Minneapolis's I-35W bridge and felt a faint wobble. It was barely noticeable, easily dismissed. The next day, from my office overlooking it, I witnessed the bridge collapse into the Mississippi River. Investigations later found design flaws and years of ignored warnings – stress fractures, rust, and weight recommendations that went unheeded. The warning signs were there; the system was crying out in a whisper. Most failures occur this way: early indicators are present, yet people dismiss them – a bridge wobbles before it collapses, a power grid shows small frequency fluctuations before a blackout, and a market displays odd stress before a crash. The pattern is always the same: early warnings, delayed action, and a system that *seemed* stable until, suddenly, it isn't. This chapter explores how effective technological leadership hinges on recognizing and acting on such weak signals through a holistic framework that bridges signals and systems thinking, global transition dynamics (GTD), cybersecurity, infrastructure resilience, and even the nuanced realm of small-signal stability. We can develop a pragmatic playbook for resilience and innovation in an interconnected world by weaving together these themes.

Signals and Systems Thinking – Weak Signals, Strong Implications

Everything in our technological universe can be viewed as a signal or a system. Signals are inputs or events that convey information – a fluctuation in voltage, a change in market price, or a patient's fever. Systems process, amplify, or suppress these signals. Weak signals are the early whispers of trouble: subtle deviations from expected patterns that indicate underlying stress or emerging disruption. They are "the first tremor before the quake," not mere noise or randomness. A minor frequency dip in a power grid, irregular transactions in finance, and a few unusual ER cases in healthcare are all weak signals that hint at far bigger events to come. Yet, counterintuitively, systems often appear stable until the moment they fail. Why? Because weak signals rarely announce themselves with fanfare, and humans are conditioned to tune out what seems insignificant. Psychologically and organizationally, we ignore faint anomalies, comfortable in normalcy until jolted by crisis. *They seem like noise.* A delayed trade or a minor network glitch is dismissed as an anomaly – until it isn't. *They get lost in stability.* We assume the grid will balance itself as it always has, and the market will rebound as it usually does – until the balance is lost

Table 12.1 Weak Signals Preceding System Failures.

Weak Signal Category	Early Indicator (signal)	Outcome when Ignored (system failure)
Structural Deficiency	Small but accumulating stresses or design flaws in infrastructure	*Bridge Collapse*: The I-35W bridge showed wobbles; ignored cracks led to 13 deaths.
Grid Instability	Minor frequency deviations, relay trips in the power network	*Blackout*: An unnoticed line contact caused the 2003 Northeast blackout, which affected 50 million people.
Financial Irregularities	Unusual liquidity shifts, spikes in defaults in markets	*Recession*: Early subprime mortgage defaults were ignored before the 2008 financial crisis.
Cyber Intrusion	Persistent but small network anomalies, unauthorized access attempts	*Cyberattack*: Suspicious login failures foreshadowed Ukraine's 2015 grid hack that cut power to 230,000.
Healthcare Anomalies	Clusters of odd symptoms or delayed lab results	*Outbreak*: Local pneumonia cases in 2019 were dismissed until COVID-19 spread globally.
Supply Chain Disruptions	Irregular shipping delays, low inventory of key parts	*Shortage*: Years of warnings about semiconductor supply issues preceded the 2021 chip crisis.
Transportation Weaknesses	Recurring aircraft maintenance issues or congestion patterns	*Accident*: Internal safety concerns about the Boeing 737 MAX were ignored until two fatal crashes occurred.
Energy System Stress	Early signs of fuel shortages or pipeline pressure drops	*Grid Failure*: Texas received years of winterization warnings before the 2021 grid collapse in a freeze.
National Security Signals	Geopolitical shifts, changes in terror chatter, or recon activity	*Attack*: Prior intelligence about planned attacks (e.g., pre-9/11 reports) was not acted on in time.

Subtle precursors in the system's normal operation presaged a major failure in each case. These early warnings were ignored or misread, illustrating how focusing only on "loud" events can blind leaders to creeping crises. The cost of inaction was enormous, from lives lost to massive economic damage. Effective leaders cultivate the ability to detect and interpret weak signals across all subsystems of their domain.

one day. *People wait for certainty*. When a weak signal grows loud enough to be undeniable, options to prevent disaster have vanished; as one analysis put it, "When proof arrives, options are gone." This blindness to weak signals has spelled catastrophe across domains.

Consider a few examples summarized in Table 12.1 to ground this in reality. These cases span infrastructure, finance, health, and security – different systems, same story. In each, a weak signal was waving red flags well before the failure but was overlooked or misinterpreted until it escalated into a crisis.

These diverse examples reinforce a critical lesson: systems rarely fail without warning. The weak signals are present, but leaders must be willing and able to listen. Developing that sensitivity is a cornerstone of technological leadership. It requires a shift from reactive management to proactive sensing. Leaders should institutionalize foresight rather than chasing the latest headline risk: gathering data on small anomalies, encouraging frontline observers to speak up about "something off," and using analytical tools to distinguish meaningful weak signals from background noise. For instance, pattern recognition artificial intelligence (AI) can trawl through sensor or market data to flag subtle deviations, but human judgment is needed to act on those flags. Crucially, a culture that rewards vigilance over complacency must be in place, where raising a concern about a wobbling bridge (or an odd network log or an unusual patient lab result) is encouraged, not dismissed. Technological leadership begins with systems thinking, seeing each organization not as static machinery but as a living system of interdependent parts, constantly sending signals. By attuning to those signals, leaders can catch the inflection points where a small deviation portends a coming shock. As the next sections show, this mindset scales from the micro (engineered stability of a single grid) to the macro (GTD of entire economies).

From Small Disturbances to Cascading Failures – Stability vs. Early Warnings

It is important to clarify the nuance between managing small disturbances *within* a system and recognizing weak signals *about* the system. Engineers often speak of small-signal stability – a system's ability to absorb minor fluctuations and remain stable. For example, an electric power grid continuously corrects slight changes in frequency or load; a well-regulated grid can handle these routine deviations, keeping the lights on without incident. This is *small-signal stability* in action: the system's feedback controls damp out little oscillations. On the other hand, weak signals are small anomalies that are *not* fully absorbed – patterns that hint the system may be drifting toward a new, undesired state. In other words, small-signal stability is about a system's immediate resilience to tiny shocks, whereas weak signals are about long-term resilience – the warnings of a brewing crisis. Both involve small changes, but one is about *staying* stable, and the other is about realizing that stability might not last.

To illustrate this distinction, consider a few parallel scenarios in different domains. In financial systems, day-to-day liquidity fluctuations are expected and are typically corrected by normal market mechanisms – this is akin to small-signal stability. However, if banks quietly begin hoarding cash or interbank lending rates creep up unexpectedly, these could be weak signals of deeper financial instability ahead (as before the 2008 crisis). In cybersecurity, a handful of failed login attempts on a network are ordinary and autonomously handled by security protocols (stability). In contrast, a pattern of *persistent*, globally distributed intrusion attempts is a weak signal of an impending coordinated attack. Hospitals routinely handle daily variations inpatient admissions (stability) in healthcare. Still, when clusters of patients with unusual symptoms start appearing across multiple clinics, it's a weak signal of an emerging outbreak. Telecom networks always correct small packets of data loss or momentary bandwidth spikes. Still, sustained congestion in certain network nodes signals a looming outage if nothing is reconfigured. Even supply chains have a baseline of minor delays that logistics can reroute around (stability). Yet, a pattern of ongoing shortages in a critical component (like semiconductor chips) broadcasts a weak signal of a larger disruption developing. In summary, *small-signal stability keeps systems functioning; weak signals warn of failure before it happens.*

For leaders, the goal is twofold: maintain small-signal stability through robust design and real-time controls, and monitor weak signals that suggest when those controls may be insufficient. These two are linked. Often, a system gives us a grace period – a series of small alerts that, if heeded, can prevent a collapse. If ignored, the system transitions through phases of degradation. In the context of complex systems and control theory, we can describe a generic sequence of system modes that any critical infrastructure or organization might experience under stress:

- **Normal Operation:** The system performs as expected within defined parameters.
- **Alert:** Early warning indicators fall outside normal bounds, though manageable.
- **Pre-disturbance:** Minor anomalies accumulate and begin to stress the system.
- **Disturbance:** A disruptive event occurs, but the system's core still holds (a partial failure).

- **Emergency:** Thresholds are exceeded; the system can no longer maintain stability, and a major failure unfolds.
- **Cascading Failure:** Initial failures propagate to interconnected components or subsystems, amplifying the crisis.
- **Systemic Reconfiguration:** Automated or manual interventions kick in to reroute, shed load, or adapt the system to contain damage.
- **Restoration:** Recovery processes restore functionality, though perhaps in a degraded form.
- **Degraded Stability:** After the event, the system operates at reduced performance with new vulnerabilities exposed.
- **New Normal:** The system settles back to its original state or, more often, into a modified equilibrium, potentially with reforms instituted to address the failure (Table 12.2).

The attached table provides the framework. What risks are being missed?

Notes: These documented patterns of system behavior differ across operational states. The exact time scales depend on circumstances, configurations, and types of disturbances. Key Points:

- ***Time Scale Variability:*** *Progression through system modes – from normal operation to cascading failures and restoration – varies based on design, redundancy, and initiating events.*
- ***Cascading Failures:*** *These can spread quickly in interconnected systems like power grids and financial networks. The time scales are general estimates and can vary in practice.*
- ***Restoration and Stability:*** *The length of these phases varies widely depending on the severity of the disturbance and the effectiveness of the response strategy.*

These time scales offer a framework, but actual events may diverge due to the dynamic and uncertain nature of the critical infrastructure systems.

This progression is a generalized template seen in everything from blackout scenarios to financial crashes to organizational crises. Critically, the early phases (Alert and Pre-Disturbance) are where weak signals live and where there is still time to act before the disturbance hits. Traditional risk management has tended to focus on the obvious crisis and the emergency response (the middle of this list) without appreciating the "softer" states leading up to it. By only planning for normal versus emergency, many organizations develop a dangerous blind spot. If things go wrong, they miss the chance to intervene in the Alert/Pre-Disturbance phase and are unprepared for Cascading Failures or Degraded states. As a result, they end up reactive, not adaptive, scrambling after the fact. The lesson for leaders is to manage the *full life cycle of system behavior*: invest in sensors and analytics for early warnings, drill on near-miss scenarios (not just full disasters), and design fail-safes that prevent a localized disturbance from snowballing. In practice, this might mean, for example, a power grid operator monitors frequency and load in real time and pays attention to "nuisance" alarms or minor equipment malfunctions that seem trivial, because those are often the first domino. It also means having systemic resilience plans: what if a failure *does* cascade? Is there a way to rapidly isolate parts of the network, reroute power or data flows, enact manual controls, etc., to prevent total collapse? The 2003 Northeast Blackout began with a single line grounding on a tree in Ohio on a hot day; software alarms that could have flagged the growing imbalance went unnoticed, and within hours, dozens of power plants tripped offline in a cascade. Conversely, modern *self-healing grid* designs include algorithms that detect anomalies and isolate the affected section in milliseconds, preventing a regional catastrophe. This adaptive capacity – to bend, not break, under stress – is the essence of resilience.

Table 12.2 System Failure and Recovery Modes Across Critical Infrastructure System Modes.

System Mode	Description	Time Scale & Examples in Critical Infrastructure
Normal State	The system operates as expected and is stable but is disrupted by discrete changes due to disturbances.	Continuous (Power: Grid stable at 50/60 Hz, Financial: Markets stable, Healthcare: Hospitals operational, Telecom: Networks running, Transportation: Traffic flow normal, Water Supply: Pressure balanced, and Military: No active threat).
Alert	Early warning signs were detected, and potential risks were identified.	Milliseconds – Seconds (Power: Voltage deviations, Financial: Market fluctuations, Healthcare: Early outbreak indicators, Telecom: Network congestion detected, Transportation: Traffic slowdown, Water Supply: Minor leaks, and Military: Increased reconnaissance).
Pre-Disturbance	Small anomalies appear, and system stress increases.	Seconds – Hours (Power: Protective relays triggered, Financial: Liquidity stress, Healthcare: Rising ER visits, Telecom: Packet loss, Transportation: Flight delays, Water Supply: Contaminant traces found, and Military: Cyber intrusions increase).
Disturbance	A disruption occurs but remains contained.	Milliseconds – Minutes (Power: Generator trip, Financial: Flash crashes, Healthcare: Hospital overload, Telecom: Network saturation, Transportation: Major accidents, Water Supply: Pump failure, and Military: Small-scale cyber or drone attack).
Emergency	The system crosses a critical threshold, and instability spreads.	Milliseconds – Hours (Power: Grid collapse, Financial: Bank runs, Healthcare: ICU surge, Telecom: Nationwide outage, Transportation: Airport shutdowns, Water Supply: Treatment plant failure, and Military: Kinetic attack or major cyber assault).
Cascading Failure	Failures propagate across interconnected components.	Milliseconds – Days (Power: Failure spreads across regions, Financial: Contagion in global markets, Healthcare: Infrastructure overwhelmed, Telecom: Interconnection failures, Transportation: Supply chain collapse, Water Supply: Region-wide water shortage, and Military: Full-scale attack or war mobilization).
Systemic Reconfiguration	The system dynamically restructures to contain damage or adapt.	Seconds – Weeks (Power: Load shedding, Financial: Bailouts, Healthcare: Emergency staffing, Telecom: Network rerouting, Transportation: Alternative logistics, Water Supply: Emergency water distribution, and Military: Retaliation or strategic shift).
Restoration	Recovery mechanisms restore function but not full capacity.	Minutes – Months (Power: Grid synchronization, Financial: Market stabilization, Healthcare: Recovery efforts, Telecom: Network repairs, Transportation: Route restoration, Water Supply: Treatment capacity restored, and Military: Ceasefire or operational shift).
Degraded Stability	The system operates at reduced performance with vulnerabilities.	Hours – Years (Power: Grid partially restored, Financial: Weak investor confidence, Healthcare: Staff burnout, Telecom: Limited bandwidth, Transportation: Reduced fleet operations, Water Supply: Contaminant risks, and Military: Geopolitical instability).
New Normal	The system returns to its original state or stabilizes in a modified form.	Days – Decades (Power: Grid redesign, Financial: New regulations, Healthcare: Public health reforms, Telecom: Infrastructure modernization, Transportation: Smart city adoption, Water Supply: Climate-adaptive systems, and Military: New strategic doctrine).

Global Transition Dynamics – A Framework for Interconnected Change

Suppose signals and systems thinking give leaders a microscope to catch early warnings within their domain. In that case, GTD provides a wide-angle lens to view how systems interact at scale. In our hyper-connected era, no critical infrastructure stands alone: the energy grid, water supply, healthcare, finance, transportation, and cyber networks are interdependent. A disturbance in one can trigger shockwaves in others. GTD is a proactive, analytical framework developed to understand and model these complex, cross-sector transitions. It leverages high-fidelity simulations, data integration, and state-of-the-art forecasting to anticipate how changes in one realm (say, energy policy) might ripple through others (economy, environment, national security). At its core, GTD rests on the concept of state variables – parameters that represent the minimal information required to describe the state of a system and predict its future behavior. This idea, rooted in R.E. Kalman's seminal work on control theory, ensures that models focus on what truly matters rather than getting lost in a sea of data. In essence, GTD asks: *What are the key variables that capture the health and trajectory of our global system?* Those could be physical measures (atmospheric CO_2, grid frequency stability), economic indicators (GDP growth, debt levels), technological metrics (AI adoption rate, cybersecurity threat level), or social factors (education index, political stability). By continuously monitoring these state variables, GTD provides status estimation – a real-time picture of where we stand – and foresight prediction – a simulation-based projection of where we are headed.

Under the hood, GTD uses a blend of model-based and data-driven techniques. For instance, it employs Kalman filters and Bayesian inference to assimilate real-time data into its models, constantly updating the state estimates as new information arrives. In a power grid context, a GTD model might include state variables like regional voltages and phase angles; as phasor measurement units (PMUs) stream in readings at 30 times a second, a Kalman filter continuously refines the grid's stability margin estimate. If a big generator trips offline, the model quickly reflects the new state and can predict how close the system is to frequency collapse, giving operators a chance to intervene. In an economic context, Bayesian methods in GTD adjust forecasts when a pandemic unexpectedly slows growth or a new technology investment boom begins. At the same time, GTD incorporates machine learning and reinforcement learning to handle global systems' nonlinear, adaptive aspects. Smart grids, for example, can use reinforcement learning agents to decide how to reroute power or activate storage when there's a sudden spike in demand or drop in supply. The common thread is adaptivity – the models learn and evolve as the world changes rather than relying on static assumptions.

GTD is especially powerful for leaders in its ability to connect domains in one analytical view. Traditional planning might look at energy, economy, climate, and security in silos. GTD breaks those silos by linking the state variables and equations of multiple sectors. A classic example: how do energy investments affect climate outcomes, economic stability, and public health? A GTD model can integrate an energy sector transition (e.g., % of renewables in the grid) with climate models (projected temperature rise), with health models (incidence of heat-related illnesses or pollution impacts), and even with geopolitical models (resource conflicts or migration flows). By synthesizing data from diverse sectors, GTD reveals that no decision operates in isolation; every choice reverberates through a web of consequences. For instance, a policy to subsidize electric vehicles isn't just about transportation – it has upstream effects on electricity demand (grid stress or innovation incentive), on oil markets (possibly lowering global oil prices, affecting economies of oil-producing nations), and on urban air quality (improving public health). GTD would track key state variables in each related domain to forecast the net outcome and any unintended side effects. In one case study, GTD analyses showed that aggressive renewable energy adoption in a region could mitigate climate risks and create jobs. Still, if not paired with grid upgrades and storage, it might paradoxically reduce resilience by making the power supply more variable. Insights like these enable *strategic foresight*: leaders can craft policies that maximize benefits while fortifying any newly revealed vulnerabilities.

The output of GTD is often a set of scenarios and transition pathways. Rather than a single prediction, it gives a landscape of "futures" – if we pull lever A (e.g., invest $X in infrastructure resilience), we likely get outcome B (reduced outage durations). Still, if we neglect C (say, cybersecurity in the grid), then outcome D (high risk of cyber-induced blackouts) enters the picture. This kind of rigorous foresight was highlighted in a recent IEEE study leveraging GTD: investments in infrastructure robustness correlated with shorter and less severe outages over decades. Likewise, GTD-enabled predictions for 2025 suggested that smart grid and AI-driven energy systems could cut outages by 20% via predictive maintenance and real-time control – essentially, using signals to fix problems *before* they cause blackouts. Such predictions aren't wild guesses; they are grounded in data from pilot projects and historical trends processed through the GTD framework. Having these dynamic maps of the future is invaluable for a technological leader. It shifts leadership from reactive crisis management to proactive navigation. By the modeled outcomes, you can justify major decisions (like overhauling a grid, funding an AI oversight body, or forging an international climate accord) and avoid "fighting the last war" while blind to new risks. In sum, GTD operationalizes systems thinking at the global scale, giving leaders tools to anticipate and shape the transitions that define our time. It embodies the mantra that *the best way to predict the future is to create it* – or at least to simulate it, understand it, and prepare for it.

Cybersecurity and Digital Resilience – Signals in Cyberspace

No discussion of signals, systems, and global interdependence is complete without addressing cybersecurity, the nervous system of our modern world. Our power grids, financial transactions, communications, and even cars are now connected via digital networks – "invisible networks that are as essential as the roads and bridges of the physical world." This connectivity brings immense benefits and equally immense vulnerabilities. A cyber threat often starts as a very faint signal: an odd network scan here, a phishing email there. Taken in isolation, these seem like minor nuisances (and often are). But at a system level, they can be an early warning of a serious attack. The WannaCry ransomware outbreak of 2017 offers a vivid example. What began as malicious code exploiting a known software flaw rapidly rippled globally – within hours, hospitals in the UK were paralyzed, corporate and government systems across Europe and Asia were locked up, and thousands of users saw their data held hostage. The fragility of our interconnected digital infrastructure was laid bare in an instant. The weak signals were there in advance: security researchers had warned of Windows' vulnerability and even released patches, but many systems remained unpatched. Organizations ignored the signal, and a cascade failure ensued in cyberspace.

Cybersecurity today must be understood as integral to systemic resilience, not a narrow IT issue. For instance, a cyberattack on a power grid can cascade into widespread physical disruption – trains halted, pipelines shut, and 911 centers knocked offline. Similarly, a hack on financial systems can erode market trust and liquidity. Thus, *cyber resilience* is about recognizing signals across the digital-physical divide. Smart leaders ask: What are the weak signals of cyber risk in my enterprise and industry? Perhaps it is an increase in phishing attempts targeting your sector, chatter on the dark web about a specific industrial control system, or new legislation that could provoke hacktivists. These might not make headlines, but they warrant action – hardening defenses, rehearsing incident response, and collaborating with peers on threat intel. Modern cybersecurity strategy is shifting from reactive defense to proactive, intelligence-driven risk management. This means not just building taller firewalls but also mapping out the interdependencies (if my data center goes down, what happens to my supply chain? If city traffic lights are hacked, what is the knock-on effect on emergency services?). It also means treating cybersecurity as a continuous adaptive process, much like an immune system, rather than a onetime deployment of tools.

Encouragingly, emerging technologies are providing new ways to strengthen cyber resilience. AI and machine learning, for example, can sift through vast streams of network logs to identify anomalies far faster than humans, spotting an attack's weak signals *before* it fully forms. AI-based systems now detect subtle patterns like a user's login behavior changing or a normally dormant account suddenly becoming active, which might indicate a breach in progress. In parallel, technologies like blockchain are being explored to secure critical data and transactions via decentralized trust, making it harder for a single breach to compromise an entire system. Importantly, cybersecurity is also an arena where collaboration is key. No company or nation can single-handedly see all threats. Sharing those early warning signals – through industry information-sharing groups or international alliances – greatly increases collective defense. For instance, when one bank detects a novel phishing email, disseminating the indicators (malicious IPs, file hashes, etc.) to others can prevent a broader outbreak. In effect, the community of defenders acts as a system, processing and amplifying signals of threat to preempt widespread failure.

Ultimately, technological leadership in cybersecurity also has a human dimension. Security expert Bruce Schneier noted that "resilience in cyberspace 'is not just about defense; it is about adaptation, collaboration, and trust.' Leaders must cultivate a culture of eternal vigilance" – not in a paranoid sense, but as an organization-wide ethos that everyone is a sensor and first responder for anomalies. This includes ethical considerations: ensuring that as we fortify our digital systems, we safeguard privacy, equity, and the most vulnerable users. Cyber resilience isn't purely technical; it's a leadership issue that blends policy (e.g., setting standards for software quality and disclosure of breaches), training (drilling teams on recovery and continuity), and even psychology (countering complacency and enabling agile decision-making under pressure). When done right, strong cybersecurity postures become a strategic enabler for innovation – organizations confident in their resilience can embrace new digital tools and services. In contrast, those plagued by breaches will naturally grow risk-averse and fall behind.

Infrastructure Resilience – Reinforcing the Physical Foundations

While digital systems capture headlines, the physical infrastructure undergirding society – our power plants and grids, water treatment facilities, transportation networks, buildings, and bridges – remains just as crucial and needs resilient leadership. The principles of signals and systems thinking apply here in full force. We've already seen how a sagging power line was the weak signal that led to the 2003 blackout. Similarly, the tragic collapse of the I-35W bridge was foreshadowed by subtle structural weaknesses. The flip side of these failures is success stories, where paying attention to signals and building resilience saved lives. In 2018, Hurricane Michael slammed Florida's Panhandle with Category 5 force, obliterating buildings left and right. Amid the wreckage, one hospital in Panama City stood firm – it had been consciously engineered with storms of that magnitude in mind. Its windows, structure, and backup systems were designed to handle extreme wind and surge. As a result, even as surrounding medical facilities failed, this hospital stayed operational throughout the disaster, treating patients when needed most.

This example is a testament to infrastructure resilience by design. The weak signals were not hurricanes themselves (which are obvious). Still, the data and science show that storms were getting stronger over time – essentially, climate trends acting as signals to engineers and planners that the "100-year storm" might now be a 20-year storm. Those who heeded that information and adapted building codes and designs fared far better.

Around the world, we see similar innovations from learning the right lessons. After generations of devastating earthquakes, Japan implemented stringent building standards with base isolation and energy-dissipating structures. The result: modern high-rises in Tokyo or Osaka that sway and flex during quakes, *absorbing* seismic energy and preventing catastrophic collapse. These buildings treat an earthquake's tremors almost like a signal to be managed – they respond in real time, a physical analogy to an automatic balancing grid. In the Netherlands, centuries of battling North Sea floods taught engineers that working *with* nature yields the best resilience. Dutch dikes and levees now incorporate overflow areas and wetlands; instead of simply building ever-higher walls, some floodwaters are given places to go that cause minimal harm and even create ecological benefits. This reflects systems thinking: recognizing that a river is part of a larger environmental system and planning accordingly. In Kenya, to counter worsening droughts (another slow-onset climate change signal), communities are deploying solar-powered wells and pumps to keep water flowing even when rain is scarce. And in cities like New York, a response to increased heat waves and intense rainstorms has been to invest in green roofs and porous pavement – relatively small interventions at the building or block level that collectively reduce urban heat islands and absorb stormwater, easing the burden on power grids and sewers. These are all pragmatic, actionable measures leaders have taken to bolster infrastructure resilience once they recognize the signals of a changing risk environment.

However, infrastructure resilience isn't just a technical or financial challenge; it's also a question of priorities and equity. A refrain in resilient design is that *it's not about what we build but what we value*. Do we value long-term safety and sustainability enough to incur higher upfront costs, possibly, or to change entrenched practices? Do we value *all* communities such that we invest in resilient infrastructure in affluent city centers and low-income neighborhoods, rural towns, and developing countries? Unfortunately, signals of risk often exist longest in marginalized areas – substandard housing that gives the first hints of climate stress or an impoverished region where a disease outbreak first brews. Yet, those areas are least equipped with resilient infrastructure. Technological leadership must also encompass social foresight, ensuring inclusive resilience planning. As one analysis asked pointedly, how do we ensure all communities benefit from resilient infrastructure, not just the wealthy?

If seawalls are built, do they protect the entire shoreline or just the tourist districts? Will the upgrades also reach remote and economically disadvantaged regions if power grids are modernized? These considerations tie back to the earlier GTD perspective – resilience is a *system property* of society. A chain is only as strong as its weakest link; a city is only as resilient as its most vulnerable population. The COVID-19 pandemic, for example, revealed that neglecting public health capacity in any segment of society can end up harming all of society. The same goes for infrastructure: a break in one under-maintained levee can flood an entire city.

From an actionable standpoint, infrastructure leaders should implement what might be called Resilience by Design and Renewal. This includes performing rigorous stress tests and audits of existing structures (analogous to banks doing stress tests for financial resilience). The weak signals of infrastructure decay – hairline cracks, unusually high maintenance frequencies, overcapacity incidents – should trigger retrofits or replacements before a failure. It also means adopting adaptive design for new projects: building structures that can be modified or upgraded as conditions change – for example, constructing bridges with the ability to add reinforcement or increase height later if flood levels rise or designing power systems that can swiftly isolate and reroute power in emergencies (the concept of "modular resilience"). Finally, it involves community engagement: people on the ground often notice weak signals first (like strange tremors or water seepage), and their knowledge can guide smarter investment if leaders listen.

Avoiding the "Golden Hammer" – Innovation with Purpose, Not Hype

As leaders scan for signals and invest in resilience, they must guard against a common pitfall in technological leadership: the "Golden Hammer" syndrome. This is the tendency to grab the trendiest new technology and treat it as the universal solution to every problem – as the proverb goes, if all you have is a hammer, everything looks like a nail. Time and again, industries have poured vast resources into a hyped technology without a clear alignment to actual needs or without respecting the nuances of the system they operate in. The result can be wasted capital, meager returns, and strategic inertia when the hype doesn't pan out. Consider the telecom industry's frenzy over 5G wireless technology. In the late 2010s, 5G was touted as the revolutionary hammer to usher in new services (from autonomous cars to immersive VR). Whoever deployed it fastest would reap huge rewards. Telecom operators have spent nearly $1 trillion in network upgrades globally since 2018, including a U.S. spectrum auction that drew over $80 billion (with one carrier, Verizon, spending $52.9 billion). But the returns have been painfully thin a few years on: the expected "killer applications" that leverage 5G's ultrafast speeds have been slow to emerge, and traditional revenue streams haven't grown proportionately. The historical return on assets for wireless carriers has stayed in the low single digits (~2–4%), barely budging with 5G and more akin to a utility company than a tech disruptor. This imbalance – massive outlay, modest payoff – led nearly half of telecom CEOs to worry that their current path (heavily 5G-centric) might render their businesses economically untenable within a decade. The 5G story illustrates how even a *genuine* innovation can become a Golden Hammer if over-applied without a clear value proposition. As one industry assessment concluded, "Even a game-changing technology must align with a clear value proposition – without that, a Golden Hammer can quickly become a very expensive nail."

The saga of blockchain technology in the finance sector provides another cautionary tale. Hailed as a trust-guaranteeing, decentralized ledger that would revolutionize everything from payments to supply chains, blockchain became a buzzword that attracted wild investment. By the late 2010s, roughly 90% of major global banks had initiated blockchain projects or pilots in some form, and surveys suggested over half expected to be running production blockchain systems by 2019. Yet those rosy projections never materialized. Fast forward a few years: the landscape is littered with stalled pilots and expensive write-offs. An estimated $25 billion spent on enterprise blockchain and related "cleantech" ventures resulted in over half of the projects failing to deliver returns. Studies found that around 90% of blockchain-based supply chain projects never made it past experimental stages into real deployment. High-profile initiatives like the Australian Securities Exchange's attempt to replace its clearing system with blockchain ended in cancellation after five years and $168 million wasted. The pattern is familiar: a technology with real potential was mishandled by trying to force fit it into too many use cases too soon, without regard to whether it was the right tool or if the ecosystem (regulations, interoperability, user readiness) was prepared. Organizations swung the blockchain hammer at every problem and found that many of those "nails" weren't nails.

The lesson for leaders is not that 5G, blockchain, AI, or any new tech is "bad" – on the contrary, they can be transformative when correctly applied to the right problem. The lesson is about discernment and alignment: technology must serve as a strategy, not vice versa. A savvy technological leader will treat hype as just another signal to be scrutinized, not a mandate to act. This means demanding hard evidence of a new tool's value and fit. What problem are we trying to solve, or which opportunity should we capture? Is this technology the best way to do so, or are we chasing it because others are? Are we prepared to handle its risks and limitations? For example, before hospitals rushed to invest in AI diagnostics (another arena prone to Golden Hammer thinking), some leaders noted the weak signals of trouble – early trials where AI missed nuanced conditions, or doctors resisted adoption. Those leaders pumped the brakes, ensuring they first had the data infrastructure and staff training to benefit from AI rather than unthinkingly implementing it everywhere. Others who did not ask those questions ended up with multimillion-dollar software that clinicians barely use.

In practice, avoiding the Golden Hammer entails a few key behaviors. One is embracing pilot projects *with purpose* – experiment, but measure results ruthlessly and be willing to terminate efforts that don't show promise. Another is maintaining a balanced portfolio of technologies: just as an investor diversifies assets, an organization should not bet its future on a single tech trend. Diversification increases the chance that at least some innovations will pay off and reduces the impact if one fails. Additionally, leaders should strengthen the feedback loops between the operational front lines and the strategic decision-makers. Often, those closer to the customer or the factory floor have the earliest insight into whether a new tech is solving problems or creating new ones (these insights are weak signals, too). A culture where data and honest feedback trump hype and optimism bias can save companies from overcommitting to a golden hammer. Finally, align technology adoption with economic reality: perform rigorous cost-benefit and life cycle analyses. Many firms now institute "gates" in their innovation process – a project must clear hurdles of demonstrated value before scaling up. This financial and strategic discipline might have prevented some 5G overspending and blockchain boondoggles. In short, technological leadership requires balancing vision with pragmatism. Yes, be bold and explore new tools – but remain grounded in the system's actual needs and constraints. Not every challenge is a nail, and not every new gadget is the hammer you need. Combining tried-and-true methods with selective innovation often yields better resilience and performance than a wholesale leap into the unknown.

Conclusion – Leadership at the Edge of the Road

Technological leadership in the 21st century means leading at the edge of the road – that tipping point where stability can tip into chaos or stagnation can tip into a breakthrough. The road ahead is not straight; it is full of sharp curves, unexpected gravel, and gusts of change. The leaders who excel lean into the curve rather than clinging blindly to how things were. They don't fight complex systems' dynamics; they learn and harness them. This chapter has argued that we equip ourselves with a robust framework for action by bridging core disciplines – from signals and systems thinking to global dynamics, cybersecurity, infrastructure engineering, and prudent innovation management. We've seen that resilience is not a passive trait but an active adaptation process. True resilience means adapting *before* impact so that the next shock doesn't break you in the first place. Whether it's a corporation adjusting its strategy in anticipation of market shifts, a city reinforcing its grid *before* the next storm, or a cybersecurity team patching vulnerabilities *before* hackers exploit them, the principle is the same: act on the signals when they are weak. Do not wait for full-blown failure to force your hand.

At the same time, leadership is about more than avoiding disaster – it is about steering toward progress. The concepts of GTD remind us that we can shape positive outcomes, not just avert negatives. By anticipating global transitions, leaders can position their organizations to seize new opportunities (like green energy jobs or AI-driven efficiencies) and to collaborate internationally on shared challenges. In a sense, the best leaders become systems integrators and knowledge synthesizers. They take inputs from engineers, data scientists, policymakers, and frontline workers – each seeing different signals – and integrate those into a coherent picture to inform decisions. They foster a culture where weak signals are elevated, not shot down, cross-silo learning is the norm, and flexibility is built into every plan. They also remain humble in the face of complex systems. As noted, *the system always wins unless you learn how it works.* No leader can outmuscle a hurricane, out-hack a hacker, or out-command a global market. But leaders can ride and channel the wave by understanding the forces at play. They "don't control the system; [they] work with it, anticipate its shifts, and position [themselves] ahead of the momentum." This mindset – proactive, vigilant, adaptive, and collaborative – defines technological leadership in our era.

As a final thought, consider this chapter part of a larger mosaic. It is deliberately positioned to bridge the technical and the strategic, the past lessons, and the future vision. In the context of this book, it serves as a connective tissue between understanding resilient systems (in previous chapters) and the actionable leadership practices that follow. This chapter lays a foundation for the subsequent discussions on strategic systems thinking and collaborative leadership in practice by integrating insights on weak signals, systemic resilience, and global foresight. Here, I hope that it provides you, the reader, with a holistic framework that transitions to those higher-level themes.

In essence, *From Weak Signals to Global Resilience* is at the heart of this narrative, anchoring our understanding of how all the disparate threads – technology, people, systems, risks, and innovations – weave together into the fabric of leadership. Equipped with this systems perspective, we can fully appreciate and integrate the lessons from the rest of the book and ultimately become the leader who "reads beyond the headlines," prepares before the crisis hits, and rides the road of technology with foresight and purpose.

Resources for Further Reading

Dynamic Programming and Control Systems

R. Bellman, *Dynamic Programming*. Princeton, NJ, USA: Princeton University Press, 1957.

D. Bertsekas, *Dynamic Programming and Optimal Control*, 4th ed. Belmont, MA, USA: Athena Scientific, 2012.

C. G. Cassandras and S. Lafortune, *Introduction to Discrete Event Systems*, 2nd ed. New York, NY, USA: Springer, 2008.

R. E. Kalman, "Mathematical description of linear dynamical Systems," *SIAM J. Control*, vol. 1, no. 2, pp. 152–192, 1963.

L. P. Kaelbling, M. L. Littman, and A. R. Cassandra, "Planning and acting in partially observable stochastic domains," *Artif. Intell.*, vol. 101, no. 1–2, pp. 99–134, 1998.

S. Meyn, *Control Techniques for Complex Systems*. Cambridge, UK: Cambridge University Press, 2007.

K. Ogata, *Modern Control Engineering*, 5th ed. Upper Saddle River, NJ, USA: Pearson, 2010.
W. B. Powell, *Reinforcement Learning and Stochastic Optimization: A Unified Framework for Sequential Decisions*. Hoboken, NJ, USA: Wiley, 2022.
M. L. Puterman, *Markov Decision Processes: Discrete Stochastic Dynamic Programming*, 2nd ed. Hoboken, NJ, USA: Wiley, 2005.
R. S. Sutton and A. G. Barto, *Reinforcement Learning: An Introduction*, 2nd ed. Cambridge, MA, USA: MIT Press, 2018.

Telecommunications and 5G

GSMA Intelligence, *The $1 Trillion 5G Investment Challenge*, 2023. [Online]. Available: https://www.gsma.com/newsroom/press-release/1-trillion-5g-investment-challenge/.
Omdia, *Global ARPU Trends in the 5G Era*, 2024. [Online]. Available: https://omdia.tech.informa.com/reports/omdia-global-arpu-trends-5g-era-2024.
PwC, *The State of 5G: Capturing More Value from Telecoms*, 2023a. [Online]. Available: https://www.pwc.com/gx/en/industries/technology/publications/state-of-5g.html.
PwC, *ROI or Die: The 5G Imperative for Telecoms*, 2023b. [Online]. Available: https://www.pwc.com/gx/en/industries/technology/publications/roi-or-die-5g.html.
PwC, *Telecom Outlook 2023-2028*, 2023c. [Online]. Available: https://www.pwc.com/gx/en/industries/technology/publications/telecom-outlook-2023-2028.html.

Blockchain and Cryptocurrency

DiCaprio, "Five Lessons from 2022's Blockchain Failures," *Traders Magazine*, Jan. 26, 2023. [Online]. Available: https://www.tradersmagazine.com/am/five-lesson-from-2022s-blockchain-failures/tradersmagazine.com.
S. Neji, "Five Reasons Why Enterprise Blockchain Deployments Failed," *#dltledgers*, 2023. [Online]. Available: https://dltledgers.com/blog/why-enterprise-blockchain-deployments-failed/dltledgers.com.
The Cryptocurrency Crash of 2022: Which Lessons for the Future? Springer, 2023. [Online]. Available: https://link.springer.com/article/10.1007/s10203-023-00345-6.
"5 Lessons From 2022 That Changed Crypto Forever," *Yahoo Finance*, 2023. [Online]. Available: https://finance.yahoo.com/news/5-lessons-2022-changed-crypto-123456789.html.
2022 was a hard year for crypto," *The World Economic Forum*, 2023. [Online]. Available: https://www.weforum.org/agenda/2023/01/2022-was-a-hard-year-for-crypto/.

Financial Technology and Digital Transformation

"Financial Firms Scaled Investments in Next-Gen Tech in 2022, Signaling New Phase of Digital Transformation," *Traders Magazine*, 2022. [Online]. Available: https://www.tradersmagazine.com/am/financial-firms-scaled-investments-in-next-gen-tech-in-2022/.
"In Digital Assets, Disintermediation Fear Lessens," *Traders Magazine*, 2022. [Online]. Available: https://www.tradersmagazine.com/am/in-digital-assets-disintermediation-fear-lessens/.

Technology Integration and Strategy

Amazon, *The AWS Growth Strategy*, 2023. [Online]. Available: https://aws.amazon.com/executive-insights/growth-strategy/.

Apple, *Lessons from Apple Pay's Market Entry*, 2023. [Online]. Available: https://www.apple.com/newsroom/2023/05/lessons-from-apple-pays-market-entry/.

Healthcare and Artificial Intelligence

CB Insights, *AI Healthcare Startup Failures*, 2023. [Online]. Available: https://www.cbinsights.com/research/report/ai-healthcare-startup-failures/.

IEEE Spectrum, *How IBM Watson Overpromised and Underdelivered on AI Health Care*, 2019.

13

Beyond the Triangle: Innovation from the Margins

Not all innovation begins with abundance. Some of the most transformative breakthroughs emerge from environments where scarcity is the only certainty. While the giants of industry break triangles with bandwidth and capital, another equally vital force pulses at the grassroots, where necessity fuels ingenuity, and resilience is forged without permission.

What happens when budgets vanish, timelines collapse, and the system offers no backup? What happens is Jugaad. Gambiarra. Shanzhai. Ingenuity by any other name.

The following chapter explores *The Frugal Code*, a philosophy and practice of innovation born in constraint but built to endure. It is not a detour from strategic leadership; it is its crucible.

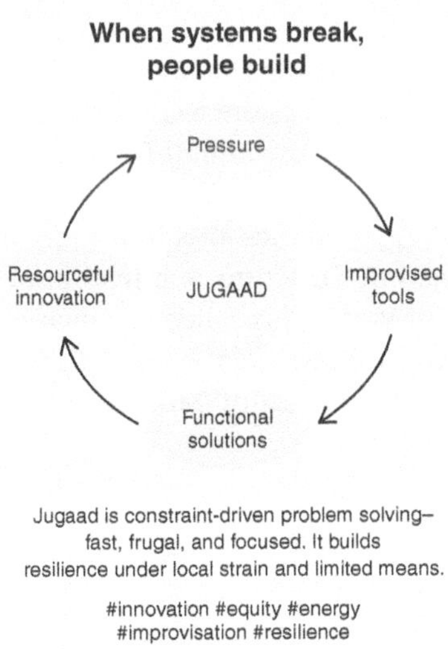

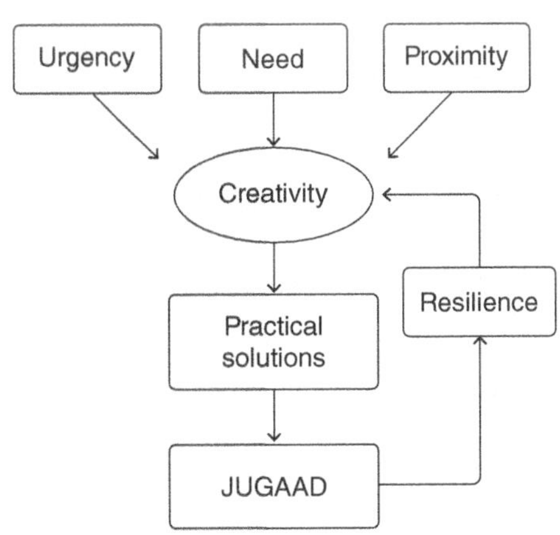

The Frugal Code: Innovating in Complexity, Scarcity, and the Real World

"No one ever bureaucratized their way to a revolution." – Massoud Amin

"Constraints are often the driving force of creativity, not its enemy."– Navi Radjou, *Jugaad Innovation*

"Resilience is not built in the boardroom. It is hammered together in the field, from broken parts, sharp minds, and unyielding purpose." – Field Note, Post-Earthquake Nepal

Introduction: The Age of Constraint, the Call to Innovation

We are living through a period of mounting complexity and systemic strain. Whether the challenge is climate volatility, geopolitical disruption, public health breakdown, collapsing infrastructure, or economic precarity, we are learning that traditional models of innovation – dependent on ideal conditions, lavish budgets, and lengthy timelines – often fail to deliver in the very moments when they are most needed.

The pandemic, natural disasters, infrastructure failures, and regional conflicts have all exposed a fundamental truth: resilience and innovation are no longer optional traits for institutions or individuals. They are survival skills. But there is a gap between knowing we need to adapt and knowing how to do so, especially in conditions of constraint. That is where frugal innovation steps in – not as a temporary workaround, but as a systemic, values-driven, and practical approach to building resilience from the ground up.

Imagine this: In a village outside Pune, a farmer retrofits a discarded fan motor into a functioning water pump. In Recife, Brazil, a teenager uses salvaged wires to restore a cracked solar panel and lights his family's home. In Kansas, a laid-off machinist builds a windmill from scrap metal and bolts it to a silo – cutting his energy bill in half.

These are not exceptional stories. They are standard examples of innovation under constraint – solutions that emerge in places where funding, infrastructure, and institutional support are absent. In India, this mindset is known as Jugaad. In Brazil, it is called Gambiarra. In China, Shanzhai. In Kenya, it parallels what is called Juakali. In early America, it was called ingenuity.

They all share the same DNA: speed, necessity, and practicality. Moreover, they point to something bigger – how resilience emerges from the bottom up.

What Is Jugaad?

Jugaad (Hindi: जुगाड़) is a colloquial term from India that translates loosely as a clever workaround, an ingenious fix, or improvisation. It reflects a practical approach to solving problems using limited resources under constraints.

In other countries:

1) **Brazil:** "Gambiarra" – quick, creative, DIY repairs in low-income or improvisational contexts.
2) **China:** "Shanzhai" – originally meaning imitation, but has evolved into grassroots hardware ecosystems and rapid design cycles.
3) **Kenya:** "Juakali" – from the Swahili for "under the hot sun," describing informal artisanship and resilience-driven production.
4) **USA:** Early American innovators like Benjamin Franklin, Thomas Edison, and Cyrus McCormick exemplified this mode of thinking – practical, iterative, and fast.

Radjou et al. [1] outlined six core principles of Jugaad innovation: frugality, flexibility, simplicity, inclusivity, adaptability, and reuse of existing resources.

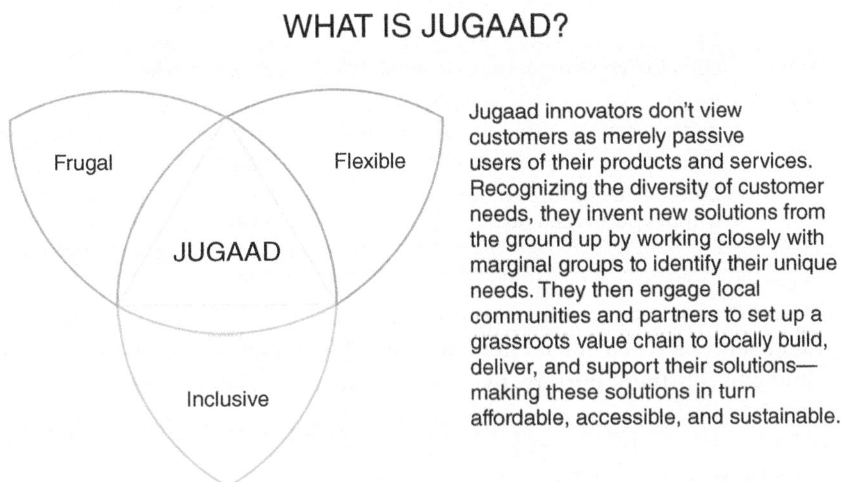

Global Examples:

India: Farmers in Punjab have engineered multipurpose ploughs from motorcycle scrap. The Tata Nano – marketed as the world's most affordable car – was built using a modular, low-cost design. Engineers at the Indian Institute of Science studied the underlying drivers of such grassroots innovations and presented findings that informed policy and academic programs. This research was recommended and integrated into the Technological Leadership Institute (TLI) graduate curriculum and has been used continuously since 2012.

Brazil: Gambiarra is widespread in urban peripheries. During COVID-19, Fiocruz rapidly developed test kits despite restricted supply chains. Community solar grids, DIY filtration systems, and repurposed medical devices have emerged from these conditions.

China: Shanzhai manufacturing began as reverse-engineered consumer electronics. Shenzhen's maker movement grew from these roots, producing agile, decentralized hardware ecosystems.

USA: Cyrus McCormick's mechanical reaper was developed from basic farm tools and feedback loops in rural Virginia, long before he moved to Chicago. It redefined productivity in American agriculture.

Puerto Rico: Following Hurricane Maria, local electricians rebuilt solar systems from remnants and restored power to isolated homes faster than official recovery efforts.

Uganda: Blood supply for rural clinics is tracked through SMS networks, developed by health workers facing communication breakdowns. The mHealth Alliance later studied these frugal platforms for global application.

Defining the Terrain: What Frugal Innovation Is – and Isn't

First, it is important to clarify the terminology to understand the power of frugal innovation. Frugal innovation is often confused with improvisation, invention, or trial and error. However, each serves a different function in the innovation landscape.

Term	Description	Use Case
Trial and Error	Experimental iteration with limited structure. Learning by doing.	Early-stage problem-solving and grassroots exploration.
Invention	Creation of something entirely new, often isolated from context or adoption.	R&D, breakthrough technologies, and academic research.
Frugal Innovation (Jugaad)	Practical, resourceful solutions created under constraint using available means.	Crisis response, emerging markets, informal economies, lean startups, and resilient systems.
Systemic Innovation	Embedded, ongoing innovation capacity within systems and institutions.	Education reform, urban development, infrastructure renewal, and public-sector transformation.

Frugal innovation is not about cutting corners or reducing quality. It is about simplifying intelligently. It is not a fallback when you do not have enough – it is a mindset and method that turns scarcity into strength, constraint into creativity, and obstacles into launchpads.

The Six Core Principles of Frugal Innovation

The concept of frugal innovation has been popularized through *Jugaad Innovation* by Navi Radjou, Jaideep Prabhu, and Simone Ahuja, which outlines six foundational principles. These are not abstract guidelines – they are applied strategies that have produced real breakthroughs in low-resource environments and global companies.

Principle	Core Question	Application	Case Example
1. Seek Opportunity in Adversity	What pain point or constraint reveals a hidden opportunity?	Reframe crisis as an innovation trigger.	3D-printed prosthetics from debris in post-quake Haiti.
2. Do More with Less	How can we deliver essential value at a fraction of the typical cost?	Focus on core function, minimize resource use.	GE's $800 ECG machine for rural India.
3. Think and Act Flexibly	What tools or knowledge can be adapted or repurposed?	Cross-functional remixing, domain-crossing.	Cuban hospitals are fixing dialysis machines with car parts.
4. Keep It Simple	What is the simplest version that still solves the problem?	Embrace minimum viable products and test early.	TATA Nano is an ultra-affordable family car.
5. Include the Margins	Who is excluded from formal systems – and what can they teach us?	Involve informal, underserved voices in codesign.	M-PESA mobile banking for Kenya's unbanked populations.
6. Follow Your Heart	What do we care about enough to solve regardless of external rewards?	Align innovation with mission, not metrics.	Aravind Eye Care's mission is to eliminate needless blindness.

These six principles serve as a blueprint for design and a deeper ethical and strategic foundation for responding to real human needs under duress.

From One-Off Hacks to Repeatable Systems

While improvisation can be clever, it is not enough. The mindset must be systematized to scale frugal innovation from isolated hacks into institutional or civic capacity. This means designing structures that support fast, low-cost experimentation under pressure and embedding frugal thinking into processes, teams, and leadership rhythms.

The "Frugal Innovation Lab" is a powerful structure for this – a small, time-boxed, cross-functional unit tasked with solving specific problems using what is available. Rather than launching million-dollar projects with abstract goals, the lab is charged with creating a working, testable prototype under tight constraints in 10 days or less. These constraints are not limitations – they are catalytic.

To go beyond improvisation, we must **build ecosystems** that allow frugal innovation to flourish systemically – **Launch a Frugal Innovation Lab:**

- **Team:** 5–10 people from different functions.
- **Challenge:** One urgent, real-world problem.
- **Resources:** Strictly limited.
- **Timeframe:** 10 days or less.
- **Output:** Minimum viable solution deployed and tested.

To spark thinking within these labs or innovation efforts, seven pragmatic questions serve as high-leverage starting points:

1) What is broken but still being used, and why hasn't it been fixed?
2) Where are people already hacking or working around the system?
3) What would we try if we had $10 and one afternoon?
4) What would a 12-year-old or a grandparent do with this problem?
5) Can we repurpose tools, space, or existing systems?
6) Where is the greatest waste of time, energy, money, or dignity?
7) What is the smallest functional version we could deploy tomorrow?

These questions are field-tested prompts that have yielded everything from community health interventions to microenterprise designs and major company pivots. They are the seeds of breakthrough thinking – not from luxury but from **limitation sharpened by empathy**.

Proof in Action: Global Case Studies of Scaled Frugal Innovation

Frugal innovation is not the domain of startups and nonprofits alone. Some of the world's largest and most admired companies have built frugal innovation into their DNA – and have sustained success because of it.

3M, for instance, built its innovation engine on two cultural practices: the 15% Time Rule (employees can use 15% of their time to pursue personal projects) and the 30% Revenue Rule (30% of annual revenue must come from products launched in the past four years). These structures produced the Post-it Note, among many others, and have enabled 3M to stay relevant and profitable across generations. Open collaboration with customers (via lead-user innovation), cross-functional teams, and in-house manufacturing of low-cost components has been instrumental.

Google borrowed from this model with its 20% Time policy, which gave rise to Gmail, Google Maps, and AdSense. The secret was not in the time itself but in the cultural permission to experiment without fear of failure.

When its engineers designed emergency ventilators from car parts already in stock, Tesla, often associated with cutting-edge engineering, used frugal innovation principles during the COVID-19 crisis. Dropbox, another tech giant, institutionalized frugal creativity through quarterly Hack Weeks, during which anyone from any department could pitch a working idea directly to leadership. Many product features and performance improvements originated from these grassroots efforts.

Outside the private sector, frugal innovation has transformed public systems. In Medellín, Colombia, instead of investing in expensive subways, the city installed escalators and cable cars to connect informal hillside neighborhoods. School attendance rose 66%, and crime plummeted. In the Philippines, emergency ambulance bikes made from bamboo drastically improved rural emergency response times. M-PESA transformed Kenya's economy by creating a frictionless mobile money system accessible through basic SMS phones.

The Leadership Model: Shifting Culture and Mindset

Frugal innovation is not just a technical process. It is a **cultural shift**.

Leadership Commitments for Resilience-Oriented Innovation

1) **Decentralized Authority:** Let local teams own rapid prototyping.
2) **Redefine ROI:** Measure impact by **lives improved per dollar** or **time saved per constraint**.
3) **Celebrate Smart Hacks:** Create storytelling moments that elevate nontraditional successes.
4) **Invest in the Invisible:** Support unglamorous fixes, retrofits, and duct-tape solutions.

Case Study: Aravind Eye Care System, India

1) Performs over 400,000 eye surgeries/year – most for free or subsidized.
2) Uses an **assembly-line process** to reduce the cost from $300 to under $30.
3) Has an embedded spiritual mission: "To eradicate needless blindness."
4) Manufactures its intraocular lenses for ~$2 (vs. $100 in the West).

This is frugal innovation **scaled** through **systems thinking**, compassion, and lean design.

Real-World Applications: How to Use This in Cities, Companies, and Crises

1. In Complex Cities

1) **Emergency Transport:** Motorcycle ambulances or bicycle-based med delivery.
2) **Water Access:** Rain-harvesting kits made from soda bottles.
3) **Citizen Data:** SMS-based reporting on trash, streetlights, or violence.

2. In Companies

- **Resource Starvation Drills:** Simulate operating with 10% of the current budget.
- **Rapid Repurpose Labs:** Monthly time-boxed sessions to reuse old tools or workflows.
- **Embedded Field Immersions:** Leaders spend three days with the lowest-level employees to learn what is broken and how they fix it.

3. In Crisis Zones

1) **Medical Workarounds:** Use of plastic bags for PPE and car batteries to power ventilators.

2) **Education Hacks:** Chalk-based outdoor classes with shared flip charts in refugee zones.
3) **Tech Repurposing:** Drones used for medicine delivery in war-torn or flooded areas.

Examples: From Boardrooms to Bamboo Bikes

Frugal and resilient innovation is not limited to under-resourced zones. Even the world's most successful companies have **institutionalized these mindsets** to stay adaptive.

3M: Innovation as a Cultural Engine – Industrializing Ingenuity

3M, known for over 60,000 products, is one of the world's most quietly effective innovation machines. **Tactics that work:**

1) **15% Time Rule:** Employees are encouraged to spend 15% of their work time on side projects of personal interest.
   Example: Post-it Notes came from this policy when scientist Art Fry repurposed a failed adhesive.
2) **30% Rule:** 30% of revenues must come from products introduced in the last four years, forcing a constant refresh.
3) **Lead-User Method:** They cocreate solutions with extreme users ahead of the curve, identifying problems before they scale.
4) **Cross-Functional Collaboration:** Scientists, marketers, and engineers codevelop solutions. Diversity of perspective is built in.
5) **Open Innovation:** 3M partners with research labs and startups, recognizing that not all genius resides internally.

> *"We do not just invest in R&D – we invest in the freedom to solve problems."*
>
> —3M Innovation Culture

Takeaway: 3M proves that frugal thinking is not about scarcity but **permission, autonomy, and structure that values ingenuity**. (Harvard Business Review)

Google: The Scaled Spirit of Jugaad

Google adopted a variation of the 3M model, known as the **20% Time**.

1) **Gmail**, **Google Maps**, and **AdSense** began as side projects from curious engineers without a formal assignment.
2) Their secret is not the time – it is the **culture of safety to experiment**.
3) Teams are celebrated for what they try, not just what works.

Takeaway: Innovation does not come from bonuses. It comes from trust, time, and psychological freedom. (Wired)

Tesla: Innovation Through Speed and Simplicity

Tesla builds its edge through **extreme vertical integration** and **rapid iteration:**

1) Engineers, software teams, and designers sit together.
2) Decisions are made fast. Prototypes are tested in real-world conditions (not just labs).
3) During the COVID crisis, they built emergency ventilators from car parts in days.

Takeaway: Even in billion-dollar sectors, frugality is a **design principle**. It creates speed, not scarcity.

Dropbox: Hack Weeks and Cross-Functionality

Dropbox runs **quarterly hack weeks** where employees work on any idea, regardless of department or hierarchy.

1) These have led to interface changes, file sync improvements, and entire product lines.
2) Success is not judged by polish, but **potential**.

Takeaway: The best ideas often come from collisions between code and conversation, between frustration and possibility.

Civic and Urban Innovation Examples – This model scales to public systems, too:

1. Medellín, Colombia

1) Once known for violence, Medellín reimagined its transit system with **escalators, cable cars, and microbuses**, tailored to hilly barrios.
2) Total cost: a fraction of a subway.
3) Outcome: School attendance rose 66%; crime dropped.

2. Nairobi, Kenya

1) M-PESA began as a pilot for microloan repayment.
2) Locals adapted it to send money home, pay bills, and even issue micro-insurance.
3) It now serves **40 million+ users**, many unbanked.

3. Philippines

1) Bamboo-based bicycle ambulances in remote villages reduced emergency response time by 70%.
2) Created by local artisans using available materials.

4. **Aravind Eye Care (India)**

1) Assembly-line eye surgery drops cost $30.
2) Spiritual mission eliminates needless blindness.
3) Manufactures its lenses for $2.

Embedding Frugal Innovation into Your Organization

To make this real in any company, agency, or system:

1. Start Small

1) Launch a Frugal Innovation Lab.
2) Run one 10-day sprint on a real internal problem.
3) Limit the budget and tech. Maximize field input and clarity.

2. Change the Questions

1) Do not ask, "How much will this cost?"
2) Ask, "How can we deliver the core value right now?"

3. Reward Attempts, Not Just Outcomes

1) Honor unfinished prototypes.
2) Celebrate stories of adaptation and recovery.

4. Use the "30/15 Rule"

1) Aim for 30% of your work from the last three years.
2) Give people 15% free time to chase bold ideas.

Beyond Invention: The Code of Resilience

Old saying: **Necessity is the mother of invention.**

Updated truth:

Constraint is the forge of resilience. Frugality is not lack – it is the lever.

In systems, true innovation emerges not when more resources are thrown at a problem but when sharper questions are asked, more voices are involved, and false complexity is removed.

Resilient innovation does not wait for the perfect moment. It **moves when nothing is ideal**, but the purpose is clear.

The Personal Dimension: Frugal Innovation in Everyday Roles

Frugal innovation is not only institutional; it is deeply human. It belongs not just in boardrooms but also in classrooms, homes, and daily life. The following table illustrates how the principles apply across different roles.

Role	Practical Application	How It Works
Engineer	Builds low-cost IoT devices from scrap electronics.	Hosts "build from leftovers" design sprints using only repurposed materials.
Teacher	Uses recycled materials for hands-on learning.	Assigns weekly challenges with limited tools to encourage problem-solving.
Business Owner	Test new service ideas at farmers' markets before investing in the full storefront.	Focuses on lean MVPs, customer testing, and rapid pivots.
Home Manager	Reuses containers, furniture, and fabrics to create home storage or activities.	Runs monthly "reinvention nights" to reimagine household use.
Executive	Creates a mini innovation fund with $500 per team for local experiments.	Measures success by time saved, friction reduced, and team creativity.
Mid-Level Manager	Old training materials are used to develop new onboarding formats.	Rotates internal team ownership of workflow improvements.

These examples demonstrate that frugal innovation is not bound by scale or sector – it is a universally applicable way of thinking and acting.

The Innovation Ladder: From Survival to Systemic Strength

The real innovation journey moves through four evolutionary stages, each deepening capacity and culture.

Stage	Mindset	Mechanism	Outcome
Survival	"We had no choice."	Improvised workarounds.	Onetime fixes.
Repeatable	"We have done this before."	Templates, shared tools.	Small-scale improvements.
Embedded	"This is how we operate."	Labs, rituals, and incentives.	Organizational transformation.
Systemic	"We design for adversity."	Feedback loops, public policy, and civic infrastructure.	Societal resilience.

The goal is not to remain reactive but to become **structurally adaptive** – able to innovate in rhythm with changing needs, not just in response to emergencies.

Why Frugal Innovation Is the Operating System of Resilience

In complex systems – whether cities, companies, or households – resilience is not merely the ability to recover. It is the ability to adapt intelligently under constraint. It is not about withstanding the storm unchanged. It is about emerging from the storm smarter, faster, more inclusive, and more focused.

Frugal innovation delivers this by forcing simplicity, inviting collaboration, accelerating learning, and anchoring innovation in human experience rather than abstraction. It is both an ethical stance and a tactical advantage.

Where bureaucracies seek control, frugal innovation seeks movement. Where traditional design asks for more, frugal thinking asks: "What is essential?"

Moreover, where many strategies falter in the face of complexity and scarcity, the frugal code rises quietly, steadily, and powerfully.

Start with Real, Not Ideal

Resilience does not begin with perfect plans. It starts with honest constraints, human ingenuity, and the will to act before everything is ready.

The frugal code is not a shortcut. It is a compass. One that turns scarcity into clarity, adversity into opportunity, and broken parts into systems that endure.

Frugal innovation is not a tool. It is a **lens**.

- When you are short on time, it teaches you to simplify.
- When you are short on resources, it teaches you to prioritize.
- When the system does not work, it teaches you to build a better one – quietly, creatively, and with conviction.

Whether you manage a home, a hospital, a company, or a classroom, **Jugaad is the code of resilience:** practical, repeatable, and human.

Closing Framework: The Frugal Innovation Ladder

Stage	Culture	Mechanism	Outcome
Survival Hacks	"We had to."	Improvisation	Onetime fix
Repeatable Genius	"We can do this again."	Sprint playbooks	Internal reuse
Organizational DNA	"This is how we think."	Labs, rituals, HR support	Continuous flow
Civic or Systemic Muscle	"We design for constraint."	Governance + feedback	Structural resilience

In summary:

- **Train People in Frugal Principles:** Short workshops. Peer storytelling. No jargon.
- **Run Adversity Labs:** Time-boxed, real-world, and team-based.

- **Measure Value-Per-Dollar or Time Saved** – not scale or polish.
- **Build a Constraint Library:** Record what worked when things failed.
- **Celebrate the Small Wins:** Post-it Notes. Bamboo bikes. Homemade dashboards.
- **Constraint is Not the End of Creativity. It is the birthplace of clarity.**

The tools are already in your hands. Now build.

Reference

1 N. Radjou, J. Prabhu, and S. Ahuja, *Jugaad Innovation*. Wiley, 2012.

Resources for Further Reading

Aravind Eye Care, www.aravind.org.
Harvard Business Review, "Creating Breakthroughs at 3M," Sept 1999.
IDEO.org, *Field Guide to Human-Centered Design*, 2015.
MIT D-Lab, "Design for Scale Curriculum."
M-PESA, "Vodafone Foundation Case Studies."
E. Pilloton, *Design Revolution*. Metropolis Books, 2009.
C.K. Prahalad, *Fortune at the Bottom of the Pyramid*, 2006.
"Tesla ventilator initiative," 2020.
UN-Habitat, "Medellín Urban Innovation Profiles."
Wired Magazine: "The 15% Solution," 1998.

14

The Evolution and Structure of the North American Power Grid

Introduction

The North American power grid is among the largest and most intricate systems ever built, serving as the backbone of modern civilization. Its development has spanned over a century, evolving from small, isolated power stations in the late 19th century to a highly interconnected network that includes the United States, Canada, and parts of Mexico. Over the years, technological breakthroughs, regulatory changes, market restructuring, and emerging challenges such as cybersecurity and climate change have shaped the grid's structure. This chapter provides a detailed examination of its evolution, including key milestones, technological advancements, legislative developments, and the rise of the Smart Grid. By understanding its past and present, we can better prepare for the future of the power system.

The Birth of Electrification (1880s–1930s)

The origins of the North American power grid can be traced to Thomas Edison's development of direct current (DC) systems. In 1882, he opened the Pearl Street Generating Station in New York City, the first commercial power station supplying electricity to homes and businesses. However, Edison's DC system had a major limitation: it could not efficiently transmit power over long distances. This restriction hindered the expansion of electrical networks beyond localized city centers.

Recognizing the need for an alternative, Nikola Tesla and George Westinghouse championed alternating current (AC), which allowed for long-distance transmission using transformers. The turning point for AC technology occurred in 1893 when Westinghouse successfully demonstrated AC-powered lighting at the Chicago World's Fair. This victory led to the construction of the Niagara Falls hydroelectric plant in 1895, which transmitted AC power over 20 miles to Buffalo, New York, solidifying AC as the dominant electrical transmission system.

As AC technology gained widespread adoption, investor-owned utilities (IOUs) emerged as major players in electricity distribution. However, electrification remained largely confined to urban centers, with rural areas left without access due to the high cost of extending power lines. By the early 20th century, Samuel Insull, a former Edison associate, promoted the state-regulated monopoly model, where Public Utility Commissions (PUCs) regulated utilities to ensure fair pricing and universal service. This model laid the foundation for the regulated power industry (Table 14.1).

Table 14.1 Key Developments in Early Electrification.

Year	Event
1882	Edison's Pearl Street Station begins operations in New York City.
1893	Westinghouse demonstrates AC at the Chicago World's Fair.
1895	Niagara Falls hydroelectric plant begins AC transmission to Buffalo.
1907	The state-regulated monopoly model is introduced.
1920s	Urban electrification expands rapidly, but rural areas remain underserved.

The New Deal and Rural Electrification (1930s–1950s)

The Great Depression of the 1930s prompted massive federal intervention in infrastructure development. One of the most transformative policies was the Rural Electrification Act (REA) of 1936, which created the Rural Electrification Administration (REA) to provide low-interest loans for rural electric cooperatives. These nonprofit cooperatives rapidly expanded electricity access, increasing rural electrification rates from 10% in 1936 to over 90% by the mid-1950s.

At the same time, large-scale hydroelectric projects reshaped the U.S. power landscape. The Tennessee Valley Authority (TVA), established in 1933, built hydroelectric dams across the Tennessee River system, providing cheap electricity and economic revitalization to the southeastern United States. Similarly, the Bonneville Power Administration (BPA) was created in 1937 to distribute power from hydroelectric dams in the Pacific Northwest.

By 1953, engineers deployed the first 345-kV transmission line, allowing for significantly longer-distance power transmission with reduced losses. This era marked the transition from localized grids to regional interconnections (Table 14.2).

Expansion and Grid Centralization (1950s–1970s)

The post-World War II economic boom drove an unprecedented increase in electricity demand. Utilities expanded their generation capacity to meet this demand, relying primarily on coal-fired and nuclear power plants. The grid evolved from fragmented local systems into a highly interconnected national network during this period. The three major interconnections that form today's grid emerged:

1) Eastern Interconnection – Covers most of the United States and Canada east of the Rocky Mountains.
2) Western Interconnection – Covers the western United States, Alberta, and British Columbia.
3) Texas Interconnection (ERCOT) – Operates independently from the rest of the United States.

Table 14.2 Federal Electrification and Hydroelectric Projects.

Year	Event
1933	The Tennessee Valley Authority (TVA) was created
1936	The Rural Electrification Act (REA) was enacted
1937	Bonneville Power Administration (BPA) was established
1953	The first 345-kV transmission line was deployed.

The first 765-kV high-voltage transmission line was introduced in 1969, enabling long-distance power transfers. However, the increased interconnectivity also introduced risks. The Northeast Blackout of 1965, which affected 30 million people, highlighted vulnerabilities in grid reliability. This event led to the formation of the North American Electric Reliability Council (NERC) in 1968, which established voluntary reliability standards.

Deregulation and Market Restructuring (1980s–2000s)

By the late 20th century, efforts to deregulate the electricity industry aimed to introduce competition. The Public Utility Regulatory Policies Act (PURPA) of 1978 allowed independent power producers (IPPs) to sell electricity, breaking IOU monopolies. The Energy Policy Act of 1992 further encouraged competition by mandating open transmission access.

Federal Energy Regulatory Commission (FERC) Order 888, issued in 1996, required utilities to open their transmission networks to third-party generators, leading to the establishment of Independent System Operators (ISOs) and Regional Transmission Organizations (RTOs).

The Smart Grid Revolution (1998–Present)

The modern electric power grid has undergone a transformative shift with the development of the Smart Grid, a vision first conceived and articulated in January 1998 by Dr. Massoud Amin while at the Electric Power Research Institute (EPRI). Recognized as the father of the smart grid, Dr. Amin pioneered the integration of automation, artificial intelligence (AI), cybersecurity, and self-healing capabilities into power systems, setting the foundation for a resilient, adaptive, and secure electric infrastructure. His vision resulted in a concerted research effort, leading to the EPRI/DOD Complex Interactive Networks/Systems Initiative (CIN/SI) (1998–2002), a groundbreaking program that established the essential building blocks of the self-healing grid.

The Genesis of the Smart Grid Concept

Before the late 1990s, the electric grid was primarily a one-way delivery system with centralized control. While utilities adopted Supervisory Control and Data Acquisition (SCADA) systems for monitoring and control, these systems lacked real-time intelligence, adaptability, and automated recovery mechanisms. The 1990s witnessed an increase in large-scale blackouts, exposing the vulnerabilities of a growing but rigid and aging power infrastructure. Recognizing these weaknesses, Dr. Amin envisioned a self-healing grid capable of detecting, isolating, and responding to faults autonomously, ensuring reliability and resilience.

In January 1998, Dr. Amin formally proposed the Smart Grid vision to the EPRI Research Advisory Council, securing \$24 million in funding – 68% from EPRI and 32% from the U.S. Department of Defense (DDR&E). This investment launched the Complex Interactive Networks/Systems Initiative (CIN/SI), a three-and-a-half-year program (1998–2002) that revolutionized power system resilience through network science, AI-driven optimization, real-time monitoring, and cybersecurity innovations. The initiative developed 24 enabling technologies, many of which have since been commercialized and deployed across modern energy systems.

Core Innovations and Technological Breakthroughs

The Smart Grid is defined by its ability to sense, analyze, and respond dynamically to cyber and physical threats, integrating advanced machine learning algorithms, distributed computing, and real-time control. The foundational work carried out under CIN/SI led to several key technological breakthroughs, including:

1) Self-Healing Networks – The concept of a self-healing grid, where sensors, AI, and automated switches detect, isolate, and reconfigure the system after a fault, drastically reducing outage times.
2) Cybersecurity for Power Systems – Recognizing the growing cyber threats, the initiative developed cyber-physical security frameworks, incorporating intrusion detection, anomaly prediction, and AI-based security monitoring.
3) Wide-Area Situational Awareness (WASA) – Advanced phasor measurement units (PMUs) and GPS-synchronized sensors were deployed to improve real-time grid visibility and prevent cascading failures.
4) Automated Demand Response (ADR) – AI-based forecasting and automated load management reduced peak demand stress, allowing consumers to participate in energy markets dynamically.
5) Smart Metering Infrastructure (SMI) – The initiative contributed to early prototypes of smart meters, which enable real-time energy consumption monitoring and demand-side management.
6) Grid Resilience Against Extreme Events – Studies on grid hardening, storm resilience, and protection against geomagnetic disturbances (GMDs) and electromagnetic pulses (EMPs) were initiated.

The CIN/SI initiative results provided an engineering and policy roadmap for transitioning from a reactive grid to a predictive, adaptive, and self-optimizing system.

National and Global Adoption

The success of the Smart Grid initiative under Dr. Amin's leadership influenced national policy and led to significant investments in grid modernization. By the mid-2000s, governments and utilities began implementing Smart Grid concepts at scale:

- 2003 Blackout and Policy Shift – The Northeast Blackout of 2003, which affected 55 million people, underscored the urgent need for a resilient, adaptive grid. Dr. Amin's research and findings directly influenced FERC and NERC policies to improve real-time monitoring and automated fault detection.
- 2007 U.S. Energy Independence and Security Act (EISA) – This legislation formally defined the Smart Grid and established funding for nationwide deployment.
- 2009 American Recovery and Reinvestment Act (ARRA) – The U.S. Department of Energy (DOE) allocated $4.5 billion for Smart Grid projects, marking the largest public investment in grid modernization.
- Global Impact – Countries such as China, Germany, and the United Kingdom adopted Smart Grid architectures inspired by EPRI's foundational work, implementing automated substations, AI-driven grid operations, and renewable energy integration.

Smart Grid Investments and Grid Modernization Trends

As of 2023, Smart Grid investments have continued to expand, focusing on decarbonization, resilience, and security. Recent infrastructure investment trends highlight the continued impact of Smart Grid technologies:

- Transmission Investments: 2022: $26.7 billion and 2023: $29.1 billion.
- Distribution Investments: 2022: $47.6 billion.

- Grid Reliability Metrics (Average Annual Outage Duration per Customer): 2017: 7.8 hours; 2018: 5.8 hours; 2019: 4.7 hours; 2020: 8.0 hours; 2021: 7.0 hours; 2022: 5.5 hours. The declining outage duration reflects the success of automated fault detection, demand response, and AI-driven grid optimization.

Dr. Massoud Amin's contributions extend beyond research and development. His work has been instrumental in:

- Educating the next generation of energy leaders – As the Director of the Technological Leadership Institute at the University of Minnesota, he has mentored engineers, policymakers, and industry leaders in energy resilience and infrastructure security.
- Shaping policy and industry standards – He has advised NERC, FERC, the DOE, and the White House, ensuring that grid modernization policies reflect cutting-edge science and engineering.
- Chairing key industry organizations – Dr. Amin has been Chairman of the IEEE Smart Grid, influencing global Smart Grid development efforts.

His pioneering work is recognized in Robert A. Dent's book, "The Foundations of Electricity: 125 Years of Electrification," which credits Dr. Amin with conceiving and articulating the Smart Grid vision. The book, published by IEEE and IEEMA, highlights his role in transforming grid resilience and security.

Global Electricity Grid Investments, Reliability, and Technological Advancements

The modern energy transition is reshaping investment priorities, with an increasing focus on smart grids, energy storage, nuclear power, and infrastructure upgrades. Despite over $6.2 trillion invested in renewables since 2004, grid resilience and reliability gaps persist. As nations adopt clean energy, investments in transmission, distribution, and smart grid technologies are critical for balancing intermittent renewable sources, securing energy supply, and maintaining stability.

Investment Trends in Energy Infrastructure

The global energy sector is witnessing an unprecedented transformation. According to BloombergNEF, over $2.08 trillion was invested in the energy transition in 2024 alone. Yet, significant gaps remain in grid expansion, nuclear energy, and energy storage, which are necessary for ensuring long-term energy security.

Key investment trends include:

- Transmission Investment: Over $14 trillion is needed by 2050 to accommodate renewable energy integration.
- Smart Grid & Storage: Growing at 15% annually, smart grids enable real-time energy pricing, demand response, and integration of distributed energy resources (DERs).
- Nuclear Resurgence: With a 90%+ capacity factor, advanced Small Modular Reactors (SMRs) and Gen IV reactors are gaining traction, supported by policy incentives (Table 14.3).

The U.S. Inflation Reduction Act (IRA) and Europe's Green Deal heavily subsidize grid modernization, storage, and next-gen nuclear, making these areas prime investment opportunities.

Grid Reliability and Outage Trends

Power outages remain a major concern despite increasing investments, especially in the United States. The System Average Interruption Duration Index (SAIDI) measures outage duration per

Table 14.3 Global Electricity Grid Investments (2022–2024).

Investment Category	Investment ($B, 2022)	Investment ($B, 2023)	Investment ($B, 2024)	Annual Growth Rate (%)
Total Energy Transition	1,980	2,050	2,080	12%
Renewables (Solar, Wind)	590	610	620	8%
Smart Grids	240	250	260	15%
Energy Storage	140	150	165	10%
Advanced Nuclear (SMRs)	75	80	90	10%
Transmission Upgrades	490	500	520	7%

Table 14.4 Electricity Grid Reliability (Outage Duration per Year).

Country	Outage Duration (Minutes, 2022)	Transmission Investment ($B, 2022)	Distribution Investment ($B, 2022)
United States	330	27.7	50.9
Japan	10	9.5	5.4
Germany	13.7	10.8	8.7
China	30	166	55

customer annually. Over 80% of major outages in the United States (2000–2023) were caused by extreme weather events. In contrast, countries like Japan and Germany experience significantly shorter outages due to advanced grid technology and predictive maintenance (Table 14.4).

Japan leads in grid reliability due to real-time monitoring, advanced automation, and predictive maintenance. In contrast, the U.S. grid faces aging infrastructure challenges despite higher investments.

The Foundations of Electricity: Technological Advancements

The IEEE publication "The Foundations of Electricity: 125 Years of Electricity" highlights key electrotechnology advancements and recognizes Dr. Massoud Amin's pivotal role in developing the Smart Self-Healing Grid.

Conclusion

The North American power grid has transformed remarkably, evolving from isolated generation hubs into a vast, digitalized infrastructure that underpins modern economies. However, this evolution brings new challenges requiring strategic investments and innovative solutions to ensure a resilient and sustainable energy future.

Investment in Infrastructure

Recent data underscores the critical need for substantial investment in the grid's infrastructure. Transmission investments have notably increased, with expenditures rising from $15.6 billion in

2012 to $21.9 billion in 2017, driven by the desire to provide clean energy and enhance grid reliability. infrastructurereportcard.org.

Distribution investments have also escalated, with major U.S. electric utilities spending about $47.6 billion in 2022. These investments are pivotal in modernizing the grid to handle emerging demands and to integrate renewable energy sources effectively.

Reliability and Outage Metrics

Despite increased investments, the grid's reliability metrics indicate areas for improvement. The SAIDI, which measures the average outage duration per customer, has shown variability over the years. For instance, the average U.S. electricity customer experienced approximately 5.8 hours of power interruptions in 2018. eia.gov.

This variability highlights the ongoing challenges in maintaining consistent service reliability across different regions and underscores the importance of targeted infrastructure enhancements.

Emerging Challenges and the Path Forward

The grid faces multifaceted challenges, including the integration of renewable energy sources, the imperative of enhancing cybersecurity, and the need to adapt to the impacts of climate change. Addressing these issues requires a holistic approach encompassing technological innovation, policy reforms, and strategic investments. Adopting advanced technologies such as AI and automation can be crucial in optimizing grid operations and improving resilience. Moreover, fostering collaboration among stakeholders, including government agencies, private sector entities, and consumers, is essential to drive the transition toward a more resilient and sustainable power grid.

In conclusion, while significant progress has been made in evolving the North American power grid, continued efforts are necessary to address the emerging challenges and capitalize on improvement opportunities. Strategic investments, innovative solutions, and collaborative initiatives will be key to ensuring a reliable, resilient, and sustainable energy future.

Resources for Further Reading

Evolution and History of the Power Grid

G. Bakke, *The Grid: The Fraying Wires Between Americans and Our Energy Future*. New York, NY: Bloomsbury USA, 2016. Discusses vulnerabilities in the aging grid and the need for modernization.

J. Cohn, *The Grid: Biography of an American Technology*. Cambridge, MA: MIT Press, 2017. This book details North American power grid history, tracing its technological evolution and management strategies.

R. A. Dent, *The Foundations of Electricity: 125 Years of Electrification*. IEEE Power & Energy Society, 2016.

R. Hirsh, *Power Loss: The Origins of Deregulation and Restructuring in the American Electric Utility System*. Cambridge, MA: MIT Press, 1999. Traces the forces leading to deregulation and restructuring in U.S. electricity markets.

P. F. Schewe, *The Grid: A Journey Through the Heart of Our Electrified World*. Washington, D.C.: Joseph Henry Press, 2007. Provides insights into the electrical grid's significance and development over time.

Regulation and Market Dynamics

Federal Energy Regulatory Commission, *Order No. 888: Promoting Wholesale Competition Through Open Access Non-discriminatory Transmission Services by Public Utilities*, 1996. [Online]. Available: https://www.ferc.gov, March 4, 2026.

Federal Energy Regulatory Commission, *Order No. 2000: Regional Transmission Organizations*, 1999. [Online]. Available: https://www.ferc.gov, March 4, 2026.

P. L. Joskow and R. Schmalensee, *Markets for Power: An Analysis of Electrical Utility Deregulation*. Cambridge, MA: MIT Press, 1983. A foundational analysis of the impact of deregulation on the power sector.

J. Lin and F. H. Magnago, *Electricity Markets: Theories and Applications*. Hoboken, NJ: Wiley-IEEE Press, 2017. A comprehensive guide to electricity market theory and optimization models.

U.S. Department of Energy, *Quadrennial Energy Review: Energy Transmission, Storage, and Distribution Infrastructure*, 2015. This federal report outlines investment needs and policy directions for modernizing the grid.

Technical and Operational Insights

J. Grainger and W. Stevenson, *Power System Analysis*. New York, NY: McGraw-Hill, 1994. A key resource on power system modeling, load flow analysis, and stability.

P. Kundur, *Power System Stability and Control*. New York, NY: McGraw-Hill, 1994. A seminal work on power system reliability and stability analysis.

J. Machowski, J. Bialek, and J. Bumby, *Power System Dynamics: Stability and Control*, 2nd ed. Hoboken, NJ: Wiley, 2008. Focuses on stability control methods for modern power systems.

A. Mazer, *Electric Power Planning for Regulated and Deregulated Markets*. Hoboken, NJ: John Wiley & Sons, 2007. Examines planning strategies under both regulated and deregulated frameworks.

M. Shahidehpour and M. Alomoush, *Restructured Electric Power Systems: Analysis of Electricity Markets with Equilibrium Models*. New York, NY: Wiley-IEEE Press, 2001. Discusses market restructuring with a focus on equilibrium models.

B. Wollenberg and A. J. Wood, *Power Generation, Operation, and Control*, 3rd ed. Hoboken, NJ: Wiley-IEEE Press, 2013. Covers power generation scheduling, market operations, and system control.

Smart Grid and Future Challenges

M. Amin, *The Self-Healing Grid: Concepts and Advancements*. EPRI, 1998. The foundational work proposing the Smart Grid concept to the EPRI Research Advisory Council.

M. Amin, *EPRI/DOD Complex Interactive Networks/Systems Initiative (CIN/SI): 1998--2002 Final Report*. EPRI, 2002. Documents the $24 million funding structure (68% from EPRI, 32% from DoD) and the development of self-healing grid technologies.

M. Amin, "Security Challenges for the Electricity Infrastructure," *IEEE Computer Magazine*, Apr. 2002. *Discusses cybersecurity risks in energy infrastructure.*

M. Amin, *Smart Grid Foundations and Security Challenges*. IEEE, 2015. This paper discusses the fundamental principles of smart grids, security risks, and integration challenges.

M. Amin and P. Schewe, "Preventing Blackouts," *Scientific American*, May 2007. A technical and policy-focused study on mitigating large-scale blackouts.

M. Amin and B. Wollenberg, "Toward a smart grid," *IEEE Power Energy Mag.*, vol. 3, no. 5, pp. 34–38, 2005. A key paper outlining the transition toward a digital, intelligent power grid.

P. Fox-Penner, *Power After Carbon: Building a Clean, Resilient Grid*. Cambridge, MA: Harvard University Press, 2020. Examines challenges related to decarbonization and resilience.

J. Momoh, *Smart Grid: Fundamentals of Design and Analysis*. Hoboken, NJ: Wiley-IEEE Press, 2012, introduces smart grid technologies, demand response, and real-time monitoring.

M. Shahidehpour and Y. Wang, *Communication and Control in Electric Power Systems: Applications of Emerging Technologies*. Hoboken, NJ: Wiley-IEEE Press, 2003. Discusses control and communication systems for improving grid reliability.

U.S. Department of Energy, *Smart Grid Investment Grant Program Report*, 2015. Details of federal smart grid deployments funded under ARRA in 2010 and later years. *(A complete list of IEEE citations will be included at the end.)*

Part IV

AI and Cybersecurity

15

Renewable Energy Strategy, Market Dynamics, and Competitive Insights

Introduction

The renewable energy sector remains a focal point of global decarbonization strategies, driven by climate policies, financial incentives, and technological breakthroughs. Businesses and communities leverage these dynamics to transition to cleaner, more resilient energy systems. This case study explores the strategies, offerings, and market positions of leading players such as Black Bear Energy and Renewable Energy Partners (REP), and key competitors including Tesla Energy, Siemens, and Sunrun. The analysis integrates verified data, comprehensive market insights, and actionable stakeholder recommendations.

Black Bear Energy: Overview

Company Profile: Founded in 2015 and headquartered in Boulder, Colorado, Black Bear Energy is a technology-enabled intermediary focusing on renewable energy solutions for commercial real estate owners, including Real Estate Investment Trusts (REITs) and Fortune 500 companies. The company's strategic focus on scalability and innovation has enabled it to implement more than 600 renewable energy projects across 20+ states.

Core Offerings

1) Solar Energy Deployment: Black Bear Energy specializes in sourcing and managing solar installations on rooftops, carports, and other commercial spaces to enhance energy efficiency and cost savings for clients.
2) Battery Energy Storage Systems (BESS): The company deploys storage systems designed for peak shaving, load shifting, and backup power, optimizing energy reliability and cost management.
3) Financial Expertise: Black Bear leverages federal and state-level tax incentives, such as the Investment Tax Credit (ITC), facilitating tax equity transfers to enhance project financial viability.

4) Cleantech Solutions: The company offers additional services to maximize energy savings, including fuel cells and energy-efficient lighting retrofits.
5) Proprietary Technology Platform: A vendor-agnostic online platform that enables competitive sourcing and streamlined management of renewable energy projects, offering clients transparency and flexibility.

Strategic Strengths

- Scalability: Proven capability to deliver large-scale projects across diverse geographies.
- Tax Credit Expertise: Unlocks financial viability through strategic use of solar tax credits and incentives.
- Client Advocacy: Focuses on customizing solutions to maximize client return on investment (ROI).

Key Case Study: REIT partnered with Black Bear Energy to install over 1 MW of solar across multiple properties, leveraging tax credits for a 25% cost reduction. The client achieved full ROI within five years while reducing carbon emissions by 40% across its portfolio [1, 2].

Renewable Energy Partners (REP): Community-Centric Leadership

Company Profile: Established in 2013 and based in Minneapolis, Minnesota, REP is a mission-driven organization focusing on equitable access to renewable energy. REP prioritizes community impact through education, workforce training, and energy solutions tailored to underserved areas.

Core Offerings

1) Solar Energy Projects: REP develops rooftop solar installations and community solar gardens to provide cost-effective energy solutions to schools, nonprofits, and local governments.
2) Microgrid Solutions: The company designs and deploys resilient microgrids that integrate solar, battery storage, and backup generators to ensure energy security in critical facilities.
3) Workforce Development: REP's training programs prepare individuals from underserved communities for careers in the renewable energy sector, creating long-term economic benefits.

Strategic Strengths

- Community Impact: Addresses energy equity by focusing on marginalized communities.
- Educational Outreach: Establishes clean energy curricula and hands-on training for students and workers.
- Localized Expertise: Builds strong relationships with local stakeholders to deliver tailored energy solutions.

Key Case Study: REP collaborated with local schools to implement a 500 kW solar array with battery storage. The project reduced energy costs by 30% while integrating renewable energy education into the school curriculum [3, 4].

Competitive Landscape

Aspect	Black Bear Energy	Renewable Energy Partners (REP)	Key Competitors
Founded	2015	2013	Tesla Energy (2003), Siemens (1847), Bloom Energy, Sunrun, etc.
Headquarters	Boulder, CO	Minneapolis, MN	Palo Alto, CA (Tesla), Munich, Germany (Siemens)
Core Market	Commercial real estate, REITs	Community and institutional markets	Residential, commercial, industrial
Key Offerings	Solar, BESS, cleantech solutions	Solar, microgrids, workforce development	AI-driven solutions, large-scale microgrids
Strategic Differentiator	Tax credit expertise, vendor-agnostic tools	Community equity, workforce training	Technology leadership, global scalability
Proven Project Size	Over 600 large-scale projects	Smaller, localized projects	Large-scale global deployments
Clients	REITs, Fortune 500 companies	Schools, communities, small businesses	Multinational corporations, governments

Analysis of Key Competitors and Strategic Lessons

Tesla Energy

Core Focus: Tesla Energy provides integrated solar and battery systems tailored for residential and commercial markets. The company's product portfolio includes the Powerwall (designed for residential energy storage), Powerpack, and Megapack (intended for utility-scale energy storage). Tesla Energy's holistic approach incorporates solar generation, storage, and vehicle-to-grid capabilities, making it a leader in renewable energy ecosystems.

Strengths: Tesla's unparalleled brand recognition and customer loyalty have positioned it as a dominant force in the renewable energy industry. The company's vertical integration strategy enables control over the entire value chain, from solar panel manufacturing to battery storage systems and energy management software. Moreover, Tesla leverages Artificial Intelligence (AI)-driven grid solutions, such as real-time energy monitoring and demand-response capabilities, to enhance the operational efficiency of its systems.

Recent Projects: Tesla recently completed a 100 MW energy storage installation in California (CA), which played a pivotal role in stabilizing the grid during peak demand periods. This project demonstrated Tesla's capability to provide scalable solutions for grid reliability while reducing reliance on traditional fossil-fuel-powered peaker plants. Tesla's virtual power plant (VPP) initiatives in Australia and the United States highlight its innovative leveraging of distributed energy resources [5, 6].

Siemens

Core Focus: Siemens is a global leader in industrial-scale energy solutions, specializing in microgrids, smart grids, and advanced automation technologies. The company's hybrid systems seamlessly integrate conventional and renewable energy sources, ensuring resilience and flexibility across diverse applications.

Strengths: Siemens boasts decades of expertise in integrating energy technologies, making it a preferred partner for large-scale projects. Its ability to combine conventional and renewable energy systems into cohesive hybrid solutions is a critical differentiator. Siemens also excels in deploying smart grid automation technologies, which optimize energy distribution and usage efficiency.

Recent Projects: One of Siemens' most notable projects involved implementing a 200 MW hybrid microgrid at an industrial park in Germany. The system integrated solar, wind, and conventional power sources, cutting energy costs by 35% while significantly reducing emissions. Siemens has also spearheaded projects in remote areas, demonstrating its ability to deliver reliable energy solutions under challenging conditions [7, 8].

Bloom Energy

Core Focus: Bloom Energy focuses on fuel cell technology, offering high-reliability energy solutions for commercial and industrial markets. Unlike traditional renewable solutions, Bloom's systems generate on-site power through solid oxide fuel cells, which operate on natural gas, biogas, or hydrogen.

Strengths: Bloom's fuel cells provide resilience, scalability, and high energy density, making them ideal for critical applications such as data centers, hospitals, and manufacturing facilities. The company's emphasis on hydrogen compatibility positions it for future growth in the evolving clean energy landscape.

Limitations: While Bloom Energy's technology is innovative, it lacks the integration of renewable energy from competitors like Tesla and Sunrun. Its reliance on natural gas raises questions about long-term sustainability compared to solar – or wind-based systems.

Sunrun

Core Focus: Sunrun specializes in residential solar and battery solutions, offering accessibility through flexible financing options such as Power Purchase Agreements (PPAs) and solar leases. The company's focus on affordability and customer engagement has made it a leader in the residential sector.

Strengths: Sunrun's success stems from its customer-centric model, which emphasizes affordability and convenience. By offering PPAs, customers can adopt solar energy with minimal upfront costs, paying only for the electricity generated. This strategy has enabled Sunrun to dominate the residential market and expand into emerging areas like energy storage and grid services.

Recent Successes: Sunrun has deployed thousands of residential solar systems across the United States, creating VPPs in regions like California. These installations allow customers to share excess energy with the grid, contributing to overall grid stability during peak demand [9, 10].

Sunnova Energy

Core Focus: Sunnova provides personalized solar and storage solutions for residential customers, emphasizing customer flexibility. The company's approach prioritizes customizable financing options, making renewable energy accessible to a wider audience.

Strengths: Sunnova's success is driven by its ability to adapt solutions to specific customer needs. The company builds strategic partnerships to expand its market penetration and enhance service delivery. Additionally, its focus on emerging markets positions it for significant growth in underrepresented regions.

Recent Projects: Sunnova has deployed residential solar and storage systems across multiple states, leveraging partnerships with homebuilders and developers to integrate renewable energy into new housing developments. The company has expanded its service offerings to include energy management systems and backup power solutions.

Additional Value Through Real-World Applications

Cross-Sector Insights from Siemens

Siemens' successful deployment of large-scale, globally distributed microgrids offers actionable lessons for REP and Black Bear Energy. By integrating renewables with industrial applications, Siemens has tapped into the lucrative small – and medium-sized enterprise (SME) market. REP and Black Bear Energy could adopt similar strategies to expand their client base, tailoring solutions for SMEs that require customized, scalable energy systems.

Lessons from Tesla Energy's Vertical Integration

Tesla's vertical integration, which encompasses manufacturing, software development, and deployment, highlights the operational efficiencies of controlling the value chain. REP and Black Bear Energy could explore partnerships or acquisitions to integrate more aspects of the value chain, enabling better cost management and higher profitability.

Sunrun's Residential Success

Sunrun's ability to dominate the residential market through flexible financing and customer engagement models provides a blueprint for expanding into underrepresented segments. REP could adapt this approach to target affordable housing projects, while Black Bear Energy might use it to engage smaller commercial clients that require lower-cost entry points.

Strategic Lessons and Opportunities

1) Unlocking Financial Innovation:
   Black Bear Energy's expertise in leveraging solar tax credits and financial mechanisms should inspire REP to explore similar strategies, including green bonds and energy-as-a-service models.
2) Technology Integration:
   AI-driven tools and predictive analytics could enhance energy management systems, providing clients with better ROI and operational efficiency.
3) Education and Workforce Development:
   REP's educational initiatives could scale nationally, creating a broader economic impact while fostering a new generation of renewable energy professionals.
4) Emerging Markets:
   Expanding into underrepresented regions, such as rural and underserved urban areas, could provide a competitive edge while addressing equity challenges in the energy transition.

Emerging Opportunities and Strategic Growth Areas

1) Advanced Energy Storage Solutions: Energy storage is a cornerstone of renewable energy's future. While Black Bear Energy and REP engage with battery storage, incorporating advanced technologies such as solid-state or second-life electric vehicle (EV) batteries could reduce costs and improve scalability. Companies like Tesla and Siemens are already experimenting with such innovations. REP could further explore storage solutions tailored to community needs, such as modular, scalable systems for small institutions.
2) Grid Modernization and Smart Grids: Grid modernization is critical for effectively integrating renewable energy sources. Both companies can leverage their expertise to participate in smart grid initiatives, enhancing grid flexibility, reducing energy losses, and providing better load management. Collaborating with utility companies to pilot advanced demand-response programs could position these firms as key players in grid innovation.
3) Decentralized Energy Models: Decentralized energy production, including peer-to-peer energy sharing through blockchain technology, presents a unique growth avenue. With its focus on community empowerment, REP could pioneer localized energy trading platforms. Black Bear Energy could adopt blockchain to streamline renewable energy credit (REC) tracking, providing greater transparency for commercial clients.
4) Renewable Energy for Underserved Markets: Expanding renewable energy access to underserved or emerging markets – both domestically and internationally – represents a massive opportunity. REP could use its community-centric model to target rural or economically disadvantaged areas, developing microgrids and localized energy hubs. Black Bear Energy could adapt its financial expertise to structure affordable, scalable solutions for such regions.
5) Integration of Artificial Intelligence (AI) and Machine Learning: AI can revolutionize energy management by optimizing energy use, predicting maintenance needs, and improving grid reliability. Black Bear Energy's proprietary platform could integrate AI tools for predictive analytics, while REP could use AI to train local operators in efficient energy system management.
6) Workforce Development and Diversity Initiatives: While REP is already a leader in workforce development, scaling these initiatives nationally or globally could differentiate the company further. Black Bear Energy could introduce training programs focused on commercial and industrial clients, emphasizing skills for tomorrow's clean energy workforce. Ensuring diversity in training programs would also address equity gaps in the sector.
7) Renewable Energy Bundling with Other Services: Both companies could explore bundling renewable energy solutions with other sustainability services, such as water conservation or green building certifications. This would provide added value to clients and tap into broader sustainability markets.
8) Carbon Offsetting and Sustainability Reporting: Large corporate clients increasingly seek sustainability metrics and carbon offset options. Black Bear Energy could expand into carbon offsetting services, while REP could help communities and institutions document and report their contributions to carbon reduction goals.

Strategic Challenges and Mitigation Strategies

While opportunities abound, the renewable energy sector faces unique challenges. Addressing these proactively can solidify leadership positions:

1) Supply Chain Constraints: The renewable energy industry is highly dependent on global supply chains, particularly for rare earth materials used in batteries and solar panels. Developing

partnerships with manufacturers or exploring alternative materials like perovskite for solar cells could mitigate risks.

2) Policy Uncertainty: Renewable energy incentives often depend on government policy, which can change. REP and Black Bear Energy must diversify their revenue streams and advocate for stable, long-term policies to reduce reliance on subsidies.
3) Public Awareness and Acceptance: Educating communities and clients about the benefits of renewable energy remains essential. REP's focus on community engagement can be a model for Black Bear Energy, which could enhance its public-facing communication efforts.
4) Technological Obsolescence: Rapid technological advancements mean that today's cutting-edge solutions may become outdated quickly. Both companies must invest in Research and Development (R&D) and stay agile to adopt emerging technologies.

Expanding Partnerships for Long-Term Impact

1) Collaborations with Academic Institutions: Partnering with universities could provide access to cutting-edge research while helping to train the next generation of renewable energy professionals. REP's relationship with schools could be expanded into collaborations with higher education institutions. Black Bear Energy could sponsor research initiatives focused on commercial renewable energy applications.
2) Cross-Sector Partnerships: Teaming up with industries outside of energy – such as technology firms or construction companies – could broaden market reach. For example, integrating solar panels into building materials (e.g., solar shingles) could open new real estate and construction opportunities.
3) Financing Innovations: Exploring innovative financing mechanisms, such as green bonds or climate-focused investment funds, could unlock additional capital. Black Bear Energy could lead the structuring of financial models for large-scale deployment, while REP could develop community-focused financial tools like crowdfunding platforms for renewable projects.

Comparative Analysis: REP vs. Competitors

Aspect	REP	Black Bear Energy	Tesla Energy	Siemens	Sunrun	Sunnova Energy
Founded	2013	2015	2004	1847	2007	2012
Headquarters	Minneapolis, MN	Boulder, CO	Palo Alto, CA	Munich, Germany	San Francisco, CA	Houston, TX
Core Market	Schools, Communities	Commercial Real Estate	Residential/ Commercial	Industrial	Residential	Residential
Key Offerings	Solar, Microgrids	Solar, Storage, Fuel Cells	Solar, Battery Storage	Microgrids, Hybrid Systems	Solar Systems	Solar, Storage
Proven Project Size	Small-scale, localized	600+ projects	100 MW+	200 MW+	Thousands	Emerging Markets
Unique Strengths	Community-driven impact	Vendor-agnostic, financial expertise	Vertical integration	Industrial expertise	Affordable Financing	Strategic partnerships

Strategic Recommendations for REP

1) Leverage Financial Mechanisms: As Black Bear Energy successfully demonstrated, REP could expand its impact by adopting strategies to unlock tax credits and green bonds. To attract private investment, REPs should explore innovative financial tools like green bonds and energy-as-a-service models.
2) Adopt AI and Data-Driven Platforms: Drawing inspiration from Black Bear Energy, REP could implement a data-driven procurement platform to enhance transparency and efficiency. Companies could use predictive analytics to optimize system performance, lower maintenance costs, and accurately forecast energy demand.
3) Expand Partnerships: Form alliances with technology providers or financial institutions to scale projects, particularly in underserved regions.
4) Focus on Hybrid Systems: Incorporating Siemens' approach to hybrid microgrids could open new opportunities for REP to serve industrial clients.
5) Expand Educational Efforts: REP's workforce training programs could scale nationally, providing long-term economic development benefits in underserved areas.
6) Address Supply Chain Challenges: Diversifying supply chains and exploring alternative materials for battery storage could mitigate risks associated with critical material shortages.

Additional Considerations and Missing Areas

1) Policy Influence: REP and its competitors could benefit from active engagement in shaping energy policy, particularly around community solar and equitable access.
2) Education and Workforce Development: Scaling REP's educational initiatives could economically benefit underserved communities.
3) Technological Integration: Exploring emerging technologies like hydrogen fuel cells and vehicle-to-grid systems could diversify REP's offerings.

In-depth Minnesota's Energy Transition: A Blueprint for Jobs and Resilience

Minnesota (MN) aimed to build a cleaner, more resilient, and economically strong energy system. This wasn't just about replacing coal with renewables – it was about creating jobs, revitalizing economies, and protecting communities from climate risks and cyberattacks.

A Vision for Change: The Next Generation Energy Act (2007) set a goal to cut emissions by 80% by 2050 while keeping energy reliable and affordable. The e21 Initiative (2015) modernized utilities to align with public needs and new technologies. The U.S. Department of Energy (DOE) selected MN to pilot smart grids, with Xcel Energy introducing advanced metering systems to improve reliability and efficiency.

Economic Growth: Renewable energy isn't just cleaner – it's an economic engine. Solar and wind create three times as many jobs per dollar invested as coal or natural gas. By 2023, the U.S. clean energy sector added 149,000 jobs, growing twice as fast as the broader labor market. These jobs ranged from installing solar panels to managing cybersecurity, with better wages and security than traditional energy industries.

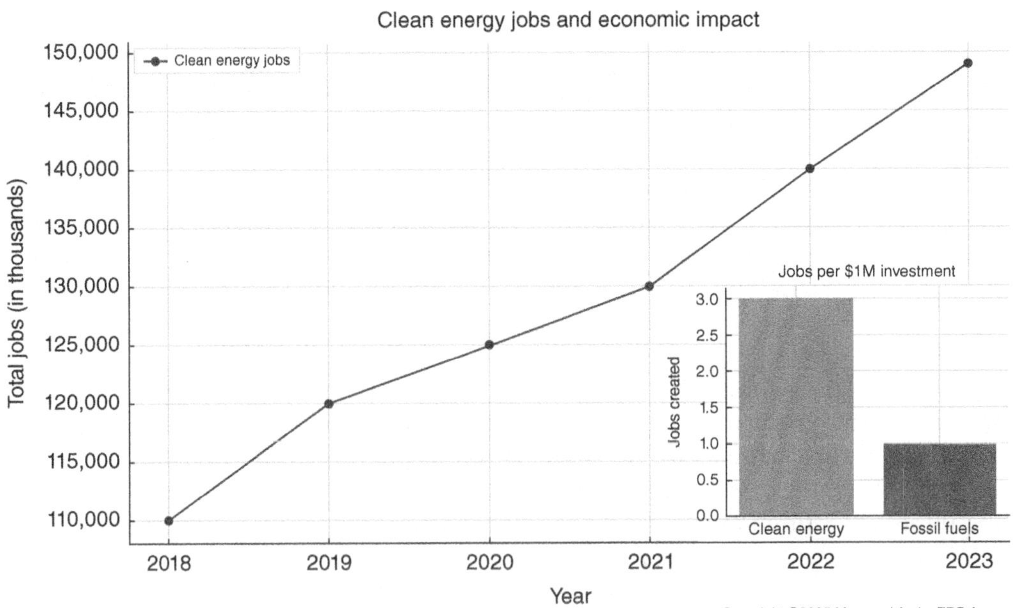

Building Resilience: Smart grids boost reliability, cut energy waste, and reduce costs for businesses and households. They also make communities more resilient to outages caused by severe weather and cyberattacks, protecting billions in economic activity.

Details: MN's clean energy shift is notable because it has no coal, oil, or gas. A decade ago, it spent $11B yearly importing energy. Now, renewables keep more money local, boosting the economy. In 2023, MN had over 62K clean energy jobs, up 4% from the previous year.

With no major fossil fuel industry, fewer entrenched interests slowed clean energy policies. Smart investments and supportive policies made MN a leader in clean energy innovation.

Other resource-limited regions also innovated. Without fossil fuels, Denmark now gets nearly 50% of its electricity from wind. With limited resources, Japan focuses on energy efficiency and renewables to meet its needs.

California has been a clean energy pioneer, implementing ambitious renewable energy targets through policies like its Renewable Portfolio Standard (RPS), which mandates 100% clean electricity by 2045. As of 2023, California derived 35% of its electricity from renewable sources such as wind, solar, and geothermal, leading the nation in installed solar capacity. The state also supports over 500,000 clean energy jobs, more than any other U.S. state, due to its larger population and economy. However, California faces challenges, including higher energy costs and grid reliability issues during extreme weather events.

Has This Energy Transition Impacted Power System Reliability? Minnesota has maintained grid reliability during its clean energy transition through key upgrades in both transmission and distribution systems.

Transmission: Utilities modernized high-voltage lines and substations to handle renewable inputs. Collaboration with Midcontinent Independent System Operator (MISO) ensures grid stability, while compliance with North American Electric Reliability Corporation (NERC) standards guarantees reliability.

Distribution: The Public Utilities Commission (PUC) requires utilities to submit grid safety and resilience plans. Minnesota has integrated 24,310 distributed energy systems (1.54 GW capacity by 2023) and deployed smart grid technologies for real-time outage detection and faster restoration.

Metrics: Due to infrastructure upgrades and automation, outage durations (System Average Interruption Duration Index (SAIDI)) and frequencies (System Average Interruption Frequency Index (SAIFI)) have improved statewide.

MN has maintained strong grid reliability during its clean energy transition. In 2023, its SAIDI was 72 minutes, far below the U.S. average of 145 minutes, reflecting investments in grid modernization, smart meters, and automated outage management. Its SAIFI was 0.92 outages per customer annually, better than the U.S. average of 1.11, showing MN's ability to integrate renewables without compromising reliability.

Globally, Japan leads with a SAIDI of 11 minutes and SAIFI of 0.06, driven by underground infrastructure and precise grid planning. Denmark, at ~20 minutes (SAIFI: 0.2), benefits from decentralized wind systems and international grid links. Norway reports a SAIDI of 93 minutes and SAIFI of 1.2, supported by hydropower resilience to extreme weather. CA lags with a SAIDI over 150 minutes and SAIFI above 1.5 due to wildfire risks, aging infrastructure, and grid strain from high renewable integration without sufficient storage.

MN's investments in technology, diversified energy, and planning prove reliability can be preserved during transitions. Modernization efforts in Japan and Denmark minimize outages, while underinvestment, as in CA, worsens challenges. Minnesota has ensured a reliable and resilient energy system while integrating renewables by modernizing infrastructure, enhancing planning, and adhering to national standards.

In summary, CA leads the United States in clean energy jobs with over 500K and gets 35% of its electricity from renewables like solar and wind. Its goals include 100% clean power by 2045, but it faces high costs and grid reliability issues.

These examples show that resource scarcity can drive clean energy success through innovation, policy, and investment. Minnesota, California, Denmark, and Japan prove clean energy leadership's economic and environmental benefits.

Overcoming Challenges: Transitioning wasn't easy. High costs and resistance from entrenched industries posed obstacles. Minnesota tackled these with practical strategies:

1) Start Small, Show Results: Projects like the North Minneapolis renewable hub powered 10,000 homes, trained 200 workers, and attracted investment.
2) Incentivize Investment: Tax breaks and grants spurred private-sector participation.
3) Focus on Training: Reskilling prepared workers for clean energy roles.
4) Track Progress: Metrics ensured accountability and guided improvements.

A Model for the Nation: MN's approach offers a roadmap:

- Invest in Innovation: Fund renewables, smart grids, and emerging technologies.
- Public–Private Partnerships: Use incentives to attract private investment.
- Local Solutions: Build energy hubs tailored to community needs.
- Measure and Adapt: Track and refine efforts based on results.

Conclusion: Minnesota shows clean energy creates jobs, strengthens economies, and builds resilience. The costs and resistance are real, but the long-term benefits outweigh them. States like Iowa and Texas, and countries like Denmark and Germany, prove that clean energy works anywhere.

Incorporating Sustainability Fitness into Technological Leadership

Bridging Sustainability and Leadership in Technology Management

The pace of technological progress is rivaled only by the urgency of environmental and societal challenges. Leaders in the technology sector are increasingly called to reconcile innovation with sustainability. Breno Nunes' Sustainability Fitness (SF) framework provides a nuanced, actionable lens through which leaders can assess and integrate sustainability into every facet of their operations. This chapter delves deep into the SF framework, blending theory with vivid real-world examples, metrics, and opportunities for collaboration that demonstrate its transformative potential.

Theoretical Foundation: Sustainability Fitness (SF)

At its core, the SF framework is a diagnostic and prescriptive tool designed to guide organizations in achieving resilience and sustainability. Developed by Breno Nunes and collaborators, SF evaluates an organization's performance across corporate, supply chain, and product levels. It is built on four interlocking components:

Essentiality and Impact Metrics

SF addresses a fundamental question: Is this product, service, or process essential? Essentiality is measured by balancing societal needs against environmental costs:

- Essential Goods: Items indispensable to societal functioning, such as renewable energy systems or medical devices.
- Nonessential Goods: Items offering limited societal value, like single-use electronics or luxury accessories, relative to their environmental footprint.

The framework also incorporates an Environmental Impact Score to quantify emissions, waste, and resource usage. Together, these metrics enable leaders to:

- Eliminate inefficiencies by identifying low-value, high-impact products.
- Redirect resources toward innovations that deliver tangible societal benefits.

System Fitness and Resilience

Modern organizations operate within dynamic socio-ecological systems. SF emphasizes:

- Balancing Resource Use: Ensuring consumption stays within planetary boundaries.
- Avoiding Collapse: Employing predictive modeling to identify risks to operations and supply chains.
- Iterative Improvement: Continuously refining strategies based on real-time insights and feedback loops.

Example Formula for System Fitness Score: A key feature is the System Fitness Score (SFS), a composite measure aggregating essentiality, environmental impact, and resilience:

$$\mathrm{SFS} = \frac{\mathrm{EI} \cdot \mathrm{SI}}{\mathrm{RI}}$$

where:

- EI is the Environmental Impact (scaled 0–1).
- SI is the Societal Impact (scaled 0–1).
- RI is the Resilience Index (scaled 0–1, based on system stability and resource availability).

This score is a comprehensive indicator of sustainability performance, enabling strategic decision-making.

Application Across Levels

SF is scalable and adaptable to:

- Corporate Strategies: Aligning policies with sustainability goals and industry standards.
- Product Innovation: Embedding sustainability into design, manufacturing, and lifecycle management.
- Regional and Global Initiatives: Driving systemic change through multi-stakeholder collaboration.

Real-World Applications: Lessons from Leaders

Apple: Building a Circular Supply Chain

Apple's journey toward a closed-loop supply chain exemplifies SF in action. By viewing disassembled iPhones not as waste but as resources, Apple demonstrates how essentiality and impact metrics guide transformative change.

- Eliminating Hazardous Materials: Apple eliminated harmful substances like mercury and polyvinyl chloride (PVC), focusing on high-essentiality materials.
- Revolutionizing Recycling: Using robots like Daisy to disassemble 200 iPhones per hour, Apple reduces e-waste and dependence on virgin resources.
- Setting Ambitious Goals: SF's commitment to achieving net-zero emissions across its supply chain by 2030 exemplifies its resilience and metric-driven strategies.

Tesla: Redefining Essentiality in Transportation

Tesla's mission to accelerate the transition to sustainable energy illustrates the integration of SF principles:

- Essentiality in Design: Tesla's EVs prioritize performance, sustainability, and societal impact.
- Digital Twins in Production: Real-time monitoring of efficiency and material use optimizes operations, reducing environmental impact.
- Energy Ecosystems: By integrating solar panels and battery storage, Tesla extends its impact beyond vehicles, aligning with broader socio-ecological goals.

Google: Data-Driven Sustainability

Google applies SF principles to its operations and innovations, using data-driven tools to address sustainability challenges:

- Efficient Data Centers: Google's data centers demonstrate operational sustainability by operating twice the energy efficiency of industry averages.
- AI for Climate Mitigation: Predictive AI optimizes renewable energy production and reduces waste.
- Carbon-Free Commitments: By 2030, Google aims to run entirely on carbon-free energy, setting an industry benchmark for environmental responsibility.

Opportunities for Collaboration

1) Corporate Applications

   Building on the examples of Apple, Tesla, and Google, organizations can:

   - Develop tailored case studies showcasing how SF strategies enhance environmental and societal value.

   - Use digital tools like digital twins to provide real-time feedback on resource use and sustainability metrics.
2) Workshops and Knowledge Sharing
   SF provides a robust foundation for workshops that:
   - Combine sustainability metrics with operational strategies.
   - Equip practitioners with actionable tools to embed resilience and essentiality into their workflows.
3) Research Synergy
   Collaborative research could explore the following:
   - Integration of SF metrics with advanced analytics.
   - Enhanced modeling of resilience and essentiality using tools like AI and machine learning.
   - Industry-specific adaptations of SF, focusing on sectors like energy, healthcare, and consumer electronics.

Integrating SF Metrics into Strategy

SF metrics provide leaders with actionable insights to:

- Prioritize investments in essential, high-impact innovations.
- Align supply chains with sustainability goals.
- Transparently communicate progress to stakeholders.

A sample Sustainability Metrics Table:

Metric	Description	Applicability	Example Scenario
Essentiality Index	Measures societal need for a product/service.	Product development, R&D	Phasing out nonessential plastics in packaging.
Environmental Impact	Quantifies emissions, waste, and resource use.	Supply chain, production	Reducing CO_2 in EV battery manufacturing.
System Fitness Score	Aggregates essentiality and environmental impact into a single metric.	Corporate sustainability strategy	Setting priorities for green innovation.

Conclusion: A Roadmap for Technological Leadership

The SF framework is a practical, powerful approach to balancing innovation with sustainability. By adopting its principles, leaders in the technology sector can:

- Drive systemic change in their organizations.
- Inspire broader industry transformation.
- Build a legacy of innovation that respects planetary and societal boundaries.

The renewable energy landscape is evolving rapidly, driven by technological advancements, financial incentives, and increasing demand for sustainable solutions. Companies like Tesla Energy, Siemens, Sunrun, and Black Bear Energy lead the way with innovative strategies, while REP's community-driven model offers a unique and impactful approach. By adopting competitors' lessons and leveraging their strengths, companies can position themselves for success in the energy transition.

Black Bear Energy and REP highlight two distinct yet complementary approaches to advancing renewable energy. Black Bear excels in scalability and financial strategies, while REP leads in community impact and workforce development. By learning from competitors like Tesla and Siemens, both companies can refine their plans to address emerging challenges and capitalize on new opportunities in the sector.

References

Market Analysis

1 Black Bear Energy, "About Us." [Online]. Available: https://blackbearenergy.com/about, accessed 6 March 2026.
2 Datanyze, "Company Profile." [Online]. Available: https://www.datanyze.com, accessed 6 March 2026.

Renewable Energy Partners

3 "REP Fact Sheet," October 2024.
4 REP Website. [Online]. Available: https://renewablenrgpartners.com/, accessed 12 January 2025.

Competitors

5 Tesla Energy, "Solar and Battery Storage." [Online]. Available: https://www.tesla.com, accessed 6 March 2026.
6 Siemens, "Microgrid Solutions." [Online]. Available: https://www.siemens.com, accessed 6 March 2026.
7 Sunrun, "Power Purchase Agreements." [Online]. Available: https://www.sunrun.com, accessed 6 March 2026.
8 Sunnova, "About Us." [Online]. Available: https://www.sunnova.com, accessed 6 March 2026.

Strategic Trends

9 Canary Media, "Solar and Tax Credits." [Online]. Available: https://canarymedia.com, accessed 6 March 2026.
10 IEA, "Renewable Energy Market Update." [Online]. Available: https://www.iea.org/reports/renewable-energy-market-update, accessed 6 March 2026.

Resources for Further Reading

R. C. Alamino and B. Nunes, "Collapse and the interplay between essentiality and impact in socio-ecological systems," *Phys. Rev. E*, vol. 107, no. 5, 054201, 2023.

B. Nunes, R. C. Alamino, D. Shaw, and D. Bennett, "Modelling sustainability performance to achieve absolute reductions in socio-ecological systems," *J. Clean. Prod.*, vol. 132, pp. 32–44, 2016.

B. Nunes, R. C. Alamino, D. Bennett, and A. Brem, "An introduction to product essentiality: Conceptualisation and measurement," *Tech. Anal. Strat. Manag.*, vol. 36, no. 1, pp. 4247–4264, 2023.

16

Opportunities in Energy Transition and Smart Infrastructure Investment

Introduction

Under Roger Sit's leadership, Sit Investment Associates has built a strong reputation for disciplined, research-driven asset management. As energy markets rapidly evolve, investors face critical decisions on infrastructure, renewables, and emerging technologies. This briefing highlights key opportunities in smart grids, energy storage, and nuclear investment – areas poised for long-term returns while balancing risk and stability.

Market Context: Energy Transition Investment Trends

- $2.08 trillion invested globally in energy transition in 2024 (BloombergNEF).
- $6.2 trillion invested in renewables since 2004, yet nuclear and grid expansion remain underfunded.
- Smart grid and storage investments are growing 15% annually, crucial for integrating renewables.
- Energy security concerns are reshaping investment priorities, with governments incentivizing grid resilience.

Why This Matters for Sit Investment Associates

- Stable, Long-Term Infrastructure Assets: Nuclear, smart grids, and energy storage offer multi-decade revenue streams, aligning with Sit Investment's fundamental, research-driven approach.
- Defensive Sector Exposure: Energy transition investments correlate less with market cycles, offering inflation protection and yield stability.
- Global Growth Drivers: The U.S. Inflation Reduction Act (IRA) and Europe's Green Deal heavily subsidize grid modernization, storage, and next-gen nuclear.

Key Investment Themes

Smart Grids & Energy Storage: The Backbone of the Transition

- U.S. Market: $250B+ invested since 2000; California and Texas leading microgrid adoption.

- Global Growth: China is deploying ultra-high-voltage (UHV) grids, and the European Union (EU) is integrating cross-border grids.
- Investment Opportunity: Smart grids enable real-time energy pricing, demand-side management, and stability for renewables.

Nuclear's Resurgence: Stable Baseload Power & Long-Term Yield

- Advanced Nuclear (SMRs & Gen IV Reactors): Lower cost, modular, shorter deployment timelines.
- High-Capacity Factor (90%+ uptime): Unlike wind and solar, nuclear does not require backup generation.
- Current Market Trends: Microsoft-backed Small Modular Reactors (SMRs) are gaining traction, and the UAE's Barakah project proves nuclear feasibility in low-cost environments.

Energy Infrastructure: Addressing the Grid Bottleneck

- Transmission Expansion Needs: $14 trillion required by 2050 to accommodate renewables.
- EV & Electrification Boom: Grid constraints pose a major investment opportunity in upgraded transmission and decentralized microgrids.
- Government Support: U.S. DOE offering loan guarantees for grid and nuclear infrastructure projects.

Strategic Recommendations for Sit Investment Associates

1) Diversified Energy Infrastructure Fund: A mix of smart grids, storage, nuclear, and renewables to balance risk and return.
2) Private Equity & Direct Investment Opportunities: High-growth startups in grid tech, SMRs, and Artificial Intelligence (AI)-driven energy management.
3) Government & Policy-Backed Investments: Leveraging tax credits, loan guarantees, and subsidies to de-risk capital allocation.

Energy Investment Landscape: Risks, Rewards, and Strategic Opportunities

The global energy transition is accelerating, with over $2.08 trillion invested in 2024 alone. Since 2004, $6.2 trillion has gone into renewable energy, yet significant gaps remain in smart grid deployment, nuclear energy, and storage infrastructure. Investment in these areas is critical for long-term stability, balancing intermittent renewables, and ensuring energy resilience. As the industry shifts, strategic capital allocation in smart grids, nuclear energy, and transmission infrastructure will be pivotal for long-term returns.

The role of smart grids and energy storage cannot be overstated. In the United States, more than $250 billion has been invested in smart grid infrastructure since 2000, with states like California and Texas leading in microgrid adoption. Globally, China and the European Union are integrating cross-border grids to enhance stability and efficiency. The market for energy storage is expanding at 15% annually, driven by increased renewable penetration and the need for grid flexibility. Without adequate storage, renewable energy remains constrained by variability, requiring backup systems that add costs and inefficiencies. Investment in smart grids enables real-time energy pricing, demand-side management, and improved stability, making renewables a more viable long-term solution.

Nuclear energy is also reemerging as a cornerstone of energy stability. The past decade saw significant declines in nuclear investment, but advanced nuclear technologies, including SMRs, are gaining momentum. SMRs offer lower costs, modular construction, and faster deployment than traditional atomic plants. Unlike wind and solar, nuclear provides stable baseload power with a 90 %+ capacity factor, eliminating the need for costly backup solutions. Countries like the United Arab Emirates (Barakah Nuclear Plant) and Canada (SMR development with Microsoft and Ontario Power Generation [OPG]) are proving nuclear's viability as a competitive alternative. While nuclear faces high upfront capital costs, its lifespan of 60–80 years, lower long-term operational costs, and minimal material requirements per TWh make it one of the most sustainable options for large-scale electricity production.

Infrastructure investment in transmission upgrades is also crucial. With over $14 trillion needed by 2050 to support electrification and renewable expansion, transmission constraints remain a bottleneck for the transition. Electric vehicle (EV) growth, grid resilience against extreme weather, and decentralized microgrid expansion demand immediate investment in transmission and distribution (T&D) networks. The U.S. Department of Energy has allocated billions in loan guarantees to support these upgrades, providing low-risk, government-backed investment opportunities.

To illustrate the financial potential and risk profiles of various energy investments, the following comprehensive table consolidates key market trends, expected returns, and sector growthbulleted rates:

Comprehensive Energy Investment Table

Investment Category	Investment ($B)	Annual Growth Rate (%)	Key Benefits	Risk Level	Expected Returns
Total Energy Transition	2,083	12	Transition to sustainable energy	Medium	Moderate
Renewables (Solar, Wind, Hydro)	6,200	8	Expanding clean energy sources	Low	Moderate
Smart Grids	250	15	Real-time pricing, grid stability	Medium	High
Energy Storage	150	10	Balancing intermittent renewables	Medium	High
Advanced Nuclear (SMRs)	80	10	Stable baseload power, long asset life	High	High
Transmission Upgrades	500	7	Supports EV & electrification growth	Low	Moderate

The risk vs. reward of energy investments is a critical consideration for investors. Smart grids and energy storage present a moderate risk with high returns, as policy incentives and infrastructure funding increasingly support these technologies. Nuclear investments, particularly SMRs, come with higher risk but offer some of the most substantial long-term returns due to their stability and extended operational lifespans. Renewable energy remains low risk but also yields moderate returns due to its dependence on market conditions and policy changes.

To further illustrate the investment landscape, the following risk vs. reward visualization helps compare various energy asset classes:

Risk vs. Reward in Energy Investment

(The chart is represented visually in the provided MS Word document, showing risk vs. reward for energy sectors.)

- X-Axis: Risk Level (Low to High)
- Y-Axis: Expected Returns (Low to High)
- Positioning:
  - Smart Grids and Storage: Moderate risk, high return
  - Renewables: Low risk, moderate return
  - Advanced Nuclear (SMRs): High risk, high return
  - Transmission Infrastructure: Low risk, moderate return

Strategic Recommendations for Sit Investment Associates

Given Sit Investment Associates' fundamental, research-driven approach, the following strategies align with long-term energy investment trends:

1) Diversified Energy Infrastructure Fund: A balanced portfolio combining smart grids, storage, nuclear, and renewables ensures stability and mitigates risk while capturing high-growth opportunities.
2) Private Equity & Direct Investment Opportunities: Investing in high-growth startups specializing in smart grid tech, SMRs, and AI-driven energy management provides early-mover advantages in emerging markets.
3) Government & Policy-Backed Investments: Leveraging tax credits, loan guarantees, and energy transition subsidies reduces capital risk and ensures predictable returns.

Summary: The energy transition is accelerating, but the market's focus on short-term renewables overlooks critical gaps in grid resilience, energy storage, and stable baseload power. Sit Investment Associates is well-positioned to capitalize on long-term, stable, yield-driven energy assets while mitigating volatility.

The global energy transition represents one of the largest investment opportunities of the century. While renewables continue to expand, the real opportunities lie in smart grids, nuclear, energy storage, and infrastructure upgrades – the critical components enabling long-term stability. Sit Investment Associates is uniquely positioned to capitalize on these opportunities, focusing on stable, yield-driven assets while managing volatility.

Sit Investment can align its strategies with the most promising and high-growth energy sectors by prioritizing grid modernization, energy resilience, and policy-backed investments. As energy demand rises and markets evolve, strategic long-term capital allocation and stable energy assets will be crucial for future-proofing investments.

Global Electricity Grid Investments and Reliability: A Comprehensive Assessment

Background: In recent decades, the United States has significantly increased its investments in electrical infrastructure, particularly in T&D systems. However, despite these efforts, the nation continues to experience frequent and prolonged power outages, primarily caused by extreme weather events and an aging grid. This brief report analyzes T&D infrastructure investments and examines power outage durations in the United States, Japan, Germany, and China (Table 16.1).

- Transmission Investment: Represents capital expenditures on the electric transmission system, covering new infrastructure and maintenance.
- Distribution Investment: Encompasses spending on electric distribution systems to deliver electricity to end-users.
- Average Outage Duration: Measured by the System Average Interruption Duration Index (SAIDI), which represents the total time an average customer experiences a non-momentary power interruption in a year.

Infrastructure Investments: The United States has experienced a significant rise in capital spending on electrical infrastructure over the past two decades. In 2003, utilities allocated about $9.1 billion to transmission systems and $19.5 billion to distribution systems. By 2023, these numbers had increased to $27.7 billion for transmission – a threefold rise – and $50.9 billion for distribution, reflecting a 160% surge. Notably, investment in overhead infrastructure, such as lines, poles, and towers, reached $17.4 billion in 2023, an 11% increase from 2022 and a 220% jump from 2003. Likewise, spending on underground lines doubled over the past 20 years, totaling $11.8 billion in 2023.

Outage Duration Trends: Despite significant investments, the reliability of the U.S. power grid has not improved accordingly. In 2021, the average electricity customer experienced approximately seven hours of power interruptions, primarily due to major weather events. Excluding these significant weather events, the average annual interruption duration remained consistently around two hours per year from 2013 to 2021.

Impact of Extreme Weather Events: Extreme weather is the main cause of significant power outages in the United States. Between 2000 and 2023, 80% of major outages – defined as incidents affecting at least 50,000 customers or interrupting service of 300 megawatts or more – were due to weather-related incidents. The frequency and severity of these events have increased, with

Table 16.1 U.S. Electric Infrastructure Investments and Outage Durations.

Year	Transmission Investment (Billion $)	Distribution Investment (Billion $)	Average Outage Duration (Hours)
2000	9.1	31.0	N/A
2010	14.1	45.0	N/A
2015	20.0	51.0	5.0
2018	35.0	N/A	5.8
2019	40.0	57.4	5.8
2022	26.7	47.6	5.5
2023	29.1	50.9	N/A

Notes: N/A indicates data is unavailable for the specified year.

climate change resulting in more intense storms, heat waves, and other adverse conditions that strain the aging grid infrastructure.

Regional Disparities: Certain regions in the United States are more susceptible to prolonged outages. For instance, Louisiana has faced serious challenges, with 47 weather-related disasters since 2014, leading to over $400 billion in damages. In contrast, states like Idaho and North Dakota have reported shorter outage durations, averaging around two hours annually. These disparities highlight the unequal impact of extreme weather and the varying resilience of regional infrastructures.

International Comparisons: Compared to other developed nations, the United States experiences more frequent and longer power outages. For example, residents in Japan face an average of only four minutes of power outages each year, indicating a highly reliable power grid. This stark contrast underscores the need for the United States to adopt best practices from countries with more resilient power systems.

Investment Trends and Reliability Metrics:

1) Global Investment Overview
   - United States: In 2022, the United States invested $27.7 billion in transmission and $50.9 billion in distribution systems. While investments have risen over the years (e.g., from $9.1 billion in transmission in 2000), the United States still faces significant reliability challenges due to aging infrastructure and extreme weather events.
   - China: China leads in global grid investment, spending $166 billion on transmission and $55 billion on distribution in 2022. These investments support its rapid renewable energy expansion.
   - Germany and Japan: Both nations invested less in absolute terms – $10.8 billion and $9.5 billion, respectively, for transmission in 2022 – but achieved far better performance due to efficient resource allocation and advanced technology.
2) Outage Duration Comparisons
   - United States: U.S. customers experienced 330 minutes of outages in 2022 (5.5 hours annually), the highest among the nations analyzed. Weather events account for 80% of major outages.
   - Japan: Reports only 10 minutes of annual outages, reflecting exceptional grid reliability and proactive maintenance practices.
   - Germany: Customers faced 13.7 minutes of outages annually in 2022, showcasing strong resilience and advanced smart grid adoption.
   - China: The average annual outage duration is 30 minutes, and its rapidly expanding grid continues to improve stability (Table 16.2).

Table 16.2 Data: Grid Investments and Outage Durations.

Country	Outage Duration (Minutes)	Transmission Investment 2022 (Billion $)	Distribution Investment 2022 (Billion $)
United States	330	27.7	50.9
Japan	10	9.5	5.4
Germany	13.7	10.8	8.7
China	30	166	55

Key Insights:

1) U.S. Challenges: Despite significant investments, the U.S. grid lags in reliability. Climate-driven extreme weather events, coupled with an aging system, exacerbate vulnerabilities. Regional disparities exist, with states like Louisiana and Texas experiencing the longest outages.
2) Global Best Practices: Japan and Germany demonstrate how smaller-scale but strategic investments, advanced technology, and robust maintenance can deliver world-class performance. For instance, Japan employs predictive maintenance to address issues before they impact consumers.
3) China's Ambition: China's aggressive investment strategy positions it as a global leader in grid expansion and renewable integration. However, rapid growth presents challenges in maintaining consistent grid reliability.

The Future of Energy: Smart Grids, Renewables, and Nuclear – A Global Perspective

Energy transition is accelerating worldwide, with investments reaching $2.08 trillion in 2024, according to BloombergNEF. This surge in funding, seen across renewable energy, grid expansion, energy storage, nuclear power, and carbon capture, is reshaping the global energy landscape. However, the debate over the most efficient, cost-effective, and sustainable energy mix remains unresolved. The role of smart grids, microgrids, and energy storage in managing intermittent renewables is becoming crucial. In contrast, nuclear energy remains a contested but potentially necessary solution for long-term stability.

The Global Investment Landscape: BloombergNEF's latest report highlights that global investment in energy transition has steadily increased since 2004, surpassing $2 trillion in 2024. Renewables account for approximately $6.2 trillion in cumulative investment over the past two decades. Yet, despite this massive funding, the total increase in global renewable power generation has been 5,500 TWh, raising concerns about efficiency compared to nuclear alternatives.

For instance, the cost of renewables over this period could have financed approximately 190 Barakah-class nuclear plants, each capable of producing 40 TWh/year, totaling 7,600 TWh/year – about 25% of global electricity demand. Moreover, nuclear plants have an operational lifespan of 60–80 years, compared to 20–30 years for solar and wind farms, making their long-term return on investment (ROI) significantly higher.

The Role of Smart Grids and Microgrids

While renewables are key to a sustainable future, their intermittent nature requires advanced grid technology to maintain energy stability. Smart grids integrate real-time data, demand forecasting, and automation to balance supply and demand efficiently. Microgrids – localized energy systems that can operate independently or connect to the main grid – are particularly useful for enhancing energy resilience in remote areas and during grid failures.

- United States: Investments in smart grids have exceeded $250 billion since 2000. California leads with microgrid adoption to address wildfire-related power outages, while Texas is investing in smart grid technology to prevent failures like the 2021 winter blackouts.

- Europe: The EU's Green Deal aims to digitize grids and integrate decentralized renewable sources. Denmark and Germany are pioneering transnational grid balancing, linking offshore wind with hydro storage from Norway.
- China: With the world's largest installed renewables capacity, China is deploying UHV transmission lines to improve grid efficiency, connecting distant solar and wind farms to industrial hubs.

Despite these advancements, grid expansion remains a costly challenge. The International Energy Agency (IEA) estimates that upgrading infrastructure to accommodate renewables will require $14 trillion by 2050. Without grid modernization, high reliance on solar and wind risks energy instability and curtailment.

The Economics of Nuclear vs. Renewables: The nuclear vs. renewables debate is heavily influenced by total system costs, rather than just Levelized Cost of Energy (LCOE). While wind and solar energy appear cheaper on paper, nuclear energy emerges as a more stable alternative when factoring in grid integration, storage, and lifecycle material use.

- Balancing Power & Backup Costs: Renewables require backup generation, usually natural gas or battery storage, adding costs not included in LCOE calculations.
- Grid Expansion: Wind and solar require extensive transmission lines, increasing system-wide investment.
- Lifespan: Nuclear plants last 60–80 years, while wind turbines and solar panels degrade within 20–30 years.
- Material Requirements: Nuclear requires 80–100 times less raw material per TWh than renewables.
- Waste Management: Solar panel and wind turbine waste will reach 88 million tons by 2050, creating a looming landfill crisis.

While nuclear energy has its challenges – high upfront costs, long development timelines, and geopolitical risks – its efficiency, longevity, and stable output make it a viable backbone for decarbonization alongside renewables.

Regional Energy Strategies

1) United States: The Biden administration's IRA has funneled $370 billion into renewables and grid infrastructure. Meanwhile, Microsoft-backed SMRs are gaining traction, though investor confidence remains low, with nuclear stocks like OPG down 50% since 2024.
2) Europe: Rising energy prices, exacerbated by the Ukraine war, have exposed the risks of over-reliance on imported energy. Manufacturers are relocating due to industrial electricity prices that are 200% higher than in the United States. Some nations, including France and the United Kingdom, are reviving nuclear projects to stabilize costs.
3) China and Russia are the only two nations aggressively expanding nuclear power, with 70+ reactors under construction. However, a lack of transparency makes it difficult to assess true costs.
4) Middle East (UAE & Saudi Arabia): The Barakah nuclear plant has been one of the few successful nuclear projects with minimal cost overruns. Saudi Arabia is eyeing nuclear power for long-term energy security.

An accurate table comparing total system costs (not just LCOE) across nuclear, renewables, and hybrid energy solutions would greatly enhance clarity, especially for investors assessing long-term viability. Many discussions focus on direct costs, but investors need a full picture of system-wide economics, risks, and returns.

Key Missing Points:

1) Investor Perspective:
   - Return on Investment (ROI): Nuclear has high upfront costs but longer lifespans (60–80 years vs. 20–30 for renewables), yielding more energy per dollar invested.
   - Risk Exposure: Renewables are vulnerable to policy shifts, while nuclear faces regulatory hurdles. Which has the lower risk-adjusted return?
   - Market Trends: Why are private investors hesitant about nuclear despite theoretical cost benefits? What would reverse this trend?
   - Break-Even Analysis: At what electricity price do nuclear and renewables achieve parity with fossil fuels?
2) System-Level Comparison:
   - Storage & Backup Costs: What's the added cost of balancing intermittent wind/solar?
   - Grid Infrastructure Needs: Nuclear energy requires stable baseload transmission, while renewable energy requires wider grid expansion. What's the capital expense of each?
   - Decommissioning & Waste Management: Which has higher end-of-life costs?
   - Scalability: Nuclear projects take 10+ years to deploy. Can modular reactors accelerate timelines?
3) Policy & Regulatory Considerations:
   - Tax Incentives & Subsidies: U.S. IRA bill supports renewables and nuclear. How do subsidies shift competitiveness?
   - Geopolitical Risks: Europe's reliance on U.S. Liquefied Natural Gas (LNG) and China's nuclear exports shift the global energy balance.

Table for Clarity:

Factor	Nuclear	Wind	Solar	Hybrid (Smart Grid + Storage)
Lifespan (years)	60–80	20–25	25–30	Varies
Energy Density (MW/km^2)	High	Low	Low	Moderate
Grid Expansion Needed?	Low	High	High	Medium
Storage/Backup Cost	None	High	High	Medium
Capital Cost ($/MW)	$6–9M	$1–2M	$1–2M	$3–5M
LCOE ($/MWh)	$80–100	$30–50	$25–40	$50–80
System Cost w/Grid & Storage ($/MWh)	$80–100	$80–150	$80–160	$70–110
Investment Risk	High (Regulatory)	Medium	Medium	Medium
Lead Time (Years)	8–12	2–4	1–3	Varies

For Investors & Sit Investment Considerations:

1) What's the Best Energy Mix for Maximizing ROI?
   - Short-Term: Wind and solar have faster deployment, benefiting from tax incentives.

- Long-Term: Nuclear offers stable returns but requires patience.
- Balanced Approach: Smart grids + modular reactors could bridge the gap.

2) What Factors Will Drive Future Investments?
   - Regulatory Changes: If nuclear licensing accelerates (SMRs), it becomes a more attractive investment.
   - Grid Modernization: Expanding smart grids will determine the feasibility of renewables.
   - Commodity Prices: If lithium and battery costs rise, nuclear becomes more cost-competitive.
3) Why Are Institutional Investors Hesitant on Nuclear?
   - Political Uncertainty: Policies shift with governments.
   - Long Payback Periods: Investors prefer projects with faster ROI.
   - Perceived Risk: Accidents like Fukushima still impact perception, despite modern reactor safety improvements.

For energy investors, a diversified portfolio incorporating renewables (short-term gains) and nuclear (long-term stability), coupled with grid modernization and energy storage, will offer the most resilient, cost-effective energy future.

The Future: Balanced Energy Systems

Neither nuclear nor renewables can solve the energy crisis alone. The most viable path forward is a hybrid model integrating smart grids, energy storage, and a mix of nuclear, renewables, and gas for flexibility.

- Investment Priorities:
  - Expand grid modernization to support renewables.
  - Scale up battery storage and hydro balancing to stabilize intermittent sources.
  - Invest in next-gen nuclear (SMRs & advanced reactors) to provide stable baseload power.
  - Improve energy efficiency in industries and buildings to reduce overall demand.

Without system-wide energy planning, nations risk higher costs, instability, and stranded assets. Smart leadership must balance innovation with pragmatism to ensure affordable, sustainable energy for the future.

Challenges and Recommendations: The aging infrastructure of the U.S. power grid presents a pressing challenge due to increasing electrical demand and the transition to clean energy. Investing in infrastructure and grid-enhancing technologies is crucial for modernizing the power system and addressing evolving energy needs. Moreover, the rise in temperatures due to climate change has caused more frequent and intense extreme weather events in recent years, placing additional stress on the grid.

In conclusion, while the United States has made significant advancements in strengthening its electrical infrastructure, the rising frequency and severity of extreme weather events, alongside an aging grid, still jeopardize the reliability of power delivery. Addressing these challenges requires a comprehensive approach, including ongoing investment in modernization, embracing advanced technologies, and proactive strategies to mitigate the effects of climate change on the power system.

Conclusions

1) Grid Modernization is Crucial: The United States must upgrade its aging infrastructure to reduce outages. Investments in resilient materials and updated transmission systems are essential, particularly in areas most affected by extreme weather.

2) Adopt Advanced Technologies: Smart grid technologies, real-time monitoring, and predictive analytics can significantly improve grid performance. Countries like Japan and Germany are already using these methods with great success.
3) Streamline Renewable Integration: Over 2.5 terawatts of renewable projects in the United States await grid connection. Simplifying permitting processes and expanding T&D systems are critical to integrating these resources effectively.
4) Focus on Climate Adaptation: Resilience must be at the forefront of investment strategies. Strengthening grids against hurricanes, wildfires, and other weather events will reduce the long-term economic and social impacts of outages.

Paths Forward: Countries can learn from each other's successes to achieve better grid performance. The United States should focus on modernizing infrastructure, adopting secure and smart technologies, and addressing permitting bottlenecks. By balancing investments in technology and resilience, nations can build grids capable of handling future energy demands.

The BloombergNEF data highlights the urgency of the energy transition and the complex trade-offs between different technologies. Renewables are growing rapidly, but require massive grid expansion and storage to maintain reliability. Nuclear remains the most energy-dense and long-lasting solution, yet it faces investment and regulatory challenges. Smart grids and microgrids offer a way to blend both approaches, ensuring flexibility and resilience.

Governments and industries must shift from isolated technology debates to integrated energy strategies. The future isn't about choosing between renewables and nuclear – it's about designing a smarter, interconnected energy system that maximizes strengths and minimizes risks.

Resources for Further Reading

Climate Central, "Surging power outages and climate change." [Online]. Available: https://www.climatecentral.org/.

Heise Online, "Germany's grid reliability metrics." [Online]. Available: https://www.heise.de/.

International Energy Agency (IEA), "Investment in transmission and distribution grids in selected countries, 2015–2021." [Online]. Available: https://www.iea.org/.

Power Technology, "China's grid investment surpasses global peers." [Online]. Available: https://www.power-technology.com/.

Reuters, "U.S. grid investments and demand surge." [Online]. Available: https://www.reuters.com/.

Trading Economics, "Electricity outage duration by country." [Online]. Available: https://tradingeconomics.com/.

U.S. Energy Information Administration (EIA), "Today in Energy: U.S. electricity transmission and distribution investments." [Online]. Available: https://www.eia.gov/.

17

Blackouts and Resilience in the Electric Grid: Vulnerabilities and Solutions

Introduction

While previous chapters examined the physical and financial underpinnings of electric power grid infrastructure, this chapter explores what happens when those systems fail – and how we can prevent such failures.

Widespread power blackouts have become a pressing concern in recent years, impacting millions of people across the globe. From North America and Europe to the Caribbean and Asia, large-scale outages between 2022 and 2025 have exposed significant weaknesses in electric grids that were once considered reliable. In April 2025, the Iberian Peninsula experienced one of Europe's worst blackouts. Spain and Portugal lost power for hours after 60% of demand vanished within five seconds, according to carbonbrief.org and reuters.com. Similarly, London's Heathrow Airport was plunged into chaos in March 2025 when a substation fire cut power, grounding over 1,300 flights and rippling delays throughout global aviation. These events underscore that electricity grids, many of which are aging and operating at their limits, are increasingly vulnerable to failures that cascade across society.

This chapter comprehensively analyzes the structural vulnerabilities that lead to systemic grid failures and the emerging strategies to enhance grid resilience. We examine the evolution, structure, and aging of the United States and global power grids, establishing how legacy design choices have created brittle systems. Next, we review recent major outages in the United States, Europe, Puerto Rico, Cuba, Saudi Arabia, and beyond (with a focus on 2022–2025 events), distilling the technical causes – from low inertia and lost generation to vegetation management failures and cyber risks – that precipitated these blackouts. We then delve into the common systemic vulnerabilities that underlie many outages: insufficient inertia in high-renewable grids, inadequate tree trimming, lack of network redundancy, obsolete equipment, cyber-physical security gaps, and poor sensing and situational awareness. In the second half of the chapter, we explore solutions and mitigation strategies. These include smart grid upgrades (advanced sensors, automation, and controls), microgrids and distributed energy resources for localized resilience, energy storage systems that buffer fluctuations, and new technologies like grid-forming inverters that stabilize low-inertia conditions. We also introduce two analytical frameworks – REBALANCE and RESONANCE – which offer a systematic approach to diagnosing grid fragility and guiding resilience improvements. Real-world applications of these frameworks are discussed, such as audits of city infrastructures, airport operations, and utility grids (including the Heathrow 2025 incident and Puerto Rico's grid collapses). We compare these approaches with traditional resilience frameworks and international best practices (from organizations like DOE, NERC, FEMA, and grid operators/ISOs), highlighting how each addresses (or overlooks) the warning signs of systemic failure. Finally, we provide recommendations for utilities, regulators, and infrastructure investors to act on the lessons learned – upgrading aging components, enforcing stringent reliability standards, investing in modernization and cybersecurity, and fostering a proactive risk management culture. The tone is technical and evidence-driven, aimed at IEEE and industry readers seeking to understand and address the complex challenge of preventing blackouts in an era of growing demands and evolving risks.

Evolution, Structure, and Aging of Power Grids

The architecture of today's power grids was largely conceived in the early to mid-20th century, and much of that legacy still defines their structure and limitations. The U.S. electric grid, often hailed as the greatest engineering achievement of the 20th century, was built as three loosely interconnected mega-systems – the Eastern Interconnection, Western Interconnection, and the Texas (ERCOT) grid – along with smaller grids for Quebec and Alaska. These networks grew organically as regional utilities linked generation plants (primarily coal, gas, hydro, and later nuclear) to population centers via high-voltage transmission lines, feeding local distribution networks. By design, early grids prioritized delivering power from central power stations outward; considerations for cross-regional support, real-time monitoring, and automated control were rudimentary. As a result, many grids today operate with 60–80-year-old infrastructure and designs that predate modern reliability practices. In the United States, for example, 70% of transmission lines and power transformers are over 25 years old and approaching the end of their design life. Key facilities – from generation plants to substations – often use decades-old equipment. Aging transformers and circuit breakers are less reliable (prone to fail under stress) and lack the digital sensors and controls that modern technology offers. This aging physical infrastructure creates latent instability: components can overheat, short, or misoperate, and their failures can propagate if not quickly isolated. Aging is a global issue as well; Europe's transmission grid, while highly meshed, includes lines and substations from the post-WWII era, and developing regions often rely on infrastructure that has seen deferred maintenance.

Compounding the age issue is that demand on the grid has vastly increased and changed in nature since those early days. Electrification of every sector, larger metropolitan loads, and new uses like data centers and electric vehicle charging put strains on equipment that may not have been anticipated in its original design. Many grids operate closer to their capacity limits, with smaller safety margins. The U.S. Department of Energy notes that peak electricity demand has risen steadily while expansion of transmission capacity has lagged, contributing to congestion and stress on critical corridors. Moreover, the contemporary push for cleaner energy has altered the generation mix: older, inertia-rich steam-turbine plants (coal and nuclear) are retiring, replaced by renewables that feed power electronics into the grid. This transition offers huge environmental benefits, but from a structural perspective, it removes some inherent stability mechanisms (like heavy spinning turbines that dampen frequency swings). NERC has warned that as conventional generators retire, "necessary reliability attributes" such as inertia and frequency response are "exiting the system," and high levels of inverter-based resources must be managed carefully to maintain stability. In short, many grids were not originally engineered for the operational realities of 2025.

Globally, electricity networks vary widely in topology and robustness. Europe's grid is highly interconnected across national borders (the Continental Europe synchronous area spans 25 countries), which generally provides resilience through mutual support, but also means a disturbance in one corner (e.g., Southeast Europe) can induce effects thousands of kilometers away if not properly mitigated. In contrast, some regions operate mostly isolated systems (e.g., Japan's eastern and western grids run at different frequencies and are connected only by small converters, limiting support during crises). Developing countries have often expanded their grids rapidly without uniform standards, leading to patchworks of modern and outdated elements. Underinvestment in maintenance and modernization is a common theme in many regions. For instance, analysts note that chronic underinvestment in generation and grid assets has left many markets with narrow reserve margins and overloaded equipment. This was evidenced by the nationwide blackout in Pakistan in January 2023, where an outdated grid control procedure led to a collapse when power was restored improperly, plunging 220 million people into darkness https://www.reuters.com/world/asia-pacific/power-surge-crashes-pakistan-grid-plunging-millions-into-darkness-2023-01-26/https://www.reuters.com/world/asia-pacific/power-surge-crashes-pakistan-grid-plunging-millions-into-darkness-2023-01-26/. In summary, the world's grids range from state-of-the-art to ramshackle, but virtually all share the challenge of aging infrastructure under rising demand and a changing resource mix. This provides the backdrop for why large failures happen – the grid is a complex, stressed system, and small failures can escalate if the conditions are ripe.

Recent Major Blackout Events (2022–2025)

North America (United States and Puerto Rico): The United States has experienced several high-profile outages in recent years, often triggered by extreme weather, exploiting underlying weaknesses. While the February 2021 Texas Freeze (Winter Storm Uri) falls just outside our 2022–2025 focus, it is emblematic: unwinterized gas plants and wind turbines in Texas tripped offline in freezing temperatures, leading to a near-total grid collapse and an estimated 4.5 million customers (over 11 million people) losing power. In late December 2022, another severe winter storm (Storm Elliott) stressed grids from the Midwest to the Southeast. Utilities in several states, including North Carolina (Duke Energy) and Tennessee (TVA), resorted to *rolling blackouts* for the first time in decades to prevent a wider collapse when demand outstripped supply during frigid conditions. Although not as widespread as Texas 2021, these controlled outages affected hundreds of thousands. They highlighted that even regulated grids with reserve margins can be overwhelmed by extreme cold if generation fails or fuel supplies falter. On the West Coast, California narrowly avoided rotating outages during record heat waves in August–September 2022; only aggressive demand response and emergency imports kept the lights on as air conditioning load hit new peaks. The near-miss underscored the tightrope between supply and demand in regions with high-renewable penetration and retiring dispatchable plants.

Puerto Rico deserves special mention due to its ongoing grid crisis. The island's electric grid has suffered repeated island-wide blackouts, most infamously after Hurricane Maria in 2017, which knocked out 100% of power to 3.4 million residents. In the period since, outages remain frequent. In April 2022, a failure at the Costa Sur power plant caused an island-wide blackout, illustrating that even in the absence of storms, the system is fragile – an aging breaker failed and triggered a cascade that left all 1.4 million customers in the dark. Then, in September 2022, Hurricane Fiona again knocked out electricity to 100% of Puerto Rico, with some areas taking up to *four weeks* to be restored. Most recently, on New Year's Eve 2024, an internal line fault collapsed, which cut power to 90% of the island (gradually restored over 48 hours). Puerto Rico's situation is extreme, where *any* disturbance (whether a storm or a single equipment failure) can escalate due to underlying stress: maintenance backlogs, lack of spinning reserves, and minimal redundancy. It showcases how *multiple failures can compound*: generation units go offline, overloaded transmission lines trip, and black-start capabilities (to self-restore power) are limited. Even as Puerto Rico sets aggressive targets for renewables by 2050, progress in strengthening its grid has been slow, and the territory remains reliant on emergency fixes. As discussed later, rebuilding efforts now emphasize distributed generation and microgrids to avoid the single-point collapse mode that characterized Maria and Fiona.

Europe (Iberian Peninsula Blackout 2025 and others): Europe's most significant recent outage struck on 28 April 2025, when a massive blackout cascaded across Spain and Portugal around midday. In what has been described as the worst blackout ever to hit the region, the entire Iberian Peninsula went dark: roughly 30 million customers lost power for hours, disrupting trains, hospitals, communications, and commerce. The disturbance was remarkably sudden – around 15 GW of generation tripped off in about five seconds, an amount equal to 60% of Spain's load at that moment, carbonbrief.org, reuters.com. This precipitous drop caused the Spanish grid to separate from the broader European network. Preliminary analyses indicate the trigger was in southwest Spain (Extremadura region), likely involving a *simultaneous loss of several large solar farms or other generators, Reuters.com*. The Iberian grid, with high solar production that day (nearly 60% of Spain's supply was solar just before the event), could not withstand the sudden imbalance. As frequency plunged and oscillations spread, automatic protections activated a full shutdown to protect equipment. Grid operators reported seeing unusually low-frequency oscillations between Iberia and the rest of Europe in the 30 minutes leading up to the collapse. en.wikipedia.org – akin to a vibration or "ringing" in the interconnected system. Similar oscillatory modes had been observed in past years (notably in 2016 and 2021) on the continental grid, but those instances did not lead to a breakup. In 2025, however, the oscillation coincided with whatever fault caused the generation drop, pushing the system past its tipping point. Restoration was painstaking; by the next morning, normal supply was mostly back, but the investigation had only begun. Authorities *ruled out* a cyberattack early on, carbonbrief.org, and despite wild speculation about solar flares or atmospheric phenomena, the focus is on grid stability with high renewables. Notably, Spain's grid operator REE stated it found at least two major incidents of generation loss that led to instability in the link with France. The event has stirred debate about whether Spain's rapid renewable deployment has outpaced the grid's capability to handle it. Critics pointed out that in February 2025, REE's parent company, Redeia, had warned of "disconnections due to high penetration of renewables without the technical capacities necessary for an adequate response," Reuters.com. In other words, without sufficient inertia, fast frequency response, or other stabilizing resources, a renewable-heavy system can be vulnerable to a domino effect – a theme we will revisit in vulnerabilities.

Elsewhere in Europe, full blackouts have been rare in recent years, but near-misses and local outages underscore the risk. In January 2021, Europe narrowly avoided a continent-wide outage when a fault in the Balkans split the synchronous grid into east and west zones for about an hour; frequency in Southeast Europe dropped to 49.74 Hz while Northwest rose to 50.6 Hz before operators shed load and restabilized the system. This incident, though outside our 2022–2025 window, prompted ENTSO-E (the European network of TSOs) to accelerate measures for handling inter-area oscillations and low-inertia conditions – efforts that will be scrutinized in light of the 2025 Iberian event. The United Kingdom experienced a notable outage in August 2019 when concurrent trips of a large offshore wind farm and a gas plant caused a frequency dip that cut power to 1 million customers; while it quickly recovered, it exposed fragilities in managing sudden generation loss. During 2022–2023, Europe also faced an *energy crisis* due to the war in

Ukraine, which tested reliability differently, including fuel scarcity. High prices and gas supply cuts did not directly cause European blackouts (thanks to emergency energy conservation and alternative supplies). Still, many countries implemented contingency plans for rotating outages in winter 2022/2023 if needed. This highlights that reliability is not just technical but is also tied to geopolitics and fuel infrastructure.

Caribbean (Cuba 2022): Cuba's electrical grid, like Puerto Rico's, has struggled with aging infrastructure and extreme events. On September 27, 2022, Hurricane Ian struck western Cuba as a Category 3 storm, and in its aftermath, the island of 11 million people lost power. The hurricane damaged the transmission network and generation plants (Cuba's grid relies on a handful of key thermal power stations), leading to a nationwide blackout. Cuban officials reported that as of the evening after the storm, the National Electric System was producing no electricity, a rare instance of an entire country's grid going to zero output. Although power was restored to much of the island within about a day as emergency crews reconnected circuits, Havana and other areas saw multi-day outages, sparking street protests by frustrated residents. The Cuban blackout illustrates how a direct hit to critical nodes can cascade: Cuba's grid was already ailing with frequent brownouts and an obsolete, overstressed infrastructure. When Ian knocked out key transmission lines and generation simultaneously, the lack of redundancy meant a nationwide failure. This event also raised security concerns – initially, some speculated (without evidence) about sabotage or cyberattacks amidst geopolitical tensions, highlighting how natural disasters and artificial threats can intersect in the public's mind during such crises.

Middle East (Saudi Arabia 2025): Even regions with substantial energy investment are not immune to blackouts. In January 2025, a large outage hit the southern provinces of Saudi Arabia (Jazan, Asir, Najran), cutting power to millions. The Saudi Electricity Co. reported that multiple power stations in those areas suddenly shut down at 16:01 on a Saturday, causing a cascading failure across the regional grid. Although service was fully restored the same day after emergency teams worked to reconnect circuits, the government's regulatory authority (SERA) ordered an investigation into the root causes and demanded recommendations to avoid similar incidents. The exact technical cause remains under review. However, a "sudden shutdown of power stations" hints at a grid protection trigger (e.g., an undervoltage or frequency event that tripped multiple plants) or a common-mode failure like a fuel supply disruption or control system issue. During heightened geopolitical tension, the incident also prompted concern about sabotage; authorities initially involved counterterrorism units until ruling out foul play. This case underlines that even a country known for energy infrastructure can face localized collapse if multiple failures coincide. It also shows the importance of *operational protocols*: a robust system should isolate disturbances, but here it appears a fault led to the loss of several generators at once, overwhelming the regional grid's ability to cope.

South Asia (Pakistan 2023, Bangladesh 2022): In January 2023, Pakistan suffered a nationwide blackout affecting an estimated 220 million people, one of the largest outages ever by population. The cause was traced to operator procedures: on a low-demand night, grid controllers in Pakistan's south disconnected transmission to save energy, but when they reconnected in the morning, a large voltage surge and frequency rise (to 50.75 Hz) occurred, tripping out lines and splitting the grid. With the network severed, power plants went into safe shutdown, and virtually the entire country lost supply for much of the day.. Following a similar nationwide outage in 2021, this event highlights issues of human factors and training, as well as equipment limits (protection relays should have coordinated better to prevent a total collapse). A few months earlier, in October 2022, Bangladesh saw 75–80% of its territory (about 140 million people) plunged into darkness when a grid failed. Reports indicated demand had slightly exceeded forecasts, and with a third of the country's gas power units out of fuel due to an ongoing energy crisis, the remaining plants couldn't support the load. The resulting imbalance caused a chain reaction outage across the national grid. Power was restored to about half the areas by that night, but it took until the next day to fully recover. For countries like Bangladesh and Pakistan, these blackouts underscore a combination of vulnerability factors: limited reserves, manual grid operations that can introduce instability, and infrastructure that cannot absorb even minor shocks. They serve as stark reminders that major blackouts are a global phenomenon, not confined to any one region or level of economic development. In 2021 alone, an analysis found that at least 350 million people worldwide were impacted by major outages, with extreme weather and *unsynchronized energy transitions* cited as key drivers alongside the traditional problems of underinvestment in renewable energy.

From these examples, several patterns emerge. Blackouts often result from converging factors rather than a single cause: an extreme weather event or technical fault occurs, and the system's latent weaknesses allow it to escalate. In Spain/Portugal 2025, the combination of high-renewable generation (low inertia) and an inter-area disturbance led to collapse. In Texas 2021 and Pakistan 2023, a mix of operational decisions and unforeseen conditions led to situations beyond the grid's engineered handling capacity. In Puerto Rico and Cuba, long-standing fragility meant any severe shock would cause total failure ("collapse by design" as one analysis termed Puerto Rico's Maria outage). These cases inform the next section's discussion of systemic vulnerabilities – essentially, the "failure modes" that keep recurring in different guises worldwide.

Systemic Vulnerabilities in Modern Grids

Despite differences in locale and trigger, major outages tend to exploit common weak links in power systems. Here, we dissect the key structural and operational vulnerabilities repeatedly implicated in blackouts. Understanding these failure modes is crucial to formulating effective countermeasures.

- Low Inertia and Frequency Instability: A power system's inertia, largely provided by the rotating mass of large generators, acts as a shock absorber for sudden imbalances between supply and demand. High inertia slows down frequency changes, giving time for controls to respond. However, as grids shift toward renewable generation – solar PV and wind, which connect via inverters with little or no physical inertia, the system becomes *less forgiving* of disturbances. Frequency can dip or spike rapidly in a low-inertia grid, triggering automated protections before human operators or slower controls can react. The 2025 Iberian blackout underscored this vulnerability: at the moment of the event, only ~16% of Spain's generation came from conventional (inertia-providing) sources like gas and nuclear. The rest was solar, wind, and hydro, which don't stabilize frequency like thermal plants. Investigators noted the grid frequency oscillated at 0.2 Hz modes before the breakup – a sign that the system was teetering on the edge with the current generation mix. Rapid phase shifts between Spain and France occurred, and within seconds of the big generation loss, the frequency deviation forced disconnection. Similarly, in the Pakistan 2023 outage, an *excess* of generation caused frequency to overshoot safe bounds (50.75 Hz) and trip the system, again an inertia issue (too little load and insufficient damping). NERC's 2023 reliability assessment explicitly flags that *retiring thermal plants and adding inverter-based resources introduce risks to frequency stability, requiring new measures to provide synthetic inertia and fast frequency response.* Without adequate inertia or equivalent controls, rate-of-change-of-frequency (RoCoF) relays can trip generators or lines in milliseconds, thinking a severe fault has occurred. Low-inertia grids must therefore carefully tune protections and add frequency support services; otherwise, they remain prone to blackouts from disturbances that a high-inertia grid would ride through.

- Vegetation Management and Weather Resilience: Many of the largest blackouts in history have started with something seemingly mundane: a tree branch contacting a power line. Vegetation encroachment on transmission lines can cause short-circuits, especially when lines sag lower in hot weather. The 2003 Northeast Blackout (50 million affected in the United States and Canada) began this way: overgrown trees in Ohio touched a sagging 345 kV line. Similarly, major outages in Europe in 2003 (Italy's 57 million-people blackout) and 2006 (Continental Europe split) were triggered by tree contacts with lines in Switzerland and Germany, respectively. Vegetation-related outages are preventable with rigorous line clearance programs. Still, they remain a concern as utilities sometimes fall behind on tree trimming due to cost or sheer scale of the network. Climate change exacerbates this by fostering more vigorous vegetation growth and introducing pests that weaken trees (making them more likely to fall). Furthermore, extreme weather like windstorms, hurricanes, wildfires, and ice storms pose direct threats: toppling trees into lines, knocking down poles, or forcing utilities to preemptively shut off power to prevent fires (as in California's Public Safety Power Shutoffs). For example, Hurricane Ian 2022 in Cuba demonstrated that an overhead network exposed to hurricane-force winds can be devastated, leading to a nationwide outage. In the continental United States, severe winter storms (e.g., the December 2022 blizzard) caused not just generation issues. Still, extensive distribution outages, such as trees and ice bringing down local lines, don't cause entire grid blackouts but leave many customers without power for days. *Vegetation mismanagement* is thus a critical vulnerability: if transmission operators don't maintain clear rights-of-way, one fallen branch can initiate a cascade. It is telling that after 2003, NERC introduced strict vegetation management standards (facilitating better inspections and imposing fines for violations), yet reports still occasionally find utilities in breach. This is an area where diligence must be relentless, especially as more frequent storms increase the number of threatening incidents.
- Lack of Redundancy and Single Points of Failure: A resilient grid has multiple pathways for power flow, so that if one element fails, others can pick up the slack (the N-1 criterion). However, many grids have bottlenecks or *single points of failure*. This is often due to historical topology: a certain city may be fed predominantly by one transmission corridor, or an entire region may rely on one or two big power plants. If those elements go down, there is insufficient alternative supply. Puerto Rico's pre-Maria grid exemplified this; it had large centralized generation in the south feeding load in the north via just a few high-voltage lines crossing the island. When those lines were all knocked out, there was no way to move power around, plunging the island into darkness. In the Heathrow Airport 2025 incident, we see a different twist on the redundancy problem. Heathrow was fed by three separate substations, which in theory are N-2 redundant (any one substation failing shouldn't shut the airport). Yet when one substation (at North Hyde) caught fire and went offline, the airport lost power completely. Why? Heathrow's internal network design did not effectively isolate and draw from the remaining live feeds. "Single-point exposure" was identified; critical systems were not sectioned such that two substations could not be covered if the third failed. A single failure propagated to a total outage despite nominal redundancy. Lack of redundancy also plagues generation capacity in some regions. During the 2022 Bangladesh blackout, over a third of the generators were out of fuel, leaving no reserve when demand spiked. reuters.com. The system had no margin to handle the loss of one more unit. Interdependence without backup is another angle – for example, multiple plants depending on the same gas pipeline (if it fails, all those plants go down, as nearly happened in New England during winter cold snaps), or multiple substations tied to the same upstream grid node. In summary, insufficient redundancy

means the grid lacks "graceful degradation" – instead of a controlled, partial shutdown under stress, it tends to all go simultaneously. This is a systemic issue that utilities and regulators must continuously address by building *network flexibility*, whether through more lines, added emergency generation, or islanding capabilities.

- Outdated Substations and Equipment Failures: A power grid is only as strong as its weakest link; often, those are aging pieces of equipment that fail unexpectedly. Substations contain transformers, breakers, relays, and control systems that orchestrate power flow and protect against faults. Many blackout investigations reveal that a protection system misoperation or equipment failure either initiated or worsened the event. For instance, the April 2022 Puerto Rico outage was traced to a failure of a 230 kV breaker at the Costa Sur plant; it failed to close properly and caused an arc that led to cascading generator trips. In the Heathrow substation fire, the old electrical switchgear or cables were likely faulty and caught fire (London Fire Brigade's investigation focused on the substation's distribution equipment), Reuters.com. Once that substation fell, Heathrow's inability to swiftly switch to backups caused the airport-wide failure. Aging transformers are another concern – if a big transformer blows (often spectacularly, with oil fires), replacement can take days or weeks, during which the load must be served via other routes. The U.S. Northeast blackout of 2003 was exacerbated by a software bug in an alarm system and *failing to isolate a line outage* – essentially a failure in the control/SCADA system, an outdated equipment issue on the software side. Many regions also have limited visibility at substations; older sites may not have real-time sensors for temperature, power flow, or breaker status. That can delay response or lead to maloperation. For example, suppose a relay (protective device) is electromechanical and not digitally networked. In that case, it might trip erroneously for a condition that no longer exists, simply because it wasn't updated with system changes. Obsolete control systems can also be targets for cyber intrusion (overlap with cyber vulnerability). In short, whether a physical component fails or a protection relay is misconfigured, outdated equipment remains a top cause of outages. Modernizing these components (using digital relays, installing health monitors on transformers, replacing old breakers) is part of the solution discussed later.
- Cybersecurity and Physical Security Risks: Modern grids are complex cyber-physical systems. While natural events and equipment failures have been the main drivers of historic blackouts, there is rising concern about malicious attacks. A coordinated cyberattack could theoretically manipulate grid controls or turn off key equipment, causing an outage. Indeed, Ukraine experienced such attacks in 2015 and 2016, where hackers took down regional distribution grids. In the United States, there have been no known large-scale cyber-induced outages yet. Still, intrusions have been detected on utility networks, and malware has been discovered (for instance, the 2020 SolarWinds hack gave adversaries potential access to some utility control systems). NERC has implemented Critical Infrastructure Protection (CIP) standards to harden the cybersecurity of the bulk power system. Still, distribution networks (outside NERC CIP) remain varied in cyber readiness. As grids get "smarter" and more networked, the attack surface grows – smart meters, IoT sensors, and remote control capabilities are potential entry points if not secured. The April 2025 Iberian blackout quickly prompted speculation of a cyber or even terrorist attack, reflecting this worry, carbonbrief.org, and reuters.com. REE had to rule out cyber intrusion as a cause publicly, carbonbrief.org. Separately, physical attacks on grid infrastructure have increased; in 2022, the United States saw multiple deliberate shootings at substations (e.g., in North Carolina and the Pacific Northwest) that caused local outages by disabling transformers. While these did not escalate to regional blackouts, they expose another vulnerability: substations often have minimal physical security (chain-link fences, few cameras), making them soft targets. An attacker with knowledge could simultaneously strike critical nodes to maximize impact. Utilities are now,

in response, adding barriers and surveillance to key substations. In short, security threats (cyber or physical) overlay all other vulnerabilities – a storm might knock out power by chance. Still, a determined actor might target the grid's known weak points. Infrastructure interdependence is also a factor: a cyberattack on a gas pipeline (as seen in the 2021 Colonial Pipeline ransomware incident) can indirectly threaten the electric grid by curtailing fuel to power plants. Or vice versa, an electric outage can hamper pipeline pumping. Therefore, grid resilience now demands an all-hazards approach, including security. A large-scale cyberattack on the grid remains a *feared but unrealized* event; nonetheless, NERC's GridEx exercises simulate such scenarios regularly to prepare operators.

- Inadequate Sensing, Monitoring, and Situational Awareness: "You can't manage what you don't measure" holds for grid operations. Several blackout reports have cited a lack of real-time visibility or delayed awareness as factors that worsened the crisis. In 2003, FirstEnergy's control center was unaware for over an hour that its lines were tripping and causing a cascade, partly due to alarm failures in its Energy Management System (EMS). Fast forward to today, and while most transmission operators have much improved control systems, some parts of the grid (especially at the distribution level or smaller municipal utilities) still lack synchronized, high-fidelity monitoring. The introduction of Phasor Measurement Units (PMUs) in the last decade – devices that provide GPS-timestamped voltage and current readings 30+ times per second – has given grid operators a powerful tool for wide-area situational awareness. PMUs can detect oscillations and angle differences that SCADA cannot. However, deployment is not yet universal. In the European 2025 blackout, PMU data in retrospect revealed the growing oscillation and phase angle drift between Spain and the rest of Europe. European. wikipedia.orgcarbonbrief.org. If that data had been integrated into automatic control schemes, it might have initiated corrective action (like activating additional reserves or triggering islanding) before the breakup. Another aspect is protective relay settings and coordination – if operators don't have a unified view, one area's protection might act without the other area knowing, leading to over-tripping. Moreover, at the distribution level, many utilities rely on customer phone calls or smart meter pings to even see that an outage has occurred. This can slow isolation and restoration. "Inadequate sensing" also refers to not detecting equipment health issues: for example, if a transformer is running unusually hot or a circuit breaker's operation time is degrading, those are signs of imminent failure that could be addressed proactively *if* someone knows. Without modern sensors, these signs are missed until a failure happens. The Northeast blackout of 2003 was partly a loss of situational awareness; likewise, during Texas 2021, there was confusion in the first hours about which generators were offline due to a lack of gas versus frozen instruments, complicating the response. Thus, insufficient real-time data and analytics can convert an avoidable event into a full-blown blackout. The goal of a "smart grid" is largely to fill this gap, installing sensors everywhere and using AI/analytics to give operators early warning of anomalies (e.g., a sudden frequency oscillation or a failing substation component) so that action can be taken before the lights go out.

These vulnerabilities – low inertia, vegetation/weather, lack of redundancy, old equipment, security gaps, and poor visibility – can act alone or in concert. In many blackout events, multiple factors compound. For instance, in Puerto Rico's grid collapse under Hurricane Maria, analysts identified how *decades of underinvestment (Burden)* and *lack of backup paths (Contingency)* meant the infrastructure had no resilience. Once the storm hit, collapse was inevitable. Similarly, Heathrow's outage was not just the substation fire (sparked by equipment failure) but also design issues (no sectionalization, so one failure took down everything). Recognizing these patterns is the first step; the next is implementing solutions to mitigate them. The following section explores technologies and practices that address these vulnerabilities, aiming to prevent blackouts or limit their scope. The subsequent section will introduce advanced frameworks for diagnosing and improving resilience, moving beyond ad-hoc fixes toward systematic strengthening.

Mitigation Strategies: Upgrading to a Smarter, More Resilient Grid

Ensuring the grid can withstand or quickly recover from disturbances requires a multifaceted approach. It ranges from deploying cutting-edge technology to adopting better operational practices and planning criteria. Here, we discuss key strategies and innovations that enhance grid resilience and reliability, mapped to the vulnerabilities they address.

1) Smart Grid Upgrades and Advanced Monitoring: Modernizing the grid with sensors, communication, and automation – the essence of the "smart grid" – directly tackles inadequate situational awareness and slow response. Wide-area monitoring systems based on PMUs are now installed on major power networks worldwide. These devices stream real-time data on grid frequency, voltage, and phase angles at different points, allowing detection of oscillations or instability in milliseconds. For example, after the 2025 Iberian event, experts highlighted that better use of PMU data could help detect inter-area oscillations and trigger controls before they escalate. In the United States, the Department of Energy's North American Synchrophasor Initiative (NASPI) led to the deployment of over 1,000 PMUs across the bulk grid, yielding tools that alerted operators to abnormalities (like the onset of a disturbance in one region affecting another). Dynamic line rating sensors are another upgrade – they monitor the actual temperature/sag of transmission lines, so operators know how much capacity is available and can avoid overloads or be warned of clearance issues (e.g., line sagging close to trees). Along with sensing is advanced control and automation: digital relays and substation automation systems can localize faults and reroute power in fractions of a second (self-healing grid). For instance, if a transmission line trips, automated systems can immediately close in a previously open line (if available) to maintain network connectivity or activate a standby generator. Distribution automation, including automatic sectionalizers and reclosers, can isolate a fallen line segment and restore power to unaffected segments within seconds, rather than waiting for a crew to intervene manually. For example, Florida Power & Light (FPL) credits its smart grid investments for quicker restoration after hurricanes, as automated feeder switches rapidly reroute power around downed lines. Another aspect is integrated data analytics: utilities employ machine learning to predict equipment failures by analyzing sensor data (transformer vibrations, breaker timing, etc.). By catching a failing component early, they can fix it in a controlled outage rather than have it blow up unexpectedly. All these upgrades create a more observant and agile grid, where operators have a detailed real-time picture and some decisions are made autonomously to contain disturbances. Under the U.S. infrastructure law, the Grid Resilience and Innovation Partnership (GRIP) programs invest billions in such technologies, aiming to strengthen the grid "against wildfires, extreme weather events, and other natural disasters" through modernization. In summary, smart grid upgrades convert a passive grid into an active, sentient system – a foundational step for resilience.
2) Vegetation Management and Weather Hardening: To address the perennial threat of trees and weather, utilities are adopting traditional and innovative measures. Aggressive vegetation management programs are being implemented with the aid of technology. Lidar surveys from helicopters or drones can map vegetation growth along thousands of miles of lines,

identifying hotspots where clearance is insufficient so crews can prioritize trimming. Some utilities use AI image analysis on drone footage to detect hazard trees (dead or leaning trees outside the right-of-way that could fall into the right-of-way). Regular trimming cycles are being shortened in high-risk areas. However, trimming alone may not suffice given the ferocity of hurricanes and wildfires. Thus, hardening measures are considered: in coastal hurricane zones, this might mean replacing wooden poles with steel or concrete, burying critical distribution lines underground (as is being done selectively in Florida and Oklahoma, for example), or installing breakaway connectors that detach lines when hit by trees to prevent pole cascade. In wildfire-prone areas, utilities have added weather stations and use predictive models to preemptively de-energize lines (Public Safety Power Shutoffs) when fire risk is extreme – a controversial but sometimes necessary tactic to avoid catastrophic fires caused by electrical faults. Long-term, microgrids (discussed below) in fire-prone communities can reduce the need for widespread shutoffs. Additionally, flood protection for substations (e.g., building floodwalls or elevating equipment) has proven vital to cope with storms. After Superstorm Sandy in 2012, many East Coast substations were rebuilt with higher flood tolerance, which paid off during subsequent hurricanes. Grid design standards are evolving to account for stronger winds and ice loads, meaning new transmission structures and lines are built with thicker margins. These efforts reduce the physical vulnerability of the grid to environmental threats. It's important to note that such hardening can be expensive. Thus, prioritization is key: often, the focus is on critical circuits (serving hospitals, downtowns, etc.) and those with a history of faults. Regulators have started to require resilience plans – for instance, Texas now mandates winterization of power plants and some grid equipment after the 2021 disaster. In vegetation and weather, success is measured in fewer storm-induced outages and quicker restoration when they do happen. While we cannot prevent hurricanes or blizzards, we can ensure the grid is not the weak link that turns a natural event into a prolonged blackout.

3) Enhanced Grid Topology and Redundancy (Network Reinforcement): One straightforward way to improve resilience is to build more network connections, eliminating *single points of failure*. This might involve new transmission lines to loop in radial areas, additional transformers or substations to provide alternative supply paths, or interties between regions that can share power during emergencies. For example, after the 2003 blackout, the U.S.-Canada task force recommended strengthening ties between the Midwest and Eastern grids and upgrading key corridors that had become bottlenecks. In Texas, discussions have emerged about increasing interconnection with the Eastern grid (ERCOT currently has only minor DC ties) to allow support during extreme events – a controversial idea, but one that could mitigate isolation. In its grid rebuild, Puerto Rico plans to integrate and tie microgrid zones together so that if one section is damaged, others can isolate and still function (more on microgrids shortly). Another facet is ensuring adequate reserve capacity and geographic diversity of generation. If a region relies on one giant plant, building smaller distributed plants or installing large-scale energy storage can provide backup if the big plant trips. Transmission operators sometimes update their N-1-1 criteria (coping with two failures in sequence), especially as extreme weather can knock out multiple elements nearly simultaneously (e.g., a storm taking down two lines). For distribution systems feeding critical loads (like airports and data centers), utilities are working with customers to create redundant feeds. Heathrow's outage taught that you can have three feeds, but if you don't have an automatic transfer scheme and sectionalized circuits, it's as good as one. So, implementing fast transfer switches and independent routing of feeders is key for such critical infrastructure. Network reinforcement is about building robustness through connectivity – a mesh is harder to break than a chain. Europe's highly meshed grid is one reason large blackouts

have been infrequent (2025 being an exception that will be carefully analyzed). Where new transmission is difficult due to rights-of-way or public opposition, non-wires alternatives like strategically placed batteries can mimic redundancy (a battery can inject power to a weak node if a line fails). The goal is to avoid any single element's failure causing widespread impact, which often means having spare capacity and pathways that kick in automatically. Redundancy isn't free, but the cost of not having it is seen in multi-billion-dollar damages from outages, Reuters .com. Regulators increasingly recognize resilience as a value in grid planning (some jurisdictions now allow utilities to earn returns on resilience investments, not just reliability).

4) Microgrids and Distributed Energy Resources (DERs): A microgrid is a self-contained network of generation and load that can operate independently (islanded) or connected to the main grid. Embracing microgrids is a paradigm shift from the purely centralized grid model. In terms of resilience, microgrids shine by limiting the scope of outages – if the main grid fails, the microgrid can isolate and keep critical facilities powered. They also reduce the load on restoration crews since parts of the system heal themselves. Around the world, we see microgrids being deployed for hospitals, military bases, campuses, and remote communities. For example, after Hurricane Maria, communities in Puerto Rico with solar + battery microgrids could restore power faster than the main grid. This has spurred investment in dozens of solar-storage microgrids on the island's schools and fire stations to serve as resilience hubs. In urban areas, microgrids can ensure continuity for critical services: airports are exploring microgrids (some U.S. airports like Pittsburgh and Denver are building their gas/solar generation and batteries to sustain operations during grid outages). A microgrid requires local generation – solar panels, wind turbines, diesel or gas generators, or fuel cells – and usually some form of storage to buffer. It also needs intelligent controls to manage the transition from grid-connected to islanded mode without interrupting power. Grid-forming inverters (discussed below) are key enablers that allow inverter-based resources to set stable voltage and frequency for the microgrid when connected. Beyond single microgrids, aggregate distributed energy resources (DERs) improve resilience by diversifying the supply. Rooftop solar on thousands of homes means a portion of the load can be met locally, even if the transmission grid is down – though without storage or microgrid capability, solar alone won't work when the grid is out (for safety, typical solar inverters shut off during outages). Pairing solar with batteries and smart inverters is crucial for resilience. DERs can also assist the main grid during emergencies; for instance, virtual power plant programs can aggregate many customer batteries and EVs to inject power or shed load to stabilize the grid. California used a fleet of residential batteries to avoid outages in 2022 by discharging them at peak stress times. Microgrids can operate at various scales: from a single building (like a high-rise with generators) to a neighborhood or town. One successful example is in Alaska, where remote towns not connected to any large grid use high-penetration wind microgrids with flywheels and intelligent controls to maintain power through Arctic storms. The key point is that distributed resources and microgrids add redundancy and autonomy to the system – they don't rely on a few central nodes and long supply lines. Instead, they create a network of many small "islands" that collectively make the grid less prone to a single failure. Regulators and grid codes are now adapting to accommodate and encourage microgrids (e.g., allowing them to disconnect and reconnect seamlessly). In a resilient grid vision, even if the main backbone goes down, vital services and communities can still have pockets of power – lights at the hospital stay on, emergency shelters have electricity, and perhaps most customers only experience a localized outage rather than a statewide blackout.

5) Energy Storage Systems: Energy storage is the ultimate multi-tool in grid resilience. Batteries (and other storage like pumped hydro or even new technologies like flow batteries) can serve

many roles: they can act as generation when supply is short, as load when there's excess (preventing over-frequency), and as a stabilizer for voltage and frequency. Large-scale lithium-ion battery farms have been deployed in places like California, Australia, and the United Kingdom in recent years, specifically to enhance reliability. One famous case is the 100 MW Hornsdale battery in South Australia: built after a statewide blackout in 2016, it has since responded to numerous grid disturbances, sometimes kicking in within fractions of a second to arrest frequency drops and avoid outages. During a 2017 Australian generator trip, the battery injected power in 100 milliseconds, far faster than conventional reserves, preventing a cascading failure. Batteries are especially valuable in low-inertia grids because they can emulate the effect of inertia by rapidly balancing load-generation mismatches (often called "synthetic inertia" when an inverter algorithm detects a frequency drop and immediately pushes out power). In the case of Iberian 2025, more battery storage might have mitigated the rate of collapse by supplying some of the lost 15 GW instantly. Storage also helps with black-start capability – after a widespread outage, batteries can help re-energize grid sections without needing a running power plant, acting as a jump-start. Many system operators are now installing batteries at critical nodes for this purpose (so that if the grid is down, they can restart key substations and then bring generators online gradually). Beyond big batteries, distributed storage (home batteries, EVs) can aggregate to support the grid through virtual power plants or microgrid mode. Energy storage can also reduce stress on lines by locally absorbing solar surges or filling in when clouds reduce solar output (smoothing the volatility that might otherwise cause grid issues). In terms of resilience metrics, adding storage has been shown to reduce the probability of load shedding in extreme conditions. For example, ERCOT (Texas) is adding several GW of battery storage to provide reserve power in both summer heat waves and winter freezes when quick-ramping support is needed. It's important to ensure storage systems themselves are reliable and have backup (e.g., a battery needs inverters and control systems that must be hardened against cyber and physical faults). Some regions are even considering vehicle-to-grid (V2G) integration, where electric cars feeding back into the grid could collectively supply significant power during an emergency (imagine a million EVs each giving 5 kW, that's 5,000 MW of capacity). While still nascent, pilot programs are exploring this, effectively turning a vulnerability (EV charging load spikes) into a strength (EVs as grid support). To summarize, energy storage adds flexibility and immediacy to grid control – it can inject or absorb energy at will, which is invaluable when the unexpected happens. As costs come down, storage deployment is accelerating, becoming a cornerstone of grid resilience strategies.

6) Grid-Forming Inverters and Virtual Inertia: One of the most exciting technical advancements addressing the low-inertia vulnerability is the development of grid-forming inverter technology. Traditional renewable energy inverters are grid-following, meaning they must sense an existing grid voltage waveform to synchronize and feed power. They don't inherently regulate frequency or voltage; they let the grid (usually dominated by big generators) handle that, pushing as much power as their MPPT (maximum power point) allows. Grid-forming inverters, by contrast, can set their own voltage and frequency reference; they can operate in islanded mode and enforce stable grid conditions much as a spinning generator does. In effect, they provide virtual inertia – if the frequency dips, a grid-forming inverter can immediately respond by drawing on stored energy (either from a battery or by momentarily reducing the kinetic energy extraction of a wind turbine) to boost the frequency. This technology is critical for future grids that may have periods of 100% power electronics-based generation (already happening in smaller systems and was achieved for a short interval on the Spanish grid in April 2023, when it ran

entirely on renewables for a day, carbonbrief.org). Without grid-forming capability, such an operation would be unstable. Manufacturers and researchers (at places like NREL and EPRI) have tested grid-forming controls on large battery inverters and wind turbine inverters. For example, pilot projects in Hawaii, Texas, and Ireland have demonstrated islands of stability created by these devices when the main grid is absent or weak. Virtual synchronous machines (VSMs) are a related concept – programming inverters to behave with the swing equations of a generator, effectively mimicking inertia and damping. The benefit for resilience is profound; renewables and storage can take over the grid support role that was once solely the domain of fossil fuel turbines. In a disturbance, grid-forming units support the system instead of tripping offline (as some old inverters do when frequency or voltage goes out of range). Think back to the Iberian blackout – if many of those solar plants had grid-forming inverters, they might have ridden through the initial issue instead of disconnecting, potentially preventing the cascade. Standards are now evolving; in 2022, the IEEE and IEC started incorporating grid-forming requirements for certain categories of resources, and ENTSO-E's grid code updates are likely to mandate more tolerant and supportive inverter behavior, given recent events. Additionally, synchronous condensers (large spinning motors not attached to a load) are being installed in some grids to provide physical inertia and voltage support as a complement to inverter-based solutions. Australia's grid operator, for instance, installed synchronous condensers in weak parts of the grid after blackouts, which stabilized inertia levels while developing more grid-forming capability. In summary, grid-forming inverters and virtual inertia represent a key technical fix to the Achilles' heel of renewable-rich grids. Turning wind turbines and batteries into active stabilizers rather than passive followers ensures that high-renewable penetration does not equate to high fragility. This is an active area of deployment – we expect that by the late 2020s, many new solar and battery systems will come standard with grid-forming functionality, fundamentally improving grid robustness.

7) Protective Relaying and System Protection Enhancements: Often overlooked in public discussion, but crucial are the improvements in system protection schemes. These automatic "safety nets" detect faults and abnormal conditions and initiate corrective actions (like tripping a line, or shedding load to rebalance frequency). In recent years, adaptive relaying has been introduced – relays that can adjust their settings based on system conditions. For instance, during a high-renewable, low-inertia period, the protection thresholds might be tuned to be a bit less sensitive to avoid over-tripping for transient swings. In contrast, in a high-inertia state, they can be more aggressive. Under-frequency load shedding (UFLS) and undervoltage load shedding (UVLS) schemes are being refined with better telemetry so that if part of the grid experiences a frequency drop, only the minimum necessary load is shed in a surgical way to arrest it. The near-miss of Europe in 2021 was partly saved by UFLS automatically dropping about 1.7 GW of load in southeastern Europe, which helped rebalance the split grids. Ensuring that these schemes keep up with the changing grid (for example, accounting for the fact that many loads are now motor drives or inverter-fed, which behave differently than pure resistive loads) is a challenge that protection engineers are working on. Another tool is system integrity protection schemes (SIPS), also known as remedial action schemes – special coordinated controls for specific high-risk scenarios. For example, if a certain intertie is overloaded and one line trips, a SIPS might automatically trip a few large industrial loads or fast-start a backup generator to prevent collapse. These schemes are *automated contingency plans*. As the grid operating envelope becomes more complex, such schemes are an important resilience layer. However, they need to be carefully designed to avoid false triggers – an infamous case was in 2020 in the United Kingdom when a lightning strike led to a perfectly sound wind farm being incorrectly tripped

by a protection scheme, contributing to an outage. Thus, newer relays emphasize "ride-through" capability – not disconnecting unless necessary – aligning with grid codes requiring generators to stay on during disturbances. Lastly, distribution system upgrades like smart fuses and sectionalizers mean that even if a storm wrecks part of the distribution network, the outage can be confined to that segment, and alternate feeds can be used for other segments. All these protection enhancements require robust communication networks (to send signals between far-flung grid elements in milliseconds). Many utilities are upgrading to fiber optic networks or private wireless for critical protection comms (replacing older microwave or leased telephone lines). The net effect is a grid that is "self-protecting" – able to take quick, precise actions to preserve its integrity, much like how a body reflexively reacts to pain to prevent injury.

In implementing these strategies, it's clear that no single technology or practice is a silver bullet. True resilience comes from layering multiple solutions, so they backstop each other. For example, a tree might still fall on a line (we can't stop every tree or storm), but if the line is sectionalized, only a small area loses power; then a local battery might keep critical loads alive; meanwhile the bulk grid, aided by other lines and generation reserves, barely notices the loss and stays stable; and perhaps an automated reconfiguration restores service quickly. That kind of multilayer defense is the goal. In the next section, we introduce frameworks that help assess whether all these measures are in place and sufficient, using a holistic view of grid and institutional resilience.

Resilience Diagnostics: The REBALANCE and RESONANCE Frameworks

While adding hardware and software solutions improves grid robustness, diagnosing resilience gaps and systematically ensuring organizational readiness are equally important. This is where structured frameworks like REBALANCE and RESONANCE come in. Developed in recent years as tools for critical infrastructure resilience assessment, these frameworks provide a comprehensive lens to evaluate not just the technical state of the grid but also institutional, procedural, and external factors that contribute to either robustness or fragility. Unlike traditional checklists or risk matrices, REBALANCE/RESONANCE adopts a dynamic systems perspective – essentially treating the grid and its operators, regulators, and stakeholders as one coupled system that can drift toward failure if not corrected.

REBALANCE is an acronym encapsulating nine core "vectors" of stress that a resilient system must manage. These are: Resilience (ability to maintain core function during disruption), Entropy (organizational/structural drift and decay over time), Burden (when stress/load exceeds system capacity), Alignment (how well public trust, governance, and goals align – legitimacy), Linkage (interdependence fragility – how failures propagate through network links), Access (openness and exposure – the "attack surface" or vulnerability to outside elements), Noise (background volatility from environment, cyber, or economic fluctuations), Contingency (the buffers, redundancy and fallback options available), and Execution (the efficacy and speed of operational response and leadership). Each vector is scored through data, incident records, expert input, etc., to gauge where the system stands. Think of these like dimensions in a multi-axis stress map – if one is badly degraded, the system is at risk. A power grid example: "Entropy" could measure aging infrastructure and deferred maintenance; "Alignment" might measure regulatory support and public trust in the utility (since loss of that could hinder responses, as seen when the public is distrustful of outage explanations); "Execution" might capture how quickly a utility mobilizes repair crews or communicates during a crisis. The idea is to identify where subtle drifts are happening – for example, maintenance backlogs growing (Entropy up), or reserve margins shrinking (Burden up) – *before* they fail. These factors are often interrelated (a drift in one accelerates problems in others). REBALANCE thus gives a structured audit of the "structural health" of the grid system.

RESONANCE complements this by looking at additional systemic blind spots and amplifiers – essentially a second layer of factors that can aggravate or mask the core vectors. The RESONANCE acronym stands for: Recovery capacity (latent) – the hidden adaptive potential before failure (e.g., unused backup resources that could be tapped, or community resilience that isn't obvious); environmental saturation – information or decision overload that creates noise (like operators inundated with alarms, causing fatigue); signal-to-noise ratio – how clearly can decision-makers distinguish true threats from background chatter (related to data quality and focus); organizational isolation – whether leadership is blind to front-line realities or early warnings (siloed communications); node failure cascade depth – how deep a failure at one node can propagate through dependencies (think inter-sector impacts: a grid outage cascading to telecoms and water systems); attack surface area – the scope of vulnerability exposures (cyber, physical entry points – large attack surface is risky); narrative manipulation – distortion of data or public perception that can lead to wrong decisions or panic (for instance, rumors or misinformation during an outage); capture by private interests – if institutions are hijacked by narrow interests that undermine public safety goals (e.g., a utility skimping on safety to maximize profit); and elasticity in time-to-adaptation – how quickly the system (technically and institutionally) can adapt and recover from disruption. These are more nuanced factors – for a grid, "attack surface" speaks to cyber/physical vulnerability size, "signal-to-noise" might be how well the operator's tools filter meaningful alerts from trivial ones, "narrative manipulation" could involve public communications during a blackout (to prevent panic or blame games that distract from recovery).

Together, REBALANCE (core health) and RESONANCE (amplifiers/blind spots) provide a comprehensive diagnostic framework for grid resilience. The approach is data-driven and proactive. Rather than waiting for a failure to investigate, this framework encourages continuous monitoring of these vectors to catch drift. For example, a REBALANCE audit might reveal that Contingency measures are insufficient (say, an audit finds that n-1-1 contingencies are not met in several grid areas, or backup generators at critical sites haven't been tested) – a clear sign that action is needed *before* an outage hits. Or it might show Alignment issues – perhaps misalignment between state policy (e.g., mandate for renewables) and grid operational reality (lagging investment in grid upgrades), which could set the stage for problems like the debate around the Spain blackout, reuters.com. By quantifying and visualizing these aspects (e.g., spider charts or vector fields of the scores), decision-makers can prioritize interventions on the most lagging fronts. An important element of REBALANCE is its focus on structural drift: the slow changes that erode resilience. As an analogy, it's like health monitoring – you want to catch the rising blood pressure and cholesterol (risk factors) before the heart attack happens.

Crucially, REBALANCE/RESONANCE frameworks are field-proven in critical infrastructure audits. They have been used in scenarios, including power grids, urban infrastructure, and organizations, to predict points of failure. For instance, in one joint audit of regional infrastructure in 2023, the framework detected deteriorating Execution, Alignment, and Entropy scores in one county, indicating mounting risk, a full five months before traditional asset risk metrics showed any issue. This early warning allowed preventative measures (investment, load reallocations, etc.) to be taken, and indeed, no service interruption occurred, whereas without intervention, a failure was likely. This example illustrates how conventional methods (like simple asset condition ratings or N-1 contingency lists) might not capture cross-domain risks. For example, an execution gap (slow decision-making) combined with entropy (aging gear) can be lethal even if each individual seems okay. The frameworks push analysts to consider such interactions. Traditional models like FEMA's all-hazards plans or NERC reliability standards often take a *reactive* or scenario-specific approach – they say "if X event happens, do Y," or ensure you meet criteria Z (like reserve margin). They don't usually measure a system's *trends toward failure*. REBALANCE asks, "What's changing now, and how far are we from collapse if that trend continues?" – a fundamentally different question. As the comparative analysis in one source notes, conventional frameworks are static snapshots, whereas REBALANCE is about dynamic monitoring. This proactive stance is invaluable given how quickly conditions can shift with technology and climate.

Let's illustrate the application of REBALANCE/RESONANCE with two case studies already introduced:

Case Study: Heathrow Airport 2025 – Latent Fragility Exposed. After the Heathrow outage, a REBALANCE audit was performed to diagnose why the failure was so crippling. It identified four critical stress vectors that contributed to the incident: Entropy (aging infrastructure and deferred upgrades – e.g., relying on old relays that didn't isolate the fault), Access (a single-point exposure with no sectionalization – the airport's power design had all eggs in one basket so to speak), Contingency (lack of fallback beyond minimal emergency lighting – there was no on-site generation ready to pick up the slack for terminals), and Linkage (the power loss cascaded into other systems: IT networks, fuel pumps, lighting, effectively paralyzing all airport functions). Each of these was measurable pre-incident – maintenance logs showed aging equipment issues (Entropy), engineering reviews had noted the single-feed risk (Access), etc., but they were *known and unprioritized*. Thus, the REBALANCE audit frames the root cause as "a substation fire" and as deeper design and governance flaws. Notably, the audit's post-event recommendations matched textbook resilience fixes: *install modular on-site generation* (like a dedicated backup power plant or large battery for the airport), *segment the electrical network of the airport so one substation failure doesn't black out all terminals*, and *develop clear failover procedures for partial power situations*. Engineers had considered these solutions before, but they were not implemented without the impetus of the audit (and the event). Essentially, Heathrow's case is a classic example of how latent fragility builds: the infrastructure was stretched and not upgraded (Entropy), no one acted on known single-point failure issues (Access, Contingency), and the interdependent nature of a modern airport meant that when power failed, everything failed (Linkage). If a REBALANCE assessment had been formally done prior, it likely would have spotlighted these vectors as out-of-tolerance, perhaps spurring preemptive corrections. After the fact, it provided a clear blueprint for remediation, presumably guiding Heathrow's upgrades. One might ask, what about the Resonance factors at Heathrow? We can surmise a couple: "Organizational Isolation" might have been at play if airport management didn't fully grasp the electrical risk (maybe the engineering teams knew, but the C-suite didn't prioritize it). "Narrative manipulation" could be relevant in how the story was told after – initially, it might have been painted as a freak accident. Still, the deeper narrative (design flaw) came out via analysis. The government's ordered probe into wider energy resilience suggests an attempt to correct the Alignment and Execution issues (ensuring lessons are learned and applied nationally), euronews.com.

Case Study: Puerto Rico's Grid – Collapse by Design and Rebuilding with Resilience. Applying REBALANCE to the Puerto Rico grid around the time of Hurricane Maria (2017) reveals how multiple vectors were maxed out, preordaining the failure. The burden was extreme: decades of underinvestment meant demand was sometimes close to outstripping the aging capacity (power plants were prone to failure, and transmission had no slack). The contingency was nil: essentially, there was no utility-scale battery storage, limited reserve generation, and no way to sectionalize – a single centralized grid with no microgrids or islandable sections. Execution and logistics faltered: after the storm, the response was slow, hindered by supply chain issues and lack of pre-positioned spares (this falls under Execution vector – the ability to adapt and respond swiftly). Even Resilience (the core meaning here being the ability to maintain service) was virtually zero for the grid, though interestingly, communities with their solar (DER) had some resilience, pointing to a contrast between centralized grid resilience vs. community resilience. Every REBALANCE vector was red for Puerto Rico: alignment (policy and regulation) was problematic, Noise (economic crisis, etc., adding to instability) was present, and so on. Maria was merely the catalyst that pushed it over the edge; as the analysis said, "engineered long before landfall". Now, with rebuilding, Puerto Rico is integrating REBALANCE principles: distributed generation (solar + storage projects, some funded by federal grants) to address Burden and Contingency; modular recovery zones and microgrids to improve Linkage and Contingency (so not everything is one big system); redundancy in supply chains (warehousing spare parts on-island, etc., to improve Execution and reduce Entropy by timely maintenance); and telemetry-based observability (adding sensors and modern SCADA for better Noise filtering and signal detection). Essentially, the rebuild is attempting to raise the scores of each vector – a holistic resilience build-out rather than just replacing poles and wires. This underscores how the framework can directly inform a reconstruction strategy: rather than purely restoring what was, it asks how to re-architect so that each resilience dimension is strengthened. Puerto Rico's journey is still ongoing, but early signs, like parts of the system handling 2022's Hurricane Fiona slightly better (quicker partial restorations), hint at improvements.

The REBALANCE/RESONANCE framework is not intended to replace engineering analyses or regulatory standards; instead, it supplements them by ensuring that no dimension of resilience is overlooked. Traditional grid planning often focuses on N-1 reliability and resource adequacy (important, but primarily technical). This framework adds layers such as institutional readiness and public interface, which are also crucial (for example, a technically solid plan can fail if the public rejects a new transmission line due to a lack of alignment or trust). Using such a diagnostic tool periodically, utilities and regulators can obtain a health report card for the grid and the surrounding enterprise. Think of it as an advanced warning system for resilience, highlighting that cyber risk (Access vector) is growing. It needs attention, or interdependence with telecom (Linkage vector) is a concern (perhaps backup communication systems need enhancement for blackout conditions).

The REBALANCE and RESONANCE frameworks provide a *comprehensive resilience audit methodology*, moving beyond siloed checks to a panoramic view of grid stability factors. Case studies at Heathrow and Puerto Rico demonstrate their practical value in both postmortem analysis and guiding future improvements. They represent the modern, systems-thinking approach needed as grids become more complex and stakes rise. The next section will briefly compare these approaches with traditional ones from organizations like DOE, NERC, FEMA, and industry best practices, to contextualize how decision-makers can incorporate both.

Comparative Frameworks and International Best Practices

Various frameworks and standards globally address grid resilience and reliability, each with a different emphasis. It's useful to compare the proactive, diagnostic approach of REBALANCE/RESONANCE with more established frameworks from entities like the U.S. Department of Energy (DOE), North American Electric Reliability Corp (NERC), the Federal Emergency Management Agency (FEMA), and grid operators (ISOs/TSOs) and note emerging best practices internationally.

Traditional approaches often fall into two categories: compliance-based reliability standards and emergency management plans. For example, NERC Reliability Standards in North America set requirements for planning and operating the bulk power system – ensure N-1 security, maintain frequency within bounds, have enough reserves, etc. These standards (and accompanying audits) have undoubtedly improved baseline reliability. They focus on known failure modes (like preventing another 2003 blackout via vegetation management rules and training requirements). However, as noted earlier, they may not fully account for *combinatorial or creeping issues* like a slow decline in maintenance effectiveness or a novel threat like a coordinated cyberattack. DOE provides guidance and technology development for grid modernization. DOE's Grid Modernization Initiative and the recent Grid Resilience State and Tribal Formula Grants (part of the 2021 infrastructure law) encourage best practices like hardening critical nodes, deploying advanced sensors, and integrating renewable and storage resources for reliability. These programs are broad and funding-driven rather than prescriptive standards. They rely on utilities to propose projects improving resilience (undergrounding lines, adding microgrids, etc.), tdworld.com. The advantage of DOE's approach is flexibility and support for innovation; the challenge is measuring outcomes – hence DOE and recipients are establishing metrics for improved outage durations, frequency, etc., to track progress.

FEMA's All-Hazards Framework typically addresses how to respond to and recover from disasters. FEMA focuses on emergency response coordination for energy, such as under the National Response Framework, where Energy is one of the critical lifelines. FEMA encourages communities to have backup power for critical facilities (e.g., hospitals required to have generators) and to plan for fuel logistics in prolonged outages. However, FEMA's guidelines mostly kick in after an event – they are about being prepared to manage the consequences (stockpiles, mutual aid, communications) rather than preventing the outage itself. That said, FEMA's newer *Building Resilient Infrastructure and Communities (BRIC)* grant program does fund mitigation, and some utilities have tapped that for projects like elevating substations or installing flood barriers, bridging the gap into prevention.

Independent System Operator (ISO)/TSO Best Practices: Grid operators themselves have developed a host of practices for reliability. For example, regional ISOs run capacity markets or resource adequacy programs to ensure enough generation is available, and ancillary service markets to procure frequency support, reserves, etc., maintaining a stable grid. The Texas blackout spurred ERCOT to create a "firm fuel" service to ensure gas plants have fuel in winter and a performance credit mechanism to incentivize reliable generation. Those are market tweaks to enhance resilience. European TSOs, under ENTSO-E, conduct annual region-wide stress tests simulating extreme conditions (e.g., a prolonged cold spell with low wind and gas shortage) to identify where bottlenecks or shortfalls could arise, then coordinate plans to address them. After the 2006 Europe disturbance, TSOs greatly improved real-time coordination, establishing systems to share data and actions quickly – this paid off in 2021 when they contained the separation event. Internationally, countries like Japan and South Korea have extremely high reliability, partly due to rigorous design standards (Japan's transmission is built to very high seismic and storm standards) and systematic drills. Singapore is often cited for best practices in power reliability – it has a fully underground transmission network (no storm exposure) and strict N-2 planning criterion, plus on-site generation mandated for critical loads like telecom exchanges.

The key difference to highlight is static vs. dynamic and cross-domain: Traditional frameworks (NERC, etc.) often categorize risks and require preparedness for each category. For instance, have a plan for a hurricane or a cyberattack. They don't necessarily track the "drift" or interaction between factors. One source said, "They do not track pressure buildup in the system and are silent until thresholds are crossed." The REBALANCE approach, in contrast, is all about spotting the buildup. It's akin to how old circuit breakers trip at a threshold, whereas modern relays might sense the rate of change and act earlier if something is accelerating toward a threshold. Another critique of some traditional frameworks is siloing: reliability standards focus on the grid, emergency management focuses on response, and cyber frameworks (like NIST SP 800-30) focus on IT risk. But a resilience event like a blackout spans all these silos – equipment fails, humans respond (or mis-respond), communications and public reaction matter. The REBALANCE framework deliberately crosses these boundaries (it considers "civic trust" alongside "buffers and redundancy"). It's essentially a superset of these specialized frameworks, tying them together.

Internationally, one emerging best practice is the concept of "Resilience by Design" in critical infrastructure. This involves, from the outset, designing systems to fail gracefully. For instance, the UK's regulator OFGEM requires electricity distribution companies to stress-test their networks against extreme scenarios (what if two primary substations fail at once?) and show plans to mitigate that. Standards for resilience are being debated: IEEE has begun working on standards for distribution reliability indices that account for resilience (long-duration, wide-area events, not just routine outages). NERC has an evolving view: its recent risk reports highlight extreme event risk, and it has formed a Reliability and Security Technical Committee looking at beyond-standard events (e.g., how to handle a solar storm that knocks out multiple transformers). Moreover, NERC now conducts a biennial Grid Security Exercise (GridEx) where utilities simulate cyber/physical attack scenarios and practice coordination with the government – this is bridging into the Execution and Alignment side by building muscle memory and trust among stakeholders.

The REBALANCE/RESONANCE approach can be seen as an attempt to integrate these best practices into a single evaluation. For a utility executive or regulator, a REBALANCE audit could highlight, for example, that while they meet NERC standards (so reliability looks good on paper), there is an Entropy issue (aging assets trending toward failure) and an Alignment issue (perhaps the utility's investments are not aligned with the state's renewable push, creating future risk). It thus provides a more actionable early warning, which conventional compliance might miss until it is too late. One comparison noted that conventional frameworks "classify and document," whereas REBALANCE "exposes and intervenes" – moving from paperwork to prevention.

Another international best practice theme is cross-sector collaboration. Events like the Heathrow outage or the Texas freeze show how the power, transportation, fuel, and communication sectors interlink. Leading countries are establishing cross-sector councils (like the UK's National Infrastructure Commission) to coordinate plans. In the United States, DOE's aforementioned Puerto Rico initiative involves DHS and HUD, recognizing that energy resilience ties into housing and security. This corresponds to the Linkage and Alignment vectors – aligning different agencies and linking solutions.

In summary, traditional reliability and emergency management frameworks provide essential foundations – minimum standards, clear roles, and resource commitments. Best practices are evolving to incorporate more extreme scenarios (NERC's changing standards, DOE funding resilience, ENTSO-E's investigations). The REBALANCE/RESONANCE framework doesn't replace those but ensures a strategic, forward-looking perspective: it's a way to check if we are merely compliant or resilient regularly. The combination of both is the ideal: use REBALANCE diagnostics to guide where we should bolster things *beyond* the minimum, and then implement through established best practices and standards changes.

Recommendations for Utilities, Regulators, and Investors

In light of the analysis above, here are key recommendations tailored to the different stakeholders responsible for keeping the lights on and improving grid resilience:

For Utilities and Grid Operators:

- Invest Aggressively in Modernization: Don't wait for failure to replace aging transformers, breakers, and control systems. Prioritize "high-consequence" equipment upgrades – those whose failure would cause widespread outages. To offset cost, leverage federal and state funding (such as DOE's grid resilience grants). Upgrade control centers with advanced situational awareness tools (PMUs, AI analytics) and ensure operators are trained to use them effectively. Consider establishing a Resilience Officer role to oversee long-term risk trends (similar to a Chief Safety Officer concept).
- Enhance Maintenance and Vegetation Programs: Treat preventive maintenance as a mission-critical activity. Use data to move from time-based to condition-based maintenance – for example, monitor transformer gas for signs of deterioration and replace it before it fails explosively. LiDAR and satellite imagery should identify hazard trees beyond the right-of-way in vegetation management and work with property owners to mitigate them. This is especially urgent as climate change may increase growth and wind damage.
- Implement Smart Grid Automation and Microgrids: Continue deploying distribution automation devices and fault location/isolation systems to reduce outage scope. Where feasible, establish microgrids for critical community assets in your service territory (hospitals, emergency response centers, etc.), possibly in partnership with those facility owners. These microgrids can save lives during major grid outages and speed up full restoration. Ensure your protection schemes and interconnection standards allow microgrids to island seamlessly when needed.
- Plan for Extreme Scenarios in Operations: Use scenario analysis beyond N-1: what if a once-in-a-century storm hits? What if multiple substations flood? Develop and drill *special operating procedures* for such events, including load shedding plans that minimize impact on critical loads (e.g., rotating outages that exempt hospitals). Conduct regular emergency drills, including black-start drills (ensuring you can re-energize the system from a complete shutdown). Incorporate lessons from events elsewhere – for example, after the Spain blackout, revisit your system's inertia and oscillation risks.
- Strengthen Cyber and Physical Security: Perform rigorous cyber risk assessments and penetration testing on your grid control networks. Implement the latest NERC CIP standards (or equivalent best practices) across IT and operational technology (OT) systems. This includes multifactor authentication, network segmentation, continuous monitoring for anomalies, and incident response plans. Similarly, harden physical security: add surveillance, alarms, and deterrents at critical substations, and coordinate with law enforcement on threat intelligence. Participate in industry exercises like GridEx to be prepared for combined cyber-physical attacks.
- Adopt Resilience Frameworks and Metrics: Consider utilizing frameworks like REBALANCE internally. For instance, as part of annual planning, score your system on those vectors – involve engineering, operations, customer service, and finance departments in assessing where stress accumulates. Use those insights to inform your strategic plans and rate case justifications.

Develop metrics such as "System Average Interruption Duration Index (SAIDI) in extreme events" or number of customers with access to backup supply to track resilience, not just normal reliability.

For Regulators and Policymakers:

- Incentivize Resilience Investments: Traditional regulation rewards cost containment and reliability, but often doesn't explicitly reward resilience (investments that pay off in low-probability events). Regulators should allow utilities to earn returns on prudent resilience investments. For example, capital expenditures for hardening or microgrids should be approved with the understanding that avoided outage costs justify the rate impact. Performance-based regulation could include resilience metrics (e.g., duration of outages after major storms) with incentives/penalties. Policymakers can also unlock funding, for example, via grants or resilience bonds. The U.S. FERC and state commissions might consider requiring utilities to file periodic "resilience plans" similar to how they file reliability plans.
- Update Standards and Building Codes: Incorporate new knowledge into binding standards. For instance, update interconnection standards to require ride-through and grid-forming capabilities for new inverter-based resources, to maintain stability with high renewables (many jurisdictions now do this for large solar/wind plants; extending it to smaller DER would be next). Strengthen tree clearance requirements and oversee enforcement through audits and fines (NERC already has this for transmission; states could do so for distribution). Building codes should mandate backup generators or storage for critical infrastructure like gas stations, cell towers, and high-rise buildings to mitigate societal impact during blackouts (some states did this after hurricanes, so ATMs, fuel pumps, etc., can run on gen-sets).
- Regional and International Coordination: Blackouts don't respect borders, so regulators should push for stronger inter-utility and interstate coordination. Support initiatives like regional mutual aid agreements for sharing equipment and crews in crises, and regional transmission planning that accounts for resilience (e.g., building interties between regions that can provide emergency support). Internationally, share best practices through bodies like the International Energy Agency (IEA) or CIGRÉ so that lessons from one country (like hardening after Fukushima in Japan, or Australia's inverter standards after their blackout) are disseminated to others. In the US, ensure federal agencies like DOE, DHS, and FEMA work hand-in-hand with states on energy security (the Feb 2022 DOE-DHS-HUD MOU for Puerto Rico is a good model, energy.gov).
- Mandate Planning for Climate Resilience: Given the increasing frequency of extreme weather, utilities must incorporate climate projections into their planning criteria. For example, if floodplain maps are changing, substation protection heights might need raising; if wildfire risk zones expand, more aggressive grid shutdown protocols or insulated cables might be required. Some regulators (e.g., in California and New York) already demand climate adaptation plans from utilities – this should become common practice. Also, diversity in energy mix and fuel security should be considered a resilience factor – for example, don't retire the last nuclear or gas plant without ensuring equivalent inertia and dispatchability from other resources or technologies.
- Support Innovation and Pilots: Policymakers should foster pilot projects for novel resilience solutions, such as community microgrids, vehicle-to-grid programs, or new storage tech, by providing regulatory sandboxes or funding support. These pilots can demonstrate viability and build business cases for wider adoption. Also, support R&D via national labs and grants.
- Promote Transparency and Public Engagement: Communicate openly about grid risks and mitigation plans. Utilities should educate customers on what is being done to prevent blackouts and

how to prepare for them (e.g., encouraging critical facilities and even households to have backup plans). Regulators can facilitate this by requiring public reporting of resilience metrics and plans. Public trust (Alignment) is crucial – if customers understand the challenges and see progress, they are more likely to support necessary investments (and cooperate during emergencies with conservation requests).

- Learn and Evolve Continuously: Treat every outage (and near-miss) as a learning opportunity. Conduct thorough postmortems not to assign blame, but to identify systemic fixes. Share lessons with the industry through forums and publications. Update operating procedures and training accordingly. Encourage a culture of resilience that empowers engineers and operators to raise concerns about vulnerabilities and pursue innovative solutions. This includes overcoming institutional silos: IT and OT teams must collaborate on cybersecurity, planners must talk with emergency responders, etc. We face an era where 100% renewable targets, electrification, and climate extremes intersect; tomorrow's grid will look different from yesterday's. Thus, regulators and utilities must be agile, revisiting standards and assumptions regularly (e.g., reevaluating whether a "once in 100-year" event might now be "once in 10 years"). Flexibility and foresight in planning are as important as the hardware upgrades themselves.

In conclusion, large-scale blackouts are not "acts of God" beyond our control – they are often the foreseeable result of identifiable weaknesses. By rigorously analyzing those weaknesses (using tools like REBALANCE/RESONANCE alongside traditional engineering studies) and investing in a portfolio of solutions – from smart grid tech and storage to stronger standards and coordinated response – we can greatly reduce the frequency and impact of blackouts. The period of 2022–2025 has sounded the alarm with events from Texas to Spain to Puerto Rico. The technical community, utilities, and governments must heed these warnings. A resilient grid is foundational to modern society's security, economy, and well-being. It is both an engineering challenge and a governance challenge. Fortunately, the knowledge and tools to achieve it are at hand – it now requires leadership, funding, and execution to re-engineer our grids for the demands of the 21st century. With proactive effort, the "day in the dark" scenarios described in this chapter can become far rarer, and when they do occur, our systems and communities will be prepared to respond and recover with minimal disruption.

Resources for Further Reading

M. Amin, "Heathrow's Wake-Up Call: Building Resilient Infrastructure in an Age of Interdependence," Case Study, Mar. 2025a.

M. Amin, "REBALANCE: Structural Recovery Vectors & RESONANCE Augmentation," *EPS Associates Working Paper/Presentation*, 2025b.

M. Amin, "Global Transition Dynamics: Mapping System Fragility," submitted to the *IEEE Control Systems Magazine*, Apr. 2025c.

Associated Press, "People in Cuba protest over the island-wide blackout caused by Hurricane Ian," *NPR*, Sep. 30, 2024.

G. Blackburn, "Heathrow defends response after substation fire grounds flights for 18 hours," *Euronews/AP*, Mar. 24, 2025.

J. Burke, "The fact these frequency oscillations mirrored those in Latvia… might suggest complex inter-area oscillations as the culprit," *LinkedIn Substack*, Apr. 2025.

Climate Central, "Weather-related power outages rising," Report, 2025.

G. Eldridge, "Understanding the April 2025 Spain Blackout: A Translation of Red Eléctrica's Briefing," *Geoff's Substack*, Apr. 30, 2025. [English summary of REE briefing on Iberian blackout].

ENTSO-E Incident Report, "Continental Europe Synchronous Area Separation on 8 January 2021," Jul. 2021.

Federal Energy Regulatory Commission (FERC) and NERC, "The February 2021 Cold Weather Outages in Texas and the South Central U.S.," Nov. 2021.

N. Ferris, "Did Spain's push for renewable energy cause its huge blackout?" *The Independent*, Apr. 30, 2025.

E. Fieser and J. Wyss, "Darkness Descends on Cuba as Hurricane Causes Blackout," *Bloomberg News*, Sep. 27, 2025.

K. Holton *et al.*, "Heathrow flights resume after closure causes global flight turmoil," *Reuters*, Mar. 22, 2025.

P. H. Jacquod, "Visualization of Pan-European Oscillation during Iberian Blackout," Apr. 2025.

LUMA Energy, "Preliminary Findings on Island-Wide Blackout of April 6, 2022," Apr. 2022. [Online]. Available: LUMA PR Release lumapr.com.

Massachusetts Institute of Technology (MIT), "The Future of the Electric Grid," MIT Energy Initiative Report, 2021 Update.

R. McKay and I. Rivera, "Puerto Rico's power grid collapses; could take two days to restore service," *Reuters*, Jan. 1, 2025.

North American Electric Reliability Corp (NERC), *2023 Long-Term Reliability Assessment*, Dec. 2023.

North American Electric Reliability Corp, "2023 ERO Reliability Risk Priorities Report," Jul. 2024.

R. Paul and S. Varadhan, "Bangladesh plunged into darkness by national grid failure," *Reuters*, Oct. 4, 2025.

"Q&A: What we do – and do not – know about the blackout in Spain and Portugal," *Carbon Brief, Apr.* 30, 2025.

Reuters, "Spain, Portugal switch back on, seek answers after biggest ever blackout," Apr. 29, 2025.

Reuters, "Power begins to return after huge outage hits Spain and Portugal," Apr. 28, 2025.

J. Runyon, "350 million worldwide experienced major power outages in 2021, report says," *Renewable Energy World*, Feb. 3, 2022. [Online]. Available: renewableenergyworld.comrenewableenergyworld .com (Accessed: May 2025).

D. Sánchez León, "Spain-Portugal blackouts: what happened, and what can Europe learn?" *The Conversation*, May 2, 2025.

Saudi Gazette, "Probe ordered into power outage in southern regions as service restored," Jan. 27, 2025.

A. Shahzad and S. Varadhan, "Power surge crashes Pakistan grid, plunging millions into darkness," *Reuters*, Jan. 26, 2025.

T&D World, "DOE Announces Nearly $474 Million in New Grid Resilience Funding," Oct. 30, 2024.

U.S. Department of Energy, "Puerto Rico Grid Recovery and Modernization."

U.S. Department of Energy, "Grid Resilience State and Tribal Formula Grants (Sec. 40101)," 2022–2024.

18

Energy Transition – Balancing Innovation, Sustainability, and Resilience

Introduction: Fragility and Opportunity in Energy Systems

In February 2022, Europe faced an energy crisis that revealed the fragility of its dependence on centralized fossil fuel systems. Russia's invasion of Ukraine disrupted natural gas supplies, exposing the vulnerabilities of energy infrastructure across the continent. Prices soared, energy rationing became a reality, and governments scrambled to secure alternative power sources. Despite decades of investment in renewables, the crisis underscored the challenges of balancing energy security, sustainability, and affordability in the global energy transition.

The energy transition is not just about replacing fossil fuels with renewable energy. It is about rethinking the systems that power our world – aligning technological innovation with equity, resilience, and environmental goals. This chapter explores the global energy transition through case studies, emerging technologies, and leadership strategies for building resilient, sustainable energy systems.

The Energy Trilemma

At the heart of the energy transition lies the "energy trilemma:" balancing three interdependent priorities:

1) **Energy Security:** Reliable access to energy underpins economic stability and national security. The 2022 European energy crisis highlighted the importance of diversifying supply chains and investing in domestic energy sources [1].
2) **Environmental Sustainability:** The energy sector accounts for 73% of global greenhouse gas emissions [2]. Achieving net-zero emissions by 2050 requires rapid decarbonization, including scaling renewables, improving energy efficiency, and developing carbon capture technologies.
3) **Affordability and Equity:** Over 730 million people globally still lack access to electricity [3]. The transition must ensure that clean energy is accessible and affordable, particularly for underserved communities in developing nations.

The interplay between these priorities varies by region, but a common thread is the need for adaptive, inclusive solutions.

Case Study: Germany's Energiewende – A Cautionary Tale: Germany's Energiewende (energy transition) offers valuable lessons on progress and pitfalls. Launched in the early 2000s, the initiative aimed to phase out nuclear and fossil fuels while scaling renewables.

Successes:

- By 2023, renewables generated 46% of Germany's electricity, up from 6% in 2000 [4].
- Investments in wind and solar created new industries and jobs, particularly in rural areas.

Challenges:

- The decision to phase out nuclear energy by 2022 increased reliance on coal and imported natural gas during shortages.
- High electricity prices sparked a public backlash and raised concerns about energy affordability.

The Energiewende highlights the importance of comprehensive planning to avoid unintended trade-offs, such as increased emissions or public opposition.

Case Study: Iceland's Geothermal Success: Compared to Germany, Iceland has nearly eliminated fossil fuels from its energy mix, thanks to its abundant geothermal and hydroelectric resources. By 2023, renewables accounted for 85% of the country's energy consumption [5].

Key Drivers of Success:

- Strategic investments in geothermal energy infrastructure.
- Public–private partnerships to fund research and deployment.
- A strong regulatory framework supporting clean energy development.

Iceland's example demonstrates the power of leveraging local resources for energy independence. However, its model may not be replicable in regions without similar geological advantages.

Technologies Driving the Energy Transition

The energy transition depends on various technologies, from established renewables to emerging innovations. Key developments include:

1) **Renewable Energy Advancements:** Over the past decade, solar power costs have dropped by 89%, while wind energy costs have fallen by 70% [6]. Emerging technologies, such as floating solar farms and offshore wind turbines, expand renewables into previously inaccessible regions.
2) **Energy Storage:** Renewables' intermittency challenges require robust storage solutions. Advances in lithium-ion, solid-state, and flow batteries are driving capacity growth. By 2023, global energy storage capacity exceeded 500 GWh, a fivefold increase from 2020 [7].
3) **Hydrogen Economy:** Green hydrogen, produced using renewable energy, offers a promising solution for decarbonizing heavy industries and transportation. The European Union allocated €8 billion in 2024 to accelerate hydrogen development [8].
4) **Smart Grids and Microgrids:** Smart grids use real-time data to optimize energy distribution, while microgrids enhance resilience by decentralizing power systems. Puerto Rico's adoption of solar microgrids after Hurricane Maria exemplifies their potential [9].

Barriers to Progress: Despite technological advancements, systemic barriers hinder the energy transition:

1) **Economic Costs:** The International Renewable Energy Agency (IRENA) estimates that achieving net-zero emissions will require $131 trillion in investments by 2050 [10]. Developing nations face trade-offs between immediate development needs and long-term sustainability goals.

2) **Geopolitical Tensions:** The transition shifts geopolitical power dynamics, with nations rich in critical minerals (e.g., lithium, cobalt) gaining influence. This raises risks of resource-driven conflicts and supply chain vulnerabilities.
3) **Technological Gaps:** While solar and wind technologies are mature, areas like hydrogen production, grid storage, and carbon capture require further breakthroughs to achieve scalability and affordability.

Leadership in the Energy Transition: Navigating the energy transition requires visionary leadership that balances innovation, equity, and resilience. Key strategies include:

1) **Innovation:** Governments and businesses must invest in research and development to advance clean energy technologies. Denmark's decades-long investment in wind energy serves as a model for policy-driven innovation [11].
2) **Equity-Centered Decision-Making:** Ensuring that underserved communities benefit from the transition is essential. For example, Morocco's program to train women as solar technicians has improved gender equity and rural energy access [12].
3) **Building Resilience Through Decentralization:** Decentralized energy systems, such as microgrids, enhance resilience to natural disasters and cyberattacks. Puerto Rico's solar microgrids now provide critical backup power for hospitals and schools [9].
4) **Promoting Public–Private Collaboration:** Partnerships between governments, industry, and academia can accelerate the transition. India's International Solar Alliance, which mobilizes funding for solar projects across 121 member countries, exemplifies collaborative leadership [13].

Strategic Pitching: Building Trust, Solving Problems, and Securing Buy-In

A great pitch is not about selling dreams but presenting a clear, structured, and credible plan to solve a real problem. Whether you are addressing investors, executives, policymakers, or military leaders, the goal is the same: to win trust by demonstrating the urgency of the problem, the clarity of the solution, the feasibility of execution, and the value of the investment. The best pitches use data, focus on outcomes, and conclude with clear next steps, ensuring the audience understands the vision and how to achieve it.

Start with Why: The Core of Any Pitch: Every pitch must begin by answering why your idea matters. As emphasized by Simon Sinek's Golden Circle framework, this approach is not just a step in the pitch process but a powerful tool to inspire and connect with your audience. Decision-makers connect more deeply when they understand the broader impact of your idea.

For example, Tesla's pitch wasn't just about selling electric cars and addressing climate change by reducing global reliance on fossil fuels. Similarly, a smart grid modernization initiative isn't merely about upgrading infrastructure; it's about ensuring energy security and reliability amid growing climate risks. You engage your audience logically and emotionally by framing your idea in terms of its broader impact.

Define the Problem with Data: After establishing the "why," define the problem clearly and back it up with data. Investors, executives, and stakeholders need to understand the urgency of the problem you are solving. A vague or overly broad problem won't resonate. Use frameworks like *Heilmeier's Catechism*, which asks:

- What are you trying to do?

- How is it done today, and what are the limitations?
- What is new in your approach?

For example, Google's early pitch addressed a specific problem: inefficient internet search engines produced poor results. Larry Page and Sergey Brin framed the issue with data, highlighting the billions of searches that failed to deliver relevant information. Similarly, when pitching utilities for smart grid modernization, you might cite $150 billion in annual economic losses due to blackouts, tying the problem to financial and operational risks.

Present a Clear and Simple Solution: Once the problem is defined, present your solution in simple and actionable terms. This straightforward approach, which doesn't require decision-makers to understand the intricacies of your technology, instills confidence in the feasibility of your solution. They need to know how your solution works, why it's better, and how it addresses the problem.

The **Quad Chart** is an excellent tool for this step, as it breaks down the pitch into four clear components: problem, solution, risks, and execution plan. For example, Airbnb's early pitch clearly outlined its solution: a platform for homeowners to rent out unused space, addressing the issue of unaffordable lodging. By focusing on a simple and scalable solution, Airbnb secured early investment, becoming a $100 billion disruptor in the hospitality industry.

Build Credibility with Proven Results: Vision alone isn't enough; credibility comes from demonstrating feasibility. Investors and decision-makers trust proven execution over lofty promises. The **Real-Win-Worth (RWW) framework** helps filter ideas by asking:

- **Is It Real?** Does it solve a real problem?
- **Can We Win?** Do we have the expertise or competitive edge to succeed?
- **Is It Worth It?** Will the financial and market returns justify the effort?

For example, Tesla didn't just sell a vision of sustainable vehicles; it demonstrated feasibility with prototypes, early partnerships (e.g., Panasonic for battery technology), and a clear roadmap. Similarly, Google's pitch included a functional prototype of its PageRank algorithm, which showcased operational success to early investors like Sequoia Capital. When pitching, emphasize concrete evidence like pilot projects, signed partnerships, or initial sales to de-risk the investment and establish trust.

Make a Specific and Justified Ask: Every pitch must include a financial or resource ask that is specific, measurable, and tied to outcomes. Generic requests like "We need $10 million to scale" lack impact. Instead, tie the ask to clear milestones. For instance, a smart grid pitch might request $5 million for two pilot projects with goals like a 50% reduction in outages and a 30% increase in renewable energy integration.

Microsoft's $1 billion partnership with OpenAI is another strong example. The pitch wasn't just about funding – it outlined how the collaboration would benefit Azure's infrastructure while advancing ethical Artificial Intelligence (AI) development. Microsoft secured trust and alignment by tying the funding to specific mutual benefits.

Close with Clear Next Steps: End your pitch with a roadmap that outlines clear, measurable next steps. A 12-month timeline with milestones demonstrates progress and creates urgency. Tesla's roadmap, for example, included launching prototypes, scaling gigafactory production, and delivering the Model S within a defined timeline. Similarly, a pitch to the Pentagon for a next-generation defense system might include completing prototype testing in six months and securing deployment contracts within a year.

The next steps must be realistic. Overpromising erodes credibility. Instead, focus on achievable goals that demonstrate progress and reduce risk.

Case Studies: Turning Ideas and Theory into Effective Action

Sequoia-Backed Companies: The $1.4 Trillion Legacy: Sequoia Capital's portfolio includes companies like Apple, Google, Airbnb, and DoorDash, creating over $1.4 trillion in market value. These companies followed structured approaches in their pitches, rooted in clarity, scalability, and alignment with market demand.

- **Example: Airbnb's Early Pitch.** Airbnb's pitch was grounded in a clear problem: travelers couldn't find affordable accommodations, while homeowners had unused space. Its pitch presented a simple solution: a platform for short-term rentals. The precise financial ask showed how a modest $20,000 investment would enable pilot launches in key cities. Within a few years, Airbnb disrupted the hospitality industry, scaling to a valuation of over $100 billion.

Lesson: Clear problem definition (lack of affordable lodging), a simple solution (connecting travelers to unused spaces), and a scalable model led to investor buy-in.

Tesla: Vision Meets Execution: Elon Musk's Tesla has become a $900 billion giant by redefining transportation and energy. In early funding rounds, Musk emphasized Tesla's unique value proposition: sustainable energy through electric vehicles (EVs). The problem was clear – traditional combustion engines contributed to climate change, and EVs were the solution. Early Tesla pitches highlighted an aggressive battery and EV production timeline aligned with consumer demand for sustainable options.

- **Key Point:** Tesla didn't just sell a vision; it demonstrated feasibility through prototypes, partnerships (e.g., with Panasonic for battery technology), and a clear roadmap. The credibility of execution, through gigafactories and early EV launches, secured funding from venture capital and public markets.

Lesson: Combine bold vision with a proven ability to execute. Investors trusted Tesla because its goals were ambitious but realistic, backed by milestones like the Model S launch.

Google: Solving a Big Problem: In its early days, Google pitched itself as a solution to a growing problem: inefficient and inaccurate internet search engines. Larry Page and Sergey Brin framed the problem using data – billions of searches produced poor results. Their solution, PageRank, was simple: a search algorithm that ranked pages based on relevance.

- **Key Ask:** Google's pitch to early investors like Sequoia was grounded in feasibility. They showed their prototype in action, proving their technology was theoretical and operational. A $100,000 check from Sequoia became one of the best Venture Capital (VC) investments in history.

Lesson: Data-driven problem definition and a clear solution resonate with decision-makers. Prototypes or functional proofs of concept build trust.

Netflix: Pivoting with Precision: Netflix's journey from Digital Versatile Disc (DVD) rentals to a streaming giant demonstrates the importance of adapting the pitch as the business evolves. Early on, Netflix pitched itself as a convenient alternative to traditional rental stores like Blockbuster, solving the problem of late fees and limited selection. By the mid-2000s, Netflix pivoted to streaming, pitching this shift as the future of entertainment.

- **Key Move:** When pitching its streaming model, Netflix justified the transition by citing market trends (growing internet speeds, consumer demand for convenience). The ask focused on content licensing and server infrastructure funding, providing clear milestones and ROI projections.

Lesson: A pitch must evolve with market shifts. Netflix's ability to pivot while maintaining clarity secured its position as a dominant player, now valued at over $150 billion.

Microsoft's AI Strategy: Strategic Partnerships and Proven Use Cases: Under Satya Nadella, Microsoft became an AI leader with tools like Azure AI and OpenAI collaborations (e.g., ChatGPT). Early AI pitches were framed around solving enterprise problems, like automating repetitive tasks or improving customer insights. Instead of selling pure technology, Microsoft tied AI solutions to business outcomes: increased efficiency, reduced costs, and scalability.

- **Key Example:** Microsoft's partnership pitch to OpenAI focused on infrastructure funding and alignment with ethical AI development. The ask was $1 billion, but the roadmap showed how Azure would benefit from the partnership, creating a symbiotic relationship.

Lesson: Tying technology to actionable, high-value use cases resonates with stakeholders. Showing mutual benefits in partnerships builds trust and momentum.

Smart Grid Innovation: Practical Deployment: A pitch to U.S. utilities for smart grid modernization is a classic example of matching the pitch to the audience. Using hard data, a problem – aging grid infrastructure prone to blackouts was defined: $150 billion in annual economic losses from power outages. The solution was smart microgrids with predictive analytics to reduce downtime.

- **Key Pitch Element:** The ask – $20 million – was tied to a pilot project in two regions, with measurable goals: a 50% reduction in outages and a 30% increase in renewable energy integration. The roadmap emphasized partnerships with local governments and energy providers to de-risk execution.

Lesson: For technical audiences like utilities, pitches must be grounded in measurable outcomes and risk mitigation.

Why These Case Studies Work: The best pitches are structured, data-driven, and actionable. Using frameworks like Heilmeier's Catechism, the Quad Chart, and RWW analysis, you can craft pitches that resonate with any audience, whether investors, government agencies, or corporate boards. This structured approach ensures your ideas are heard, funded, and implemented. Each case study highlights a few universal principles:

1) **Data-Driven Problem Definition:** Whether Google solves inefficient searches or Tesla addresses climate change, data must urgently support the problem.
2) **Clear, Simple Solutions:** Avoid jargon and focus on how the solution works and why it's better.
3) **Proven Execution:** Investors and stakeholders trust milestones, pilot results, and partnerships more than promises.
4) **Tailored Asks:** The funding request must be specific, realistic, and aligned with clear goals.
5) **Momentum with Next Steps:** A roadmap with measurable milestones creates urgency and confidence.

In summary, pitching is about winning trust and securing buy-in from investors, executives, policymakers, or military leaders. It requires structure, clarity, and proof, not speculation. The best pitches define the problem, present a clear solution, establish credibility, justify the ask, and outline the next steps.

- The **starting point** is explaining why your idea matters. Simon Sinek's *Golden Circle* emphasizes starting with "why" before moving to "how" and "what." Decision-makers respond to purpose-driven ideas. A cybersecurity firm isn't selling AI; it's preventing billion-dollar losses from cyberattacks. A grid modernization initiative isn't about upgrading infrastructure; it's ensuring energy security and reliability as climate risks grow. Framing the pitch around the broader impact builds engagement.

- Hard data must back the problem definition. Investors don't fund ideas because they sound interesting; they fund solutions to urgent, proven problems. *Heilmeier's Catechism* provides a structured way to define the problem: *What are you trying to do? How is it done today? What are the limitations?* If briefing defense agencies, use real attack statistics and vulnerabilities in existing infrastructure. If pitching utilities, reference blackout data, regulatory deadlines, and cost inefficiencies.
- The **solution must be clear and simple**. A common mistake is overloading the pitch with technical details. Decision-makers don't need to understand algorithms or engineering specs – they need to know how the solution works and why it's better than alternatives. The **Quad Chart** forces clarity by breaking the solution into four sections: problem, solution, risks, and execution plan. It ensures concise, high-impact communication.
- Credibility and feasibility come next. Investors and executives don't bet on ideas – they bet on execution. The **RWW framework** filters out weak ideas by asking three questions: *Is it real? Can we win? Is it worth it?* This is where many pitches fail. Saying that technology is "disruptive" isn't enough; it must be proven. A startup with a signed pilot agreement with Siemens has traction. A defense contractor with successful field tests has credibility. Investors and decision-makers prioritize risk mitigation and execution probability over potential.
- The **ask must be specific and justified**. Vague funding requests weaken credibility. Instead of saying, *"We need $10 million to scale,"* specify how the funds will be used: *"$5 million will fund three pilot projects, regulatory approval, and customer acquisition for the next 12 months."* The **Quad Chart's risk and execution quadrants** ensure realistic execution plans for funding requests.
- Closing with **clear next steps** builds urgency and momentum. A well-structured pitch ends with a 12-month roadmap that defines measurable goals. If pitching a new AI-driven energy system, the next steps might be a six-month pilot with a regional grid operator, followed by a full rollout in three markets within a year. If presenting a next-gen defense system to Pentagon officials, the timeline might include six-month prototype testing and securing deployment contracts within a year.
- **Pitch length depends on the audience.** Venture capitalists expect a 10-minute pitch with sharp, data-driven slides. Corporate and government stakeholders, who operate in more structured environments, may allow 20–30 minutes but still expect efficiency. Regardless of format, **clarity and conciseness** are non-negotiable. Overpromising is a critical mistake. Realistic projections, backed by execution confidence, win trust.

The best pitches follow a structured progression: define the problem, present the solution, establish credibility, justify the ask, and outline clear next steps. Using frameworks like Heilmeier's Catechism, the Quad Chart, and RWW analysis keeps pitches focused and actionable. Whether securing funding, influencing policy, or driving industry adoption, structured, evidence-backed communication differentiates ideas that get funded and those that get ignored.

Adapting to Your Audience: The structure and length of a pitch depend on the audience. Venture capitalists expect short, sharp presentations – usually 10 minutes or less – with high-impact visuals and concise messaging. Corporate boards or government agencies may allow 20–30 minutes, but still expect efficiency and focus.

Tailor your pitch to address your audience's priorities. For the Pentagon, emphasize national security, operational readiness, and cost-effectiveness. For utilities, focus on reliability, cost savings, and regulatory compliance. For venture capitalists, highlight scalability, market potential, and early traction. Always use data and proven results to build credibility and avoid overpromising to preserve trust. Tailor your pitch based on the audience, for example:

- **For High-Stakes Briefings (e.g., Pentagon):** Emphasize national security implications, risk mitigation, and operational readiness. Use the **Quad Chart** to condense key details into actionable summaries.
- **For Utilities and Infrastructure Projects:** Focus on reliability, cost savings, and regulatory compliance. Back the pitch with relevant case studies (e.g., smart grid pilots in Puerto Rico after Hurricane Maria, backed by case studies like Tesla's microgrids).
- **For VC Investors:** Highlight scalability, market potential, and early traction. Use data (e.g., growth rates or Total Addressable Market [TAM]) to show the opportunity is real.
- **For Corporate Boards:** Align the pitch with organizational goals like efficiency, profitability, or Environmental, Social, and Governance (ESG) targets. Show how the proposal supports the long-term strategy.

By following these principles and integrating case studies, you can deliver clear, credible, and actionable pitches, ensuring your ideas are funded and executed effectively.

The Road Ahead: The energy transition is fraught with challenges but offers unprecedented opportunities for innovation, economic growth, and climate resilience:

- **Resilience:** Decentralized and adaptive energy systems are critical to managing climate impacts and natural disasters.
- **Global Equity:** Bridging the energy access gap remains a pressing issue. Innovative off-grid solutions, such as mini-grids, can empower remote communities.
- **Job Creation:** In 2023, the renewable energy sector employed over 12 million people globally, with significant growth in solar, wind, and battery manufacturing [14].

The transition is a technological and societal transformation that demands bold leadership and collaboration across sectors and geographies.

Conclusion: Powering a Sustainable Future

The energy transition is humanity's greatest challenge – and its greatest opportunity. Leaders can drive progress toward a sustainable and inclusive future by investing in clean energy innovation, fostering equity, and building resilient systems. Today's decisions will shape tomorrow's energy systems – and the world.

References

1 International Energy Agency (IEA), "Energy Security Amid Global Crises," *World Energy Outlook*, 2022.
2 IPCC, "Global Greenhouse Gas Emissions by Sector," *Sixth Assessment Report*, 2023.
3 World Bank, "Energy Access Statistics," 2023.
4 German Federal Ministry for Economic Affairs and Climate Action, "Energiewende Progress Report," 2023.
5 Iceland National Energy Authority, "Geothermal Energy in Iceland," 2023.
6 International Renewable Energy Agency (IRENA), "Global Renewables Outlook," 2023.

7 U.S. Department of Energy, "Battery Storage Growth and Trends," 2023.
8 European Union, "Hydrogen Strategy and Funding Commitments," 2024.
9 Puerto Rico Energy Bureau, "Solar Microgrids: A Post-Hurricane Maria Initiative," 2023.
10 IRENA, "Investments for a Net-Zero Future," 2023.
11 Danish Energy Agency, "Wind Energy Development in Denmark," 2023.
12 World Bank, "Morocco's Solar Technician Training Program," 2023.
13 International Solar Alliance, "Scaling Solar Energy Across Borders," 2024.
14 IRENA, "Renewable Energy Employment Report," 2023.

19

Artificial Intelligence – A Double-Edged Sword

The Promise and the Peril

On a crisp morning in October 2022, a hospital in Tokyo achieved a significant milestone. Using an artificial intelligence (AI)-driven diagnostic tool, medical professionals identified an aggressive form of early-stage cancer that had eluded detection by radiologists, saving a patient's life. Just a week later, the same AI system failed catastrophically, misdiagnosing another patient and leading to unnecessary surgery. These incidents illustrate the double-edged nature of AI: a technology with unparalleled potential to improve lives but also profound risks when its limitations are overlooked.

AI systems are increasingly integrated into critical decision-making processes, offering transformative healthcare, finance, transportation, and energy capabilities. Yet, their deployment requires thoughtful oversight, ethical foresight, and careful calibration to avoid catastrophic errors. As AI becomes a defining force of the 21st century, it demands leadership to balance its extraordinary opportunities with its inherent challenges.

The Evolution of Artificial Intelligence

The journey of AI began in the mid-20th century with Alan Turing's seminal 1950 question, "Can machines think?" Since then, AI has evolved through three distinct eras:

1) **Rule-Based Systems (1950s–1980s):** Early AI systems were programmed with explicit rules and logical reasoning, excelling in constrained environments like chess or logic puzzles. These systems were robust but inflexible, relying entirely on predefined algorithms.
2) **Machine Learning (1990s–2010s):** The advent of machine learning algorithms allowed AI to analyze large datasets and improve its performance iteratively. This era saw significant advances in spam filters, recommendation engines, and fraud detection systems.
3) **Deep Learning (2010s–Present):** Leveraging neural networks and vast computing power, deep learning has enabled AI to perform tasks such as natural language processing, facial recognition, and autonomous driving. According to the International Data Corporation (IDC), by 2024, AI technologies will contribute over $500 billion annually to the global economy [1].

AI Across Key Sectors: Opportunities and Risks

AI is transforming industries, offering unmatched efficiencies while introducing significant risks. Its dual impact is evident across sectors.

1) **Healthcare:** AI has revolutionized diagnostics, as demonstrated by DeepMind's AlphaFold, which solved the decades-old protein-folding problem in 2021. This breakthrough accelerated research into treatments for diseases like Alzheimer's and Parkinson's and reduced drug development timelines by years [2].
   However, the risks are equally notable. In 2023, studies revealed that diagnostic AI tools trained predominantly on light-skinned individuals underperformed when diagnosing darker-skinned patients, highlighting systemic biases in training data [3].
2) **Finance:** AI is widely used for fraud detection, high-frequency trading, and customer service automation. For instance, JPMorgan Chase's COiN platform reduced the time required to review legal contracts from 360,000 hours annually to mere seconds [4].
   Yet, the financial sector also illustrates the perils of over-reliance on AI, as seen in 2022 when algorithmic trading amplified market volatility, triggering substantial losses for institutional investors [5].
3) **Transportation:** Autonomous vehicles powered by AI reduce accidents and optimize traffic systems. Waymo, for example, logged over 10 million miles in autonomous testing without a single fatality as of 2023 [6].
   However, incidents involving Tesla's autopilot system – such as a fatal collision in 2023 – underscore the challenges of deploying AI in uncontrolled environments and the need for rigorous testing and oversight [7].
4) **Energy and Climate:** AI models transform energy management by predicting demand, optimizing grid performance, and integrating renewable sources. Google DeepMind's collaboration with the U.K. National Grid reduced energy forecasting errors by 50%, facilitating renewable energy adoption [8].
   Yet, AI's reliance on data centers, which consume vast amounts of energy, poses a paradoxical challenge for its sustainability applications [9].

AI as a Double-Edged Sword

While AI offers transformative potential, it also amplifies existing societal challenges and introduces new ones:

- **Bias and Inequity:** AI systems often reflect the biases inherent in their training datasets. For example, in 2018, Amazon scrapped an AI-powered hiring tool after it was found to systematically discriminate against women by penalizing resumes containing gender-specific keywords [10].
- **Misinformation:** Generative AI tools like deepfakes can create compelling yet false content, eroding trust in information. During the 2024 U.S. elections, a deepfake video of a candidate caused widespread confusion before being debunked [11].
- **Job Displacement:** While AI creates opportunities in fields like software development and data analysis, it also threatens manufacturing, retail, and customer service jobs. The World Economic Forum estimated that 85 million jobs could be displaced globally by 2030, even as 97 million new roles emerge [12].

Leadership in the Age of AI

Harnessing AI responsibly requires visionary leadership prioritizing ethics, transparency, and collaboration. Leaders must address three key imperatives:

1) **Foster Ethical AI Development:** Establish guidelines for fairness, accountability, and transparency for AI systems. The European Union's AI Act regulates AI applications based on their potential harm and is a model for balancing innovation with safety [13].
2) **Promote Workforce Reskilling:** Organizations must invest in reskilling programs to address job displacement. Singapore's SkillsFuture initiative equips citizens with digital competencies and provides a replicable framework for workforce adaptation [14].
3) **Advance Global Cooperation:** Given AI's global implications, international collaboration is essential. Initiatives like the United Nations' Global Digital Compact aim to create shared ethical standards and promote equitable access to AI technologies [15].

Looking Ahead: The Future of AI

AI is a transformative force capable of solving humanity's most significant challenges while amplifying existing inequities. Its trajectory will depend on how leaders navigate their risks and opportunities.

- **AI for Good:** Initiatives like Microsoft's AI for Earth program use AI to combat deforestation, monitor biodiversity, and optimize water resource management [16].
- **Superintelligence and Existential Risks:** Thought leaders such as Stephen Hawking and Elon Musk have warned about the potential dangers of superintelligent AI systems surpassing human control. Robust safeguards and transparent regulations are critical to mitigate these risks [17].
- **Global Inequality:** Without deliberate efforts to democratize AI, developing nations risk falling further behind. Equitable access to AI tools and resources must remain a global priority [18].

Example of Generative AI: Revolutionizing Business Across Functions

Technological advances have promised efficiency, speed, and quality for decades, but rarely at all. However, the rise of generative AI is breaking that paradigm, creating opportunities for businesses to achieve these goals simultaneously. Across industries, generative AI is unlocking new ways to solve problems, streamline processes, and create value. From reimagining research and development (R&D) to optimizing supply chains and improving customer interactions, the transformative potential of this technology is reshaping industries.

This chapter examines how generative AI delivers on its promises, supported by real-world case studies, accurate data, and actionable lessons for leaders.

The Breadth of Generative AI Applications

Generative AI's potential spans all major business functions, offering value-creating opportunities. Below is an expanded table showcasing its applications across functions and industries.

Below is a table showcasing its applications across functions and industries:

Function	Use Cases	Examples
R&D	Intellectual property analysis, rapid prototyping, consumer product iteration, material optimization	Generative AI analyzes patent landscapes, accelerates prototyping, and suggests alternative materials for cost-effectiveness and sustainability.
Commercial	Real-time sales advisory, automated A/B testing for marketing, customer sentiment analysis	Tools like Salesforce Einstein enable sales teams to prioritize leads, while AI-driven marketing optimizes real-time campaigns.
Supply Chain	Warehouse design simulation, route optimization, real-time risk management	AI-driven systems like UPS ORION optimize delivery routes, saving costs, reducing emissions, and improving efficiency.
Manufacturing	Predictive maintenance, defect detection, production planning	Siemens uses AI to predict equipment failures, reducing unplanned downtime and cutting maintenance costs.
Procurement	Fraud detection, contract analytics, supplier performance evaluation	AI identifies anomalies in procurement patterns, helping companies like Unilever prevent fraud and optimize supplier relationships.
Corporate Functions	Meeting summaries, HR automation, financial reporting	IBM Watson automates HR tasks such as onboarding, employee training, and financial analysis, saving time and improving accuracy.

Generative AI in Action: Case Studies Across Industries

R&D Transformation: Pfizer's Drug Discovery

Drug discovery has traditionally been time-intensive and costly, requiring years of research and millions in investment. Pfizer embraced generative AI to analyze vast datasets of chemical compounds, identify potential drug candidates, and predict their effectiveness. By streamlining these processes, Pfizer reduced the time to identify viable candidates by 30%, accelerating the development timeline and cutting costs.

Impact: Pfizer saved millions in R&D expenses and delivered life-saving medications to the market faster.

Lesson for Leaders: Use generative AI to analyze complex data sets and identify patterns faster than human analysis allows. This enables quicker innovation and reduced costs.

Supply Chain Optimization: UPS ORION

UPS revolutionized its supply chain with the On-Road Integrated Optimization and Navigation (ORION) system, a generative AI-driven platform that calculates the most efficient delivery routes in real-time. ORION considers factors like traffic, weather, and package volume to optimize routes for its drivers.

The system saved UPS over $400 million annually, reduced fuel consumption, and cut emissions by 100,000 metric tons annually. Beyond cost savings, ORION improved delivery times and customer satisfaction.

Impact: UPS achieved financial and environmental benefits through AI-driven supply chain optimization.

Lesson for Leaders: AI can optimize logistical operations to maximize cost savings, reduce environmental impact, and improve service quality.

Reference: "UPS Saves $400M Annually with AI Routing," *Forbes*, Jul. 11, 2023.

Manufacturing Excellence: Siemens' Predictive Maintenance

Downtime in manufacturing can be a massive drain on productivity and profits. Siemens adopted generative AI to predict equipment failures before they occurred. Using sensors and machine learning algorithms, Siemens proactively monitored machinery health and scheduled maintenance.

The result? Unplanned downtime was reduced by 50%, and maintenance costs decreased by 30%. Generative AI also helped identify the root causes of defects, further enhancing production quality.

Impact: Generative AI ensured operational continuity and reduced costs.

Lesson for Leaders: Predictive maintenance powered by AI minimizes disruptions and extends the lifespan of critical equipment.

Commercial Breakthrough: Coca-Cola's AI-Powered Campaigns

Coca-Cola leveraged generative AI to analyze customer preferences and design hyper-targeted marketing campaigns. By automating A/B testing, the company identified the most effective content and messaging for different markets, leading to a 35% increase in engagement rates in key regions.

Impact: Coca-Cola's AI-driven campaigns improved customer targeting and boosted revenue growth.

Lesson for Leaders: Generative AI can deliver personalized experiences at scale, driving customer loyalty and sales.

Fraud Detection in Procurement: Unilever

Unilever implemented generative AI to monitor procurement transactions and detect irregularities. The system analyzed supplier invoices for anomalies, identifying fraudulent activities and potential contract violations. This proactive approach prevented over $20 million in losses.

Impact: Unilever enhanced its financial security and strengthened supplier relationships.

Lesson for Leaders: Use AI to ensure transparency and accountability in procurement processes, minimizing risks and losses.

Generative AI in the Paper and Packaging Industry

In the chapter and packaging industry, generative AI drives innovation in R&D, supply chains, and sustainability efforts.

Case Example: Georgia-Pacific

Georgia-Pacific, a leader in paper and packaging, implemented generative AI to optimize operator efficiency and reduce waste. By capturing the knowledge of experienced operators, AI-enabled troubleshooting systems minimize machine downtime and off-quality production.

Impact: Georgia-Pacific improved productivity and reduced costs while maintaining high product quality.

Lesson for Leaders: Generative AI can institutionalize operational expertise, making processes more efficient and consistent.

Reference: "Georgia-Pacific Optimizes Operator Efficiency Using Generative AI," *AWS Case Study*.

Key Takeaways for Leaders

Generative AI delivers real, measurable benefits when deployed strategically. Here's how leaders can start:

1) **Adopt Early, Scale Strategically:** Begin with pilot projects to test AI applications in your organization before scaling across functions.
2) **Invest in Data Infrastructure:** Clean, accessible data is essential for AI to deliver value.
3) **Foster Cross-Functional Collaboration:** Bring teams from R&D, supply chain, marketing, and other areas to identify high-impact AI use cases.
4) **Prioritize Training and Upskilling:** Equip employees with the skills to understand and implement AI tools effectively.
5) **Measure ROI Continuously:** Track AI's financial, operational, and sustainability impacts to ensure ongoing improvements and alignment with organizational goals.

Conclusion

AI represents the defining technological challenge and opportunity of our time. Its dual capacity to advance human progress and exacerbate societal risks places a tremendous responsibility on leaders, innovators, and policymakers. By fostering ethical practices, prioritizing collaboration, and investing in inclusive solutions, we can ensure that AI is a tool for progress rather than harm. The story of AI is still being written, and our choices today will shape not only the future of technology but the future of humanity itself.

Generative AI is not just a tool for efficiency – it's a strategic enabler that redefines how businesses operate. The possibilities are immense, from reducing R&D timelines at Pfizer to optimizing supply chains at UPS and enhancing marketing campaigns at Coca-Cola. In the paper and packaging industry, companies like Georgia-Pacific use AI to tackle complex challenges while reducing costs and improving quality.

For leaders, the message is clear: Generative AI isn't the future – it's the present. Organizations that embrace this technology now will build resilient, innovative, and sustainable operations capable of thriving in an increasingly competitive market.

References

1 IDC, "Worldwide Artificial Intelligence Spending Guide," International Data Corporation, 2024.

2 J. Jumper *et al.*, "Highly accurate protein structure prediction with AlphaFold," *Nature*, vol. 596, no. 7873, pp. 583–589, 2021.

3 Z. Obermeyer *et al.*, "Racial bias in healthcare algorithms," *Science*, vol. 366, no. 6464, pp. 447–453, 2019.
4 JPMorgan Chase, "How AI Is Transforming Legal Operations," 2022.
5 A. Smith, "Algorithmic Trading: Risks and Failures," *Financial Times*, 2022.
6 Waymo, "Autonomous Vehicle Safety Report," 2023.
7 NHTSA, *Tesla Autopilot Crash Investigations.* National Highway Traffic Safety Administration, 2023.
8 Google DeepMind, "AI for Energy Optimization with UK National Grid," 2023.
9 Greenpeace, "Clicking Clean: Energy Consumption of Data Centers," 2023.
10 Reuters, "Amazon Scraps AI Hiring Tool That Showed Bias Against Women," 2018.
11 Washington Post, "Deepfake Threats in the 2024 Elections," 2024.
12 World Economic Forum, "The Future of Jobs Report," 2023.
13 European Union, "The AI Act Regulatory Framework," 2024.
14 Singapore Government, "SkillsFuture: Transforming Workforce Development," 2023.
15 United Nations, "Global Digital Compact," 2023.
16 Microsoft AI for Earth, "Environmental Solutions Through AI," 2023.
17 S. Hawking, "The Risks of AI," *BBC Future*, 2017.
18 World Bank, "AI Access in Developing Nations," 2023.

Resources for Further Reading

"Coca-Cola Enhances Campaigns with AI," *AdWeek*, Oct. 25, 2023. [Online]. Available: AdWeek.
"*Generative AI in the Packaging and Paper Industry*," McKinsey & Company, November 1, 2024. [Online]. Available: McKinsey.
"Georgia-Pacific Optimizes Operator Efficiency Using Generative AI," *AWS Case Study*. [Online]. Available: AWS.
"Pfizer Speeds Drug Discovery with AI," *Nature Biotechnology*, Mar. 12, 2023. [Online]. Available: Nature Biotechnology.
"Siemens Predictive Maintenance Cuts Downtime by 50%," *Industry Week*, Sept. 15, 2023. [Online]. Available: Industry Week.
"Unilever Detects Procurement Fraud with AI," *The Guardian*, Apr. 19, 2023. [Online]. Available: The Guardian.
"UPS Saves $400M Annually with AI Routing," *Forbes*, Jul. 11, 2023. [Online]. Available: Forbes.

Part V

Regional Collaboration and Sustainable Development

20

Cybersecurity – Protecting the Backbone of Modern Society

Introduction: The Day the Pipeline Stopped

On May 7, 2021, the United States experienced a wake-up call to the vulnerabilities of its critical infrastructure. Hackers had infiltrated Colonial Pipeline's IT systems using ransomware, encrypting critical operational data and demanding $4.4 million in cryptocurrency. As a result, the company was forced to shut down its 5,500-mile pipeline, which supplies nearly 45% of the fuel consumed along the U.S. East Coast.

The attack's ripple effects were immediate and far-reaching: fuel shortages, price surges, and a loss of public confidence in the nation's ability to safeguard essential services. Panic buying exacerbated the crisis, with long lines forming at gas stations. Ultimately, Colonial Pipeline paid the ransom, but the damage to economic stability and public trust was already done. This incident underscored an unsettling reality: cyberattacks on critical infrastructure can cause as much disruption as physical attacks in the modern era.

Cybersecurity is no longer merely a technical concern but a strategic imperative central to national security, economic stability, and societal resilience. This chapter examines the evolving threat landscape, the vulnerabilities of interconnected systems, and the leadership required to navigate this complex domain.

The Evolving Cyber Threat Landscape

By 2024, cybercrime will have become one of the most significant global risks, with projected annual costs of $10.5 trillion by 2025, according to Cybersecurity Ventures [1]. Threats have grown in scale and sophistication, targeting corporations, governments, hospitals, and individuals. Key trends include:

1) **Ransomware Epidemic:** Ransomware attacks have increased in frequency and severity. By 2022, such attacks occurred every 11 seconds globally, with average payouts exceeding $800,000 [2]. High-profile incidents like the Colonial Pipeline hack demonstrate the cascading impacts on economies and communities.
2) **Nation-State Cyber Operations:** Cyber capabilities have become tools of geopolitical competition. In 2020, the SolarWinds breach, attributed to Russian operatives, infiltrated U.S. federal agencies and private companies, exposing the fragility of global supply chains [3].

3) **Zero-Day Exploits:** Unpatched software vulnerabilities, known as zero-day exploits, have become increasingly common. These vulnerabilities allow attackers to bypass even the most robust defenses before patches can be deployed [4].
4) **Phishing and Social Engineering:** Cybercriminals increasingly exploit human vulnerabilities through phishing emails, voice scams, and other social engineering tactics. Over 90% of successful cyberattacks in 2023 originated from human error [5].

Critical Infrastructure at Risk

Cyberattacks on critical infrastructure, such as power grids, water systems, healthcare facilities, and transportation networks, pose significant risks to societal stability. Two case studies highlight the vulnerabilities and potential solutions.

Case Study 1: The Ukrainian Power Grid Attack

In December 2015, hackers launched a coordinated cyberattack on Ukraine's power grid, cutting electricity to 225,000 people during a freezing winter. The attackers, later linked to Russian intelligence, used spear-phishing emails to access operational systems. Once inside, they remotely disabled equipment and switched off substations, making recovery efforts more difficult by overwriting firmware [6].

This attack was a chilling demonstration of how cyber threats can disrupt physical systems. Ukraine's response focused on resilience, including investments in decentralized energy systems, advanced cybersecurity protocols, and international partnerships. The event served as a global warning about the fragility of critical infrastructure in the face of advanced cyber threats.

Case Study 2: Healthcare Under Siege

In 2021, ransomware attacks targeted healthcare systems worldwide, with devastating consequences. Ireland's Health Service Executive (HSE) was disabled for weeks, disrupting patient care, delaying treatments, and jeopardizing lives. The attack underscored the consequences of underinvestment in cybersecurity for essential services [7].

Ireland's experience highlighted the need for proactive investments in incident response capabilities, robust backup systems, and employee training to mitigate human error. In response, HSE accelerated its cybersecurity initiatives, creating dedicated teams to monitor and respond to threats in real-time.

Why Cybersecurity Fails

Despite advancements in technology, cybersecurity often fails due to systemic weaknesses:

1) **Human Error:** Overreliance on human judgment remains a significant vulnerability. Weak passwords, poor phishing awareness, and failure to follow security protocols enable attackers to penetrate networks. For example, according to IBM Security [8], phishing attacks accounted for 36% of breaches in 2023.

2) **Legacy Systems:** Many organizations rely on outdated software and hardware, leaving them vulnerable to known exploits. The WannaCry ransomware attack in 2017 exploited such vulnerabilities, affecting organizations worldwide, including hospitals and transportation systems [9].
3) **Resource Constraints:** Smaller organizations often lack the budgets and expertise to implement robust cybersecurity measures, making them easy targets for attackers.
4) **Insufficient Collaboration:** Public and private sectors frequently fail to share threat intelligence effectively. This lack of coordination creates gaps in defenses, particularly against sophisticated nation-state actors.

Building Cyber Resilience

Cyber resilience extends beyond preventing breaches to ensuring that organizations can withstand, recover from, and adapt to cyberattacks. Key strategies for building resilience include:

1) Proactive Threat Management
   - Regularly conduct vulnerability assessments and penetration testing.
   - Implement multifactor authentication and encryption protocols to protect sensitive data.
   - Use AI and machine learning to detect anomalies and predict potential threats [10].
2) Incident Response Planning
   - Develop and regularly update a detailed incident response plan.
   - Conduct tabletop exercises to simulate breaches and test readiness.
   - Maintain robust data backup systems to ensure business continuity during attacks [11].
3) **Zero-Trust Architecture:** Adopt zero-trust models that verify every user and device, reducing insider threats and unauthorized access [12].
4) **Cross-Sector Collaboration:** Foster partnerships between governments, private companies, and academia to share knowledge and resources. The U.S. Cybersecurity and Infrastructure Security Agency (CISA) exemplifies collaboration, offering guidelines and threat intelligence to critical sectors [13].

Leadership in Cybersecurity

Leadership is crucial in navigating the complex and evolving domain of cybersecurity. Effective leaders must prioritize proactive measures, foster a culture of security, and align technical solutions with strategic objectives. Key leadership principles include:

1) **Transparency and Communication:** Clear, honest communication during crises builds public trust and minimizes reputational damage.
2) **Ethical Decision-Making:** Leaders must weigh short-term actions, such as paying ransoms, against long-term societal consequences. Transparency and consultation with law enforcement are critical [14].
3) **Investing in Workforce Development:** Fund training and certification programs to address the global shortage of cybersecurity professionals.
4) **Fostering Innovation:** Encourage the development of next-generation cybersecurity technologies, such as quantum-proof encryption and blockchain-based systems for data integrity [15].

Conclusion: Defending the Backbone of Society

The Colonial Pipeline attack, the Ukrainian power grid hack, and countless other incidents demonstrate that cybersecurity is a technical challenge and a societal imperative. As threats evolve, so must our defenses. Leaders prioritizing resilience, collaboration, and proactive measures can safeguard critical systems and help shape a secure, interconnected future.

References

1 Cybersecurity Ventures, "Cybercrime to Cost the World $10.5 Trillion Annually by 2025," 2023.
2 IBM X-Force, "Cost of a Data Breach Report 2023," IBM Security, 2023.
3 FireEye, "SolarWinds Supply Chain Attack Overview," 2021.
4 CISA, "Understanding Zero-Day Exploits," Cybersecurity and Infrastructure Security Agency, 2023.
5 Verizon, "Data Breach Investigations Report," 2023.
6 ICS-CERT, *Ukraine Power Grid Cyberattack Analysis*. U.S. Department of Homeland Security, 2016.
7 WHO, "Cybersecurity and Healthcare Systems," 2022.
8 IBM X-Force, "2023 Threat Intelligence Index," 2023.
9 Europol, *WannaCry: Lessons Learned*. European Union Agency for Law Enforcement Cooperation, 2018.
10 Google Cloud, "AI for Cybersecurity Threat Detection," 2023.
11 NIST, *Cybersecurity Incident Response Planning*. National Institute of Standards and Technology, 2022.
12 Forrester Research, "Zero Trust Adoption Framework," 2023.
13 CISA, *Cybersecurity Performance Goals*. Cybersecurity and Infrastructure Security Agency, 2023.
14 Colonial Pipeline Co., "Incident Report on the 2021 Ransomware Attack," 2021.
15 Quantum Alliance Initiative, "Post-Quantum Cryptography Standards," Hudson Institute, 2023.

21

Cybersecurity in the Age of Interdependence

Modern society's critical infrastructures – from power grids and water systems to transportation, healthcare, and finance – are increasingly interconnected and digitized. This interdependence provides efficiency and intelligence, but it also creates complex vulnerabilities. A disruption in one system can ripple across others, especially when caused by cyberattacks. Indeed, ubiquitous networking and "smart" technologies have made critical infrastructure "more complex, more interdependent and, therefore, more fragile." Cybersecurity has thus emerged as a strategic cornerstone of systems resilience: ensuring that the digital components controlling infrastructure are protected is now essential to maintaining operational continuity, public safety, and national security. One expert noted that our reliance on connected infrastructure has reached a point where a cyber incident can "cause cascading second-order effects on other sectors, leading to a large-scale catastrophe." This chapter examines the strategic role of cybersecurity in an age of digital interdependence, the rising threat landscape, frameworks for bolstering cyber resilience, instructive case studies, leadership strategies, and metrics for measuring preparedness.

Cybersecurity and Interconnected Infrastructure Systems

Today's infrastructure systems are cyber-physical and deeply interwoven. Smart grids use advanced sensors and Artificial Intelligence (AI)-enabled controls to balance electricity supply and demand; intelligent transportation systems synchronize traffic lights and vehicle telemetry; hospitals and emergency services rely on networked devices; and virtually all critical services depend on cloud computing and communications networks. In such an ecosystem, cybersecurity is not merely an Information Technology (IT) concern but a strategic function integral to system reliability. If the digital control layer is compromised, physical operations can halt. For example, a coordinated cyberattack on the electric grid could turn off power and disrupt communications, banking, fuel distribution, and healthcare – a chain reaction of failures. Government and industry leaders increasingly recognize this risk; the World Economic Forum's *Global Risks Report 2024* ranks cyber insecurity as a top global risk, warning that malware, ransomware, deepfakes, and other cyber threats "threaten supply chains, financial stability, and democracy." In other words, cyber incidents can have systemic impacts akin to – or even surpassing – traditional disasters.

Crucially, cybersecurity's role in these systems goes beyond preventing data theft; it is about ensuring the operational resilience of critical services. A breach in one node can propagate through connected networks. In 2015, a cyberattack on Ukraine's power grid temporarily turned off electricity for hundreds of thousands of people, demonstrating that cyberwarfare can directly impact the physical world. Analysts later observed that "the often dimly understood interdependencies" between infrastructure sectors mean a "blow to one…could cause cascading…effects on other sectors" https://www.cigionline.org/articles/danger-critical-infrastructure-interdependency/. This interdependence elevates cybersecurity to a board-level concern. Executives now discuss cyber risks in the same breath as financial or safety risks. Surveys indicate a growing awareness in the C-suite that cyber resilience is a pillar of overall business resilience. For instance, over one-third of organizations worldwide experienced a material cyber incident in the past year, and an overwhelming majority of business leaders expect such incidents to increase. The lesson is clear: protecting interconnected systems requires an integrated approach to security, bridging technical measures with organizational strategy.

One key aspect of this strategic role is building security and resilience by design. Infrastructure operators are embedding cybersecurity from the ground up – in system architectures, control system configurations, and even the artificial intelligence algorithms that manage resources. There is also a shift toward Zero Trust architectures in critical networks, where no user or device is implicitly trusted, to limit the lateral spread of attacks. In AI-enabled networks, where algorithms autonomously adjust infrastructure in real-time, verifying the integrity of AI models and data streams becomes part of cybersecurity's remit. Leaders in sectors like energy and transportation are beginning to mandate rigorous security testing of any new "smart" components before deployment. Such foresight is essential in the age of interdependence: it is far more effective to incorporate strong cybersecurity into the design of smart grids or Internet of Things (IoT)-enabled healthcare devices than to retrofit protections after deployment.

Finally, cybersecurity serves a trust function among interconnected partners. Critical services are often delivered by a web of organizations – public agencies, private contractors, cloud providers, and software vendors – all dependent on each other's security practices. A weakness at one entity can jeopardize the whole ecosystem. Thus, maintaining trust through robust cybersecurity (validated by audits, certifications, or information sharing about threats) underpins the collaborative resilience of interdependent infrastructure. Therefore, the strategic role of cybersecurity is to act as the glue that holds together the safety and reliability of a complex, networked society. It transforms a collection of linked systems into a resilient whole, able to absorb shocks and continue functioning even under attack.

Rising Threats in an Era of Digital Interdependence

As interdependence grows, so do the threats that seek to exploit it. Modern cyber adversaries range from profit-driven criminal gangs to sophisticated nation-state units, all too aware that hitting one target can have outsized cascading effects. Several threat trends have escalated in recent years:

- Supply Chain Attacks: One compromised supplier can open a backdoor into hundreds or thousands of customers. The *SolarWinds* software breach in 2020–2021 starkly illustrated this risk. Attackers (later identified as Russia's foreign intelligence service) infiltrated SolarWinds, a network management software vendor, and inserted malicious code into a routine update. When pushed out, that tainted update was downloaded by roughly 18,000 organizations – including multiple U.S. federal agencies and Fortune 500 companies – unknowingly giving the adversary remote access to their networks. This "one hack, infects many" technique was "one of the most widespread and sophisticated hacking campaigns ever" against government and industry. Its discovery forced organizations worldwide to "completely reassess their threat profiles, risk assessments, and defensive postures." Analysts observed a sharp rise in copycat supply chain exploits in the aftermath. Mandiant found that supply chain compromise accounted for 17% of intrusions in 2021 – a huge jump from less than 1% in 2020 – largely due to the SolarWinds campaign and its fallout. The lesson is sobering: trust can be a weapon in an interconnected digital supply chain. Whether through malicious code in software updates, compromised hardware components, or backdoors in third-party services, supply chain attacks leverage interdependence itself as the attack vector. Organizations have responded by scrutinizing their suppliers' security, deploying *software bills of materials (SBOMs)* to map component dependencies, and adopting frameworks (e.g., National Institute of Standards and Technology [NIST]'s updated guidance) emphasizing supply chain risk management.

- Ransomware and Cyber Extortion: Ransomware has evolved into a rampant, multi-billion-dollar criminal enterprise that preys on interdependent networks. By encrypting data and disrupting operations, ransomware attacks harm the immediate victim and paralyze dependent services. In May 2021, the *Colonial Pipeline*, which supplies ~45% of fuel to the U.S. East Coast, was hit by a ransomware attack that forced a six-day shutdown of fuel distribution. The attack locked up IT systems (even though the pipeline's operational controls were not directly hacked) and the company halted operations as a precaution, leading to gasoline shortages and public panic buying across multiple states. Images of cars lining up at gas stations revealed how a cyberattack can trigger a real-world crisis. Colonial Pipeline's CEO authorized a ransom payment of approximately $4.4 million in cryptocurrency to regain system control. However, U.S. authorities later recovered a portion. "The largest publicly disclosed cyber attack against U.S. critical infrastructure" showed the systemic ransomware threat. It also had cascading effects: fuel logistics companies, airlines, and consumers were indirectly impacted. The broader statistics on ransomware are alarming – by 2023, an estimated 72% of organizations worldwide had been affected by a ransomware incident. Criminal groups now often engage in "double extortion," stealing sensitive data before encrypting systems to pressure victims with both data leaks and operational paralysis. In our interconnected era, a ransomware outbreak at a managed service provider can propagate encryption malware to client networks, or a hit on a hospital system can disrupt regional healthcare. The epidemic has prompted governments to establish joint task forces and improve coordination, as seen with the U.S. forming a Joint Ransomware Task Force in 2022 https://www.cisa.gov/news-events/news/attack-colonial-pipeline-what-weve-learned-what-weve-done-over-past-two-years. It has also accelerated the adoption of best practices like segmented networks (to contain malware spread), offline backups, and incident response (IR) drills focused on swift recovery.
- Nation-State Cyber Operations: Advanced Persistent Threat (APT) groups backed by nation-states are leveraging cyber tools for strategic effect, often aiming at the critical infrastructure of rival nations. These adversaries have the resources and patience to conduct stealthy, prolonged campaigns. The Russian-linked "Sandworm" group, for instance, has repeatedly struck Ukraine's grid and government networks, not for financial gain but to degrade essential services. In December 2015, Sandworm operatives compromised Ukrainian electricity distribution companies, using malicious firmware and remote control commands to open circuit breakers and cut power – an unprecedented cyber-induced blackout affecting ~225,000 customers for several hours. The U.S. government later attributed this operation to Russian state actors https://www.cisa.gov/news-events/ics-alerts/ir-alert-h-16-056-01. A year later, in 2016, another cyberattack (using a malware framework known as *Industroyer*) disrupted part of Kyiv's transmission grid. And during the 2022 invasion, Ukrainian infrastructure faced a blitz of cyber assaults alongside kinetic strikes. While Ukraine's defenders were able to limit the impact of many cyber attempts,

the campaign revealed ever-more sophisticated Operational Technology (OT) attack techniques. A late-2022 incident saw attackers using "living off the land" techniques at the OT level to trip breakers at a Ukrainian substation, timed to coincide with physical missile strikes. Mandiant analysts noted this represented "the latest evolution in Russia's cyber-physical attack capability," showing an ability to rapidly develop new OT attack tools and potentially reuse them against other countries' industrial systems. Beyond Ukraine, other nation-state campaigns have targeted industrial control systems (e.g., the Stuxnet operation on Iran's nuclear centrifuges, or attempts to penetrate U.S. energy grid networks) and enterprise supply chains (as seen in Chinese APT targeting telecom and cloud service providers). State-sponsored cyberattacks often seek to quietly preposition access in critical networks, effectively planting "cyber landmines" that could be detonated to disrupt a crisis. The interdependence of systems means these threats pose *strategic risks*: a cyberattack could be used as a force multiplier in warfare or as a coercive tool in peacetime, with potentially cascading outages across energy, communications, and other lifeline sectors.

- Cascading Vulnerabilities: Even absent targeted attacks, systemic vulnerabilities in widely used software/hardware can pose a catastrophic risk in an interdependent environment. The *Log4Shell* vulnerability (disclosed in late 2021) in a common open-source logging library is a prime example – it affected hundreds of millions of devices and applications globally, forcing emergency patching in organizations worldwide. Similarly, weaknesses in core Internet infrastructure (like Domain Name System [DNS] or certificate authorities) or popular IoT devices can be exploited to disrupt large swaths of dependent services. The 2016 Mirai botnet attack, which hijacked insecure IoT cameras and Digital Video Recorders (DVRs), dramatically took down major Internet services on the U.S. East Coast by overwhelming DNS provider Dyn – effectively a cascading failure originating from compromised "smart" gadgets. In the age of interdependence, organizations must secure themselves and remain vigilant about flaws in the broader ecosystem that could spill over into their operations.

The threat landscape is vast, complex, and rapidly evolving. Threat actors are innovating to take advantage of connectivity and interdependence: whether by infiltrating the software supply chain, holding critical data hostage, conducting covert sabotage, or exploiting ubiquitous technologies. The cascading nature of these threats blurs traditional boundaries between sectors and even between corporate and national security. A cyberattack might start as an IT incident but quickly escalate to a societal crisis if it knocks out essential utilities or services. This reality demands that leaders adopt a posture of constant vigilance and adaptability. As discussed next, a range of frameworks and standards has emerged to help organizations systematically bolster their cyber resilience in the face of such threats.

Frameworks and Standards for Resilience Planning

Organizations are turning to established cybersecurity frameworks (CSFs) and standards to navigate the complex risk environment as roadmaps for resilience. These frameworks provide structured approaches to identify risks, implement controls, and ensure continuous improvement. Leading examples include the U.S. NIST CSF, the international ISO/IEC (International Organization for Standardization/International Electrotechnical Commission) 27001 standard, industry-specific standards (such as the North American Electric Reliability Corporation Critical Infrastructure Protection (NERC CIP) for the electric grid), and various guidelines from bodies like IEEE. Applying these frameworks to resilience planning helps organizations move from ad hoc defenses to a managed, systematic security posture.

NIST Cybersecurity Framework (CSF): Originally released in 2014 and updated through 2024, the NIST CSF has become a de facto foundation for cybersecurity risk management, especially in critical infrastructure sectors. It organizes security activities into five (now six) core functions – Identify, Protect, Detect, Respond, Recover, and (new in 2.0) Govern. This structured model guides organizations to map their assets and risks (Identify), deploy protective safeguards, establish detection and response capabilities, and plan for recovery and continuity. The latest NIST CSF 2.0 emphasizes governance and supply chain risk management, reflecting lessons from incidents like SolarWinds. It also aligns more closely with international standards to ease global adoption. One reason for CSF's wide uptake is its flexibility – it is "a flexible set of guidelines and best practices" adaptable to organizations of all sizes. By 2023, surveys showed that NIST CSF was the U.S. companies' most commonly used framework to measure cybersecurity program maturity and prioritize improvements. Many organizations use CSF's tiered maturity model to assess their current and target state, thereby integrating resilience goals into their strategic plan. For example, if the CSF assessment shows weaknesses in the Respond/Recover functions, a company might invest in IR training and backup systems as part of its resilience planning. The U.S. government has also promoted NIST CSF in critical infrastructure (e.g., through sector-specific guidance and the Cybersecurity and Infrastructure Security Agency [CISA] Cybersecurity Performance Goals, which map to CSF), further cementing it as a cornerstone of resilience efforts.

ISO/IEC 27001: ISO 27001 is a globally recognized standard for establishing an Information Security Management System (ISMS). Unlike NIST CSF's more voluntary guidance, ISO 27001 is often pursued for certification to demonstrate an organization's commitment to security best practices. The standard requires a risk-based approach: organizations must systematically assess information security risks, implement appropriate controls (referencing an extensive catalog in ISO 27002), and undergo regular audits. A strength of ISO/IEC 27001 is its broad applicability across industries and its integration of people, process, and technology controls. In 2022, ISO 27001 was updated (ISO/IEC 27001:2022) to introduce 11 new controls addressing modern challenges, such as Threat Intelligence, Secure Coding, and Cloud Services Security. It restructured its control domains to align with today's technologies. This update underscores how standards evolve to cover emerging aspects of resilience (for instance, "Information and Communication Technology (ICT) readiness for business continuity" is now explicitly included in advisera.com). Many multinational organizations favor ISO 27001 because it provides a common language for security and resilience that is recognized worldwide. As a result, ISO 27001 has "broad support in the cybersecurity industry," particularly for companies seeking a formal certification of their security program. At the same time, NIST CSF is often used as a practical tool for maturity assessment and improvement. Importantly, implementing ISO 27001 helps institutionalize resilience: it mandates policies for IR, backup, disaster recovery, and legal compliance – all key to sustaining operations under duress. An ISO-certified organization is expected to have documented plans and procedures for responding to incidents and continuing critical activities, which directly feed into resilience planning.

Industry and Domain Standards (IEEE, NERC, etc.): In addition to general frameworks, specialized guidelines target the security of specific types of infrastructure. For example, the NERC CIP standards are mandatory in the electric utility sector and outline requirements for securing bulk electric system cyber assets, including contingency planning and incident reporting, thereby enhancing grid resilience. IEEE also contributed domain-specific standards and guides. For instance, IEEE has published guidance for Distributed Energy Resources (DER) cybersecurity (IEEE 1547.3) to ensure that, as renewable energy resources and microgrids are integrated, they do not become weak links in grid security. IEEE is working on frameworks for operational resilience in cloud computing (IEEE P3454) and has convened conferences on system-of-systems security and resilience. These efforts signal recognition that resilience must be engineered at every level – from microcontrollers in field devices to enterprise cloud architectures. Organizations should stay abreast of relevant standards in their field (for example, hospitals may follow the Health Information Trust Alliance [HITRUST] CSF for healthcare, or manufacturers might look to ISA/IEC 62443 for industrial control system security) and incorporate those guidelines into their resilience planning. Adopting industry standards also aids regulatory compliance and cyber insurance eligibility, creating additional incentives for robust implementation.

Frameworks for Cyber Resilience: Traditional security frameworks are increasingly being extended or supplemented with explicit cyber resilience frameworks (CRFs). Notably, NIST's Special Publication 800-160 Vol. 2 (2019) provides principles for "Developing Cyber-Resilient Systems," focusing on engineering systems that can withstand and recover from cyberattacks. The U.S. Department of Homeland Security (DHS) and Carnegie Mellon's Computer Emergency Response Team (CERT) developed the Cyber Resilience Review (CRR), a voluntary assessment tool aligned with NIST CSF that evaluates an organization's resilience across domains like incident management, service continuity, and supply chain defense. Likewise, the World Economic Forum 2022 introduced a CRF and Cyber Resilience Index (CRI) to help organizations quantify and benchmark their resilience maturity. The CRI provides a scoring mechanism against best practices, encouraging a shift from just "Are we secure?" to "How quickly can we recover and adapt when (not if) an attack occurs?" weforum.org. These resilience-focused frameworks complement security controls with redundancy, fail-safe design, crisis communications, and continuous delivery of essential services under degraded conditions.

In practical terms, what does applying these frameworks look like? It often starts with a risk assessment and gap analysis: using NIST CSF or ISO 27001 as a checklist to identify areas that need strengthening. Then, organizations develop a resilience plan – prioritizing critical assets and processes, mapping dependencies, and ensuring plans are in place to keep those running during an incident. Frameworks guide each step. A bank, for instance, might use ISO 27001 to ensure it has strong access controls and encryption (preventive security), NIST CSF to build robust detection/response, and then CRR to evaluate how it would maintain customer services if its Automated Teller Machine (ATM) network were cyberattacked. The interplay of standards ensures no critical aspect is overlooked: governance (assigning clear roles and responsibilities), technical controls (firewalls, monitoring, backups), human factors (training and IR teams), and continuous improvement (audits, drills, lessons learned) all feature in these frameworks.

Moreover, embracing frameworks and standards helps break down silos. It gives executives, engineers, and operations teams a common vocabulary to discuss cybersecurity resilience. This is especially valuable in large, interdependent enterprises or cross-sector collaborations. When a power utility and a cloud provider align to, say, NIST CSF or ISO 27001, it becomes easier to assess each other's posture and work together on joint risk mitigations.

In summary, frameworks and standards act as force multipliers for resilience planning. They encapsulate hard-won knowledge from industry and government about what works and provide a roadmap to strengthen cybersecurity in alignment with business objectives systematically. By following these best practices, organizations can significantly reduce the likelihood of successful attacks and, critically, ensure they can contain the damage and recover quickly if an attack occurs. The next section will delve into real-world case studies highlighting how failures and successes in applying such practices have played out, offering further insight into resilience lessons for leaders.

Case Studies of Cyber Incidents and Resilience Lessons

Examining high-profile cyber incidents in recent years provides concrete lessons on interdependence and resilience. This section reviews three illustrative case studies – the Colonial Pipeline ransomware attack, the SolarWinds supply chain compromise, and the Ukrainian power grid attacks – analyzing their impact and what they teach about building cyber-resilient systems.

Case Study 1: Colonial Pipeline Ransomware Attack (2021)

Background: Colonial Pipeline is one of the largest fuel pipeline operators in the United States, transporting gasoline, diesel, and jet fuel from the Gulf Coast up the Eastern Seaboard. On May 7, 2021, Colonial Pipeline Co. fell victim to a ransomware attack (attributed to the criminal group DarkSide) that encrypted data on its IT network. Although the attackers did not directly take over pipeline controls, the company preemptively halted pipeline operations to prevent the malware from spreading to critical OT systems. This led to a complete shutdown of fuel delivery for the East Coast for several days.

Impact: The immediate effect was a disruption of about 2.5 million barrels of fuel daily. Fuel shortages were reported in North Carolina and Virginia within a day, as distribution hubs ran low. The incident sparked public panic, with images of "snaking lines of cars at gas stations across the eastern seaboard" and panic buying of fuel dominating news cycles. Colonial Pipeline's shutdown represented a rare instance of a cyberattack creating *instant, tangible economic effects* for average consumers. The U.S. federal government issued emergency declarations to allow alternative fuel transport by road. Under intense pressure, the company opted to pay the ransom of approximately $4.4 million in Bitcoin to obtain a decryption key and restore its systems. Within a week, pipeline service was back to normal, though not before causing a spike in fuel prices. Notably, the U.S. Department of Justice later traced and recovered about $2.3 million of the ransom, exploiting a cryptocurrency wallet vulnerability. This action demonstrated law enforcement's growing capabilities in cybercrime response.

Resilience Lessons: Colonial Pipeline was a wake-up call for critical infrastructure operators about the real-world stakes of cybersecurity. First, it highlighted the interdependence between IT and OT systems. Even though Colonial's OT was unharmed, the company lacked separation and confidence in the integrity of its networks, forcing a total shutdown to be safe. This underscores the need for network segmentation and zero-trust principles in industrial environments, so that the compromise of business IT systems does not automatically threaten control systems. Second, the incident revealed gaps in incident preparedness and response. Colonial Pipeline reportedly did not have a detailed cyber crisis plan or a tested scenario for a ransomware event of this scale. The decision to pay the ransom was made under duress to restore operations quickly, but it was later criticized. Organizations can learn from this to develop playbooks for ransomware, including decision frameworks for ransom payment (in consultation with authorities), communication plans to stakeholders and the public, and rapid restoration strategies (such as maintained backups and fallback manual operations where possible).

On the positive side, Colonial spurred significant government and industry collaboration post-incident. Within two months, the U.S. Transportation Security Administration (TSA) issued new security directives for pipeline operators, mandating 24-hour incident reporting, vulnerability assessments, and contingency planning. CISA, the U.S. Cybersecurity and Infrastructure Security Agency, intensified its outreach, bringing together pipeline operators in joint sessions to "strengthen security practices to safeguard OT networks." One result was the expansion of programs like *CyberSentry*, which places sensors in critical OT systems to provide early warning of cyber threats. The broader lesson is that cross-sector collaboration is key: sharing threat intelligence and best practices can uplift the security of an entire sector. Colonial Pipeline's CEO later testified to Congress about the need for a public-private partnership to prevent such attacks.

From a leadership perspective, Colonial underscored that cyber incidents can become "kitchen table issues" – affecting the daily lives of citizens – and thus cyber risk must be treated as enterprise risk. In the aftermath, boards and executives of critical infrastructure companies were prompted to ask tough questions: Do we have visibility into our cyber risks? Are we investing enough in prevention and resilience? How would we keep operating if our digital systems were knocked out? Those who had seen cybersecurity as a technical back-office problem were forced to recognize it as a core element of business continuity planning. The Colonial Pipeline case reinforces the importance of preparedness, network resilience, and rapid IR capabilities in any interconnected infrastructure organization.

Case Study 2: SolarWinds Supply Chain Breach (2020)

Background: In December 2020, cybersecurity firm FireEye and U.S. agencies discovered a sophisticated supply chain attack that had been underway for months. The incident centered on SolarWinds Orion, a popular IT monitoring platform used by over 33,000 organizations worldwide. Attackers – later identified as Russia's SVR (APT29) – compromised SolarWinds' software build process and inserted a backdoor (dubbed "SUNBURST") into Orion software updates released between March and June 2020, gao.gov. Any client that installed these Trojanized updates unknowingly opened a door for the attackers. Because many large enterprises and government agencies used SolarWinds Orion, the scale of potential compromise was vast. An estimated 18,000 organizations received the infected update from gao.gov, including the U.S. Departments of Homeland Security, Treasury, State, major telecom companies, and numerous others.

Impact: The SolarWinds breach is often cited as "one of the most widespread and sophisticated hacking campaigns" in history. However, the adversaries did not activate their backdoor on all 18,000 victims. They selectively targeted about 100 high-value organizations for follow-on exploitation, focusing on those yielding intelligence (espionage appeared to be the primary motive). Victims reported attackers moving laterally within networks, stealing authentication tokens, and exfiltrating sensitive data (such as emails and source code). The campaign went undetected for at least eight to nine months. Ultimately, FireEye's breach disclosure (the company realized its Red Team tools were stolen) led to uncovering the SolarWinds connection. The incident had a sweeping impact on the cybersecurity community and policymakers. Thousands of organizations scrambled to check for indicators of compromise and patch systems. The U.S. government imposed sanctions on Russia and launched multiple investigations. Remediation for affected entities was complex, essentially eradicating and rebuilding parts of their network infrastructure to be sure the attackers were out. The monetary cost is hard to quantify, but it easily runs into tens of millions in IR efforts across the public and private sectors. The blow to trust was more intangible: trust in software updates, vendor security, and one's ability to detect stealthy nation-state actors.

Resilience Lessons: SolarWinds taught the world how vulnerable the software supply chain can be and how an attack on one vendor can indirectly compromise thousands of others. One clear lesson is the need for "Zero Trust" not just within an enterprise, but also toward components and code supplied by others. Organizations are increasingly validating and monitoring software behavior rather than assuming vendor software is clean. Technical measures like digital code signing, integrity checking, and behavior anomaly detection have gained emphasis (had SolarWinds Orion's behavior been closely monitored, the unusual outbound communications of the backdoor might have raised alarms sooner). The attack also accelerated calls for software producers to adopt stricter security in their development pipelines – what some call a "Secure by Design" approach. The concept of a Software Bill of Materials (SBOMs) – a manifest of third-party components in a software product – went from academic discussion to mainstream recommendation in the wake of SolarWinds. In May 2021, a U.S. Presidential Executive Order on cybersecurity mandated that vendors selling to the government provide SBOMs and adhere to secure development practices. This policy move directly responded to SolarWinds, aiming to "increase software supply chain security" and resilience by giving customers transparency into what they are running.

SolarWinds underscores the importance of defense-in-depth and rapid detection from a resilience planning standpoint. Even with the compromise of a trusted software, organizations with robust monitoring might catch anomalies (for example, unusual authentication patterns or data flows). Some targeted organizations detected and thwarted the attackers in the early stages through vigilant security operations. Those that fared better had assumed breach – operating under the mentality that any system, even trusted tools, could be compromised – and thus they limited the blast radius of any single component's failure. This could mean segregating network management tools on isolated networks, or employing strict outbound network filtering (the SolarWinds malware had to communicate to the attackers' servers – a connection that might have been blocked in a locked-down environment).

Another lesson is the power of information sharing and public-private collaboration in responding to widespread incidents. The investigative piecing of the SolarWinds attack involved private firms (like FireEye, Microsoft), government agencies (National Security Agency [NSA], CISA, Federal Bureau of Investigation [FBI]), and others openly sharing details and indicators. This collaborative response likely shortened the dwell time of the adversary and provided actionable information to many potential victims quickly. In planning for cyber resilience, establishing trust relationships and communication channels (e.g., membership in an Information Sharing and Analysis Center (ISAC) or a national CERT) is invaluable. It enables faster warning and more coordinated containment when a large-scale campaign unfolds.

At a strategic level, the SolarWinds saga drove home that nation-state cyber threats require a collective defense. No single company, even one with significant security savvy, could single-handedly resist an attack of this caliber indefinitely. It took a coalition to expose and counter it. This has meant pushing for community-wide exercises and threat intelligence fusion for critical infrastructure sectors to improve readiness for the next big attack. Investments in resilience might include diversifying critical software (to avoid monocultures where one compromise has universal impact) and planning for operating if major IT management systems go down (e.g., having contingency plans for network monitoring or identity management if those tools are compromised).

The SolarWinds breach highlights the need for supply chain resilience as part of a cybersecurity strategy. Organizations must scrutinize and secure the links that bind them to external software and services. Resilience planning extends beyond one's walls to encompass vendors' and partners' reliability and transparency. "Trust but verify" has perhaps never been more apt: trust your vendors' products to do their job, but have independent means to verify they haven't been tampered with.

Case Study 3: Cyberattacks on the Ukrainian Power Grid (2015–2016)

Background: Ukraine has been at the frontlines of cyber-physical conflict for years, serving as a testing ground for nation-state cyberattacks on critical infrastructure. Two landmark attacks in December 2015 and December 2016 marked the first-known instances of hackers causing deliberate electrical blackouts. In 2015, attackers believed to be the Sandworm group (linked to Russia's GRU) compromised the IT networks of three regional electric power distribution companies in Ukraine. They used spear-phishing to gain entry, then deployed malware (BlackEnergy3 and KillDisk) to facilitate the attack. On December 23, 2015, the attackers remotely took control of distribution substation breakers (via hijacked Supervisory Control and Data Acquisition [SCADA] systems). They opened them, resulting in power outages for around 225,000 customers in western Ukraine for several hours. Operators watched helplessly as cursors moved on their screens, executing unauthorized commands – a nightmare scenario for grid engineers. Additionally, the attackers sabotaged systems by overwriting the firmware of critical devices and wiping servers, slowing the recovery.

A year later, on December 17, 2016, another cyberattack targeted a transmission-level substation north of Kyiv, using a more automated and sophisticated malware called *Industroyer* (or CrashOverride). This malware could directly send commands to substation devices (using legitimate grid control protocols), essentially acting as a cyber-weapon tuned to electric grid operations. The 2016 attack caused a smaller outage (around 1/5 the size of 2015's) and for a shorter duration, partly due to quicker manual recovery by Ukrainian grid operators.

Impact: Though temporary, these cyber-induced outages had significant psychological and strategic implications. They demonstrated that advanced actors could remotely infiltrate and disrupt power infrastructure, crossing a boundary many had only theorized about. In 2015, power was restored within 3–6 hours by manually switching breakers (Ukraine's grid had retained some manual operation capability – a saving grace). The 2016 incident's impact was mitigated by the fact that the grid's protection systems tripped to isolate the fault. There were no known long-term physical damages to equipment in either case (attackers in 2015 tried to turn off backup power supplies of control centers to complicate recovery). These incidents primarily affected Ukraine, but they sent shockwaves worldwide. Governments in Europe and North America urgently reassessed their grid security postures. If a similar attack were carried out on a less prepared grid, the effects could be more severe and harder to overcome.

Resilience Lessons: Ukraine's grid attacks provided a trove of lessons on both vulnerabilities and effective resilience practices. One key takeaway is the value of maintaining manual operations capability and training as a resilience measure. Ukrainian technicians could switch to manual mode to restore power, dispatching personnel to sites to flip switches, which ended the outages. In highly digitized systems, there is a tendency to remove manual fallback options; Ukraine showed that keeping them (and practicing their use) can be a lifesaver when digital controls are compromised.

Another lesson is the importance of cyber hygiene and monitoring in industrial networks. Investigations found that the adversaries had been present in Ukraine's power companies' IT networks for months, and they pivoted into the OT environment partly by using Virtual Private Network (VPN) access and stolen credentials. Improved network segmentation between corporate and control networks, strict access controls, and unidirectional gateways for monitoring could have blunted the attack. With assistance from international partners, Ukraine subsequently strengthened its grid defenses, deploying more advanced detection systems. Observers noted that by 2022, Ukrainian grid operators had developed a high degree of vigilance and IR skill, having "familiarity with Russian tactics, vigilant cyber hygiene, and the skilled use of manual override controls" – all contributing to their resilience.

The attacks also underscore the necessity of cross-sector and international collaboration in resilience. Following the 2015 blackout, a U.S. DHS (ICS-CERT), U.S. Department of Energy (DOE), FBI, and NERC team traveled to Ukraine to assist and learn from the incident. The information from Ukraine was rapidly disseminated through security alerts and mitigation recommendations to other utilities worldwide. This global sharing helped others shore up defenses (for example, electric utilities in the United States conducted emergency drills and patched known vulnerabilities like those exploited by BlackEnergy malware). It's a strong example of how an incident in one country can inform resilience elsewhere, effectively turning a real attack into a training scenario for everyone else. In the years since, U.S. and European grid operators have run periodic grid cybersecurity exercises (like the biennial GridEx in North America) simulating similar multifaceted attacks, and many have upgraded their IR playbooks.

Strategically, the Ukraine case studies illuminated the concept of combined cyber and physical warfare – cyberattacks were timed with political/military objectives. Critical infrastructure operators worldwide thus must consider concurrent threats (a cyberattack happening alongside a natural disaster or a physical attack). Resilience planning should account for compound scenarios. For example, how can we respond if a cyberattack occurs during a major storm or vice versa? Ukraine's experience in 2022 (facing missiles and malware together) showed the value of preplanned procedures and redundancies. One reported improvisation was using SpaceX Starlink satellite links to maintain communications when terrestrial networks were damaged, sgp.fas.org – a creative resilience tactic to deal with degraded infrastructure.

Ukraine's grid attacks taught critical infrastructure communities to prepare for the unprecedented. Cyberattacks can cause real-world harm, but their effects can be mitigated with foresight, training, and international support. The incidents reinforced the need to invest in robust IR, system redundancy, and adaptive recovery strategies. Every country or operator that depends on networked control systems noted: the question is not *if* such attacks are possible – they are – but *when* they might face a similar crisis and whether they will be ready to respond.

These case studies collectively highlight that technological interdependence magnifies the impact of cyberattacks and provides insights for resilience. Colonial Pipeline showed that even an IT-focused attack (ransomware) can halt physical services; SolarWinds showed that trust in third-party software must be managed; Ukraine showed that determined adversaries can cause physical blackouts and that strong defense and recovery are feasible. In each case, organizations and leaders had to adapt and learn. The following section will outline a proactive roadmap for leadership to build on these lessons and strengthen cyber resilience across interdependent systems.

Leadership Roadmap: Building Resilience in an Interdependent Era

Achieving cybersecurity resilience in the age of interdependence is not solely a technical endeavor – it is fundamentally a leadership challenge. Leaders at the board and executive levels must champion and integrate cybersecurity into their organizations' deoxyribonucleic acid (DNA), making it a sustained priority aligned with business goals and ethical responsibilities. This section outlines a roadmap for leaders, emphasizing foresight, investment, collaboration, incident readiness, and a culture of accountability. These elements serve as guideposts for steering complex enterprises toward greater cyber resilience.

1) Foresight and Risk Anticipation: In a rapidly evolving threat landscape, reactive postures are no longer sufficient. Leaders must adopt strategic foresight, anticipating future threat scenarios and preparing in advance. This involves staying informed through cyber threat intelligence (tracking emerging adversary tactics, vulnerabilities, geopolitical risks) and leveraging that insight in decision-making. Scenario planning and war-gaming should become routine at the executive level: for example, envisioning how a supply chain attack could disrupt operations or how a prolonged cloud outage might impact services. Leaders cultivate an anticipatory mindset by asking "How might we be attacked?" and stress-testing their systems. One cybersecurity CEO said, "Staying ahead requires more than just reacting – it demands foresight, adaptability, and a commitment to constant learning." Concretely, this could mean conducting regular tabletop exercises that include not just IT staff but also business executives, so that in a crisis, those leaders have muscle memory of the decisions and communications required. Foresight also means keeping an eye on technological change – for instance, understanding the security implications of deploying AI, 5G, or IoT at scale – and baking security considerations into those innovation plans from the outset.
2) Investment in Resilience and Innovation: Leadership must ensure that cybersecurity and resilience are adequately resourced. This goes beyond IT budgets; it includes investing in people, processes, and technologies that bolster defense and enable rapid recovery. The economics are compelling: organizations with robust security controls and IR capabilities tend to reduce the cost of breaches significantly. A 2023 analysis found that companies with an IR team and regular testing saved on average $1.49 million per breach compared to those without. Such figures prove that cyber investments pay off by averting larger losses. Leaders should view spending on cybersecurity not as a discretionary IT cost, but as a strategic investment in operational continuity and risk reduction, akin to insurance or safety programs. This includes modernizing legacy systems that are insecure by design, funding network upgrades to enable zero-trust architectures, and implementing resilient infrastructure (for example, backup power and communications for critical facilities). It also means investing in innovation for security: leveraging advanced tools like AI-driven threat detection, automation for patch management, and network analytics to handle the scale of modern threats. Many organizations are increasing their budgets here; a global survey indicated that 86% plan to boost cybersecurity spending by at least 10% following recent high-profile attacks. However, investment must be smart and tied to a clear strategy and risk model. Leaders should demand metrics on how additional funding improves specific resilience outcomes (e.g., faster IR time, higher percentage of systems with automated backups, etc.). In sum, sustained funding and support from the top enable security teams to execute multiyear resilience initiatives, rather than patching holes in an ad hoc fashion.
3) Cross-Sector Collaboration and Information Sharing: No organization is an island in an interdependent world. Leaders should actively engage in collaborative defense efforts, both within their sector and across sectors. This can take many forms: participating in industry ISACs to exchange threat intelligence, forging partnerships with government agencies (like law enforcement or cybersecurity agencies) for threat briefings and incident support, and working with suppliers and customers on joint security initiatives. The Colonial Pipeline incident clarified that only a "cohesive collaboration across government" and industry would scale to meet the threat. Indeed, in its aftermath, initiatives such as the Joint Cyber Defense Collaborative (JCDC) were established to "bring together industry, government, and international partners" to address ransomware and other threats collectively. Leaders should encourage their organizations to consume threat information and contribute, sharing anonymized incident details or indicators

when they suffer attacks, so that others can be warned. Such transparency can be challenging culturally, but it significantly raises the communal resilience. Cross-sector drills and exchanges are also invaluable; for example, a power grid company might coordinate with telecom providers and local governments to rehearse responses to a combined power and communications outage scenario caused by a cyberattack. By building these relationships in peacetime, during a crisis, there is already trust and clear lines of communication. In practice, this could mean having signed agreements or memoranda of understanding in place for IR assistance, or seconding staff to joint analytic teams (as some financial institutions do with government cyber centers). The outcome is a network of defenders that can react faster and more effectively than any lone organization.

4) Incident Response Maturity and Drills: Since no defense is foolproof, leaders must ensure their organizations are ready to respond and recover when incidents occur. IR planning and capability is a core pillar of cyber resilience. Executive leadership should mandate the creation of a robust IR plan that spells out roles, communication channels, and recovery steps for various incident types (data breach, ransomware, system outage, etc.). More importantly, they should ensure this plan is regularly tested and refined. Studies show a clear benefit: organizations that thoroughly test their IR plans (through drills or simulations) handle breaches more efficiently and minimize damage. One security expert emphasized that "there's a big difference between how companies envision their response versus what happens when you try to play it out. It has to be practiced. Doing so saves time and costs." Leadership should be involved in these exercises – for instance, practicing the decision-making around taking systems offline, making a public announcement, or engaging with law enforcement. This ensures that decisions can be made swiftly and confidently in a real crisis. IR maturity also includes establishing relationships with external IR firms or consultants ahead of time (e.g., having a retainer contract), so that expert help is a phone call away if an incident escalates beyond internal capacity. Finally, post-incident reviews are a leadership responsibility: after any significant incident or drill, executives should lead a debrief on lessons learned and drive the implementation of improvements (whether that's fixing a technical gap or adjusting a policy). The organization continuously hardens its resilience by treating every incident as an opportunity to improve.
5) Ethical Responsibility and Cyber Accountability: With great connectivity comes great responsibility. Leaders today must recognize that cybersecurity is not just a technical necessity but also an ethical obligation to customers, employees, and society. When organizations collect and digitize massive amounts of data or control critical services, they protect those assets from harm. Fulfilling this duty is part of corporate social responsibility in the digital age. As one analysis put it, businesses have a "moral obligation to protect their customers and business partners… they are expected to have adequate cybersecurity and respond promptly and decisively in the event of a cyber incident." Failing to do so can be seen as negligence – for example, if known vulnerabilities are left unpatched and lead to a breach, or if a company hides a violation for months (as in the notorious case of Equifax 2017), thereby increasing harm to users. Ethical leadership in cybersecurity means prioritizing transparency, fairness, and trust. This could manifest as promptly disclosing incidents to those affected, even if not legally required, or choosing not to pay ransom to criminals (a dilemma that pits quick restoration against potentially encouraging more crime). It also includes respecting privacy and implementing security measures that safeguard user data as rigorously as the company's crown jewels. By fostering an ethical culture, leaders ensure that teams don't cut corners on security for convenience or profit and empower employees at all levels to speak up about risks. Moreover, boards should incorporate cyber risk into their governance – for instance, having a cybersecurity or technology committee, and

linking executive accountability (and compensation) to good cybersecurity practices. In essence, leaders set the tone that security is everyone's responsibility and that protecting the interconnected ecosystem is a core value, not just a checkbox.

6) Continuous Improvement and Metrics: Leadership must embrace cybersecurity resilience as an ongoing journey, not a onetime project. The environment, threats, and business operations all change, so resilience strategies must evolve in tandem. Leaders should adopt a continuous improvement mindset, guided by clear metrics and feedback loops to manage this. This means regularly measuring cybersecurity readiness – using tools and indices as discussed in the next section (e.g., tracking an internal cyber resilience scorecard, benchmarking via frameworks). Metrics might include patching timelines, phishing test success rates, incident detection time (Mean Time to Detect), response time (Mean Time to Recover), and compliance levels with frameworks (such as what percentage of NIST CSF subcategories are at a desired maturity). By reviewing these metrics periodically (at least quarterly at the board level), leaders can identify where more resources or attention are needed and verify that investments yield results. These metrics mustn't be used punitively, but rather to celebrate progress and spotlight areas for growth. Many organizations also undergo independent cyber resilience assessments or certifications to validate their posture. Achieving improvements in such assessments year-over-year can be a strategic goal that leadership sets. Additionally, staying current means updating policies and plans when new threats emerge – for instance, incorporating ransomware-specific playbooks in the last few years, or adjusting remote work security policies post-2020. A learning organization will adapt its controls and training frequently, and its leaders will support this agility rather than seeing security as static. In practical terms, leadership can institute quarterly cyber risk reviews and annual external audits and ensure that any incident or near-miss triggers an update to practices. This commitment to iterate and improve keeps an organization resilient despite the unpredictable nature of cyber threats.

In conclusion, leadership is the linchpin of cyber resilience. Leaders can significantly fortify their organizations by taking a proactive stance, looking ahead to risks, empowering collaboration, funding robust defenses, honing response skills, upholding ethics, and measuring progress. The interdependent nature of modern systems means that leadership actions (or inaction) in one organization can have far-reaching consequences. Those at the helm must thus navigate technical complexities and human, economic, and societal dimensions of cybersecurity. The next section will delve into how leaders can gauge their organization's preparedness using various metrics and tools, tying together the strategic initiatives described here with concrete ways to track and validate resilience.

Measuring Cybersecurity Readiness: Metrics and Tools

Leaders and practitioners need effective ways to measure readiness to manage and improve cybersecurity resilience. Unlike traditional engineering domains, cybersecurity lacks a single silver bullet metric, but a combination of indices, scoring systems, and assessment tools can provide a meaningful picture. Organizations can identify gaps and track progress over time by quantifying security aspects, from vulnerability levels to response capabilities. This section discusses several key metrics and tools today, including the Common Vulnerability Scoring System (CVSS), the MITRE ATT&CK (Adversarial Tactics, Techniques, and Common Knowledge) framework, CRIs, and other emerging measures.

Vulnerability Severity – CVSS: One fundamental aspect of readiness is understanding the severity of technical vulnerabilities in systems. The CVSS is a widely used standard for scoring the severity of software flaws on a scale from 0 to 10 (low to critical). CVSS provides a standardized way to communicate how bad a vulnerability could be, considering factors like how easy it is to exploit and its impact on confidentiality, integrity, and availability. "CVSS has long been a prime method for accurately assessing vulnerability severity," offering a common language for organizations to prioritize patching. For example, a vulnerability with a CVSS score of 9.8 (out of 10) might allow an unauthenticated attacker to fully take over a system remotely – a top priority to fix. In contrast, a score of 4.0 might denote a less urgent issue. By tallying the number of high-CVSS vulnerabilities in their environment and tracking how quickly they remediate them, organizations gain a metric for technical exposure and response efficiency. In recent years, CVSS has evolved (with version 4.0 released in 2023) to improve precision by adding new metric groups for threat and environment. This allows more context-aware scoring, for instance, factoring in if exploit code is actively circulating (which would increase urgency). CVSS can be used as a resilience metric to compute an "average residual risk" of systems or to ensure that no critical vulnerabilities remain unpatched beyond a certain time. Many compliance regimes now include requirements like "remediate CVSS 7.0+ vulnerabilities within 30 days," embedding this metric into governance. However, CVSS alone doesn't capture everything (it doesn't measure the effectiveness of detection or response), so it should be combined with other indicators for a full picture.

Threat Coverage and Detection Capability – MITRE ATT&CK: Another way to gauge readiness is by evaluating how well an organization can detect and defend against known adversary tactics. The MITRE ATT&CK framework is a comprehensive knowledge base of cyber adversary tactics and techniques, based on real-world observations. It's often visualized as a matrix with techniques mapped under tactics (stages of an attack). Security teams use ATT&CK to assess their visibility and controls against each method. For instance, can we detect if an attacker uses credential dumping? Do we have a way to block lateral movement via Remote Desktop Protocol? Organizations can find gaps by mapping defensive tools and telemetry to the ATT&CK techniques they cover. ATT&CK has effectively become a measuring stick for threat coverage. It's also used in threat intelligence to categorize adversary behaviors. A significant trend is that many organizations have embraced ATT&CK as a common language to evaluate security posture: "almost half of organizations surveyed in 2022 were using MITRE ATT&CK to strengthen their defenses, while 41% used it occasionally," and over 60% of security leaders considered it very important to their strategy wiz.io. This widespread adoption means that ATT&CK can serve as a benchmark; for example, a company might strive to have detection analytics for 80% of the techniques relevant to its threat profile, then measure improvement toward that goal. MITRE ATT&CK evaluations (run by MITRE Engenuity) also test popular security products against simulated attack sequences, producing data that organizations can use to gauge which controls perform well. By using ATT&CK as a metric framework, organizations move beyond generic "we have a SOC (Security Operations Center)" statements to detailed evidence: e.g., "We can detect 9 of the 11 techniques typically used by ransomware groups – and we have plans to address the remaining two." This coverage-driven metric is very actionable and directly tied to real attack scenarios, making it invaluable for resilience planning.

Cyber Resilience Indexes and Maturity Scores: At a higher organizational level, indices that aggregate multiple factors into a single resilience score are emerging. For instance, the World Economic Forum's CRI provides a quantitative benchmark of an organization's cyber resilience by measuring performance against best practices. weforum.org. Such indexes often consider a range of domains: governance, defense measures, response readiness, business continuity, and supply chain security, among others. The CRI, combined with a CRF, lets an organization assess where it stands (e.g., 75 out of 100) and track improvement over time. weforum.org. Likewise, consultancies and industry groups have developed cyber resilience benchmarks – Aon's Cyber Resilience Report or Accenture's Security Index are examples – that allow leaders to compare their preparedness against peers. One practical metric is the "Mean Time to Recover (MTTR)" from cyber incidents, which indicates how quickly core services can be restored after disruption. A shorter MTTR suggests stronger resilience. Another is "downtime hours avoided" due to resilient architecture (for example, measuring how often failover systems prevented an outage). Organizations might also keep a resilience scorecard that blends technical metrics (like patching rates, IR times) with governance metrics (like training completion, audit findings). Any index or composite score aims to distill complex readiness data into a form executives can grasp and use to drive decisions. For public infrastructure, national-level cyber readiness indexes attempt to gauge a country's overall resilience by looking at policy, education, and investment indicators.

Drill and Exercise Performance: Cyber exercise or drill performance is a direct measure of readiness. Many organizations conduct regular simulations (from tabletop discussions to full operational "red team vs blue team" exercises). Scoring these exercises – e.g., how quickly was the attack detected? Were all required actions taken? Where did communications slow down? – can yield metrics such as "exercise incident detection time" vs. "real incident detection time." Improvement in these metrics over successive drills is a tangible sign of increased readiness. Some sectors hold annual sector-wide exercises (like finance's Quantum Dawn or energy's GridEx); participation in these and lessons learned can be benchmarked. Also, tracking the number of unscheduled downtime events caused by cyber incidents per year is a blunt but important measure – the fewer and shorter such events, the more resilient the operations.

Compliance and Audit Scores: While compliance is not equal to security, passing security audits and meeting standards is a baseline indicator of a mature program. Metrics here include audit findings counts, percentage of the critical conclusions addressed by due dates, and maintenance of certifications (like ISO/IEC 27001 certification status). Suppose an organization can show a year-over-year reduction in repeat audit findings or expansion of its certified scope, which suggests progress in institutionalizing security controls. Some regulatory regimes are introducing explicit cyber resilience metrics; for instance, the European Union's NIS2 directive will likely require certain uptime or incident notification metrics to be reported.

Emerging Measures – Cyber Stress Tests: Inspired by financial stress tests, some organizations and researchers are exploring cyber stress testing, where various worst-case scenarios are modeled (like a simultaneous attack on multiple critical systems) to see if resilience thresholds are exceeded. This can yield quantitative results – e.g., in a model, if X system and Y system fail, the probability of maintaining service is Z%. While somewhat experimental, such modeling can guide investments to where they yield the biggest risk reduction.

External Ratings: Another trend is services that provide outsider cyber risk scores (SecurityScorecard, BitSight, etc.) by scanning an organization's Internet-facing footprint and checking for vulnerabilities, misconfigurations, or past breaches. Insurers and partners use these scores (often letter grades or 0–100) to gauge how well an entity might withstand cyber threats. While not fully indicative of internal resilience, they motivate organizations to fix externally visible issues. An upward trend in an external rating can complement internal metrics to show improvement.

In implementing metrics, it's crucial to remember that no single metric is comprehensive. Organizations benefit from a *portfolio of metrics* – some leading indicators (like patching cadence, which can predict future risk) and some lagging indicators (like breach impact, which shows outcomes). The key is that metrics should drive action. If a metric indicates a shortfall (e.g., too many critical vulnerabilities open, or low detection coverage in one ATT&CK tactic), it should trigger management attention and remediation efforts. Over time, the data helps to allocate resources to where they matter most and to verify whether past resilience investments are paying off.

For example, an organization might set targets such as maintaining an average CVSS score below four on all Internet-facing assets, detecting and responding to simulated attacks (exercises) within one hour, achieving a CRI of "Advanced" (the next tier up) by next year, and ensuring all critical suppliers meet certain security ratings. These targets provide concrete checkpoints on the journey to resilience.

Measuring cybersecurity readiness involves technical, procedural, and organizational lenses. CVSS scores reveal how exposed you are to known technical risks; ATT&CK coverage shows how prepared you are to counter adversary techniques; resilience indices and exercises demonstrate how well you can keep running amid disruptions. By combining these, leaders gain a 360-degree view of preparedness. The quantification of cyber risk and resilience, though an evolving discipline, is improving, giving organizations tools to baseline themselves and strive for measurable enhancements. The final part of this chapter provides actionable recommendations synthesizing all the insights discussed – a checklist for leaders aiming to build truly cyber-resilient organizations.

Conclusion and Recommendations

The age of interdependence demands a new caliber of cybersecurity leadership. As this chapter has detailed, the stakes are immense: cyberattacks can cascade through interconnected networks, disrupt essential services, and inflict societal harm. Yet, with strategic foresight, robust frameworks, collaborative effort, and continuous improvement, organizations can bolster their resilience against even the most daunting threats. Cybersecurity in critical systems is not just about preventing breaches – it's about ensuring that our digitally enabled society can withstand and rapidly recover from cyber shocks.

Leaders should integrate cybersecurity into the core of organizational strategy, treating it as fundamental to resilience and innovation. Doing so protects the enterprise's mission and stakeholders and contributes to the stability of the larger networks we rely on. In closing, we distill five to seven actionable recommendations for leaders who aspire to build and govern cyber-resilient organizations:

- Embed Cyber Risk into Enterprise Risk Management: Treat cybersecurity on par with financial, operational, and safety risks. Establish a top-level risk ownership for cyber (e.g., a board committee and a chief information security officer reporting into leadership). Regularly review cyber risks in enterprise risk registers and ensure they are factored into strategic decisions (mergers, new technology adoption, etc.), acknowledging that cyber risk is business risk.
- Adopt and Tailor Resilience Frameworks: Leverage established frameworks (NIST CSF, ISO/IEC 27001, etc.) to create a structured cybersecurity program. Conduct a maturity assessment and address gaps by setting a multiyear roadmap with clear milestones. Emphasize not just protection, but detection, response, and recovery capabilities. Tailor these frameworks to your context – identify your "crown jewels" and ensure resilience plans specifically safeguard those critical assets and processes.
- Invest in People and Skills Development: Recognize that a skilled workforce is your first line of defense. Invest in continuous training – from technical cybersecurity staff to executive leadership

and boards – so everyone knows their role in maintaining security. Develop specialized skills through exercises and hands-on practice (e.g., red team/blue team drills). Build a culture that encourages reporting anomalies and near misses without blame. A cyber-aware and prepared team can prevent or contain incidents far more effectively than technology alone.

- Strengthen Technology Foundations (Zero Trust and Beyond): Modernize your IT and OT environments with resilience in mind. Implement Zero-Trust principles (verify everything, limit access privileges) across networks to compartmentalize potential breaches. Ensure robust data protection (encryption, backups) and use automation for prompt patch management. Leverage advanced analytics and threat intelligence platforms for real-time situational awareness of threats. Regularly test backup restoration and failover systems to guarantee they work when needed. Aim for an architecture where no single point of failure – whether a system or a supplier – can take down critical operations.
- Foster Partnerships and Community Defense: Don't go it alone. Join industry cyber forums or ISACs to exchange threat intel and best practices. Collaborate with government cybersecurity agencies for early warnings and incident support (e.g., participate in programs like CISA's JCDC or equivalent). Work with supply chain partners to raise the security baseline – this could mean contractual security requirements, joint incident drills with key vendors, or sharing assessment results. By building trust and communication channels with peers and authorities, you ensure that when an incident happens, you are not isolated – you have a support network to lean on.
- Plan, Practice, and Refine Incident Response: Develop a detailed IR and business continuity plan that covers a variety of scenarios (ransomware, data breach, system outage, etc.). Crucially, these plans should be practiced through regular drills that involve technical teams, management, and communications staff. Treat simulations as learning opportunities to improve coordination and speed. Ensure backup systems and manual workarounds are documented and tested for when digital systems fail. After any exercise or incident, conduct a postmortem and implement improvements. An organization that practices crisis response will handle real crises far better, reducing downtime and damage.
- Lead with Transparency and Accountability: Set the tone that cybersecurity and ethics go hand-in-hand. Be forthright about cyber risks and incidents with stakeholders – timely disclosure and honest communication build trust (and often stave off worse fallout). Hold your organization accountable through metrics and audits, and likewise hold yourself accountable – for instance, tie a portion of executives' performance evaluations to cybersecurity outcomes (such as meeting resilience goals). Demonstrate through actions that security is a core value: for example, don't delay critical security updates for convenience, and ensure compliance requirements are met even when they are hard. Leading by example in prioritizing security will permeate the organization and encourage every employee to do their part.

By implementing these recommendations, leaders can significantly enhance their cyber resilience. In doing so, they protect their enterprise's value and mission and contribute to the stability and trust in the interconnected systems underpinning modern life. In an era where digital and physical realms are inextricably linked, such leadership in cybersecurity is a vital component of technological stewardship and societal responsibility. The challenges are great, but as this chapter has shown, the knowledge, tools, and strategies to confront them are at hand – it remains for visionary leaders to apply them with determination and foresight.

Resources for Further Reading

Center for Internet Security (CIS), "The SolarWinds Cyber-Attack: What You Need to Know," Dec. 2020. [Online]. Available: https://www.cisecurity.org/solarwinds.

International Organization for Standardization (ISO), "ISO/IEC 27001: What's New in IT Security?" Oct. 2022. [Online]. Available: https://www.iso.org/contents/news/2022/10/new-iso-iec-27001.html.

International Organization for Standardization (ISO) and International Electrotechnical Commission (IEC), "ISO/IEC 27001:2022 – Information Security, Cybersecurity and Privacy Protection – Information Security Management Systems – Requirements," Oct. 2022. [Online]. Available: https://www.iso.org/standard/27001.

MITRE Corporation, "SolarWinds Compromise, Campaign C0024," [Online]. Available: https://attack.mitre.org/campaigns/C0024/.

MITRE Corporation, "Get Started | MITRE ATT&CK®," [Online]. Available: https://attack.mitre.org/resources/.

MITRE Corporation, "MITRE ATT&CK®," [Online]. Available: https://attack.mitre.org/.

National Institute of Standards and Technology (NIST), "NIST Cybersecurity Framework 2.0: Resource & Overview Guide," NIST SP 1299, Feb. 2024a. [Online]. Available: pubs.nist.gov/nistpubs/SpecialPublications/NIST.SP.1299.pdf.

National Institute of Standards and Technology (NIST), "The NIST Cybersecurity Framework (CSF) 2.0," NIST CSWP 29, Feb. 2024b. [Online]. Available: https://csrc.nist.gov/pubs/cswp/29/the-nist-cybersecurity-framework-csf-20/final.

National Institute of Standards and Technology (NIST), "NIST Cybersecurity Framework 2.0: Using the CSF Tiers – A Quick-Start Guide," NIST SP 1302, Oct. 2024c.

U.S. Congress, House of Representatives, Committee on Homeland Security, "Cyber Threats in the Pipeline: Lessons from the Federal Response to the Colonial Pipeline Ransomware Attack," Hearing, June 15, 2021. [Online]. Available: https://www.congress.gov/117/chrg/CHRG-117hhrg45310/CHRG-117hhrg45310.pdf.

U.S. Congress, Senate, Committee on Homeland Security and Governmental Affairs, "Threats to Critical Infrastructure: Examining the Colonial Pipeline Cyberattack," Hearing, June 8, 2021. [Online]. Available: https://www.govinfo.gov/content/pkg/CHRG-117shrg46569/pdf/CHRG-117shrg46569.pdf.

U.S. Department of Energy, Office of Cybersecurity, Energy Security, and Emergency Response (CESER), "Colonial Pipeline Cyber Incident," May 2021. [Online]. Available: https://www.energy.gov/ceser/colonial-pipeline-cyber-incident.

U.S. Department of Homeland Security, Cybersecurity and Infrastructure Security Agency (CISA), "Remediating Networks Affected by the SolarWinds and Active Directory/M365 Compromise," Dec. 2020. [Online]. Available: https://www./news-events/news/remediating-networks-affected-solarwinds-and-active-directorym365-compromise.

U.S. Department of Homeland Security, Cybersecurity and Infrastructure Security Agency (CISA), "The Attack on Colonial Pipeline: What We've Learned & What We've Done Over the Past Two Years," May 2023. [Online]. Available: https://www./news-events/news/attack-colonial-pipeline-what-weve-learned-what-weve-done-over-past-two-years.

U.S. Government Accountability Office (GAO), "SolarWinds Cyberattack Demands Significant Federal and Private-Sector Response," May 2021a. [Online]. Available: https://www.gao.gov/blog/solarwinds-cyberattack-demands-significant-federal-and-private-sector-response-infographic.

U.S. Government Accountability Office (GAO), "Colonial Pipeline Cyberattack Highlights Need for Better Federal and Private Sector Preparedness," May 2021b. [Online]. Available: https://www.gao.gov/blog/colonial-pipeline-cyberattack-highlights-need-better-federal-and-private-sector-preparedness-infographic.

U.S. Government Accountability Office (GAO), "Cybersecurity: Federal Response to SolarWinds and Microsoft Exchange Incidents," GAO-22-104746, Oct. 2021c. [Online]. Available: https://www.gao.gov/products/gao-22-104746.

World Economic Forum, "Global Cybersecurity Outlook 2024," Jan. 2024. [Online]. Available: https://www.weforum.org/publications/global-cybersecurity-outlook-2024/.

World Economic Forum, "Global Cybersecurity Outlook 2025," Jan. 2025. [Online]. Available: https://www.weforum.org/publications/global-cybersecurity-outlook-2025/.

22

Bridging the Digital Divide – Equity in the Age of Connectivity

Introduction: The Challenge of Inclusion in a Digital World

The Internet is often described as the great equalizer, offering unprecedented access to knowledge, services, and opportunities. Yet, for billions of people, the promise of digital inclusion remains unfulfilled. The digital divide – a gap between those with access to digital technologies and those without – reinforces global inequalities, limiting economic potential, healthcare access, and education opportunities.

Bridging this divide requires visionary leadership, cross-sector collaboration, and innovative solutions addressing exclusion's root causes. This chapter examines the barriers to digital inclusion, the strategies being implemented worldwide, and actionable frameworks for creating equitable access to the tools of the digital age.

The Reality of the Digital Divide

While global Internet penetration reached 73% in 2024, this figure hides stark disparities:

- **Geographical Divide:** Only 40% of sub-Saharan Africa's population is online, compared to 95% in North America and Europe [1].
- **Urban–Rural Divide:** Rural communities, particularly in developing nations, often lack broadband infrastructure.
- **Economic Divide:** High costs for devices and connectivity exclude low-income households from digital access.
- **Gender Divide:** Women in developing countries are 23% less likely to have access to the Internet than men, which limits their economic and educational opportunities [2].

These inequities have significant implications. Without digital access, marginalized populations face limited economic participation, education, and healthcare opportunities, widening existing socioeconomic gaps.

Case Study: India's Aadhaar Program

India's Aadhaar program highlights the potential of digital tools to drive inclusion. Launched in 2009, Aadhaar is the world's largest biometric identification system, providing unique Digital Identity (ID) numbers to over 1.3 billion residents by 2023 [3].

Key Impacts:

- **Financial Inclusion:** Aadhaar enabled millions of previously unbanked individuals to access financial services, reducing corruption and improving subsidy distribution.
- **Healthcare:** Digitized health records linked to Aadhaar streamlined medical services, particularly in rural areas.
- **Economic Empowerment:** Entrepreneurs used Aadhaar-linked credit systems to secure loans and expand businesses.

Challenges: Despite its success, Aadhaar faced criticism for privacy concerns, data breaches, and exclusion errors. These issues underscore the importance of safeguarding user rights in digital inclusion initiatives.

Case Study: Starlink and the Role of Satellite Internet

Satellite Internet offers a scalable solution for bridging connectivity gaps in underserved regions. SpaceX's Starlink, a low-Earth orbit satellite constellation, has connected remote areas worldwide, including rural Alaska and conflict-affected Ukraine.

Advantages:

- **Rapid Deployment:** Unlike fiber-optic infrastructure, satellite Internet can be deployed quickly in remote areas.
- **Global Reach:** Starlink's coverage spans oceans, mountains, and deserts, making it ideal for regions lacking traditional infrastructure.

Challenges:

- **Affordability:** At $110 per month, Starlink's service remains inaccessible for low-income households.
- **Environmental Concerns:** The proliferation of satellites raises concerns about orbital debris and sustainability [4].

Starlink's success highlights the need for public–private collaboration to make satellite Internet accessible and affordable for underserved populations.

Regional Challenges and Opportunities

Africa:

- Mobile broadband dominates connectivity, with innovations like M-Pesa revolutionizing financial services.
- Barriers include high data costs and limited infrastructure investment [5].

Latin America:

- Urban centers are well-connected, but rural areas face significant gaps.
- Initiatives like the World Bank's Digital Inclusion Program have piloted telemedicine and e-learning solutions with mixed results.

Southeast Asia:

- Countries like Indonesia have invested in large-scale fiber-optic networks like the Palapa Ring, but population density strains urban networks.
- Governments are accelerating 5G deployment to meet rising demand [6].

Leadership in Bridging the Divide

Effective leadership is critical to addressing the digital divide. Key priorities include:

1) **Policy and Regulation:** Governments must treat broadband as a public utility, similar to electricity and water. Policies that incentivize private investment in underserved areas are essential.
2) **Public–Private Partnerships (PPPs):** Collaboration between governments, tech companies, and Non-Governmental Organizations (NGOs) can accelerate deployment and reduce costs. For instance, Google's Project Loon used high-altitude balloons to provide Internet access during emergencies [7].
3) **Digital Literacy:** Access alone is insufficient without the skills to navigate digital tools. Community-driven education initiatives are essential, particularly for women and marginalized groups.
4) **Ethical Considerations:** Privacy, security, and equitable access must guide digital initiatives. Safeguards against data misuse and exclusion are critical for trust and sustainability.

Case Study: Kenya's Community-Driven Approach

Kenya has become a leader in using mobile technology to address the digital divide. The M-Pesa platform allows users to send and receive money via mobile phones, bypassing traditional banking systems. This innovation has expanded economic opportunities for millions, particularly in rural areas.

Key Strategies:

- **Community Ownership:** Local leaders deployed and promoted M-Pesa, ensuring cultural acceptance.
- **Scalable Models:** M-Pesa's low-cost structure allowed it to scale rapidly, reaching over 30 million users by 2024.

Kenya's approach demonstrates the power of leveraging local resources and partnerships to drive digital inclusion [8].

Framework for Bridging the Digital Divide

To create equitable access, leaders can adopt the following framework:

1) **Assess Needs and Barriers:** Conduct comprehensive assessments to identify gaps in infrastructure, affordability, and literacy.
2) **Develop Inclusive Policies:** Create policies prioritizing marginalized groups, incentivizing investment, and protecting user rights.
3) **Leverage Innovative Technologies:** Utilize solutions like satellite Internet, Internet of Things (IoT), and Artificial Intelligence (AI) to address unique regional challenges.

4) **Build Partnerships:** Foster collaboration across governments, businesses, and civil society to pool resources and expertise.
5) **Monitor and Adapt:** Regularly evaluate the impact of digital inclusion initiatives and adjust strategies based on feedback and outcomes.

Bridging the Digital Divide – Empowering a Connected Future

Introduction: The Divide in a Digital World

In 2024, 73% of the global population – 5.8 billion people – have Internet access, marking a significant leap in global connectivity over the last two decades. Yet, this figure conceals stark disparities. While urban centers in North America and Europe enjoy near-universal access, rural regions in sub-Saharan Africa, Latin America, and Southeast Asia often remain disconnected. For billions, the promise of the digital age – education, healthcare, and economic opportunities – remains unfulfilled [1].

The digital divide, the gap between those with access to digital technologies and those without, perpetuates global inequality. Addressing this challenge is not just a technological endeavor but an ethical imperative that requires leadership, collaboration, and innovation. This chapter examines the roots of the digital divide, explores case studies, and presents actionable strategies for fostering digital inclusion.

Understanding the Digital Divide

The digital divide manifests in three primary dimensions:

1) **Geographic Divide:** Rural and remote areas are often underserved due to infrastructure costs. For example, 95% of North America has broadband coverage, but only 40% of sub-Saharan Africa is connected [2].
2) **Economic Divide:** High device costs, data plans, and connectivity costs exclude low-income households. The average monthly broadband price in sub-Saharan Africa is $45, which is unaffordable for most families [3].
3) **Skills Divide:** Even when infrastructure exists, a lack of digital literacy prevents marginalized communities from leveraging technology effectively. Globally, women are 20% less likely than men to use the Internet [4].

These disparities compound existing inequalities, limiting access to education, healthcare, and economic mobility.

Case Study: India's Aadhaar System

India's Aadhaar initiative, the world's largest biometric ID system, demonstrates how digital infrastructure can empower underserved populations. Launched in 2009, Aadhaar assigns residents a unique identification number based on biometric and demographic data. By 2023, over 1.3 billion Indians had registered [5].

Impact:

- **Financial Inclusion:** Aadhaar enabled millions to open bank accounts and access government subsidies directly, reducing fraud and inefficiency.

- **Healthcare Access:** The system streamlined the delivery of medical benefits, particularly in rural areas.
- **Economic Empowerment:** Digital IDs allow entrepreneurs to secure credit, fostering grassroots economic growth.

Challenges: Despite its successes, Aadhaar faced criticism for data privacy concerns and exclusion errors, highlighting the need for robust governance and accountability.

Case Study: Starlink and Satellite Connectivity

Satellite Internet is emerging as a solution for remote and underserved regions. SpaceX's Starlink system, a constellation of low-Earth orbit satellites, has brought high-speed Internet to rural Alaska, isolated Pacific islands, and conflict zones like Ukraine [6].

Advantages:

- **Rapid Deployment:** Unlike fiber-optic cables, satellites can provide connectivity within weeks.
- **Global Reach:** Starlink offers coverage in regions with impractical traditional broadband infrastructure.
- **Scalability:** Its modular design supports continuous expansion.

Challenges:

- **Affordability:** At $110 per month, Starlink remains inaccessible for many low-income households.
- **Environmental Concerns:** The proliferation of satellites raises questions about orbital debris and long-term sustainability.

Starlink exemplifies how private-sector innovation can complement public efforts to close the digital divide.

Regional Insights: The Divide Across Continents

1) **Africa:** Mobile broadband dominates connectivity in Africa, with over 80% of Internet users accessing it via smartphones. Innovations like M-Pesa, a mobile payment platform, showcase the potential for leapfrogging traditional infrastructure. However, affordability and digital literacy remain significant barriers [7].
2) **Latin America:** While urban centers have high connectivity rates, rural areas often lack basic infrastructure. Initiatives like the World Bank's Digital Inclusion Program aim to bridge these gaps, but progress is slow due to weak governance and underfunded local projects [8].
3) **Southeast Asia:** Governments in Southeast Asia have embraced PPPs. Indonesia's Palapa Ring Project, a network of undersea fiber-optic cables, has connected 514 districts, though high population density in urban areas strains networks [9].

Strategies for Bridging the Divide

1) **Universal Broadband Access:** Governments must treat broadband as an essential utility. Key steps include subsidizing connectivity costs for low-income households and investing in

public infrastructure. For example, South Korea's universal broadband initiative has achieved near-total coverage [10].

2) **Public–Private Collaboration:** Partnerships between governments, telecom providers, and tech companies can accelerate deployment. Google's Project Loon, which used high-altitude balloons to provide Internet in disaster zones, exemplifies innovative collaboration [11].
3) **Digital Literacy Programs:** Access is meaningless without the skills to use technology effectively. Programs targeting women and marginalized communities, like the United Nations (UN) Women's EQUALS Global Partnership for Gender Equality in the Digital Age (EQUALS) initiative, are essential for inclusive growth [12].
4) **Ethical Frameworks:** Digital inclusion projects must prioritize privacy, data security, and community consultation. Missteps in privacy protection, as seen in India's Aadhaar program, can erode trust and undermine progress.

The Role of Emerging Technologies

Emerging technologies can play a pivotal role in closing the digital divide:

1) **Artificial Intelligence:** AI-powered tools can analyze gaps in connectivity, optimize resource allocation, and personalize digital literacy programs. For example, the AI platform LearnCloud tailors educational content for underserved populations [13].
2) **Blockchain:** Blockchain-based systems offer secure, decentralized solutions for identity verification, financial inclusion, and data sharing. Projects like the UN's Building Blocks have used blockchain to distribute aid in refugee camps [14].
3) **5G Networks:** The rollout of 5G networks promises faster, more reliable connectivity, particularly in urban centers. However, ensuring equitable access to 5G in rural areas requires significant investment [15].

Leadership in Digital Inclusion

Leaders play a critical role in bridging the digital divide. Effective strategies include:

1) **Inclusive Policymaking:** Governments must engage diverse stakeholders, including marginalized communities, in policymaking processes. For example, Kenya's Information and Communication Technology (ICT) Master Plan involved extensive public consultation to align digital goals with local needs [16].
2) **Investing in Local Solutions:** Tailoring solutions to regional contexts ensures scalability and sustainability. In Bangladesh, Grameenphone's community-based approach has successfully expanded rural connectivity [17].
3) **Promoting Global Cooperation:** Bridging the divide requires collaboration across borders – initiatives like the International Telecommunication Union's (ITU's) Connect 2030 Agenda foster international partnerships to promote universal connectivity [18].

Conclusion: Toward a Connected Future

The digital divide is one of the 21st century's defining challenges. Addressing it requires more than technological solutions – bold leadership, innovative collaboration, and a commitment to equity. By investing in connectivity, fostering digital literacy, and ensuring inclusive access, leaders can empower billions and unlock the full potential of the digital age.

Conclusion: A Shared Responsibility

Bridging the digital divide is a technological challenge and a moral and economic imperative. As digital connectivity becomes increasingly essential for education, healthcare, and economic participation, ensuring equitable access will define the inclusivity of the 21st century.

Leadership in this domain demands foresight, collaboration, and a commitment to equity. By prioritizing inclusive policies, investing in innovation, and empowering communities, we can build a digitally connected world that leaves no one behind.

References

Global Connectivity and Digital Inclusion

1 Internet World Stats, *Global Internet Penetration Report*, 2024.
2 Internet World Stats, *Global Internet Penetration Data*, 2024.
3 World Bank, *Digital Inclusion in Emerging Economies*, 2023.
4 World Bank, *Digital Inclusion Program in Latin America*, 2023.
5 International Telecommunication Union (ITU), *Measuring Digital Development: Facts and Figures*, 2024.
6 International Telecommunication Union (ITU), *Connect 2030 Agenda*, 2024.
7 Korean Ministry of Science and ICT, *Universal Broadband Report*, 2023.

Gender Equity and the Digital Divide

8 International Telecommunication Union (ITU), *Bridging the Gender Digital Divide*, 2023.
9 UN Women, *EQUALS: Closing the Gender Digital Divide*, 2023.
10 UN Women, *EQUALS Initiative Progress Report*, 2023.

National Infrastructure Projects

11 Indonesia Ministry of Communication, *Palapa Ring Project Report*, 2023.
12 Indonesian Ministry of Communication, *Palapa Ring Project Report*, 2023.
13 Kenyan Ministry of ICT, *ICT Master Plan*, 2023.

Frontier Connectivity Solutions

14 SpaceX, *Starlink Overview and Impact Report*, 2023.
15 SpaceX, *Starlink Coverage and Deployment*, 2023.
16 Google, *Project Loon Case Studies*, 2023.
17 Google, *Project Loon Overview*, 2023.
18 Ericsson, *5G Networks: Bridging Connectivity Gaps*, 2023.

Resources for Further Reading

Frontier Connectivity Solutions

Grameenphone, *Community-Based Connectivity Solutions*, 2023.

Digital Identity and Financial Inclusion

Aadhaar, *Transforming India Through Digital Identity*. Unique Identification Authority of India, 2023.
M-Pesa, *Financial Inclusion Through Mobile Technology*, 2023.
Safaricom, *M-Pesa Impact Analysis Report*, 2024.
Unique Identification Authority of India (UIDAI), *Aadhaar: Transforming India Through Digital Identity*, 2023.

Emerging Technologies for Inclusion

LearnCloud, *AI-Powered Educational Tools for Underserved Populations*, 2023.
UN World Food Programme, *Building Blocks: Blockchain for Aid Distribution*, 2023.

23

Crisis Leadership – Navigating Uncertainty with Resilience

Crises are inflection points that test not only the systems we rely on, but the leaders entrusted with guiding them. From hurricanes and earthquakes to pandemics, cyberattacks, and energy failures, the unpredictability of crisis events demands more than procedural response. It demands leadership that is fast, clear, adaptive, and grounded in public trust. The following chapter distills lessons from global crises into an integrated, narrative framework for understanding and applying crisis leadership.

When Systems Break, Leadership Is the First Line of Defense

In 2005, Hurricane Katrina made landfall along the Gulf Coast of the United States, exposing both the physical vulnerability of infrastructure and the institutional weakness of government coordination. The storm displaced hundreds of thousands and led to over 1,800 deaths, but it was the failure of leadership – not just wind and water – that turned a natural disaster into a humanitarian catastrophe. Despite clear meteorological warnings, Federal Emergency Management Agency (FEMA) failed to preposition emergency resources. Communication broke down between federal, state, and local governments. Evacuation plans were inadequate, especially for the elderly, disabled, and economically marginalized. Trust eroded as citizens waited for aid that never came.

By contrast, six years later, New Zealand faced its own calamity when the 2011 Christchurch earthquake struck, killing 185 people and leveling much of the city's infrastructure. Yet, the national and local response was swift, coordinated, and transparent. Prime Minister John Key addressed the nation daily. Local communities organized grassroots support like the Student Volunteer Army. Decision-makers incorporated feedback from those most affected and adjusted recovery plans accordingly. While the loss was tragic, the response became a model of resilience.

From Pandemic to Policy Failure: COVID-19 as a Global Leadership Audit

The COVID-19 pandemic exposed global fault lines in crisis governance. Some nations navigated the storm with resolve and coherence; others floundered amid political fragmentation, public mistrust, and communication failures. New Zealand, under Jacinda Ardern, pursued a "go hard, go early" containment strategy that combined swift lockdowns with empathetic public messaging.

Germany's Angela Merkel relied on science-backed transparency to foster compliance with health protocols.

Meanwhile, vaccine nationalism and unequal global access undermined collective progress. In the United States, Operation Warp Speed achieved remarkable technical success in vaccine development through unprecedented public–private collaboration, yet struggled with equitable distribution and public trust in early phases. The lesson is clear: crisis leadership succeeds when it pairs scientific rigor with empathetic, inclusive governance.

Decentralization, Technology, and Trust: The Puerto Rico and Estonia Lessons

Not all crises are acute. Some, like energy transitions or cyber threats, emerge through chronic underinvestment and systemic fragility. When Hurricane Maria devastated Puerto Rico in 2017, it was not the storm alone but the brittle, centralized energy infrastructure that collapsed. Millions went without electricity for months. But this catastrophe became a turning point. Local leaders partnered with Non-Governmental Organizations (NGOs), engineers, and communities to redesign energy systems around microgrids and renewables, prioritizing resilience over dependence.

Similarly, after a series of coordinated cyberattacks in 2007 that disabled Estonia's digital infrastructure, the nation undertook a comprehensive cyber resilience campaign. With support from North Atlantic Treaty Organization (NATO), Estonia created a digitally integrated yet secure society that today serves as a global model for e-governance.

Case in Contrast: The Texas Power Crisis of 2021

In February 2021, a historic winter storm froze large swaths of Texas, triggering a cascade of failures across the state's electrical grid. The state's energy system, isolated from national grids and unprepared for cold extremes, faltered. Millions lost power, heat, and water. A lack of communication, inadequate planning, and regulatory blind spots exacerbated the human toll.

This was not a failure of engineering but a failure of foresight and leadership. Despite prior warnings, key stakeholders failed to winterize infrastructure. Leadership must be willing to make uncomfortable investments in resilience long before they are tested.

Chapter 16: Leadership in Crisis – Navigating Complexity in the Age of Interdependence

Introduction: The Test of Leadership in Crisis

Crises have a unique way of testing the limits of human systems, exposing vulnerabilities, challenging assumptions, and requiring immediate action under extreme uncertainty. Whether it's a natural disaster, a public health emergency, or a cyberattack, these events demand more than operational efficiency; they require visionary leadership capable of fostering resilience, adaptability, and cooperation.

This chapter explores key lessons from notable global crises, focusing on leadership strategies that transcend the immediate moment to rebuild stronger and more resilient systems. By examining examples like New Zealand's response to earthquakes, FEMA's post-Katrina transformation, and the global COVID-19 vaccine rollout, this chapter provides actionable frameworks for leaders navigating complex, high-stakes situations.

The Role of Leadership in Crisis

Effective leadership in crisis involves more than managing resources and implementing policies. It requires:

1) **Decisiveness Under Uncertainty:** Leaders must act swiftly, even with incomplete information.
2) **Transparent Communication:** Clear, consistent messaging builds trust and minimizes misinformation.
3) **Adaptability:** As crises evolve, leaders must be willing to pivot strategies.
4) **Collaboration Across Sectors:** Coordinating efforts among governments, private sectors, and communities is essential.

Case Study: New Zealand's Earthquake Recovery

On February 22, 2011, Christchurch, New Zealand, was struck by a 6.3 magnitude earthquake, killing 185 people and causing widespread destruction. The response, led by Prime Minister John Key and local leaders, became a model for community-driven recovery.

Key Leadership Actions:

- **Establishment of CERA:** The Canterbury Earthquake Recovery Authority (CERA) integrated government, private sector, and community efforts, streamlining decision-making.
- **Community Empowerment:** Grassroots organizations, such as the Student Volunteer Army, were mobilized to provide direct aid, fostering local engagement and resilience.
- **Transparent Communication:** Daily briefings from government officials reduced uncertainty and misinformation.

Outcome: The recovery emphasized not just rebuilding infrastructure but restoring trust and social cohesion. The process demonstrated the power of participatory leadership in creating a resilient society [1].

Case Study: FEMA's Post-Katrina Transformation

Hurricane Katrina in 2005 exposed severe deficiencies in FEMA's crisis response capabilities, from inadequate resource allocation to poor coordination with state and local authorities. Images of stranded residents in New Orleans symbolized systemic failure.

Under the leadership of Administrator Craig Fugate, FEMA underwent a comprehensive transformation post-Katrina. Fugate introduced a "whole community" approach prioritizing inclusivity, preparedness, and collaboration.

Key Changes:

- **Prepositioning Resources:** Supplies were distributed to high-risk areas before disasters occurred.

- **Inclusive Planning:** Vulnerable populations, including the elderly and disabled, were explicitly considered in response plans.
- **Simulation and Training:** Large-scale exercises tested readiness and improved agency coordination.

Result: These reforms were tested during Hurricane Sandy in 2012. FEMA's improved coordination and proactive resource deployment significantly reduced response times and minimized damage [2].

Case Study: Global COVID-19 Vaccine Rollout

The COVID-19 pandemic presented one of modern history's most complex global crises. The rapid development and distribution of vaccines became a defining test of leadership, requiring coordination across governments, private sectors, and international organizations.

Successful Strategies:

- **Operation Warp Speed (USA):** A collaboration between the U.S. government, pharmaceutical companies, and academia accelerated vaccine development, reducing timelines from years to months.
- **Digital Health Systems (Rwanda):** Leveraging a robust digital infrastructure, Rwanda ensured equitable vaccine distribution and achieved one of the highest vaccination rates in Africa.
- **Unified Leadership (New Zealand):** Prime Minister Jacinda Ardern's clear, science-driven communication fostered public trust and compliance with vaccination campaigns.

Challenges:

- **Vaccine Hoarding:** Wealthy nations purchased large quantities of vaccines, leaving low-income countries reliant on donations.
- **Logistical Bottlenecks:** Weak supply chains in developing nations delayed vaccine delivery.

Key Takeaway: The global vaccine rollout underscored the importance of collaboration, transparency, and equity in crisis response. It also highlighted the need for investment in health infrastructure to prepare for future pandemics [3].

Leadership Frameworks for Crisis Management

From these case studies, key principles of effective crisis leadership emerge:

1) **Anticipation and Preparedness:** Use scenario planning and stress testing to identify vulnerabilities. For example, the World Economic Forum's Global Risk Report provides frameworks for assessing systemic risks [4].
2) **Transparent Communication:** Trust is the foundation of effective crisis response. Leaders must communicate consistently and honestly to manage public expectations.
3) **Adaptive Decision-Making:** Flexibility is critical as crises evolve. During the Christchurch earthquake recovery, New Zealand's leaders adjusted priorities based on community feedback.
4) **Collaboration Across Sectors:** Public–private partnerships can amplify resources and expertise. Operation Warp Speed exemplifies how collaboration accelerates outcomes.
5) **Focus on Equity:** Crisis responses must address the needs of marginalized populations to ensure comprehensive recovery.

Emerging Challenges and Innovations

As crises become more complex and interdependent, leaders must adapt to new realities:

1) **Climate-Driven Crises:** Rising sea levels and extreme weather events are increasing the frequency and severity of natural disasters. Leaders must integrate climate resilience into crisis planning.
2) **Cybersecurity Threats:** Cyberattacks on critical infrastructure, like the Colonial Pipeline ransomware incident, demonstrate the growing importance of digital resilience in crisis management [5].
3) **Global Interdependence:** Supply chain disruptions during the COVID-19 pandemic highlighted the vulnerabilities of globalized systems. Leaders must balance efficiency with redundancy.
4) **Artificial Intelligence (AI) and Predictive Analytics:** Tools like AI-powered risk modeling and real-time data analysis can enhance crisis response. For example, BlueDot's AI system detected the COVID-19 outbreak before it was officially reported [6].

Principles of Crisis Leadership: A Synthesis

From these cases, five principles emerge:

1) **Foresight and Preparedness:** Effective leaders conduct scenario planning, invest in stress testing, and preposition assets. Singapore's national pandemic plan, updated after severe acute respiratory syndrome (SARS), enabled early and effective COVID-19 response.
2) **Transparency and Trust-Building:** Clear, consistent communication – especially when uncertainty is high – builds public compliance and morale. Angela Merkel's science-driven addresses are a model.
3) **Inclusion and Equity:** Crises magnify inequalities. Plans that overlook marginalized populations compound harm. FEMA's Katrina response failed in part because it ignored the realities of vulnerable communities.
4) **Decentralization and Community Engagement:** Resilience scales locally. The Student Volunteer Army in Christchurch, the solar microgrids in Puerto Rico, and Estonia's digital civic engagement all point to the power of community.
5) **Collaboration Across Sectors:** No single entity can address systemic crises alone. Operation Warp Speed, despite flaws, showed how cross-sector collaboration accelerates innovation.

A Framework for Action

Leaders can translate these principles into action through a five-phase approach:

1) **Anticipate and Plan:** Use data modeling, simulations, and risk mapping.
2) **Engage Broadly:** Establish coalitions among government, private sector, and civil society.
3) **Communicate Relentlessly:** Own the narrative, acknowledge uncertainty, and be transparent.
4) **Act and Adapt:** Deploy rapid response but revise plans as new information emerges.
5) **Recover and Institutionalize Learning:** Conduct postcrisis reviews to codify best practices.

Conclusion: Resilience Is a Leadership Choice

Crises are inevitable, but catastrophic outcomes are not. The difference lies in leadership that prepares, includes, adapts, and communicates. Whether the shock is sudden or creeping, whether the risk is physical, biological, or digital, the task of leadership is to respond not only to what is visible but to anticipate what is coming. When done well, crisis leadership does more than save lives – it builds trust, advances equity, and strengthens the foundations of a resilient future.

Crises reveal leaders' true capabilities. Those who succeed are not necessarily the most well-resourced but the most adaptive, transparent, and collaborative. Whether rebuilding after an earthquake, transforming a flawed agency, or navigating a global pandemic, effective crisis leadership is defined by its ability to turn challenges into opportunities for systemic improvement.

By adopting the frameworks and principles outlined in this chapter, leaders can build resilient systems capable of withstanding future crises. In an age of uncertainty, the ability to anticipate, adapt, and collaborate will determine survival and the potential for thriving in a complex world.

References

1 Canterbury Earthquake Recovery Authority (CERA), "Final Report on Earthquake Recovery," New Zealand Government, 2016. [Online]. Available: https://www.cera.govt.nz.

2 FEMA, "Lessons Learned from Hurricane Sandy," Federal Emergency Management Agency, 2012. [Online]. Available: https://www.fema.gov.

3 WHO, "Global Vaccine Equity Progress Report," World Health Organization, 2023. [Online]. Available: https://www.who.int.

4 World Economic Forum, "Global Risk Report 2024," World Economic Forum, 2024. [Online]. Available: https://www.weforum.org.

5 U.S. Department of Homeland Security, "Colonial Pipeline Ransomware Incident Report," DHS, 2021. [Online]. Available: https://www.dhs.gov.

6 BlueDot, "*AI-Powered Outbreak Detection: COVID-19 Case Study*," BlueDot, 2020. [Online]. Available: https://www.bluedot.global.

Resources for Further Reading

J. Ardern, "Pandemic Leadership: Lessons from New Zealand," 2021. [Online]. Available: https://www.beehive.govt.nz.

Canterbury Earthquake Recovery Authority (CERA), "Christchurch Earthquake Recovery Report," New Zealand Government, 2016. [Online]. Available: https://www.cera.govt.nz.

FEMA, "Hurricane Katrina After-Action Report," *Federal Emergency Management Agency*, 2006. [Online]. Available: https://www.fema.gov.

International Atomic Energy Agency (IAEA), "Fukushima Daiichi Accident Report," 2015. [Online]. Available: https://www.iaea.org.

A. Merkel, "Science-Driven Leadership in Crisis," Federal Republic of Germany, 2020. [Online]. Available: https://www.bundesregierung.de.

National Academies of Sciences, Engineering, and Medicine, "Lessons from Operation Warp Speed," 2022. [Online]. Available: https://www.nationalacademies.org.

Singapore Ministry of Health, "COVID-19 Preparedness Report," 2021. [Online]. Available: https://www.moh.gov.sg.

U.S. Department of Health and Human Services, "Operation Warp Speed Overview," 2021. [Online]. Available: https://www.hhs.gov.

Part VI

Long-Range Leadership, Pivotal Technologies, Governance, and Lessons for the Future

24

Building Collaborative Ecosystems – Harnessing Collective Potential

Introduction: Why Collaborative Leadership Matters

The 21st century is marked by deeply interwoven global challenges: climate change, supply chain fragility, cyber-physical threats, systemic inequities, and rapid technological disruptions. These crises are not confined to a single sector, discipline, or nation. They are transboundary, complex, and accelerating. Addressing them requires more than technical solutions – it demands collaborative leadership to build purpose-driven ecosystems that span traditional boundaries.

From my experience leading transformative efforts – such as the Electric Power Research Institute (EPRI)/Department of Defense (DoD) Complex Interactive Networks/Systems Initiative (CIN/SI), and the post-9/11 Infrastructure Security Initiative (ISI) – I have come to understand that collaboration, if carefully architected, is a catalytic force for innovation, resilience, and equitable growth. These initiatives taught me that aligning stakeholders across diverse interests – academia, government, industry, and civil society – is possible if we ground our efforts in trust, transparency, and a shared mission.

This chapter offers a blueprint for building and sustaining collaborative ecosystems. Through real-world case studies – ranging from global frameworks like the Paris Agreement, to innovation engines like the Defense Advanced Research Projects Agency (DARPA) and the Intelligence Advanced Research Projects Activity (IARPA), to local models like Barcelona's smart city transformation – it articulates the principles, tools, and leadership strategies required to thrive in a world that demands systems thinking and collective action.

The Power of Collaborative Ecosystems

Collaborative ecosystems are strategic networks that harness the comparative strengths of distinct actors – governments, private sector entities, research institutions, communities, and nongovernmental organizations – to address complex challenges. Their design is complementary: no single participant can solve the problem alone, but they unlock breakthroughs that exceed the sum of their parts.

Core Benefits:

1) **Resource Efficiency:** Shared investment and pooled expertise reduce duplication and increase Return on Investment (ROI) on innovation.

2) **Innovation Acceleration:** Cross-pollination of ideas across sectors fosters novel approaches to entrenched problems.
3) **Risk Distribution:** Ecosystems buffer systemic risk through diversified participation, enhancing resilience.
4) **Scalable Impact:** Coordinated collaboration enables large-scale implementation, often spanning geographies and sectors.

Case Study: CIN/SI – From Fragmented Systems to Unified Resilience

In 1998, I launched the EPRI/DoD CIN/SI, an ambitious program that brought together over 28 universities, the U.S. military, EPRI stakeholders, and industry leaders to examine and address cascading failures in interdependent infrastructures – power grids, water systems, and communications networks.

Structure and Achievements:

- **Participants:** 100+ researchers, 240 graduate students, and postdocs.
- **Investment:** Over $20 million shared between DoD (36%) and EPRI (64%).
- **Deliverables:** 24 core technologies, 4,000+ research products synthesized in a 155-page final report [1].

Key Tools:

- **Probabilistic Risk Assessment (PRA):** Simulated cascading failures, revealing weaknesses later confirmed during the 2003 Northeast blackout [2].
- **Red Team Exercises:** Penetration testing on utilities nationwide uncovered Information Technology–Operational Technology (IT-OT) vulnerabilities – a precursor to issues seen in the Colonial Pipeline cyberattack [3].
- **Self-Healing Grid Technologies:** Enabled dynamic rerouting and autonomous fault isolation – now foundational to smart grid resilience.
- **Dynamic Risk Models:** Supported long-term planning under evolving threat conditions.
- **Knowledge Frameworks:** Disseminated learnings across stakeholder organizations through workshops and interoperable tools.

CIN/SI's success lay in its systemic orientation and rigorous cross-sector coordination. It demonstrated how collaborative ecosystems, with the right governance and trust mechanisms, could rapidly translate research into applied resilience.

Transition to Post-9/11 Infrastructure Security (ISI)

After the September 11 attacks, I was asked to establish and lead EPRI's post-9/11 ISI, building on the CIN/SI foundation. Complemented by the revamped Enterprise Information Security (EIS) Year 2000 (Y2K) programs, this effort focused on critical infrastructure vulnerabilities and recovery strategies.

Notable Outcomes:

- **Red Team Testing:** Deployed at scale across North America, identifying key failure points in grid and telecom systems.

- **Recovery Transformers:** Modular, rapidly deployable high-voltage transformers to shorten restoration time during catastrophic outages.
- **Secure Networks:** Air-gapped, encrypted communication systems hardened against cyber and electromagnetic pulse (EMP) threats [4].

The ISI effort predated and shaped the resilience strategies used in response to the 2003 blackout and subsequent emerging threats.

The Paris Agreement: A Global Collaboration Milestone

The 2015 Paris Agreement remains the most comprehensive international collaboration to combat climate change. Signed by 196 countries, it commits the world to limiting global temperature rise to below 2 °C, aiming for 1.5 °C.

Design Strengths:

1) **Shared Vision:** Unified climate targets across geopolitical divides [1].
2) **Flexibility:** Nationally Determined Contributions (NDCs) allow local customization while maintaining global accountability [2].
3) **Inclusive Engagement:** Scientists, Non-Governmental Organizations (NGOs), cities, and corporations – like Microsoft and Apple – contributed commitments [4].
4) **Transparency Mechanisms:** Periodic reviews and reporting systems enforce accountability [3].

Persistent Gaps:

- Unequal funding for vulnerable countries.
- Nonbinding enforcement mechanisms.
  These challenges indicate the need for more adaptive and equitable frameworks within global ecosystems.

DARPA and IARPA: Innovation Through Mission-Driven Collaboration

DARPA and IARPA are examples of structured, innovation-focused ecosystems. Their success rests on bold missions, interdisciplinary teams, and tolerance for failure.

DARPA's Ecosystem Model:

- **Clear Missions:** Global Positioning System (GPS), Advanced Research Projects Agency Network (ARPANET), stealth tech, autonomous vehicles – all emerged from focused problem statements [5].
- **Risk Acceptance:** Failures are essential to radical innovation [6].
- **Public–Private Collaboration:** From academia to startups, partnerships foster commercialization pipelines [7].

While smaller, IARPA follows similar principles in predictive analytics, Artificial Intelligence (AI), and cybersecurity, demonstrating that even highly classified domains benefit from collaborative openness within defined boundaries.

Smart Urbanism: Barcelona as a Living Lab

Barcelona's transformation into a "smart city" reveals how local ecosystems can drive global innovation. By integrating technology firms, city planners, academics, and citizens, Barcelona improved sustainability, services, and economic inclusion.

Key Achievements:

1) **IoT Integration:** Reduced electricity use by 15% through demand optimization [8].
2) **Participatory Governance:** Citizens contributed to project design, increasing public trust [9].
3) **Economic Uplift:** Attracted over €2 billion in smart city investment by 2023 [10].

Barcelona's iterative, feedback-driven model exemplifies adaptive collaboration and places civic empowerment at the heart of ecosystem success.

Principles of Effective Collaborative Ecosystems

From global treaties to municipal innovation, several principles recur:

1) **Shared Vision:** An articulated mission catalyzes alignment.
2) **Stakeholder Mapping:** Identify and integrate actors based on influence, interest, and contribution potential [11].
3) **Trust and Transparency:** Frequent, open communication and equitable information sharing are essential [3, 12].
4) **Adaptability:** Ecosystems must evolve with context – technological, political, or environmental [10].
5) **Inclusive Leadership:** Empowering underrepresented voices strengthens relevance and equity.

Leadership Framework for Collaboration

Effective ecosystem leadership requires intentional design. Leaders must:

1) **Set Clear, Measurable Objectives:** Ground collaboration in well-defined outcomes.
2) **Build Multidisciplinary Teams:** Encourage intersections across engineering, policy, sociology, and business.
3) **Foster Continuous Communication:** Use collaboration platforms and in person engagement to sustain momentum.
4) **Deploy Enabling Technologies:** Tools like blockchain, AI, and digital twins enhance visibility and accountability.
5) **Evaluate and Adjust:** Use data to track progress, adjust approaches, and share lessons learned.

Tools and Tactics for Ecosystem Builders

1) **Public–Private Partnerships (PPPs):** Blend funding and expertise, as seen in DARPA and ISA [5, 13].
2) **Co-creation Workshops:** Facilitate collaborative problem-solving and design thinking.
3) **Scenario Planning:** Anticipate multiple futures; stress test strategies against uncertainty.
4) **Digital Infrastructure:** Real-time platforms enhance collaboration across time zones and silos.

Actionable Insights for Leaders

- **Invest in Collaborative Hubs:** Institutions like the Masdar Institute fuel innovation by convening stakeholders [14].
- **Institutionalize Incentives:** From tax credits to fast-track grants, incentives align market behavior with mission-driven collaboration.
- **Empower Local Voices:** Ensure representation and buy-in from impacted communities.
- **Scale What Works:** Replicate successful pilots using modular, open-source frameworks.

Challenges to Collaboration

1) **Funding Fragmentation:** Ecosystem longevity requires sustained investment.
2) **Misaligned Incentives:** Metrics must reinforce, not hinder, cross-sectoral success.
3) **Political and Cultural Friction:** Bridging norms across institutions and borders takes time and trust.
4) **Short-Termism:** Long-term challenges (e.g., climate resilience) require governance models that transcend election cycles or market pressures.

Conclusion: A Blueprint for Collective Resilience

The collaborative ecosystems profiled here – from national security research to climate policy, from city design to critical infrastructure protection – prove that transformation is possible when we lead beyond silos. Visionary leaders must now institutionalize collaboration as a core competency, not a side initiative.

To navigate 21st-century uncertainty, we must build inclusive, adaptive, and outcome-oriented coalitions. The world needs leaders who design with systems, partner with purpose, and lead with humility. The time for action is now.

References

1 M. Amin, *Complex Interactive Networks/Systems Initiative (CIN/SI): Final Summary Report*, Electric Power Research Institute (EPRI), Palo Alto, CA, Mar. 2003. [Online]. Available: https://massoud-amin.umn.edu/research/publications.

2 U.S.-Canada Power System Outage Task Force, *Final Report on the August 14, 2003 Blackout in the United States and Canada*, Apr. 2004. [Online]. Available: https://www.energy.gov/sites/default/files/oeprod/DocumentsandMedia/BlackoutFinal-Web.pdf.

3 Cybersecurity and Infrastructure Security Agency (CISA), "Colonial Pipeline Cyber Incident," U.S. Dept. of Homeland Security, 2021. [Online]. Available: https://www.cisa.gov/news-events/alerts/2021/05/11/ongoing-cybersecurity-threats.

4 Microsoft Corp., "Microsoft will be carbon negative by 2030," Jan. 2020. [Online]. Available: https://www.microsoft.com/en-us/sustainability/emissions-impact-dashboard.

5 DARPA, *The DARPA Model for Transformative Innovation*, Defense Advanced Research Projects Agency, Arlington, VA, 2022. [Online]. Available: https://www.darpa.mil/work-with-us/the-darpa-model.

6 National Academy of Sciences, *Interdisciplinary Research and DARPA's Success*, Washington, D.C., 2023. [Online]. Available: https://www.nationalacademies.org/news/2023/01/darpa-model-and-cross-sector-innovation.

7 J. Gertner, *The Idea Factory: Bell Labs and the Great Age of American Innovation*. New York, NY: Penguin Books, 2022.

8 Barcelona City Council, "Barcelona Smart City Projects," 2023. [Online]. Available: https://ajuntament.barcelona.cat/digital/en/smart-city.

9 P. Townsend, "Citizen Engagement in Smart City Development," *Smart Cities World*, May 2022. [Online]. Available: https://www.smartcitiesworld.net.

10 European Commission, "Barcelona: Leading the Smart City Agenda," *EU Smart Cities Marketplace*, 2023. [Online]. Available: https://smart-cities-marketplace.ec.europa.eu.

11 R. Mitchell, B. Agle, and D. Wood, "Toward a theory of stakeholder identification and salience: Defining the principle of who and what counts," *J. Bus. Ethics*, vol. 112, no. 1, pp. 45–56, 2022.

12 DARPA, "Risk-Tolerant Innovation Strategies," DARPA, 2022. [Online]. Available: https://www.darpa.mil/about-us/innovation.

13 International Solar Alliance (ISA), "Global Solar Energy Roadmap," 2023. [Online]. Available: https://isolaralliance.org.

14 Masdar Institute, "Renewable Energy Innovation at Masdar," 2023. [Online]. Available: https://www.masdar.ae.

Resources for Further Reading

International Energy Agency (IEA), "Global Energy System Resilience and Redundancy," 2021. [Online]. Available: https://www.iea.org/reports/resilience-in-energy-systems.

International Energy Agency (IEA), "Update on Nationally Determined Contributions (NDCs)," *IEA Climate Tracker*, Apr. 2023. [Online]. Available: https://www.iea.org/reports/climate-pledges.

United Nations Framework Convention on Climate Change (UNFCCC), "The Paris Agreement," 2015. [Online]. Available: https://unfccc.int/process-and-meetings/the-paris-agreement/the-paris-agreement.

U.S. Department of Defense and EPRI, *Post-9/11 Infrastructure Security Initiative – Confidential Findings Summary*. Electric Power Research Institute, 2003. [Selected findings publicly released].

U.S. Department of Energy, "Smart Grid Investment and Modernization Outcomes," Office of Electricity, 2023. [Online]. Available: https://www.energy.gov/oe/technology-development/smart-grid.

World Resources Institute (WRI), "Enhancing Accountability in the Paris Agreement," 2023. [Online]. Available: https://www.wri.org/initiatives/paris-agreement-tracker.

25

Collaborative Leadership Across Regions – Building Ecosystems for Global Impact

Introduction: The Importance of Regional Collaboration

Effective collaboration is crucial for addressing global challenges, including climate change, energy transition, urbanization, and digital transformation. Yet, the approaches to collaboration vary significantly across regions due to differences in resources, priorities, and challenges. By examining these differences, leaders can extract valuable lessons from successful practices while addressing persistent challenges to build adaptive, resilient ecosystems.

This chapter integrates regional assessments of collaborative ecosystems with key insights into leadership styles that foster transformative impact. Case studies from North America, Europe, Asia, Africa, South America, and Australia illustrate the diverse global strategies and highlight actionable insights for fostering collaboration.

Regional Collaborative Ecosystems

North America: Innovation and Resilience in Energy and Infrastructure

North America exemplifies the interplay of technological advancements, economic strength, and shared challenges, particularly in the areas of energy and infrastructure resilience. However, disparities in priorities and fragmented systems hinder unified progress.

Energy Transition

In 2023, the United States derived 79% of its energy from fossil fuels, reflecting its persistent reliance on traditional energy sources. In contrast, Canada leads the continent in renewable energy production, with hydropower contributing over 60% of its energy mix. Mexico, heavily reliant on oil exports, has shown promise in solar and wind energy, driven by government reforms. Yet, challenges like grid fragmentation – exemplified by Texas's 2021 grid failure that left millions without power – underscore the need for a cohesive regional approach.

Ambitious projects, such as the U.S.'s offshore wind farms, which target 30 GW by 2030, and large-scale storage facilities like California's Moss Landing Energy Storage Project, demonstrate innovation in renewable energy. Cross-border integration among Canada, the United States, and Mexico remains essential for enhancing regional energy security and efficiency.

Infrastructure Modernization

Aging infrastructure is a critical challenge. In the United States, 43% of roads are in poor condition, and over 45,000 bridges are structurally deficient. According to the American Society of Civil Engineers (ASCE), a funding gap of $2.6 trillion is projected by 2029. Innovative solutions, such as the Internet of Things (IoT)-enabled traffic systems in Los Angeles and Maryland's Purple Line public-private partnership (PPP), showcase how technology and new financing models can drive progress.

Resilient design solutions, such as flood barriers, have been implemented in New York City to mitigate climate risks, while California's seismic retrofits protect against earthquakes. These examples highlight the importance of integrating resilience into modernization efforts.

Europe: Leadership in Data Privacy and Urban Sustainability

Europe stands out for its commitment to ethical governance and sustainable development, offering global leadership in collaborative innovation.

Data Privacy and Governance

The General Data Protection Regulation (GDPR) sets a global benchmark for data privacy, compelling organizations worldwide to adopt compliant practices. While fostering consumer trust, GDPR compliance often presents financial and logistical challenges for smaller enterprises. The regulation's implementation underscores Europe's leadership in prioritizing ethical data governance alongside technological advancement.

Urban Innovation

European cities excel in integrating urban planning with sustainability. Copenhagen's green mobility initiatives reduced car traffic by 25%, while Amsterdam's smart grid systems achieved a 15% reduction in energy consumption. Despite these successes, challenges remain in scaling and coordinating smart city initiatives across administrative boundaries. The collaborative leadership of municipal governments, private companies, and community organizations has proven essential in overcoming these barriers.

Asia: Rapid Urbanization and Technological Pioneering

Asia combines economic ascent with technological innovation, but disparities in adoption highlight the need for inclusive development strategies.

5G Leadership

China leads the world with 2.7 million 5G base stations, enabling advancements in healthcare, logistics, and autonomous vehicles. South Korea's deployment of 5G technologies supports urban solutions, such as smart transportation systems. However, connectivity gaps in rural areas limit the broader impact of these technologies. Addressing these disparities requires targeted infrastructure investments and regional cooperation.

Manufacturing Automation

Japan's adoption of robotics has reduced production costs by 15%, enhancing global competitiveness. Similarly, India's thriving startup ecosystem fosters innovation in Artificial Intelligence (AI) and IoT, driving regional growth. Balancing automation with job creation and environmental sustainability remains a critical challenge for both economies. Governments, industries, and educational institutions must collaborate to ensure equitable progress.

Africa: Decentralized Solutions for Inclusion

Africa's decentralized solutions' ingenuity addresses critical gaps in energy, finance, and agriculture. Despite resource constraints, scalable models effectively meet local needs, offering lessons for global implementation.

Energy Access

Off-grid solar systems, such as those developed by M-KOPA, have electrified over one million households, reducing reliance on kerosene and improving air quality. However, underdeveloped energy infrastructure in rural areas constrains broader economic growth. Expanding these solutions requires investment in scalable technologies and supportive regulatory frameworks.

Mobile Banking

Platforms like M-Pesa enable financial inclusion for over 50 million users, transforming access to banking services. However, reliable Internet connectivity and robust regulatory frameworks are essential for expanding these systems sustainably. Collaborative efforts among governments, technology providers, and financial institutions are critical to addressing these challenges.

South America: Balancing Growth and Sustainability

South America's rich natural resources fuel economic growth, but environmental challenges highlight the need for sustainable practices.

Sustainable Agriculture

Brazil's precision farming techniques increased agricultural yields while reducing water usage by 25%. Chile's sustainable wine production has set industry benchmarks. However, soil degradation and deforestation pose significant threats to the long-term viability. Regional collaboration in research, technology sharing, and policy enforcement is essential to address these challenges.

Tech Entrepreneurship

Startups like Nubank in Brazil drive fintech innovation, attracting international investment. Similarly, Colombia's burgeoning e-commerce sector reflects regional entrepreneurial strength. Infrastructure deficiencies and regulatory hurdles remain barriers to scalability, underscoring the importance of supportive ecosystems.

Australia: Climate Resilience and Decentralized Innovation

Australia's investments in climate resilience offer valuable lessons for global adaptation strategies.

Climate Resilience

Australia leads in per capita solar adoption and has developed advanced wildfire prediction systems and water management tools. These technologies mitigate risks and highlight the importance of cross-sector collaboration in addressing climate challenges. Collaborative leadership has played a crucial role in integrating these innovations into public policy and community planning.

Collaborative Leadership in Practice

Leadership in collaborative ecosystems relies on styles tailored to regional needs:

1) **Collaborative Leadership** promotes inclusivity and shared accountability, which are essential for Tesla's microgrid deployment in Puerto Rico.
2) **Transformational Leadership:** Inspires systemic innovation, as demonstrated by Google's Objectives and Key Results (OKRs).
3) **Servant Leadership:** Prioritizes community needs, as seen in Habitat for Humanity's housing solutions.
4) **Adaptive Leadership:** Balances flexibility and innovation, as shown by General Electric's (GE's) Industrial IoT transformation.
5) **Systems Thinking Leadership:** Addresses interdependencies, critical for Australia's climate resilience strategies.

Case Studies in Collaborative Leadership in Action

Leadership often falters not because of a lack of ideas but due to disjointed efforts and a failure to align around shared goals. Uniting diverse stakeholders toward common objectives is paramount in complex systems. Collaborative leadership thrives in these environments, fostering inclusivity, shared accountability, and the convergence of expertise. This chapter examines the transformative impact of collaborative leadership, illustrating its principles through verified case studies and actionable insights.

Driving Innovation Through Unified Leadership

Google: Objectives and Key Results (OKRs): In the early 2000s, Google faced the challenge of managing rapid growth while avoiding organizational silos. John Doerr introduced the OKRs framework to address this. This system aligned leadership across diverse teams by setting unified, measurable goals, allowing for clarity and focus in a fast-paced environment.

During the development of Google's AI platform, OKRs played a critical role in integrating ethical considerations into engineering and user experience processes. The framework ensured that innovation was upheld and advanced responsibly. Google's success demonstrates how transformational and collaborative leadership, combined with clear accountability, can focus efforts and achieve breakthroughs.

The Struggle and Success of Digital Transformation

General Electric: Industrial IoT and Predix: Under Jeff Immelt's leadership, GE embarked on a bold digital transformation to become a leader in the Industrial IoT. Immelt's vision required aligning traditionally independent engineering, IT, and operations teams to build the Predix platform, enabling predictive analytics for industrial operations.

Although the transformation faced resistance and complexity, the collaborative approach eventually positioned GE as a pioneer in industrial IoT. This case highlights the power of adaptive leadership to address large-scale changes, bridging human and technical challenges to achieve shared objectives.

Leadership in Crisis Recovery

Hurricane Maria: Puerto Rico's Energy Rebuild: In 2017, Hurricane Maria devastated Puerto Rico, leaving millions without power. The collapse of the island's centralized grid underscored its vulnerability. Recovery required collaboration across government agencies, nonprofits, and private companies.

Tesla deployed solar microgrids with battery storage, addressing immediate energy needs while creating sustainable, decentralized solutions. Local leaders ensured that these efforts aligned with community needs, showcasing the synergy between servant and participatory leadership. This initiative demonstrated how technology and inclusive decision-making could transform crisis recovery.

Housing Through Collaboration

Habitat for Humanity: Inclusive Leadership for Impact: Habitat for Humanity addresses housing insecurity by fostering collaboration among local volunteers, future homeowners, and corporate partners. This inclusive leadership model delivers affordable housing and builds community ownership.

By 2020, Habitat had served over 5.3 million people globally through its housing solutions. The organization's participative and servant leadership approaches demonstrate how collaboration can drive large-scale social impact, emphasizing shared goals and accountability.

Multidisciplinary Leadership in Healthcare

Mayo Clinic: Collaborative Care for Complex Cases: The Mayo Clinic's multidisciplinary leadership model integrates expertise from various specialties to deliver tailored patient care. This approach ensures comprehensive treatment plans, enhancing outcomes in complex cases.

By valuing diverse perspectives and fostering collaboration, the Mayo Clinic has become a global leader in healthcare innovation. This case highlights the importance of collaborative leadership in environments that require precision and adaptability.

Key Leadership Styles for Success

1) **Collaborative Leadership:** Fosters inclusivity and shared accountability. Examples include Tesla's microgrid deployment and Habitat for Humanity's housing projects.

2) **Transformational Leadership:** Inspires systemic change and innovation, as seen in Google's OKRs and GE's digital transformation.
3) **Servant Leadership prioritizes community needs and ethical decision-making, as** evident in Tesla's work in Puerto Rico.
4) **Participative Leadership:** This approach encourages diverse input and co-creation, as demonstrated by Habitat for Humanity and the Mayo Clinic.
5) **Adaptive Leadership:** Empowers teams to navigate uncertainty and innovate, exemplified by GE's IoT initiatives.

Why Collaborative Leadership Matters

These stories reveal a universal truth: leadership succeeds when it unites diverse teams with a clear purpose. Collaborative leadership ensures resources, expertise, and vision converge to achieve meaningful outcomes. Without alignment, even the best ideas can falter, leaving opportunities unrealized.

Conclusion: Leadership for Complex Systems

Global challenges require integrated ecosystems driven by inclusive, innovative leadership. These case studies demonstrate that nations can establish networks that foster sustainability, equity, and resilience by leveraging their regional strengths.

Collaborative leadership exemplifies the future of impactful leadership in complex systems. Leaders can navigate uncertainty and achieve sustainable progress by aligning stakeholders, fostering inclusivity, and emphasizing shared goals. This approach ensures that innovation and impact are both ethical and equitable.

Resources for Further Reading

American Society of Civil Engineers (ASCE), "Infrastructure Report Card," 2022.

Australian Renewable Energy Agency (ARENA), "Solar Energy Adoption," 2023.

Brazilian Ministry of Agriculture, "Precision Farming Initiatives," 2023.

California Independent System Operator (CAISO), "Energy Imbalance Market Annual Report," 2023.

Copenhagen City Council, "Green Mobility Initiatives," 2023.

"Collaborative Leadership in Action: Case Studies & Insights," Voltage Control, 2022. [Online]. Available: https://voltagecontrol.com/articles/successful-collaborative-leadership-in-action-case-studies-and-real-world-examples/.

J. Doerr, *Measure What Matters*. Portfolio Penguin, 2018a.

J. Doerr, *Measure What Matters: How Google, Bono, and the Gates Foundation Rock the World with OKRs*, Portfolio Penguin, 2018b. [Online]. Available: https://www.amazon.com/Measure-What-Matters-Google-Foundation/dp/0525536221.

S. Girod and L. Duke, IMD Business School, 2019a.

S. Girod and L. Duke, "*Digital Transformation at GE: Shifting Minds for Agility*," IMD Business School, 2019b. [Online]. Available: https://www.imd.org/research-knowledge/strategy/case-studies/digital-transformation-at-ge-shifting-minds-for-agility/.

Habitat for Humanity, Annual Report, 2020a.

Habitat for Humanity, "Homes, Communities, Hope, You | FY2020 Annual Report," 2020b. [Online]. Available: https://www.habitat.org/multimedia/annual-report-2020/.

International Energy Agency (IEA), "Hydropower in Canada," 2023.

Japan Robotics Association, "Manufacturing Automation Trends," 2023.

M-KOPA Solar, "Off-Grid Energy Solutions," 2023.

Moss Landing Energy Storage, "Battery Energy Storage Project," 2023.

North American Electric Reliability Corporation (NERC), "Interconnections and Regional Energy Security," 2023.

Safaricom, "M-Pesa: Transforming Lives," 2022.

S. Sweeney, *New Labor Forum*, 2018a.

S. Sweeney, "Puerto Rico's energy future: Keeping power with the people," *New Labor Forum*, vol. 27, no. 2, pp. 18-25, Apr. 2018b. [Online]. Available: https://newlaborforum.cuny.edu/2024/04/11/puerto-ricos-energy-future-keeping-power-with-the-people/.

26

Global Leadership Frameworks – Aligning Regional Efforts for Global Progress

Introduction: The Need for Unified Leadership

The challenges of the 21st century – climate change, geopolitical instability, economic inequality, and technological disruption – require leadership frameworks that can align regional, national, and institutional efforts into cohesive strategies. Without unified global approaches, these interconnected challenges are exacerbated by inefficiencies, duplication of efforts, and competing interests, ultimately hindering progress. Effective global leadership frameworks must emphasize practical, actionable collaboration while respecting regional differences and leveraging localized expertise.

This chapter examines key components of successful global leadership frameworks. By analyzing multilateral agreements, collaborative networks, and technological innovations, this chapter offers actionable guidance for building systems that are scalable, equitable, and resilient.

Lessons from Global Leadership Initiatives

The Paris Agreement: Balancing Diversity and Unity

The Paris Agreement is a landmark example of global cooperation, uniting nearly 200 nations to limit global warming to below 2 °C. The agreement balances global and regional priorities through mechanisms such as Nationally Determined Contributions (NDCs), which allow nations to tailor their climate commitments to their unique circumstances.

Key Mechanisms:

- **NDCs:** Enable nations to address their specific needs while contributing to global objectives.
- **Transparency Frameworks:** Require periodic reporting and verification of progress, fostering accountability and trust.
- **Financial Support:** Provide funding to developing nations to bridge equity gaps in climate adaptation and mitigation.

Lessons Learned:

- Clear, measurable goals that respect local autonomy enhance participation.
- Transparent reporting systems build trust and promote compliance.
- Persistent challenges include securing sufficient funding and addressing noncompliance from major emitters.

The United Nations Sustainable Development Goals (SDGs): Comprehensive and Inclusive Framework

The SDGs are an ambitious global framework adopted in 2015. They encompass 17 goals that address poverty, inequality, education, and environmental sustainability. The SDGs' comprehensive approach ensures that all sectors are interconnected, fostering systemic solutions to global challenges.

Key Success Factors:

- **Interconnected Goals:** Recognize the interdependence of issues such as education, gender equality, and climate change.
- **Broad Stakeholder Engagement:** Mobilize governments, private enterprises, and communities to collectively address challenges.
- **Monitoring Systems:** The Global Indicator Framework tracks progress, identifies gaps, and guides policy adjustments.

Lessons Learned:

- A systems-thinking approach is vital for addressing complex global challenges.
- Regional collaboration ensures equitable progress across diverse economic and political contexts.
- Implementation remains uneven, with developing nations requiring greater support to meet targets.

The Belt and Road Initiative (BRI): Leveraging Economic Connectivity

China's BRI illustrates how large-scale economic and infrastructure projects can foster global connectivity. The BRI has strengthened ties between Asia, Europe, and Africa by investing in transportation, energy, and trade systems.

Key Success Factors:

- **Infrastructure Investment:** Multi-billion-dollar projects, including railways, ports, and energy grids, create shared economic value.
- **Technology Integration:** Leveraging AI and blockchain improves transparency and operational efficiency in cross-border trade.
- **Regional Alignment:** Collaborative projects align the interests of participating nations, fostering interdependence and mutual benefit.

Lessons Learned:

- Infrastructure-led strategies can enhance economic interconnectivity but must include environmental and debt sustainability safeguards.
- Effective global frameworks require equitable terms for developing nations to prevent economic over-dependence or exploitation.

Principles for Global Leadership Frameworks

From these initiatives, several actionable principles emerge for creating effective global leadership frameworks:

1) **Shared Vision and Objectives:**
   Establish clear, measurable goals that are universally understood and accepted by all parties involved. For instance, the SDGs provide a framework that aligns global and local priorities under a common vision.
2) **Inclusive Stakeholder Engagement:**
   Actively involve governments, private sectors, NGOs, and marginalized communities. The Paris Agreement's multi-stakeholder approach enhanced both legitimacy and effectiveness.
3) **Scalable and Adaptive Structures:**
   Design frameworks that evolve in response to changing circumstances. For example, the periodic review of NDCs under the Paris Agreement ensures adaptability over time.
4) **Transparency and Accountability:**
   Implement systems that ensure open reporting and evaluation of progress. Blockchain-based solutions can enhance transparency in areas like climate finance and international trade.
5) **Balance Autonomy with Coordination:**
   Respect regional differences while aligning efforts toward shared objectives. As the Paris Agreement shows, flexible commitments facilitate participation without undermining collective goals.

Tools and Strategies for Global Leadership

Building effective global frameworks requires pragmatic tools and strategies that leaders can readily apply.

Digital Platforms for Collaboration

Technologies such as AI and blockchain can support real-time monitoring and transparency. For instance, the Climate Action Tracker evaluates and reports on global climate commitments, highlighting performance gaps and areas for improvement.

Public–Private Partnerships (PPPs)

Combining government resources with private-sector innovation can accelerate progress. DARPA's partnership model showcases the transformative potential of public–private collaboration in advancing technologies such as GPS and the Internet.

Regional Hubs for Global Challenges

Establishing centers of excellence at the regional level can localize global initiatives, thereby enhancing their effectiveness. For example, Africa's Kigali Cooling Efficiency Program focuses on sustainable solutions tailored to the continent's needs.

Data-Driven Decision-Making

Big data tools can inform policy decisions, allocate resources efficiently, and measure outcomes. The Global Indicator Framework of the SDGs exemplifies how robust data systems can guide strategic action and accountability.

Leadership Styles for Global Collaboration

Effective leadership in global collaboration hinges on adopting styles that drive unity and innovation:

1) **Transformational Leadership:** Inspire a shared vision that galvanizes stakeholders to work toward collective goals.
2) **Collaborative Leadership:** Cultivate partnerships and prioritize mutual benefits, fostering trust and long-term engagement.
3) **Systems-Thinking Leadership:** Approach problems holistically, recognizing interdependencies and optimizing solutions across sectors.
4) **Adaptive Leadership:** Remain flexible in changing conditions and ensure frameworks are resilient to future disruptions.
5) **Servant Leadership:** Focus on addressing the needs of vulnerable and underrepresented populations to ensure equitable participation and representation.

Conclusion: Leadership for a Connected World

Addressing global challenges requires frameworks that align diverse regional efforts under unified, strategic approaches. Initiatives such as the Paris Agreement, the SDGs, and the BRI demonstrate the importance of striking a balance between local autonomy and global coordination. Leaders can foster scalable and equitable systems by applying principles such as inclusivity, adaptability, and transparency. Integrating digital tools, PPPs, and regional hubs further ensures that solutions are pragmatic and actionable.

In an interconnected world, leadership must embrace collaboration as the cornerstone of progress. Effective frameworks address immediate challenges and lay the foundation for sustainable, long-term global resilience.

Resources for Further Reading

Belt and Road Initiative, "Infrastructure Development Across Asia," 2023.
Climate Action Tracker, "Evaluating Global Climate Commitments," 2023.
DARPA, "Public-Private Innovation Models," 2023.
Kigali Cooling Efficiency Program, "Sustainable Cooling Solutions," 2023.
United Nations Framework Convention on Climate Change (UNFCCC), "Paris Agreement," 2015.
United Nations, "Sustainable Development Goals," 2015.

27

Regional Dynamics – Transforming Challenges into Opportunities

Introduction: Regional Leadership in a Global Context

While global frameworks provide overarching strategies, regional dynamics ultimately shape the implementation of these objectives. Each region faces unique challenges, from resource allocation and infrastructure deficits to sociopolitical constraints. These challenges, however, often present opportunities for innovation and regional leadership. Effective regional strategies are critical for driving sustainability, resilience, and equity in global progress.

This chapter examines how various regions transform their distinct circumstances into drivers of growth and development. Through detailed case studies, we analyze regional successes in renewable energy, urban mobility, and digital transformation, offering actionable insights for leaders to replicate and adapt these strategies.

Regional Leadership Case Studies

North America: Leading in Innovation and Resilience

Renewable Energy Transition

North America is at the forefront of renewable energy adoption, driven by advancements in solar, wind, and energy storage. Despite this progress, outdated infrastructure and energy inequities continue to challenge the region.

- **Texas Wind Power:** With 37 GW of installed capacity, Texas leads North America in wind energy production. However, the 2021 grid failure exposed vulnerabilities in integrating renewable energy with traditional systems.
- **Canada's Hydropower Leadership:** Over 60% of Canada's electricity is generated from hydropower, exemplifying the country's strength in renewable energy.

Lessons Learned

Investments in grid modernization and cross-border energy partnerships are crucial for integrating renewable sources and enhancing reliability. Regional collaboration can harmonize the integration of renewable energy, addressing vulnerabilities exposed by isolated systems.

AI in Healthcare

North America is also revolutionizing healthcare through AI applications, which enhance diagnostics, predictive analytics, and telemedicine.

- **Case Study: IBM Watson:** By integrating AI, IBM Watson increased cancer diagnostic accuracy by 30%.
- **Teladoc Health:** During the COVID-19 pandemic, Teladoc Health expanded rural healthcare access by 150%.

Lessons Learned

To bridge healthcare disparities, AI-driven innovations must be complemented by initiatives that ensure equitable access. Adopting robust data privacy frameworks modeled on Europe's General Data Protection Regulation (GDPR) can bolster trust in AI systems.

Europe: Sustainability and Ethical Governance

Circular Economies

Europe is leading the way in developing circular economy frameworks, with a focus on resource efficiency and sustainable production.

- **The Netherlands' Circular Economy Plan:** Aims to achieve 100% circularity by 2050, reducing resource dependency and waste.

Lessons Learned

Circular economies require strong regulatory support and collaboration between the public and private sectors. Coordination across regions, such as EU-wide recycling initiatives, enhances scalability and standardization.

Data Privacy and Smart Cities

Europe's GDPR has set a global benchmark for ethical data use, while cities like Copenhagen are integrating sustainability with technology.

- **Copenhagen's Green Initiatives:** The city's carbon-neutral metro system has reduced emissions by 70%.

Lessons Learned

Ethical leadership fosters public trust, while smart city technologies must prioritize inclusivity and citizen engagement to maintain trust. Regulatory frameworks, such as GDPR, demonstrate how governance can drive both innovation and accountability.

Asia: Urbanization and Digital Connectivity

Smart City Development

Asia's rapid urbanization has spurred innovations in smart city infrastructure, particularly in economies such as Singapore and South Korea.

- **Singapore's Smart Nation Initiative:** Integrates IoT, AI, and big data to optimize urban living, from healthcare to transportation.

Lessons Learned

Governance frameworks must ensure equitable access to smart city solutions. Regional collaborations, such as the ASEAN Smart Cities Network, are essential for knowledge sharing and scalability.

5G Leadership

Asia also leads the 5G revolution, enabling transformative applications in healthcare, logistics, and autonomous vehicles.

- **China's 5G Infrastructure:** With 2.7 million base stations, China's 5G network supports real-time manufacturing and urban planning connectivity.

Lessons Learned

Investments in 5G infrastructure must prioritize long-term sustainability and equitable access, particularly in underserved rural areas. Public–private partnerships are key to accelerating adoption.

Africa: Decentralized Innovation and Resilience

Mobile Banking Revolution

Africa's mobile banking sector exemplifies how decentralized technologies can leapfrog traditional infrastructure gaps.

- **Kenya's M-Pesa:** Integrates mobile banking with renewable energy financing, empowering over 50 million users.

Lessons Learned

Tailored solutions leveraging mobile technology can scale rapidly, providing financial inclusion in underserved regions. Combining decentralized energy systems with financial tools further enhances accessibility.

Renewable Energy Access

Off-grid solar solutions address Africa's energy challenges, particularly in rural areas.

- **M-KOPA Solar:** It has provided affordable solar systems to over one million households, reducing reliance on kerosene and improving air quality.

Lessons Learned

Public–private partnerships are essential for scaling off-grid energy systems. Localized solutions must align with broader regional development goals.

South America: Sustainability and Agricultural Innovation

Climate-Smart Agriculture

South America addresses food security and environmental degradation through the adoption of precision agriculture and sustainable practices.

- **Brazil's Low-Carbon Agriculture Plan:** Integrates satellite monitoring and AI to optimize farming practices and reduce water usage by 25%.

Lessons Learned

Agtech solutions must strike a balance between productivity and environmental conservation. Partnerships between governments, tech companies, and farmers are essential for scalability.

Tech Entrepreneurship

The region's startups are driving innovation in fintech and e-commerce.

- **Nubank in Brazil:** Has attracted international investment, expanding financial access to underserved communities.

Lessons Learned

Fostering entrepreneurial ecosystems requires regulatory support and access to capital. Aligning tech innovation with social impact maximizes regional benefits.

Australia: Climate Resilience and Local Innovation

Climate Adaptation Technologies

Australia's frequent wildfires and water scarcity have driven investments in climate resilience.

- **Advanced Wildfire Prediction Systems:** Have minimized environmental and economic impacts.

Lessons Learned

Regional solutions must address immediate challenges while contributing to global frameworks. Collaborating with indigenous communities ensures the development of culturally informed adaptation strategies.

Effective Leadership Styles for Regional Collaboration

To address these diverse challenges, effective leadership styles are critical:

1) **Adaptive Leadership:** Adjust strategies to fit local contexts, recognizing that solutions must be tailored to each region's unique challenges.
2) **Collaborative Leadership:** Build partnerships that pool resources and expertise, as seen in cross-border energy policies in North America.
3) **Servant Leadership:** Prioritize the needs of marginalized communities, ensuring that development initiatives are inclusive and equitable.
4) **Systems-Thinking Leadership:** Approach challenges holistically, integrating interconnected solutions across sectors.

Conclusion: Regional Foundations for Global Progress

Regional leadership is the cornerstone of global progress. By addressing unique challenges with tailored solutions, regions contribute to the collective goals of sustainability, resilience, and equity. From North America's renewable energy initiatives to Africa's mobile innovation and Asia's smart city advancements, localized efforts have the potential to drive transformative change.

Leaders must build on these regional strengths, fostering cross-border collaboration to create a more interconnected and sustainable world. By adopting adaptive, collaborative, and inclusive leadership styles, they can ensure that regional innovations contribute to global progress.

Resources for Further Reading

Australian Climate Council, "Wildfire Adaptation Strategies," 2023.
Brazil Ministry of Agriculture, "Low-Carbon Agriculture Plan," 2023.
Canadian Hydropower Association, "Hydropower and Canada's Energy Mix," 2023.
China 5G Authority, "5G Infrastructure Report," 2023.
Copenhagen City Council, "Green Metro System," 2023.
Dutch Ministry of Infrastructure, "Circular Economy 2050," 2023.
IBM Watson, "AI in Healthcare," 2023.
M-KOPA Solar, "Off-Grid Energy Solutions," 2023.
Nubank, "Financial Inclusion through Fintech," 2023.
Safaricom, "M-Pesa Case Study," 2023.
Singapore Smart Nation, "IoT in Urban Development," 2023.
Texas Renewable Energy, "Wind Power Statistics," 2023.

28

Regional Pathways to Sustainable Development – Strategic Approaches and Solutions

Introduction: Regional Strategies for a Global Future

Sustainable development is a shared global objective, yet its path is not uniform. Each region faces unique geographical, cultural, and socioeconomic challenges that demand tailored strategies. When effective, these strategies offer lessons that transcend borders, enabling nations to learn from each other and implement context-specific solutions. This chapter explores how regions have pioneered innovative approaches to sustainable development, focusing on how these solutions have evolved from local ingenuity and resilience to impactful, globally relevant models.

Effective leadership is pivotal in creating, managing, and scaling these strategies. Leaders who understand the nuances of their regions while maintaining a vision for the global implications of their work are critical to achieving lasting progress. Using adaptive, collaborative, and transformational leadership styles, they can navigate complexity and inspire actionable solutions.

Case Study: Technological Disruption in the U.S. Labor Market: Lessons from History for AI

The history of technological disruption in the U.S. labor market reveals critical patterns that offer valuable lessons for understanding the future impact of artificial intelligence (AI). Over the last century, the U.S. labor market underwent two profound structural shifts driven by general-purpose technologies (GPTs) such as steam power, electricity, and digital computing.

From 1880 to 1960, the share of workers in agriculture plummeted, dropping from 41% to just 6%. This was due to advances in mechanization and agricultural productivity. From 1960 to 1980, the second major transformation occurred as workers transitioned out of blue-collar factory jobs and into office-based roles, including administrative support and white-collar professions. Blue-collar employment fell by 10% points during this period, continuing to decline steadily into the 21st century.

These shifts were driven by the disruptive power of GPTs, which redefined industries and reshaped the labor market over decades. The rise of AI, as the authors argue, has the potential to function as a GPT on the scale of steam power or electricity. Early evidence already suggests that AI is contributing to labor market disruptions in several notable ways:

1) **Labor Market Polarization is Declining:** While the 2000s saw a rise in low- and high-paying jobs and a hollowing out of middle-paid occupations, this pattern has shifted. Low- and middle-paid occupations have declined sharply, while high-paying knowledge work has grown.

2) **Growth in Low-Wage Service Jobs Has Stalled:** Historically resilient areas of employment, such as low-wage service jobs, are now seeing limited growth. Technological advancements, particularly in automation and robotics, are impacting these sectors.
3) **STEM Employment is Rising Rapidly:** The employment share of Science, Technology, Engineering, and Mathematics (STEM) occupations has grown by more than 50% since 2010, driven by the proliferation of software development and computer-related roles.
4) **Decline in Retail Jobs:** Retail sales employment has decreased by 25% in the past decade, largely due to advancements in e-commerce and automation in retail operations.

These patterns illustrate both the challenges and opportunities presented by AI. For example, while AI is expected to reduce back-office administrative roles, it will likely increase demand for strategic decision-making and complex problem-solving in managerial and professional roles. In this way, AI is not merely automating existing jobs but amplifying the expectations of knowledge workers.

Why This Analysis Matters in the Context of Workforce Resilience: The examples of technological disruption provided by David Deming, Christopher Ong, and Lawrence H. Summers offer a powerful historical framework for understanding how today's workforce can adapt to the rapid rise of AI. As with previous GPTs, AI is not simply replacing jobs but restructuring the labor market in ways requiring individuals and organizations to adapt continuously.

This analysis supports several key recommendations in this chapter:

- **Upskilling and Reskilling Workers:** The historical shift from agriculture to manufacturing and from manufacturing to office work demonstrates the importance of developing new skills as technological capabilities evolve. Today, this means focusing on STEM skills, data literacy, and critical thinking.
- **Fostering Adaptability:** The slowing pace of occupational churn since 1990 underscores the importance of proactive, rather than reactive, workforce planning. Investments in education, training, and adaptable organizational practices are vital for resilience.
- **Leveraging Human Creativity and Decision-Making:** As AI takes on routine tasks, the demand for knowledge workers who can analyze complex problems and generate innovative solutions will likely grow. Leaders must focus on cultivating these capabilities within their teams.

Building Resilience for the AI-Driven Economy: As the authors note, historical patterns of technological disruption demonstrate both the risks and opportunities of innovation. By understanding the trajectory of past GPTs, such as steam power and electricity, we can better prepare for the workforce disruptions caused by AI. Chapters 25, 26, and 9 explore how individuals, organizations, and policymakers can respond to these shifts, offering strategies to ensure technological change leads to broad-based prosperity.

Africa: Unlocking Development Through Local Innovation

Africa's journey toward sustainable development highlights its ability to leverage resources and creativity to overcome significant structural challenges. The continent has long faced energy scarcity, agricultural vulnerability, and financial exclusion, yet it continues to produce innovative, scalable solutions.

Renewable Energy for Underserved Communities

In the remote villages of Rwanda, where access to electricity was once considered an unattainable luxury, the implementation of solar mini-grids has transformed daily life. These systems, powered by renewable energy, illuminate homes, enable small businesses to thrive, and improve access to education. The electrification of over 300,000 rural households is a testament to the power of community-driven initiatives combined with modern technology.

The leadership driving these projects is marked by adaptability and a focus on inclusivity. Local leaders have worked closely with international organizations to understand the unique needs of these communities, ensuring that the solutions provided align with cultural practices and economic realities.

Agriculture and Climate-Resilient Practices

Traditional farming methods in Nigeria's arid regions struggled to withstand erratic rainfall and prolonged droughts. However, introducing drought-resistant crop varieties has revolutionized agricultural practices, boosting yields by 20% and reducing food insecurity. Farmers have become active participants in the process, contributing local knowledge to enhance the success of these initiatives.

Transformational leadership has been critical here. Leaders have fostered a collective sense of ownership and pride in the progress achieved by inspiring communities to embrace new techniques and technologies.

Fintech for Financial Inclusion

In Ghana, the financial inclusion story is one of ingenuity meeting necessity. Mobile credit platforms such as FarmDrive have empowered small-scale farmers who were previously excluded from traditional banking systems. These platforms provide access to microloans tailored to the agricultural calendar, enabling farmers to invest in seeds, equipment, and fertilizers.

Collaborative leadership, where fintech entrepreneurs partner with local governments and international investors, has driven this success. Leaders have created ecosystems where technology meets grassroots needs, ensuring financial tools are accessible to even the most remote populations.

South America: Bridging Natural Wealth and Economic Inequality

South America's rich biodiversity and vast natural resources present opportunities and challenges. The region's path toward sustainable development is shaped by its need to balance economic growth with environmental conservation.

Climate-Smart Agriculture

Brazil's Low-Carbon Agriculture Plan exemplifies this balance. By integrating satellite monitoring with precision farming, the initiative has reduced water usage by 25% while enhancing productivity. Farmers now utilize data to make informed decisions, optimizing crop yields while minimizing environmental impact.

Effective leadership in this context has been characterized by systems thinking. By understanding the interconnectedness of agriculture, technology, and climate change, leaders have developed strategies that address multiple challenges simultaneously.

Green Infrastructure Development

In Medellín, Colombia, urban reforestation projects have transformed the city into a model of green infrastructure. Air pollution has dropped by 11%, and public health outcomes have improved. The city's leadership adopted a citizen-centric approach, involving residents in the planning and implementation of these projects.

Servant leadership has been at the core of Medellín's transformation. By prioritizing the well-being of citizens and engaging them as active participants, leaders have created a sense of community ownership that ensures the longevity of these initiatives.

Renewable Energy Expansion

Chile's Atacama Desert, with its abundant sunshine, has become a hub for solar energy. Today, 30% of the country's electricity is generated from solar power, a remarkable achievement that positions Chile as a global leader in renewable energy.

Transformational leadership has driven this success. Leaders with a clear vision for a sustainable future have inspired investment and innovation, turning a barren desert into a source of national pride and progress.

Asia: Driving Growth Through Technological Leadership

Asia's rapid urbanization and technological advancements make it a global innovation powerhouse. However, this growth presents challenges such as resource management, environmental degradation, and equitable development.

Urban Mobility and Connectivity

Expanding metro systems in cities like Delhi and Bengaluru have significantly reduced traffic congestion and improved urban productivity in India. Commuters save weekly hours, and carbon emissions have decreased as more people shift from private vehicles to public transit.

Adaptive leadership has been essential in navigating the complexities of urban planning. Leaders have worked to integrate infrastructure projects with broader economic and environmental goals, ensuring sustainable outcomes.

Water Management in Arid Regions

Israel's mastery of desalination technology and drip irrigation is a model for regions grappling with water scarcity. By supplying 85% of its drinking water through desalination, Israel has turned a potential crisis into a competitive advantage.

Visionary leadership has played a crucial role here. By prioritizing long-term planning and technological investment, leaders have ensured water security for future generations while exporting their expertise to other nations.

Digital Transformation for Inclusive Growth

In the Philippines, digital literacy programs have enabled small businesses to access global markets. As a result, economic participation has increased by 15%, lifting countless families out of poverty.

Collaborative leadership has been vital. Leaders have created ecosystems that empower individuals with the skills needed to thrive in a digital economy by bringing together educators, tech companies, and local governments.

North America: Innovating for Resilience

North America's approach to sustainable development is defined by its technological leadership and focus on resilience in the face of environmental and infrastructure challenges.

Resilient Energy Systems

Canada's investment in hydrogen energy exemplifies its commitment to diversifying its energy portfolio. By 2030, hydrogen is expected to account for 20% of the country's energy production, reducing reliance on fossil fuels and enhancing energy security.

Systems thinking leadership has been instrumental in aligning public policy, private investment, and technological innovation to achieve this vision.

Infrastructure Modernization

The U.S. Bipartisan Infrastructure Law has allocated $550 billion to modernize transportation systems, reduce emissions, and enhance resilience against extreme weather. Projects like high-speed rail and smart highways are reshaping how Americans move and connect.

Pragmatic leadership, focusing on actionable solutions and measurable outcomes, has driven these efforts. Leaders have prioritized collaboration across political and institutional boundaries to accelerate progress.

Regenerative Agriculture

The regenerative farming movement in the U.S. Midwest has sequestered over 5 million tons of carbon annually while improving soil health. These practices combat climate change and ensure food security for future generations.

Servant leadership, emphasizing land stewardship, has inspired farmers to adopt sustainable practices that benefit their communities and the planet.

Effective Leadership Styles for Sustainable Development

- **Transformational Leadership:** Inspiring innovation and fostering a shared vision for sustainable futures.
- **Collaborative Leadership:** Building partnerships across sectors to pool resources and expertise.
- **Adaptive Leadership:** Tailoring solutions to local contexts while aligning with global goals.
- **Systems-Thinking Leadership:** Addressing interconnected challenges with holistic strategies.
- **Servant Leadership:** Prioritizing the needs of communities and the environment for equitable, inclusive progress.

Conclusion: Regional Lessons for Global Progress

The diversity of strategies highlighted in this chapter underscores a universal truth: sustainable development requires a combination of local ingenuity and global collaboration. By embracing effective leadership styles and fostering partnerships across regions, leaders can transform challenges into opportunities and build a resilient, equitable world.

Resources for Further Reading

Brazilian Ministry of Agriculture, “Low-Carbon Agriculture Plan.”
Chilean Energy Ministry, “Solar Energy in the Atacama Desert.”
FarmDrive, “Mobile Financial Services for Farmers.”
Israeli Water Authority, “Desalination and Drip Irrigation.”
Medellín Environmental Authority, “Urban Reforestation Initiatives.”
Nigerian Agricultural Ministry, “Drought-Resistant Crops in Nigeria.”
Rwanda Energy Group, “Solar Mini-Grid Projects.”
U.S. Department of Transportation, “Bipartisan Infrastructure Law.”

29

Regional Innovations in Sustainable Development – Emerging Strategies and Case Studies

Introduction: Pioneering Regional Approaches to Sustainability

Sustainability is an evolving narrative shaped by each region's unique cultural, geographical, and socioeconomic contexts. Around the world, countries are finding innovative ways to address environmental, social, and economic challenges. While often localized, these efforts contribute to broader global solutions, offering practical insights and replicable models. This chapter presents a comprehensive view of how sustainable development is being driven by regional leadership and ingenuity. It draws from detailed case studies across Asia-Pacific, Europe, China, Russia, and North America, highlighting the strategies, tools, and leadership approaches that have produced measurable results.

Asia-Pacific: Urban Innovations for Sustainable Growth

The Asian Development Bank (ADB) has launched numerous projects targeting the vulnerabilities of urban environments across the Asia-Pacific. The Urban Climate Change Resilience Trust Fund stands out for its scale and impact. The initiative has funded over 25 projects across 20 cities that have improved public transportation networks, upgraded flood defenses, and constructed resilient housing for climate-affected communities. These efforts have reached more than two million people. Leadership in these projects demonstrated adaptability, responding dynamically to challenges like flooding, energy insecurity, and urban sprawl. Local partnerships were central to implementation, with municipalities, NGOs, and citizens actively defining and refining solutions. This collaboration ensured that interventions reflected the community's needs and were sustainable over time.

Europe: Enhancing Regional Attractiveness and Innovation

In Abruzzo, Italy, policymakers implemented a comprehensive regional sustainability monitoring framework grounded in locally developed indicators. These metrics, covering sectors such as biodiversity, energy, waste management, and transportation, allowed for the continuous evaluation and improvement of development strategies. This approach, developed in collaboration with research institutions and local stakeholders, ensured that regional planning remained evidence-based and adaptable to emerging trends. The system allowed for real-time adjustments to policy and prioritized transparency and public participation. Abruzzo's leadership used this framework to effectively integrate environmental priorities into economic and social development agendas.

In Appenzell Ausserrhoden, Switzerland, the challenge of rural decline and economic restructuring prompted a transdisciplinary initiative. Local leaders collaborated with researchers, farmers, and civic organizations to codevelop regional strategies emphasizing ecological sustainability and cultural heritage. Among the initiatives were promoting sustainable agricultural practices, local food systems, and ecotourism. Community workshops were used to design programs that ensured stakeholder buy-in and regional relevance. The resulting initiatives were grounded in the region's identity and reflected a shared vision for sustainable rural prosperity.

East Germany offers a distinct model centered around the role of research institutions in driving regional innovation. In regions impacted by postindustrial decline, Fraunhofer Institutes have acted as innovation intermediaries, linking universities, businesses, and municipal governments. One initiative led to the development and deployment of passive house prototypes that reduced energy use by up to 30%. These efforts extended beyond technical implementation; they included vocational training, public engagement, and policy advocacy to institutionalize energy-efficient construction practices.

China: Scaling Sustainability from Rural to Urban

In Lincang, a mountainous area in Yunnan Province, sustainability efforts focused on balancing ecological conservation with rural development. Authorities implemented a regional sustainability index identifying gaps in agricultural productivity, ecosystem services, and income generation. The index informed a series of targeted investments in eco-agriculture, forest management, and renewable energy. Over 1.5 million residents have benefited from improved infrastructure and higher agricultural yields. Local leadership adopted a strategic approach, coordinating efforts across different administrative levels and engaging local cooperatives and academic experts to guide implementation.

Dongtan was conceptualized as one of the world's first fully sustainable cities on Chongming Island near Shanghai. While its original timeline was delayed, Dongtan's planning process had a lasting influence on urban development in China. The city's master plan included energy self-sufficiency, integrated transit networks, high-density green building clusters, and on-site waste processing systems. The design emphasized resource cycles and low-carbon urban form, integrating land-use planning with energy and water systems. Though the full vision was not realized, lessons from Dongtan informed later eco-city developments.

In Hangzhou, the development of the Hangzhou Metropolitan Circle represents a successful fusion of digital innovation and sustainable urban planning. Over the past decade, Hangzhou has invested heavily in digital infrastructure, including artificial intelligence systems that manage traffic, utilities, and emergency response. These systems have reduced energy use and improved public service efficiency. The city's digital economy has grown by 25%, partly driven by policies that align economic development with environmental targets. This model has been recognized for linking smart governance with sustainability outcomes.

Russia: Sustainable Land Use and Education

Russia's Comprehensive Plan for Rural Development, launched in 2019, sought to address long-standing land use and infrastructure inequalities. By 2023, the program had invested more than $20 billion in clean water systems, transportation upgrades, and housing across 1,000 rural villages. These efforts reached approximately three million residents. A hallmark of the plan was its collaborative structure, integrating the efforts of federal ministries, regional authorities, private investors, and agricultural cooperatives. By simultaneously addressing legal, technical, and social barriers, the plan delivered tangible improvements in rural quality of life.

At Ural Federal University, sustainability education has become a cornerstone of technical training and community engagement. Engineering students must participate in applied research projects, many involving wastewater treatment, renewable energy integration, and energy efficiency audits for local industries. Over 500 students participate annually, and projects often result in pilot deployments. These initiatives are led by faculty who combine mentoring with systems-level project design, preparing graduates to contribute meaningfully to regional and national sustainability challenges.

North America: Infrastructure and Economic Resilience

The Hartford-Springfield Knowledge Corridor, spanning Connecticut and Massachusetts, exemplifies regional infrastructure planning and economic development alignment. The initiative was launched to revitalize an economically struggling corridor by linking investments in public transit with housing, workforce training, and innovation hubs. Key projects included the expansion of high-speed rail, transit-oriented development near station areas, and creating cross-state economic zones to attract businesses and retain skilled workers. The effort united the state governments, universities, metropolitan planning organizations, and community groups. Its success was recognizing the interdependence of transportation, labor markets, and land use.

Conclusion: Embracing Regional Innovations for Sustainable Futures

The regional case studies presented in this chapter illustrate the diversity of effective pathways toward sustainability. Whether grounded in rural revitalization, digital innovation, or transdisciplinary collaboration, these efforts share common traits: they are locally informed, systems-oriented, and driven by inclusive leadership. They highlight that sustainability is not a one-size-fits-all endeavor but a process of continuous learning, adaptation, and innovation rooted in specific contexts. By studying these examples, leaders elsewhere can draw lessons for creating sustainable, resilient, and inclusive futures.

Resources for Further Reading

Asian Development Bank, "Urban Climate Change Resilience Trust Fund," 2023.
Q. Cheng *et al.*, "Sustainability interactions in mountainous areas," *Environ. Sci. Pollut. Res.*, vol. 31, 2024.
Chinese Urban Planning Institute, "Dongtan Eco-City Design," 2023.
Fraunhofer Society, "Energy-Efficient Housing Prototypes," 2024.
Hangzhou Smart City Development Report, "Digital Economy Metrics," 2023.
NADO, "Aligning Strategies to Maximize Impact: Transportation and Economic Development," 2023.
Russian Ministry of Agriculture, "Comprehensive Plan for Rural Development," 2023.
SpringerLink, "Constructing Regional Development Strategies: A Case Study Approach," 2024.
UNECE, "Supporting Sustainability: The Case Study of the Abruzzo Regional Indicators," 2023.
Ural Federal University, "Industrial Wastewater Purification Projects," 2023.

30

Regional Pathways to Resilience – Navigating Environmental and Economic Challenges

Introduction: The Imperative for Regional Resilience

In an era marked by climate change, geopolitical uncertainty, and economic disruptions, resilience is no longer a luxury but an existential necessity. As regions globally confront unique challenges, they deploy innovative strategies to navigate risks while fostering long-term stability. This chapter examines how regional approaches to resilience integrate environmental sustainability, economic vitality, and social well-being.

The stories of North America, Europe, Asia-Pacific, Africa, South America, and Australia reveal a tapestry of resilience strategies shaped by local contexts. These examples demonstrate how effective leadership, tailored interventions, and technology-enabled solutions safeguard communities and systems against future shocks.

North America: Balancing Renewable Energy Transition and Grid Reliability

The transition to renewable energy in North America has brought both progress and vulnerabilities. California's ambitious renewable energy policies, such as the Renewable Portfolio Standard (RPS), mandate 60% renewable energy by 2030. However, the 2020 rolling blackouts highlighted the fragility of aging grid infrastructure.

Case Study: California's Renewable Energy Challenges

California's investment in solar, wind, and energy storage has been transformative, reducing greenhouse gas emissions by 35% since 2000. Yet, the state's energy grid, designed for a centralized fossil-fuel-based system, struggled to manage the variability of renewable sources during extreme heat waves. Technologies like microgrids and Artificial Intelligence (AI)-enabled predictive maintenance are now being deployed to address these gaps.

Key Leadership Style: Systems Leadership

California's systems leaders emphasize interconnectedness, balancing immediate energy needs with long-term goals. They work with utilities, regulators, and local governments to build resilience into the grid.

Insight

Modernizing grid systems with advanced technologies and collaborative planning ensures reliability while supporting renewable energy transitions.

Europe: Redesigning Systems for Resilience

Europe has been a global leader in adopting resilience-focused frameworks, addressing energy security and resource efficiency through circular economies and diversification strategies.

Case Study: The Netherlands and Circular Economies

Amsterdam's Circular Innovation Program has integrated recycling and reuse into urban systems, reducing waste by 50% in the past decade. The program ensures resource efficiency while fostering economic resilience by aligning waste management with economic production.

Case Study: Energy Security in Germany

Following the disruption of natural gas supplies in 2022, Germany accelerated renewable energy adoption, increasing wind and solar capacity by 20% within two years. Energy storage investments and policies to diversify energy sources have strengthened Germany's resilience.

Key Leadership Style: Transformational Leadership

Transformational leaders in Europe inspire collective action, mobilizing industries, governments, and communities toward systemic change.

Insight

Resilience in Europe demonstrates the value of aligning resource efficiency with energy security to mitigate environmental and economic risks.

Asia-Pacific: Innovations in Disaster Risk Management

The Asia-Pacific region faces some of the world's most severe climate risks, from typhoons in Japan to droughts in India. Regional resilience strategies often integrate cutting-edge technology with traditional practices.

Case Study: Disaster Resilience in Japan

Japan's investments in shock-absorbing buildings, early warning systems, and seawalls have minimized casualties during earthquakes and tsunamis. For instance, the $12 billion Fukushima seawall project has significantly reduced tsunami risk for millions of residents.

Case Study: India's Climate-Resilient Agriculture

India's National Innovations on Climate Resilient Agriculture (NICRA) initiative promotes water conservation, crop diversification, and climate-smart farming. Over 10 million farmers have adopted NICRA's practices, increasing yields by 20% in drought-prone regions.

Key Leadership Style: Resilient Leadership

Resilient leaders in the Asia-Pacific region combine foresight with action, enabling communities to adapt to evolving risks.

Insight

Integrating technological innovations with community-driven practices enhances disaster preparedness and climate resilience.

Africa: Building Resilience Through Decentralized Systems

The dual challenges of economic underdevelopment and climate vulnerability shape Africa's resilience strategies. Decentralized energy systems and nature-based solutions are transforming local economies.

Case Study: Off-Grid Solar in Kenya

Kenya's M-KOPA Solar platform provides affordable energy to over 1 million households, reducing reliance on centralized infrastructure. With pay-as-you-go financing models, M-KOPA has driven a 60% increase in energy access in rural areas.

Case Study: The Great Green Wall Initiative in West Africa

The Great Green Wall combats desertification through reforestation and land restoration. Stretching 8,000 kilometers across the Sahel, the initiative has restored 20 million hectares of degraded land and created 350,000 jobs.

Key Leadership Style: Inclusive Leadership

Inclusive leaders empower communities by prioritizing equity and local participation in decision-making processes.

Insight

Decentralized systems and nature-based solutions strengthen resilience while promoting equitable development.

South America: Resilience in Vulnerable Regions

South America's diverse geography poses unique resilience challenges, from coastal cities threatened by rising seas to drought-prone agricultural regions.

Case Study: Chile's Coastal Adaptation

Chile has invested in mangrove restoration and sea defense systems, protecting 2 million coastal residents from rising sea levels and storm surges. These measures have reduced flood risks by 40% in vulnerable areas.

Case Study: Drought-Resilient Agriculture in Brazil

Brazil's precision irrigation programs have reduced water consumption by 30% while improving yields in arid regions. The use of drought-resistant crops has also enhanced food security.

Key Leadership Style: Visionary Leadership

Visionary leaders in South America adopt forward-thinking approaches to address both current risks and future challenges.

Insight

Combining natural and engineered solutions enhances resilience in vulnerable regions, balancing environmental sustainability with economic development.

Australia and Oceania: Balancing Innovation and Resilience

Australia and Pacific Island nations are on the frontlines of climate change, facing rising sea levels, wildfires, and water scarcity. Their resilience strategies often rely on advanced technologies and community engagement.

Case Study: Wildfire Management in Australia

Australia's Fire Prediction and Resource Allocation System (FIRAS) uses AI and satellite data to predict wildfire behavior and optimize resource deployment. This system has reduced wildfire response times by 30%, saving lives and property.

Case Study: Water Management in the Murray-Darling Basin

The Murray-Darling Basin Plan balances agricultural needs with ecosystem health, ensuring water availability for 2 million people and supporting 30% of Australia's food production.

Key Leadership Style: Data-Driven Leadership

Data-driven leaders leverage advanced analytics to inform decision-making, ensuring efficient resource allocation.

Insight

Advanced technologies, coupled with community participation, strengthen resilience against climate risks.

Lessons for Building Regional Resilience

The strategies explored in this chapter reveal common themes and actionable lessons for resilience building:

1) **Localized Solutions:** Tailoring resilience strategies to regional contexts ensures effectiveness and community buy-in.

2) **Integrated Systems:** Addressing interdependencies across sectors, such as energy and water, enhances adaptive capacity.
3) **Technology as an Enabler:** Digital tools like AI and satellite monitoring amplify resilience efforts.
4) **Equitable Approaches:** Inclusive strategies promote social stability and broad-based benefits.
5) **Continuous Adaptation:** Regularly updating plans based on emerging risks fosters long-term resilience.

Conclusion: Pathways to a Resilient Future

Building resilience at the regional level is critical to addressing global challenges. From North America's grid modernization to Africa's off-grid solutions, these stories demonstrate the power of tailored approaches and visionary leadership. By embracing these lessons, policymakers and practitioners can create adaptive systems that ensure sustainability, stability, and equity in an uncertain future.

Resources for Further Reading

Australian Department of Industry, "Fire Prediction and Resource Allocation System," 2023.
California Independent System Operator, "Energy Imbalance Market Annual Report," 2023.
Circular Economy Europe, "Amsterdam's Circular Innovation Program," 2023.
Great Green Wall Initiative, "Combating Desertification Across the Sahel," 2023.
Indian Council of Agricultural Research, "NICRA: Climate-Resilient Agriculture," 2023.
Japan Meteorological Agency, "Disaster Preparedness Frameworks," 2023.
Miami-Dade County, "Resilience Planning: Sea-Level Rise Mitigation Strategies," 2023.
Ministerio del Medio Ambiente, "Coastal Adaptation Strategies in Chile," 2023.
M-KOPA Solar, "Affordable Energy Access in Africa," 2023.

31

Governance and Resilience – Systems of Adaptation and Accountability

Introduction: Governance as a Pillar of Resilience

Governance is the backbone of societal resilience, providing the frameworks and mechanisms to navigate crises and adapt to evolving challenges. In today's interconnected world, where climate risks, economic disruptions, and technological shifts frequently intersect, governance systems must do more than maintain order – they must actively foster resilience. From centralized disaster response frameworks in Japan to decentralized water management systems in Kenya, effective governance provides the foundation for societal stability and adaptive capacity.

This chapter explores governance models worldwide that demonstrate how transparency, accountability, and collaboration enable resilience. It also examines instances where governance failures, such as the 2021 Texas power crisis, serve as cautionary tales for policymakers and leaders.

Centralized vs. Decentralized Governance Models

Governance structures shape how systems respond to crises and adapt to long-term challenges. Depending on the context, both centralized and decentralized models have distinct advantages.

Case Study: Centralized Governance in Japan

Japan's disaster preparedness framework is a global benchmark for centralized governance. Managed by the Cabinet Office, Japan's system integrates real-time disaster monitoring with nationwide response protocols. For example, during the 2011 Tōhoku earthquake and tsunami, centralized coordination ensured the rapid deployment of relief efforts and the activation of advanced early warning systems, which saved thousands of lives.

Case Study: Decentralized Water Management in Kenya

Water Resource User Associations (WRUAs) empower local communities to manage water resources in Kenya. These decentralized governance structures have improved water access in arid regions by allowing communities to tailor solutions to their unique needs. The WRUAs have also reduced resource conflicts by fostering local accountability and collaboration.

Key Leadership Style: Adaptive Leadership

Adaptive leaders excel in aligning governance structures to the specific challenges they face. In Japan, this meant leveraging advanced technology for disaster management, while in Kenya, it involved empowering local stakeholders to cocreate solutions.

Insight

Both centralized and decentralized models have strengths. The key to resilience is tailoring governance systems to align with the nature of the challenge and the needs of affected communities.

Collaborative Governance Frameworks

Collaboration is essential for addressing complex challenges that span jurisdictions and sectors. Collaborative governance integrates government institutions, private sectors, and civil society to achieve shared goals.

Case Study: Denmark's Energiewende

Denmark's Energiewende initiative has successfully transitioned the country to renewable energy, with wind power now supplying nearly 50% of electricity needs. This success was achieved through a collaborative governance model that combined public–private partnerships, community ownership of energy projects, and transparent policymaking.

Case Study: The U.S.–Mexico Border Infrastructure Fund

Cross-border collaboration between the United States and Mexico has successfully financed water treatment plants and transportation infrastructure, improving public health and economic ties. This fund exemplifies how international cooperation enhances regional resilience by integrating resources and aligning objectives.

Key Leadership Style: Transformational Leadership

Transformational leaders inspire multi-stakeholder collaboration, mobilizing diverse resources to address shared challenges.

Insight

Collaboration enhances resource efficiency and fosters stakeholder trust, critical for addressing transboundary issues such as energy security and public health.

Governance for Climate Resilience

Effective governance for climate resilience must integrate scientific data, enforce accountability, and prioritize equity.

Case Study: The European Green Deal

The European Green Deal aims for climate neutrality by 2050. It combines regulatory tools like carbon pricing with funding mechanisms to support member states in transitioning to renewable energy. This comprehensive approach ensures accountability while incentivizing innovation.

Case Study: Australia's Climate Resilience Strategy

Australia has invested heavily in wildfire prediction systems and water management technologies, including artificial intelligence (AI-enabled) monitoring and adaptive water allocation frameworks for the Murray–Darling Basin. These measures have mitigated disaster risks and protected critical ecosystems like the Great Barrier Reef.

Key Leadership Style: Ethical Leadership

Ethical leaders prioritize long-term environmental sustainability while ensuring that governance frameworks are inclusive and equitable.

Insight

Governance frameworks addressing climate resilience must balance regulatory oversight with localized innovation to achieve sustainable outcomes.

Challenges in Governance and Resilience

Governance failures often expose vulnerabilities, leading to cascading disruptions. Identifying and addressing these gaps is critical for building robust systems.

1) **Inadequate Crisis Preparedness**
   The 2021 Texas power crisis revealed systemic vulnerabilities in energy governance. A lack of weatherization standards for power infrastructure led to widespread blackouts during a winter storm, affecting over 4.5 million people and causing an estimated $195 billion in damages.
2) **Navigating Competing Priorities**
   Brazil's Low-Carbon Agriculture Plan has significantly reduced emissions while maintaining agricultural productivity. However, inconsistent implementation across regions highlights the challenge of balancing economic growth with environmental goals.
3) **Equity and Inclusivity in Governance**
   Governance failures often exacerbate inequities. For example, Africa's reliance on international funding for renewable energy projects limits local ownership and decision-making, even in successful initiatives like Kenya's decentralized solar systems.

Tools and Strategies for Resilient Governance

1) **Multi-stakeholder Engagement**
   Governance systems that engage diverse stakeholders are more likely to achieve broad-based resilience.

Example: Singapore's Urban Redevelopment Authority integrates public consultation into urban planning, ensuring that policies align with societal needs.

2) **Transparent Decision-Making**
Transparency builds trust and accountability.
Example: The European Union's General Data Protection Regulation (GDPR) sets a global standard for ethical governance in digital systems.
3) **Leveraging Technology for Resilience**
Governments increasingly rely on technology to enhance governance.
Example: Digital twins simulate urban planning scenarios, improving disaster readiness in cities like Barcelona. Blockchain ensures supply chain transparency, particularly in regions reliant on global trade.
4) **Metrics for Performance Assessment**
Governance systems must be evaluated through clear metrics.
Example: FEMA's National Incident Management System (NIMS) includes standardized evaluation protocols to assess disaster response effectiveness.

Lessons in Governance for Resilience

1) **Customized Approaches:** Tailor governance systems to address specific challenges, whether centralized, decentralized, or hybrid.
2) **Equity and Inclusion:** Ensure governance frameworks prioritize marginalized populations to promote fairness and stability.
3) **Technological Integration:** Leverage digital tools to enhance decision-making and streamline resource allocation.
4) **Long-Term Vision:** Balance immediate needs with sustainable development goals to ensure long-term resilience.

Conclusion: Governance as a Driver of Resilience

Governance is the keystone of societal resilience, shaping how communities respond to crises and adapt to emerging challenges. From Denmark's renewable energy leadership to Kenya's community-driven water management, these case studies demonstrate the power of well-designed governance systems to build sustainable, adaptive, and inclusive societies. As the world faces increasingly complex risks, leaders must prioritize transparency, collaboration, and ethical foresight to ensure a resilient future.

Resources for Further Reading

Australian Department of Industry, "Fire Prediction and Resource Allocation System," 2023.
Cabinet Office of Japan, "Japan's Disaster Preparedness Framework," 2023.
Danish Energy Agency, "Denmark's Wind Energy Framework," 2023.
European Commission, "The European Green Deal," 2023.
FEMA, *National Incident Management System (NIMS)*. Federal Emergency Management Agency, 2023.
Kenyan Ministry of Water Resources, "Water Resource User Associations (WRUAs)," 2023.
Texas Tribune, "Lessons from the 2021 Texas Power Crisis," 2023.
Urban Redevelopment Authority, "Singapore's Urban Planning and Sustainability," 2023.

32

Identifying and Assessing Emerging and Pivotal Technologies – A Practical Guide

Introduction: The Transformative Power of Technology

Emerging and pivotal technologies have consistently reshaped industries, economies, and societies. These technologies offer immense potential for innovation and growth, from artificial intelligence (AI) and quantum computing to renewable energy solutions and biotechnology. However, their adoption and integration require a nuanced understanding of their development stages, transformative potential, and societal implications.

This chapter explores tools and frameworks for assessing emerging and pivotal technologies, highlighting real-world examples of their transformative impact. It also examines the leadership styles that drive successful technology adoption, emphasizing the importance of foresight, collaboration, and ethical stewardship.

Defining Emerging and Pivotal Technologies

1) **Emerging Technologies**
   Emerging technologies are in their early stages of development, with significant potential for innovation but limited commercialization. Examples include blockchain, advanced robotics, and gene-editing tools like CRISPR [1, 2].
2) **Pivotal Technologies**
   Pivotal technologies drive systemic change, enabling shifts across industries and societal structures. These include quantum computing, AI, and renewable energy innovations like green hydrogen and solid-state batteries [3, 4].

Key Leadership Style: Visionary Leadership

Visionary leaders are essential for identifying technologies that align with long-term goals. Their ability to anticipate trends and articulate a clear vision inspires stakeholders and drives strategic investments.

Insight

Understanding the distinction between emerging and pivotal technologies enables leaders to prioritize resources effectively, balancing innovation risks with potential rewards.

Tools for Assessing Emerging and Pivotal Technologies

Technology Foresight

Technology foresight is a structured approach to identifying future opportunities and challenges.

- **Step 1: Environmental Scanning**
  Use PESTLE analysis (Political, Economic, Social, Technological, Legal, Environmental) to assess external factors influencing technology adoption.
  Example: A logistics company identified AI applications for global supply chains through PESTLE analysis, focusing on automation and optimization under diverse regulatory conditions.
- **Step 2: Scenario Building**
  Develop scenarios to evaluate technological impacts under various conditions, such as regulatory changes or market fluctuations.
  Example: A smart grid initiative used scenario planning to model renewable energy adoption in different regulatory environments, identifying key risks and opportunities.
- **Step 3: Strategic Alignment**
  Align technology investments with organizational goals.
  Example: A pharmaceutical company integrated foresight findings into its R&D roadmap, accelerating mRNA vaccine development during the COVID-19 pandemic.

Leadership Style: Strategic Leadership

Strategic leaders excel in using foresight tools to align technological initiatives with broader organizational goals, ensuring long-term sustainability and resilience.

Filters for Identifying Pivotal Technologies

Filters help leaders focus on transformative technologies by evaluating their potential across key dimensions:

- **Scalability:** Can the technology impact multiple industries or geographies?
  Example: Autonomous vehicles revolutionize logistics, healthcare (ambulances), and urban planning.
- **Adoption Readiness:** Is the technology accessible and cost-effective?
  Example: Small Modular Reactors (SMRs) require significant initial investments but offer scalable energy solutions.
- **Market Potential:** Does the technology address critical challenges or create substantial value?
  Example: Quantum computing optimizes supply chains, solving problems previously deemed intractable.
- **Synergy with Other Technologies:** Can it integrate with existing systems to amplify impact?
  Example: AI and Internet of Things (IoT) enable smart cities by combining data analytics with real-time decision-making capabilities.

Leadership Style: Analytical Leadership

Analytical leaders use data-driven approaches to evaluate technologies systematically, ensuring that decisions are grounded in evidence rather than speculation.

Scenario Planning

Scenario planning evaluates potential futures to identify risks and opportunities associated with technology adoption.

- **Step 1: Define Key Variables**
  Identify factors influencing the technology's success, such as regulatory frameworks, infrastructure readiness, and market trends.
- **Step 2: Develop Scenarios**
  To evaluate potential outcomes, create "best case," "worst case," and "base case" scenarios.
  Example: A global e-commerce firm used scenario planning to assess blockchain's impact on supply chain transparency, anticipating regulatory and operational hurdles.
- **Step 3: Test Strategies**
  Evaluate organizational strategies under each scenario to refine implementation plans.

Leadership Style: Adaptive Leadership

Adaptive leaders excel in dynamic environments by leveraging scenario planning to navigate uncertainties and make informed decisions.

Real-World Applications

1) **Quantum Computing in Supply Chains**
   - **Emerging Potential:** Quantum chips enable logistics, encryption, and materials science breakthroughs.
   - **Pivotal Attributes:** Quantum computing's scalability across industries and the potential to revolutionize computational problems position it as a transformative force.
2) **AI in Healthcare**
   - **Tools Applied:** Scenario planning and Technology Readiness Levels (TRLs) assessed AI-based diagnostics.
   - **Outcome:** AI accelerated decision-making and reduced diagnostic errors by 25% during the COVID-19 pandemic.
3) **SMRs for Decentralized Energy**
   - **Framework Used:** Filters and PESTLE analysis evaluated scalability, regulatory challenges, and market potential.
   - **Impact:** SMRs are pivotal for energy resilience, especially in underserved regions.

Key Leadership Styles for Technology Assessment

1) **Visionary Leadership:** Inspires innovation by anticipating future trends and aligning them with long-term objectives.
2) **Strategic Leadership:** Ensures technology investments are purposeful and aligned with organizational goals.
3) **Analytical Leadership:** Evaluates technologies systematically, ensuring data-driven decisions.
4) **Adaptive Leadership:** Navigate uncertainty using foresight tools and scenario planning.

Lessons for Leaders

1) **Customize Tools to Context:** Tailor filters, foresight, and scenario planning to specific industries or organizational needs.
2) **Adopt a Holistic View:** Combine multiple assessment tools to evaluate technologies from internal and external perspectives.
3) **Invest in Education:** Equip teams with the skills and knowledge to apply assessment tools effectively.

Insight

Successful technology leadership requires a balance of vision, strategy, and adaptability supported by robust analytical frameworks.

Conclusion: Enabling Informed Decision-Making

Emerging and pivotal technologies hold the potential to transform industries and societies. Organizations can navigate uncertainties and unlock innovation by leveraging structured assessment tools and aligning leadership styles with the demands of technology adoption. As industries face increasing complexity, leaders who combine vision, strategy, and ethical foresight will define the next generation of technological progress.

References

1 National Science Foundation, "Emerging Technology Trends," 2023.
2 L. Carlson, "Pivotal technology frameworks," *Res. Technol. Manag.*, vol. 50, no. 6, pp. 34–42, 2024.
3 NASA, "Technology Readiness Levels (TRLs)," 2022.
4 Google Quantum Research, "Quantum Chip Advances," 2024.

Resources for Further Reading

Energy Resilience Solutions, 2024.
MIT Sloan School of Management, "Management of Technology Tools and Frameworks," 2022.
J. Smith and R. Brown, "AI in telemedicine: Lessons from COVID-19," *Healthcare Innov. J.*, vol. 11, no. 3, pp. 21–27, 2023.

33

Leadership in Pivotal Technologies – A Framework for the Future

Introduction: The Imperative of Leadership

Leadership is pivotal in ensuring these innovations translate into meaningful progress in a world driven by relentless technological advancement. Technology's journey – from its inception as an emerging idea to its adoption as a transformative force – requires more than scientific breakthroughs and market readiness. It demands leaders who can navigate complexity, anticipate risks, and inspire collaborative ecosystems to harness technology's full potential.

This chapter explores the principles of effective leadership in the context of pivotal technologies, examining real-world applications and the leadership styles that enable transformative outcomes. From quantum computing to renewable energy, the path to technological adoption is fraught with challenges, including ethical dilemmas, market disruptions, and geopolitical considerations. Leaders who succeed in this space embrace strategic foresight, build collaborative ecosystems, and integrate ethical stewardship into their decision-making.

The Role of Leadership in Pivotal Technologies

Technological leadership is distinct from traditional management. While management ensures stability and efficiency, leadership in pivotal technologies drives evolution, fostering innovation while managing risks. It requires a unique combination of skills: visionary thinking to anticipate what's possible, strategic foresight to plan for the unknown, and adaptability to pivot in response to unexpected challenges.

In 2021, for example, a healthcare organization faced mounting pressure to reduce costs and improve patient outcomes. By deploying an artificial intelligence (AI)-driven diagnostic tool, the organization achieved a 25% reduction in diagnostic errors and cut average patient wait times by 40%. However, the success of this initiative relied not only on technical expertise but also on the leadership that bridged the gap between clinicians, data scientists, and policymakers. The organization's leaders ensured compliance with data privacy regulations and addressed ethical concerns, demonstrating that leadership in pivotal technologies extends beyond technical adoption to encompass governance and trust-building.

Strategic Foresight: Preparing for the Unknown

Strategic foresight is at the core of successful technological leadership – anticipating future trends, disruptions, and opportunities. This requires more than predictive analytics or trend monitoring;

it demands a deep understanding of interconnected systems and the foresight to navigate potential outcomes.

One of the most compelling examples of strategic foresight is seen in the logistics industry's adoption of autonomous delivery systems. A multinational logistics firm used scenario planning to evaluate how autonomous vehicles would impact operations under varying regulatory, technological, and economic conditions. By developing multiple scenarios, the company identified potential barriers, such as resistance from local governments and infrastructure gaps, and crafted strategies to address them. This proactive approach allowed the firm to stay ahead of competitors while minimizing risks associated with early adoption.

Leaders who excel in strategic foresight embrace complexity, recognize interdependencies, and use scenario planning and roadmapping tools to align technological investments with long-term goals.

Ethical Decision-Making: Balancing Progress and Responsibility

Emerging and pivotal technologies often outpace existing regulatory frameworks, creating ethical dilemmas requiring careful navigation. Leaders in this space must go beyond compliance to actively consider the societal impacts of their decisions.

Consider the example of a renewable energy firm developing wind farms in rural communities. While the project promised economic benefits and environmental gains, it also faced opposition from residents concerned about land use and noise pollution. The firm's leaders engaged the community through town hall meetings, incorporating their feedback into the project's design. This approach resolved conflicts, fostered trust, and demonstrated a commitment to ethical stewardship.

Ethical leadership in pivotal technologies involves aligning technological advancements with societal values, conducting rigorous impact assessments, and prioritizing transparency in decision-making processes.

Building Collaborative Ecosystems

No single organization can realize the full potential of pivotal technologies in isolation. Collaboration across sectors, disciplines, and geographies is essential for scaling innovation and addressing complex challenges.

Denmark's wind energy initiative offers a powerful example of collaborative leadership. The country transformed its energy landscape through public–private partnerships, with wind power now accounting for nearly 50% of its electricity supply. This success was not achieved through technical innovation alone but through the collaborative efforts of government agencies, private companies, and research institutions.

Effective leaders in this space act as facilitators, bringing together diverse stakeholders to create ecosystems that leverage collective resources and expertise. By fostering trust, aligning incentives, and ensuring shared accountability, they enable the success of initiatives that would be impossible to achieve independently.

Adaptive Leadership: Responding to Change

The dynamic nature of pivotal technologies demands leaders who can adapt to changing circumstances. Adaptive leaders remain flexible, learning and evolving as conditions shift, whether responding to unforeseen challenges or seizing new opportunities.

The rollout of fifth-generation (5G) networks illustrates the importance of adaptability. Early adopters faced significant hurdles, from infrastructure delays to public concerns about health and privacy. Adaptive leaders navigated these challenges by iterating their strategies, addressing stakeholder concerns, and leveraging emerging insights to optimize deployment. This iterative approach mitigated risks and accelerated adoption in key markets.

Adaptive leadership thrives on continuous learning, open-mindedness, and a willingness to challenge assumptions. This ability to pivot ensures organizations remain resilient in the face of uncertainty.

Case Studies: Lessons in Leadership

1) **Quantum Computing in Defense**
   Quantum computing offers transformative potential for encryption and materials science, but its development is fraught with high costs and geopolitical tensions. A defense contractor demonstrated visionary leadership by investing in quantum research while collaborating with governments to establish ethical guidelines. This approach balances innovation with societal responsibility, ensuring advancements serve national and global security interests.
2) **AI in Healthcare**
   During the COVID-19 pandemic, a hospital system integrated AI-driven diagnostic tools to manage patient flow and optimize resources. Strategic foresight enabled the organization to anticipate regulatory hurdles, while adaptive leadership ensured rapid implementation despite logistical challenges.
3) **Renewable Energy in Developing Regions**
   A nonprofit organization worked to bring solar power to rural African communities. The initiative achieved scalable and sustainable outcomes by building collaborative ecosystems that included local governments, international donors, and private companies. Ethical leadership ensured that the project prioritized local needs and avoided unintended consequences.

Key Leadership Styles for Pivotal Technologies

1) **Visionary Leadership:** Inspires innovation by articulating a clear vision and aligning stakeholders around long-term goals.
2) **Strategic Leadership:** Balances immediate objectives with future opportunities, using tools like scenario planning and foresight.
3) **Ethical Leadership:** Ensures that technological advancements align with societal values and minimize harm.
4) **Adaptive Leadership:** Navigates uncertainty through continuous learning and flexibility.
5) **Collaborative Leadership:** Builds ecosystems that leverage diverse resources and expertise.

Conclusion: Leading the Future

Leadership in pivotal technologies is not about managing the status quo but shaping the future. By combining vision, strategy, ethics, adaptability, and collaboration, leaders can navigate the complexities of technological transformation while ensuring that progress serves the greater good.

The lessons and examples in this chapter demonstrate that successful leadership in this space is as much about people and values as technology. In a rapidly evolving world, leaders who embrace this balance will drive meaningful change and create a sustainable future.

Resources for Further Reading

Danish Energy Agency, "Denmark's Wind Energy Framework," 2023.
Google Quantum Research, "Quantum Chip Advances," 2024.
"AI in telemedicine," *Healthcare Innov. J.*, vol. 11, no. 3, pp. 21–27, 2023.
MIT Sloan School of Management, "Leadership and Ethics in Technology," 2022.
National Science Foundation, "Emerging Technology Trends," 2023.

34

Thriving in Complexity – Lessons for Resilience and Leadership

Leadership, systems, and life are all about navigating complexity. From the flicker of a kerosene lamp in Tabriz to global efforts in critical infrastructure, aerospace, and leadership, this journey has been a testament to the transformative power of resilience. As the chapters of this book have shown, no system, leader, or plan can eliminate uncertainty. Still, we create systems that adapt, recover, and thrive by understanding, modeling, and managing complexity.

The book began with a prologue emphasizing the urgency of building resilience and fostering technological leadership. Each chapter has explored facets of this journey, from the evolution of smart grids to the principles of governance, the integration of pivotal technologies, and the pathways for global collaboration. In this final chapter, we combine those lessons, connecting the book's title, themes, and chapters into a cohesive vision for the future.

Understanding Complexity

In every challenge, decision, and opportunity, we encounter systems. These systems generate signals that, when interpreted, reveal their health and direction. Success begins with learning to read those signals:

- **What Signals Are You Sensing?** Systems communicate through trends, anomalies, and feedback – both obvious, like power grid performance metrics, and subtle, like shifting societal norms or team morale.
- **What Can You Measure?** Whether uptime, emissions reductions, or financial outcomes, metrics must connect to meaningful goals.
- **What Can't You Measure?** Trust, resilience, and integrity are hard to quantify but essential to the success of any system.

Systems' physical, digital, or human resilience relies on a leader's ability to interpret signals and act accordingly. Reflecting on earlier chapters, we see examples of how understanding these dynamics led to transformative outcomes – from securing North America's grid post-9/11 to advancing smart cities in Europe.

Modeling for Resilience

Signals provide data, but modeling helps us understand and improve system behavior. Effective leaders use models to predict outcomes, test scenarios, and guide decision-making.

- **Data-Driven Models:** Historical and real-time data identify patterns and optimize responses. For instance, smart grids use predictive analytics to prevent outages.
- **First-Principles Models:** Built from fundamental dynamics, these models simulate behavior under stress. The design of Black Hawk helicopters exemplifies how first-principles modeling ensures stability in extreme conditions.
- **Iterative Design:** No model is perfect. Continuously refine models with new data to adapt to changing circumstances.

These principles were echoed in Chapter 27's exploration of sustainable development and Chapter 9's analysis of energy transitions. Whether assessing renewable energy integration or disaster risk management, using foresight tools and modeling frameworks has consistently proven pivotal.

Managing Toward Resilience

Leadership bridges the gap between understanding systems and achieving outcomes. Managing resilience means balancing short-term responses with long-term goals. It requires clarity, adaptability, and action:

1) **Define Clear Goals:** Success starts with clarity. For smart grids, goals like efficiency and emissions reductions provided measurable benchmarks.
2) **Monitor and Adapt:** Feedback loops enable rapid responses to disruptions, from cyberattacks to natural disasters.
3) **Act Decisively:** Waiting for perfect data is a trap. Leaders must make informed decisions and adjust as needed.
4) **Handle Disruptions:** Stabilize first, then optimize. This principle guided efforts during Desert Shield and in post-9/11 grid modernization.

The lessons from earlier chapters underscore the importance of proactive, strategic management. For example, Chapter 8 highlighted how effective governance frameworks drive resilience in climate adaptation and energy security.

Leadership Styles for Thriving in Complexity

As this book has shown, leadership styles are not one-size-fits-all. The most effective leaders adapt their approach based on context, challenges, and objectives. Let's revisit key leadership styles defined in earlier chapters:

- **Transformational Leadership** inspires change by aligning organizational goals with a compelling vision. As explored in Chapter 29, this style is essential for driving innovation in pivotal technologies.
- **Adaptive Leadership:** This approach embraces flexibility and continuous learning. As detailed in Chapter 9, it proved invaluable in navigating uncertainty during the energy transitions.
- **Servant Leadership:** Prioritizes the needs of the team and stakeholders. Chapter 7's focus on regional resilience demonstrated how servant leadership fosters collaboration and trust.
- **Strategic Leadership:** Balances short-term efficiency with long-term goals. The governance models in Chapter 8 exemplified the power of strategic foresight in addressing global challenges.

The integration of these styles enables leaders to navigate complexity effectively. They create systems that thrive under the harshest conditions by fostering trust, inspiring innovation, and adapting to change.

Practical Steps for Leaders

As a culmination of the book's lessons, here are actionable steps for leaders:

1) **Measure What Matters:** Focus on metrics that align with strategic goals and drive meaningful outcomes.
2) **Model for Foresight:** Use scenario planning, risk assessments, and iterative design to anticipate and address challenges.
3) **Manage with Resilience:** Act decisively, learn from disruptions, and focus on long-term objectives.
4) **Foster Collaboration:** Build ecosystems that leverage diverse expertise and align stakeholders around shared goals.
5) **Invest in Innovation:** As emphasized in Chapter 30, prioritize R&D and technological advancements to address emerging challenges.
6) **Lead Ethically:** Ensure equity, integrity, and sustainability principles guide decisions.

Reflections and Closing Thoughts

From the kerosene-lit villages outside of Tabriz to leading global initiatives in resilience, this journey has underscored one enduring truth: systems matter because people matter. When systems fail, people suffer. When systems endure, people thrive.

The book's title, *Navigating Complexity: Leadership, Systems, and Resilience,* reflects this central theme. It is a call to action for leaders to embrace the challenges of complexity with courage, clarity, and compassion. By understanding systems, modeling for foresight, and managing with resilience, we can build a future that is not only sustainable but also equitable and prosperous.

Leadership isn't about eliminating uncertainty; it's about thriving within it. Progress isn't about perfection – it's about persistence. As we move forward, let us commit to building systems – and lives – that adapt, recover, and flourish.

Let's get to work.

Resources for Further Reading

M. Amin, *Toward a Secure and Smart Self-Healing Grid*. EPRI Research Advisory Committee, 1998.
California Independent System Operator (CAISO), "Energy Imbalance Market Annual Report," 2023.
DARPA, "Innovation Ecosystems for Defense Applications," 2023.
European Commission, "European Green Deal," 2023.
FEMA, "National Incident Management System (NIMS)," 2023.
Inter-American Development Bank, "Rural Electrification in Latin America," 2023.
MIT Sloan School of Management, "Leadership in Technological Innovation," 2023.
National Science Foundation, "Emerging Technology Trends," 2023.

35

Stability and Freedom: Structural Foundations of Resilient Societies

Summary

Stable societies are built on four fundamental pillars: legitimacy, trust, the rule of law, and human dignity. These elements determine whether institutions endure stress or collapse under it. Historical and comparative research indicates that when these foundations weaken, political systems become brittle, economic coordination falters, public health deteriorates, and security becomes unstable [1–4]. Drawing on historical texts, contemporary governance indicators, and empirical case studies, this paper examines the architecture that sustains free societies, the frameworks that shape civic behavior, the manifestations of systemic erosion, and the policy pathways for restoration. It integrates classical sources, such as Madison and Tocqueville [1, 2], with modern institutional economics [5, 6], trust research [7–9], and global governance metrics [9–11], offering both theoretical depth and policy relevance. We find that civic norms, shared factual reality, collective attention, social memory, and long-term institutional time horizons mediate the strength of these pillars. The analysis concludes that wealth, technology, and military strength cannot secure liberty if legitimacy, trust, law, and dignity decay. Lincoln's warning about "the silent artillery of time" remains both empirically and normatively urgent [12].

Introduction

The stability of societies depends less on visible power than on an underlying invisible architecture. Constitutions, markets, armies, and technologies all operate within structures of legitimacy, trust, law, and shared dignity. When these foundations are sound, nations can withstand economic shocks, pandemics, political crises, and external threats; when they erode, decline is often silent and cumulative until it becomes sudden and visible [1–4]. We define stability as the capacity of a political community to maintain social order, resolve conflicts peacefully, and adapt through change without collapsing. Four core concepts – legitimacy (the recognized right to govern), trust (social and institutional confidence), the rule of law (the supremacy of fair, enforceable rules), and dignity (inviolable respect for individual worth) – emerge as the structural foundations of such resilience. Where legitimacy is absent, coercion replaces consent and instability follows. For example, a recent global survey found that majorities in many democracies are dissatisfied with how their system is working, undercutting leaders' perceived right to rule [13]. By contrast, when people believe in the legitimacy of institutions, they grant a "reservoir of support" that buffers regimes against shocks. Trust is closely intertwined: it functions as a form of social capital that lowers transaction costs and enables cooperation, whereas pervasive distrust breeds defection and conflict. In the United States, public trust in the federal government has fallen from roughly 75% in the 1960s to just 22% today – near historic lows after years of polarization and policy failures [8]. Likewise, the rule of law provides the backbone tying power to principle. It restrains abuses and reassures citizens that disputes will be resolved fairly, underpinning both stability and economic development [14]. Finally, dignity is essential: societies that honor the equal worth of each person foster inclusion and loyalty, whereas systematic indignities – corruption, brutality, and extreme inequality – corrode the social contract. The Universal Declaration of Human Rights opens by declaring that recognition of "human dignity is the foundation of freedom, justice and peace in the world" [14]. Indeed, much of modern political unrest, from populist backlashes in wealthy democracies to revolutions in the developing world, reflects a demand for dignity from those who feel excluded or disrespected.

Across historical periods and political systems, the strength of these foundations has often determined whether states endure under pressure or collapse. Madison wrote in Federalist No. 51 that "if men were angels, no government would be necessary" [1], underscoring the need for structures to channel human behavior. Tocqueville observed that American democracy drew resilience not only from its institutions but from citizens' shared habits of heart and active associations [2]. Modern institutional economists have similarly demonstrated that weak legal and normative structures help explain persistent underdevelopment even in resource-rich states [6]. Empirical studies show that high-trust societies govern more effectively, weather crises with less coercion, and recover more quickly [5, 7, 9]. Organisation for Economic Co-operation and Development (OECD) data reveal that countries with higher public trust in government also exhibit better fiscal performance, higher compliance, and greater resilience [9, 11, 15].

The urgency of this inquiry is clear. Across many democracies, indicators of institutional trust, rule of law, and civic cohesion have declined over the past decade [8, 15, 16]. Pew surveys show that public trust in the U.S. federal government remains near historic lows [8]. At the same time, the World Justice Project reports that global rule-of-law scores have dropped for seven consecutive years [17]. Information ecosystems have fragmented, shared facts have weakened, and political polarization has intensified across advanced economies [10, 16, 18, 19]. These trends make the invisible pillars of free societies both more visible and more fragile.

This paper systematically examines these dynamics. It begins by analyzing the four foundational pillars – legitimacy, trust, the rule of law, and dignity. It then explores the broader civic frameworks through which norms, shared facts, collective attention, institutional memory, and time horizons shape social stability. Next, manifestations of erosion are examined across core systems, including democracy, markets, public health, security, and civic life, drawing on contemporary cases from the United States, Europe, and the Global South. The paper concludes with strategies for restoration, integrating comparative evidence and emphasizing both sound institutional design and revitalized civic capacity.

Methodology

This research employs a mixed-methods approach combining historical analysis, comparative political economy, and contemporary empirical data. First, classical political texts – including Madison's *Federalist Papers* [1] and Tocqueville's *Democracy in America* [2] – provide normative and structural foundations for understanding legitimacy, trust, law, and civic culture. Second, the study integrates insights from modern institutional economics [5, 6], social capital theory [7, 9], and governance studies [11, 15], focusing on the mechanisms through which institutions generate stability or fragility.

Empirically, the analysis draws on data from multiple international sources, including the OECD's trust in government indicators [9], the World Justice Project's Rule of Law Index [17], Pew Research Center longitudinal surveys of public trust [8], the Varieties of Democracy (V-Dem) dataset tracking democracy indicators [20], WHO studies on information during pandemics [10], and the World Bank's Worldwide Governance Indicators [21]. Case studies complement these – for example, Poland's democratic backsliding and partial reversal (2015–2023), the long-term decline of U.S. institutional trust (1964–present), Chile's 2019 social unrest, and Taiwan's recent digital civic engagement reforms – chosen to illustrate structural dynamics in varied contexts.

The methodology is interdisciplinary and multimethod. We triangulate qualitative historical reasoning (to capture complex causal mechanisms and long-term processes) with quantitative institutional data (to identify broad patterns). This mixed approach allows us to move from abstract principles to concrete evidence. For instance, historical examples suggest that broad-based legitimacy and the rule of law are preconditions for stability. We examine current cases where legitimacy is low (e.g., autocracies facing unrest) vs. where it is robust (consolidated democracies), and look at data on public trust and stability metrics. We remain attentive to context – recognizing that expressions of legitimacy or trust differ in a tribal society vs. an industrialized democracy – but we also seek general structural patterns.

We explicitly consider time and change. Resilience is not a static end-state but a capacity to adapt over time. Thus, our methodology includes historical comparisons (e.g., how societies recovered from past crises of legitimacy or trust) and analysis of longitudinal trends (such as the decades-long decline in social capital in the United States). By including sources such as Abraham Lincoln's 1838 address and other historical accounts, we remind ourselves that concerns over institutional decay are not new – and that learning from the past is part of our evidence base. In summary, we construct a theoretical scaffold, map contemporary data onto it, and then examine narrative case evidence to understand how these structural foundations operate in practice. This combination of conceptual depth and empirical breadth is intended to yield academically robust and globally relevant conclusions.

Foundations

Legitimacy

Legitimacy is the belief among citizens that institutions and leaders exercise authority lawfully, fairly, and in the public interest. Without legitimacy, consent gives way to coercion, and political systems become brittle and unstable. Madison emphasized that the durability of republican government rests on a constitutional design that "obliges it to control itself" [1]. Tocqueville observed that American democracy rested not just on formal rules but on a widespread perception of their fairness [2]. Empirically, legitimacy correlates strongly with regime stability and public compliance with laws and policies [6, 15, 20].

Today, warning signs abound. The World Justice Project finds that checks on government power have weakened in the majority of countries, with global rule-of-law scores declining for seven consecutive years [17]. In several cases, declining legitimacy has preceded outright institutional breakdown. For example, in Hungary and Venezuela, public disillusionment with unfair governance set the stage for democratic erosion before legal crises became evident [20]. Many postcolonial states also faced a "crisis of legitimacy" after independence: arbitrary borders and transplanted institutions left citizens feeling little loyalty to the new regimes. In such contexts, local strongmen or sectarian identities often supplanted national authority, fueling chronic instability. Rebuilding legitimacy has required inclusive dialogues (giving citizens a voice in writing new constitutions) and power-sharing arrangements to reflect diverse groups – steps that broaden the base of consent and allegiance.

Trust

Trust functions as civic infrastructure. Fukuyama defined trust as "the expectation of regular, honest, and cooperative behavior, based on commonly shared norms" [5]. High-trust societies govern more effectively, coordinate economic activity more easily, and maintain civic peace with lower enforcement costs [5, 7, 9]. Conversely, pervasive distrust raises transaction costs and invites corruption and conflict. OECD data show that trust in national governments averaged only about 39% in 2023 across member countries, down from 45% in 2007 [11]. In the United States, public confidence in the federal government plummeted from 77% in 1964 to around 22% in recent years [8]. Such figures signal a crisis of confidence across democracies.

Trust operates on multiple levels – horizontal trust among citizens and vertical trust between citizens and institutions – and both are mutually reinforcing. In high-trust settings, people can engage in transactions and collective endeavors without elaborate safeguards, relying on shared norms of honesty and integrity. By contrast, in low-trust environments, cooperation extends only to close circles (family or clan) and everything beyond requires costly enforcement, stunting broader social and economic development. For example, communities with dense networks of social capital can self-organize disaster relief or public health campaigns, whereas communities with fragmented social capital struggle to mobilize. Crucially, political trust (confidence that authorities are competent and act in the public interest) underpins compliance. During the COVID-19 pandemic, societies with higher trust in government and science achieved greater voluntary uptake of health measures, whereas low-trust contexts saw resistance and conspiracy theories undermine the response [11, 22]. Trust can take years to build through reliable performance and fairness, but it can be lost quickly through scandals or failures and is arduous to rebuild once it has been broken. The widespread mistrust evident today correlates with rising voter volatility, street protests, and susceptibility to demagogues. Put simply, when people do not trust that institutions will treat them fairly or effectively, they either withdraw from civic life or seek alternative channels (sometimes extremist) for their grievances. Building and maintaining trust – through honest communication, delivering on promises, and visibly curbing corruption – is therefore central to societal resilience.

Rule of Law

The rule of law binds power to principle. It means that clear, public, and stable laws – applied equally to all, including those in power – form the basis of governance. This principle provides a predictable framework in which trust and legitimacy can flourish. Tocqueville remarked that "in America, the law is king" [2], capturing how a robust legal order can supplant arbitrary rule by individuals. Political economists have identified the rule of law as a cornerstone of development and stability: a reliable legal system reduces uncertainty, enforces contracts, protects rights, and thereby encourages investment and cooperation [14]. When the legal order is perceived as corrupt or biased, citizens lose faith in the system's fairness; conversely, when even rulers are bound by law, people develop a stake in the system's preservation.

The link between the rule of law and resilience is evident in both democratic and autocratic systems. In democracies, the rule of law ensures that established rules govern the political process itself – leaders can be removed via constitutional procedures, and minority rights are protected from the tyranny of the majority. This legal scaffolding prevents crises from spiraling into violence. In newer or weaker democracies, erosion of the rule of law often yields "illiberal democracy," where elections occur but elected leaders undermine courts, concentrate power, and erode checks and balances. Hungary and Poland have exemplified such backsliding in the past decade, as ruling parties have captured judicial institutions and media, leading to internal division and conflict with broader democratic norms. Studies conclude that, beyond formal constitutional design, what truly matters is whether officials and citizens internalize a culture of legality – one that respects the rules, exercises restraint, and upholds principles of civility and mutual toleration [23]. In essence, the lived culture of politics is as important to the resilience of the rule of law as the written statutes. In fragile states or autocracies, the absence of the rule of law often manifests as predatory governance and rampant corruption, fueling public anger and unrest. The Arab Spring uprisings of 2011, for instance, were triggered in part by outrage at lawless authoritarian abuses; protesters demanded "dignity" in the form of justice under law. Successful transitions from authoritarian rule (for example, South Africa after apartheid) typically involve major legal reforms – such as new constitutions, retrained judges and police, and anti-corruption agencies – to institutionalize the rule of law and signal that even powerful actors are now constrained by higher principles. Empirical measures, such as the WJP Rule of Law Index, consistently show a strong association between high rule-of-law scores and various measures of stability, including lower violence, less corruption, higher income levels, and greater investor confidence [17]. The rule of law also underpins human dignity by securing equal rights and fairness. Ultimately, democracy and the rule of law are mutually reinforcing: undermining legal constraints in the name of popular will, or vice versa, imperils both liberty and stability.

Dignity

Dignity connects individuals to the civic whole. Tocqueville noted that democratic life depends on citizens feeling they have a stake in collective decision-making [2]. When dignity is systematically denied – through exclusion, humiliation, or injustice – people often disengage from society or adopt oppositional identities. Ensuring basic dignity through equal rights and inclusion is not only an ethical duty but also a practical imperative for legitimacy. Citizens who feel respected by their institutions are far less likely to turn to violence or extremist alternatives. Conversely, pervasive indignities like corruption that belittle the average person, police brutality that violates basic rights, or extreme inequality that erodes self-worth all corrode the social fabric. As one scholar observed, many modern upheavals represent a "demand for dignity" by groups who feel ignored or disrespected. International norms echo this insight: the foundation of the postwar human rights framework is that recognizing human dignity is essential to freedom, justice, and peace [14]. Empirically, OECD studies find that perceptions of procedural fairness and having a voice strongly predict whether people voluntarily comply with public policies and remain civically engaged [9, 15]. In short, dignity is the social bedrock upon which trust and legitimacy ultimately stand.

Frameworks

Institutions do not operate in a vacuum; they are sustained or undermined by broader civic frameworks. These frameworks shape how legitimacy, trust, law, and dignity are experienced and maintained in daily life. Five elements stand out: civic norms, shared factual reality, collective attention, social memory, and institutional time horizons.

Civic Norms

Rights and laws written into constitutions require complementary civic habits to give them force. Tocqueville suggested that "the health of a democratic society may be measured by the quality of functions performed by private citizens" [2]. Likewise, Madison's system of checks and balances was designed to channel human ambition toward accountability rather than destruction [1]. Where informal norms of restraint, fair play, and tolerance are strong, political conflicts tend to be self-limiting; where such norms weaken, polarization and institutional hardball take hold [7, 8, 16]. Recent democratic erosions illustrate this vividly: when politicians treat rivals as enemies to be jailed, or when segments of the populace refuse to accept electoral outcomes, formal rules alone may not be enough to save the system. A comparative study of rule-of-law resilience in Europe concluded that, beyond having laws on paper, stability depends on whether officials actually respect the spirit of those laws and uphold principles of civility and mutual toleration [23]. In essence, the lived culture of politics can protect or undermine formal institutions.

Civic norms also affect everyday social trust and collective action. Communities steeped in norms of reciprocity – neighbors helping neighbors, honest dealings in business, active volunteerism – build dense networks of social capital that can rapidly mobilize in a crisis. For example, local civic groups spontaneously organized relief efforts in high-trust communities after disasters during the COVID-19 pandemic. By contrast, a culture of cynicism or every person for themselves leaves society brittle when collective action is needed. Nurturing positive civic norms through education, leadership example, and community programs is thus part and parcel of strengthening society's structural foundations. This might include civic education curricula that emphasize democratic values, public ceremonies or rituals that reinforce a sense of shared identity, and cross-partisan dialogues that rebuild some common understanding in polarized societies.

Shared Facts

A common base of factual knowledge and a social commitment to truth are essential for effective collective decision-making. Shared facts facilitate constructive public discourse and informed policymaking. When societies lose a shared factual foundation – when groups cannot even agree on basic realities, such as election results, public health data, or economic figures – deliberation fragments and cooperation falters. We are living through an era of what has been termed "Truth Decay," marked by the diminishing role of facts in public life and the splintering of shared reality [24]. The proliferation of misinformation and algorithm-driven echo chambers means that segments of society now inhabit vastly different informational universes. This directly undermines trust: those immersed in conspiratorial falsehoods come to view institutions as malevolent, while those in the fact-based community lose trust in compatriots they see as misled. As Daniel Kahneman observed, people often "assess the relative importance of issues by the ease with which they are retrieved from memory" – and modern media exploits this by saturating attention with sensational content [19]. Outrage and novelty crowd out sober analysis, making it harder to sustain focus on slow-burning challenges, such as infrastructure or climate change. Authoritarian actors exploit the attention economy's biases by seeding distractions or disinformation to divert public attention from their abuses.

The consequences of losing shared facts are severe. In the public health sphere, repeated misinformation about vaccines led significant minorities in various countries to distrust medical experts and reject life-saving COVID-19 vaccines, hampering pandemic control [10]. In the political realm, baseless but incessant claims of election fraud convinced millions of Americans to doubt the legitimacy of the 2020 U.S. presidential vote – culminating in a violent assault on the Capitol by a mob falsely believing that the election was "stolen." When different groups hold incompatible versions of reality, the legitimacy of the overall system suffers (since one side does not accept the other's victories as valid) and the rule of law is endangered (as people take extralegal action based on fabricated premises). In essence, a breakdown of common facts is a direct threat to resilience, akin to a society losing its shared language.

Countering this fragmentation requires multipronged efforts. Independent, high-quality journalism and public broadcasting can provide citizens with verified information that they trust. Civic education and media literacy programs can equip the next generation to evaluate sources and resist manipulation critically. There are positive examples: Finland has implemented comprehensive media literacy education and remains notably resilient against disinformation relative to other European Union (EU) countries. Meanwhile, governments and technology platforms are exploring ways to curb the most corrosive aspects of the attention economy – for instance, by adjusting algorithms to dampen the spread of virulent falsehoods or improving content moderation (while safeguarding legitimate debate). Evidence shows that transparent communication by authorities can enhance public trust. When governments openly share data and explain policies based on evidence, citizens are more likely to believe that "government uses the best available evidence" in decisions. Yet currently, only about 41% of people express such confidence on average [25]. Rebuilding a culture of truth is essential: norms that value honesty and expertise must be reinforced, and deliberate purveyors of falsehoods should face social and political costs. This cultural shift may be one of the most challenging tasks, but without it, the other pillars – legitimacy, trust, the rule of law, and dignity – rest on shaky ground.

Attention

Collective attention is both scarce and contested. Modern media ecosystems often divert public attention toward outrage and spectacle rather than fostering collective problem-solving. In a society inundated with information, trivial or sensational events can overshadow sober deliberation on structural issues. Barry Schwartz's work on the paradox of choice demonstrates that an overload of options or stimuli can lead to paralysis rather than empowerment [18]. In governance, fragmented or fickle public attention translates into short political time horizons and difficulty sustaining long-term initiatives [10, 19]. Policymakers struggle to invest in preventive infrastructure or gradual reforms if citizens and the media are continually distracted by the crisis of the day. Moreover, the quest for constant engagement in the attention economy often rewards extremism and conflict over moderation and consensus, further polarizing society.

The distortion of priorities caused by attention deficits undermines resilience. Challenges such as climate adaptation, demographic shifts, or education reform require sustained focus over years – yet democratic systems geared toward immediate gratification struggle to address these issues. Authoritarians can exploit short attention spans by creating diversions or blaming scapegoats, thereby avoiding accountability for underlying problems. Kahneman's insight, "what you see is all there is," underscores how easily public perceptions can be manipulated when attention is selective [19]. Addressing this requires institutions that can plan for the long term (such as independent agencies or cross-party commissions insulated from daily political pressures) and a media landscape that elevates important long-range issues. Some countries have attempted to institutionalize longer time horizons – for example, through future generations commissions or strategic foresight units in government. Ultimately, a society that cannot concentrate on its most consequential challenges will find its stability eroded by neglect.

Memory

Institutional memory encodes lessons learned from the past. Tocqueville warned that democratic societies are prone to "forget their ancestors," meaning they often fail to heed historical lessons [2]. Institutions preserve memory through laws, archives, educational curricula, and civic rituals that transmit values and knowledge across generations. When these mechanisms break down, societies must relearn past lessons at great cost. Comparative studies of post-authoritarian transitions show that countries with strong archives, intact legal records, and a continuous rule-of-law tradition tend to be more resilient, having internalized safeguards against repeating past mistakes [6, 20]. By contrast, nations with disrupted memory – where history is suppressed or distorted – may swing from one crisis to the next without the benefit of institutional learning.

Memory is not only about documents and history books; it also lives in culture and norms. For example, Germany's postwar resilience owes much to a deliberate effort at *Vergangenheitsbewältigung* (coming to terms with the past), which built a strong societal commitment to democracy and human rights after the horrors of the 20th century. In societies that have experienced civil conflict or authoritarian rule, truth commissions and memorials are often used to embed lessons of "never again" into collective memory. If those efforts are neglected, collective amnesia can set in – opening the door for old patterns of division or abuse to resurface. Thus, maintaining institutional memory through education, transparent record-keeping, and public remembrance is a key component of structural resilience.

Time Horizons

Time is a structural factor often overlooked. Societies that squander their institutional time – by constantly shifting policies or failing to plan beyond the next election – lose the capacity to act effectively on long-term issues. OECD research shows that countries with more consistent policy planning cycles and future-oriented governance (such as Finland or New Zealand) maintain higher trust levels and achieve more effective outcomes [11, 15]. Conversely, nations with volatile or myopic policy horizons tend to experience strategic drift, especially during crises [20]. For instance, a key contributor to the 2008 financial crisis was the short-termism in both private markets and public oversight – immediate profit-taking and political expediency took precedence over prudence, allowing systemic risks to accumulate.

Embedding longer time horizons into governance can improve resilience. This can involve multiyear budgeting, cross-party agreements on strategic priorities, or institutions mandated to consider the interests of future generations (such as Wales's pioneering Future Generations Commissioner). When leaders and citizens adopt a long-term perspective, they are more likely to invest in preventive measures, accept short-term sacrifices for long-term benefits, and avoid the constant shifting from one urgent problem to another. Building this forward-looking orientation is challenging in fast-paced political systems, but without it, societies risk being perpetually unprepared for foreseeable challenges.

Manifestations

When foundations and frameworks weaken, symptoms appear across core systems of society. We can observe these manifestations empirically in both the United States and around the world – in the functioning of democracy, the stability of markets, public health outcomes, security, and the tenor of civic life.

Democracy

Elections are ultimately tests of legitimacy. In recent years, public confidence in electoral processes has wavered. For instance, in the United States, only about half of citizens were confident in the accuracy of the vote count during the contentious 2020 election. Confidence rebounded slightly by 2024 but remained below pre-2016 levels [8]. In Poland, sweeping judicial reforms starting in 2015 undermined checks on the ruling party, contributing to significant democratic backsliding. Yet by 2023, mass civic mobilization and a change in government began to reverse some of these authoritarian trends, illustrating both the fragility and the potential resilience of democratic institutions [20]. Globally, the erosion is widespread: the V-Dem project reports that liberal democratic indicators declined in over 70 countries between 2013 and 2023 [20]. Moreover, as noted earlier, a 2025 survey found that a median of 64% of people in advanced economies are dissatisfied with how democracy is working in their country [13]. This discontent provides fertile ground for populist or anti-system movements, further stressing democratic norms.

Markets

Markets rest on trust and predictability. Fukuyama described trust as the "lubricant" of economic activity, enabling complex financial and commercial systems to function [5]. When trust evaporates, markets can seize up with astonishing speed. The 2008 global financial crisis illustrated this dynamic: after the failure of major institutions, credit markets froze and panic spread until unprecedented government guarantees restored a degree of confidence. One emblematic incident came when the Reserve Primary Fund "broke the buck" in September 2008, triggering a run on money market funds and a systemic liquidity crunch that required federal intervention [26]. More recently, in March 2023, the collapse of Silicon Valley Bank led to a quarter of its deposits being withdrawn in a single day, prompting broader fears of contagion in the banking sector [27]. In both cases, formal regulations and legal mechanisms proved insufficient without swift action to shore up trust.

The breakdown of ethical norms and legal oversight also contributed to these crises. Leading up to 2008, the pursuit of short-term profits over long-term stability, combined with lax regulatory enforcement, created hidden risks that eventually imploded, eroding public trust in financial elites and regulators. In the aftermath, many countries experienced surges of populist anger, as citizens sought to punish those perceived as having betrayed the public's trust. This illustrates how market failures can lead to political instability. Conversely, where the rule of law is strong and corporate governance is transparent, markets are more stable. Investors have confidence that contracts will be honored and malfeasance will be punished, thereby reducing risk premiums and fostering growth. In sum, economic prosperity and stability cannot be sustained without the underpinning of trust and legal integrity.

Public Health

Public health depends on trust in expertise, institutions, and the social contract. This became starkly apparent during the COVID-19 pandemic. Where citizens trusted health authorities and governments, they were more willing to follow guidance on masks, distancing, and vaccines; where trust was low, misinformation flourished, and compliance faltered. The World Health Organization documented how an "infodemic" of false information undermined vaccination efforts, particularly in communities already skeptical of authority [10]. Comparative studies show a strong correlation between trust in science and vaccine uptake across OECD countries [11, 22].

In high-trust countries like Denmark, vaccination rates exceeded 80%, whereas in lower-trust settings, they lagged far behind, contributing to higher mortality rates. Beyond pandemics, trust influences routine public health outcomes – whether people heed boil-water advisories, comply with tuberculosis treatment, or support funding for health programs. When institutions communicate transparently and consistently, and when citizens believe that leaders are acting in the public interest, health interventions are more effective, and societies are more resilient against disease.

Security

Law enforcement and national security fundamentally rely on public cooperation and legitimacy. Police and security agencies can only be truly effective if they have consent and trust from the communities they serve. When authority is perceived as unbiased and lawful, people are more likely to comply with laws and assist in maintaining order (for example, by reporting crimes or providing information). However, when authority is perceived as corrupt, discriminatory, or arbitrary, public cooperation plummets and enforcement costs soar [6, 17, 21]. Even a well-equipped security force will struggle to maintain order if citizens view it as an illegitimate occupier rather than a protector.

In societies with strong legal legitimacy, police and military forces can lean on voluntary compliance and community partnership. Consider community policing models, where trust between officers and residents leads to a shared responsibility for safety. In contrast, in places where the rule of law is weak or politicized, security forces often face hostility or apathy from the public, leading regimes to rely more heavily on coercion and violence to maintain control. This creates a vicious cycle: heavy-handed tactics further erode legitimacy, fueling unrest and making genuine security even harder to achieve. Thus, investing in the legitimacy of security institutions – through professionalization, accountability mechanisms, and community engagement – is essential to enduring stability.

Civic Life

Civic life is a mirror of dignity and inclusion in society. Tocqueville famously observed that voluntary associations were a bedrock of American democracy, cultivating habits of cooperation and public-mindedness [2]. When social trust and civility prevail, public discourse focuses on persuasion and problem-solving; when they break down, discourse degrades into hostility and zero-sum competition for power. We see symptoms of this degradation in many democracies: town halls and school board meetings turning acrimonious, social media amplifying hate speech, and declining participation in mainstream civic organizations. OECD surveys link citizens' sense of procedural fairness and social inclusion to their willingness to engage civically – people who feel heard and respected are far more likely to volunteer, vote, or otherwise contribute to public life [9, 15]. Conversely, those who perceive institutions as distant or biased often retreat from civic participation, creating a void that extremists or demagogues can fill.

Societies that fail to provide dignity and inclusion will eventually face backlash. Chile's 2019 protests, for example, were sparked by a minor increase in transit fares but quickly expanded into a broader revolt against longstanding inequalities and indignities – including high living costs and elites perceived as out of touch [28]. The intensity of the unrest revealed accumulated "dignity deficits" among large segments of the population. The aftermath saw Chile embark on a process to rewrite its constitution in an attempt to address citizens' demands for a more equitable and respectful social contract. This underscores that seemingly stable civic life can mask deep frustrations that erupt unless dignity is protected. Rebuilding civic life where it has frayed involves creating forums for dialogue across differences, emphasizing civic education and mutual respect, and ensuring that all groups in society feel seen and valued in the nation's public narrative.

Restoration and Policy Implications

Restoring structural resilience requires deliberate reinforcement of the four pillars and their supporting civic frameworks. This involves both domestic initiatives (e.g., within the United States) and international or comparative approaches, since many challenges are shared across countries.

Institutional Design

Institutions must be transparent, predictable, and firmly rooted in the rule of law. Reforms that strengthen judicial independence, electoral integrity, and legislative oversight are central to bolstering legitimacy and accountability [1, 6, 17]. In the United States, this could include further investment in secure and accessible election infrastructure, protections for nonpartisan election administration, and tighter limits on emergency powers to prevent executive overreach [8, 16]. Comparative evidence shows that increasing transparency can yield dividends: for instance, Estonia's pioneering e-governance systems and open data initiatives have significantly reduced opportunities for corruption and boosted public trust in government [29]. Digital tools that openly track government spending or decision-making make it harder for abuses to hide in the shadows, thereby reinforcing institutional legitimacy.

Participatory Governance

Beyond formal checks and balances, involving citizens directly in governance can deepen legitimacy and trust. One model is Taiwan's vTaiwan platform, which uses a deliberative online process to crowdsource citizen input on complex policy issues – striking a balance between open participation and structured discussion that yields actionable consensus [30]. Another example is participatory budgeting, first popularized in Brazil, where residents directly decide portions of municipal budgets. In many cities across the Global South, well-designed participatory budgeting has improved public trust and the alignment of spending with community priorities [31]. The key is genuine empowerment: when citizens see that their voices can influence policy, their trust in institutions and willingness to engage both increase.

Legal Reform and Rule of Law

Strengthening the rule of law requires sustained investment in the integrity and capacity of legal institutions. This includes adequately resourcing courts, insulating judges from partisan pressure, improving legal education, and enforcing anti-corruption measures. Countries that managed to stabilize fragile democracies often did so by bulwarking their judiciaries. For example, postapartheid South Africa's constitution created robust protections for judicial independence, which helped the country navigate political crises in subsequent decades [20], [21]. Conversely, when judicial independence erodes – as it did in Hungary through the 2010s – the decline in rule of law is accompanied by broader democratic regression [20]. International bodies can also assist: European Union mechanisms to monitor and sanction rule-of-law breaches (such as withholding funds) have at times slowed the backsliding in member states like Poland. Ultimately, fostering a culture of rule of law domestically – where no one is above the law, and the law itself is viewed as legitimate – remains the cornerstone of preventing tyranny and maintaining public trust.

Civic Capacity and Trust-Building

Rebuilding trust requires governments to deliver results competently and fairly and to communicate honestly. OECD studies show that trust increases when citizens perceive public services to be effective and see that officials are transparent about both successes and failures [9, 11, 15]. Initiatives to improve media literacy and civic education can also strengthen the informational foundations of trust. Finland's national media literacy curriculum, for example, has been credited with fostering a citizenry more resilient to misinformation, thereby enhancing long-term trust in institutions. Wales's innovative Future Generations framework – which requires policymakers to consider impacts on the young and unborn – has sought to build public confidence that government is acting with foresight and in the common interest [11, 32]. In the United States, targeted investments in civic education, local journalism, and participatory budgeting are being explored as ways to renew social capital, improve shared facts, and reconnect communities to governance [8, 16]. Such efforts illustrate that trust and good governance can be mutually reinforcing: as one improves, it tends to pull the other up as well.

Global Cooperation

Many structural challenges do not stop at national borders – including misinformation, financial contagion, pandemics, and climate change. Thus, enhancing resilience also requires international cooperation. Standards for information integrity (such as cross-national efforts to combat disinformation), global health surveillance systems, and coordinated economic safety nets are all examples of collective measures that can buttress national foundations. Institutions like the OECD, the World Health Organization, and the World Bank provide forums and frameworks for such shared learning and action [9, 10, 21]. By pooling knowledge on what works and by jointly addressing transnational threats, countries can reinforce each other's legitimacy and capacity. For instance, during the COVID-19 crisis, international collaborations on vaccine development and distribution (though imperfect) showed the potential of cooperative approaches in protecting global public goods. Looking forward, a concerted emphasis on democratic norms and rule-of-law promotion in foreign policy – through peer networks, assistance programs, and international law – could help counteract autocratization trends worldwide. In a deeply interconnected world, no society's foundations are entirely its own; global solidarity is part of the solution to fortifying freedom and stability.

Conclusion

A resilient society is not a gift of fate but an ongoing achievement – the product of sound structures tended vigilantly over time. We have explored how legitimacy, trust, rule of law, and dignity form the invisible pillars supporting stable freedom. Lincoln warned of "the silent artillery of time," which strikes at the foundations of nations more surely than any foreign foe [12]. His warning endures. Wealth, technology, or military power cannot secure freedom if legitimacy, trust, law, and dignity decay from within. Nations that invest in these invisible structures – through transparent institutions, shared civic frameworks, and inclusive governance – can weather storms that would otherwise break them. Those who neglect these pillars may appear stable until the moment they collapse.

The cracks in the temple of liberty are visible. The path to repair is also clear. The remaining question is whether societies will act with foresight and unity to reinforce their foundations before time's silent artillery does its worst.

Methods Appendix

Data Sources – OECD trust in government indicators (2017, 2023); World Justice Project Rule of Law Index (2015–2023); Pew Research Center public trust and democracy surveys (1958–2025); Varieties of Democracy (V-Dem) dataset (2010–2023); WHO COVID-19 "Infodemic" reports (2020–2022); World Bank Worldwide Governance Indicators (1996–2023).

Indicators – Public trust in national government; Rule of Law Index scores; V-Dem Liberal Democracy Index; COVID-19 vaccination uptake rates; judicial independence measures; electoral integrity indices; corruption perception scores.

Limitations include variations in measurement methodologies across datasets, potential time lags between institutional changes and observed outcomes, incomplete data coverage for some developing countries, and the challenge of attributing causality in complex social systems.

References

1 J. Madison, Federalist No. 51, 1788. Accessed: Feb. 2, 2026 [Online]. Available: https://avalon.law.yale.edu/18th_century/fed51.asp.

2 A. de Tocqueville, *Democracy in America.* New York: Vintage, 1945.

3 U.S. Constitution, 1787. Accessed: Feb. 2, 2026 [Online]. Available: https://constitution.congress.gov/constitution/.

4 The Bill of Rights, 1791. Accessed: Feb. 2, 2026 [Online]. Available: https://constitution.congress.gov/browse/essay/bill-of-rights/.

5 F. Fukuyama, *Trust: The Social Virtues and the Creation of Prosperity.* New York: Free Press, 1995.

6 D. Acemoglu and J. Robinson, *Why Nations Fail: The Origins of Power, Prosperity, and Poverty.* New York: Crown, 2012.

7 R. Putnam, *Bowling Alone: The Collapse and Revival of American Community.* New York: Simon & Schuster, 2000.

8 Pew Research Center, *Public Trust in Government: 1958–2023.* Washington, D.C., 2023 Accessed: Feb. 2, 2026 [Online]. Available: https://www.pewresearch.org/politics/2023/09/19/public-trust-in-government-1958-2023/.

9 OECD, *Trust and Public Policy: How Better Governance Can Help Rebuild Public Trust.* Paris: OECD, 2017.

10 WHO, *Infodemic Management: WHO Information Network for Epidemics.* Geneva, Switzerland: World Health Organization, 2022 Accessed: Feb. 2, 2026 [Online]. Available: https://www.who.int/teams/risk-communication/infodemic-management.

11 OECD, *Governance at a Glance 2023.* Paris: OECD, 2023.

12 A. Lincoln, *Address Before the Young Men's Lyceum of Springfield*. Illinois: Abraham Lincoln Online / Northern Illinois University, 1838 Accessed: Feb. 2, 2026 [Online]. Available: https://www.abrahamlincolnonline.org/lincoln/speeches/lyceum.htm.

13 Pew Research Center, "Dissatisfaction with democracy remains widespread in many nations," Jun. 30, 2025. Accessed: Feb. 2, 2026 [Online]. Available: https://www.pewresearch.org/global/2025/06/30/dissatisfaction-with-democracy-remains-widespread-in-many-nations/.

14 United Nations General Assembly, *Universal Declaration of Human Rights*. Paris: United Nations, 1948.

15 OECD, *How's Life? 2020: Measuring Well-being*. Paris: OECD, 2020.

16 Pew Research Center, *Political Polarization in Advanced Economies*. Washington, D.C.: Pew Research Center, 2022 Accessed: Feb. 2, 2026 [Online]. Available: https://www.pewresearch.org/global/2022/06/22/political-polarization-in-advanced-economies/.

17 World Justice Project, *Rule of Law Index 2023*. Washington, D.C.: WJP, 2023.

18 B. Schwartz, *The Paradox of Choice*. New York: HarperCollins, 2004.

19 D. Kahneman, *Thinking, Fast and Slow*. New York: Farrar, Straus and Giroux, 2011.

20 Varieties of Democracy (V-Dem), *V-Dem Dataset v13*. Gothenburg: University of Gothenburg, 2023.

21 World Bank, *Worldwide Governance Indicators*. Washington, D.C.: World Bank, 2023 Accessed: Feb. 2, 2026 [Online]. Available: https://info.worldbank.org/governance/wgi/.

22 OECD, *Trust and COVID-19 Vaccine Uptake*. Paris: Organisation for Economic Co-operation and Development, 2022 Accessed: Feb. 2, 2026 [Online]. Available: https://www.oecd.org/coronavirus/policy-responses/trust-and-covid-19-vaccine-uptake-61f2c9f5/.

23 M. Skóra, "A Decade of Rule of Law Backsliding: Lessons Learnt for the Next EU Legislative Period," *Review of Democracy* (CEU), Jun. 11, 2024. Accessed: Feb. 2, 2026 [Online]. Available: https://revdem.ceu.edu/2024/06/11/a-decade-of-rule-of-law-backsliding/.

24 J. Kavanagh and M. D. Rich, *Truth Decay: An Initial Exploration of the Diminishing Role of Facts and Analysis in American Public Life*. Santa Monica, CA: RAND Corporation, 2018.

25 "OECD Trust Survey 2023: Key Findings," OECD Press Release, Nov. 2023.

26 Financial Crisis Inquiry Commission, *Final Report of the National Commission on the Causes of the Financial and Economic Crisis in the United States*. Washington, D.C.: U.S. Government Printing Office, 2011 Accessed: Feb. 2, 2026 [Online]. Available: https://www.govinfo.gov/content/pkg/GPO-FCIC/pdf/GPO-FCIC.pdf.

27 U.S. Federal Reserve, *Review of the Federal Reserve's Supervision and Regulation of Silicon Valley Bank*. Washington, D.C.: Board of Governors of the Federal Reserve System, 2023 Accessed: Feb. 2, 2026 [Online]. Available: https://www.federalreserve.gov/publications/review-of-the-federal-reserves-supervision-and-regulation-of-silicon-valley-bank.htm.

28 CEPAL, *Social Unrest in Chile: Structural Causes and Responses*. Santiago, Chile: United Nations (ECLAC/CEPAL), 2020 Accessed: Feb. 2, 2026 [Online]. Available: https://www.cepal.org/en/publications/45727-social-unrest-chile-structural-causes-and-responses.

29 Transparency International, *Estonia: Digital Governance and Anti-Corruption*. Berlin, Germany, Transparency International, 2021 Accessed: Feb. 2, 2026 [Online]. Available: https://www.transparency.org/en/publications/estonia-digital-governance-and-anti-corruption.

30 vTaiwan, *Open Government and Digital Deliberation*. Taipei, Taiwan: vTaiwan Project / Taiwan Government, 2021 Accessed: Feb. 2, 2026 [Online]. Available: https://info.vtaiwan.tw/.

31 A. Fung and E. Wright, *Deepening Democracy: Institutional Innovations in Empowered Participatory Governance*. London: Verso, 2003.

32 Wales Future Generations Commissioner, *Well-being of Future Generations Act Progress Report*. Cardiff, UK: Office of the Future Generations Commissioner for Wales, 2021 Accessed: Feb. 2, 2026 [Online]. Available: https://www.futuregenerations.wales/publications/.

Part VII

Technological Leadership

36

The Costs of Failure – Lessons from Leadership Collapse

Overview

Technological progress often masks systemic fragility. Leadership failure – whether ethical, operational, or strategic – can produce cascading consequences that reverberate across industries, ecosystems, and nations. This chapter analyzes five emblematic cases of leadership breakdowns, each offering unique insights into what went wrong, the true cost of failure, and what resilient, ethically anchored leadership could have prevented.

Case Study 1: Boeing 737 MAX – Engineering Compromised by Financial Pressure

Background

In 2011, Boeing launched the 737 MAX to compete with Airbus's fuel-efficient A320neo. To reduce redesign costs and accelerate time to market, Boeing opted to retrofit the existing 737 platform rather than develop a new airframe. This decision initiated a series of compromises with deadly consequences.

Failure Mechanism

- **MCAS System Flaws:** The Maneuvering Characteristics Augmentation System (MCAS) was introduced to compensate for aerodynamic instability due to larger engines. Critically, it relied on a single angle-of-attack (AoA) sensor.
- **Pilot Training Neglected:** Boeing lobbied to avoid requiring simulator training to make the MAX more attractive to airlines. MCAS was not disclosed in flight manuals.
- **Corporate Culture Shift:** Following the 1997 merger with McDonnell Douglas, Boeing leadership increasingly prioritized financial performance over engineering excellence.

Outcomes

- **Crashes:** Lion Air Flight 610 (October 29, 2018) and Ethiopian Airlines Flight 302 (March 10, 2019) crashed, killing 346 people.
- **Grounding:** The MAX fleet was grounded worldwide for 20 months (March 2019–November 2020).

- **Financial Losses:**
  - **Boeing's Direct Costs:** $20+ billion (including $2.5B DOJ settlement, customer compensation, production halt)
  - **Stock Decline:** Market cap fell from $238B (February 2019) to $98B (March 2020)
- **Leadership Accountability:**
  - CEO Dennis Muilenburg was ousted in December 2019.
  - Congressional reports concluded Boeing "gambled with public safety."

Leadership Lessons

1) Short-term gains undermined system safety.
2) Suppressing dissent and sidelining engineers eroded institutional knowledge.
3) Failure to disclose critical system behavior to pilots violated ethical norms.

Case Study 2: Facebook and Cambridge Analytica – Ethical Lapses in Data Governance

Background

In 2015, academic Aleksandr Kogan collected Facebook user data via a personality quiz app, later passing data on ~87 million users to political consultancy Cambridge Analytica, in violation of Facebook's policies.

Failure Mechanism

- **Inadequate Oversight:** Facebook failed to audit third-party data practices rigorously.
- **Reactive Governance:** Facebook delayed disclosing the breach to regulators and users.
- **Profit-Driven Culture:** Leadership prioritized user growth and engagement over privacy, despite known risks.

Outcomes

- **Public Exposure:** March 2018 reports by *The Guardian* and *The New York Times* revealed the breach.
- **Market Impact:**
  - Stock fell ~19% in two weeks, erasing over $100 billion in market value.
- **Regulatory Repercussions:**
  - $5 billion FTC fine (largest privacy penalty in history, July 2019).
  - GDPR enforcement in Europe, sparking a global wave of data privacy legislation.
- **Trust Erosion:**
  - Pew Research (2019): 44% of U.S. users aged 18–29 deleted the app after the scandal.

Leadership Lessons

1) **Ethical foresight must guide platform design.**
2) **Transparency in crises matters more than reputational containment.**
3) **Data rights are civil rights in the digital age.**

Case Study 3: Enron – Accounting Fraud and the Illusion of Innovation

Background

Once lauded as the most innovative company in America, Enron used special purpose entities (SPEs) to hide debt and inflate earnings. Leadership engineered a façade of profitability that unraveled catastrophically.

Failure Mechanism

- **Fraudulent Accounting:** Enron used mark-to-market accounting to project hypothetical future profits as current income.
- **Conflicted Oversight:** Auditor Arthur Andersen facilitated the fraud and shredded evidence during the SEC investigation.
- **Leadership Malfeasance:** Executives, including CEO Jeff Skilling and Chairman Ken Lay, knowingly misled stakeholders.

Outcomes

- **Collapse:** Enron filed for bankruptcy on December 2, 2001 – the largest in U.S. history at that time.
- **Job and Retirement Losses:**
  - 20,000 employees lost jobs.
  - $1.2 billion in employee retirement plans were wiped out.
- **Regulatory Fallout:**
  - Passage of the Sarbanes-Oxley Act (2002), requiring executive accountability and stricter auditor independence.

Leadership Lessons

1) **Culture of deception metastasizes across institutions.**
2) **Risk externalization cannot substitute for value creation.**
3) **Regulatory capture and unchecked hubris erode market integrity.**

Case Study 4: Fukushima Daiichi – Crisis Management Failures in Complex Systems

Background

On March 11, 2011, a magnitude 9.0 earthquake and subsequent tsunami devastated Japan's Tōhoku region. At the Fukushima Daiichi Nuclear Power Plant, backup power systems failed, leading to reactor meltdowns.

Failure Mechanism

- **Design Flaws:** Backup diesel generators were located in basements vulnerable to flooding.
- **Overconfidence:** TEPCO (Tokyo Electric Power Company) downplayed tsunami risks despite historical warnings.

- **Information Suppression:** Initial response was opaque; the government and TEPCO delayed releasing radiation data.

Outcomes

- **Human and Economic Cost:**
  - Over 154,000 people evacuated.
  - Estimated cleanup cost: $200–250 billion USD (as of 2024).
- **Trust Collapse:**
  - Public confidence in nuclear power plummeted.
  - Germany announced its nuclear exit by 2022.

Leadership Lessons

1) **Redundancy and physical resilience are non-negotiable.**
2) **Crisis response requires transparency and trust-building.**
3) **"Black swan" risks demand proactive scenario planning, not probability denial.**

Case Study 5: Wells Fargo – Misaligned Incentives and Systemic Ethical Failure

Background

From 2002 to 2016, Wells Fargo employees opened over 3.5 million unauthorized accounts to meet aggressive sales targets under the "Gr-eight" cross-sell campaign.

Failure Mechanism

- **Pressure Culture:** Front-line staff were pushed to meet quotas under threat of termination.
- **Lack of Whistleblower Protection:** Internal complaints were ignored; some whistleblowers were fired.
- **Delayed Accountability:** Senior leadership denied responsibility despite knowledge of widespread misconduct.

Outcomes

- **Financial and Legal Penalties:**
  - Over $3 billion in fines (U.S. Department of Justice [DOJ], U.S. Securities and Exchange Commission [SEC], and Consumer Financial Protection Bureau [CFPB] settlements).
- **Leadership Fallout:**
  - CEO John Stumpf resigned in 2016; banned from banking in 2020.
- **Reputational Damage:**
  - Trust index dropped to 23% among U.S. customers in 2017 (Gallup).
- **Cultural Reckoning:** 5,300 employees were terminated over several years.

Leadership Lessons

1) **Incentives shape behavior – ethics must precede metrics.**
2) **Culture audits are strategic necessities, not HR formalities.**
3) **Reputation is a system-wide asset: once compromised, difficult to restore.**

Cross-Cutting Themes and Failure Archetypes

Archetype	Symptoms	Root Cause	Prevention Strategy
Financialization of Vision	Innovation used as a financial façade	Profit maximization over stakeholder value	Reinforce mission alignment and ethical key performance indicators (KPIs)
Opaque Complexity	Incomprehensible systems to insiders	Insufficient feedback loops	Regular independent audits and red teaming
Suppressed Dissent	Fear of retaliation for raising alarms	Culture of denial	Whistleblower channels and reward mechanisms
Reactive Ethics	Crisis-first approach to accountability	Weak board governance	Ethics embedded in product, process, policy
Redundancy Neglect	Single-point failures tolerated	Optimized fragility	Scenario planning and resilient design reviews

Conclusion: The Hidden Cost of Collapse

Failures of leadership are rarely sudden – they are the outcome of eroded trust, ignored signals, and decisions made in defiance of complexity. Whether the result is a grounded airplane, breached democracy, collapsed bank, or contaminated ecosystem, the cost is more than monetary. It is human.

Resilient leadership requires more than vision. It demands:

- Early warning systems for ethics and engineering
- Cultures that reward dissent and learning
- Systemic design for trust, redundancy, and transparency

Failure, when studied rigorously, becomes the seedbed of reform. Leadership that endures is leadership that learns – before, during, and after the storm.

37

Planetary Intelligence – Ecological Systems, Climate Resilience, and the Leadership We Need

The Earth Is Not a Stakeholder – It Is the System

In the summer of 2023, the world crossed a sobering threshold: for the first time in recorded history, global average temperatures stayed above the 1.5 °C warming limit for more than 12 consecutive months. The Amazon began emitting more carbon than it absorbed. Arctic permafrost, long considered dormant, released methane in explosive bursts. Floods in Libya wiped out entire cities. Canada's wildfires produced more carbon in one season than the nation's economy had in decades. The Anthropocene did not begin with a bang but with smoke, fire, silence, and retreat.

The Earth does not negotiate. It integrates. It remembers. And it reacts with the full force of its entangled systems. The question before us is not whether we can reduce emissions or deploy renewable technologies. It is whether we can lead as though the Earth is not a stakeholder to be served but the entire system in which stakeholders emerge.

Climate as a Complex Adaptive System

Climate is not simply a collection of weather patterns or CO_2 levels. It is a distributed, nonlinear, self-regulating system – what ecologist James Lovelock once conceptualized as "Gaia": a planetary-scale system whose components (biosphere, atmosphere, hydrosphere, cryosphere, and lithosphere) coevolve dynamically.

Earth's Feedback Architecture

- **Carbon Cycle Feedbacks:** Soil and forest systems regulate CO_2 but degrade under thermal stress.
- **Albedo Shifts:** Ice melt reduces reflectivity, accelerating heat absorption.
- **Methane Release:** Melting permafrost stores 1,400+ Gt of carbon (NOAA, 2022), with methane 84× more potent than CO_2 over 20 years.

Crossing Planetary Boundaries

The Stockholm Resilience Centre identifies **nine planetary boundaries**. As of 2024, we have transgressed six:

Boundary	Status (2024)	Safe Threshold
Climate Change	Breached	<350 ppm CO_2
Biodiversity Loss	Breached	<10 extinctions/million species/year
Biogeochemical Flows (N, P)	Breached	Sustainable nutrient cycles
Land-System Change	Breached	<15% converted forest cover
Freshwater Use	Breached (seasonal)	<4,000 km^3/year
Novel Entities (Plastics, PFAS)	Breached	Undefined threshold
Ocean Acidification	Approaching	≥80% preindustrial saturation
Ozone Depletion	Safe (recovering)	>276 DU (Dobson Units)
Atmospheric Aerosols	Uncertain	Undefined

Each breached boundary is not an isolated risk, but a pressure point on the global life-support system. Leadership in this context must move from **extraction** to **integration**.

The Ecological Infrastructure of Civilization

Civilizations do not survive in spite of nature; they survive because of the biosphere's ecological services: pollination, freshwater provisioning, carbon sequestration, disease regulation, and oceanic stability. These services have been treated as externalities. But in reality, they are existential inputs.

Value of Nature's Services

- Estimated Annual Value: $125–150 trillion [1].
- Amazon Basin alone regulates **20%** of the Earth's freshwater and **absorbs 2.2 billion tons** of CO_2 annually [2].
- Wetlands provide $14,785/ha in annual services vs. the $2,000/ha they are sold for in development [3].

This mismatch between ecological value and economic visibility is not just a market failure – it is a moral blindness embedded in decision-making architectures.

Technological Systems Must Mirror Ecological Systems

To lead through the climate crisis, technological systems must emulate ecological systems in three core dimensions:

Resilience over Efficiency

- **Natural Systems:** Redundant, diverse, decentralized.
- **Conventional Systems:** Optimized, centralized, brittle under stress.

Example

- U.S. power grids in Texas (2021 freeze) and California (2020 fires) failed under climate-induced extremes.
- **Microgrids** and **vehicle-to-grid systems** are decentralized, resilient alternatives gaining traction.

Circularity over Linearity

- Natural systems operate on **closed-loop cycles** (carbon, nitrogen, and water).
- Current systems run on **linear extract–produce–waste models**.

Scalable Circular Tech Examples

- **Ecovative:** Mycelium-based biodegradable packaging
- **CarbonCure:** Captures CO_2 in concrete curing
- **Gr3n:** PET chemical recycling through microwave-induced depolymerization

Regenerative Intelligence

- Ecosystems self-correct through **feedback, adaptation, and diversity**.
- Artificial intelligence (AI) and Internet of Things (IoT) systems must transition from monitoring to **adaptive governance**.

Example

- **Precision agriculture** with real-time AI soil analytics reduces water and fertilizer usage by up to **60%** [4].
- **Digital twin forests** track biomass and CO_2 absorption across seasons, improving reforestation strategies.

Climate Technologies: Hype vs. Transformation

The Hype Funnel

Many so-called green technologies (direct air capture, hydrogen) promise miracles but remain **cost-intensive**, **energy-intensive**, and **unscalable** without state subsidy.

Direct Air Capture Reality (2024)

- Cost: $600–$1,200 per ton CO_2 (IEA)
- Global capacity: ~10,000 tons/year vs. 36 billion tons emitted

High-Impact, Scalable Interventions

Technology	Cost/Ton CO_2 Avoided	Readiness	Leadership Leverage
Grid-scale storage (Li-ion)	$120–200	TRL 8–9	Regional utilities
Agroforestry systems	$5–15	TRL 7	National ministries
Heat pumps	$20–50	TRL 9	Urban governments
Smart heating, ventilation, and air conditioning (HVAC) and insulation	$10–25	TRL 9	Building codes
Synthetic protein	$100–300	TRL 6–7	Food sector leaders
Rice methane abatement	<$10	TRL 8	Southeast Asia G20

Real leadership is not defined by "moonshot funding" but by the **mobilization of institutional alignment, behavioral adoption, and systemic resilience**.

The Geoethical Imperative

Intergenerational Risk Transfer

We are not only externalizing environmental costs but **transferring risks across time**. This violates the core principle of sustainable development: meeting the needs of the present **without compromising the ability of future generations** to meet theirs.

Quantitative Forecasts

- By 2050, 1.2 billion people will be at risk of displacement due to climate [5]
- Annual global damages will reach \$3.8 trillion under 3 °C warming [6]
- Fisheries will collapse by 2048 under current acidification trends (Science, Vol 314)

Just Transition as Leadership Litmus Test

A leadership model that succeeds must deliver:

- **Energy Justice:** Equitable access and pricing for low-income communities
- **Ecological Restoration:** Rewilding, biodiversity corridors, and carbon sinks
- **Historical Accountability:** Developed nations funding \$100B+/year for global south transitions (still unmet, UNFCCC COP28)

Strategic Leadership for Planetary Stewardship

System-Based Decision Architecture

- **Foresight:** Use horizon scanning and scenario modeling (e.g., Shell's Sky 2050 and WEF's Risk Interconnection Map)
- **Feedback Loops:** Integrate satellite and ground sensors into climate-aware governance
- **Nonlinear Key Performance Indicators (KPIs):** Incorporate ecosystem thresholds and irreversible tipping points into Environmental, Social, and Governance (ESG) scoring

Multilevel Governance

Leadership is now layered:

- **Local:** Climate-resilient zoning and community energy
- **Corporate:** Decarbonization pathways and biodiversity accounting
- **National:** Carbon border adjustments and circular economy laws
- **Transnational:** Loss and damage finance and climate liability courts

Cultural Narratives

No transformation sustains without a shift in meaning. Leadership must reframe:

- From **climate as constraint → climate as compass**
- From **tech as fix → tech as harmonizer**
- From **humans as dominators → humans as participants in biotic systems**

The Covenant with the Future

There are no sidelines in the climate system. Every decision is embedded. Every delay is systemic. Every innovation carries not just carbon implications, but ethical ones. Leadership in the 21st century is not defined by charisma or credentials, but by climate consciousness, ecological humility, and the courage to govern with planetary feedback in mind.

We must design sentient systems – not only intelligent, but wise; not only powerful, but regenerative. To lead now is to make the Earth think through us – to weave technology, biology, governance, and justice into one continuous feedback loop of resilience.

References

1 R. Costanza *et al.*, "Changes in the global value of ecosystem services," *Global Environ. Change*, vol. 26, pp. 152–158, 2014. Accessed: Feb. 2, 2026. Available: https://www.sciencedirect.com/science/article/pii/S0959378014000685.

2 World Wide Fund for Nature (WWF), *Living Planet Report 2022: Building a Nature-Positive Society*. Gland, Switzerland: WWF, 2022 Accessed: Feb. 2, 2026. Available: https://livingplanet.panda.org/en-GB/.

3 The Economics of Ecosystems and Biodiversity (TEEB), *The Economics of Ecosystems and Biodiversity: Mainstreaming the Economics of Nature—A Synthesis of the Approach, Conclusions and Recommendations*, 2011. Accessed: Feb. 2, 2026. Available: https://teebweb.org/publications/teeb-for/policymakers/.

4 J. Deere, *Sustainability Report 2023*. Moline, IL, USA: Deere & Company, 2023 Accessed: Feb. 2, 2026. Available: https://www.deere.com/en/our-company/sustainability/.

5 International Organization for Migration (IOM), *World Migration Report 2022*. Geneva, Switzerland: IOM, 2021 Accessed: Feb. 2, 2026. Available: https://worldmigrationreport.iom.int/wmr-2022.

6 Swiss Re Institute, *Biodiversity and Ecosystem Services: A Business Case for Re/Insurance*. Zurich, Switzerland: Swiss Re, 2023 Accessed: Feb. 2, 2026. Available: https://www.swissre.com/institute/research/topics-and-risk-dialogues/biodiversity-and-ecosystems.html.

38

What You Do, and How You Tell, Matters: From Building Systems to Shaping Minds, the Real Legacy Lies in Purpose, Integrity, and Clarity

When people ask what I've learned from four decades of building complex systems – power grids, defense networks, security, and resilience protocols – the answer is rarely technical. It's this: what you make is not just a project but a story about your values. And if no one can understand that story – if it's unclear, impersonal, or dishonest – it won't carry. It won't last. This piece is about both halves of that truth: what we do, and how we tell it. One without the other fails. But together, they create the only legacy worth leaving.

Part 1:
What Comes After the Work **(reflections on engineering, ethics, and legacy)**

I spent 40 years working on complex systems – power grids, defense networks, energy infrastructure, and resilience frameworks. They were built to last. Some did. Some didn't. Over time, the questions I asked myself changed. It was no longer enough to ask, "Did it succeed?" I began to ask, "Did it matter?" A dear friend and colleague of mine, Mr. Michael Wright, often put it pointedly: "Do you know what you do, does?" That simple, strange question never let go of me. It forced me to look beyond the immediate success metrics and examine my work's deeper impact. I saw good people burn out and strong leaders drift off course. Not from a lack of effort or intelligence – but from forgetting who the work was for. In chasing deadlines, budgets, and technical feats, we risk losing sight of the people and purpose at the heart of our projects. In my later years, I slowed down and started writing – not official reports or strategic plans, but honest reflections on what I saw and what I learned the hard way. I wrote books, essays, poems, and stories. Some were technical, others personal. All of them were true. They were not written to impress, but to clear the air and distill what mattered in those four decades of building systems. Through that process, a few principles emerged clearly, like bedrock under shifting sands:

- If your work doesn't help someone stand up, it's not the work we need.
- If your system can't carry the people at the edge, it's already failed.
- It wasn't built to last if your voice disappears in silence when things get hard or when many go quiet.

These statements became guideposts for me. They may sound simple, but each was paid for in experience – often in mistakes and crises. I want to explore these ideas in depth in the following pages, drawing on real historical cases and my journey. This is not a chronological memoir of my career; it is a thematic exploration of responsibility, ethics, and integrity in the things we build. It is written for anyone – leaders, engineers, policy-makers, students, or concerned citizens – who wonders about the impact of what they do on other people, on society, and the future. Together, we will examine where things have gone wrong, where they have gone right, and what lessons we can carry forward. I write in the spirit of a nonfiction treatise, blending lived experience with deep research. My goal is not to preach (I have no halo to polish), but to share stories and evidence that might illuminate a better path.

From Success to Significance: Did It Succeed vs. Did It Matter?

Early in my career, I was obsessed with making projects succeed. Success meant delivering on requirements – the bridge held up traffic, the power plant came online, and the software system passed its tests. We cheered when the thing we built "worked." However, as years passed, I encountered projects that technically succeeded, yet they left me and others with gnawing doubts. We had to ask: What was the point of this success? Who benefited? A system can meet its specifications and still fail in the bigger picture.

One stark historical example of this distinction is the Manhattan Project, which successfully achieved its goal during World War II: creating the first atomic bombs. By purely technical standards, it was a triumph of physics and engineering. The bombs worked as designed. But after the mushroom clouds faded, many of the scientists grappled with whether that success mattered in a good way – or whether it was, in fact, a grave moral failure. J. Robert Oppenheimer, the scientific director of the project, famously reflected on the devastation of Hiroshima and Nagasaki by saying, "I feel I have blood on my hands." He had led the team that accomplished its mission, yet the human cost of that accomplishment haunted him. In the years that followed, Oppenheimer advocated for international nuclear arms control and opposed further developments like the hydrogen bomb – actions which resulted in him being ostracized and stripped of his security clearance.

Even more telling is the story of one of Oppenheimer's colleagues, Joseph Rotblat. Rotblat was a Polish physicist on the Manhattan Project who had initially joined to help stop Nazi Germany. But in late 1944, he learned that the project's leadership viewed the bomb not merely as a deterrent to Nazi aggression, which was by then waning, but as a geopolitical tool against the Soviet Union. Rotblat was alarmed. In his conscience, the work no longer aligned with its original defensive purpose. He made the extraordinary decision to leave the Manhattan Project in 1944, the only scientist to do so. His departure was met with suspicion (the U.S. government even falsely accused him of espionage to discredit his protest), but Rotblat held to his principles. He spent the rest of his life campaigning against nuclear weapons, helping to found the Pugwash Conferences on Science and World Affairs – efforts that earned him a share of the Nobel Peace Prize in 1995. Reflecting on the dawn of the atomic age, Rotblat said, "Science became identified with death and destruction." Unlike many of his peers, he asked not just "Can we build it?" but "Should we?" and "What will it do to the world?" He walked away when he concluded that his work was contributing to potential injustice and suffering. Rotblat's story starkly contrasts with Oppenheimer's. Oppenheimer stayed and saw the project through, later wrestling with remorse; Rotblat decided that if the work no longer helped humanity stand up, it was not work worth doing, and he acted on that belief. His question was effectively: Do you know what your work does to the world? And once he answered it, he chose significance over success.

This lesson transcends nuclear science. We face it in technology, engineering, and business today. In Silicon Valley, for instance, some tech professionals have begun to ask whether the products they build truly help people or merely addict them, surveil them, or even harm them. In 2018, Google was in a dilemma akin to a mini-Manhattan Project moment. The company was contracting with the U.S. Department of Defense on "Project Maven," an effort to use artificial intelligence to analyze drone surveillance footage – technology that could improve targeting for lethal drone strikes. By technical metrics, Google's artificial intelligence (AI) prowess made the project a likely success. However, thousands of Google's employees questioned whether this "success" would have a positive impact or make the company complicit in taking lives. In an unprecedented internal revolt, over 4,000 Google employees signed a petition urging the company to cancel the military project, and about a dozen employees even resigned from their jobs in protest. They raised ethical concerns that went to the heart of the purpose: they didn't want their work in AI to be used for killing or surveillance without oversight. Faced with this pressure, Google's leadership ultimately announced that it would honor its current contract but not renew it, and the company established new AI ethics guidelines pledging not to build technologies for use in weapons. This episode illustrated how workers redirected a company's definition of success – from simply landing a lucrative contract – toward a broader notion of significance and responsibility. They asked, in effect, "What does our work do, and does it align with our values?" and forced their leaders to answer that question.

The difference between "Did it work?" and "Did it matter?" is the difference between output and outcome. A power plant's output might be gigawatts of electricity generated (a success). Still, the outcome could be lights on in schools and hospitals – or, if misused, pollution that sickens a community. A social-media platform's output might be a slick app with millions of users (success). Still, the outcome might be a radical change in how those users consume information, possibly undermining social cohesion or mental health. With experience, I learned to pursue significance over mere success. This means defining the purpose of a project from the outset in human terms: who will be better off if we succeed? How will we know that our success made a positive difference? Those questions must be asked early and often. Otherwise, we risk pouring our energy into building monuments that might stand tall but cast dark shadows.

Who Is It For? Remembering the People at the Center

One reason projects lose their meaning is that leaders and teams can drift away from the original purpose and audience of the work. I've seen leaders start inspired to serve society, yet over time become fixated on internal goals – market share, promotions, and legacy – to the point where they forget the people they intended to help. I've seen engineers (myself included) get so deep into the technical challenges that we lose sight of the end user. This drift is often gradual and unintentional, but it can lead to systems that, in the end, serve themselves or their creators' egos more than the public. To prevent that, we must keep asking: Who is this for?

Consider the realm of civil infrastructure – systems explicitly built to serve communities. If any domain should never lose sight of who it's for, it is public infrastructure. Yet even here, that focus can fade, sometimes with disastrous consequences. The Flint water crisis is a heartbreaking example. Flint, Michigan, was in economic distress, and officials were pressured to cut costs. In 2014, a state-appointed emergency manager (who had authority over the city's decisions) switched Flint's water supply from the established source (treated water from Detroit's system) to the local Flint River to save money. It was supposed to be a temporary, budget-friendly fix. But in the rush to implement this change cheaply, the basic question "who is this for?" – in other words, "what will this do to the people drinking the water?" – was tragically neglected. The river water was not properly treated with corrosion inhibitors, and it began to leach lead from Flint's aging pipes into the drinking water. Citizens immediately complained that the water from their taps looked and tasted odd, even foul. They saw rashes, hair loss, and illnesses. But for months and months, their voices were downplayed or ignored by those in charge. Officials assured residents that everything was fine, even as evidence showed that it was not. In reality, tens of thousands of Flint residents – including children – were being exposed to dangerous levels of lead, a potent neurotoxin. The water supply also saw a spike in bacteria; an outbreak of Legionnaires' disease (a severe pneumonia) killed at least 12 people and sickened dozens more, likely due to the compromised water system. General Motors, which had a plant in Flint, noticed that the water was corroding auto parts and stopped using it in their factory – yet the city's people were told to keep drinking it. It eventually came out that some in the state government knew about the problems early on but chose to delay action, prioritizing cost savings and image management over citizens' health. Internal communications revealed that financial cost, not public health, remained the "primary decision driver" even as the crisis worsened. Here was a textbook case of a system failing the people it was meant to protect. The water system was poisoning the community it was built to serve. The engineers and managers involved had solved the technical problem of sourcing water more cheaply – a "success" on paper – but they had utterly lost sight of the fundamental purpose of a public water supply: to deliver safe, clean water to the public. In Flint, the work did not help people stand up; it knocked them down. The system did not carry the people to the edge; it pushed them over. And when residents raised their voices, many in power fell into an appalling silence or denial, at least initially. Only through the tenacity of concerned parents, local activists, and outside scientists (like Professor Marc Edwards from Virginia Tech and Dr. Mona Hanna-Attisha, a local pediatrician who found elevated lead in children's blood) did the truth finally come out and the water source get changed back after 18 months of damage. The lessons from Flint are numerous: technical competence cannot substitute for moral competence, and efficiency cannot trump safety. If officials ever say, "We met the budget and timeline, so it's a success," while citizens are lining up for bottled water and children are getting sick, something has gone wrong about who the work was for.

Another sobering case: Hurricane Katrina and the New Orleans levees. New Orleans sits below sea level, protected by an elaborate levee system and floodwalls built by the U.S. Army Corps of Engineers over decades. Those defenses are life-or-death infrastructure for the people behind them – arguably some of the most essential systems one can build for a community. Yet, the system catastrophically failed when Hurricane Katrina struck in August 2005. Over 50 levee breaches occurred around New Orleans, and roughly 80% of the city was flooded. In some neighborhoods, like the Lower Ninth Ward (a historically Black, low-income community), floodwaters reached 10 to 15 feet high, submerging houses up to the rooftops. People who could not evacuate – people experiencing poverty, older people, and the vulnerable – were trapped in attics or on roofs for days. Over 1,500 people in Louisiana lost their lives, many by drowning. The devastation left hundreds of thousands homeless; entire communities were uprooted. Why did this happen? In the immediate aftermath, some federal officials claimed that the hurricane was too strong, exceeding what anyone could have anticipated. "I don't think anybody anticipated the breach of the levees," President Bush said at the time. But that wasn't true – experts had warned for years that New Orleans' levees were vulnerable to a major hurricane. And more pointedly, investigations revealed that many of the levees did not perform even to their design level. The system was supposedly designed to handle a Category 3 hurricane, yet some floodwalls gave way at water levels far below what they should have held. In 2006, the Army Corps of Engineers produced a 6,000-page report taking responsibility for many of the failures: they admitted the levees failed due to "flawed and outdated engineering practices" in their design and construction. Decades of piecemeal planning, budget shortfalls, and compromises had left a patchwork system that was not truly up to the task of protecting the people in a worst-case scenario. In essence, those levees were a system that wasn't built to last when the real test came. The people at the edge – the people in the lowest-lying, poorest areas – paid the highest price for those flaws. There is also a social dimension: after the fact, a government task force characterized the disaster as a case of environmental injustice, noting that Flint's poor, predominantly African American population (in the water crisis) did "not enjoy the same degree of protection" as other communities. Similarly, in New Orleans, it was the least privileged who suffered the worst from the infrastructure failure. In both cases, the systems we built betrayed our most vulnerable.

The overarching lesson is that public systems must be designed with the most vulnerable in mind, not the most convenient average scenario. If a levee system can't protect the lowest-income neighborhood as well as the richest, it has failed its moral purpose. If a power grid crumples in a cold snap and leaves those on the margins freezing in the dark (as happened in Texas during the 2021 winter storm), that grid was not truly "resilient" in the human sense. Resilience means the ability to withstand shocks while protecting those who rely on you. Too often, we declare a system resilient because it meets a specification or standard – but if that standard was set without considering the needs of those at the edge cases, the declaration is an illusion. Real resilience includes social resilience: ensuring that every segment of the community can weather the storm.

As an engineer, I gradually learned that technical specs like "1% annual chance flood" or "N-1 grid reliability" are not just numbers – they are ethical choices about whom we are willing to let suffer how often. We cannot forget the faces and lives behind the metrics. Who is it for? It's for all of us, especially those who have the least cushion when systems break. Keeping the people at the center isn't just a feel-good notion; it leads to better engineering. New Orleans' new levee system (upgraded after Katrina at a cost of over $20 billion) is now built to a higher standard – but even so, experts still question if it's enough for the next big storm. The work of remembering who it's for is never done; it requires continuous vigilance, scientific honesty about risks, and the courage to invest in safety margins that protect everyone, not just the comfortable.

Integrity and Voice: When the Alarms Go Unheeded

The third principle I listed – "It wasn't built to last if your voice disappears in silence when things get hard or when many go quiet" – speaks to the importance of integrity and speaking up, especially in dire moments. A system might look robust on the outside, but if internally its culture silences truth-tellers and dissenters, it is fragile at the core. Eventually, reality catches up, and often with catastrophic results. In my career, I witnessed how the absence or presence of ethical courage could tip the balance between disaster and safety. It's not enough for a few individuals to know the right thing; they must also be empowered to say it and act on it.

Perhaps no story illustrates this better than the Space Shuttle Challenger disaster. By January 1986, NASA's Space Shuttle program had chalked up successes and was pushing to increase launch frequency. The Challenger's mission that month included a much-publicized plan to send a schoolteacher (Christa McAuliffe) into space – a symbol of making spaceflight accessible. But behind the scenes, some engineers were deeply worried about the Shuttle's solid rocket boosters. Specifically, a team at the contractor Morton Thiokol had for years been tracking a dangerous flaw in the booster O-ring seals. These rubber O-rings were critical gaskets meant to prevent hot combustion gases from leaking at the joints between booster segments. If the O-rings failed, it could cause exactly what ultimately happened to Challenger – an explosion of the fuel tank. The engineers had seen O-ring damage on multiple previous flights and had repeatedly raised alarms that, in cold weather, the O-rings might not seal at all. On the night before the launch,

January 27, 1986, the predicted temperature for the Florida launch was far lower than any previous shuttle launch. Roger Boisjoly, a Thiokol engineer, and his colleagues were beside themselves. All their data indicated that launching in such cold could be extremely dangerous, because the O-rings could stiffen and fail. They urgently recommended a postponement. Initially, Thiokol management backed their engineering team and told NASA, "We recommend not to launch." This was a moment where the voices of concern were sounding the alarm loud and clear. But then the room went quiet – or rather, too noisy with pressure. NASA managers, facing schedule pressures and perhaps overconfident from past successes, pushed back hard. According to later investigations, one NASA manager responded to the no-launch recommendation with anger, effectively challenging the engineers to prove beyond a doubt that the O-rings would fail, rather than heeding their cautious advice. Under this intense pressure, Thiokol executives convened privately and overruled their engineers. One executive famously told a concerned engineer to "take off your engineering hat and put on your management hat" – meaning, stop thinking about safety and start thinking about appeasing the customer and the schedule. They reversed the recommendation and gave NASA the green light. The next morning, Challenger lifted off and, 73 seconds later, disintegrated as the O-ring in the right booster failed due to the cold, releasing flames into the fuel tank. The shuttle and its crew of seven were lost. It was one of the most devastating and public technological failures in history, and it was entirely preventable. A subsequent Presidential commission concluded that organizational culture was as much to blame as the technical flaw. NASA had a culture that silenced dissent and normalized deviance – each prior near-miss had been downplayed, warning signals lost in the noise of a "can-do" culture. As one analysis put it, the "nation's smartest minds" were "blind to the subtleties of engineering complexity" and had "overlooked a litany of clear warnings," seduced by the myth that they could do the impossible routinely. The Challenger case seared into my mind that if those in charge do not listen – truly listen – to the voices urging caution, then all the technical excellence in the world can't save you from catastrophe. Silence and acquiescence can kill. It's an ethical failure as much as an engineering one. Indeed, the first canon of the Engineering Code of Ethics is to hold paramount the safety, health, and welfare of the public. In the Challenger saga, that canon was violated not by intent to harm, but by a lack of moral courage to delay a launch and disappoint political masters. When Boisjoly later recounted the incident, he expressed deep regret that he didn't fight even harder and go public to stop the launch. He and a few others knew the risk – but organizational dynamics effectively stifled their voices.

You might think such a failure could never happen again, but similar patterns have recurred. In 2003, the Space Shuttle Columbia broke apart on re-entry. Again, NASA's culture of not heeding engineers' concerns was scrutinized – some engineers suspected damage to Columbia's wing during launch. Still, managers discouraged aggressive investigation, with one manager famously saying, "In my opinion, not much we can do about it" (a fatalistic attitude that shut down further debate). The pattern is sadly common across industries. In the Boeing 737 MAX crisis of 2018–2019, we saw a modern echo of Challenger's lessons – this time in the corporate sector. Boeing had been one of the world's most respected airplane manufacturers, famed for its engineering-led culture and commitment to safety. But over the years, especially after a 1997 merger with McDonnell Douglas, Boeing's internal culture shifted dramatically. Control shifted from engineers to business executives, who are fixated on stock prices and market competition. The company's emphasis tilted away from "safety first" (the mantra one expects in aerospace) to "shareholder value first." This culture change set the stage for the development of the 737 MAX aircraft, a new variant of an older 737 design equipped with larger engines. To save time and money, Boeing made a series of decisions that in hindsight proved calamitous. Rather than design a new airframe from scratch (which would be expensive and require extensive retraining of pilots), they retrofitted the 737 design to accommodate the bigger engines. This made the plane prone to pitching up in certain flight conditions. Instead of a robust hardware solution to that aerodynamic issue, Boeing installed a software fix – the MCAS system – that would push the nose down automatically if the plane's sensors detected a dangerous climb. Boeing also assured airlines that pilots of existing 737s wouldn't need simulator training to fly the MAX, so they downplayed the differences. Boeing even omitted mention of the new MCAS system from pilot manuals to avoid raising alarms or training requirements. All of these decisions were driven by market and profit goals: Airbus had a new model coming out, Boeing wanted to respond quickly and cheaply, and a promise to airlines (like Southwest) of no extra pilot training became a selling point. Internally, some Boeing engineers did raise concerns. They requested additional testing of the MCAS or noted that its reliance on a single sensor posed a safety risk. But in case after case, those voices were overruled or muted. Engineers were explicitly told not to suggest design changes that would slow the schedule or add cost. A finance-driven management culture exerted subtle and not-so-subtle pressure: one engineer later reported being warned "very directly and in threatening ways" that if he kept pressing for safety-related improvements that might delay the project, his job could be at risk. Some employees later confessed that they didn't feel the plane was safe – a few even told their families not to fly on a 737 MAX – yet they did not scream those fears from the rooftops; they stayed quiet and hoped for the best. Regulators, too, were kept in the dark or cowed; Boeing had lobbied for a regulatory setup where it could essentially self-certify much of its plane, and many FAA officials had become too cozy with the company. In 2018 and 2019, two 737 MAX airliners crashed (Lion Air Flight 610 and Ethiopian Airlines Flight 302), both under similar circumstances: the MCAS software, fed by a faulty sensor, commanded the planes into nose dives that the pilots could not recover from. A total of 346 people died in those two crashes. In the aftermath, investigations laid bare the trail of bad decisions and unheard warnings. A House of Representatives report in 2020 was scathing, citing "a culture of concealment" at Boeing and a failure of oversight. One software developer at Boeing wrote, "Today, safety doesn't come first – money comes first, and safety's only utility in this company is in helping to keep the money coming." It's a chilling statement, and an utterly damning one for an engineering firm. The 737 MAX was grounded worldwide for over a year while fixes were implemented, costing Boeing tens of billions of dollars in losses, not to mention the incalculable cost in reputation. The root cause was not simply a technical mistake; it was the silencing of voices and a lapse of professional integrity. Boeing had, in earlier decades, lived by the credo that "everyone is responsible for safety" and that any engineer could halt a test or flight if they spotted a concern. But by the 2010s, that voice had been muffled. The few who did object internally were told to be team players or that management "had everything under control." The result was a system that was not built to last, because it was built on a shaky moral foundation. It only took a couple of failures to expose it.

Stories like Challenger and the 737 MAX are tragic. Still, they also yield heroes – people who, like Roger Boisjoly or like the handful of Boeing engineers who did voice concerns (albeit unsuccessfully), at least tried to honor their ethical duty. We also have examples where voices of integrity did manage to avert disaster or bring truth to light. Think of whistleblowers who exposed corporate and governmental wrongdoing – their voices often come at great personal cost, but they prevent greater harm. For instance, Frances Haugen, a former Facebook employee, in 2021 leaked internal documents showing that the company knew its platforms could cause societal harm (from inciting violence to harming teen mental health) but often put profit over corrective measures. By speaking out, she sparked a global conversation on social-media ethics. In the world of finance, Sherron Watkins famously warned Enron's CEO of the firm's accounting fraud. When ignored, her memo later helped investigators piece together the truth, thus blowing open one of the biggest corporate scandals ever. And in government science, consider Dr. James Hansen of NASA, who spoke up to Congress in 1988 about the reality of climate change when it was still a contentious topic – he felt a moral obligation not to stay silent about the data he was seeing. In each of these cases, "what you do" includes what you decide to do when you realize something is wrong. Will you have the courage to use your voice, or will you bury the truth? Not everyone is in a position to become a public whistleblower, nor is every issue one of imminent public danger. But within teams and organizations, the principle remains: a system where people are afraid to speak up about risks or unethical practices is a system courting failure. Conversely, organizations that encourage ethical dialogue and empower even junior members to raise a red flag tend to catch problems early and adapt. Psychological safety – the sense that you won't be punished or humiliated for speaking up with a concern – is often cited as a key factor in high-reliability organizations (aviation, nuclear power operations, etc.). As I reflect on projects I led or was part of, I realize the ones I am most proud of aren't those that went perfectly from the start – rather, they're the ones where someone caught a flaw or a blind spot in time because they felt safe enough to speak up, and we heeded them. Those are the times the system showed true resilience: not in never having issues, but in being able to confront and fix them transparently.

In sum, integrity is the invisible infrastructure that holds any complex endeavor together. Contracts, specs, and org charts mean nothing if people cannot trust each other to tell the truth. When things get hard – when budgets are on the line, when a delivery date looms, or when admitting a mistake could be embarrassing – that's exactly when integrity is tested. Those are the moments when doing the right thing might mean pressing pause, owning up to a flaw, or pushing back on pressures. It's hard. It can feel like "many go quiet" in those moments, especially if careers or reputations seem to be at stake. But if everyone goes quiet, the result may be far worse than a missed deadline or a blown contract – it could be lives lost, public trust shattered, and a legacy of regret. I often think back to that phrase: "Do you know what you do, does?" It reminds me that what we do isn't just a line on a résumé or a product on a shelf – it has a real impact on people's lives. Knowing what it does means also knowing when it could harm, and taking responsibility to prevent that.

Building with Purpose, Ethics, and Humanity

After all these years and all these lessons, you might expect a person to become cynical or to give up on building big things. But I am not pessimistic. Cautionary tales aside, I still believe in building. We need to develop better and smarter than ever – whether it's infrastructure for a sustainable planet, technologies that improve quality of life, or institutions that promote justice and knowledge. The challenges we face in the world (climate change, inequality, energy transition, digital transformation, and more) demand ambitious work. But we must approach that work from a different place than we often have in the past. As I sometimes say to myself, I now choose to work "lower to the ground, closer to the point." Lower to the ground means grounded in reality, humility, and the daily lives of people (no more ivory towers or lofty promises detached from consequences). Closer to the point means always asking, "What is the point of this project? What core purpose am I serving, and is this the best way to serve it?"

Let's unpack that with a positive vision. What does it look like to do work that "helps someone stand up," that "carries those at the edge," and that amplifies voices instead of silencing them? It looks like a city infrastructure project that doesn't just aim to beautify downtown but also creates jobs for marginalized workers and adds affordable housing so that the whole community rises together. It looks like an energy system designed to be green and reliable, but also deliberately set up to ensure the poorest households benefit from lower costs and aren't the first to lose power in an outage. It looks like a technology company that bakes ethics into its design process from day one. For example, a team developing an AI product will have a "red team" or ethicists on board to question how the AI could be misused or whom it might unintentionally hurt. They'll have the power to veto launches until issues are addressed. It looks like leaders who foster a culture of agency and voice at every level: a military general who encourages junior officers to question orders that seem off, a hospital administrator who empowers nurses to speak up if a procedure might harm a patient, or a CEO who publicly rewards an employee who pointed out a flaw in a product rather than burying the news.

There are real instances of this kind of leadership. For example, after the Challenger and Columbia losses, NASA did try to reform its culture. It created channels for anonymous safety reporting and brought in outsiders to challenge internal groupthink. Some credit these changes for the safer missions that followed. In the nuclear power industry, after the Three Mile Island accident in 1979, the industry globally set up the World Association of Nuclear Operators (WANO) to share safety information transparently across companies and countries, embodying a principle that "an accident anywhere is an accident everywhere" – which is a way of saying: we're all in this together, and we must all speak up and learn from each other for the common good. In civil engineering, professional societies now emphasize "ethical resiliency" – meaning not just designing structures to resist loads, but designing organizations that resist pressures to cut corners. Engineering curricula have started to include ethics courses and case studies (like the ones we've discussed) so that new engineers enter the field aware that their calculations and drawings live in a social context with moral stakes. When I see young engineers at a place like Flint working alongside community members to test water and demand action, it gives me hope that the next generation is embracing a renewed sense of responsibility.

To truly build with purpose and ethics, we also need memory and foresight. Memory, in that we honor the lessons of past failures and sacrifices – we don't shrug off history as someone else's problem. Foresight, in that we think about long-term implications and future generations. A bridge built today should be sound not just for us, but for our children; a policy implemented now should be judged by its impact on those yet to be born. There is a concept in some Indigenous cultures about considering the "seven generations" ahead in any decision. Modern projects rarely have that kind of time horizon in mind, but what if we tried? At the very least, it would reframe decisions about "did it matter" on a longer timeline. Will this matter in 50 years in a good way? If not, why are we doing it? Questions like that can be inconvenient to short-term interests, but they are essential to sustainability.

Throughout this journey, I have also learned about justice, dignity, and trust. We touched on environmental injustice in Flint and New Orleans – those are reminders that the ethical dimension of work includes fairness and respect. Who bears the risks and who gets the benefits? If your work consistently marginalizes or harms a certain group, it's failing the test of justice. Dignity comes into play in how we treat those affected by our projects – do we bulldoze over communities or do we engage them with respect and give them a voice in shaping solutions? A resilient system is not one imposed on people; it's one built with people's participation and buy-in. And trust – perhaps the most fragile and precious commodity – is the currency that makes any complex system function. When trust is broken (as happened with Boeing's passengers, Flint's citizens, or the American public after certain scandals), it is incredibly hard to rebuild. Thus, maintaining integrity and transparency is not just morally right, but pragmatically wise: it preserves trust. Trust, in turn, feeds back into resilience because people who trust each other cooperate better in crises.

In writing all this, I am mindful of not presenting myself as having always known these things. On the contrary, much of what I've shared came out of my own mistakes or slow realizations. I've been the young engineer gung-ho about a clever design who didn't listen enough to an older mentor saying, "Think about the people who will use this." I've been the middle manager under pressure who perhaps kept quiet one day when I should have spoken up about an unrealistic deadline that might compromise safety. I've felt the sting of seeing a project I was involved in fall short of its promise for reasons that, in hindsight, were avoidable had we been more honest up front. But I've also experienced the redemption of course-correcting – of pausing a project to reevaluate its goals, of openly admitting to stakeholders "we got it wrong" and then working to fix it. Those moments are hard, but they are the ones that mattered. They are when I felt the work meant something beyond a paycheck or a plaque.

So, what comes after the work? In the title of this piece – What Comes After the Work – I'm hinting at legacy and impact. When the immediate work is done, what echoes does it leave? Memory is one: will people remember it as something that helped them or hurt them, something that lifted people or let them down? Conscience is another: can you, the builder or leader, sleep at night knowing what your work has done? For Rotblat, leaving the bomb project gave him a lifetime of peace advocating for nuclear disarmament; for Oppenheimer, staying meant carrying a burden of guilt that he tried to alleviate later. When I retire (and I nearly have), I want to know that I tried to build things that mattered in a good way, and that I encouraged others to do the same. I've come to realize that how we do the work – with what ethics and concern – is as much the legacy as what we built. A bridge will one day collapse or be replaced, a program will be updated or sunset, but the values we practice and impart can persist in the people we mentor and the culture we leave behind.

I want to close by turning the questions outward, to you, the reader, who might be an engineer, a leader, a student, or simply someone curious about how systems shape society. What are you building? Who is it for? What does it do? These simple questions can launch profound discussions. They are questions Mr. Wright effectively posed to me with his phrasing, and they became a refrain in my mind. Try bringing these questions into your next team meeting, or design review, or personal career reflection. They cut through jargon and get to the ethical heart of the matter. If you find the answers unsatisfying or uncomfortable, that's a sign to rethink and realign your work. If the answers inspire you – if you can say, "I'm building X for this group of people, and it will help them by doing Y" – then you have a north star of purpose to guide you through the challenges.

In my journey, I've met many others who reached a point of reckoning and asked these questions. Some changed careers, some reformed their organizations from within, and some started new initiatives focused on social good. It's a hard path, but a rewarding one. It has a way of rekindling the energy and dignity in work, preventing burnout because you know why you labor. It's the antidote to drifting aimlessly. And collectively, if more of us take this approach, it could shift whole industries toward more ethical, sustainable practices – what you might call a movement from success to significance.

I invite you to be part of that movement. Whether you're coding software, managing a factory, teaching students, or crafting policy, you have a sphere of influence. You can insist on clarity of purpose (#Clarity) and uphold integrity (#Integrity). You can champion foresight (#Foresight) by considering future generations, and responsibility (#Responsibility) by owning the impacts of your work. You can remember justice (#Justice) and voice (#Voice) – ensuring the marginalized are heard and treated fairly. You can commit to craft (#Craft) and excellence, but not at the expense of ethics (#Ethics). And in doing so, you build trust (#Trust) with those around you and those you ultimately serve.

This treatise has been a blend of personal witness and external evidence. I've cited cases of failure and success not to cast blame, but to uncover patterns: the pattern that technical problems are often ethical problems in disguise, that the soft factors of agency and culture decide the hard outcomes of systems, and that personal courage (or its absence) can reverberate far beyond one's immediate job. I have written it to "clear the air" – to share what I've come to believe holds after stripping away illusions.

I still choose to be a builder. But I now know that the true measure of "built to last" is not just steel and concrete – it's whether the work carries forward our values when we are gone. A bridge stands for decades, but what does it stand for? Let it stand for connection, for opportunity, for inclusion. A codebase might run for years; let it run in service of knowledge and empowerment, not division or exploitation. A policy might shape society; let it shape it toward justice and resilience.

If our work helps even one person stand up who would have been knocked down, if it brings those at the edges into the fold instead of leaving them behind, and if it speaks truth to power instead of falling meekly silent, then it is work worth doing. It is work that matters. In my career, I had to learn this the long way. My hope in writing this is to shorten that learning curve for others and to affirm for those who may already feel this way that you are not alone. There is a community of us who care about these things – #Agency, #Clarity, #Dignity, #Ethics, #Purpose, #Resilience, #Voice, and all those values I tagged at the outset, not as buzzwords but as beacons.

The question that never lets go – "Do you know what you do, does?" – is now firmly in your hands. May you answer it with honesty. May you answer it with action. And may you, at the end of your endeavors, take pride in what comes after your work: the lives touched, the futures made brighter, and the knowledge that you did what was right and true.

Thank you for reading, and thank you even more for doing.

Resources for Further Reading

Byers, Dylan. "Google says it will not renew controversial Pentagon contract." CNN (CNNMoney), June 1, 2018. (Reports on Google's decision, under employee pressure, to discontinue its work on Project Maven, an AI project with the U.S. military. Cites that 4,000 Google employees petitioned against building "warfare technology" and about a dozen resigned in protest, leading Google to establish new AI ethics principles and avoid future weapons-related work.)

Iskander, George. "The Manhattan Project Shows Scientists' Moral and Ethical Responsibilities." Scientific American, March 2, 2022. (Discusses J. Robert Oppenheimer's remorse – "I feel I have blood on my hands." – and Joseph Rotblat's ethical choice to leave the Manhattan Project upon learning of its ulterior aims, as well as Rotblat's quote "Science became identified with death and destruction.")

Prentice, Robert. "Ethical Lessons Learned from the Challenger Disaster." Ethics Unwrapped, University of Texas, July 1, 2024. (Details NASA's normalization of risk and the silencing of engineers' warnings leading up to the 1986 Challenger shuttle explosion, violating the paramountcy of public safety.)

Prentice, Robert. "Engineering Ethics and the Boeing Scandal." Ethics Unwrapped, University of Texas, Jan. 25, 2022. (Analyzes the Boeing 737 MAX case: how Boeing's shift from an engineering-driven, safety-first culture to a profit-driven culture after 1997 led to design shortcuts, suppressed safety concerns, and two fatal crashes killing 346 people. Highlights internal pressure on engineers and the abandonment of the NSPE ethical canon to prioritize public safety.)

Pruitt, Sarah. “How Levee Failures Made Hurricane Katrina a Bigger Disaster.” History.com, updated May 28, 2025. (Describes the 2005 failure of New Orleans’ levee system during Hurricane Katrina. Notes that 80% of the city was flooded and an estimated 1,500+ people died in Louisiana. Later investigations showed some levees failed below their design thresholds, and the Army Corps of Engineers acknowledged flawed engineering practices. Emphasizes that the poorest communities, like the Lower Ninth Ward, were hardest hit, and that long-ignored warnings and underinvestment in robust flood protection exacerbated the disaster.)

Ray, Michael. “Flint water crisis.” Encyclopædia Britannica, last updated June 28, 2025. (Summary of the Flint, Michigan water crisis (2014–2015), in which cost-cutting measures led to lead-contaminated drinking water and a Legionnaires’ disease outbreak. Tens of thousands were exposed to toxic water, and at least 12 died. Chronicles how officials ignored residents’ complaints and scientific evidence, prioritizing cost over health until the crisis could no longer be denied.)

39

Telling a Story That Sticks (How Narrative Clarity Serves Action and Change)

A charity once tested two appeals for a life-saving cause. One letter overflowed with statistics – percentages of people in need, charts of declining resources. The other told the story of a single girl: her name, her daily walk for water, her hopes. Donors who read about the girl gave far more than those who saw only numbers. The facts were dire, but a simple story moved hearts and wallets. Something in our minds responded to her plight in a way spreadsheets couldn't touch. Why? Because human beings are wired for story.

From ancient campfires to crowded conference rooms, stories have been our most natural way of understanding the world. We live through stories. Psychologists observe that people and cultures define themselves through narratives – we are the stories we tell about ourselves. Historian Yuval Noah Harari bluntly says, "Storytelling is our superpower." It allowed early humans, weak on their own, to cooperate in tribes and nations. We invented shared myths – about gods, about countries, about money – and believed them together, enabling cooperation on a massive scale. In Harari's words, fictional stories became the "glue" that made large-scale human society possible. Our capacity to believe stories, be they legends or ideals, lets us build cathedrals, launch revolutions, and organize communities. We're the only species on Earth that can do this. Storytelling made us human.

Modern science confirms that our brains crave narrative. A clear story engages more of the brain than raw data does. In one neuroscience study, listening to a well-told tale caused the neurons of listeners to light up in patterns mirroring the storyteller's – a phenomenon called "neural coupling." As the story builds, our brain releases dopamine, a chemical that rewards us for paying attention and helps fix events in memory. This is why an emotional tale or a dramatic twist can stay with you for years while a list of facts fades away.

Experiments at Harvard Business School showed that after a day, people forgot 73% of a statistic's details, but only about 33% of a story's details. In other words, stories stick. "People are more likely to recall information over a longer period when it's wrapped in an anecdote as opposed to statistics," the researchers concluded. We remember narratives better because a story gives context and texture to facts. Our brains prefer it – as one professor said, humans don't communicate in numbers, but in natural language and story. Even in professional settings, a narrative cuts through the distraction of data overload and holds attention. We lean in, wanting to know what happens next.

Good stories captivate us because they speak to something fundamental in our minds. Over the years, storytellers and researchers have noted certain elements that make a narrative compelling. These aren't gimmicks – they reflect how we emotionally engage with information. Conflict, character, emotion, and structure are the four most important aspects.

Conflict: The Hook of Tension

Every great story, from a childhood fable to a market analysis, is driven by conflict. Conflict is the engine of narrative – the struggle or problem that must be resolved. Without it, there is no suspense, no reason to care. "Conflict is the beating heart of story," one storytelling expert writes. You're left with a sequence of flat and disconnected events without any dilemma or tension. Conflict can be as grand as a war between kingdoms or as humble as a person overcoming self-doubt. It might be a tough question or an unexpected problem in a business context. The key is that conflict introduces tension. It raises a question: How will this situation be resolved? That question keeps the audience invested. Communication research has found that when an audience is presented with a problem or mystery, their curiosity spikes – they get a "pleasantly uncomfortable" urge to know the answer. In Will Storr's words, "the more context we learn about a mystery, the more anxious we become to solve it." Our reward circuits light up while waiting for a conclusion, almost as if anticipating a treat. This is why a well-crafted opening – "Despite an increased budget, sales fell 4%. Why?" – can pull people to the edge of their seats. Conflict and questions make us active listeners, not passive ones. We start guessing outcomes and mentally participating. In short, conflict drives engagement. It gives your story stakes and your audience a reason to care, whether narrating a hero's journey or explaining a dip in quarterly earnings.

Characters and Empathy: People in the Story

If conflict is the engine, characters are the vehicle that carries us through the story. Humans are social creatures – we relate to people, not abstract ideas. That's why even the most data-driven presentation benefits from a face or name to care about. A well-developed character, whether fictional or real, creates a bridge of empathy to the audience. We see the world through their eyes. We root for them to succeed or fear they won't. This emotional connection is powerful. Scientists have found that when we engage with a character-driven story, our brains release oxytocin, a hormone associated with trust and empathy. In one study, people who read a narrative about a child with cancer had a surge of oxytocin – and the higher their oxytocin levels, the more they were willing to donate to help others in need. A character makes the stakes personal. For example, instead of saying "Millions suffer from disease X," a story might introduce us to one patient – their name, daily struggle, and hopes. Suddenly, the abstract suffering has a human face. As writer Pat Morden noted, stories connect with us "at a gut level," stirring emotions that motivate action more effectively than any fact sheet. We pay greater attention to vivid, relatable details than to dry aggregates. This is not a weakness in our thinking; it's a feature of how we're built. Our ancestors survived by observing and remembering the experiences of others. We're tuned to listen when someone says, "Let me tell you about a person I met..." Crafting any narrative – a speech, an article, or a report – helps to put a person in the story. It could be a customer, an employee, a historical figure, or even yourself. Give the audience someone to care about. If they can see the situation through that character's eyes and feel what's at stake for them, they will care about the outcome. And once they care, they'll stay hooked.

Emotional Resonance: Stirring Feelings That Last

Stories make us feel – and that is a source of their power. Facts can inform, but emotions inspire. Think of the moments from stories that have stayed with you: maybe a scene that made you cry, laugh, or gasp in fear. These feelings are the imprint of the experience. Emotional resonance means the story's feelings echo in the reader or listener. It's not about manipulating people's emotions unfairly; it's about finding the genuine heart of the narrative. If you're telling a true story of a hardship overcome, it's natural that it may be sad or triumphant – let the audience feel that. Emotion is vital for making an impact. Psychologists note that decisions are often driven by feelings more than by rational analysis. A story that generates empathy, joy, or even righteous anger can spur someone to act when a list of reasons might not. In fundraising, for example, one famous experiment showed that a single gripping story prompted more generosity than pages of statistics – because compassion is triggered by emotion, not numbers alone. Emotion also makes information memorable. Neurologically, when something emotionally significant happens, our brains mark it as important. That dopamine rush during an exciting or moving event helps encode the memory. Advertisers have long recognized this: a touching or humorous ad sticks in your mind and keeps you recalling the product. One marketing analysis found that ads with above-average emotional response drove substantially higher sales growth than those focusing only on rational content.

The lesson for any storyteller is to find the feeling in your message. Are you celebrating a success? Framing a daunting challenge? Even in a technical report, you can convey wonder at a discovery or urgency about a problem. Use detail and language that evokes senses and feelings – the relief of a solution, the tension of a risk, the pride in a job well done. When people feel the story, they won't forget it, which may move them to change or help. As author and writing coach William Zinsser put it, "the secret of good writing is to strip every sentence to its cleanest components," – but that doesn't remove emotion, it reveals it. Simplicity can be deeply emotional. A plain, honest sentence like "We had hoped, but we lost him," in the right context, can break a reader's heart. You don't need melodrama or flowery language; the story's truth, told with clarity, will carry its power.

Structure: A Journey with a Beginning, Middle, and End

A great story is well-shaped. Even the most fascinating characters and conflicts can fall flat if the tale is haphazardly told. That's where structure comes in – the deliberate arrangement of events and revelations to guide the audience through an experience. At the most basic level, structure means having a clear beginning, middle, and end. In the beginning, we set the stage: introduce the setting, the characters, and the conflict. In the middle, tension rises: obstacles appear, complications twist the plot, and questions demand answers. Ultimately, we reach a resolution: the dispute is resolved (for better or worse), and we see the outcome and what it means. This classical narrative arc has endured from ancient myths to modern novels because it resonates with how we experience life's challenges. As Will Storr emphasizes, a well-structured story maintains engagement – it doesn't meander aimlessly. Each part of the story leads to the next logically or causally, keeping the audience oriented and interested. Surprises and twists are placed at the right moments to jolt attention or reframe the context, but they fit into an overall flow toward resolution. A satisfying narrative builds. Consider a mystery novel: the opening hooks us with a crime, the middle complicates with red herrings and clues, the climax unveils the culprit, and the finale ties up loose ends.

Without that progression, the mystery's tension would never catch on or fizzle. The same principles apply beyond fiction. If you're presenting, don't just dump information in arbitrary order – structure it like a story. For example, set up a problem that matters (beginning), explore why it's challenging and what's been tried (middle), then reveal a solution or insight (end). Even a short anecdote in a speech benefits from a mini-structure: context, conflict, and resolution. Audiences find stories satisfying when they feel a sense of trajectory. It doesn't have to be a happy ending, but it should feel like we went somewhere. A clear structure also makes the message coherent. Research on communication uses the term "coherence" – the integrity of the story's parts. Each idea should build on the previous. If details appear, they eventually connect to the main point. This spares the audience the cognitive work of puzzling how things relate; instead, they can relax into the narrative flow. In practical terms, you might outline your communication in story form: Here's the situation. Here's the challenge we faced. Here's how we tackled it – the struggles and insights. And here's what happened in the end. Such a structure invites listeners to follow along step by step. They are far more likely to stay engaged than if you present conclusions out of the blue or wander through tangents. Structure matters, as every storyteller knows. It turns a list of events into an experience and, ultimately, drives the lesson or message home with impact.

Having explored why stories grip us, we should turn to what stories do. A compelling narrative doesn't just entertain or inform – it inspires action. Stories have changed people's minds and even altered the course of history. At the start, that little girl's story moved donors to open their wallets; on a larger scale, narratives have sparked movements and shaped cultures.

From Story to Action: How Narratives Change Behavior

In 1852, Harriet Beecher Stowe published Uncle Tom's Cabin, a novel about the cruel realities of slavery. The book galvanized antislavery sentiment in the United States. Legend states that when Stowe later met President Abraham Lincoln during the Civil War, Lincoln greeted her by saying, "So you're the little woman who wrote the book that started this great war." Whether or not Lincoln's quip is apocryphal, it was widely repeated because it rang true: Stowe's story had pricked the conscience of a nation. It personalized the suffering of enslaved people in a way political pamphlets hadn't. Readers could see Eliza fleeing across the ice with her child in her arms; they could weep for Tom's fate. That emotional experience steeled many against the institution of slavery. In Stowe's vivid narrative, abstract moral arguments became human and urgent. The novel's impact was so great that it's often cited as an instigating factor in the Civil War. This is an extreme example, but it shows a general truth: stories influence behavior. Depending on the tale, they can spur people to empathy and action, or conversely, to hatred and violence. Throughout history, effective leaders have used storytelling to unite people behind a cause or vision.

On a smaller scale, businesses and organizations find that story-driven communication gets results. Executives who want to rally their employees or pitch to investors often begin with a narrative. They know a story will be remembered long after bullet points are forgotten. For instance, a company founder might share a personal anecdote about how they struggled with a problem their product now solves. That origin story can inspire employees with a sense of purpose and authenticity that no mission statement could achieve. Venture capitalists often say they invest in a founder's story as much as in the product – they want to be sold on the journey, the passion, the inevitability of success through overcoming obstacles. Silicon Valley icon Richard Branson once remarked, "entrepreneurs who cannot tell a story will never be successful." He understood that a dry business plan rarely fires people up, but a story can. One venture capitalist has said the most underrated skill in business is storytelling. A well-told narrative can turn a presentation into a movement, a brand into a legend.

The persuasive power of storytelling is also well documented in psychology and marketing research. People are more likely to donate to charity when they hear about an individual in need rather than statistics about millions. Participants were given $5 in one experiment and shown one of two appeals for a hunger relief charity. Some saw facts and figures about food shortages;

others read about a single starving young girl named Rokia. On average, those who read Rokia's story donated more than twice as much as those who saw the statistical appeal. Remarkably, even adding statistics to Rokia's story reduced the effectiveness – the numbers seemed to blunt the emotional impact of her narrative. We empathize with one person's story more deeply than with any abstraction. This is known in behavioral science as the "identifiable victim effect." It's not rational in a strict sense (one life moves us more than thousands), but it's human. Stories create identification, and identification drives action.

Chemistry is involved here, too. Dr. Paul Zak, a neuroeconomist, conducted experiments where people watched short films or read stories, some of which were designed to be very moving (for example, a father and son dealing with the son's terminal illness). He found that when a character-driven story engaged people, their brains released oxytocin – the hormone a mother produces when bonding with her baby. The amount of oxytocin predicted how likely they were to help others afterward, such as donating a reward payment to charity. In other words, a story that evoked empathy in someone primed them to do something compassionate. We often think we make decisions logically, but as fundraising expert Tom Ahern said, giving is an "act prompted by empathy, desire, pleasure, anger, and a host of other emotions." Stories are uniquely suited to stir those emotions. They connect with us "at a gut level," making us want to act, not because someone presented an argument, but because we feel the urge to help.

Beyond charity, narratives sway opinions in fields from politics to public health. Advertisers leverage storytelling to make products appealing (the classic "Nike revolution" ads of the 1980s didn't list shoe features – they told a story of overcoming odds). Educators use stories to make lessons stick. Even in data-heavy fields, like finance or science, a well-placed anecdote or metaphor can illuminate the meaning behind the numbers, leading to that aha moment where an audience finally gets it. Thomas Graeber's 2023 study on leadership, titled "Memorable Leaders Don't Just Spout Statistics, They Tell Stories," found that a week after a presentation, people recalled the story-based content far better than pure data points. The implication is clear: if you want your message to endure and possibly change behavior, embed it in a narrative.

All this said, it's important to remember that the truth and clarity of a story also determine its impact. A misleading story can spread misinformation as easily as a truth can spread understanding. Harari notes that while storytelling unites, it can also deceive or divide. Propaganda is essentially a story told to manipulate public opinion. So, there is an ethical dimension: we should strive to tell honest stories and use their power for good ends. The most compelling stories in history – those that changed the world for the better – have had a core of truth and authenticity. They revealed something real about the human condition, injustice, or hope, and that truth fueled action.

Assuming we have a truthful, engaging story, there remains a final ingredient to making it truly effective: clear and simple writing. All the powerful elements of storytelling – conflict, characters, emotion, structure – can be undermined if delivered in a muddled manner. If the audience struggles to understand you, the story's spell breaks. Great storytellers throughout history, from classical orators to modern writers, have understood that simplicity and clarity are essential. As we turn to the craft of writing, we'll see why stripping away clutter and speaking is not a limitation, but a superpower of its own.

The Art of Clarity: Stripping Away Clutter for Impact

Imagine Lincoln at Gettysburg, dedicating the battlefield in 1863. The speaker before him, Edward Everett, was a famed orator who delivered a two-hour address full of classical allusions and flowery rhetoric. Lincoln spoke in plain, spare language for about two minutes – 10 sentences in total. Which speech do we remember today? Everett himself recognized what happened. The next day, he wrote to Lincoln, admiring the President's brevity and power: "I should be glad, if I could flatter myself that I came as near to the central idea of the occasion in two hours, as you did in two minutes." Lincoln's Gettysburg Address, at just 272 words, became one of the most famous speeches in American history. Its clarity and directness – "of the people, by the people, for the people" – cut straight to the heart. This illustrates a timeless principle: less is more when less is said with sincerity and precision.

Good writing omits needless words. This mantra comes from William Strunk Jr., who in 1918 wrote The Elements of Style, a little book that has guided generations of writers. Strunk's rule 13 was: "Omit needless words." He explained: "Vigorous writing is concise. A sentence should contain no unnecessary words, a paragraph no unnecessary sentences, for the same reason that a drawing should have no unnecessary lines and a machine no unnecessary parts." Every word that doesn't serve a purpose weakens the whole. Clutter and verbosity act like weeds in a garden, choking the beauty of the ideas. The writer and teacher William Zinsser put it vividly: "Clutter is the disease of American writing. We are a society strangling in unnecessary words, circular constructions, pompous frills, and meaningless jargon." That was his diagnosis in On Writing Well, and it rings true today whenever we read a corporate memo or academic paper that feels almost designed to confuse. Zinsser's cure was ruthless simplicity: prune the clutter, clear the weeds. He advised writers to examine every sentence and ask: Is this needed? Is there a shorter, clearer way to say this? This doesn't mean all writing should be staccato or robotic – you can be concise and still have style and voice. But it does mean respecting the reader's attention. Every unnecessary word is a tiny tax on the reader's patience. Enough of them, and you bankrupt their willingness to continue.

Clarity in writing is not just about brevity; it's about honesty and directness. George Orwell, famous for his crisp prose, believed that clear language was tied to clear thought – and that political jargon and euphemism were often used to hide ugly truths. "Good prose is like a windowpane," Orwell wrote. The reader should be able to see straight through the words to the meaning, without decorative smudges blurring the view. If the prose draws attention to itself, it's like muddy glass – the content gets obscured. Orwell practiced what he preached. In an era of grandiloquent politicians, he wrote newspaper columns and essays in plain English, conveying complex ideas in a way any literate person could grasp. He also formulated some famous guidelines for writers: never use a long word where a short one will do, never use a foreign phrase or jargon if there's an everyday English equivalent, cut out unnecessary words, use the active voice, and so on. Orwell even demonstrated by example, taking a well-known passage from the Bible and translating it into pretentious "modern English" to show how ridiculous and unclear it became. His point was that simplicity is a virtue – not because readers are dumb, but because clarity is a sign of respect. It shows you have thought through what you want to say and aren't trying to dress it up or muddle it.

Clarity also means keeping your logic front and center. Peggy Noonan, a speechwriter for Ronald Reagan and an essayist known for her elegant yet accessible style, advises that "the most moving thing in a speech is its logic." By logic, she means the clear progression of ideas – the because behind each statement. People often assume that flowery language or dramatic flourishes make a speech impactful. Still, Noonan argues the opposite: what truly moves listeners is a good idea plainly stated and built step by step. "It's not the sentimental exhortations," she writes, "it's the logic behind your case." This is a crucial insight. Emotional storytelling and clear logic are not contradictory; in fact, they work best together. A story gives flesh to the logic, and logic gives backbone to the story.

When you present a narrative, make sure the audience isn't lost about why you're telling them this. Each part of the story should reinforce the core message or argument. If a detail doesn't serve the point, consider removing it, no matter how much you love it. The process of editing for clarity can be painful – you might cut out clever phrases or interesting tangents – but it forges a stronger piece. In the end, your readers or listeners will thank you because you're making it easy for them to follow and remember. As Strunk and White summed up, "clear thinking becomes clear writing," and one can't happen without the other.

Being clear and simple doesn't mean being dull. On the contrary, simplicity can be profound. Think of some of the most quotable lines in literature or speeches: "All men are created equal." "Less is more." "I have a dream." They resonate precisely because they are straightforward and uncluttered. There is a certain poetry in plainness. Ernest Hemingway, a master of spare prose, proved that a few short, ordinary words can evoke enormous emotion. In A Farewell to Arms, he wrote: "The world breaks everyone, and afterward many are strong at the broken places." The sentence is simple in structure and vocabulary, yet it carries the weight of hard-earned truth. Hemingway's style – often subject, verb, object, with minimal adjectives – forces each word to bear significance. Readers find their images and feelings in the blanks he leaves. This is powerful: it invites engagement. Peggy Noonan's best speeches did the same. They weren't loaded with academic terms or convoluted sentences; they spoke to everyday people in language that rang true. When Reagan addressed the nation after the Challenger shuttle disaster, Noonan helped craft his words to the grieving families and schoolchildren watching. The speech was gentle, clear, and direct: "The future doesn't belong to the fainthearted; it belongs to the brave." Not fancy, but strong – and memorable to this day.

In practice, achieving clarity means revising. The first draft is often for getting your ideas out; the second (and third, and fourth...) is for pruning and organizing. Read your work aloud. Do the sentences flow, or do you trip over excess verbiage? Would someone unfamiliar with the topic understand this easily? Have you defined jargon or, better yet. Sometimes, a metaphor can clarify – a concrete image can illuminate an abstract point – but beware of mixed metaphors or clichés that will confuse or bore the audience. Aim for fresh, precise expressions. If you find you've said "at this point," can it just be "now"? If you wrote "implement a solution," would "solve" do? Small edits like that, multiplied, make a big difference. The goal is not to dumb down your content; it's to smarten up your delivery.

Let's return to the combination of storytelling and clarity. They are not opposing forces – together they make each other stronger. A story gives meaning to facts, and clear writing gives the story credibility and punch. You might have the most thrilling tale to tell, but if you bury it in lazy, bloated prose, it will fail to captivate. Conversely, you might have a very straightforward message to deliver. Still, if you couch it in a narrative with vivid examples, it will land with far greater impact than a sterile exposition. The greatest communicators have always balanced the two. Winston Churchill, for example, was a superb storyteller in his speeches, often painting a scene or a character (he once described the Soviet Union as "a riddle, wrapped in a mystery, inside an enigma"). Yet, he was also plainspoken and even blunt when needed ("Blood, toil, tears, and sweat"). The mix of imagery and directness rallied a nation at war.

We live in an age of information overload and rampant distraction. That makes storytelling and clarity more important than ever. A well-told narrative cuts through because it speaks to the ancient part of our minds that listens for meaning. A clearly expressed idea stands out because it shines amidst the fog of jargon and noise. When you put the two together – a compelling story told in a clean, sincere style – you create something formidable. You can educate and inspire at the same time. You can engage both heart and mind.

Telling Your Own Story, Clearly and Honestly

In the end, all of these principles come down to a simple practice: tell the truth in a story form, and don't waste words doing it. Whether you're writing a LinkedIn post, giving a keynote, or drafting a chapter of a book, you can apply what we've covered. Start by asking: What's the real story here? Identify the conflict or question at the core of your message – that's your hook. Identify who the "characters" are – maybe it's you, your team, a customer, a historical figure – and bring their perspective in. Let yourself feel the emotions in the story (excitement, frustration, hope, fear) and convey those authentically. Structure your piece so it has a logical flow – perhaps situation, complication, resolution. And then, crucially, edit for clarity.

As you edit, imagine a person from your intended audience reading or listening. Will they grasp each point without strain? If there's a complex concept, can you swap in a down-to-earth analogy? If technical details are necessary, can you at least provide context so they see why it matters? Remember that clarity isn't just about short sentences – it's also about context. A good story gradually gives context before big reveals, so the audience isn't lost. As Will Storr observed, context increases curiosity up to a point. We want to understand, so we keep listening. But too little context, and we're confused; too much irrelevant context, and we're bored. It's a balance you learn by practicing and by getting feedback. Don't be afraid to show a draft of your writing to a colleague or friend and ask, "Did this hold your interest? Was anything unclear or unnecessary?" Fresh eyes often catch things we miss.

To make your narrative resonate, consider adding concrete, specific details. Specificity is a friend to both storytelling and clarity. Instead of saying "Our project had some delays," say "Our project launch was delayed by three weeks when the supplier suddenly went bankrupt." Instead of "She was very angry," say "She slammed the report on the table, her voice shaking." Specific images and facts give the audience something real to latch onto. They also lend credibility – you appear as a truthful observer, not a spinner of vague generalities. Mark Twain once advised writers, "Don't say the old lady screamed. Bring her on and let her scream." In other words, show, don't just tell. Let the evidence or action appear in the story whenever possible so that the audience can conclude naturally.

It's also helpful to loop back to your key message at the end of your narrative. Stories can take on lives of their own, so as the storyteller, you gently steer them to the point. In a persuasive piece, you might conclude by explicitly stating the lesson or the call to action. In a more reflective or artistic piece, you might end with an image or line that subtly underscores the theme. Either way, ensure the ending feels conclusive and ties to the purpose you began with. A good ending often echoes the beginning in some way – it answers the question you posed or resolves the tension you set up, providing a sense of closure.

Finally, infuse your voice into the narrative, even as you keep it clear. Clarity doesn't mean blandness. Your choice of which story to tell, which details to highlight, and what you feel about it – all that creates voice. Write as one person to another, as if you're having a conversation with the reader. Peggy Noonan suggests imagining you're talking to a single intelligent friend rather than "delivering oratory to thousands." That helps keep your tone genuine and your language natural. If a bit of humor or a personal anecdote fits, include it. Those touches make your communication human. Ernest Hemingway, though an advocate for sparse prose, had a voice so distinctive that "Hemingwayesque" is an adjective. He achieved that not by being wordy but by being truthful and selective – choosing words that felt real to the experiences he described, and leaving out what didn't matter.

In sum, the essence of strong communication is storytelling with purpose and simplicity. We've seen how stories engage our identity, brains, and emotions, and how they can drive people to remember and act. We've also seen that without clarity, even a great story can falter – but with clarity, even a simple message can soar. When you merge the two, you get something magical: a narrative that is compelling and crystal-clear. It invites the audience on a journey and ensures they never lose the path. It resonates with their emotions and remains anchored in meaning.

Think back to that little girl carrying water, and the donors who gave after hearing about her. That story was powerful not because it was embellished – it was powerful because it was true and specific, and it let people see one child's world. It had conflict (a daily struggle for survival), character (a relatable girl), emotion (the pathos of her situation), and an implicit structure (problem in the beginning, hope for a solution at the end). And it was told in plain language that anyone could understand. No big words or complex graphs; no moralizing lecture. Just an honest narrative that spoke for itself.

You have stories like that to tell – we all do. They might be stories of problems solved at work, lessons learned in life, challenges overcome by a team, or visions for a better future. The key is to tell them intentionally and cleanly. If you do, you will find your words carry farther and sink deeper. You will see eyes light up in understanding, or hear feedback like, "I connected with what you wrote." That is one of the most gratifying feelings as a communicator: not just to be heard, but to have struck a chord with someone.

So, the next time you face a blank page or a silent auditorium, remember the principles we've explored. Start with a story – even a small one – to ground your message in human experience. Use conflict to draw interest, and let characters give your message a face. Welcome emotion; don't shy from it, harness it. Organize your narrative so it flows and builds to a point. Then, ruthlessly clarify. Cut the fluff, drop the jargon, and say what you mean as plainly as possible. Be guided by that Hemingway-Noonan-Orwell ideal of strong, straight talk – writing that is simple and true.

In doing so, you honor your audience's time and intelligence. You allow your ideas to shine without barriers. And you carry on a tradition as old as our species and as vital as ever: telling a good story, well. It's a craft and an art, one you can spend a lifetime refining. But even at its most basic, when done with heart and clarity, it has the power to change minds and touch souls. That is the ultimate reward of mastering storytelling and writing with purpose – you not only inform or persuade, but you also connect. And in that connection, real impact happens.

In conclusion, at the end of the day, whether we are writing code, laying concrete, teaching, or leading, we are storytellers and builders. And the world needs us to do both with courage and clarity. The systems we design and the narratives we share should help people stand – not just technically function. They should invite connection, not confusion. That is the deep craft of our time: not just to succeed, but to matter. And to do that, we must know what our work does – and say so clearly, truthfully, and well.

Epilogue: The Story Beneath the System

If the system is what we build, and the story is how we make sense of it, then our greatest responsibility is alignment. Systems that serve no human center fail, even if they run. Stories that stir emotion without truth mislead, even if they inspire. What the world needs now is neither efficiency nor eloquence alone. It requires builders who can think narratively and speak ethically. It needs leaders who ask not only, "What are we doing?" but also, "Who will this help stand up?"

That is the deep link between ethics and clarity. Between the grid and the page. Between what we do and how we tell. The world listens. Let us speak with precision. Let us build with care. Let us remember that in both story and system, the most enduring work is work that honors the human at its core.

Resources for Further Reading

S. Arrowsmith, "Pressure point: The crucial role of conflict in data storytelling," *Greenbook, Feb.*, vol. 15, 2024 Accessed: Feb. 2, 2026. Available: https://www.greenbook.org/insights/grow-your-insights-business/pressure-point-the-crucial-role-of-conflict-in-data-storytelling.

T. Graeber, "Looking to leave a mark? Memorable leaders don't just spout statistics, they tell stories," *Harvard Business School Working Knowledge*, 2023 Accessed: Feb. 2, 2026. Available: https://www.library.hbs.edu/working-knowledge/looking-to-leave-mark-memorable-leaders-tell-stories-dont-spout-statistics.

T. Graeber, "Stories, statistics, and memory," *The Quarterly Journal of Economics*, vol. 139, no. 4, pp. 2181–2233, 2024 Accessed: Feb. 2, 2026. Available: https://academic.oup.com/qje/article/139/4/2181/7691253.

Y. N. Harari, *Sapiens: A Brief History of Humankind*. London, U.K: Penguin Random House (Vintage), 2014 Accessed: Feb. 2, 2026. Available: https://www.penguin.co.uk/books/419795/sapiens-by-harari-yuval-noah/9781529913934.

NeuroLeadership Institute, "The neuroscience of story listening," *Jul.*, vol. 5, 2022. Available: https://www.neuroleadership.com/articles/the-neuroscience-of-story-listening.

D. A. Small and G. Loewenstein, "Helping a victim or helping the victim: Altruism and identifiability," *J. Risk Uncertainty*, vol. 26, pp. 5–16, 2003 Accessed: Feb. 2, 2026. Available: https://link.springer.com/article/10.1023/A:1022299422219.

G. J. Stephens, L. J. Silbert, and U. Hasson, "Speaker–listener neural coupling underlies successful communication," *Proc. Natl. Acad. Sci. U.S.A.*, vol. 107, no. 32, pp. 14425–14430, 2010 Accessed: Feb. 2, 2026. Available: https://europepmc.org/article/PMC/2922522.

W. Storr, *The Science of Storytelling: Why Stories Make Us Human, and How to Tell Them Better*. London, U.K: HarperCollins, 2019 Accessed: Feb. 2, 2026. Available: https://www.harpercollins.com.au/9780008276959/the-science-of-storytelling-why-stories-make-us-human-and-how-to-tell-them-better/.

W. Strunk Jr., and E. B. White, *The Elements of Style*, 4th ed. New York, NY, USA: Longman, 1999 Accessed: Feb. 2, 2026. Available: https://en.wikipedia.org/wiki/The_Elements_of_Style.

P. J. Zak, "Why your brain loves good storytelling," *Harvard Business Review*, 2014 Accessed: Feb. 2, 2026. Available: https://hbr.org/2014/10/why-your-brain-loves-good-storytelling.

W. Zinsser, "Simplicity," in *On Writing Well*, Ch. 2. Accessed: Feb. 2, 2026.

40

Leapfrogging the Legacy – Innovation Under Constraint in the Global South

From Absence to Advantage

In the remote highlands of Rwanda, a drone launches from a red clay runway, carrying blood transfusion supplies to a rural clinic inaccessible by road during the rainy season. Operated by Zipline, a logistics company headquartered in Silicon Valley, the system now runs hundreds of deliveries per day, not in California, but across Rwanda and Ghana. In Nairobi, Kenya, farmers access microloans and crop insurance through mobile platforms that utilize only SMS technology. In Bangladesh, schoolchildren attend digital classrooms via solar-powered tablets, circumventing unreliable grids.

These are not fringe cases. They are signals of a new paradigm: leapfrogging innovation – a form of transformation born not from surplus but from scarcity, not from the weight of legacy systems, but from their absence.

The Logic of Leapfrogging

Leapfrogging is not merely technological substitution. It is **discontinuous development** – an innovation trajectory that **skips intermediary stages** common in the Global North.

Classical vs. Leapfrogging Development

Dimension	Traditional Path (Global North)	Leapfrogging Path (Global South)
Power Access	Fossil > Grid > Renewables	Off-grid solar > Microgrids
Finance	Banks > ATMs > Online Banking	Mobile money (M-Pesa, bKash)
Education	Industrial model > Online learning management system (LMS)	Community-led microlearning & EdTech
Health Delivery	Centralized hospitals	Distributed diagnostics + telemedicine
Transportation	Private cars > Public transit reform	Motorbikes + app-based shared mobility

In essence, leapfrogging bypasses legacy infrastructure, regulatory inertia, and sunk cost traps, offering a more agile entry point into the technological frontier.

The Enabling Conditions

While often perceived as disadvantaged, leapfrogging nations possess unique enabling constraints – structural conditions that force innovation.

Infrastructure Absence

- **Africa:** 600 million people lack access to electricity; 900 million lack clean cooking solutions (IEA, 2026)
- **Opportunity:** The absence of a central grid made it possible to **deploy solar microgrids** at speed and scale:
  - **M-KOPA (Kenya):** Over 1 million solar home systems deployed
  - **SELCO (India):** Powers 500,000+ homes and micro-enterprises

Demographic Dividend

- **Sub-Saharan Africa:** Median age = 18.6 (UN, 2024)
- **South Asia:** 1 billion + under age 30
- **Implication:** Young, mobile-first populations are early adopters of digital-native services and less attached to analog systems.

Institutional Flexibility

- Emerging states can **design regulations concurrently** with new technology.
- Example: Rwanda's Civil Aviation Authority became the **first in the world** to regulate drone delivery at a national scale (2016).

Case Studies in Leapfrogging Innovation

Kenya: Mobile Finance Revolution

- **M-Pesa**, launched by Safaricom in 2007, allowed mobile-based money transfers without a bank account.
- By 2024:
  - 98% of Kenyan households use mobile money
  - Contributed to **2% annual GDP growth** (World Bank)
  - Inspired replications in 20+ countries

Leadership Insight: Kenya's Central Bank resisted early IMF and World Bank pressure to impose banking regulations on M-Pesa. The decision to permit technological experimentation before regulation sparked one of the most inclusive financial revolutions globally.

Bangladesh: Solar Home Systems (SHS)

- Grameen Shakti (a sister org of Grameen Bank) financed the deployment of **4 million SHS** between 2003 and 2018.
- 85% of these were installed in rural areas without grid access.
- Integrated **microfinance + modular tech + local capacity building**.

Leadership Insight: By training over 2,000 women as SHS technicians and entrepreneurs, the program effectively blended **gender equity, technological diffusion, and local ownership**.

Rwanda: Data-Driven Public Health

- Zipline's drone network has conducted over 750,000 medical deliveries across Rwanda and Ghana.
- Reduced vaccine spoilage by 67%, and blood delivery time from 4 hours to 30 minutes.
- Drone data integrated into national AI-driven logistics planning platforms.

Leadership Insight: Rwanda's top-down, systems-oriented governance enabled full regulatory integration, with cross-ministerial cooperation among key sectors, including health, aviation, and Information and Communication Technology (ICT).

India: Digital Identity Infrastructure

- Aadhaar, the world's largest biometric ID system (1.4+ billion enrollees), enables:
  - Direct cash transfers to 300 million beneficiaries
  - Paperless KYC for banking, SIM cards
  - Integrated with India Stack, a digital infrastructure backbone enabling APIs for governance, finance, and data

Leadership Insight: India leapfrogged legacy welfare systems and banking bottlenecks, enabling pro-poor, low-cost digital state capacity.

Constraints as Catalysts: The Innovation Under Constraint Model

Constraint	Systemic Innovation Triggered
No power grid	Solar microgrids, pay-as-you-go solar
No formal banking	Mobile finance, agent networks
Low literacy	Icon-based user interface (UI), voice-based AI assistants
Remote geography	Drones, low-Earth-orbit satellite broadband
Poor roads	Motorcycle-based mobility logistics

Constraints force **functional creativity**. They eliminate redundancy, drive cross-sector collaboration, and enhance the functionality of technologies. In leapfrogging economies, **innovation is not a luxury** – it is a survival imperative.

System-Level Risks and Friction Points

Leapfrogging is not linear, nor is it without hazards. The following risks demand adaptive leadership:

Digital Colonialism

- Platforms (e.g., Meta's Free Basics) offered free internet access but controlled content and throttled the open web.
- Leapfrogging can create data dependency on foreign-owned platforms without reciprocal benefit-sharing.

Informality and Data Gaps

- Informal economies (up to 90% in some African countries) generate few traceable metrics.
- Without robust data, AI and public policy risk encoding blind spots.

Regulatory Catch-Up

- Technological speed often outpaces policy formulation.
- **Example:** Motorcycle ride-hailing apps in Nigeria and Indonesia were banned after uncontrolled proliferation.

The Architecture of Leapfrog Leadership

Effective leapfrog leadership is not about mimicking Silicon Valley; it is about creating contextually intelligent, structurally inclusive ecosystems that foster innovation.

Institutional Design Principles

- **Sandboxes Before Statutes:** Pilot new technologies under controlled environments (e.g., Ghana's fintech sandbox).
- **Modular Infrastructure:** Design public systems as interoperable platforms (India Stack and Open Network for Digital Commerce).
- **Public–Private Foresight Coalitions:** Integrate long-term planning across academia, startups, and civil society.

Metrics That Matter

Leapfrogging success should not be measured by gross domestic product (GDP) or a number of unicorns, but by:

- Access equity (rural vs. urban tech penetration)
- Gender parity in tech usage
- Local value capture (patents and talent retention)
- Resilience gain per dollar invested

The Ethical Edge: Avoiding the Pitfalls of the North

Leapfrogging nations have the opportunity to design ethical, inclusive, and ecologically aligned systems from the start.

- Do not build *surveillance capitalism* under new names.
- Do not replicate *carbon-intensive pathways* in pursuit of short-term gains.
- Do not offload *e-waste or carbon credit obligations* onto domestic populations.

  Instead, leapfrog into:

- Digital rights frameworks embedded in tech design (e.g., Kenya's Data Protection Act, 2019)
- Circular economy models from inception (e.g., Nigeria's e-waste laws)
- Green industrial policy (e.g., Morocco's solar manufacturing ecosystem; Vietnam's electric vehicle startups)

The Future May Already Be Distributed

The next Silicon Valley may not look like Silicon Valley. It may emerge in Lagos, Dhaka, Kigali, or Medellín – not despite their constraints, but because of them. Leapfrogging is no longer just a strategy for catching up; it has become a key approach for surpassing competitors. It is a strategy for redesigning the future – from the bottom up, from the outside in, and from the margins to the center.

If the industrial giants of the North defined the 20th century, the 21st may be remembered for the agile, adaptive, and deeply human innovation ecosystems of the Global South. The challenge is not just to innovate, but to lead systems that learn, adapt, and regenerate – without repeating the extractive patterns of those who came before.

Reference

International Energy Agency (IEA), International Renewable Energy Agency (IRENA), United Nations Statistics Division (UNSD), World Bank Group, and World Health Organization (WHO), *"Tracking SDG7: The Energy Progress Report 2024,"* Washington, DC, World Bank, Jun 2024. Available: https://sdgs.un.org/publications/tracking-sdg7-energy-progress-report-2024-58022 (accessed Feb. 2, 2026).

41

The Currency of Trust: Why Our Future Depends on What We Cannot See

Introduction

I have seen cities go dark and markets freeze. In those moments of crisis, the true measure of strength was simple: Could people still rely on one another? Could they stand steady when fear pressed in? When a massive regional power blackout struck, the electrical grid failed instantly – yet neighbors stepped up to direct traffic and share resources. When a financial panic hit and credit markets ground to a halt, algorithms and balance sheets were useless – only emergency action and public faith in institutions kept the economy from collapsing. The systems that failed were technical. The systems that held were human. Each time, it became clear that our most vital infrastructure is not steel or software; it is trust. Trust is the invisible currency that powers our grids, markets, democracies, and families. When trust collapses, no technology or wealth can save a society. This book explores how trust is built and broken, why it has become the fragile thread holding our world together, and how we can fortify this hidden grid before it frays beyond repair.

We live in an age when trust is under strain everywhere. Global surveys paint a stark picture: only about 40% of people worldwide expect their families to be better off in five years – a steep drop in optimism compared to a few years ago [1]. Economic anxiety, disinformation, and polarization are eroding the glue that holds communities together. Public trust in governments and institutions is near historic lows in many countries. For example, in the United States, only around 20% of citizens now trust the federal government to do what is right – down from 77% in the 1960s, reflecting decades of deepening cynicism [2]. A worldwide "trust recession" is unfolding, driven by leadership failures, inequality, and rapid technological change. The World Economic Forum warns that the erosion of social cohesion and societal polarization – essentially a collapse of trust – ranks among the top global risks for the coming decade [1]. In an unraveling world of climate threats, cyberattacks, and geopolitical rifts, trust is both more precious and more precarious than ever.

Yet amid the uncertainty, there is hope: trust can be rebuilt. Throughout history and across societies, trust has proven to be a renewable resource – if we choose to renew it. Communities that nurture trust and social bonds have shown remarkable resilience in the face of disasters and turmoil. Leaders who earn trust through integrity and empathy have guided nations through dark hours. In the chapters ahead, we will examine trust as the cornerstone of effective systems (from power grids to supply chains), as the key to ethical leadership and social stability, as the linchpin of resilient communities, and as the unwritten code passed across generations. Through storytelling, personal experiences, and analysis of real cases on every continent, we will see why our future depends on what we cannot see – the currency of trust – and how investing in it can secure a brighter, more resilient world for all.

The Currency of Trust: The Invisible Fuel of Systems

Trust is the currency that powers every system we rely on. In the modern world, nearly everything runs on trust. We take for granted that flipping a light switch will bring electricity, that money in the bank will be available when needed, or that a posted letter will reach its destination. Beneath each of these expectations lies a bedrock of trust in people, institutions, and processes we cannot personally verify at all times. Economist Kenneth Arrow once observed that "virtually every commercial transaction has within itself an element of trust" [3]. Indeed, trust is the invisible fuel that keeps commerce, technology, and society functioning. When it falters, systems grind to a halt – sometimes overnight.

Consider the global financial system, which essentially runs on trust. Money itself is a promise, valuable only because people believe in its value. In the 2008 global economic crisis, that belief was shaken to the core. Banks stopped trusting one another's solvency and the interbank lending that greases the economy froze up. As one analysis noted at the time, "banks don't fully trust each other... there's just an environment of distrust right now, and that's the core of this entire crisis" [4]. Interest rates for overnight loans between banks spiked to record levels as each institution feared the next might collapse. Credit – whose Latin root *credere* means "to believe" – evaporated. It took massive government interventions to halt the downward spiral. The lesson was stark: when trust vanishes, so does liquidity; markets seize up, and even enormous wealth cannot substitute for the missing trust. What holds a complex financial network together is the shared belief that obligations will be honored. Once that faith is lost, the network unravels. The 2008 meltdown proved that elaborate financial engineering is no match for a fundamental loss of confidence. In the end, trust *is* the capital that matters most.

The same is true for other critical systems. Energy grids, supply chains, and communications networks all require layers of trust to function smoothly. Our electric power grid is often called the most complex machine ever built – thousands of power plants and millions of miles of lines working in unison. But the real glue is the trust between all actors: that grid operators will keep supply and demand balanced, that maintenance crews will do their jobs, that customers will pay their bills, and that regulators will enforce safety. When that trust is absent or broken, even the best technology can falter. For instance, during widespread power outages, what often determines whether society stays calm or descends into chaos is public trust. In one stark comparison, New York City's blackout of 1965 (as well as a later outage in 2003) was met with relative calm and community cooperation, while the blackout of 1977 – coming amid fiscal crisis and frayed social trust – resulted in an eruption of looting and arson across dozens of neighborhoods [5]. The technical cause in each case was similar (a grid failure), but the social aftermath diverged dramatically. Trust was the variable: in 1977, many New Yorkers had lost faith in city leadership and economic justice, and when the lights went out, civic order collapsed. In 1965 and 2003, by contrast, a stronger residue of trust and social capital helped people pull together despite the darkness. These episodes reveal that the true resilience of infrastructure lies in human relationships. A power grid can fail safely if people trust each other – or become a catastrophe if they do not.

Look at any efficient supply chain, and you will see trust holding it together. Whether it's a farmer trusting a distributor to pay on delivery, or a factory trusting its parts supplier to meet quality standards, each link relies on faith in the next. The COVID-19 pandemic provided a dramatic illustration: at the outset, many nations and states engaged in frantic bidding wars for masks and medical supplies, often out of distrust that others would share fairly. Global supply chains are strained under protectionism and suspicion. In contrast, regions that coordinated and trusted each other's data – for example, scientists sharing vaccine research openly – achieved faster breakthroughs.

In normal times, too, businesses depend on reliable partners and a stable rule of law to enforce contracts. When trust in those foundations erodes, companies either withdraw or incur steep costs protecting themselves. Economists have found a strong correlation between societal trust levels and economic prosperity [3]. High-trust societies tend to have lower transaction costs, smoother business cooperation, and more investment. Low-trust environments, on the other hand, are forced to spend enormous effort on verification, security, and litigation, which acts like sand in the gears of commerce. In this way, trust is a form of capital – social capital – that boosts efficiency and growth. A handshake deal in a high-trust community can be more valuable than a legal contract in a low-trust one.

Zooming in further, even our families and personal lives run on trust as currency. From the time we are children, we learn to trust (or distrust) the world around us, and those early experiences form our "human code" for life. A child who grows up in a trusting, supportive environment gains confidence to explore and collaborate; one raised amid broken promises and betrayal may carry deep skepticism. In workplaces, teams with high trust among members consistently outperform those mired in office politics and fear. Marriages and friendships thrive on trust and crumble when it is gone. In communities, neighbors who trust each other can create informal safety nets – watching each other's homes and helping with childcare – that no government program can replicate. During disasters, these trust-fueled networks become literal lifelines. Sociologists have documented how communities with strong interpersonal trust and civic engagement cope far better with crises. People share resources, evacuate the vulnerable, and rebuild together. In contrast, communities divided by suspicion may see "everyone for themselves" behavior that makes the crisis far worse. Trust, at its heart, is a mutual belief in reliability and goodwill. It allows us to overcome the inherent uncertainty in any interaction. Without that belief, cooperation collapses to its lowest denominator – contracts and coercion – which are costly and brittle substitutes.

In short, trust is the unseen currency spent in every human exchange. It is the confidence that others – known or unknown – will do the right thing. It greases the everyday machinery of life, from buying groceries with paper money to electing leaders to managing a public health crisis. We notice trust only when it's missing: when a scam makes us wary of strangers, when a failed promise leaves us jaded, and when a breach of integrity makes headlines. Only then do we realize how much of our world runs on assumptions of good faith. As we will explore, those assumptions are being tested today as never before. But first, we must understand how fragile this currency can be, and what happens to societies when the bank of trust runs dry.

The Fragile Thread: Trust and Survival in an Unraveling World

History shows that civilizations do not usually fall from a lack of power or wealth – they fall when the thread of trust binding them breaks. Great powers have been defeated by internal rot long before external enemies overran them. Ancient Rome had formidable legions and vast treasure, yet a decay of civic virtue and trust marked its decline. As corruption spread and the social contract frayed, ordinary Romans lost faith in their leaders and the unity of the empire. By the end, many communities did not bother to defend Rome against barbarian invaders – the allegiance to Rome had already died in their hearts. In the same vein, the Soviet Union did not collapse in 1991 because it ran out of tanks or missiles. It collapsed when its people ceased to believe in the system. A popular dark joke in the USSR captured this loss of trust: "They pretended to pay us, and we pretended to work." Decades of propaganda, economic stagnation, and broken promises had eroded any trust the Soviet citizens had in their government. When reformist leaders tried to introduce openness (glasnost) and restructuring (perestroika), it was too late – the foundation of trust was gone, and the union fell apart with shocking speed. These examples underline a vital point: when trust dies, systems of any size – empires, governments, companies – become hollow. They may linger on for a time, but they are one shock away from collapse because no one is truly invested in their success.

Today, that fragile thread of trust is under strain across the world. We see the warning signs in our headlines and our communities. In many democracies, public trust in elected leaders and institutions has hit record lows, fueling a populist wave of anger. People who feel lied to or left behind by elites become susceptible to demagogues who promise easy fixes while further poisoning the well of trust. For instance, surveys in country after country show deep skepticism that governments are working in the people's interest – a view reinforced by high-profile scandals of corruption and abuse of power. Each scandal is not just a single breach; it accumulates into a narrative that "the system is rigged" or "politicians cannot be trusted," driving citizens into cynicism or even onto the streets in protest. In authoritarian states, the trust problem takes a different form – an enforced facade of trust maintained by fear. Because dissent is punished, people may profess loyalty but privately doubt the regime. Such top-down "trust" is brittle: when cracks appear, they can quickly widen into collapse, as seen in Eastern Europe's revolutions. Genuine trust cannot be commanded; it must be earned.

One profound contemporary challenge is how technology and media are disrupting trust. We live in the age of the "information apocalypse," where it is increasingly difficult to know what information is real. Social media platforms amplify rumors and fake news, while artificial intelligence (AI)-generated deepfakes can mimic faces and voices with alarming realism. In this environment, the basic trust that is necessary for a common reality starts to erode. If citizens cannot even agree on what is true – for example, whether an election was free and fair, or whether a vaccine is safe – society loses the ability to solve problems collectively. A vivid example emerged during the COVID-19 pandemic: in some communities, rampant misinformation led large segments of the public to distrust health authorities and reject lifesaving vaccines or masks. In the United States and elsewhere, political polarization and low trust in government hampered the pandemic response, contributing to higher death tolls. Studies found that regions with higher trust in government and science had greater compliance with public health measures and ultimately better health outcomes. In contrast, in the low-trust areas, people were less willing to follow guidelines or believe official information [6]. In essence, when trust in leadership and expertise falters, even well-intended policies fall flat. The pandemic was a stress test that exposed how vital social trust is to managing collective threats.

Our growing dependence on digital technology has also introduced new trust vulnerabilities. Cyberattacks and data breaches regularly undermine trust in institutions' ability to protect sensitive information. When a major company loses millions of customer records to hackers, or a city's infrastructure is paralyzed by ransomware, the public's faith in the security of the digital systems we rely on takes a hit. Each incident plants seeds of doubt: Is my hospital's network safe from hackers? Will the power grid stay up if targeted? Can I trust that the water treatment plant won't be compromised? While engineers race to patch technical weaknesses, the psychological impact of these breaches is cumulative. We risk entering a climate of "zero trust" by default – a mindset where people assume every digital interaction is suspect. Such skepticism, if unchecked, can stifle adoption of beneficial innovations (for instance, people may reject smart city systems or e-health records out of fear) and can drive a retreat into isolated, analog alternatives. Building a secure cyber future, thus, isn't only about better firewalls; it's about maintaining public trust that our critical systems won't betray us.

Perhaps no issue illustrates the global trust deficit more than the climate crisis. Climate change is a collective action problem of unprecedented scale – solving it demands trust and cooperation among nations, industries, and communities worldwide. Yet trust has been in short supply. Developing countries, for example, have been told to trust that wealthy nations will deliver financial support for green development and adaptation to climate impacts. In 2009, richer countries pledged to mobilize $100 billion per year by 2020 to help poorer nations cope with climate change. That promise was not kept on time. By 2020, the reported funding was around $83 billion, and much of it was in the form of loans rather than grants – far less useful to vulnerable economies [7]. This failure has frayed the trust of developing nations, who understandably feel wronged. "Rich countries are treating poorer countries with contempt," said Nafkote Dabi, a climate policy lead at Oxfam, in 2023, warning that broken promises are "fatally undermining" climate negotiations [7]. The result is a dangerous stalemate: low-income countries grow hesitant to commit to emissions targets or land protections if they can't trust the promised support, while wealthier countries become wary of contributing more without seeing reciprocal action. It's a vicious cycle of mistrust that hampers our ability to face a truly existential threat. Only by restoring trust – through verifiable actions, transparency, and fairness – can this cycle be broken. The same holds within nations: citizens must trust that climate policies are equitable and not just burdensome, or they will resist change. We've seen instances of public backlash (like the "Yellow Vest" protests in France against fuel taxes) when people felt climate initiatives violated a social trust by unfairly harming certain groups.

The fragility of trust is evident even in seemingly stable systems. Modern infrastructure often gives an illusion of invincibility – until a crisis hits. When natural disasters strike – a massive hurricane, a regional blackout, a pandemic – they reveal the underlying condition of social trust. Communities with strong trust and social networks tend to respond effectively: neighbors check on neighbors, volunteers mobilize, and there is faith that help will come. In such communities, people often self-organize to fill gaps, demonstrating resilience. Research has shown that social capital (the networks and trust within a community) is a top predictor of disaster recovery success [8]. After Japan's 2011 earthquake and tsunami, for example, villages with higher levels of civic engagement and trust had significantly lower mortality rates and recovered faster than those with weaker social ties [8]. By contrast, where trust is low, crises can spiral into social breakdown. We have witnessed instances where rumors incite panic buying, or residents refuse to evacuate because they don't believe officials. During Hurricane Katrina in 2005, the slow and chaotic government response, combined with pre-existing distrust, especially among poorer communities in New Orleans, led to a collapse of order in parts of the city. Misinformation and distrust hampered coordination, and it took years for trust in institutions even partially to mend after that failure. Simply put, *resilience is not just about robust physical systems, but about robust social trust*. A community might have strong levees or stockpiles, but if people don't trust the authorities or each other, those assets can be squandered or underused.

On the global stage, the absence of trust between nations can mean the difference between peace and conflict. Treaties and alliances are only as strong as the credibility of the parties involved. The post-World War II international order – institutions like the United Nations, NATO, the European Union – was built on the hard-earned trust that cooperation was better than war. But in recent years, cracks have appeared. Longstanding allies question each other's commitments (as seen when trust wavered in transatlantic relations), and rivals exploit disinformation to sow distrust. The risk is that without a basic thread of trust, even imperfect, the world could slide into a new era of fragmentation and rivalry, undermining progress on issues that require unity (from pandemics to terrorism to climate). Indeed, the theme of the World Economic Forum's 2023 summit was "Cooperation in a Fragmented World," reflecting the urgent need to rebuild trust and cooperation in the face of fragmentation [1].

Looking ahead, we face a fork in the road. Down one path, societies continue to splinter as trust in each other and institutions crumbles. In this scenario, problems multiply: democracies gridlock or backslide into authoritarianism because social trust – the willingness to compromise and see others as legitimate – disappears. Economies stagnate as investors and consumers retreat out of fear. Communities balkanize, each withdrawing behind gates or into online echo chambers. Technology, instead of connecting us, becomes a battleground of misinformation and surveillance, further eroding trust. In the worst case, the loss of confidence could even lead to violence and conflict, as groups demonize each other and no longer believe in peaceful solutions. This grim future is not a sci-fi dystopia; elements of it are already visible wherever trust has collapsed.

But down the other path, we recognize the value of trust and act decisively to restore it. This path is harder but far more hopeful. It involves leaders being accountable and transparent, demonstrating ethics that earn back public confidence. It affects communities actively investing in social bonds – fostering dialogue across divides, building inclusive institutions, and remembering that trust is built slowly, through consistent honesty and empathy. It involves redesigning systems – from policing to social media algorithms – with trust and fairness in mind, so they don't inadvertently breed cynicism or inequality. Perhaps most importantly, it involves each of us making daily choices that strengthen trust: keeping our word, listening to others' concerns, and showing up for our neighbors. Trust, after all, is contagious. Small acts of trustworthiness can ripple outward, inspiring others to do the same.

The thread of trust may be frayed, but it has not snapped. Around the world, countless efforts are underway to mend and reinforce it. In cities, programs bring police officers and community members together to rebuild trust through dialogue and collaboration. In business, "stakeholder capitalism" is on the rise – a model where companies aim to build trust by serving not just shareholders but employees, customers, and communities ethically. In governance, some nations are experimenting with participatory budgeting or citizen assemblies to give the public a real voice, thereby strengthening trust in decision-making. Even on the global stage, youth movements and NGOs are working across borders to foster trust where governments have failed – for example, connecting young climate activists from north and south to form bonds of solidarity. These are the strands being woven to repair the fragile thread. The question is whether we can scale them up fast enough.

Trust: The Hidden Grid of the 21st Century

Beneath all the visible grids and networks of our modern age lies an invisible one: the grid of trust. We often think of infrastructure in terms of concrete, wires, code, and policies. But an electrical grid or an information network is ultimately a framework that depends on people doing their part reliably. Trust is the *unseen circuitry* that links all the components together. Although it doesn't appear in any schematic diagram, its failure can cause the whole system to crash. The 21st century has brought astounding technological connectivity – we can transact, communicate, and travel across the globe faster than ever. Ironically, this hyper-connectivity makes the trust grid even more critical. With so many interdependencies, a single failure can cascade widely. The only way complex systems don't "break all at once" (as one grid expert said in a post-9/11 briefing [8]) is if there are buffers – and trust provides many of those buffers.

Think of trust as the bandwidth in a societal system. High trust means information and cooperation flow quickly, like data on a high-speed connection. Low trust is like narrow bandwidth – everything slows down with additional verification, oversight, and friction.

In high-trust organizations, a project can move forward with a handshake or a few emails because each party trusts the other to follow through. In low-trust settings, even minor tasks require lengthy contracts, approvals, and micromanagement, bogging down progress. Trust is also akin to voltage stability in the power grid: it keeps the current of civic life flowing steadily, preventing small disturbances from turning into blackouts. When trust is high, a society can absorb shocks – people give each other the benefit of the doubt and pull together in emergencies. But if trust is already low, even a minor shock (a rumor, a temporary outage, an election result) can trigger disproportionate turmoil. In that sense, trust is the safety margin that prevents systemic failures. A society with no trust margin is operating one incident away from crisis at all times.

Moreover, trust acts as the operating system for collaboration. Just as computers need a common OS to coordinate processes, humans need trust to coordinate at scale. An example of trust as a hidden infrastructure can be seen in disaster response. After a major earthquake or hurricane, official responders may be overwhelmed. It's often community networks – churches, local businesses, volunteer groups – that improvise solutions in the first critical hours. These networks succeed because they run on trust and local knowledge. I recall working on community resilience workshops in a city where we mapped "heat zones" of trust: we asked residents, "*Who would you turn to if the system fails?*" The answers (the neighbor with a generator, the corner store owner who extends credit, and the old church that still has a landline) revealed an invisible web of relationships that was the true first-responder system. When we showed city officials these community trust maps, it was a revelation. The city's formal plans had overlooked the informal trust grid that holds things together when formal systems falter. As one community organizer succinctly put it, "Relationships are the real infrastructure."

Even in cutting-edge fields like AI, the hidden grid of trust looms large. Developers can create astonishing AI systems, but if users don't trust them – if the public fears the algorithms are biased, or that AI decisions are unaccountable – adoption will stall and backlash will grow. That's why concepts like "ethical AI" and "trustworthy AI" have become prominent: engineers recognize that without building trust into the design (through transparency, fairness, and human oversight), the technology will fail to deliver its promise. Similarly, in the energy transition to renewables and smart grids, success depends not just on solar panels and batteries, but on trust: consumers must trust the grid to remain reliable as we integrate new energy sources, and communities must trust that the burdens and benefits of clean energy will be shared fairly. If people fear the lights will go out or jobs will be lost unfairly, they will resist change. Trust is thus the bridge between present and future – it allows societies to undertake ambitious transformations (be it digitalization, decarbonization, or globalization) with confidence rather than fear.

Because trust is largely invisible, it can be woefully undervalued by decision-makers focused on tangibles. A company might slash employee benefits to cut costs, not realizing it is chipping away at workers' trust and thus long-term productivity. A government might roll out a policy without community input, saving time upfront but breeding mistrust that will cost dearly in enforcement and compliance problems later. These are examples of "efficiency" that backfires by weakening the hidden grid. Truly forward-thinking leadership, in contrast, treats trust as a critical asset to invest in. For instance, some police departments are shifting from pure enforcement metrics to "trust metrics" – measuring community sentiment and engagement – on the understanding that public safety ultimately flows from public trust. In business, the rise of corporate ethics and sustainability efforts reflects an understanding that trust from consumers and society is a competitive advantage that drives everything from brand loyalty to the ability to attract talent.

Investing in trust can take many forms. It might mean corporations embracing transparency – sharing data on their supply chains or environmental impact – even if not legally required, to earn public confidence. It can mean governments practicing radical honesty with citizens about challenges and mistakes, instead of spinning or covering up, thereby building credibility over time. It certainly means leaders being willing to stand in truth even when it costs them, because nothing builds trust like visible integrity, and nothing destroys it faster than deceit. As an example, in 1982, the pharmaceutical company Johnson & Johnson faced a crisis when Tylenol bottles were found to be tampered with poison. They chose full transparency and a nationwide recall at great expense, putting customer safety first. That courageous response is credited with not only saving the brand but also setting a gold standard for corporate trustworthiness in a crisis. It sent a message: we value your trust more than our short-term profit. And customers remembered.

On a societal level, building the trust grid often comes down to inclusion and fairness. People trust a system when they feel they have a voice in it and a stake in its outcomes. This underscores the importance of ethical governance and reducing inequality. Data shows that societies with high inequality tend to suffer lower levels of interpersonal and institutional trust [3]. When large segments of the population feel left out or exploited, trust in the social order erodes. By contrast, more egalitarian societies – for example, many in Northern Europe – exhibit both high trust and strong economic performance. In surveys, over 60% of people in countries like Norway, Sweden, and Denmark agree that "most people can be trusted." In contrast, in highly unequal societies such as Brazil or Colombia, that figure can drop to below 10% [3]. The Nordic nations, which rank among the world's highest in social trust, also consistently rank high in happiness and low in corruption, suggesting a virtuous cycle where trust and equality reinforce each other. This doesn't mean homogeneous societies automatically trust more – rather, it means that when people perceive the system as just and everyone as playing by common rules, trust flourishes even among diverse groups. By ensuring access to basic services, reducing extreme disparities, and honoring the social contract with all citizens, governments can cultivate the sense that "we're all in this together," which is the bedrock of trust.

We must also recognize that trust is an intergenerational contract. The decisions one generation makes set the context for the next. If young people grow up watching leaders lie or evade responsibility, why would they trust any institution as adults? If they see society discount their future – for instance, by wrecking the environment or running up unpayable debts – the unwritten pact between generations frays. On the other hand, when a generation rises to a challenge in a way that inspires the next, trust transcends time. Think of the trust placed by one generation of freedom fighters in the idea of democracy, which they passed to their children; or the trust that built institutions like the United Nations after World War II, rooted in the hope of "never again." We today owe a debt of trust to future generations: to steward a world they can live in and to model the values that will hold it together. In practical terms, that means tackling long-term problems with honesty and courage now, not kicking cans down the road. It means teaching and exemplifying ethical behavior. It also means amplifying youth voices in decision-making today, as a sign that we trust and respect those who will inherit the consequences of our actions. When we do right by the future, we reinforce the bond of trust across time – a "human code" of honor that can guide humanity even through dark corridors of history.

Ultimately, strengthening the hidden grid of trust comes down to leadership – not just from those at the top, but leadership from every seat. Each of us is a node in the trust network, capable of either strengthening or weakening it. Leaders in formal roles (CEOs, presidents, mayors) bear a heavy responsibility to set the tone. If they hoard power and veil the truth, they may achieve short-term control but at the cost of eroding the trust that underpins long-term stability. By contrast, leaders who act as custodians of trust – who prioritize truth, transparency, and the common good – leave a legacy far more durable than any monument. They understand that trust, once broken, is hard to rebuild, so it must be protected vigilantly. In the words of one leadership maxim: *"Leaders who hoard power lose it. Leaders who protect trust leave something that endures."* This ethos applies equally to community leaders, managers, educators, and parents. When people consistently experience fairness, compassion, and reliability from those in charge, it reinforces a broader culture of trust.

In this age of disruption, we have seen how brittle the old definitions of strength are. Military might, economic size, and technological prowess – none guarantee resilience if trust is absent. The mightiest power grid fails if people don't cooperate to conserve energy in a crisis. The strongest economy falters if consumers and banks panic and withdraw trust. The most advanced democracy can teeter if citizens refuse to trust election results or each other's basic decency. The future, it turns out, will not be won by the society with the biggest arsenal or the fastest supercomputers. It will be won by those who can still be trusted when everything else breaks. Trust is the factor that turns disparate individuals into a collective capable of overcoming challenges. It is what allows us to take that leap of faith to cooperate, innovate, and sacrifice when needed.

The hidden grid of trust carries more weight than concrete, steel, or code. Every decision we make, every interaction, is either building it up or wearing it down. The choices might be as small as being honest in a business deal or as large as a nation owning up to a historical injustice – but each one matters. If we treat trust as our most precious infrastructure and nurture it with the same care that we give to our roads and bridges, we will create systems that endure. If we squander it, we risk a future of constant breakdowns and fragmentation.

Conclusion: Rebuilding Trust, Rebuilding the Future

In the end, trust is the ground beneath every step we take as a civilization. It is the silent pact that turns chaos into cooperation and adversity into collective action. Break that ground, and nothing lasting can be built. Keep it strong, and even in ruin, a path to recovery can be found. The currency of trust is unique: the more we spend it in good faith, the richer we become. Unlike wealth or technology, trust is not zero-sum – it multiplies with use, forging resilience in families, teams, and nations.

The journey we have taken through these chapters leads to a clear conclusion: trust is survival. Our survival in crises, our prosperity in peacetime, and our ability to tackle global challenges – all hinge on rebuilding and sustaining trust. In a world unraveling in many ways, trust is the fragile thread stitching it together and the only thread capable of mending it. Each of us has a role in strengthening that weave. We strengthen it by expecting honesty and accountability from our leaders and ourselves. We strengthen it by resisting the temptation to retreat into tribalism or cynicism and instead by extending a hand across divides. We strengthen it by remembering that behind every system – electrical, economic, or political – are human beings, and it is ultimately our willingness to trust each other that determines if those systems stand or fall.

If there is one message to carry forward, it is a hopeful one: trust can be rebuilt. We know this because it has been done before. After eras of conflict, nations have reconciled and built trust anew – as Germany and France did after 1945, laying the foundation for a peaceful Europe. After periods of corruption and abuse, institutions have been reformed and public trust earned back – as seen in various anti-corruption successes from Singapore to Georgia, where determined leadership broke the cycle of cynicism. Communities torn by disaster or violence have found healing through truth and reconciliation, whether through South Africa's Truth and Reconciliation Commission or grassroots peace circles in war-torn villages. These stories show that even when trust has been gravely damaged, it is not beyond repair. It requires truth-telling, courage, and time, but the thread can be reforged.

For general readers, policymakers, business leaders, and young professionals alike, the implications are clear. Trust is not a "soft" sentiment to consider after the real work is done – it *is* the real work. A policy that ignores trust, an innovation introduced without trust, or a strategy that undercuts trust is ultimately self-defeating. Conversely, an initiative that builds trust can transform what is possible. This applies whether you are leading a multinational corporation, governing a city, or simply trying to support your neighbors. The specific tools will differ – it might be adopting a community-driven planning process for a new project, implementing ethical guidelines for AI, practicing radical candor with your constituents, or mentoring youth in your neighborhood – but the end goal is the same: to weave a stronger fabric of trust.

As we look to the future, we must ask ourselves: what legacy will we leave? If we continue on a path of fractured trust, we risk leaving a world of walls – walls between people, walls between nations, and walls around hearts. But if we choose to invest in trust, we can build a world of bridges. Bridges that allow energy and ideas to flow, that enable joint solutions to shared problems, and that connect people in empathy. The currency of trust, though invisible, might be the most valuable asset humanity has in the 21st century. By integrating trust with resilience, ethics, and our human code of values, we create a bulwark against whatever storms may come.

In an era obsessed with innovation and disruption, perhaps the boldest innovation is to reimagine strength not as dominance or dominance, but as the ability to earn and hold trust. It is a humble kind of strength – quiet, often unseen – but as we have argued, it underlies all others. Trust is the hidden grid that lights our way. If we make trust our priority – in our systems, our leadership, our communities – we will find that we possess an endless energy source for renewal and progress. With trust, even broken nations and broken systems can find the will to rise again. Without it, even our greatest achievements are built on sand.

The future will belong to those who can be trusted when everything else breaks. Let us strive to be those people and to cultivate those leaders. Let us design our technologies and policies with the wisdom that relationships are infrastructure and conscience is core to resilience. The challenges of our time – be it climate change, technological upheaval, or social division – demand nothing less than a resolution to rebuild trust at every level. Each act of trust we foster is a strand in the rope that will hold us together.

Our future truly depends on what we cannot see. It depends on trust – the currency that carries our hopes and the thread that ties our fate. If we value it, protect it, and grow it, there is no storm we cannot weather, no grid we cannot keep alive. Trust is the line that holds. May we never let it go.

References

1 J. Myers, "Trust barometer: 4 steps to rebuilding trust," *World Economic Forum*, 18 Jan. 2023. Accessed Aug. 18, 2025. [Online]. Available: https://www.weforum.org/stories/2023/01/edelman-trust-barometer-rebuilding-trust/.

2 J. Davidson, "American trust in government near 'historic lows,' Pew finds," *The Washington Post*, 09 Jun. 2022.

3 E. Ortiz-Ospina, M. Roser, and P. Arriagada, "Trust," *Our World in Data*, Apr. 2024. Accessed Aug. 18, 2025. [Online]. Available: https://ourworldindata.org/trust.

4 N. Irwin, "Mutual distrust freezes lending among banks," *The Washington Post*, 30 Sep. 2008.

5 "New York City blackout of 1977," *Wikipedia*. Accessed Aug. 18, 2025. [Online]. Available: https://en.wikipedia.org/wiki/New_York_City_blackout_of_1977.

6 A. Devine *et al.*, "Trust in government and COVID-19 vaccine hesitancy," *Nat. Commun.*, vol. 13, no. 1, p. 5209, 2022.

7 Oxfam International, "Rich countries' continued failure to honor their $100 billion climate finance promise threatens negotiations and undermines climate action," Press Release, 5 Jun. 2023.

8 D. P. Aldrich, *Building Resilience: Social Capital in Post-Disaster Recovery*. Chicago, IL: University of Chicago Press, 2012.

42

Beyond the Horizon – Leadership, Foresight, and Institutional Transformation Toward 2050

Prologue: Time as Infrastructure

On the wall of the Long Now Foundation in San Francisco stands a clock designed to tick for 10,000 years. Its hands move once per year. Its chime rings once per millennium. The clock was not built to tell time. It was built to reshape how we think about time – to push our decisions beyond quarterly earnings, political terms, and election cycles into the realm of intergenerational stewardship.

The 2040–2050 decade is not a distant abstraction. It is a strategic inflection point – a temporal hinge where cascading crises and compounded innovations will either catalyze resilient futures or collapse fragile systems. The institutions that survive – and thrive – will be those designed not for control, but for complexity; not for stability, but for continuous adaptation.

The Forces Shaping 2040–2050

Demographic Shifts

- Global population projected to reach 9.7 billion by 2050.
- Africa will account for more than half of global population growth; Nigeria alone will surpass the United States in population.
- Aging in the Global North: By 2040, one in three Europeans will be aged 60 or older.

Implications: Aging economies will require systems for care, retraining, and wealth redistribution. Youth-driven economies will require investment in education, civic platforms, and meaningful employment pathways.

Climate Compounding Effects

- IPCC Sixth Assessment: Extreme weather events will triple in frequency by 2040 if warming exceeds 2 °C.
- Coastal Displacement: Over 300 million people will be at risk of annual flooding by 2050.
- Global GDP at Risk: Estimated 11–14% loss under moderate emission scenarios.

Implications: Institutions will face climate-related liquidity risks, supply chain disruptions, and climate justice-related unrest. Leadership must account for irreversible ecological thresholds and integrate climate scenario planning into financial, operational, and legal systems.

Technological Convergence

By 2040, the convergence of artificial intelligence, synthetic biology, quantum computing, neurotechnology, and ambient intelligence systems will redefine the structure of institutions.

Domain	Forecasted Capabilities by 2040	Strategic Challenge
AI (AGI potential)	Autonomous research, legal, and military agents	Alignment, oversight, and value encoding
Synthetic Biology	CRISPR 3.0, biofabrication at scale	Biosecurity and ecological release
Quantum Computing	Post-quantum encryption and materials modeling	Cryptographic resilience and ethical access
Neurotechnology	Direct neural interfaces and memory editing	Cognitive sovereignty and psychological risk
Smart Environments	Sensor-embedded architecture and adaptive cities	Data privacy and behavioral autonomy

Implications: Institutions must evolve **governance frameworks not just for technologies, but for emergent capabilities** – those not yet fully understood, but inevitable.

Leadership for Nonlinear Futures

From Strategic Planning to Strategic Foresight

The traditional strategic planning cycle – assess, plan, and execute – is **ill-suited to nonlinear risk**. In its place must arise **strategic foresight:** the disciplined exploration of alternative futures.

Toolkits for 2040–2050 Leadership:

- **Scenario matrices** (Shell's TINA and Waves Scenarios)
- **Cross-impact analysis** (Leontief matrix and probabilistic modeling)
- **Backcasting** from normative visions (e.g., carbon neutrality and post-scarcity)
- **Futures wheel mapping** (First-, second-, and third-order consequence mapping)

Case Example: Finland's Parliament Committee for the Future has institutionalized foresight in national policy, evaluating AI futures, green transition paths, and demographic risks as part of its **annual legislative planning**.

Leadership Archetypes for the Next Era

Archetype	Core Capacity	Risk If Absent
Sensemaker	Synthesizes complex signals into actionable insight	Misdiagnosis of emergent shifts
Custodian	Embeds intergenerational ethics and ecological integrity	Intertemporal harm and collapse
Catalyst	Mobilizes ecosystems of innovation	Institutional inertia and stagnation
Recomposer	Redesigns structures in real-time	Misfit institutions and talent drain
Bridgebuilder	Integrates diverse epistemologies and cultures	Polarization and governance failure

Effective leadership toward 2050 requires **polymaths over specialists**, **translators over technocrats**, and **architects over administrators**.

Institutional Transformation: Design for Evolution

From Hierarchies to Fractal Organizations

Future-resilient institutions will be:

- **Modular:** Capable of self-contained decision cycles
- **Fractal:** Capable of scaling values and logic across layers
- **Interoperable:** Plug-and-play interfaces across public–private data sectors

Example: Taiwan's Digital Ministry, led by Audrey Tang, operates as a fluid structure, with no hierarchy, open-source codebases, and citizen-led policymaking through the vTaiwan platforms. Its agility during COVID-19 showcased the power of **fractal governance**.

From KPIs to Dynamic Fitness Functions

Linear key performance indicators (KPIs) (e.g., revenue and output) fail under conditions of volatility. Institutions must develop **fitness functions** – metrics that adjust in response to feedback loops and environmental context.

Metric Examples:

- **Resilience Yield:** Speed of system recovery post-disruption
- **Integrity Index:** Stakeholder trust and internal ethical coherence
- **Diversity Flux:** Rate of inflow and influence of diverse perspectives
- **Ecological Ratio:** Resource input vs. ecosystem restoration output

From Compliance to Regenerative Governance

Instead of viewing regulation as a constraint, future institutions will build **regenerative governance models** – systems that generate systemic value across dimensions.

Governance Level	Current Mode	Future Mode
Corporate	ESG reporting	Integrated impact accounting (ecological + civic)
National	Rule enforcement	Civic codesign + scenario-based regulation
Global	Treaty frameworks	Polycentric, distributed consensus systems

Example: The Earth System Governance Project is piloting "climate constitutions" for legal personhood of ecosystems – a step toward **planetary citizenship regimes**.

Strategic Investments for 2040–2050 Readiness

To lead into the next epoch, institutions must invest in five **capability stacks:**

1) **Foresight Infrastructure**
   - National Futures Labs
   - Corporate foresight units with direct board access
   - Scenario repositories shared across sectors

2) **Cognitive Diversity Engines**
   - Strategic inclusion of divergent perspectives
   - Rotation of decision roles among multidisciplinary actors
   - Bias audits for artificial intelligence (AI) and human decision systems
3) **Rapid Prototyping Capacity**
   - Innovation sandboxes across policy, finance, and urban systems
   - Failsafe budgets and incentives for experimentation
   - "Fail forward" documentation embedded in learning systems
4) **Planetary Integration Tools**
   - Ecological accounting integrated with Enterprise Resource Planning (ERP) systems
   - Real-time feedback from sensors into policy dashboards
   - Citizen science pipelines for data collection
5) **Ethical Calibration Loops**
   - Annual ethical stress tests for algorithms and leaders
   - Participatory audit mechanisms (e.g., Citizens' Assemblies)
   - Intergenerational councils for value oversight

The Horizon and the Mandate

2040–2050 is not just a projection. It is a proving ground. It will reveal whether we have **upgraded our institutions** to match the complexity of our crises, whether our foresight was backed by action, and whether we designed not just for prosperity, but for **permanence with dignity**.

Leadership into 2050 will require:

- **Systems thinking as default, not luxury**
- **Temporal Agility** – knowing when to act short-, medium-, and long-term
- **Boundary-Spanning Ethics** – beyond nations, generations, and species
- **Institutional Humility** – willingness to be redesigned, even dismantled

Stewardship as the Final Metric

The question for future leaders is not whether they were right, or rich, or fast. It is whether the systems they touched became more resilient, more just, more wise.

2040–2050 will not be remembered for predictions that came true, but for transformations that came to life. This is not a time to manage what is, but to architect what must be. Not a time to scale what works, but to **seed what matters** – even if the harvest belongs to generations yet unborn.

The future is not something we forecast. It is something we **earn** through foresight, fidelity, and the fierce insistence that tomorrow must be more humane than today.

Reference

United Nations Department of Economic and Social Affairs, Population Division, *"World Population Prospects 2024: Summary of Results," UN DESA/POP/2024/TR/NO. 9*, New York, NY, USA, United Nations, 2024 (accessed Feb. 2, 2026). Available: https://population.un.org/wpp/Publications/.

Afterthoughts: Designing the Future Through Technological Leadership

The Compass: Why This Book Exists

Technological Leadership for Resilience and Progress began not with a call to innovation, but with a call to conscience.

From the book's *Prelude*, grounded in personal history and systemic failures witnessed in war zones, blackout corridors, and boardrooms, a fundamental premise emerges: technology without stewardship becomes a hazard; leadership without systems thinking becomes peril.

This work serves as a navigation tool across time, disciplines, and crises, offering guidance to those responsible for managing infrastructures, institutions, and intelligence architectures through the most volatile century in human history.

The Terrain: What the Book Maps

1) **The Foundations of Technological Leadership**
   Chapters 1 through 4 establish the lexicon and architecture of resilient leadership. Leadership is reframed not as charisma or control, but as a disciplined approach to navigating complexity. Technological leadership requires both anticipation and integration – the ability to sense weak signals, make ethical choices under uncertainty, and orchestrate systems of action.
   "Leadership is not only the capacity to respond to crisis; it is the ability to evolve before crisis demands it."
2) **Strategy, Structure, and Systemic Design**
   Chapters 5 through 18 move into implementation. Strategic planning tools – scenario modeling, real options analysis, the Balanced Scorecard, system dynamics, and technology readiness levels (TRLs) – are situated within real-world contexts:
   - Blackouts at Heathrow
   - COVID-19 ventilator pivots
   - Cascading infrastructure failures in Puerto Rico and Texas

   Ethical foresight, **operational resilience**, and **cross-sector design logic** emerge as unifying themes.
3) **Sectoral Deep Dives and Transformational Levers**
   Chapters 19 through 34 offer a tour across critical sectors:
   - Cybersecurity as public infrastructure (Chapters 22–23)
   - Artificial intelligence (AI) ethics, data sovereignty, and digital dignity (Chapters 24–25)

- Smart cities and infrastructure integration (Chapters 8–27)
- Climate adaptation, innovation ecosystems, and systemic risk (Chapters 9–31)

These chapters present not just problems, but playbooks:

- **Modular governance for AI**
- **Cyber-physical system modeling frameworks**
- **Ecosystem design maps for urban transformation**

4) **The Lenses and Tools**

Throughout the book, it integrates a *Lens Map* – a way of seeing that adapts to context:

- **Ethical lens** for decisions under asymmetry
- **Systems lens** for multiorder feedback understanding
- **Temporal lens** for actions with delayed consequences
- **Geospatial lens** for global-local alignment
- **Resilience lens** for integrity under volatility

Each chapter adds another facet to this compass, enabling readers to triangulate action with foresight and integrity.

The Inflection: What the New Chapters Add

Chapter 36: The Cost of Failure

This chapter distills the anatomy of collapse, encompassing cases such as those of Boeing, Facebook, Enron, Fukushima, and Wells Fargo, demonstrating that failure is not sudden but systemic. It emerges from:

- Prioritization of efficiency over resilience
- Ethical blindness reinforced by financial incentives
- Suppression of dissent and delay in adaptive response

Failure is treated not as a cautionary tale but as a strategic teacher. The case studies map the symptoms, root causes, and preventable trajectories of organizational collapse.

"Collapse does not announce itself – it accumulates."

Chapter 37: Planetary Intelligence

This chapter reframes climate not as an environmental issue, but as an infrastructural condition of civilization itself. By integrating planetary boundaries, ecological accounting, and regenerative design principles, the chapter establishes a model for technological systems that learn from and reinforce biological systems.

It advances three design imperatives:

- Build for resilience, not efficiency
- Model for circularity, not linear throughput
- Govern for regeneration, not minimization

Technologies examined include precision agriculture, ecological digital twins, circular manufacturing, and post-carbon infrastructures.

"The Earth is not a stakeholder to be served – it is the system in which stakeholders emerge."

Chapter 38: Leapfrogging Economies

This chapter challenges linear development narratives by documenting innovation born from constraint. In Kenya, Bangladesh, Rwanda, and India, leaders bypassed 20th-century infrastructure models to create agile, inclusive systems:

- Mobile finance as public infrastructure (M-Pesa)
- Solar home systems with gendered entrepreneurship (Grameen Shakti)
- Biometric digital governance and open application programming interfaces (APIs) (India Stack)

Leapfrogging is positioned not as catch-up but as design freedom – the ability to build new systems unencumbered by legacy inertia.

"Constraint is not a barrier to innovation – it is its crucible."

Chapter 39: The 2040–2050 Horizon

The final chapter extends the foresight arc to address demographic upheaval, ecological tipping points, and technological convergence. It introduces future architecture for institutions, including:

- Fractal governance models
- Dynamic fitness metrics over static key performance indicators (KPIs)
- Ethical calibration loops and intergenerational councils
- Planetary integration tools (e.g., ecological accounting and participatory AI)

It culminates in a leadership model rooted not in control, but in custodianship, sensemaking, and systemic imagination.

"Leadership in the future will be judged not by what it built, but by what it sustained."

The Unifying Ethos: Leadership as Stewardship

Across the 39 chapters, a consistent set of convictions crystallizes:

- That leadership is design work – the design of choices, values, and infrastructures under uncertainty.
- That resilience is not merely the capacity to bounce back, but the ability to reconfigure meaningfully.
- That technological progress, decoupled from ethical integrity and systems integration, amplifies fragility.
- That we are in a time where **every institution is a technology of consequence** and must be treated as such.

Final Reflection: What This Book Asks of the Reader

This is not a book to be passively consumed. It is a decision engine, a map for designers, engineers, public servants, entrepreneurs, and visionaries tasked with making the 21st century a habitable place.

It offers no prescriptions, only principled scaffolds. It tells no myths of genius innovators, only systems in flux and the leaders who chose to intervene with integrity.

The book ultimately asks:

- Will we design systems that reward foresight, not just feedback?
- Will we educate leaders fluent in interdependence, not just independence?
- Will we build institutions that can sense, reflect, and adapt, not just scale?

If the answer is yes, then the future is not just something we survive, but something we steward with dignity.

Prologue: Final Thoughts on Leadership That Lasts: Navigating Complexity with Integrity and Vision

In an era defined by rapid technological advancements, global interdependencies, and unforeseen disruptions, effective leadership transcends the traditional boundaries of resource management and strategic execution. It requires the capacity to synthesize vast amounts of information, comprehend interconnected systems, and inspire teams to navigate uncertainty with clarity and purpose. This chapter explores the essence of modern leadership through a reflective and practical lens, integrating foundational principles, actionable insights, and real-world applications that converge to create leaders who thrive in complexity while remaining grounded in values and ethics.

Cognitive Endurance: Leadership at the Edge of Complexity

The pressure on system leaders has reached unsustainable levels. Leaders in sectors from energy to national security must manage complexity, pace, and ambiguity at scales unmatched in any prior era. This is about structural overload.

In 2023, over 62% of U.S. infrastructure executives reported "decision fatigue" as their primary leadership constraint. Cyber response teams often operate at burnout levels, with turnover in Security Operations Center (SOC) units exceeding 45% annually. In the federal emergency response system, the average tenure for incident commanders declined from 3.2 years in 2015 to 1.8 years in 2023. The signal is clear: our institutions are losing their steadiest hands.

Cognitive overload disrupts pattern recognition, compresses time horizons, and erodes moral clarity. None of these are technical failures. They are leadership design failures – avoidable if engineered correctly.

Systems resilience must include executive recovery time, defined decision thresholds, role rotation, and deep redundancy. The most durable organizations train for stress before it arrives and embed slack into their structures without apology. Leadership today is about staying functional under conditions designed to break you.

The Evolution of Leadership: From Stability to Complexity

Historically, leadership was framed by the relative stability of its time. Industrial Revolution-era leaders, such as Andrew Carnegie and John D. Rockefeller, focused on managing large-scale operations and optimizing resources to achieve maximum efficiency. Today, the landscape is drastically different. Leaders contend with cybersecurity threats, climate change, and the transformative impacts of artificial intelligence (AI). Volatility, uncertainty, complexity, and ambiguity (VUCA) have reshaped leadership expectations, demanding adaptability and a forward-thinking mindset.

The shift underscores that while principles such as integrity, vision, and resilience remain timeless, the tools and strategies leaders employ must evolve to meet the demands of a rapidly changing world.

The Leadership Compass: Core Principles for Navigating Complexity

1) **Integrity as a Guiding Star:** Integrity is the foundation of sustainable leadership. Acting honestly and fairly builds trust, the essential currency of effective leadership. Decisions rooted in integrity foster lasting relationships and a resilient organizational culture.
2) **Vision Beyond the Horizon:** Visionary leaders anticipate the next step and the trajectory of change. They inspire teams to align their efforts toward a shared, transformative purpose, cultivating a sense of direction and possibility.
3) **Resilience in Adversity:** Resilience transcends individual grit; it is a systemic capability. Leaders must instill adaptability and perseverance within their organizations to navigate challenges and thrive in uncertain times.
4) **Empathy and Inclusion:** Empathy enables leaders to understand and value diverse perspectives. Inclusion fosters collaboration and innovation, ensuring every voice contributes to collective success.
5) **Commitment to Ethical Innovation:** Leaders are responsible for aligning innovation with ethical standards in an age of rapid technological progress. This commitment ensures that advancements prioritize societal well-being alongside organizational goals.

Reflections from a Life Review: Leadership Lessons for the Modern Era

Reflecting on over four decades of leadership experience across academia, industry, and policy, certain lessons are universally applicable. These lessons are grounded in successes and failures, the relentless pursuit of truth, and the pursuit of meaningful impact.

The Power of Reflection: Reflection is the bridge between experience and wisdom. It transforms events into insights, enabling leaders to evaluate their decisions and recalibrate their actions. A 2024 study in *The Journals of Gerontology* found that reflective practices enhanced decision-making and reduced stress, corroborating the value of introspection in organizational behavior [1].

The Importance of Clarity: Amid ambiguity, clarity is a leader's compass. Distilling complexity into actionable insights requires analytical rigor and effective communication, fostering team understanding and alignment [2].

Systems Thinking: Every decision reverberates through interconnected systems. Leaders must anticipate these ripple effects to design holistic and sustainable solutions that effectively address them. Systems thinking transforms isolated challenges into opportunities for integrated progress [3].

Case Studies: Leadership in Action

The following case studies illustrate the application of leadership principles in real-world contexts, highlighting the interplay between foresight, ethics, and innovation.

Case Study 1: Transforming Public Safety with AI at the University of Minnesota

By 2020, rising campus crime and inefficiencies in traditional surveillance prompted the University of Minnesota to integrate AI-powered video analytics and predictive policing. This initiative reduced crime rates by 20% and response times by 50%, exemplifying data-driven and ethical innovation [4, 5].

Case Study 2: Addressing the Digital Divide in Public Schools

Digital inequities in underserved communities demanded systemic intervention. A comprehensive initiative expanded broadband access, distributed 5,000 devices, and trained teachers, resulting in a 15% increase in student engagement and a 10% improvement in standardized test scores. This effort showcased servant leadership and systems thinking [6, 7].

Case Study 3: Kurt Manufacturing's Automation Revolution

Facing inefficiencies in manual surface finishing, Kurt Manufacturing piloted a UR10 robotic arm, reducing cycle times by 75% and labor costs by $200,000 annually. Transformational leadership drove process optimization, resulting in a 105% return on investment (ROI) within the first year [8].

Leadership Tools: Bridging Theory and Practice

Leadership requires not only principles but also the application of practical tools:

- **Scenario Planning:** Prepares leaders for multiple future scenarios, enhancing adaptability.
- **Cost–Benefit Analysis (CBA):** Evaluates financial and nonfinancial implications of decisions.
- **Heilmeier's Catechism:** Guides clarity and feasibility in innovation projects.
- **Geographic Information Systems (GIS):** Provides actionable insights for resource allocation and predictive modeling.

Leadership Styles: What Works and Why

The case studies highlight the effectiveness of various leadership styles:

1) **Adaptive Leadership:** Fosters flexibility and collaboration in navigating complex and uncertain situations.
2) **Transformational Leadership:** Inspires innovation and aligns teams with a shared purpose.
3) **Servant Leadership:** Prioritizes equity and inclusivity, building trust and cohesion.
4) **Data-Driven Leadership:** Anchors decisions in evidence, ensuring accountability and precision.

Leadership That Gets Results Earns Respect and Builds Resilience

Leadership isn't about titles or knowing every technical detail – it's about values, decisions, and actions that earn respect and drive meaningful results. Effective leaders inspire through vision, empower through trust, and build systems that endure under pressure. This blend of purpose, resilience, and adaptability defines leadership as something that makes an impact.

Vision anchors decisions and motivates teams. Without a clear purpose, actions become reactive, and innovation stalls. Mary Barra's leadership at General Motors exemplifies how bold visions drive transformation. By committing to an all-electric future, she aligned GM with global decarbonization goals and positioned it as a leader in sustainability. Leaders who articulate and pursue clear goals inspire confidence, innovation, and long-term success.

Resilience depends on preparation and systems. Puerto Rico's adoption of solar microgrids after Hurricane Maria demonstrated the power of decentralized energy systems to recover and thrive after disruption. Resilient leaders and organizations anticipate challenges and design strategies to withstand them. Adaptability, built into systems, ensures survival and growth during times of crisis.

Ethical decision-making builds trust and amplifies impact. When faced with crises, decisions grounded in values strengthen relationships and credibility. Moderna's rapid development of its mRNA vaccine demonstrated how striking a balance between urgency and scientific rigor can save lives and open doors to further innovation. Ethical leadership ensures that short-term actions support long-term trust and progress.

Innovation thrives in cultures of experimentation. Organizations that make creativity a core value adapt quickly to change. Ford's pivot during the COVID-19 pandemic to produce ventilators demonstrated how resourcefulness and agility can save lives and strengthen a brand. Leaders who support calculated risks and treat failures as learning opportunities cultivate environments where innovation flourishes.

Collaboration creates stronger solutions. Big challenges often require partnerships that span multiple teams, industries, and borders. Moderna's global alliances accelerated vaccine production, proving that collaboration drives faster and more effective outcomes. Leaders who foster partnerships amplify impact and resilience, addressing problems no single entity could solve alone.

Data-driven decision-making reduces uncertainty. Leaders who trust data make faster, smarter choices. Investments in smart grids following the 2003 blackout demonstrated how real-time data can prevent cascading failures and enhance infrastructure resilience. In complex environments, data provides clarity and supports informed decision-making and decisive action.

Psychological safety drives performance. Teams that feel secure sharing ideas, admitting mistakes, and taking risks perform better. Google's Project Aristotle identified psychological safety as the top factor in team effectiveness. Leaders who create supportive environments unlock innovation and foster trust.

Diversity strengthens teams and innovation. Inclusion brings fresh ideas, fosters adaptability, and enhances outcomes. IAMOT's diverse conferences highlighted how representation in leadership drives innovation in technology. Leaders who value diversity build smarter, more effective organizations.

Continuous learning ensures relevance. IBM's pivot to AI and cloud computing required upskilling thousands of employees and revitalizing its market position. Leaders who embrace growth and flexibility prepare their teams and organizations for the future. Lifelong learning is a cornerstone of sustained leadership.

Effective leadership balances autonomy and accountability. Empowering teams while holding them responsible for results builds trust and ownership. The Management of Technology (MOT) program taught students to solve real-world challenges independently, equipping them to tackle complex problems confidently.

Transparency and decisive action inspire trust. Clear communication about challenges and decisions builds credibility. Satya Nadella's cultural transformation at Microsoft demonstrated how openness and decisive leadership rebuild trust and focus.

Balancing innovation with calculated risk drives progress. Tesla's bold push for autonomous vehicles combined ambition with rigorous testing, ensuring safety while advancing technology. Leaders who manage risk strategically enable sustainable innovation.

Humility strengthens leadership. Leaders who listen, admit when they are wrong, and remain open to new ideas foster collaboration and trust. Jacinda Ardern's empathetic response to the Christchurch attack united a nation, showing how humility inspires loyalty and respect.

Legacy comes from enduring systems. Real leadership builds frameworks that outlast individuals. The MOT program's emphasis on practical solutions and continuous learning has empowered alumni to drive lasting change across industries. Strong systems ensure progress continues long after leaders have moved on.

Gratitude strengthens teams. Recognizing contributions builds morale and loyalty. Starbucks' Howard Schultz often credited employees for the company's success, fostering trust and commitment. Gratitude creates stronger relationships and motivates teams to excel.

Leadership is about purpose, resilience, and action. It requires clarity of vision, the courage to take risks, and the humility to keep learning. Leaders who build trust, foster inclusion, and focus on achieving meaningful outcomes create teams and systems that thrive in uncertain environments. The lessons of visionary leaders like Mary Barra, the resilience of systems in Puerto Rico, and adaptable organizations like IBM remind us that leadership isn't about avoiding failure – it's about turning challenges into opportunities and building legacies that last.

This leadership framework provides practical steps for driving impact, fostering innovation, and navigating complexity. Leaders can inspire, endure, and elevate their teams to meet future demands by staying grounded in values and committed to progress.

References

1 R. N. Butler, "The life review: An interpretation of reminiscence in the aged," *Psychiatry*, vol. 26, pp. 65–76, 1963.

2 Jobs, S., "*Stanford Commencement Address*," online. 2005.

3 P. M. Senge, *The Fifth Discipline: The Art and Practice of the Learning Organization*. Doubleday, 1990.

4 G. Jacobs, F. van Houdt, and G. Coons, "Studying surveillance AI-cologies in public safety: How AI is in the world and the world in AI," *Surveillance & Society*, vol. 22, no. 2, pp. 188–208, 2024.

5 Grumet, J., *Hospice and Reflection*, *Hospice Quarterly*, 2023.

6 "Digital Equity Initiatives in Education," *Journal of Educational Policy*, vol. 28, pp. 89–101, 2020.

7 "Broadband Expansion Strategies," *Infrastructure Policy Review*, vol. 18, pp. 67–80, 2021.

8 "Automation in Manufacturing," *Journal of Robotics*, vol. 12, pp. 45–56, 2021.

Appendix A

Leadership Tools

This appendix consolidates the full suite of analytical tools, strategic frameworks, and applied methods referenced throughout the book. To maintain clarity and momentum in the narrative, we've moved the detailed explanation of tools from Chapters 4A to 4E into this appendix, preserving their original sequence while making them easier to reference in real-world settings.

Each tool is thoroughly documented, including its origin, use case, limitations, and practical value. The catalog is grouped into three tables:

- **Table A.2: Fundamental Tools** – widely applicable framing, scanning, and aligning methods.
- **Table A.3: Advanced Tools** – in-depth tools for innovation strategy, resilience, life cycle planning, and risk modeling.
- **Table A.4: Hybrid Tools** – frameworks that integrate multiple tools or span across fundamental and advanced use cases.

Practitioner Readiness Checklist

Before selecting and applying the tools, leaders are strongly encouraged to use the Practitioner Readiness Checklist (Table A.1). It ensures strategic fit, systems awareness, stakeholder alignment, and operational capacity. The checklist draws from the principles outlined in Chapters 1 and 4 and reflects lessons from decades of infrastructure modernization, risk management, and technological leadership across sectors.

All tools in this appendix are cross-referenced in the main chapters and available here for deeper exploration or operational deployment. These tools include the following and considerably more, as discussed in this appendix. A PowerPoint file discussing selected tools accompanies this book and can be downloaded. The link to the PPT is available on the About the Companion Website page (www.wiley.com/go/amin) in the front matter, and on the back cover.

Table A.1 Practitioner Readiness Checklist.

#	Readiness Category	Key Question(s)	Criteria for "Ready"	Notes/Action Needed
1	Clarity of Purpose	Do you have a clearly defined "Why"?	Vision documented; team alignment confirmed	Use the Golden Circle and Catechism
2	Stakeholder Alignment	Are stakeholders aligned with the mission?	Mapped, briefed, and roles understood	Use the Stakeholder Mapping Matrix
3	Problem Framing	Is the challenge scoped correctly?	Framed as a system, not a symptom	Use Heilmeier's Catechism, SWOT
4	Data Availability	Do you have access to validated data?	Data sources confirmed and vetted	List uncertainties and gaps
5	Systems View Applied	Have you mapped dependencies and loops?	Boundaries and dynamics identified	Use Cross-Impact + GTD
6	Scenario Planning	Are multiple futures explored?	Scenarios built and stress-tested	Use Scenario Matrix, Futures Wheel
7	Strategic Fit	Do strategies match internal/external conditions?	Alignment via SWOT/TOWS verified	Revisit SWOT, TOWS
8	Lifecycle Awareness	Have you assessed the tech maturity?	TRL mapped; S-curve placement identified	Use S-Curve and TRL
9	Cost-Risk Integration	Are risk and cost models integrated?	LCCA and options analysis complete	Combine with Risk Matrix
10	Technology Mapping	Are roadmaps built for needed technologies?	Timeline, milestones, and dependencies mapped	Use Technology Roadmap
11	Execution Planning	Is there a phased implementation plan?	Resources, KPIs, and timelines set	Use Gantt format if needed
12	Capacity & Skills Check	Do you have the right team and skills?	Talent, gaps, and partners mapped	Use Reskilling, Gap Analysis
13	Resilience Strategy	Are systems designed for stress recovery?	Redundancies and adaptive capacity are modeled	Use Resilience Audit, GTD
14	Feedback Mechanisms	Are learning loops embedded?	Review processes and recalibration plans in place	Define KPI governance
15	Narrative Coherence	Can strategy be communicated effectively?	Clear storyline for internal and external use	Use Storytelling Tools

16	Ethics & Responsibility	Are societal and ethical impacts considered?	ESG and bias impacts reviewed	Apply GTD and Ethics Review tools
17	Tool Familiarity	Does the team understand the tools?	Guidance reviewed, ownership assigned	Reference this appendix
18	Monitoring Plan	Are tracking metrics defined?	Dashboards and KPI reviews are integrated	Scorecard-based systems
19	External Dependencies	Are macro-level risks accounted for?	Scenarios and contingencies are defined	Use Horizon Scanning
20	Adaptability Under Pressure	Can the team pivot under VUCA conditions?	Simulations and adaptive planning are complete	Conduct VUCA resilience drills
21	Post-Implementation Review	Are lessons captured for future use?	Learning cycle structured and resourced	Build feedback loops
22	Tool Selection Matrix	Are tools selected to match the challenge type and phase?	Rationalized and mapped to the problem phase	Start with Tables A.–A.4

Table A.2 Fundamental Tools for Strategic Clarity and Leadership Readiness.

Tool Name	Creator(s)	Year	Primary Purpose	Application in the Book	Strengths	Limitations
Heilmeier's Catechism	George H. Heilmeier	1975	Framework for evaluating research and innovation proposals	Used in Chapter 4A to assess early-stage initiatives	Clarifies purpose, cost, risk, and metrics	May oversimplify complex systems
SWOT Analysis	Albert Humphreys (popularized)	1960s	Strategic situational analysis	Chapter 4A, systems scanning and stakeholder reviews	Universally applicable, simple	Static lacks system dynamics
PESTLE Analysis	Derived from business strategy tools	1980s	Environmental scanning	Chapter 4A: external trend analysis	Structured macro scan	Lacks organizational context
The Golden Circle	Simon Sinek	2009	Clarifies purpose, process, and product	Chapter 1 and 4A; leadership alignment	Sharpens internal motivation	May oversimplify strategic complexity
5 Whys Technique	Sakichi Toyoda	1930s	Root cause analysis	Chapter 4C: Infrastructure Failure Review	Uncovers root issues quickly	Can lead to premature conclusions
SMART Goals	Peter Drucker (popularized)	1950s–1981	Goal setting	Chapter 4A; implementation tracking	Aligns objectives with performance	Not system-aware
Systems Mapping	General systems theory	1950s–present	Identifying system components and flows	Chapter 4A and 5; complexity modeling	Visual, cross-disciplinary	Labor-intensive
STARR (Stability, Transformation, Adaptation, Recovery, and Resilience)	Massoud Amin	2008	Framework for evaluating resilience strategies	Used throughout Chapters 5–13	Rich temporal view of change	May require extensive modeling

Table A.3 Advanced Tools for Strategic Foresight and Evaluation.

Tool Name	Creator(s)	Year	Primary Purpose	Application in the Book	Strengths	Limitations
Scenario Planning Matrix	Shell/GBN/Global Business Network	1970s–1990s	Explore uncertainty using divergent futures	Chapter 4B, national and enterprise planning	Expands vision, stress-tests plans	Complex; requires facilitation
Technology Readiness Levels (TRLs)	NASA	1970s	Assess technology maturity	Chapter 4B and Chapter 6C	Clear 9-level scale	Not cross-sector compatible
Cost-Benefit Analysis	Jules Dupuit (France)	1844	Economic decision-making framework	Chapter 4B: Infrastructure Investment	Quantifies trade-offs	Misses the social/ethical impact
Life Cycle Analysis	EPA/Industrial Ecology	1990s	Full impact analysis over time	Chapter 4B: Technology Roadmaps	System-wide scope	Data-intensive
Multi-Criteria Decision Analysis (MCDA)	Keeney and Raiffa	1976	Complex decision prioritization	Chapter 4C: infrastructure and energy trade-offs	Handles multiple variables	Can be opaque to stakeholders
Global Transition Dynamics (GTD)	Amin and Starr	2003	Model for managing major transitions	Chapter 5 and Appendix	Integrative, resilience-driven	Requires data + context
Portfolio Mapping	Bruce Henderson (BCG)	1970s	Strategic resource allocation	Chapter 4B and 25	Aids visual prioritization	Too reductive for systems
Risk–Impact Matrix	Common business practice	1980s	Prioritize risks based on impact and likelihood	Chapter 4D: applied in the Heathrow case	Easy communication	Ignores interdependence

Table A.4 Hybrid or Integrative Tools Used Across Contexts.

Tool Name	Creator(s)	Year	Primary Purpose	Application in the Book	Strengths	Limitations
Red Teaming + Scenario + Risk Matrix	US DoD/Multiple	1990s	Stress-test strategy from adversarial POV	Used in Chapter 4C and the Heathrow case	Challenges assumptions	Requires a culture of candor
Technology Roadmapping + TRLs + SWOT	Composite	2000s	Align tech strategy with maturity and context	Chapter 4B	Combining time, value, and readiness	Labor-intensive
Delphi Method + MCDA	RAND Corporation + Keeney/Raiffa	1960s–1970s	Forecasts with expert consensus and multiple factors	Chapter 4B–4C	Consensus + prioritization	Time-consuming
Resilience Scorecards + Systems Mapping + GTD	Amin, Starr, et al.	2003–2024	Evaluate dynamic resilience across systems	Chapters 5, 13, and the GTD appendix	Cross-sector resilience	Needs continuous data
Innovation Diamond + Golden Circle	Design Thinking Field	2000s	Align creativity with mission and need	Chapter 4A–4C	Vision-to-execution clarity	Subjective unless grounded

Tools, Methods, and Templates

Leadership and Foresight Tools

This appendix compiles all the foresight, planning, and decision-making tools used across Chapters 1 and 4 throughout the book. It is structured in three tables:

- **Table A2:** Fundamental Tools
- **Table A3:** Advanced Tools
- **Table A4:** Hybrid or Integrative Tools

These tools are presented in the order they originally appeared, for coherence and pedagogical continuity.

Basic Tools for Strategic Leadership and Decision-Making

This chapter highlights the basic skills important for successful strategic management and decision-making. These tools have been utilized in leadership development, business strategy, R&D assessment, and policy planning, among other areas. The emphasis is on tools that help organizations understand systems and make better decisions. Every tool is explained: what it is, what it does, its benefits, weaknesses, and how it can be used in real life.

The Golden Circle: Unlocking the Power of Purpose-Driven Leadership

Simon Sinek introduced this concept in his transformative 2009 TED Talk, "How Great Leaders Inspire Action," and further explored it in his book *Start with Why*; the Golden Circle has become a cornerstone framework for purpose-driven leadership and strategic planning. At its essence, the Golden Circle reshapes how leaders and organizations think about motivation, alignment, and impact. Its deceptively simple structure – "Why," "How," and "What" – holds profound implications for how ideas are communicated, teams are inspired, and businesses thrive.

Origins and Evolution of the Golden Circle

Simon Sinek's Golden Circle emerged from years of research into why some individuals and organizations achieve extraordinary success while others stagnate. Sinek observed that successful leaders and companies tend to think differently; instead of starting with "What" (the products or services they offer) or "How" (the processes and strategies they employ), they focus on "Why" – the underlying purpose, cause, or belief that drives everything they do. This inside–out approach, he argued, not only distinguishes great organizations but also fosters deep emotional connections with employees, customers, and stakeholders.

The Golden Circle was inspired by Sinek's study of iconic leaders and companies, including Martin Luther King Jr., the Wright Brothers, and Apple. In their unique ways, these entities captured the hearts and minds of their audiences not through what they did but through the clarity of their purpose. For Sinek, this insight provided a replicable framework that could transform leadership, marketing, and organizational strategy.

The Golden Circle resonated with audiences worldwide. Sinek's TED Talk has since amassed over 60 million views, making it one of the most-watched talks in TED history. More importantly, the framework has become a practical tool for leaders striving to create purpose-driven cultures and enduring success.

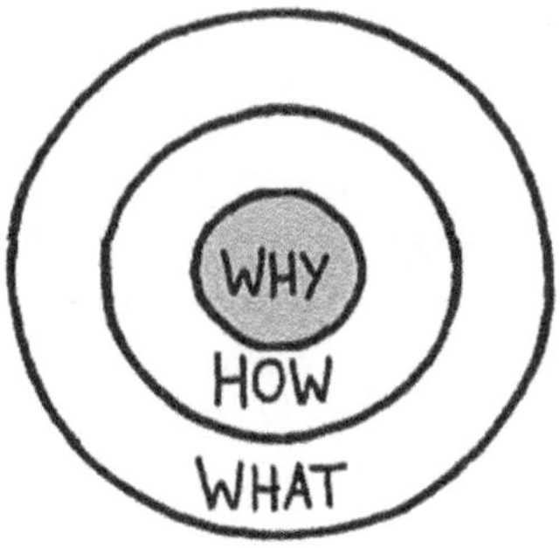

Figure: Golden Circle (Simon Sinek): A framework for identifying purpose, focusing on "Why" (mission), "How" (process), and "What" (outcomes). It is used to create alignment and motivation across organizations. *Reference*: Simon Sinek, *Start with Why: How Great Leaders Inspire Everyone to Take Action*, Portfolio, 2009. [This is copyrighted by Simon Sinek (see: https://www.nist.gov/image/golden-circlejpg). You may want to reach out to Sinek's team and see if you can get a license for the graphic with attribution. NIST had permission and credited him in this way: **Credit:** Copyright 2013, Simon Sinek, Inc. Used with permission from www.startwithwhy.com.]

The Framework in Action

At the heart of the Golden Circle is the idea that people don't buy what you do; they buy why you do it. Sinek uses three concentric circles to illustrate this principle. The innermost circle represents "Why" – the purpose or belief that inspires action. Surrounding it is "How" – the methods or processes used to bring the purpose to life. The outermost circle is "What" – the tangible products, services, or results an organization delivers.

This structure forces leaders to rethink their communication and strategy. Starting with "Why" ensures that decisions are anchored in purpose, which drives the authenticity and consistency needed to inspire trust. The "How" provides the roadmap for operationalizing purpose, while the "What" serves as the measurable output.

Lessons from the Golden Circle

The Golden Circle underscores that clarity of purpose is not just a theoretical advantage but a practical necessity. Organizations that operate with a clear "Why" achieve better alignment among their teams, foster stronger customer loyalty, and often outperform their competitors – purpose, when authentic and well-communicated, becomes a magnet for talent and investment.

However, the framework also highlights potential pitfalls. Focusing solely on "Why" without delivering on the "How" and "What" can lead to unfulfilled promises and erode credibility. Moreover, the "Why" must resonate deeply – it cannot be contrived or superficial, as today's stakeholders are adept at identifying insincerity.

Case Study: Steve Jobs and the Apple Renaissance

Few leaders have exemplified the Golden Circle as effectively as Steve Jobs during his second tenure at Apple. By 1997, when Jobs returned to the company he cofounded, Apple was teetering on the brink of collapse. Years of fragmented leadership and unfocused product development had left it struggling to compete against Microsoft and other industry giants.

Jobs's turnaround strategy was grounded in rediscovering Apple's "Why." He articulated it as a commitment to challenging the status quo and empowering individuals through beautifully designed, user-friendly technology. This purpose, encapsulated in the now-famous slogan "Think Different," became the rallying cry for employees and customers alike.

Jobs didn't stop at redefining the company's purpose. He aligned Apple's operations and product development with this vision, focusing resources on a few transformative projects. The 1998 release of the iMac was a testament to Apple's "How" and "What." Its revolutionary design and seamless user experience embodied the company's core belief in innovation and simplicity. Subsequent products like the iPod, iPhone, and iPad solidified Apple's position as a global leader in technology, driving it to become the world's first trillion-dollar company.

The success of Apple's revival demonstrates the power of aligning purpose with action. Jobs's ability to start with "Why" and execute with precision across the "How" and "What" created a legacy that reshaped not only Apple but the entire tech industry.

Applications and Impact

The Golden Circle's relevance extends far beyond technology. Companies like Southwest Airlines, Patagonia, and Tesla have also embraced purpose-driven strategies to achieve competitive advantage and cultural cohesion. For Tesla, the "Why" of accelerating the world's transition to sustainable energy informs everything from its product design to its marketing campaigns, fostering a loyal customer base despite premium price points.

Even in noncorporate contexts, the Golden Circle has proven transformative. Nonprofit organizations and social movements have used the framework to galvanize support and drive impact. For example, Martin Luther King Jr.'s "I Have a Dream" speech resonates not because it detailed a plan of action but because it articulated a vision of justice and equality that inspired millions.

The Risks and Limitations

While the Golden Circle offers a powerful lens for leadership and strategy, it has limitations. Defining a compelling "Why" can be challenging, particularly for legacy organizations or those in highly commoditized industries. Without tangible actions to support it, a "Why" risks being dismissed as empty rhetoric. Additionally, organizations that overemphasize purpose at the expense of execution may struggle to meet stakeholder expectations.

For these reasons, leaders must balance inspiration with accountability. Purpose must be backed by robust processes ("How") and measurable outcomes ("What"). When purpose aligns with practice, it drives emotional engagement and tangible success.

A Blueprint for the Future

The Golden Circle remains one of the most influential frameworks for leaders seeking to inspire and innovate in an increasingly complex world. Its simplicity belies its transformative potential, offering a roadmap for aligning purpose with action in an era where consumers and employees demand authenticity, transparency, and impact; starting with "Why" has never been more critical.

As Steve Jobs demonstrated at Apple, the Golden Circle is not merely a theory but a tool for action. When leaders articulate and live their "Why," they build organizations that transcend short-term gains and leave lasting legacies.

References

1 S. Sinek, *Start with Why: How Great Leaders Inspire Everyone to Take Action*. Penguin Group, 2009.
2 W. Isaacson, *Steve Jobs*. Simon & Schuster, 2011.
3 C. Gallo, *The Innovation Secrets of Steve Jobs: Insanely Different Principles for Breakthrough Success*. McGraw-Hill Education, 2014.
4 *Apple Reports Fourth Quarter Results*. Apple Inc, 2007.
5 TED, "Simon Sinek: How Great Leaders Inspire Action," 2009.

Heilmeier's Catechism: A Framework for Defining and Evaluating Innovation

Heilmeier's Catechism is a guiding question set by George H. Heilmeier, former Director of the Defense Advanced Research Projects Agency (DARPA). This framework is widely used in technology development, R&D project evaluation, and strategic decision-making to assess the feasibility and potential impact of new ideas. Its emphasis on clarity, practicality, and foresight makes it an invaluable tool for engineers, policymakers, and executives navigating innovation challenges.

Origins and Evolution of Heilmeier's Catechism

George Heilmeier, an electrical engineer and inventor credited with developing liquid crystal display (LCD) technology, created his catechism during his tenure at DARPA in the 1970s. Heilmeier recognized the need for a systematic method to evaluate the value and viability of high-risk, high-reward research projects. His questions were designed to cut through jargon and compel innovators to articulate their ideas regarding real-world impact, practicality, and potential risks.

Originally intended for internal DARPA use, the catechism has since been widely adopted across sectors, including defense, healthcare, academia, and corporate innovation. It is particularly valued for its ability to balance visionary thinking with pragmatic scrutiny.

The Framework in Action

Heilmeier's Catechism is composed of the following foundational questions:

1) **What are you trying to do?**
   Articulate goals in simple, clear terms, avoiding technical jargon.
2) **How is it done today, and what are the limitations?**
   Identify existing methods, their shortcomings, and the gaps your solution addresses.
3) **What is new in your approach, and why do you think it will be successful?**
   Highlight the innovative aspects of your idea and its potential advantages.
4) **Who cares?**
   Define the target audience or beneficiaries of your work.
5) **What are the risks?**
   Acknowledge potential challenges and uncertainties.
6) **What are the midterm and final "exams" to check for success?**
   Specify metrics and criteria for evaluating progress and outcomes.

These deceptively simple questions demand thoughtful, precise answers that distill complex ideas into actionable insights. By focusing on the "big picture" and the operational details, the catechism ensures that innovation efforts are grounded in reality while aiming for transformative impact.

Lessons from Heilmeier's Catechism

The catechism's power lies in bridging the gap between visionary ideas and practical implementation. By forcing innovators to answer fundamental questions about their projects, it reduces the risk of investing in ideas that are either infeasible or misaligned with organizational goals. It also encourages accountability by requiring clear metrics for success.

However, its strength is also a limitation: the focus on clarity and simplicity may inadvertently discourage the exploration of highly abstract or speculative ideas that could yield breakthroughs in the long term. Organizations must pair the catechism with complementary tools that encourage creativity and out-of-the-box thinking to address this.

Case Study: DARPA and the Internet

One of the most significant applications of Heilmeier's Catechism was its use at DARPA in developing ARPANET, the precursor to the modern Internet. In the late 1960s and early 1970s, DARPA faced the challenge of enabling secure and reliable communication across geographically dispersed networks.

Using Heilmeier's principles, DARPA researchers framed their objectives in clear, actionable terms. For instance:

- **What are you trying to do?** Enable computers to communicate over long distances in a secure, decentralized manner.
- **How is it done today, and what are the limitations?** Communication relied on dedicated telephone lines, which were expensive and vulnerable to disruption.
- **What is new in your approach, and why do you think it will be successful?** Packet-switching technology, which allows data to be broken into packets and transmitted independently, was a novel concept that promised greater efficiency and resilience.

The outcome of this project laid the foundation for the Internet, transforming global communication and commerce. Heilmeier's Catechism played a critical role in ensuring that ARPANET's ambitious goals were matched with practical, well-defined steps.

Applications and Impact

Heilmeier's Catechism is now a staple in innovation management across industries. In the corporate world, it has been used by companies like Google to evaluate moonshot projects under its X Development division. In academia, research proposals often include Heilmeier-inspired justifications to secure funding.

Moreover, the catechism's emphasis on defining success metrics has influenced public policy and nonprofit initiatives, ensuring that taxpayer or donor resources are allocated effectively. Its versatility and rigor make it applicable to virtually any field where innovation and uncertainty intersect.

The Risks and Limitations

While Heilmeier's Catechism is invaluable, it is not a one-size-fits-all solution. Its focus on practicality can sometimes stifle exploratory research that lacks immediate applications. Additionally, the emphasis on defining metrics and success criteria may inadvertently prioritize measurable outcomes over qualitative or long-term benefits.

Organizations must recognize these limitations and use the catechism as part of a broader innovation toolkit. For example, pairing it with frameworks like scenario planning or design thinking can foster a more holistic approach to managing uncertainty and complexity.

A Blueprint for the Future

Heilmeier's Catechism remains a timeless framework for navigating the complexities of innovation. Its structured approach to evaluating ideas ensures that resources are invested wisely, risks are managed proactively, and goals are aligned with stakeholder needs. As global challenges become increasingly interconnected, the catechism's emphasis on clarity, accountability, and impact will only grow in importance.

The enduring legacy of George Heilmeier lies not only in his contributions to technology but also in his ability to distill the essence of innovation into a simple yet profound set of questions. By embracing these principles, leaders and organizations can chart a path toward sustainable progress in an ever-changing world.

References

1 G. Heilmeier, *Innovation at DARPA*. Defense Advanced Research Projects Agency Archives, 1976.
2 J. Abbate, *Inventing the Internet*. MIT Press, 2000.
3 National Academy of Engineering, "The Life and Legacy of George Heilmeier," 2012.
4 J. Gertner, *The Idea Factory: Bell Labs and the Great Age of American Innovation*. Penguin Group, 2012.
5 DARPA, "50 Years of Innovation: A History of DARPA's Contributions to Technology," 2020.

Quad Charts: A Tool for High-Stakes Communication and Decision-Making

Quad charts are an indispensable tool for presenting complex ideas succinctly in high-stakes environments, particularly in the defense, aerospace, and research sectors. Designed for clarity and efficiency, these four-quadrant frameworks distill key information into a single page, enabling decision-makers to grasp essential details quickly and make informed judgments. Over the years, quad charts have become a staple for briefing senior leadership at the Pentagon, NASA, and corporate boards.

Origins and Evolution of Quad Charts

The use of quad charts emerged in the defense sector during the mid-20th century as a response to the increasing complexity of military and R&D projects. With limited time to review voluminous reports, senior decision-makers required concise summaries that highlighted the core aspects of a proposal or project. Quad charts met this need by dividing information into four quadrants, each focused on a distinct aspect of the topic at hand. This format became a standard tool in environments like DARPA, the Department of Defense (DoD), and later, in the private sector.

The format's utility extends beyond military applications. Its adaptability has made it a preferred tool in industries like healthcare, technology, and education, where leaders must distill dense information into actionable insights.

The Structure of a Quad Chart

A typical quad chart comprises four quadrants, each dedicated to a specific category of information. While the exact content varies based on the context, a standard quad chart used in defense and R&D typically includes:

1) **Problem Statement or Objective (Top Left):**
   This quadrant articulates the problem being addressed or the primary goal of the project. It establishes the context and stakes, ensuring that the decision-maker understands the "why" behind the initiative.
2) **Proposed Solution or Approach (Top Right):**
   Here, the focus shifts to the "how." This quadrant summarizes the proposed solution, outlining the key methods, technologies, or strategies that differentiate the approach from existing ones.
3) **Benefits and Impact (Bottom Left):**
   This section highlights the expected outcomes, emphasizing the value the project will deliver. For defense applications, this might include enhanced capabilities or cost savings. For R&D, it could focus on innovation potential or strategic advantages.
4) **Risks and Challenges (Bottom Right):**
   Recognizing potential obstacles, this quadrant outlines key risks, uncertainties, and mitigation strategies. It demonstrates due diligence and preparedness, fostering confidence in the project's feasibility.

Lessons from Using Quad Charts Effectively

Quad charts are more than just a formatting tool; they are a discipline in structured thinking and concise communication. By forcing presenters to distill their ideas into four essential components, the charts eliminate extraneous details and ensure focus on the most critical aspects.

This simplicity is its strength – and its limitation. While quad charts excel at providing high-level overviews, they may oversimplify complex issues or omit nuanced details. To address this, presenters often accompany quad charts with supplemental materials, such as detailed reports or appendices.

Case Study: The Pentagon and DARPA Briefings

Quad charts have been a cornerstone of briefing formats at the Pentagon and DARPA. For instance, when DARPA-funded researchers present new technologies, they use quad charts to communicate their projects' relevance to national security.

A hypothetical example might look like this:

- **Problem Statement:** Adversarial advancements in hypersonic missile technology pose a threat to U.S. strategic superiority.
- **Proposed Solution:** Develop a next-generation, counter-hypersonic defense system using AI-enabled radar and interceptors.
- **Benefits:** Enhanced protection of critical assets, preservation of strategic deterrence, and a technological edge over adversaries.
- **Risks:** Technical feasibility challenges, cost overruns, and timeline uncertainties.

The structure ensures that high-ranking officials, often managing multiple projects simultaneously, can quickly grasp the core message, ask pertinent questions, and make decisions without wading through excessive detail.

Applications in Other Sectors

The versatility of quad charts has led to their adoption in diverse fields:

- **Corporate Strategy:** Used to present business cases, mergers, or product development initiatives to C-suite executives.
- **Healthcare:** Employed to evaluate clinical trial proposals or new treatment strategies.
- **Academia:** Utilized in grant applications and research proposals to summarize objectives, methods, and expected impacts.

In each case, the ability to synthesize complex ideas into an easily digestible format makes quad charts a powerful communication tool.

Capabilities and Limitations

Capabilities:

- **Efficiency:** Quad charts provide decision-makers with all critical information at a glance, saving time in high-pressure environments.
- **Clarity:** The structured format ensures that key points are not buried under extraneous details.
- **Versatility:** The tool can be adapted to various contexts, from technical presentations to strategic planning.

Limitations:

- **Oversimplification:** Important nuances may be lost when compressing information into a four-quadrant format.
- **Reliance on Accompanying Materials:** Decision-makers may need additional context or data to fully understand the implications of the proposal.
- **Risk of Overstandardization:** Excessive reliance on the format can lead to uniformity that stifles creative thinking.

The Future of Quad Charts

As decision-making environments become increasingly complex, the need for concise and effective communication tools like quad charts will continue to grow. Advances in data visualization and artificial intelligence (AI)-driven analytics may enhance the tool's capabilities, enabling real-time updates and dynamic reconfigurations. However, the fundamental discipline of clarity and brevity will remain its core strength.

References

1 DARPA Briefing Standards, "Defense Advanced Research Projects Agency Archives," 2020.
2 J. Gertner, *The Idea Factory: Bell Labs and the Great Age of American Innovation*. Penguin Group, 2012.
3 U.S. Department of Defense, "Standard Operating Procedures for R&D Proposals," 2018.
4 National Academy of Sciences, "Communicating Complex Ideas in High-Stakes Environments," 2015.

5 C. Christensen, *The Innovator's Dilemma: When New Technologies Cause Great Firms to Fail.* Harvard Business Review Press, 1997.

SWOT Analysis: A Comprehensive Examination

Introduction: The Origins and Significance of SWOT Analysis

SWOT analysis – an acronym for Strengths, Weaknesses, Opportunities, and Threats – is one of the most widely used strategic planning tools across industries. It emerged in the 1960s from research conducted by Albert Humphrey and his team at the Stanford Research Institute, who were working to identify why corporate planning often failed. The SWOT framework evolved as a simple yet powerful tool to help organizations better understand internal and external factors influencing their success.

Today, SWOT Analysis remains a cornerstone of strategic decision-making, used by businesses, governments, nonprofits, and academic institutions. Its appeal lies in its versatility: it can be applied to a variety of contexts, from launching a new product to assessing organizational health or exploring market opportunities.

Framework and Methodology

SWOT Analysis is a structured process for evaluating an organization's internal capabilities and external environment. It is divided into two key dimensions:

1) **Internal Analysis:**
   - **Strengths:** Internal attributes or capabilities that give the organization an advantage over competitors.
   - **Weaknesses:** Internal limitations or vulnerabilities that hinder organizational performance.
2) **External Analysis:**
   - **Opportunities:** External factors or trends that the organization can leverage for growth or improvement.
   - **Threats:** External challenges or risks that could jeopardize success.

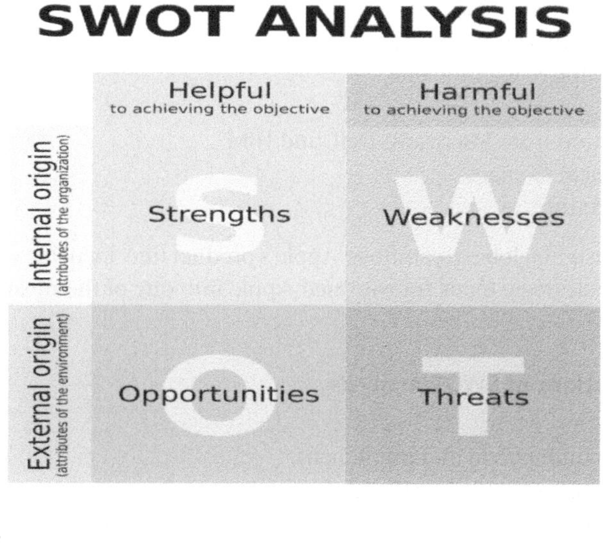

Figure: SWOT analysis.

By examining these dimensions, organizations can develop actionable strategies that capitalize on strengths, address weaknesses, seize opportunities, and mitigate threats.

Application: Practical Steps

The process of conducting a SWOT analysis typically involves the following steps:

1) **Identify the Objective:** Define the purpose of the analysis (e.g., entering a new market, evaluating organizational performance).
2) **Gather Data:** Collect qualitative and quantitative information about internal operations and the external environment.
3) **Categorize Factors:** Organize findings into the four SWOT categories.
4) **Prioritize:** Determine the most critical factors within each category.
5) **Develop Strategies:** Create actionable plans that align with the findings.

Case Study: SWOT Analysis in Action

Apple Inc.

When Steve Jobs returned to Apple in 1997, the company was on the brink of bankruptcy. Jobs used a SWOT analysis to reposition Apple as a leader in innovation. Here's how the framework is applied:

1) **Strengths:**
   - Brand loyalty and a dedicated user base.
   - Skilled design and engineering teams.
   - Established reputation for innovation.
2) **Weaknesses:**
   - Narrow product portfolio.
   - High production costs.
   - Declining financial performance.
3) **Opportunities:**
   - Growing demand for portable devices.
   - Emerging markets in Asia and Europe.
   - Potential for disruptive technologies.
4) **Threats:**
   - Intense competition from Microsoft, Dell, and IBM.
   - Rapid technological changes.
   - Economic uncertainty.

Armed with these insights, Jobs streamlined Apple's product line, focusing on the iMac, iPod, and later the iPhone. This strategic focus transformed Apple into one of the most valuable companies in the world.

Strengths and Limitations of SWOT Analysis

Strengths:

- **Simplicity:** Easy to understand and implement.
- **Flexibility:** Applicable across industries and organizational scales.
- **Comprehensive:** Encourages holistic thinking by integrating internal and external factors.

Limitations:

- **Subjectivity:** Results depend on the quality of input and participant perspectives.
- **Static Nature:** Provides a snapshot in time, requiring frequent updates in dynamic environments.
- **Oversimplification:** May overlook complex interdependencies or nuances.

Enhancing SWOT Analysis

To address its limitations, organizations often pair SWOT analysis with complementary tools like:

- **PESTLE Analysis:** To explore external macro-environmental factors in greater depth.
- **Porter's Five Forces:** To assess competitive dynamics.
- **Scenario Planning:** To evaluate potential future developments.

Relevance in Today's World

SWOT analysis continues to thrive in a rapidly changing global landscape. For instance, it has been instrumental in:

1) **Digital Transformation:** Companies like Netflix used SWOT to transition from DVD rentals to streaming, identifying digital distribution as an opportunity while addressing the threat of declining physical media.
2) **Sustainability Initiatives:** Organizations like Unilever employ SWOT to integrate sustainability into their business models, leveraging opportunities in green markets while mitigating regulatory threats.
3) **Crisis Management:** During the COVID-19 pandemic, healthcare organizations applied SWOT to adapt to supply chain disruptions and evolving patient needs.

Lessons for Graduate Students and Practitioners

- **Clarity:** Focus on actionable insights rather than exhaustive lists of factors.
- **Contextual Relevance:** Tailor the analysis to the specific industry, market, or organizational context.
- **Iterative Approach:** Update the analysis regularly to reflect changing conditions.

Applications Across Industries

1) **Corporate Strategy and Business Development:** SWOT is a fundamental tool in crafting business strategies. For example, Apple has consistently leveraged its strengths in product innovation and brand loyalty to capture opportunities in emerging markets like India while addressing weaknesses such as supply chain vulnerabilities.
2) **Public Policy and Urban Planning:** City governments use SWOT analysis to identify strengths (e.g., skilled workforce), weaknesses (e.g., infrastructure gaps), opportunities (e.g., smart city initiatives), and threats (e.g., climate change) when designing urban development plans.
3) **Healthcare:** Hospitals and healthcare providers rely on SWOT to improve patient outcomes by aligning internal resources (skilled personnel, cutting-edge equipment) with external opportunities (government incentives) while mitigating threats like regulatory compliance and insurance constraints.

4) **Education and Training:** Universities conduct SWOT analyses to enhance programs, attract diverse student populations, and address funding limitations. For instance, Harvard Business School evaluated its strengths in global alumni networks to capitalize on opportunities in digital education post-pandemic.

Case Study: SWOT at Tesla, Inc.

Tesla's strategic success offers a clear example of SWOT in action.

- **Strengths:** Dominance in electric vehicle (EV) technology, innovative battery designs, and brand reputation as a sustainability leader.
- **Weaknesses:** Dependence on rare earth metals, high production costs, and operational inefficiencies in scaling manufacturing.
- **Opportunities:** Rising global demand for EVs, government incentives for renewable energy, and partnerships in charging infrastructure.
- **Threats:** Growing competition from established automakers (e.g., Ford and GM's EV lines) and supply chain disruptions exacerbated by geopolitical tensions.

By addressing its weaknesses (e.g., streamlining production processes) and capitalizing on opportunities (e.g., expansion into energy storage markets), Tesla has solidified its leadership position in the EV sector.

Benefits of SWOT Analysis

1) **Accessibility:** The simplicity of the framework allows individuals and organizations at any scale to apply it effectively.
2) **Customizability:** SWOT can be tailored to suit any context – be it a multinational corporation or a community nonprofit.
3) **Holistic Viewpoint:** By balancing internal and external factors, SWOT ensures a well-rounded strategic perspective.
4) **Actionable Insights:** The framework often uncovers immediate, practical actions while highlighting long-term opportunities for growth.

Limitations and Criticisms

While SWOT is a versatile tool, it has its limitations:

1) **Subjectivity:** The outcomes of a SWOT analysis depend on the perspectives of those conducting it, potentially leading to biased or incomplete assessments.
2) **Static Nature:** SWOT provides a snapshot in time but lacks mechanisms for dynamic monitoring or adaptation.
3) **Oversimplification:** By categorizing factors into four quadrants, SWOT may overlook nuanced interdependencies or long-term systemic risks.

Overcoming Limitations: Combining SWOT with other frameworks like Political, Economic, Social, Technological, Legal, and Environmental (PESTLE) analysis or scenario planning adds depth to the analysis, ensuring a more comprehensive strategic outlook.

Lessons Learned

1) **Collaborative Implementation:** The effectiveness of SWOT increases when diverse teams contribute to the analysis, incorporating varied perspectives.

2) **Regular Updates:** To remain relevant, SWOT should be revisited periodically to reflect evolving internal and external conditions.
3) **Integration with Decision-Making Tools:** Embedding SWOT insights into broader strategic planning processes ensures actionable results and alignment with organizational goals.

Conclusion: Relevance in Modern Leadership

SWOT analysis remains a vital tool for navigating complexity and uncertainty in strategic planning. SWOT analysis remains an invaluable tool for today's leaders, offering clarity and direction amidst complexity. Its universal applicability across industries ensures its continued relevance, from tech giants navigating innovation to policymakers addressing global challenges.

By blending SWOT with advanced methodologies and incorporating real-time data, modern organizations can unlock their full potential to drive impactful decisions.

Citations

1 A. Humphrey, *SWOT Analysis for Organizational Success*. Stanford Research Institute, 1960.
2 "Tesla's strategic SWOT analysis," *Har. Bus. Rev.*, vol. 97, no. 3, pp. 45–55, 2019.
3 J. Kotler, *Strategic Management and SWOT Applications*. Wiley Publishing, 2018.
4 P. Drucker, "Management for the Future: Applying SWOT," *J. Bus. Strat.*, vol. 33, no. 2, pp. 23–30, 2016.

Porter's Five Forces Framework

The Origins of a Game-Changing Strategy Tool

In the late 1970s, Michael E. Porter, a young professor at Harvard Business School, sought to address a significant gap in the field of business strategy. At the time, most frameworks for strategic planning focused inward, emphasizing a company's resources and internal processes. Porter, however, believed that organizations needed a systematic way to evaluate the external forces shaping their industries. This conviction led him to develop the Five Forces framework, which he first introduced in his groundbreaking book *Competitive Strategy: Techniques for Analyzing Industries and Competitors*, published in 1980.

Porter's academic rigor and sharp intellect were evident even in his early career. A Rhodes Scholar with degrees from Princeton and Harvard, Porter brought a multidisciplinary perspective to business strategy. He combined insights from economics, industrial organization theory, and corporate strategy to create a model that has since become one of the most widely taught and applied tools in business schools and boardrooms worldwide. His framework offered a fresh perspective: industries were not merely arenas for competition but dynamic ecosystems where external pressures shaped profitability and growth.

The Five Forces Unpacked

Porter identified five key forces that determine the competitive intensity and, ultimately, the profitability of any industry. These forces act as both constraints and opportunities for businesses, and understanding them is critical for crafting strategies that ensure long-term success.

The first force, the **threat of new entrants**, examines the barriers that protect established companies from new competitors. High barriers, such as significant capital requirements or strong brand loyalty, make it harder for newcomers to disrupt the market. The second force, the **bargaining power of suppliers**, assesses the influence suppliers have over prices and availability of key inputs. A small number of powerful suppliers can significantly impact profitability. The third force, the **bargaining power of buyers**, shifts the focus to customers and their ability to demand lower prices or higher quality. Industries dominated by well-informed or concentrated buyers often face intense pricing pressure.

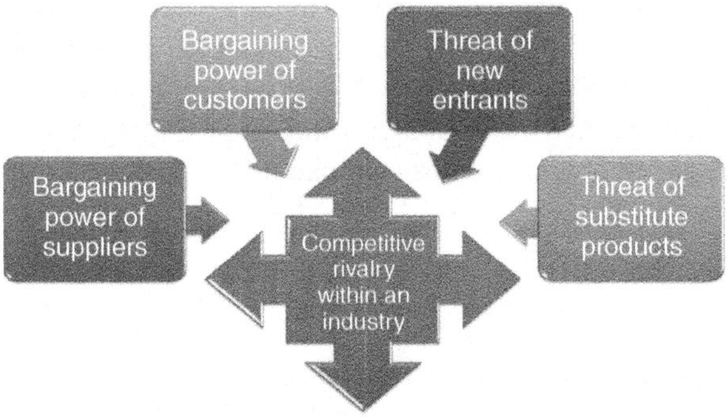

Figure: Porter's Five Forces Framework: Evaluates competitive pressures in an industry: supplier power, buyer power, competitive rivalry, threat of substitution, and threat of new entrants. *Reference*: Michael E. Porter, *Competitive Strategy: Techniques for Analyzing Industries and Competitors*, Free Press, 1980 https://commons.wikimedia.org/wiki/File:Porters_five_forces.PNG, Creative Commons Attribution-Share Alike 3.0 Unported].

The fourth force, the **threat of substitutes**, explores the risks posed by alternative products or services that fulfill the same customer needs. Finally, the **intensity of rivalry among existing competitors** captures the level of competition within the industry, which can lead to price wars, advertising battles, and reduced profitability. Together, these forces create a detailed map of the competitive landscape, enabling businesses to identify threats and opportunities.

Case Study: Steve Jobs, Apple, and the iPhone Revolution

One of the most compelling applications of Porter's Five Forces can be seen in Apple's strategic decisions during the launch and evolution of the iPhone. In 2007, Steve Jobs unveiled a device that not only disrupted the telecommunications industry but also redefined consumer expectations for mobile technology. Jobs, a master strategist, intuitively grasped many of the insights that Porter's framework formalized.

Apple faced significant **barriers to entry**, which Jobs turned into competitive advantages. The company's deep investments in research and development created proprietary technologies, such as the multitouch interface and the A-series processors, that competitors struggled to replicate. Furthermore, Apple's strong brand loyalty and seamless ecosystem of devices and services made it difficult for new entrants to carve out market share.

In terms of **supplier power**, Apple exercised control by diversifying its supply chain and negotiating exclusive contracts for critical components, such as the Retina display. This not only secured a consistent supply but also limited suppliers' ability to dictate terms. On the **buyer power** front, Apple reduced consumer leverage by offering a highly differentiated product. Customers were willing to pay a premium for the iPhone's superior design, functionality, and user experience, mitigating price sensitivity.

The **threat of substitutes**, such as feature phones and early smartphones from competitors, was effectively neutralized by Apple's innovation. The iPhone's integration with the App Store, iCloud, and iTunes created a user experience that alternatives could not match. Lastly, Apple managed **industry rivalry** by positioning the iPhone as a luxury product, avoiding the price wars that plagued other manufacturers. Instead of competing on cost, Apple competed on innovation and user experience.

Capabilities and Strengths of the Framework

Porter's Five Forces remains a cornerstone of strategic planning because of its versatility and depth. It provides leaders with a structured approach to analyze their industries, uncover hidden risks, and identify untapped opportunities. By focusing on external pressures, the framework helps businesses anticipate market shifts and craft strategies that align with broader industry dynamics. It has been applied across sectors, from technology and healthcare to manufacturing and retail, demonstrating its universal relevance.

Limitations and Evolving Contexts

While the Five Forces framework is immensely valuable, it is not without its limitations. Critics argue that it assumes industries are static, with clearly defined boundaries. In today's interconnected and rapidly evolving markets, such as those influenced by digital transformation and globalization, these boundaries often blur. Moreover, the framework does not explicitly account for the role of government regulation, social trends, or technological disruptions, which can significantly impact industry dynamics. However, these limitations do not diminish the tool's utility; rather, they highlight the need for complementary analyses.

Lessons for Modern Leaders

Porter's Five Forces teaches leaders to look beyond their organizations and understand the broader forces shaping their industries. It emphasizes the importance of external awareness and strategic foresight in navigating competitive landscapes. For modern leaders, the framework serves as a reminder that success requires more than internal efficiency; it demands adaptability to external pressures and proactive positioning in the face of change.

Citations

1 M. E. Porter, *Competitive Strategy: Techniques for Analyzing Industries and Competitors*. Free Press, 1980.

2 W. Isaacson, *Steve Jobs*. Simon & Schuster, 2011.

3 P. Ghemawat, "Competition and Business Strategy in Historical Perspective," *Bus. Hist. Rev.*, vol. 76, no. 1, pp. 37–74, 2002.

Value Chain Analysis: Unpacking Competitive Advantage

Introduction: The Origins of Value Chain Analysis

Value Chain Analysis (VCA) was introduced by Michael Porter in his seminal book *Competitive Advantage: Creating and Sustaining Superior Performance* (1985). As a professor at Harvard Business School, Porter developed the framework to provide a systematic way for organizations to analyze their internal activities and determine how they create value for customers. At its core, VCA helps organizations identify competitive advantages by dissecting their operations into discrete, interconnected activities.

Unlike traditional cost accounting, which primarily focuses on financial efficiency, VCA emphasizes the value delivered to customers through each activity. By understanding which activities contribute most to customer satisfaction and profitability, companies can optimize their processes, reduce waste, and create products or services that outperform competitors. VCA has become a staple in industries ranging from manufacturing and logistics to healthcare and technology.

Framework and Components

Porter's value chain is divided into two broad categories:

1) **Primary Activities:** Directly involved in creating and delivering a product or service.
   - **Inbound Logistics:** Receiving, storing, and managing inputs like raw materials.
   - **Operations:** Transforming inputs into final products or services.
   - **Outbound Logistics:** Distributing finished goods to customers.
   - **Marketing and Sales:** Communicating the product's value and generating demand.
   - **Service:** Supporting customers post-sale, including maintenance and troubleshooting.
2) **Support Activities:** Enable the primary activities to function efficiently.
   - **Firm Infrastructure:** Administrative and financial systems supporting overall operations.
   - **Human Resource Management (HRM):** Recruitment, training, and workforce management.
   - **Technology Development:** Research, innovation, and IT systems driving product or process improvements.
   - **Procurement:** Sourcing and acquiring inputs required for operations.

The interactions between these activities determine the total value delivered to customers, which can then be compared to the costs incurred to establish profit margins.

Applications Across Industries

1) **Manufacturing:** VCA is integral to optimizing operations in manufacturing. For example, Toyota uses lean manufacturing principles to streamline operations and minimize waste in its inbound logistics and production processes. This focus on efficiency enables Toyota to deliver high-quality vehicles at competitive prices.
2) **Technology and Software:** Tech companies like Apple rely on VCA to differentiate themselves. Apple's superior design (a result of technology development) and its tightly controlled retail experience (outbound logistics and marketing) create a brand synonymous with innovation and quality.

3) **Healthcare:** Hospitals use VCA to improve patient outcomes while reducing costs. For instance, Cleveland Clinic evaluates inbound logistics (patient intake), operations (diagnostic processes), and service (posttreatment care) to optimize patient flow and resource allocation.
4) **Retail and E-Commerce:** Amazon employs VCA to dominate e-commerce through efficient inbound logistics (fulfillment centers), operations (automated warehousing), and outbound logistics (Prime delivery). Its investment in technology development, such as predictive analytics and AI, further enhances customer satisfaction.

Case Study: Amazon's Value Chain Dominance

Amazon exemplifies the transformative potential of VCA.

- **Inbound Logistics:** Amazon has a vast network of suppliers and warehouses, supported by sophisticated supply chain management software that tracks inventory in real time.
- **Operations:** Automation through robotics and AI streamlines order processing, ensuring speed and accuracy.
- **Outbound Logistics:** With its extensive delivery network, including Amazon Prime, the company guarantees fast, reliable shipping, further strengthening customer loyalty.
- **Marketing and Sales:** Amazon leverages data analytics to personalize recommendations and offers, driving repeat purchases.
- **Service:** Exceptional customer support and seamless return processes ensure long-term trust and satisfaction.

By optimizing every link in its value chain, Amazon has achieved unparalleled efficiency and scalability, setting a benchmark for the retail industry.

Benefits of Value Chain Analysis

1) **Enhanced Efficiency:** By identifying bottlenecks or redundant processes, organizations can streamline workflows and reduce costs.
2) **Customer-Centric Strategies:** Focusing on activities that directly impact customer satisfaction ensures long-term loyalty and competitive differentiation.
3) **Competitive Benchmarking:** VCA enables companies to compare their value chains with industry leaders, identifying areas for improvement.
4) **Cross-Functional Collaboration:** The framework encourages departments to align their efforts, breaking silos and fostering a unified organizational vision.

Limitations and Criticisms

While powerful, VCA has its limitations:

1) **Resource-Intensive:** Conducting a thorough VCA requires significant time and expertise, particularly for large organizations with complex operations.
2) **Dynamic Market Conditions:** Static value chains may become obsolete in rapidly changing industries, necessitating frequent updates to remain relevant.
3) **Subjectivity:** Determining the value contribution of certain activities can be subjective, leading to biased conclusions.

Lessons Learned

1) **Holistic Integration:** VCA is most effective when combined with complementary tools like SWOT analysis or PESTLE to provide a broader strategic context.
2) **Focus on Differentiation:** Organizations should prioritize activities that create unique value for customers rather than merely cutting costs.
3) **Adaptability:** Given the dynamic nature of global markets, value chains must be revisited regularly to reflect new technologies, customer preferences, and competitive pressures.

Conclusion: Relevance in Modern Leadership

VCA remains a critical tool for leaders aiming to build resilient, customer-focused organizations. By dissecting operations into manageable components, it provides clarity and direction, helping businesses adapt to challenges and seize opportunities.

In an era defined by technological disruption and global interconnectivity, VCA offers a structured approach to navigating complexity while delivering superior value to stakeholders.

Citations (IEEE Format)

1 M. Porter, *Competitive Advantage: Creating and Sustaining Superior Performance*. New York, NY: Free Press, 1985.
2 "Value chain applications in E-commerce: Amazon case study," *J. Bus. Strat.*, vol. 45, no. 4, pp. 33–47, 2020.
3 T. Ohno, *Toyota Production System: Beyond Large-Scale Production*. Productivity Press, 1988.
4 "Healthcare value chains: Improving patient outcomes," *Har. Bus. Rev.*, vol. 95, no. 6, pp. 21–29, 2018.

Theory of Change

Introduction to the Theory of Change

The **Theory of Change (ToC)** is a systematic methodology used to map the pathways that lead from specific actions or interventions to long-term outcomes. First formalized in the 1990s by organizations like the Aspen Institute, ToC emerged as a response to the need for clear frameworks in evaluating the effectiveness of social programs and interventions. Today, it is widely used in fields such as international development, public policy, corporate social responsibility (CSR), and strategic project management.

ToC focuses on defining the causal relationships between inputs, activities, outputs, outcomes, and long-term impacts. It emphasizes the importance of measurable milestones and critical assumptions that underpin success, making it both a planning and evaluation tool.

Core Components of the Theory of Change

1) **Inputs:** Resources, time, and expertise invested in the initiative.
2) **Activities:** Actions or interventions implemented to achieve goals.

3) **Outputs:** Immediate, tangible results of activities.
4) **Outcomes:** Intermediate changes in behavior, knowledge, or conditions resulting from the outputs.
5) **Impact:** Long-term, systemic changes aligned with the overarching vision.

ToC often incorporates **critical assumptions** – hypotheses about conditions or factors outside the organization's control that must hold true for the initiative to succeed. This aspect encourages organizations to identify potential risks and mitigate them early in the process.

Developing a Theory of Change Framework

The process of creating a ToC involves:

1) **Define the Long-Term Goal:** Establish the ultimate objective, such as reducing carbon emissions, increasing literacy rates, or improving employee retention.
2) **Map Pathways to Impact:** Identify the intermediate outcomes necessary to achieve the goal and work backward to define outputs and activities.
3) **Articulate Assumptions:** Specify the conditions required for each step of the process to succeed.
4) **Identify Indicators:** Develop measurable metrics to track progress at each stage.
5) **Visualize the Framework:** Create a diagram or flowchart that clearly illustrates the connections between inputs, activities, outputs, outcomes, and impact.

Applications in Industry and Development

ToC has found widespread application across sectors due to its flexibility and emphasis on evaluation. Its use spans from guiding social programs in developing nations to enhancing strategic planning in multinational corporations.

- **International Development:** Organizations like UNICEF and the World Bank use ToC to plan and evaluate initiatives targeting poverty alleviation, education access, and healthcare improvement. By clarifying the causal pathways, ToC enables these organizations to align resources with outcomes effectively.
- **Corporate Social Responsibility (CSR):** Companies like Patagonia employ ToC to ensure their sustainability initiatives, such as reducing carbon footprints, align with long-term environmental goals.
- **Strategic Change in Organizations:** ToC is increasingly applied in business contexts to guide transformations like digital adoption, diversity initiatives, and culture change. It provides clarity on how specific interventions lead to measurable organizational improvements.

Case Study: The Gates Foundation's Education Strategy

The Gates Foundation utilized the ToC to design its education initiatives aimed at increasing high school graduation rates in underserved communities across the United States. The framework enabled the foundation to:

1) Identify root causes of dropout rates, such as lack of access to quality teaching and socioeconomic challenges.

2) Define clear intermediate outcomes, including teacher training programs and the introduction of personalized learning platforms.
3) Monitor progress using indicators like attendance rates, teacher retention, and student test scores.
4) Adapt strategies based on real-time evaluation to ensure alignment with the desired impact.

Advantages of the Theory of Change

1) **Clarity and Focus:** ToC forces organizations to explicitly connect their activities to long-term goals, ensuring alignment and accountability.
2) **Transparency:** By articulating assumptions, ToC exposes risks and external dependencies, fostering better decision-making.
3) **Evaluation-Driven:** ToC emphasizes measurable outcomes, making it easier to assess the success of initiatives.
4) **Flexibility:** It is adaptable across domains, from environmental conservation to technological innovation.

Limitations and Challenges

1) **Complexity:** Large-scale initiatives with multiple variables can make ToC diagrams overly intricate, reducing their utility.
2) **Data Dependency:** Effective ToC relies on accurate data collection and analysis, which can be resource-intensive.
3) **Overemphasis on Linear Progression:** ToC may struggle to accommodate the nonlinear dynamics of complex systems, particularly in unpredictable environments.

Practical Tips for Implementation

1) **Engage Stakeholders:** Collaborate with stakeholders early to ensure the ToC reflects diverse perspectives and priorities.
2) **Iterate Regularly:** Treat the ToC as a living document, revising it as new information becomes available.
3) **Visualize Clearly:** Use software like Kumu, Visio, or Miro to create accessible and visually compelling ToC diagrams.
4) **Integrate with Other Tools:** Combine ToC with scenario planning or cost–benefit analysis (CBA) to address uncertainty and resource allocation.

Relevance in Today's World

In an era of increased accountability and demand for evidence-based decision-making, the ToC has gained prominence as a critical planning tool. Its ability to foster clarity and alignment makes it indispensable in addressing complex challenges like climate change, social equity, and digital transformation.

Citations in IEEE Format

1 J. P. Connell and A. C. Kubisch, *Applying a Theory of Change Approach to the Evaluation of Comprehensive Community Initiatives*. Aspen Institute Roundtable on Comprehensive Community Initiatives for Children and Families, 1998.

2 L. Dhillon and S. Vaca, "Refining theories of change," *Eval. J. Australas.*, vol. 12, no. 1, pp. 4–14, 2012.
3 The Bill & Melinda Gates Foundation, "Education Strategy Overview," Foundation Reports, 2020.
4 P. Rogers, *Theory of Change: Methodological Briefs*. Florence: UNICEF Office of Research – Innocenti, 2014.

S-Curves for Technology Growth

Introduction: Understanding S-Curves

The **S-Curve model** is a foundational concept in technology and market growth, illustrating the lifecycle of technologies, innovations, and industries. Developed from studies on innovation diffusion and economic growth, the S-Curve describes how technologies follow a predictable trajectory, from slow initial adoption to rapid growth and eventual saturation. Its origins trace back to research by Everett Rogers on the "Diffusion of Innovations" in the 1960s, which was later adapted to study technological development and market penetration.

The S-Curve's characteristic shape represents cumulative progress over time: a slow initial phase of learning and experimentation (the early adopters), followed by a period of rapid growth (mainstream adoption), and eventually a plateau (market saturation or technological maturity).

Core Phases of the S-Curve

1) **Emergence (Slow Start):** This phase involves invention, development, and early adoption. Progress is slow due to high costs, technical uncertainties, and limited market awareness.
2) **Accelerated Growth:** In this phase, technological breakthroughs, cost reductions, and wider adoption drive exponential growth. Competitors enter the market, and adoption by the majority begins.
3) **Maturity and Saturation:** Growth slows as the technology reaches market saturation or physical performance limits. Innovation diminishes, and incremental improvements dominate.
4) **Decline or Disruption:** Technologies may enter a decline phase as newer innovations (on their own S-Curve) disrupt and replace them.

Applications Across Industries

The S-Curve framework is invaluable for strategic planning, R&D management, and market forecasting. Its applications span diverse industries:

- **Technology Development:** Used to anticipate when investments in a technology will yield diminishing returns and when to shift to new innovations.
- **Market Strategy:** Helps businesses determine when to scale operations or diversify product lines to maintain growth.
- **Energy Transition:** Applied to assess the adoption trajectories of renewable energy technologies like solar panels and wind turbines.

Case Study: Apple's iPhone and the Smartphone Market

Apple's iPhone provides a textbook example of the S-Curve in action. When introduced in 2007, the iPhone's adoption followed a typical S-Curve trajectory:

1) **Emergence:** The first iPhone was revolutionary but expensive, with limited apps and features compared to its successors. It appealed mainly to early adopters willing to experiment with cutting-edge technology.
2) **Accelerated Growth:** Subsequent iPhone models improved usability, expanded app ecosystems, and lowered costs. This led to rapid global adoption, propelling Apple to become one of the most valuable companies.
3) **Maturity:** By the mid-2010s, the smartphone market began to saturate. Incremental improvements, such as better cameras and faster processors, replaced groundbreaking innovations. Market growth plateaued, and Apple shifted its focus to services and ecosystem integration.
4) **Disruption Avoidance:** Recognizing the potential for market saturation, Apple has invested heavily in complementary technologies (e.g., wearables, augmented reality (AR)) to ensure sustained relevance.

Advantages of S-Curves

1) **Strategic Timing:** Identifies when to invest in or exit technologies based on their growth phase.
2) **Resource Allocation:** Helps prioritize R&D spending by signaling when a technology is reaching diminishing returns.
3) **Competitive Edge:** Anticipates disruption by tracking emerging technologies that could replace incumbents.

Challenges and Limitations

1) **Nonlinearity:** Not all technologies follow a predictable S-Curve due to external disruptions, policy changes, or unforeseen breakthroughs.
2) **Early Identification:** It can be difficult to determine if a technology is still in its early phase or already accelerating.
3) **Disruption Blind Spots:** The S-Curve focuses on individual technologies but may miss external disruptors that reshape industries.

Practical Implementation

1) **Performance Benchmarking:** Measure the progress of a technology against industry benchmarks to identify its position on the curve.
2) **Scenario Planning:** Combine the S-Curve with scenario planning to assess the impact of external factors like regulation or competitor innovations.
3) **Portfolio Management:** Use the S-Curve to maintain a balanced portfolio of emerging, growing, and mature technologies.

Emerging Trends in S-Curve Applications

- **Artificial Intelligence:** The adoption of AI technologies is currently in the accelerated growth phase, with rapid improvements in capabilities and widespread integration across sectors.
- **Electric Vehicles (EVs):** EVs have transitioned from emergence to accelerated growth as battery costs decline and infrastructure improves.
- **Quantum Computing:** Still in the emergence phase, with high costs and technical hurdles, quantum computing holds transformative potential for the next decade.

Relevance in Today's Context

The S-Curve model is more relevant than ever in an era of rapid technological advancement and market volatility. For organizations navigating digital transformation or sustainability transitions, understanding S-Curve dynamics enables strategic foresight and proactive decision-making.

Citations

1 E. M. Rogers, *Diffusion of Innovations*, 5th ed. New York: Free Press, 2003.
2 R. N. Foster, *Innovation: The Attacker's Advantage*. Summit Books, 1986.
3 C. M. Christensen, *The Innovator's Dilemma: When New Technologies Cause Great Firms to Fail*. Boston: Harvard Business Review Press, 1997.
4 G. Berkhout, "Adopting a technology S-curve framework for energy transition analysis," *Energy Pol. J.*, vol. 48, pp. 584–595, 2017.
5 Meeker, M., "Internet Trends Report," *Bond Capital Annual Reports*, 2021.

Lean Startup Methodology

Introduction: Origins of the Lean Startup Approach

The **Lean Startup Methodology**, pioneered by Eric Ries in his seminal book *The Lean Startup* (2011), revolutionized how businesses approach innovation, product development, and market entry. Drawing inspiration from Toyota's Lean Manufacturing principles, which emphasize waste reduction and iterative improvement, Ries adapted these concepts for startups navigating uncertainty. The methodology aims to help entrepreneurs build scalable businesses with minimal risk and resources by focusing on continuous feedback, rapid experimentation, and validated learning.

The Lean Startup Methodology challenges traditional approaches that rely on extensive upfront planning and large capital investments before market testing. Instead, it advocates for agility, adaptability, and customer-centric innovation, making it a cornerstone of modern entrepreneurship and corporate innovation initiatives.

Core Principles of Lean Startup

1) **Build-Measure-Learn Feedback Loop:**
   - The heart of the methodology, this iterative loop involves quickly building a *Minimum Viable Product (MVP)*, measuring its performance, and learning from customer feedback. This cycle minimizes waste and ensures the product aligns with user needs.
2) **Validated Learning:**
   - Entrepreneurs test hypotheses about customers, markets, and products through real-world experiments. Metrics, not assumptions, drive decision-making.
3) **Pivot or Persevere:**
   - If an experiment reveals that the initial approach isn't working, the startup pivots – adapting its strategy – or perseveres if the direction proves viable.
4) **Continuous Deployment:**
   - By releasing incremental updates frequently, teams can test features in real time and gather immediate feedback.

5) **Innovation Accounting:**
   - A structured approach to measuring progress in uncertain conditions. It emphasizes actionable metrics (e.g., customer acquisition, retention) over vanity metrics like page views.

Applications Across Industries

The Lean Startup Methodology isn't limited to tech startups; it has found widespread application in various fields:

- **Corporate Innovation:** Large corporations like GE and Intuit have adopted lean principles to foster intrapreneurship and accelerate product cycles.
- **Healthcare:** Startups in med-tech use MVPs to test new devices with minimal regulatory risk and resource expenditure.
- **Government and Nonprofits:** Governments employ lean principles for designing citizen-centered services with limited budgets.

Case Study: Dropbox's MVP Strategy

Dropbox, now a leader in cloud storage, initially started with an MVP that was not a fully functional product but a simple explainer video. The video demonstrated the product's core functionality and invited potential users to sign up. This approach enabled Dropbox to:

1) Gauge market interest without incurring high development costs.
2) Collect early feedback to refine its feature set.
3) Build a user base before investing in full-scale development.

The overwhelming response validated Dropbox's assumptions about market demand, allowing the company to secure funding and scale quickly.

Advantages of Lean Startup Methodology

1) **Resource Efficiency:** By focusing only on essential features through MVPs, startups conserve time and money.
2) **Risk Mitigation:** Continuous testing and adaptation reduce the likelihood of catastrophic product failures.
3) **Customer-Centric Innovation:** Frequent feedback loops ensure products meet real market needs.
4) **Scalability:** Lean principles adapt well to companies of all sizes, from fledgling startups to multinational corporations.

Challenges and Limitations

1) **Misinterpretation of MVP:** Many startups equate an MVP with a bare-bones product, neglecting quality or usability, which can alienate early adopters.
2) **Metrics Overload:** Teams may become bogged down in measuring too many metrics without focusing on actionable insights.
3) **Market Context:** In industries with long development cycles (e.g., pharmaceuticals), lean principles may not be fully applicable.

Practical Implementation

1) **Hypothesis-Driven Development:**
   - Start with clear hypotheses about customer needs and test them through MVPs.
2) **Customer Development:**
   - Engage directly with users to gather insights and refine the product.
3) **A/B Testing:**
   - Compare different iterations of features or designs to determine what resonates best with users.

Emerging Trends in Lean Startup

- **AI-Powered MVPs:** Startups increasingly use AI to simulate user feedback and optimize features before physical testing.
- **Lean and Agile Hybrid Models:** Many organizations combine Lean Startup principles with Agile project management for greater adaptability.
- **Corporate Startups:** Lean principles are being institutionalized in corporate incubators and accelerators to drive internal innovation.

Relevance in Today's Context

In today's fast-paced markets, where consumer preferences shift rapidly and technological advancements disrupt industries, the Lean Startup Methodology is more relevant than ever. It empowers organizations to navigate uncertainty with agility, fostering a culture of experimentation and learning. From Silicon Valley tech giants to emerging social enterprises, lean principles continue to reshape how businesses innovate and thrive.

Citations

1 E. Ries, *The Lean Startup: How Today's Entrepreneurs Use Continuous Innovation to Create Radically Successful Businesses*. Crown Business, 2011.
2 S. Blank, *The Four Steps to the Epiphany: Successful Strategies for Startups That Win*. Wiley, 2013.
3 S. Thomke, "Experimentation matters: Unlocking the potential of new technologies for innovation," *Harv. Bus. Rev.*, vol. 81, no. 4, pp. 67–75, 2003.
4 A. Osterwalder, *Business Model Generation: A Handbook for Visionaries, Game Changers, and Challengers*. Wiley, 2010.
5 J. Gans, "Do entrepreneurs need a lean startup mindset?" *MIT Sloan Manag. Rev.*, vol. 59, no. 1, pp. 1–7, 2017.

Developing the Next Tool: Technology Roadmapping

Introduction to Technology Roadmapping

Technology roadmapping is a strategic planning tool designed to align an organization's technological capabilities with its business objectives. Originating in the late 1970s at Motorola, this method was developed to address the rapid pace of innovation and the need for structured foresight in product development. It has since evolved into a widely used framework across industries, enabling organizations to anticipate technological trends, allocate resources effectively, and maintain a competitive edge.

Core Concepts of Technology Roadmapping

A technology roadmap is essentially a visual representation that links an organization's goals to its technological initiatives over a defined timeline. It integrates market trends, customer needs, and internal capabilities to outline a cohesive strategy for technology development. Roadmaps are typically organized into layers, including market drivers, product features, technology requirements, and resource allocation.

Key components include:

- **Market and Business Drivers:** Identifying external forces such as customer demands, regulatory requirements, and competitive pressures.
- **Technology Capabilities:** Assessing existing technologies and identifying gaps that need to be filled.
- **Milestones and Timelines:** Setting clear objectives and tracking progress over time.
- **Resource Planning:** Allocating human, financial, and technical resources efficiently.

Types of Technology Roadmaps

1) **Product Roadmaps:** Focused on aligning technology development with product timelines and market launches.
2) **Strategic Roadmaps:** High-level plans that guide organizational priorities and investments in technology.
3) **Program and Project Roadmaps:** Detailed plans for specific R&D projects or technology deployments.

Applications in Industry

Technology roadmapping is widely used in industries like aerospace, pharmaceuticals, automotive, and IT. For instance, semiconductor companies like Intel use roadmaps to guide R&D investments in chip design, ensuring alignment with market trends and production capabilities.

Case Study: Tesla's Battery Development Tesla's technology roadmap for battery innovation demonstrates how this tool can drive competitive advantage. By setting long-term goals to reduce battery costs and increase energy density, Tesla aligned its R&D efforts with market demands for affordable EVs. The roadmap provided a clear strategy for investments in manufacturing facilities, such as the Gigafactory, and partnerships with suppliers.

Advantages of Technology Roadmapping

1) **Alignment:** Ensures that technological efforts are consistent with business objectives.
2) **Prioritization:** Helps organizations focus resources on high-impact projects.

3) **Risk Management:** Identifies potential bottlenecks and external threats early in the planning process.
4) **Flexibility:** Encourages adaptability by updating roadmaps as market conditions change.

Limitations

While technology roadmapping is a powerful tool, it is not without challenges:

- **Complexity:** Developing and maintaining roadmaps for large organizations can be resource-intensive.
- **Uncertainty:** Rapidly changing market conditions and technological disruptions may render roadmaps obsolete.
- **Interdisciplinary Coordination:** Requires collaboration across departments, which can be difficult in siloed organizations.

Steps to Create a Technology Roadmap

1) **Define Objectives:** Establish clear business goals and identify key market drivers.
2) **Assess Current Capabilities:** Evaluate the organization's existing technology assets and gaps.
3) **Identify Milestones:** Break down the roadmap into actionable steps with measurable outcomes.
4) **Engage Stakeholders:** Involve cross-functional teams to ensure alignment and buy-in.
5) **Monitor and Update:** Continuously revise the roadmap based on performance metrics and market feedback.

Practical Applications and Recommendations

Technology roadmapping is particularly useful for organizations facing rapid technological change or high levels of competition. To maximize its effectiveness:

- Use software tools like Roadmunk or Aha! to streamline the roadmap creation process.
- Incorporate scenario planning to account for uncertainties in the market.
- Regularly review and update the roadmap to reflect changes in organizational priorities or external conditions.

Citations in IEEE Format

1 R. Phaal, C. J. Farrukh, and D. R. Probert, "Technology roadmapping: Linking technology resources to business objectives," *Int. J. Technol. Manag.*, vol. 26, no. 1, pp. 2–19, 2003.

2 M. L. Garcia and O. H. Bray, *Fundamentals of Technology Roadmapping*. Sandia National Laboratories, 1997.

3 Tesla, Inc, *Battery Innovation and Manufacturing Strategy*. Tesla Investor Day Presentation, 2023.

Business Model Canvas: A Blueprint for Innovation and Strategy

Origins of the Business Model Canvas

In 2004, Alexander Osterwalder, a Swiss business theorist and entrepreneur, revolutionized the way businesses conceptualize and execute their strategies. Dissatisfied with the fragmented and overly complex approaches to business planning, Osterwalder developed a simple, visual framework that would allow organizations to articulate, analyze, and innovate their business models. The result was the *Business Model Canvas* (BMC), a nine-block framework that quickly gained global recognition. It was popularized in his 2010 book, *Business Model Generation*, coauthored with Yves Pigneur.

Osterwalder's academic and entrepreneurial background heavily influenced his work. With a Ph.D. from the University of Lausanne focused on business model innovation, he brought a rigorous yet practical perspective to his creation. His goal was to create a universally applicable tool – one that could guide startups, multinationals, and nonprofits alike in designing scalable, effective business strategies. Today, the BMC is taught in business schools worldwide and is a go-to tool for innovation and strategic alignment.

Breaking Down the Business Model Canvas

The BMC comprises nine interconnected building blocks, each representing a critical component of a business:

1) **Customer Segments:** Defines the groups of people or organizations a business aims to serve. From niche markets to mass audiences, understanding customer needs is foundational.
2) **Value Propositions:** Articulates the unique value a company delivers to its customers, whether through innovation, cost reduction, or exceptional service.
3) **Channels:** Details how a business delivers value to its customers, whether through digital platforms, retail stores, or direct sales.
4) **Customer Relationships:** Outlines how a company interacts with its customers, from automated systems to personalized service.
5) **Revenue Streams:** Identifies the ways in which a business earns income, such as subscriptions, licensing, or one-time sales.
6) **Key Resources:** Highlights the assets – human, physical, intellectual, and financial – that enable the business to operate.
7) **Key Activities:** Focuses on the critical tasks and operations that drive success, from product design to marketing.
8) **Key Partnerships:** Identifies external entities that provide essential resources, services, or market access.
9) **Cost Structure:** Analyzes the major costs associated with operating the business, ensuring alignment with revenue streams and overall strategy.

The beauty of the BMC lies in its simplicity. By visually mapping out these elements, organizations gain a clear, holistic view of their business models, enabling them to identify inefficiencies, explore new opportunities, and ensure strategic coherence.

Case Study: Steve Jobs and Apple's Ecosystem

The BMC provides a lens through which we can analyze Apple's extraordinary success under Steve Jobs. Upon his return to Apple in 1997, Jobs embarked on a journey to revitalize the company by designing a highly integrated ecosystem of products and services.

1) **Customer Segments:** Jobs understood the importance of targeting both the tech-savvy and the everyday consumer. Apple's customer base ranged from creative professionals to families, unified by a desire for intuitive, beautifully designed products.

2) **Value Propositions:** The hallmark of Apple's value proposition was simplicity and seamless integration. Products like the iPod, iPhone, and Mac offered not just functionality but an elevated user experience.
3) **Channels:** Apple revolutionized retail with its Apple Stores, providing a space where customers could explore and experience products firsthand. The App Store further extended this channel into the digital realm.
4) **Customer Relationships:** Apple cultivated loyalty through its Genius Bar, personalized product setups, and regular software updates, creating long-term relationships with customers.
5) **Revenue Streams:** By diversifying income sources – hardware sales, software licensing, and App Store revenues – Apple built a resilient financial model.
6) **Key Resources:** The company's intellectual property, brand equity, and talented workforce were unmatched. Apple's control over its supply chain and proprietary technologies provided a strong competitive edge.
7) **Key Activities:** Innovation and design remained at the core of Apple's operations. The company also excelled in marketing, creating iconic campaigns like "Think Different."
8) **Key Partnerships:** Apple's relationships with component suppliers, app developers, and telecommunication companies were critical in delivering value to its customers.
9) **Cost Structure:** Apple's investments in R&D, manufacturing, and marketing were offset by high profit margins, particularly on its flagship products.

The BMC captures how Jobs transformed Apple into a vertically integrated powerhouse, demonstrating how every building block interconnects to create a cohesive strategy.

Capabilities and Strengths of the Business Model Canvas

The BMC's primary strength is its adaptability. It can be used to design, analyze, or iterate business models across industries. Startups employ it to validate their ideas, while established firms use it to explore new growth avenues. Its visual format makes it an accessible and collaborative tool, fostering innovation and alignment within teams.

Limitations and Challenges

Despite its versatility, the BMC is not without its limitations. Critics argue that it oversimplifies complex business dynamics, offering a snapshot rather than a comprehensive analysis. It does not explicitly address external factors like competitive forces, regulatory changes, or technological disruptions, which are critical to strategic planning. However, these shortcomings can be mitigated by using the BMC in conjunction with tools like Porter's Five Forces or PESTLE analysis.

Lessons for Modern Leaders

The BMC teaches leaders to think holistically. By mapping out each component, they gain a clear understanding of how decisions in one area affect the entire system. It encourages adaptability, enabling organizations to pivot in response to market changes. Most importantly, it fosters a customer-centric approach, reminding leaders that all strategies must ultimately deliver value to their target audience.

Citations

1 A. Osterwalder and Y. Pigneur, *Business Model Generation: A Handbook for Visionaries, Game Changers, and Challengers*. Wiley, 2010.
2 W. Isaacson, *Steve Jobs*. Simon & Schuster, 2011.
3 H. Chesbrough, *Open Business Models: How to Thrive in the New Innovation Landscape*. Harvard Business Review Press, 2006.

PESTLE Analysis: A Strategic Lens for Macro-environmental Factors

Origins and Evolution of PESTLE Analysis

The PESTLE analysis framework emerged in the mid-20th century as a strategic tool to help organizations understand the macro-environmental factors influencing their operations. The original version, often called Political, Economic, Social, and Technological (PEST), was expanded to include Legal and Environmental factors to reflect evolving complexities in global business and policy landscapes.

Developed by strategic planners and management consultants, PESTLE has since become a cornerstone of strategic analysis, particularly in industries exposed to external uncertainties like energy, healthcare, and technology. Its primary purpose is to provide a structured way of examining external forces that could affect organizational performance, helping decision-makers anticipate challenges and opportunities.

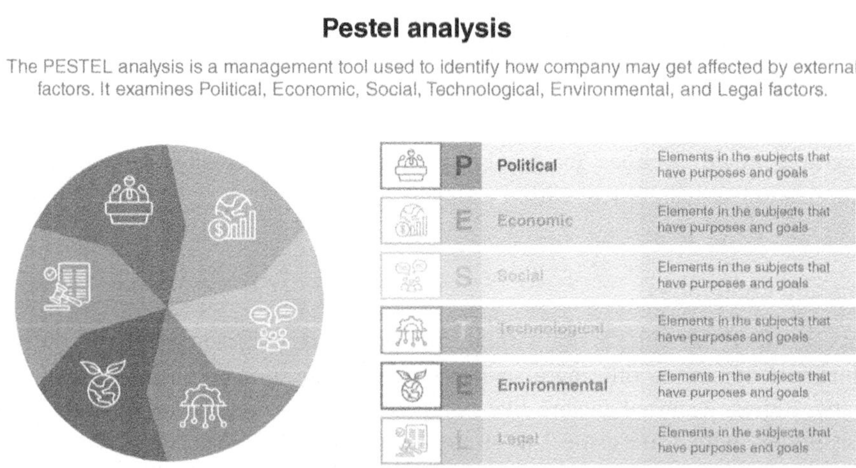

Figure: PESTLE Analysis: Analyzes external factors – Political, Economic, Social, Technological, Legal, and Environmental – that impact decisions. [*Reference*: Francis J. Aguilar, *Scanning the Business Environment*, Macmillan, 1967. https://www.vecteezy.com/free-vector/pestle-analysis].

Methodology: The Six Pillars of PESTLE

PESTLE analysis examines six critical dimensions:

1) **Political:** Analyzes government policies, regulations, and political stability. For instance, trade tariffs, international agreements, and tax policies can significantly impact industries like manufacturing and technology.

2) **Economic:** Evaluates economic conditions such as inflation, interest rates, exchange rates, and unemployment. These factors shape market demand and operational costs.
3) **Social:** Assesses demographic trends, cultural norms, consumer behavior, and workforce dynamics. This aspect ensures strategies align with societal expectations and market needs.
4) **Technological:** Examines technological advancements, innovation trends, and digital disruptions. This factor is crucial in industries undergoing rapid transformation, such as AI and renewable energy.
5) **Legal:** Covers laws and regulations, including labor laws, intellectual property rights, and industry-specific compliance requirements.
6) **Environmental:** Focuses on sustainability issues like climate change, resource scarcity, and environmental regulations, increasingly vital in the age of Environmental, Social, and Governance (ESG) priorities.

Applications in Practice

PESTLE is widely used in strategic planning, risk management, and market entry analysis. It helps organizations identify external risks and opportunities, ensuring that strategies are adaptable and resilient.

Case Study: Renewable Energy Adoption in Emerging Markets

In the renewable energy sector, PESTLE analysis is instrumental in understanding barriers and enablers:

- **Political:** Government incentives for renewable energy projects, such as feed-in tariffs or subsidies, play a critical role in adoption.
- **Economic:** The cost of solar and wind technology versus fossil fuels impacts investment decisions.
- **Social:** Public acceptance of renewable projects can determine their success, especially in regions with cultural resistance to change.
- **Technological:** Advances in battery storage and grid integration affect project feasibility.
- **Legal:** Permitting processes and land acquisition laws often dictate project timelines.
- **Environmental:** Renewable projects are often evaluated against their ecological impact, such as biodiversity concerns.

Strengths and Limitations

Strengths:

- **Holistic Perspective:** PESTLE provides a comprehensive understanding of the external environment.
- **Versatility:** Applicable across industries and for diverse strategic objectives.
- **Forward-Looking:** Encourages proactive rather than reactive decision-making.

Limitations:

- **Time-Intensive:** Requires extensive data collection and analysis.
- **Dynamic Nature:** Factors can change rapidly, necessitating frequent updates.
- **Subjectivity:** Interpretations of PESTLE findings can vary, depending on analysts' expertise.

Modern Relevance

In today's interconnected world, PESTLE analysis remains indispensable. From multinational corporations navigating global markets to startups seeking to scale in new regions, the framework enables leaders to anticipate changes and align strategies with external realities. Moreover, PESTLE's emphasis on sustainability and legal compliance resonates with current ESG imperatives, making it more relevant than ever.

Citations

1 G. Johnson, K. Scholes, and R. Whittington, *Exploring Corporate Strategy: Text and Cases.* Prentice Hall, 2011.
2 F. J. Aguilar, *Scanning the Business Environment.* Macmillan, 1967.
3 A. Clarke and M. Fuller, "Strategic foresight: How PESTLE analysis supports organizational planning," *J. Bus. Strat.*, vol. 36, no. 5, pp. 24–33, 2020.
4 Renewable Energy World. "Leveraging PESTLE for Strategic Growth in Renewable Markets." Accessed January 2025.

Scenario Planning: Navigating Uncertainty Through Strategic Foresight

Origins and Historical Context of Scenario Planning

Scenario planning originated during the 1940s, developed by military strategists to anticipate a range of potential future events during World War II. Later, in the 1970s, it was popularized in the corporate world by Pierre Wack and his colleagues at Royal Dutch Shell. Wack applied scenario planning to forecast disruptions in the global oil market, such as the 1973 OPEC oil embargo. The tool's success in helping Shell navigate volatile energy markets catapulted it into the mainstream as an essential method for strategic foresight.

Unlike forecasting, which often extrapolates current trends, scenario planning involves crafting diverse, plausible future scenarios to explore the implications of uncertainty. This approach helps organizations prepare for multiple outcomes rather than betting on a single predicted future.

Methodology: The Core Elements of Scenario Planning

Scenario planning typically follows a structured process:

1) **Define the Scope and Objectives:**
   - Identify the central question or decision to be addressed (e.g., "How should our organization adapt to climate change policies over the next 20 years?").
   - Establish the time horizon and key focus areas, such as market trends, technological advancements, or geopolitical developments.
2) **Identify Key Drivers of Change:**
   - Use tools like PESTLE analysis to pinpoint factors that could influence the future.
   - Group drivers into "knowns" (predictable trends) and "unknowns" (uncertainties with high impact).
3) **Develop Plausible Scenarios:**
   - Create a matrix using two critical uncertainties as axes. For example, in energy markets, uncertainties might include "global adoption of renewable energy" and "rate of technological innovation."

- o Populate the matrix with four plausible scenarios, exploring the extremes of these uncertainties (e.g., rapid adoption with high innovation vs. slow adoption with low innovation).

4) **Analyze Implications:**
   - o Assess how each scenario might impact the organization, its stakeholders, and broader systems.
   - o Identify potential strategies to capitalize on opportunities or mitigate risks in each scenario.
5) **Develop Actionable Strategies:**
   - o Use the insights from scenario analysis to create flexible, robust strategies.
   - o Establish "signposts" or indicators to monitor which scenario is unfolding, enabling timely adjustments.

Applications in Practice

Scenario planning is widely used across industries and policy domains:

1) **Energy Sector:**
   - o Royal Dutch Shell continues to use scenario planning to address energy transitions, carbon neutrality, and geopolitical risks. For example, its *Sky* scenario explores a pathway to a net-zero emissions future, while the *Mountain* and *Oceans* scenarios examine slower, fragmented transitions.
2) **Healthcare:**
   - o Scenario planning has been critical in managing public health crises. For instance, during the COVID-19 pandemic, governments and healthcare organizations used scenarios to anticipate vaccine rollout challenges and hospital capacity issues.
3) **Technology:**
   - o In the tech sector, companies like IBM and Google use scenario planning to navigate rapid innovation cycles and regulatory shifts. Scenarios help them explore the implications of emerging technologies like quantum computing and AI on their business models.

Case Study: Royal Dutch Shell's Strategic Pivot

Shell's use of scenario planning during the 1973 oil crisis remains a hallmark example. By envisioning scenarios where oil prices would skyrocket due to geopolitical tensions, Shell preemptively diversified its supply chain and renegotiated supplier contracts. As a result, the company weathered the crisis more effectively than its competitors, securing its place as a leader in the global energy market.

Strengths and Limitations

Strengths:

- **Flexibility:** Encourages the development of strategies that are resilient across a range of possible futures.
- **Enhanced Decision-Making:** Promotes long-term thinking and prepares leaders to navigate uncertainties with confidence.
- **Collaboration:** Engages diverse stakeholders in imagining the future, fostering buy-in and shared understanding.

Limitations:

- **Resource-Intensive:** Requires significant time and expertise to develop meaningful scenarios.
- **Subjectivity:** The selection of key drivers and scenario narratives can introduce biases.
- **Unpredictable Black Swans:** While useful for managing uncertainties, scenario planning may miss low-probability, high-impact events.

Modern Relevance

Scenario planning has gained renewed importance in the face of global challenges such as climate change, geopolitical instability, and technological disruption. Organizations increasingly use this tool to anticipate risks and opportunities in volatile, uncertain, complex, and ambiguous (VUCA) environments. Additionally, advancements in computational modeling and AI have enhanced scenario planning's precision and scalability.

Lessons for Today's Leaders

Scenario planning emphasizes adaptability over certainty. In an era where traditional linear forecasting often fails, the ability to think in terms of multiple plausible futures equips leaders to navigate complexity effectively. By embracing uncertainty as an opportunity rather than a threat, organizations can build resilience and maintain strategic agility.

Citations

1 P. Wack, "Scenarios: Uncharted waters ahead," *Harv. Bus. Rev.*, vol. 63, no. 5, pp. 73–89, 1985.

2 P. J. H. Schoemaker, "*Scenario planning: A tool for strategic thinking*," *Sloan Manag. Rev.*, vol. 36, no. 2, pp. 25–40, Winter 1995.

3 A. Wilkinson and R. Kupers, "Living in the futures," *Harv. Bus. Rev.*, vol. 91, no. 5, pp. 118–127, 2013.

4 P. Schwartz, *The Art of the Long View: Planning for the Future in an Uncertain World.* Currency Doubleday, 1991.

Balanced Scorecard: Aligning Strategy with Execution

Origins and Historical Context of the Balanced Scorecard

The Balanced Scorecard (BSC) was introduced in the early 1990s by Dr. Robert Kaplan and Dr. David Norton as a performance measurement framework. Dissatisfied with the traditional reliance on financial metrics to gauge organizational success, Kaplan and Norton sought a more holistic approach. Their groundbreaking paper, "The BSC – Measures That Drive Performance," published in *Harvard Business Review* (1992), proposed a system that integrated nonfinancial metrics into strategic management.

The initial motivation behind the BSC was simple yet transformative: to provide organizations with a comprehensive view of their performance by balancing financial outcomes with nonfinancial drivers. Over time, it evolved into a strategic management tool, enabling organizations to align their business activities with vision and strategy, improve internal and external communication, and monitor organizational performance against strategic goals.

Framework and Methodology

The BSC framework emphasizes four key perspectives, though organizations can adapt or expand these to suit their needs:

1) **Financial Perspective:**
   - Focuses on financial performance indicators, such as profitability, revenue growth, cost management, and shareholder value.
   - **Key Questions:** How do we look to shareholders? Are we achieving financial success?
2) **Customer Perspective:**
   - Measures customer satisfaction, retention, and market share.
   - **Key Questions:** How do customers perceive us? Are we meeting their needs?
3) **Internal Processes Perspective:**
   - Assesses operational efficiency, innovation, and the quality of internal business processes.
   - **Key Questions:** What must we excel at internally to deliver value to customers and shareholders?
4) **Learning and Growth Perspective:**
   - Focuses on organizational capacity, employee training, corporate culture, and innovation.
   - **Key Questions:** How can we continue to improve and create value?

Kaplan and Norton stressed that these perspectives should be interconnected and aligned with the organization's overarching strategy. This alignment is achieved through strategic objectives, performance measures, targets, and initiatives.

Example of a Balanced Scorecard Objective: For an EV company like Tesla:

- **Financial:** Increase quarterly revenues by 20%.
- **Customer:** Improve customer satisfaction scores by 15% through enhanced service.
- **Internal Processes:** Optimize production lines to reduce assembly time by 10%.
- **Learning and Growth:** Train 100% of employees on new battery technologies.

Implementation Steps

1) **Clarify Vision and Strategy:**
   - Translate high-level strategic goals into actionable objectives across the four perspectives.
2) **Identify Key Metrics:**
   - Choose specific, measurable indicators that track progress toward objectives.
3) **Set Targets:**
   - Define realistic yet challenging goals for each metric.
4) **Develop Initiatives:**
   - Outline projects and activities needed to achieve targets.
5) **Monitor and Review:**
   - Regularly track performance, analyze gaps, and adjust strategies as needed.

Applications in Practice

The BSC is widely used across industries and sectors:

1) **Corporate Strategy:**
   - Companies like IBM have used the BSC to transition from hardware-focused strategies to service and software-based models. By linking performance measures to strategic goals, IBM realigned its workforce and resources to achieve sustainable growth.

2) **Healthcare:**
   - The Cleveland Clinic implemented the BSC to enhance patient outcomes and streamline operations. The clinic became a global leader in healthcare delivery by focusing on patient satisfaction and internal efficiency.
3) **Government and Nonprofits:**
   - Municipal governments, such as the City of Charlotte, North Carolina, have used the BSC to improve transparency and align services with community needs. Metrics tracked include citizen satisfaction and resource allocation efficiency.

Strengths and Limitations

Strengths:
- **Strategic Alignment:** Ensures that day-to-day activities align with long-term organizational goals.
- **Holistic Perspective:** Provides a multidimensional view of performance, balancing short-term financial metrics with long-term value creation.
- **Flexibility:** Can be tailored to fit organizations of any size, type, or industry.

Limitations:
- **Complexity:** Designing and implementing a BSC requires significant effort and expertise.
- **Overemphasis on Metrics:** Poorly chosen metrics or excessive focus on data collection can lead to misaligned priorities.
- **Cultural Resistance:** Employees may resist adoption if the system is perceived as burdensome or overly rigid.

Modern Relevance

The BSC remains a critical tool for modern organizations, especially in digital transformation and sustainability. Many organizations now incorporate additional perspectives, such as environmental impact or social responsibility, reflecting the growing importance of ESG metrics.

For instance, companies like Unilever have adapted the BSC to track progress toward their sustainability goals, integrating measures like carbon footprint reduction and supply chain transparency alongside traditional financial metrics.

Lessons for Leaders

The BSC demonstrates that what gets measured gets managed. Leaders can foster a culture of accountability, innovation, and continuous improvement by focusing on multiple performance dimensions. It encourages organizations to think beyond quarterly profits and invest in the long-term drivers of success, such as employee development and customer loyalty.

Citations

1 R. S. Kaplan and D. P. Norton, "The balanced scorecard – measures that drive performance," *Harv. Bus. Rev.*, vol. 70, no. 1, pp. 71–79, 1992.

2 R. S. Kaplan and D. P. Norton, *The Strategy-Focused Organization: How Balanced Scorecard Companies Thrive in the New Business Environment*. Harvard Business Review Press, 2000.

3 P. R. Niven, *Balanced Scorecard Step-by-Step: Maximizing Performance and Maintaining Results.* John Wiley & Sons, 2002.
4 Case Study: "IBM's Balanced Scorecard Transition." *MIT Sloan Management Review*, 2006.
5 "Balanced scorecard in healthcare: The Cleveland clinic's approach." *Healthc. Q.*, vol. 15, no. 4, 2012.

Design Thinking

Introduction: The Origins and Philosophy of Design Thinking

Design Thinking, a problem-solving approach grounded in empathy and creativity, has its roots in the work of design practitioners like Herbert Simon, who outlined the concept in *The Sciences of the Artificial* (1969). The methodology gained broader recognition through Stanford University's Hasso Plattner Institute of Design (known as the d.school) and the global design consultancy IDEO, spearheaded by Tim Brown. Design Thinking's fundamental premise is that human-centric solutions are more innovative, effective, and impactful when they address users' real needs and experiences.

While originally employed in product and industrial design, Design Thinking has evolved into a universal methodology applicable across industries, from technology to healthcare, education, and government services. Its iterative process emphasizes understanding user needs, generating creative solutions, prototyping, and testing – ensuring that innovation is rooted in practicality and usability.

The Five Stages of Design Thinking

1) **Empathize:**
   - The foundation of Design Thinking is empathy. Practitioners immerse themselves in the user's context through interviews, observation, and surveys to deeply understand their experiences, challenges, and needs.
   - **Tools:** Journey maps, ethnographic studies, and user personas.
2) **Define:**
   - Insights gathered in the empathize phase are synthesized to articulate the core problem or opportunity. This step involves framing a problem statement that guides the ideation process.
   - **Tools:** Problem statements, *How Might We* questions, and affinity diagrams.
3) **Ideate:**
   - Teams brainstorm diverse ideas and solutions without judgment, focusing on creativity and out-of-the-box thinking. This phase encourages expansive ideation followed by narrowing down feasible concepts.
   - **Tools:** Brainstorming, mind mapping, and SCAMPER techniques.
4) **Prototype:**
   - Selected ideas are translated into tangible prototypes – mock-ups, models, or interactive simulations. The goal is to visualize solutions quickly and cost-effectively.
   - **Tools:** Wireframes, role-playing, and 3D models.
5) **Test:**
   - Prototypes are tested with real users to gather feedback, uncover flaws, and refine the solution iteratively. This phase often loops back to earlier stages, fostering continuous improvement.
   - **Tools:** Usability testing, A/B testing, and iterative development.

Applications Across Industries

Design Thinking has transcended its origins in design to influence a wide array of fields:

1) **Healthcare:**
   - The Cleveland Clinic uses Design Thinking to enhance patient experiences, redesigning everything from waiting rooms to surgical procedures based on patient and caregiver input.
2) **Education:**
   - The K12 Lab at Stanford's d.school employs Design Thinking to empower educators to develop creative curriculums tailored to diverse student needs.
3) **Business Innovation:**
   - IBM has integrated Design Thinking into its product development processes, resulting in a reported 300% return on investment by improving product usability and customer satisfaction.
4) **Government Services:**
   - The United States Digital Service applies Design Thinking to simplify citizen interactions with government systems, such as improving healthcare.gov's interface.

Case Study: Airbnb's Transformation Through Design Thinking

Airbnb's adoption of Design Thinking in 2009 marked a turning point in its growth trajectory. Facing stagnant user growth, cofounders Joe Gebbia and Brian Chesky shifted their focus from coding features to understanding user pain points. Through interviews and on-the-ground research, they discovered that low-quality photos of rental listings deterred bookings.

Using this insight, Airbnb developed a simple yet transformative solution: sending professional photographers to hosts' homes to capture appealing images. This empathic approach increased bookings and highlighted the power of customer-centric innovation. Airbnb's valuation skyrocketed from $10 million to over $30 billion within a decade, cementing its place as a disruptor in the hospitality industry.

Advantages of Design Thinking

1) **Empathy-Driven Innovation:**
   - By prioritizing user needs, Design Thinking generates solutions that are not only innovative but also practical and relevant.
2) **Iterative and Flexible:**
   - Its nonlinear process encourages adaptability, enabling teams to refine solutions dynamically.
3) **Collaboration-Oriented:**
   - Cross-functional collaboration fosters diverse perspectives and holistic problem-solving.
4) **Scalable:**
   - Design Thinking can be applied to challenges of any scale, from designing a product to overhauling entire systems.

Challenges and Limitations

1) **Time-Intensive:**
   - The iterative nature of Design Thinking can extend timelines, particularly when extensive testing and prototyping are required.

2) **Overreliance on Creativity:**
   - Without structured guidance, teams may struggle to balance creativity with practicality.
3) **Bias in User Insights:**
   - The empathize phase relies heavily on qualitative data, which can introduce biases if not rigorously analyzed.

Emerging Trends in Design Thinking

1) **Digital Transformation:**
   - Tools like Figma and Miro enhance the prototyping and ideation stages, enabling remote collaboration in design teams.
2) **AI Integration:**
   - AI-driven tools now assist in generating user personas, simulating prototypes, and predicting user behavior.
3) **Systems Thinking Fusion:**
   - Combining Design Thinking with systems thinking provides a broader perspective, addressing interconnected challenges in urban planning and climate adaptation.

Relevance in Today's Context

As global challenges become more complex, the need for human-centric innovation has never been greater. Design Thinking's focus on empathy, creativity, and iterative improvement equips leaders and practitioners with a robust toolkit for addressing problems in an increasingly interconnected world. Whether improving customer experiences, designing sustainable systems, or fostering social change, Design Thinking remains a vital methodology.

Citations in IEEE Format

1 T. Brown, *Change by Design: How Design Thinking Creates New Alternatives for Business and Society*. Harper Business, 2009.
2 H. A. Simon, *The Sciences of the Artificial*, 3rd ed. MIT Press, 1996.
3 R. Martin, *The Design of Business: Why Design Thinking is the Next Competitive Advantage*. Harvard Business Press, 2009.
4 J. Kolko, "Design thinking comes of age," *Harv. Bus. Rev.*, vol. 93, no. 9, pp. 66–71, 2015.
5 Dam, R. F., and Siang, T. Y., "What is Design Thinking and Why is it so Popular?" *Interaction Design Foundation*, 2020.

Innovation Diffusion Models

Introduction and Background

The concept of innovation diffusion, which focuses on understanding how new ideas, products, or technologies spread through a population, has been central to fields as diverse as marketing, technology adoption, and social change. First developed by Everett M. Rogers in his seminal work *Diffusion of Innovations* (1962), the framework was influenced by earlier studies in sociology, including

Gabriel Tarde's laws of imitation and Bryce Ryan and Neal Gross's research on hybrid corn adoption in rural Iowa. Rogers consolidated these ideas into a comprehensive model that was applied across industries and disciplines.

At its core, the innovation diffusion model examines the factors influencing the adoption of new ideas or products over time. It identifies key players in the adoption process, stages of adoption, and contextual factors that either accelerate or hinder diffusion. Understanding this process is crucial for businesses, policymakers, and researchers aiming to introduce transformative innovations or achieve widespread behavioral change.

Key Components of the Model

1) **Adopter Categories**
   Rogers classified adopters into five groups based on their willingness to adopt innovations:
   - **Innovators (2.5%):** Risk-takers and the first to adopt new ideas, often motivated by curiosity.
   - **Early Adopters (13.5%):** Visionary leaders who recognize the potential of innovations and influence others.
   - **Early Majority (34%):** Pragmatic users who adopt once they see proven benefits.
   - **Late Majority (34%):** Skeptical individuals who require significant social proof.
   - **Laggards (16%):** Traditionalists who resist change and adopt only when necessary.
2) **The S-Curve of Adoption**
   Adoption rates typically follow an S-curve pattern:
   - **Introduction Phase:** Slow adoption by innovators and early adopters.
   - **Growth Phase:** Rapid adoption as the early and late majority embrace the innovation.
   - **Maturity Phase:** Adoption slows as the market becomes saturated.
3) **The Innovation-Decision Process**
   This process outlines the stages individuals go through before adopting an innovation:
   - **Knowledge:** Awareness of the innovation.
   - **Persuasion:** Forming an attitude toward the innovation.
   - **Decision:** Choosing to adopt or reject the innovation.
   - **Implementation:** Using the innovation.
   - **Confirmation:** Reinforcing the decision to adopt.
4) **Attributes of Innovations**
   Certain characteristics of an innovation influence its adoption rate:
   - **Relative Advantage:** The degree to which the innovation is perceived as better than existing solutions.
   - **Compatibility:** Align with potential adopters' values, experiences, and needs.
   - **Complexity:** Ease of understanding and use.
   - **Trialability:** The ability to test the innovation before full adoption.
   - **Observability:** The visibility of its benefits to others.

Applications and Case Studies

1) **Technology Adoption in Consumer Markets**
   - The diffusion of smartphones provides a textbook example of Rogers's model. Apple's iPhone attracted innovators and early adopters in 2007 due to its groundbreaking interface and design. Over time, as competitors entered the market and prices fell, adoption spread to the early and late majority, culminating in near-universal smartphone use today.

2) **Public Health Interventions**
   - In the 1970s, the World Health Organization (WHO) used the innovation diffusion framework to promote the adoption of oral rehydration therapy (ORT) in developing countries. By engaging early adopters in rural communities and demonstrating its effectiveness, the campaign gradually convinced skeptical populations, leading to widespread use and significant reductions in child mortality rates.
3) **Energy Transitions**
   - The adoption of renewable energy technologies like solar panels follows the innovation diffusion curve. Innovators adopted solar technology decades ago, driven by environmental concerns and government incentives. Early adopters and the early majority followed as costs decreased and efficiency improved, driving the industry toward broader acceptance.

Strengths and Limitations

Strengths

- **Interdisciplinary Applications:** The model is versatile and applicable to diverse domains, such as marketing, education, healthcare, and sustainability.
- **Actionable Insights:** Organizations can effectively tailor strategies to target specific groups by identifying key adopter categories.
- **Predictive Power:** The S-curve and decision-making process provide a framework for forecasting adoption trends.

Limitations

- **Simplification of Human Behavior:** The model assumes rational decision-making, which may not account for emotional, cultural, or systemic barriers.
- **Static Categorization:** Adopter categories can be fluid, as individuals may behave differently depending on the innovation.
- **Focus on Adoption Over Sustained Use:** The model emphasizes initial adoption rather than long-term integration and impact.

Pragmatic Use in Technological Leadership

For leaders in technology and innovation, the diffusion model offers practical tools to strategize rollouts, target influential adopters, and anticipate challenges. Key strategies include:

- **Leveraging Early Adopters:** Engaging this group to demonstrate credibility and build momentum.
- **Reducing Complexity:** Simplifying innovations through user-friendly design and comprehensive training programs.
- **Enhancing Observability:** Showcasing success stories and tangible benefits to convince skeptical populations.

Relevance Today

In today's fast-paced, interconnected world, innovation diffusion occurs faster than ever before. Social media and digital platforms amplify observability and trialability, accelerating adoption cycles. However, Rogers's model's fundamental principles remain relevant, guiding leaders in navigating adoption dynamics across new contexts such as AI, renewable energy, and public health.

Citations

1 E. M. Rogers, *Diffusion of Innovations*. Free Press, 1962.
2 G. Tarde, *The Laws of Imitation*. Henry Holt and Company, 1903.
3 Ryan, B., & Gross, N. C. (1943). "The Diffusion of Hybrid Seed Corn in Two Iowa Communities." *Rural Sociology*.
4 World Health Organization (1978). "Promotion of Oral Rehydration Therapy."
5 G. Parker and M. Van Alstyne, *Platform Revolution*. W.W. Norton & Company, 2016.

A Few Additional Foundational Tools, Concisely

Real-Win-Worth (RWW) Framework

What It Is: A tool for evaluating the prospects of business opportunities or projects from the standpoint of their feasibility.

Purpose: To determine whether a project is doable (Real), can outperform the competitors (Win), and is worthwhile (Worth).

The inputs needed are market research, competitor analysis, and an assessment of the project's potential costs and benefits.

How to Use It:

- **Objective:** Does the project appear doable in the sense that it can be done technically and in practice?
- **Win:** Analyze the competition and the project's position in the market.
- **Worth:** Establish the financial and strategic returns that can be obtained from the project.

Applications: New product development, market entry strategies, project funding, and investment appraisal.

- **Strengths:** Offers a sequential methodology for identifying and selecting projects with a high potential for success.
- **Limitations:** Based on the availability of market information and the organization's internal processes.
- **Case Study:** This is how Procter & Gamble (P&G) uses real-win-worth (RWW) to assess new consumer products, their market feasibility, and profitability.

Stress-Testing and What-If Analysis

What It Is: A simulation approach to determine how systems respond to various stress conditions.

Purpose: To understand and recognize risks and potential challenges that may occur.

Inputs Needed: System models, historical data, and possible stress events.

How to Use It:

- Enumerate the sources of stress, such as economic downturns, supply chain disruptions, or cyberattacks.
- Recreate the system through artificial simulation in response to these stressors.
- Identify measures to manage the risks that have been highlighted.

Applications: Financial systems, critical infrastructure, and corporate risk management.

- **Strengths:** Stresses the system's weaknesses and assists in developing preventive strategies.
- **Limitations:** Works only as well as the scenarios and assumptions incorporated into the model.
- **Case Study:** The Fed exercises what are called stress tests on banks every year to check on the condition of financial systems during economic crises.

Technology Readiness Levels (TRLs)

What It Is: A scale of nine levels developed by NASA to measure the maturity of a technology.

Purpose: To determine the advancement of technologies from initial research (TRL 1) to full implementation (TRL 9).

Inputs Needed: R&D outcomes, prototype testing, and validation results.

How to Use:

Assess the technology and place it on the TRL scale.

Identify gaps and prioritize investment or development actions.

- **Applications:** Innovation management, R&D planning, and defense projects.
- **Strengths:** Provides a clear and systematic framework for assessing technology maturity.
- **Limitations:** Does not consider market adoption or societal impact.
- **Example:** NASA and DARPA apply TRLs to evaluate emerging technologies such as robotics in space exploration and defense.

The Delphi Method

What It Is: A systematic way to build expert consensus through iterative surveys.

Purpose: To predict future developments and support strategic planning.

Inputs Needed: Subject matter experts, structured surveys, and iterative feedback loops.

How to Use:

Select a panel of subject matter experts.

Conduct successive surveys, refining insights based on prior responses.

Compile results to identify trends or priorities.

- **Applications:** Strategic foresight, technology forecasting, and policy analysis.
- **Strengths:** Captures diverse perspectives and eliminates individual biases.
- **Limitations:** Time-consuming and reliant on expert availability.
- **Example:** The RAND Corporation used the Delphi Method to predict advancements in military technologies.

Strategic Foresight and Technology Roadmapping, Just to Get You Started (with Detailed Version in the Next Chapter)

What It Is: The integration of foresight techniques and roadmapping for identifying, evaluating, and planning future technologies.

Purpose: To ensure that technological advancement aligns with strategic planning goals in the long run.

Inputs Needed: Trend evaluation, stakeholders' opinions, resource availability, and future projections.

How to Use It:

- Identify future trends and technological changes through foresight exercises.
- Develop a technological roadmap with a timeline, goals/objectives, and resources needed.
- Incorporate the roadmap into the organizational strategy to guide R&D investments.

Applications: Management of innovation, planning of R&D, and strategies for sustaining national competitiveness.

- **Strengths:** Both a predictive tool and a strategic planning instrument; encourages interdisciplinary work; creates a working vision.
- **Limitations:** Becomes outdated quickly if not regularly updated in rapidly evolving environments.
- **Case Study:** NASA uses roadmapping technology to identify investments in space exploration technologies.

Cost–Benefit and Risk Interdependency Models

What It Is: An approach to understanding interconnected systems' trade-offs, risks, and repercussions.

Purpose: To balance cost, risk, and system robustness.

Inputs Needed: Cost information, risk factors, interconnection structure, and performance indicators.

How to Use It:

- Identify the components of the system and their interrelationships.
- Assess each component's costs, risks, and impacts on other elements.
- Optimize resources and minimize vulnerabilities.

Applications: Infrastructure development, cybersecurity, and energy grid reliability.

- **Strengths:** Depicts risk propagation and interconnectedness, identifying areas for resource investment.
- **Limitations:** Requires high-level modeling capabilities and quality data.
- **Case Study:** This model was applied in the United Kingdom to determine investments in SMRs to improve grid reliability while mitigating associated risks.

Porter's Value Chain Analysis for Strategic Innovation

What It Is: A framework that helps identify activities that create or do not create value for an organization.

Purpose: To identify where to innovate and where to gain a competitive edge in the value chain.

Inputs Needed: Organizational activity details, costs, structures, and market knowledge.

How to Use It:

- Identify the primary and support activities in the value chain.
- Evaluate each activity for its value creation and competitive advantage.
- Identify where to innovate, reduce costs, or improve quality.

Applications: Business model design, operational redesign, and competitive strategy.

- **Strengths:** Identifies innovation opportunities within the organization.
- **Limitations:** Does not consider external market or legal factors.

- **Case Study:** Amazon's supply chain management uses VCA to enhance customer satisfaction and reduce costs.

Business Model Stress Testing and Simulation (Detailed Version Upcoming in the Next Chapter on Advanced Tools)

What It Is: Business model stress testing and simulation is a structured method for evaluating the flexibility and durability of a company's business model under various challenging scenarios. It identifies vulnerabilities, validates strengths, and ensures the business can adapt to unforeseen economic, operational, or technological changes. This approach allows leaders to forecast risks, test responses, and refine strategies in controlled environments before real-world disruptions occur.

Why It Matters: Markets are unpredictable, and businesses operate in environments where shocks – such as economic downturns, supply chain disruptions, or technological shifts – can expose weaknesses. Stress testing enables leaders to anticipate problems, build resilience, and ensure their operations remain viable under pressure. For instance, during the 2008 financial crisis, companies that had tested their models for downturn scenarios were better equipped to weather economic uncertainty, while others faltered.

How to Use It

1) **Define the Business Model:** Start by mapping out the core components – value proposition, customer segments, revenue streams, key partnerships, and cost structures – using tools like the *BMC*.
2) **Identify Stress Scenarios:** Develop hypothetical but plausible stress conditions. Examples include:
   a) A 30% drop in market demand due to economic recession.
   b) Disruption in a key supply chain component for 60 days.
   c) Regulatory changes increase operational costs by 25%.
3) **Simulate Outcomes:** Use models or software to simulate how the business would perform under these scenarios. Quantify the impacts on cash flow, profit margins, and customer retention.
4) **Analyze Weak Points:** Identify where the business model cracks – cost overruns, inability to scale, or dependency on a single market or supplier.
5) **Refine for Resilience:** Develop mitigation strategies, such as diversifying revenue streams, strengthening supplier networks, or improving cash reserves.

Applications

- **Market Entry Strategies:** Companies stress-test new business models to anticipate challenges in unfamiliar markets. For instance, an energy startup entering a developing region may simulate currency devaluations or infrastructure limitations to adapt pricing and distribution plans.
- **Mergers and Acquisitions:** Before merging, firms simulate how the combined model would respond to challenges like increased competition, cultural misalignment, or cost integration.

Case Example: Netflix is a prime example of business model adaptability through stress testing. Originally a DVD rental service, Netflix simulated scenarios of shrinking DVD demand as broadband internet improved. Their simulations indicated streaming services would dominate the market, prompting Netflix to shift its model entirely. By preparing for the decline of its existing model, Netflix avoided stagnation and became a global streaming leader.

Strengths and Limitations

- **Strengths:** Stress testing offers a forward-looking assessment that enhances preparedness and agility. By simulating multiple paths, businesses can proactively address risks.
- **Limitations:** The accuracy of the outcomes relies heavily on realistic assumptions and high-quality data. Models that exclude unforeseen risks may provide a false sense of security.

References:

- Osterwalder, A., *Business Model Generation: A Handbook for Visionaries, Game Changers, and Challengers,* Wiley, 2010.
- McKinsey & Company, "Stress testing business models for resilience," 2022.

Summary Table: Foundational Tools for Strategic Leadership.

Tool	Purpose	Strengths	Limitations	Applications
Golden Circle	Align purpose, processes, and outcomes	Promotes clarity and alignment	Vague without operational plans	Leadership, branding, team alignment
Heilmeier's Catechism	Evaluate R&D projects	Clear, simple, and flexible	Lacks market and social impact analysis	R&D planning, technology roadmaps
Quad Chart	Summarize project data	Easy to understand and communicate	Oversimplifies complex issues	Proposals, briefings, decision-making
Porter's Five Forces	Analyze the competitive environment	Systematic and structured analysis	Static does not capture market dynamics	Market entry, competitive strategy
Technology Readiness Levels (TRLs)	Assess technology maturity	A clear framework for development stages	Ignores market adoption and social impact	R&D planning, defense, innovation management
Business Model Canvas	Plan business models and operations	Simple, versatile, widely applicable	May lack detailed financial insights	Business strategy, startups, project analysis
Delphi Method	Build consensus for predictions	Reduces individual biases, diverse insights	Time-consuming, requires expert availability	Strategic foresight, policy analysis

References

S. Sinek, *Start with Why: How Great Leaders Inspire Everyone to Take Action.* Penguin, 2009.

G. Heilmeier, *DARPA Guidelines for Project Planning and Assessment.* DARPA, 1975.

Department of Defense (DoD), *Quad Chart Guidelines,* 2015.

M. Porter, *Competitive Strategy: Techniques for Analyzing Industries and Competitors.* Free Press, 1980.

M. Porter, *Competitive Advantage: Creating and Sustaining Superior Performance.* Free Press, 1985.

NASA, *Technology Readiness Guidelines.* NASA, 1992.

A. Osterwalder, *Business Model Generation: A Handbook for Visionaries, Game Changers, and Challengers.* Wiley, 2010.

RAND Corporation, *Delphi-Based Decision-Making Frameworks.* RAND, 2021.

Further Concepts: Strategic Leadership and Decision-Making Instruments

Introduction

This chapter builds on the essential tools by presenting the reader with advanced methodologies that help the user deal with the issues in interconnected systems, large-scale transitions, and decisions in unpredictable environments. These tools help leaders forecast, respond, and innovate effectively using systems thinking and advanced modeling approaches. While the tools in Appendix A are foundational, the tools in this chapter are dynamic and interactive and require sophisticated application and practice. In this chapter, we discuss a bit more **advanced methods, mostly systems:**

1) **Integrated Resource Planning (IRP):** Balances supply-side and demand-side resources for efficient energy planning. *Reference*: U.S. Energy Policy Act, 1992.
2) **Design Thinking: A Human-Centered Approach to Innovation**

Sustainability, Security, and Resilience

3) **Lifecycle Analysis:** Evaluate environmental impacts across a product's lifecycle. *Reference*: ISO 14040, *Environmental Management – Life Cycle Assessment Principles and Framework*, 2006.
4) **Resilience Metrics and Analysis:** Quantifies system robustness and recovery capabilities. *Reference*: Holling, C. S., *Annual Review of Ecology and Systematics*, 1973.

Systems and Engineering Approaches

5) **Systems Engineering:** Designs and manages complex systems with lifecycle integration. *Reference*: NASA and DoD, *Systems Engineering Handbook*, 1990s.
6) **Engineering Systems Framework:** Examines systems' technical, social, and economic dimensions. *Reference*: MIT Engineering Systems Division, 1990s.
7) **Causal Loop Diagramming for Complex Systems:** Visualizes reinforcing and balancing feedback loops to clarify system dynamics. *Reference*: Peter Senge, *The Fifth Discipline*, 1990.
8) **Total Cost of Ownership (TCO) Analysis:** This evaluates systems' initial and ongoing costs to determine their long-term value. *Reference*: Andrew Ng, foundational work on lifecycle costing, 1980s.

Decision-Making Tools

1) **Strategic Decision-Making Under Uncertainty:** Combines probabilistic models and risk assessments for unpredictable conditions. *Reference*: Avi Gilboa and Itzhak Gilboa, *Making Better Decisions*, 2009.
2) **Monte Carlo Simulations for Risk and Sensitivity Analysis:** Models system uncertainty and variability through random sampling. *Reference*: Stanislaw Ulam, *Monte Carlo Method*, 1940s.
3) **Dynamic Systems Modeling and Analysis:** Simulates interactions within systems to predict outcomes and test interventions. *Reference*: Jay W. Forrester, *Industrial Dynamics*, 1961.
4) **Global Transition Dynamics (GTD):** Leverages AI and real-time data to address interconnected global challenges. *Reference*: C. Starr and M. Amin, 2003, 2023, and 2024.

Resilience Metrics, Risk Assessment, and Advanced Tools

Introduction

The modern landscape of resilience and risk management requires tools that adapt to growing complexities and uncertainties across industries. In addition to traditional approaches like resilience metrics and risk assessment, advanced methodologies like Criticality, Accessibility, Recuperability, Vulnerability, Effect, and Recognizability (CARVER), Delphi-Supporting Hypothesis, and Risk Prioritization (D-SHARP), and others provide precision, depth, and adaptability. These tools are instrumental in tackling challenges across the defense, energy, healthcare, and technology sectors.

This expanded narrative delves deeper into the principles, methodologies, and applications of resilience metrics, risk assessment, CARVER, and D-SHARP, underscoring their critical relevance in today's risk-prone environment.

Resilience Metrics and Analysis

Resilience metrics quantify a system's ability to anticipate, withstand, and recover from disruptions. Grounded in ecological research by C.S. Holling, this framework has evolved to incorporate engineering, organizational, and infrastructure systems. It now includes technical measures and social dimensions, emphasizing holistic resilience planning.

Applications and Advanced Metrics

1) **Energy Systems:**
   - Enhanced metrics for microgrids combine system redundancy with renewable energy adoption to ensure energy reliability during disruptions.
   - **Case Study:** Hurricane Harvey (2017) demonstrated how Texas energy grids, bolstered by resilience planning, ensured power continuity in critical regions.
2) **Healthcare Resilience:**
   - Metrics evaluate surge capacity, supply chain redundancy, and adaptability during crises like COVID-19, reducing systemic failures in patient care.
3) **Smart Infrastructure:**
   - IoT-enabled resilience metrics allow predictive monitoring of bridges, dams, and urban systems, identifying potential disruptions in real time.

Risk Assessment and Management (RAM)

RAM remains foundational to strategic planning, offering a structured approach to identifying, analyzing, and mitigating risks. Modern advancements incorporate AI/machine learning (ML) to refine risk detection and prediction, especially in dynamic environments.

Applications of Advanced RAM

1) **Finance:** Value-at-risk (VaR) models integrate market simulations, predicting the impacts of global fluctuations on asset portfolios.
2) **Aerospace:** Real-time risk analytics in space missions factor in equipment degradation, environmental variables, and human behavior to preempt mission failure.
3) **Defense:** Geopolitical risk assessments leverage big data and sentiment analysis to forecast instability and guide strategic operations.

CARVER: A Tactical Approach to Risk Assessment

Origins and Purpose

CARVER was developed by the U.S. military as a decision-support tool to evaluate and prioritize targets. Initially used in warfare, its adaptability makes it applicable across critical infrastructure, cybersecurity, and supply chain management.

Core Components

1) **Criticality:** How vital is the system to mission success or organizational continuity?
2) **Accessibility:** How easily can the system be compromised?
3) **Recuperability:** How quickly can the system recover after a disruption?
4) **Vulnerability:** What is the likelihood of the system being attacked or failing?
5) **Effect:** What are the consequences of disruption, ranging from operational delays to societal impacts?
6) **Recognizability:** How easily can an adversary identify the system's vulnerabilities?

Applications

1) **Cybersecurity:** CARVER evaluates risks to IT systems, identifying weak access points and prioritizing defensive measures.
2) **Critical Infrastructure:** Water treatment facilities and power plants use CARVER to assess physical and cyber risks, enabling layered defenses.
3) **Supply Chain Resilience:** CARVER identifies chokepoints in global logistics, ensuring smoother operations in crises.

Case Study: The Department of Homeland Security used CARVER during Hurricane Katrina to evaluate which infrastructure systems (levees, hospitals) needed immediate attention and resource allocation, saving lives and restoring functionality faster.

D-SHARP: A Precision Tool for Decision-Making

Overview

D-SHARP integrates Delphi method insights with probabilistic risk assessment (PRA) to support high-stakes decision-making. Its unique focus on hypothesis-driven risk evaluation makes it invaluable for the energy, aerospace, and intelligence sectors.

Core Methodology

1) **Delphi Insights:** Experts provide iterative feedback, refining risk hypotheses based on collective expertise.
2) **Risk Modeling:** Bayesian and Monte Carlo simulations quantify risks, identifying the most probable and impactful outcomes.
3) **Prioritization:** Ranked risks enable decision-makers to focus resources effectively, optimizing mitigation strategies.

Applications

1) **Energy Transition:** Evaluating risks of adopting renewable technologies, including intermittency and geopolitical supply dependencies.
2) **Aerospace:** Assessing risks of equipment failure and launch delays in next-gen rocket systems.
3) **National Security:** Predicting risks of cyberterrorism by integrating threat intelligence and expert consensus.

Case Study: NASA employed D-SHARP to evaluate the Artemis I lunar mission risks, prioritizing redundant system designs to mitigate launch delays and ensure mission success.

Integration with Resilience and Risk Management

1) **Lifecycle Risk Monitoring**
   Modern organizations integrate resilience metrics and advanced tools like CARVER and D-SHARP into continuous lifecycle assessments. These processes address evolving risks at every stage, from design to decommissioning.
2) **Emerging Technologies**
   AI/ML models now analyze resilience and risk data in real time, automating decision-making processes for rapid and accurate responses.

Strengths and Limitations

Strengths

- **Comprehensive Analysis:** Tools like CARVER and D-SHARP complement traditional methods, providing multifaceted risk evaluations.
- **Scalability:** Applicable across industries, from tactical military planning to global energy transitions.
- **Precision:** Advanced probabilistic models improve the accuracy of risk prioritization and mitigation strategies.

Limitations

- **Data Dependency:** High-quality data is critical; incomplete datasets can compromise outcomes.
- **Complexity:** The technical sophistication of these tools demands expert interpretation.

Lessons Learned

1) **Proactive Planning:** Resilience metrics combined with RAM tools like CARVER enable organizations to effectively prepare for known and unforeseen risks.
2) **Strategic Prioritization:** Tools like D-SHARP focus attention and resources on the most impactful risks, avoiding overextension.
3) **Systemic Integration:** Integrating resilience and risk tools into organizational frameworks ensures a culture of continuous improvement.

Conclusion

As systems grow increasingly interconnected and complex, traditional risk assessments must evolve. Advanced tools like CARVER and D-SHARP provide precision, flexibility, and actionable insights that modern industries and governments require. By integrating these methods with resilience frameworks, organizations can safeguard against disruptions and thrive in an era defined by uncertainty.

Citations

1 Holling, C.S., "Resilience and Stability of Ecological Systems," *Annual Review of Ecology and Systematics*, 1973.
2 Department of Homeland Security, "Risk Analysis for Critical Infrastructure: Lessons from Hurricane Katrina," *DHS Infrastructure Reports*, 2006.
3 NASA, "Artemis I Mission Risk Assessments," *NASA Technical Briefings*, 2022.
4 U.S. Department of Defense, "CARVER Methodology for Risk Prioritization," *DOD Strategy Reports*, 2018.
5 National Institute of Standards and Technology (NIST), *Cybersecurity Resilience Metrics*. NIST Publications, 2021.

Resilience Metrics and Analysis: Measuring a System's Ability to Recover from Disruptions

Introduction

Resilience metrics have become indispensable in systems engineering, organizational management, and infrastructure development. These metrics provide a quantitative and qualitative understanding of a system's ability to withstand disruptions, adapt to changing conditions, and recover functionality in adversity. Rooted in the ecological theories of C.S. Holling, resilience analysis has expanded to address critical challenges in energy systems, healthcare, transportation, and cybersecurity. In today's complex and interconnected world, resilience metrics provide a roadmap for designing robust systems and minimizing downtime.

Background and Development

The concept of resilience first emerged in ecological studies in the 1970s. C.S. Holling introduced it as a measure of an ecosystem's capacity to absorb disturbances and reorganize without losing its core functions. Over time, these principles were adapted for engineering and organizational applications, emphasizing system redundancy, robustness, and recovery capabilities. Institutions like the Department of Homeland Security and the National Institute of Standards and Technology (NIST) have refined resilience metrics to ensure their applicability in critical infrastructure and industrial contexts.

Core Components of Resilience Metrics

1) **Anticipation:** The ability to predict and prepare for disruptions. Advanced weather modeling, for example, helps utility companies anticipate extreme weather events.
2) **Absorption:** The capacity to absorb shocks without significant performance degradation. For instance, redundant systems in power grids ensure continued functionality during localized outages.
3) **Adaptation:** The ability to adjust and reconfigure operations during disruptions. Healthcare systems, for instance, adapted by expanding ICU capacity during the COVID-19 pandemic.

4) **Recovery** is the speed and efficiency with which a system regains its original functionality. Rapid restoration of telecommunications after natural disasters is a hallmark of resilient networks.

Applications Across Industries

- **Energy Systems:** The Texas energy grid, impacted by extreme weather in 2021, highlighted the importance of incorporating resilience metrics into grid planning. Microgrids with renewable energy sources and storage capabilities have demonstrated superior resilience in similar contexts.
- **Healthcare:** Hospitals now use resilience metrics to evaluate their preparedness for crises, such as pandemics or cyberattacks on electronic medical records.
- **Urban Planning:** Cities like Rotterdam have adopted resilience strategies to protect against rising sea levels, designing flood barriers and adaptive infrastructure.

Case Study: The Rockefeller Foundation's 100 Resilient Cities initiative has supported municipalities worldwide in implementing resilience frameworks that address physical and social vulnerabilities.

Strengths and Limitations

- **Strengths:**
  - Provides a structured approach to identifying vulnerabilities.
  - Facilitates cross-sector collaboration by establishing common resilience goals.
  - Incorporates adaptive strategies to handle future uncertainties.
- **Limitations:**
  - Requires extensive data collection, which may not always be available.
  - Complexity in implementation due to the interdisciplinary nature of resilience planning.

Lessons Learned:

The integration of resilience metrics ensures that systems are robust, adaptable, and efficient in responding to disruptions. Key takeaways include the importance of stakeholder engagement, real-time monitoring, and investment in redundancy.

Risk Assessment and Management: Identifying, Analyzing, and Mitigating Risks in Systems and Strategies

Introduction: RAM is the cornerstone of industry decision-making. RAM minimizes vulnerabilities and ensures long-term operational success by systematically identifying potential risks, analyzing their likelihood and impact, and developing mitigation strategies. In an era of uncertainty, from geopolitical tensions to climate change, managing risk is a critical skill for leaders in technological and organizational domains.

Historical Development: The practice of risk management dates back to the 19th century, and its roots are in actuarial science and financial analysis. Over time, its principles were adopted by fields like engineering and public health, especially as large-scale infrastructure projects and public safety initiatives demanded structured frameworks. RAM methodologies were formalized in the 20th century, with tools like Failure Mode and Effects Analysis (FMEA) and Probabilistic Risk Assessment (PRA) gaining prominence.

Core Steps in Risk Assessment

1) **Risk Identification:** Cataloging potential risks, such as technical failures, natural disasters, or cyberattacks.
2) **Risk Analysis:** Quantifying each identified risk's probability and potential impact.
3) **Risk Evaluation:** Prioritizing risks based on their significance and the resources required for mitigation.
4) **Risk Treatment:** Implementing strategies to avoid, reduce, or transfer risks.
5) **Monitoring and Review:** Continuously assessing the effectiveness of risk management plans.

Applications Across Sectors

- **Defense:** Risk assessments guide strategic decisions in combat operations, infrastructure protection, and cybersecurity. The U.S. DoD's CARVER tool (detailed earlier) is a key example of prioritizing risks in high-stakes scenarios.
- **Energy:** Risk assessments for nuclear power plants focus on identifying operational failures, external threats, and potential environmental impacts.
- **Healthcare:** Hospitals use RAM to prepare for pandemics, ensuring adequate medical supplies, staff readiness, and cybersecurity for patient records.

Advanced Tools and Techniques

- **Monte Carlo Simulations:** These probabilistic models provide insights into risk variability, helping organizations plan for worst-case scenarios.
- **D-SHARP Methodology:** Combines expert-driven insights with advanced computational models to prioritize risks and develop mitigation strategies.
- **Artificial Intelligence:** ML algorithms now predict risks in supply chains, natural disasters, and market fluctuations, enabling proactive decision-making.

Case Study: After the 2011 Fukushima Daiichi nuclear disaster, Japan overhauled its risk assessment protocols for nuclear facilities. Probabilistic models now evaluate risks from tsunamis, earthquakes, and operational failures, ensuring improved safety standards.

Strengths and Limitations

- **Strengths:**
  - Provides actionable insights for resource allocation and mitigation planning.
  - Adapts to dynamic environments with ongoing risk monitoring.
  - Encourages collaboration across stakeholders to address systemic vulnerabilities.
- **Limitations:**
  - Heavily dependent on the accuracy of input data.
  - Can become overly complex, delaying decision-making processes.

Lessons Learned

Effective risk management requires balancing quantitative and qualitative approaches, fostering a culture of vigilance, and leveraging cutting-edge technologies for real-time analysis. Integrating RAM into strategic decision-making ensures that organizations remain resilient in an uncertain world.

Conclusion: Resilience metrics and risk assessment are deeply interconnected tools that provide comprehensive frameworks for addressing vulnerabilities in systems and strategies. From measuring recovery capabilities to prioritizing risk mitigation, these methodologies are indispensable for leaders in both the public and private sectors. By integrating advanced tools like CARVER and Monte Carlo simulations, organizations can enhance their ability to anticipate disruptions, adapt to challenges, and thrive in a rapidly changing world.

Citations

1 C. S. Holling, "Resilience and stability of ecological systems," *Annu. Rev. Ecol. Syst.*, 1973.
2 Department of Homeland Security, "Resilience Metrics for Critical Infrastructure," *DHS Reports*, 2020.
3 Fukushima Nuclear Risk Assessment Overhaul, *International Atomic Energy Agency Reports*, 2012.
4 Monte Carlo simulation applications, *Risk Manag. J.*, 2021.
5 National Institute of Standards and Technology (NIST), *Advanced Risk Management Practices.* NIST Publications, 2022.

Strategic Decision-Making Under Uncertainty: Navigating Unpredictable Environments

Introduction

Strategic decision-making under uncertainty is a critical discipline in leadership and management, especially in today's VUCA world. Leaders and organizations are frequently required to make high-stakes decisions without having complete information or clear visibility into the future. This method involves frameworks, tools, and approaches designed to mitigate risks, leverage opportunities, and make informed decisions even when outcomes are unpredictable.

Historical Context

The need for structured decision-making under uncertainty became prominent during World War II when military leaders had to make quick strategic choices based on incomplete intelligence. Postwar, the approach was adapted by businesses, particularly during periods of economic volatility like the oil crises of the 1970s. With the advent of globalization, technological innovation, and complex interdependencies among industries, decision-making under uncertainty has become a cornerstone of strategic management.

One of the early pioneers in this field, Herbert A. Simon, introduced the concept of "bounded rationality," emphasizing that human decision-making is limited by the information available, cognitive biases, and time constraints. His work laid the groundwork for modern decision science and earned him the Nobel Prize in Economics in 1978.

Core Components of Strategic Decision-Making Under Uncertainty

Strategic decision-making under uncertainty typically involves several interrelated steps and frameworks:

1) **Framing the Problem**
   Clearly defining the decision to be made, including its scope, objectives, and constraints. This ensures that efforts are focused and relevant variables are considered.
2) **Scenario Planning**
   Developing multiple plausible future scenarios to identify potential risks and opportunities. For example, energy companies often use this method to plan for market shifts, regulatory changes, or technological breakthroughs.
3) **Decision Trees and Probabilistic Analysis**
   Using structured visualizations to map out decision paths and their associated probabilities. Decision trees, for instance, break down complex decisions into manageable components, assigning probabilities and outcomes to each branch.
4) **Sensitivity Analysis**
   Testing how changes in key variables impact outcomes. This approach helps organizations identify the most critical factors influencing a decision.
5) **Real Options Analysis**
   Borrowed from financial theory, this technique evaluates the value of keeping options open when making investment decisions under uncertainty.
6) **Monte Carlo Simulations**
   Using computational models to simulate thousands of potential outcomes, providing insights into variability and risk. These simulations are especially useful in finance, energy, and supply chain management.

Applications Across Sectors

- **Energy:** Oil and gas companies use scenario planning to navigate fluctuating prices, geopolitical tensions, and the transition to renewable energy.
- **Healthcare:** Hospitals employ sensitivity analysis to prepare for varying patient volumes and resource needs during pandemics.
- **Technology:** Tech firms like Google use decision trees to evaluate investment opportunities in emerging markets and technologies.
- **Defense:** Military leaders rely on probabilistic analysis to evaluate risks in operational planning, balancing mission success with personnel safety.

Case Study: Unilever and Scenario Planning

Unilever has been a leader in using scenario planning to address long-term uncertainties related to climate change, market dynamics, and resource scarcity. By developing detailed scenarios of possible futures, the company has created adaptive strategies to ensure sustainable growth while mitigating risks. For example, its "Future Foods" initiative aligns with potential consumer preferences and regulatory environment shifts.

Strengths and Limitations

- **Strengths:**
  - Encourages flexibility and adaptability in decision-making.
  - Reduces the impact of cognitive biases by relying on structured frameworks.
  - Promotes proactive rather than reactive planning.
- **Limitations:**
  - Relies heavily on the quality and availability of data.
  - Can be resource-intensive in terms of time and expertise.
  - Scenarios and simulations may oversimplify complex interdependencies.

Lessons Learned

1) **Bias Mitigation:** Decision-makers must be aware of biases, such as overconfidence or anchoring, which can skew judgment.
2) **Collaboration:** Involving diverse perspectives enhances the robustness of decision frameworks.
3) **Iterative Processes:** Revisiting assumptions and models regularly ensures relevance in dynamic environments.

Future Trends in Strategic Decision-Making Under Uncertainty

Emerging technologies like AI and ML are enhancing the capabilities of strategic decision-making tools. AI-powered algorithms can process vast amounts of data to identify patterns, predict outcomes, and recommend optimal courses of action. Similarly, advancements in data visualization tools are making complex analyses more accessible to nontechnical stakeholders.

Another trend is the integration of ethical considerations into decision-making frameworks. For example, businesses now evaluate decisions on financial outcomes and their societal and environmental impacts, aligning with the principles of ESG criteria.

Conclusion: Strategic decision-making under uncertainty is no longer an optional skill but an essential capability for leaders navigating today's complex and unpredictable world. Organizations can turn uncertainty into a competitive advantage by employing structured frameworks, leveraging advanced tools, and fostering a culture of adaptability. Whether planning for climate risks, market disruptions, or technological advancements, this discipline empowers leaders to make informed, resilient, and impactful decisions.

Citations

1 H. A. Simon, *The Sciences of the Artificial*. MIT Press, 1969.
2 P. J. H. Schoemaker, "Scenario planning: A tool for strategic thinking," *Sloan Manag. Rev.*, vol. 36, no. 2, pp. 25–40, Winter 1995.
3 Unilever, "Future Foods Initiative," Unilever Reports, 2021.
4 S. Makridakis and N. N. Taleb, "*Living in a world of low levels of predictability*," *Int. J. Forecast.*, vol. 25, no. 4, pp. 840–844, Oct. 2009.
5 G. Koller, *Risk Assessment and Decision Making in Business and Industry*. CRC Press, 2005.
6 D. Kahneman, *Thinking, Fast and Slow*. Farrar, Straus and Giroux, 2011.

Integrated Resource Planning (IRP): Balancing Supply and Demand in Resource-Constrained Systems

Introduction

Integrated Resource Planning (IRP) is a comprehensive methodology for balancing supply and demand within systems that face resource constraints. It was first developed in the energy sector during the late 1970s and 1980s to address the growing complexity of energy production, distribution, and consumption. Over time, IRP has evolved into a powerful framework used across industries like utilities, water management, and infrastructure planning.

The essence of IRP lies in its holistic approach to planning. Rather than focusing solely on supply-side solutions (e.g., building new power plants), IRP incorporates demand-side management strategies, such as energy efficiency programs and conservation efforts. This dual focus ensures long-term planning aligns with economic, environmental, and societal goals.

Historical Context

IRP emerged as a response to the energy crises of the 1970s, highlighting the vulnerabilities of resource-intensive systems. Traditional planning methods – centered on building additional capacity to meet demand – proved inadequate in addressing challenges like fuel scarcity, rising costs, and environmental concerns. By the 1980s, state regulatory commissions in the United States began mandating IRP for electric utilities to ensure more sustainable and cost-effective energy planning.

One of the early adopters of IRP was the Pacific Gas and Electric Company (PG&E), which integrated demand-side measures into its planning processes to optimize resource utilization. This pioneering work demonstrated the potential of IRP to achieve significant cost savings and reduce environmental impact.

Core Principles of IRP

IRP operates on several key principles:

1) **Holistic Assessment**
   IRP evaluates supply- and demand-side resources to identify cost-effective and sustainable solutions.
2) **Long-Term Perspective**
   Planning horizons typically span 10–30 years, allowing organizations to anticipate future challenges like population growth, technological advancements, and climate change.
3) **Stakeholder Engagement**
   Effective IRP involves collaboration with regulators, consumers, and other stakeholders to ensure that planning decisions reflect diverse perspectives and priorities.
4) **Flexibility and Adaptability**
   IRP incorporates scenario analysis and sensitivity testing to prepare for uncertainties like economic fluctuations or policy changes.
5) **Sustainability and Equity**
   Modern IRP emphasizes minimizing environmental impacts and promoting equitable access to resources, aligning with broader sustainability goals.

IRP Process

The IRP process typically includes the following steps:

1) **Data Collection and Analysis**
   Planners gather detailed data on current supply resources, demand trends, and technological advancements, which form the foundation for modeling future scenarios.
2) **Demand Forecasting**
   Advanced statistical and computational models predict future resource demand under various conditions, accounting for economic growth, policy shifts, and consumer behavior.
3) **Resource Options Evaluation**
   Potential supply-side (e.g., renewable energy, natural gas) and demand-side (e.g., energy efficiency programs, demand response) options are assessed for feasibility, cost, and environmental impact.
4) **Scenario Planning**
   Planners develop multiple scenarios to explore how different strategies perform under varying conditions, such as policy changes or extreme weather events.

5) **Optimization and Selection**
   Using optimization algorithms, planners identify the portfolio of resources that best meets demand while minimizing costs and environmental impacts.
6) **Implementation and Monitoring**
   Once the plan is approved, resources are deployed, and their performance is monitored to ensure alignment with the original objectives.

Applications of IRP

Energy Sector

IRP is most widely applied in the energy sector, where it helps utilities balance power generation with customer demand. For example:

- **Case Study: Georgia Power**
  Georgia Power's IRP integrated renewable energy projects, such as solar farms, alongside demand-side measures like energy efficiency rebates. This approach reduced reliance on coal, cutting carbon emissions and saving costs.

Water Management

In water-scarcity regions, IRP balances supply-side interventions (e.g., reservoirs, desalination plants) with demand-side interventions (e.g., water conservation programs).

- **Case Study: California Drought Management**
  The California Department of Water Resources employed IRP to manage water distribution during droughts, prioritizing conservation efforts and reallocating water to critical areas.

Infrastructure Planning

IRP informs long-term infrastructure investments, ensuring that roads, bridges, and public transit systems meet future demand while optimizing resource use.

- **Case Study: Singapore Urban Planning**
  Singapore's Urban Redevelopment Authority uses IRP principles to integrate land-use planning with transportation and energy systems, creating a sustainable urban environment.

Strengths and Limitations of IRP

- **Strengths**
  - Promotes cost-effectiveness by integrating supply- and demand-side solutions.
  - Encourages sustainable practices by considering environmental impacts.
  - Builds resilience by preparing for multiple future scenarios.
  - Facilitates stakeholder collaboration, enhancing transparency and public trust.
- **Limitations**
  - Data-intensive, requiring extensive information on resource availability, usage patterns, and technologies.
  - Complex modeling and analysis may necessitate advanced computational tools and expertise.
  - Uncertainty in long-term forecasts can affect the reliability of planning outcomes.
  - Regulatory and political challenges may hinder implementation.

Future Trends and Innovations in IRP

Advancements in digital technologies are revolutionizing IRP. Smart grids, for instance, enable real-time monitoring and optimization of energy systems, enhancing the accuracy of demand forecasts and resource allocation. Similarly, AI and ML algorithms are improving scenario modeling and decision-making processes.

Another emerging trend is the integration of equity considerations into IRP. Utilities and policymakers are increasingly focused on ensuring that vulnerable populations have access to affordable and reliable resources, aligning with the principles of ESG frameworks.

Conclusion

IRP remains a cornerstone of sustainable resource management. Its holistic, long-term approach equips organizations to navigate the complexities of resource constraints while balancing economic, environmental, and societal goals. By integrating supply-side innovations with demand-side strategies, IRP ensures that systems remain resilient, cost-effective, and equitable in the face of uncertainty. As technologies and societal priorities evolve, IRP will play a vital role in shaping sustainable futures across industries.

Citations

1 S. Stoft, *Power System Economics: Designing Markets for Electricity*. IEEE Press, 2002.
2 California Energy Commission. "Integrated Resource Planning Overview," *California Energy Reports*, 2021.
3 U.S. Department of Energy. "Integrated Resource Planning: Best Practices for Electric Utilities," 2019.
4 Singapore Urban Redevelopment Authority. "Master Planning for Sustainable Development," *Annual Review*, 2020.
5 M. W. Rosegrant, "Water resources planning: Integrated approaches," *J. Water Manag.*, 2018.
6 D. Koutsoyiannis, "Energy and water resource synergies in planning," *Hydrol. Sci. J.*, 2017.

Design Thinking: A Human-Centered Approach to Innovation

Origins and Historical Context of Design Thinking

Design Thinking traces its origins to the 1960s as a problem-solving methodology emerging from architecture and engineering. The term gained prominence in the 1980s, popularized by David Kelley, founder of IDEO, a global design consultancy firm. Kelley's work at IDEO, alongside Stanford University's d.school (the Hasso Plattner Institute of Design), propelled Design Thinking into the mainstream as a structured framework for creative problem-solving. It was originally developed to bridge the gap between creativity and practical application, providing a repeatable process to tackle complex, ambiguous challenges.

Tim Brown, CEO of IDEO, further expanded the scope of Design Thinking, positioning it as a tool for product design and organizational transformation. His book, *Change by Design*, became a seminal text, highlighting how companies like Apple, Google, and P&G used Design Thinking to innovate and create user-centric solutions.

The Framework and Methodology

Design Thinking is grounded in empathy and iteration. It emphasizes understanding the human perspective, brainstorming creative solutions, and refining them through feedback. The process is often broken down into five key stages:

1) **Empathize:**
   - Gain a deep understanding of user needs through observation, interviews, and engagement.
   - **Example Tools:** Empathy maps, shadowing, and contextual inquiry.
2) **Define:**
   - Synthesize findings into a clear problem statement or design challenge.
   - **Example:** Apple's redefinition of user needs during the development of the iPod as "1,000 songs in your pocket."
3) **Ideate:**
   - Generate a wide range of creative ideas without judgment through brainstorming or co-creation sessions.
   - **Techniques:** Mind mapping, "How Might We" questions, and SCAMPER (Substitute, Combine, Adapt, Modify, Put to another use, Eliminate, and Rearrange).
4) **Prototype:**
   - Build simple, low-fidelity representations of the solution to test assumptions and concepts.
   - **Example:** IDEO created physical mockups to test hospital layouts for efficiency and user experience.
5) **Test:**
   - Solicit feedback from real users to refine prototypes and improve solutions.
   - Iterative adjustments ensure the solution evolves to meet user needs effectively.

Applications in Practice

1) **Corporate Innovation:**
   - P&G employed Design Thinking to develop Swiffer, revolutionizing home cleaning. By observing customers struggle with traditional mops, P&G teams empathized with user frustrations and prototyped solutions that addressed ease of use and convenience.
2) **Healthcare:**
   - Kaiser Permanente used Design Thinking to improve patient experience in emergency rooms. The hospital redesigned workflows by mapping patient journeys and engaging frontline staff in ideation to reduce wait times and improve care delivery.
3) **Education:**
   - Stanford d.school's K–12 Lab Network equips educators with Design Thinking tools to foster classroom creativity. By prototyping teaching methods, educators tailor learning experiences to diverse student needs.
4) **Technology:**
   - Airbnb embraced Design Thinking to transition from a struggling startup to a market leader. Founders Brian Chesky and Joe Gebbia created user-centered prototypes, interviewing hosts and guests to refine the platform for a seamless experience.

Strengths and Limitations

Strengths:

- **Empathy-Driven:** Prioritizes user needs, ensuring solutions are relevant and impactful.
- **Iterative:** Encourages flexibility, allowing ideas to evolve based on real-world feedback.
- **Collaborative:** Fosters teamwork across disciplines, breaking down silos in organizations.

Limitations:

- **Time-Intensive:** Requires significant investment in user research and prototyping.
- **Ambiguity in Metrics:** Success often depends on subjective user feedback, which can complicate evaluation.
- **Cultural Resistance:** Organizations with hierarchical or rigid structures may resist the open-ended, iterative nature of Design Thinking.

Modern Relevance and Emerging Trends

Design Thinking remains highly relevant as organizations strive to navigate uncertainty and drive innovation. Its principles align with the demands of the digital age, where user experience (UX) and customer-centricity are paramount.

Emerging trends in Design Thinking include:

- **Integration with AI:**
  - Companies like IBM use Design Thinking alongside AI tools to create predictive solutions in healthcare and retail.
- **Social Innovation:**
  - Nonprofits like the Bill & Melinda Gates Foundation employ Design Thinking to address global challenges, such as improving sanitation in developing countries.
- **Remote Collaboration:**
  - Tools like Miro and MURAL enable teams to conduct Design Thinking workshops virtually, expanding accessibility.

Lessons for Leaders

Design Thinking offers leaders a powerful tool to foster creativity, align teams, and solve problems that defy linear approaches. By embedding empathy, experimentation, and inclusivity into organizational processes, leaders can build cultures of innovation that drive sustainable growth.

For technological leaders, the methodology bridges engineering precision and user-centered design, ensuring that solutions are functional, desirable, and impactful. It exemplifies the shift from a product-first to a user-first mindset, a hallmark of successful organizations in today's competitive landscape.

Citations

1 T. Brown, *Change by Design: How Design Thinking Creates New Alternatives for Business and Society*. Harper Business, 2009.

2 T. Kelley, *The Art of Innovation: Lessons in Creativity from IDEO, America's Leading Design Firm*. Crown Business, 2001.

3 Dam, R. F., & Siang, T. Y. "What Is Design Thinking and Why Is It So Popular?" *Interaction Design Foundation*, 2020.

4 Case Study, "How airbnb used design thinking to disrupt the hospitality industry," *Harv. Bus. Rev.*, 2017.
5 "Design thinking in healthcare: Kaiser Permanente's journey." *Healthc. Q.*, vol. 16, no. 2, 2015.

Lifecycle Analysis: Evaluating Sustainability from Cradle to Grave

Origins and Development of Lifecycle Analysis (LCA)

Lifecycle Analysis (LCA) was born from the growing awareness of environmental challenges in the 1960s and 1970s. Early studies, such as Coca-Cola's 1969 internal research comparing the environmental impact of glass and plastic bottles, marked the beginning of lifecycle thinking. These efforts laid the groundwork for a systematic approach to evaluating environmental impacts. The oil crisis of the 1970s and subsequent debates on resource conservation further propelled the adoption of LCA. The methodology gained formal recognition in the 1990s when the ISO established standards for LCA (ISO 14040 series), providing a globally accepted framework for its application.

Today, LCA is a critical tool in sustainability science. It evaluates the environmental, social, and economic impacts of products, processes, and systems across their entire lifecycle – from raw material extraction to production, use, and disposal.

The Framework and Methodology of LCA

LCA is structured around four main phases, each of which provides a comprehensive view of a system's impact:

1) **Goal and Scope Definition:**
   - Establishes the purpose of the analysis, system boundaries, and functional unit (e.g., one unit of a product or one megawatt-hour of energy).
   - **Example:** An automotive company might set a goal to compare the environmental footprint of electric versus internal combustion engine vehicles.
2) **Inventory Analysis:**
   - Compiles data on all inputs (materials, energy) and outputs (emissions, waste) across each lifecycle stage.
   - **Example:** For a smartphone, this includes mining raw materials like lithium and cobalt, energy use during production, and e-waste generated at the end of life.
3) **Impact Assessment:**
   - Quantifies potential impacts across categories such as global warming potential, water use, air pollution, and toxicity.
   - **Example:** Measuring the carbon footprint of producing one kilowatt-hour of electricity from coal versus solar power.
4) **Interpretation:**
   - Synthesizes findings to guide decision-making, identify hotspots, and recommend improvements.
   - **Example:** A company might use LCA results to redesign a product, replacing energy-intensive materials with more sustainable alternatives.

Applications of Lifecycle Analysis

LCA is versatile, with applications across numerous sectors. Its insights drive sustainability in product design, corporate strategy, policymaking, and more.

1) **Renewable Energy:**
   - Wind, solar, and hydropower projects use LCA to measure environmental benefits compared to fossil fuels. For instance, Tesla employs LCA to optimize the lifecycle of its lithium-ion batteries, addressing energy use during production and facilitating recycling at end-of-life.
2) **Consumer Goods:**
   - Companies like Unilever and Nestlé conduct LCAs on packaging to reduce waste and transition to biodegradable materials. Coca-Cola's switch from glass to PET bottles was informed by lifecycle comparisons of environmental impact.
3) **Construction:**
   - The building industry integrates LCA to evaluate materials like concrete, steel, and timber. For example, adopting cross-laminated timber instead of concrete reduces embodied carbon emissions significantly.
4) **Public Policy:**
   - Governments rely on LCA to shape policies like carbon pricing and renewable energy subsidies. The European Union's Environmental Product Declarations (EPDs) mandate LCA for products marketed in Europe.

Strengths and Limitations

Strengths:

- **Comprehensive Insights:** LCA evaluates a product's entire lifecycle, preventing the shift of environmental burdens from one stage to another (e.g., from production to disposal).
- **Quantitative Rigor:** Its standardized methodology ensures reproducibility and comparability across studies.
- **Policy Influence:** LCA findings often inform legislation, sustainability standards, and corporate ESG strategies.

Limitations:

- **Data Intensity:** Collecting accurate lifecycle data across global supply chains can be challenging and resource-intensive.
- **Scope Creep:** Poorly defined boundaries may yield misleading results or fail to capture indirect impacts.
- **Subjectivity in Weighting:** Impact categories (e.g., carbon emissions vs. water use) may carry different significance depending on regional or organizational priorities.

Case Studies of Lifecycle Analysis in Action

1) **Apple's iPhone:** Apple leverages LCA to quantify and reduce the environmental impact of its products. Each iPhone undergoes a thorough LCA, examining raw material extraction (e.g., mining rare earth elements), manufacturing energy use, and recycling options. By incorporating recycled aluminum and responsibly sourcing rare earth elements, Apple has reduced its carbon footprint while advancing its commitment to sustainability.
2) **The Danish Wind Energy Sector:** Denmark uses LCA to evaluate wind turbines, a cornerstone of its renewable energy strategy. By analyzing material use, energy inputs, and decommissioning, Danish manufacturers like Vestas have optimized their processes to achieve energy payback periods as low as six months – meaning turbines produce more energy than their manufacturing required within half a year of operation.
3) **The European Union's Circular Economy Directive:** The EU applies LCA principles to transition industries toward circular economy models. Policies require manufacturers to design

for recyclability and account for the environmental costs of materials across their lifecycles. For example, automotive producers must meet stringent end-of-life recycling targets, reducing waste and promoting resource efficiency.

Practical Relevance and Emerging Trends

LCA has evolved to address contemporary challenges like climate change, resource scarcity, and social equity. Modern advancements in LCA include:

1) **Integration with Artificial Intelligence:** AI streamlines inventory analysis by automating data collection and identifying patterns in large datasets. For instance, AI-powered tools help track emissions in global supply chains.
2) **Dynamic LCA:** Incorporates real-time data to account for variables like energy grid fluctuations or changing market conditions, offering more accurate impact assessments.
3) **Social LCA:** Extends traditional LCA to evaluate social impacts, such as labor conditions and community well-being. This approach aligns with broader ESG reporting frameworks.
4) **Regional Customization:** Location-specific LCA models account for differences in energy sources, waste management practices, and economic conditions, providing more nuanced insights.

Lessons for Leaders

LCA equips leaders with a powerful tool to make informed decisions that balance profitability with sustainability. By identifying environmental hotspots and quantifying trade-offs, LCA enables organizations to set tangible goals, enhance brand reputation, and comply with regulatory requirements.

For technological leaders, LCA provides a framework to integrate sustainability into product development and innovation processes. It fosters accountability, aligning engineering practices with societal demands for environmental stewardship. In an era of rising stakeholder expectations, mastering LCA is not just a competitive advantage – it is a strategic imperative.

Citations

1 ISO 14040:2006, *Environmental Management – Life Cycle Assessment – Principles and Framework.* International Organization for Standardization.
2 T. Brown, "Lifecycle thinking in product design," *J. Sustain. Des.*, vol. 12, no. 3, 2018.
3 "Tesla's Environmental Impact Reports," *Tesla Inc.*, 2021.
4 "LCA and the Circular Economy in the EU," *European Commission Reports*, 2020.
5 Case Study: "The Carbon Payback of Wind Energy," *Danish Energy Agency*, 2022.
6 Apple. "Environmental Responsibility Report," *Apple Inc.*, 2022.

Systems Engineering (Part 1 of 3)

Introduction: A Holistic Approach to Complexity

Systems engineering emerged as a formal discipline during the mid-20th century, a response to the growing complexity of technological and organizational challenges. With its origins in World War II and major projects like the Manhattan Project, it became indispensable for managing the interplay between interdisciplinary teams, intricate designs, and evolving user needs. Institutions like Bell Telephone Laboratories, NASA, and the DoD spearheaded the refinement of systems engineering, which has since become foundational across industries, including aerospace, energy, healthcare, and transportation.

At its core, systems engineering is a holistic methodology that integrates technical, organizational, and societal factors. It seeks to optimize not just individual components but the entire system, ensuring harmony and functionality across its lifecycle. By focusing on interdependence and systemic interactions, systems engineering bridges technical precision with real-world applications.

Core Concepts and Frameworks

Requirements Analysis: Defining Success

The cornerstone of any successful system is a clear understanding of its purpose and requirements. Systems engineering begins with **requirements analysis**, a process of identifying user needs, constraints, and desired outcomes. This involves engaging stakeholders to ensure all perspectives are considered.

For example, during the development of the Apollo Program, NASA prioritized objectives like astronaut safety, precision lunar landings, and mission reliability. Systems engineers translated these broad goals into actionable requirements, laying the groundwork for detailed designs and mission success [1].

Lifecycle Thinking: From Cradle to Grave

Lifecycle thinking extends systems engineering beyond design and implementation to encompass the entire lifespan of a system – from inception to decommissioning. This comprehensive approach ensures sustainability, adaptability, and efficiency.

For instance, modern infrastructure projects incorporate LCA to assess environmental impacts, maintenance requirements, and eventual disposal. Tesla employs lifecycle thinking to refine its EV production, optimizing raw material use, energy efficiency, and recycling processes [2].

Risk Management: Anticipating the Unexpected

Risk management is integral to systems engineering, addressing potential failures and uncertainties. Engineers use probabilistic models, failure mode analysis, and redundancy planning to mitigate risks.

The Apollo 13 mission exemplifies this principle. When an oxygen tank exploded mid-flight, NASA's robust risk management protocols enabled the crew's safe return. Redundant systems and contingency planning proved lifesaving, underscoring the value of foresight [3].

Cost–Benefit Analysis: Balancing Resources and Outcomes

Systems engineering incorporates **CBA** to evaluate the trade-offs between resource investments and expected outcomes. This quantitative tool helps decision-makers prioritize initiatives with the greatest impact.

The U.S. Department of Energy used CBA to justify investments in smart grid technologies. By comparing upfront costs with long-term benefits – reduced outages, enhanced cybersecurity, and energy efficiency – policymakers secured funding for transformative infrastructure [4].

Applications Across Sectors

Aerospace: Reaching Beyond Earth

In aerospace, systems engineering is the backbone of ambitious projects like satellite deployments and interplanetary missions. NASA's Artemis program, aiming to return humans to the Moon, leverages advanced systems engineering to integrate hardware, software, and operations across international teams [5].

Healthcare: Enhancing Patient Outcomes

Healthcare systems rely on systems engineering to optimize workflows, reduce errors, and improve patient care. For example, Kaiser Permanente uses systems engineering to streamline patient scheduling, integrate electronic health records, and enhance operational efficiency [6].

Energy: Transitioning to Sustainability

In the energy sector, systems engineering guides the integration of renewable technologies, grid modernization, and resilience planning. Projects like Copenhagen's transition to a carbon-neutral city exemplify the application of systems thinking to balance economic growth with environmental sustainability [7].

Transportation: Smarter Mobility

Autonomous vehicles and smart cities depend on systems engineering to manage the interaction of sensors, algorithms, and infrastructure. By simulating real-world scenarios, engineers optimize safety, efficiency, and user experience [8].

Emerging Trends and Challenges

Digital Twins: Simulating Reality

Digital twins – virtual replicas of physical systems – are revolutionizing systems engineering. These models allow engineers to simulate operations, predict failures, and refine designs in real time.

In aviation, GE's digital twin technology monitors jet engines, enabling predictive maintenance and minimizing downtime [9].

Artificial Intelligence (AI) and Machine Learning (ML)

AI and ML are enhancing systems engineering by automating data analysis, identifying patterns, and optimizing complex systems. For example, AI-driven supply chain models reduce inefficiencies, ensuring timely delivery of resources [10].

Sustainability and Resilience

As global challenges like climate change intensify, systems engineering must prioritize sustainability and resilience. Advanced frameworks now incorporate environmental metrics, stakeholder engagement, and adaptive planning to address these imperatives.

Relevance for Graduate Students and Professionals

For students and professionals, systems engineering offers a versatile toolkit for addressing multifaceted problems. Mastery of its principles enables practitioners to:

1) **Navigate Complexity:** Manage interdisciplinary teams and integrate diverse perspectives.

2) **Innovate Responsibly:** Balance technological advancement with societal and environmental considerations.
3) **Drive Impact:** Deliver solutions that are efficient, scalable, and sustainable.

Case Study: Graduate students at MIT's Systems Design and Management program apply systems engineering to real-world challenges, from autonomous vehicle development to pandemic response planning. Their projects exemplify the practical, transformative potential of the discipline [11].

Conclusion

Systems engineering is not just a methodology – it is a mindset for understanding and solving complex, interconnected challenges. Its holistic approach, rigorous frameworks, and interdisciplinary applications make it indispensable in the 21st century. Whether designing spacecraft, modernizing energy grids, or revolutionizing healthcare, systems engineering empowers individuals and organizations to achieve sustainable success.

As the field evolves, emerging technologies like digital twins, AI, and advanced computational tools promise to expand its reach and capabilities. For those embarking on careers in engineering, management, or policy, mastering systems engineering offers a pathway to making meaningful contributions in an increasingly interconnected world.

Citations

1 Apollo Program: NASA Historical Reference Collection.
2 Tesla Environmental Impact Report 2023.
3 G. Kranz, *Failure Is Not an Option*. Simon & Schuster, 2000.
4 U.S. Department of Energy, "Smart Grid Investment Report," 2021.
5 NASA Artemis Program Overview, 2023.
6 "Healthcare Optimization at Kaiser Permanente," Systems Engineering Journal, 2020.
7 Copenhagen Energy Transition Report, 2022.
8 *Autonomous Vehicle Systems Engineering*. SAE International, 2021.
9 "Digital Twins for Jet Engine Optimization," GE Aviation White Paper, 2022.
10 "AI in supply chain management," *IEEE Trans. Eng. Manag.*, 2023.
11 *MIT Systems Design and Management Program*. Case Studies Archive, 2023.

Systems Engineering: Core Areas (Part 2 of 3)

Stakeholder and Requirements Analysis: Aligning Vision with Needs

The Foundation of Systems Success

Every great system begins with a conversation – between engineers, end users, investors, regulators, and a host of other stakeholders. Stakeholder and requirements analysis is the foundational step in systems engineering, where the diverse and often conflicting needs of all involved parties are identified and reconciled. The goal is to ensure that the final product or system addresses these needs effectively and measurably, without losing sight of overarching objectives.

The roots of this component trace back to the post-World War II era, when the complexity of military projects required clear communication between developers and operators. Early systems engineering pioneers, like those at Bell Labs, emphasized the importance of translating vague expectations into specific, actionable requirements. Over time, this process evolved into a structured discipline with tools like requirements traceability matrices and stakeholder prioritization models.

A Case in Practice: The London Underground Jubilee Line Extension

When extending the Jubilee Line of the London Underground, stakeholders ranged from city planners and commuters to environmental groups and local businesses. Initial consultations revealed competing priorities: faster commuter travel, minimal disruption to local businesses, and reduced environmental impact. Through rigorous stakeholder analysis, engineers developed a set of requirements that balanced these needs, resulting in a project that improved connectivity while minimizing ecological and social disruption.

Relevance Today

In today's landscape of rapid technological advancement, stakeholder and requirements analysis remains critical. For example, in the development of autonomous vehicles, stakeholders include regulatory bodies, insurance companies, technology firms, and the general public. Each has unique concerns – ranging from safety and liability to affordability and usability. Systems engineers must integrate these perspectives into comprehensive, measurable requirements that guide the development process.

Challenges and Lessons Learned

Stakeholder analysis is as much an art as it is a science. Engineers must navigate conflicting priorities, manage expectations, and mitigate risks of scope creep. Effective communication, empathy, and adaptability are key. The Apollo program exemplified how a clear set of requirements – rigorously aligned with the mission's goal of landing humans on the moon – can lead to groundbreaking achievements.

Architectural Design: The Blueprint for Integration

The Skeleton of Complex Systems

Once requirements are defined, the next step is designing the system's architecture – its skeleton. This involves breaking the system into subsystems or modules, each with defined functions, interfaces, and performance criteria. The architectural design ensures that these modules interact seamlessly, creating a cohesive and efficient whole.

Architectural design became prominent in the 1960s during the rise of large-scale engineering projects, such as NASA's Apollo program and the Interstate Highway System. These projects required meticulous planning to manage interdependencies and avoid catastrophic failures.

A Modern Example: Smart Grids

In modern energy systems, architectural design is critical for creating smart grids that integrate renewable energy sources, battery storage, and traditional power plants. Engineers must define how data flows between sensors, control systems, and operators to optimize performance while maintaining reliability.

Methodologies and Tools

- **Systems Modeling Language (SysML):** A visual modeling language used to create system architecture diagrams that illustrate component interactions and dependencies.
- **Digital Twins:** Virtual replicas of physical systems used to simulate and refine architectural designs.

Challenges and Lessons

A poorly designed architecture can result in inefficiencies, incompatibilities, and failure. Boeing's 787 Dreamliner program demonstrated how overreliance on a decentralized supply chain without robust architectural integration led to costly delays. Effective architectural design requires a balance of creativity, precision, and collaboration.

Model-Based Systems Engineering (MBSE): Digital Transformation of Design

The Shift to Digital

Model-based systems engineering (MBSE) represents a paradigm shift from document-driven processes to model-driven approaches. It leverages digital tools to visualize, simulate, and refine systems throughout their lifecycle. MBSE has become essential in industries like aerospace and automotive, where complexity demands precision and agility.

Origins and Growth

MBSE emerged in the 1990s as a response to the limitations of traditional methods, which relied heavily on static documents. Early adopters like Lockheed Martin and NASA championed MBSE to streamline the development of spacecraft and defense systems. Today, tools like MATLAB, Simulink, and Cameo Systems Modeler are industry standards.

Applications and Impact

- **Automotive:** MBSE enables manufacturers to test autonomous driving algorithms in virtual environments, reducing the need for costly physical prototypes.
- **Aerospace:** NASA's Artemis program uses MBSE to model interactions between spacecraft, landers, and support systems for lunar missions.

Limitations and Considerations

While MBSE enhances efficiency, it requires significant upfront investment in software, training, and data management. Moreover, maintaining the accuracy of models over a system's lifecycle is a continuous challenge.

Lifecycle Management: From Cradle to Grave

Sustaining Systems Through Time

Lifecycle management ensures that systems remain functional, efficient, and relevant from their inception to decommissioning. This component is particularly critical in sectors like infrastructure, energy, and defense, where systems often operate for decades.

Historical Roots

The concept of lifecycle management gained prominence in the 1970s with the advent of complex industrial systems, such as nuclear power plants. Early pioneers emphasized the importance of maintenance planning and end-of-life strategies to ensure safety and cost-effectiveness.

Case Study: The Airbus A380

The Airbus A380 program illustrates the importance of lifecycle management. While the aircraft was an engineering marvel, its high operating costs and limited market demand led to its early discontinuation. Better LCA during the design phase might have mitigated these issues.

Emerging Trends

Today, lifecycle management incorporates sustainability metrics, such as carbon footprints and circular economy principles, to align with global ESG goals. For instance, renewable energy projects now include decommissioning plans to recycle wind turbine blades and solar panels.

Verification and Validation (V&V): Ensuring Performance and Reliability

Testing the System's Limits

Verification and validation (V&V) are critical steps in systems engineering, ensuring that systems perform as intended under real-world conditions. Verification confirms that components meet design specifications, while validation ensures that the overall system fulfills stakeholder needs.

Aerospace Pioneers

The Apollo 11 mission exemplified rigorous V&V practices. Every subsystem underwent exhaustive testing to ensure reliability under extreme conditions. This attention to detail was instrumental in the mission's success.

Modern Applications

- **Pharmaceuticals:** V&V is used to ensure drug safety and efficacy through clinical trials.
- **IT Systems:** In cybersecurity, V&V identifies vulnerabilities and ensures compliance with regulatory standards.

Challenges and Evolving Practices

As systems grow more complex, V&V must evolve. Advanced simulation tools and AI-driven testing are increasingly used to predict and prevent failures. However, maintaining a balance between thorough testing and time-to-market pressures remains a challenge.

Case Study Expansion for Core Areas

1) **Stakeholder and Requirements Analysis: SpaceX Falcon Heavy**
   SpaceX's Falcon Heavy rocket is a compelling case study in stakeholder-driven design. The requirements gathered from stakeholders – NASA, commercial clients, and even internal teams – demanded a reusable, cost-effective heavy-lift launch vehicle. The result? A system that addressed the dual needs of lowering costs and ensuring reliability for high-stakes missions, while also incorporating real-time feedback from operational launches like Falcon 9.
   - **Tool Example:** IBM's Rational Dynamic Object-Oriented Requirements System (DOORS) for managing stakeholder requirements and ensuring traceability from concept to delivery.
2) **Architectural Design: Boeing 787 Dreamliner**
   The Dreamliner highlights the importance of robust architectural design in global supply chains. Boeing outsourced the production of major components to different countries. This decentralized model was innovative but led to integration delays due to insufficient

architectural coordination. Enhanced use of digital twins and collaborative tools like Dassault Systèmes' CATIA could have mitigated these risks.

- **Tool Example:** SysML, Digital Twins, and Integrated Product Development tools to optimize system architecture and interdependencies.

3) **MBSE in Practice: NASA's Artemis Program**
NASA's Artemis program heavily relies on MBSE to design systems for lunar exploration. Engineers use virtual models to simulate spacecraft interactions, assess mission risks, and adapt designs dynamically. MBSE enables rapid iterations, integrating cross-disciplinary inputs in a way that documents alone cannot.
   - **Tool Example:** Cameo Systems Modeler, MATLAB, and Simulink for creating and validating models.
4) **Lifecycle Management: Offshore Wind Farms**
Offshore wind farms exemplify lifecycle management challenges. Engineers must plan for equipment wear, environmental impact, and recycling from installation logistics to decommissioning. Tools like SimaPro and GAMS help assess sustainability across the lifecycle.
   - **Real Example:** The Hornsea Project in the United Kingdom, the world's largest offshore wind farm, integrates lifecycle tools to evaluate cost, performance, and environmental impact over 25+ years.
5) **Verification and Validation (V&V): Tesla Autopilot**
Tesla's Autopilot system demonstrates the necessity of robust V&V in emerging technologies. Engineers conduct extensive simulations and real-world tests to validate algorithms under diverse driving conditions. Failures in V&V, such as accidents due to edge cases in AI models, highlight the stakes of this process.
   - **Tool Examples: ANSYS, CARLA (a** simulator for autonomous driving), and cloud-based platforms for iterative testing.

Expanded Practical Tools

1) **Trade Studies and Decision Matrices**
A trade study evaluates multiple design options against key criteria like cost, performance, and risk. In defense systems, choosing between different radar technologies involves weighing technical specs against budget constraints.
   - **Tool Example:** Pugh Matrix for comparative analysis; MATLAB and GAMS for quantitative trade-offs.
2) **Design for Sustainability**
The need for sustainable engineering is paramount. Design approaches like **Lifecycle Sustainability Assessment (LCSA)** evaluate environmental, social, and economic factors in system design.
   - **Tool Example:** SimaPro and OpenLCA for assessing lifecycle sustainability.
   - **Real Case:** Apple used 100% recycled aluminum in its MacBook line, which began as an LCSA-driven initiative.
3) **Risk Analysis and Management**
Risk management frameworks such as **Fault Tree Analysis (FTA)** and **FMEA** are critical in industries like aerospace and energy.
   - **Real Case:** The Challenger disaster underscores the importance of preemptive risk analysis in identifying potential failure points.
   - **Tool Example:** RiskWatch and ReliaSoft for risk quantification and mitigation strategies.

4) **Integrated Resource Planning (IRP)**
IRP balances supply and demand in energy planning, ensuring efficient resource allocation.
   - **Real Case:** California's energy grid modernizations, integrating renewables through IRP frameworks.
   - **Tool Example:** HOMER Energy for microgrid design and GAMS for utility-scale modeling.

Additional Core Topics to Include

1) **Design Thinking for Systems**
While design thinking is traditionally associated with product design, it has applications in systems engineering, focusing on end-user needs and iterative problem-solving.
   - **Real Case:** IDEO's work with Ford on human-centered design for vehicle interfaces.
2) **Philosophy of Technology in Systems Design**
Understanding the ethical and societal implications of technology is crucial for responsible innovation. This is particularly relevant in AI and autonomous systems.
   - **Real Case:** Google's AI principles guide the responsible deployment of ML models.
3) **Modal Aspects in Systems Design**
The modal analysis evaluates a system's response to dynamic conditions. This approach is widely used in structural engineering and acoustics.
   - **Real Case:** Modal analysis of skyscrapers to predict performance during earthquakes.

Additional Case Studies and Lessons

1) **Systems Engineering in Public Health: COVID-19 Response**
The rapid development and distribution of mRNA vaccines (e.g., Pfizer and Moderna) relied on systems engineering principles to streamline manufacturing, logistics, and deployment. MBSE and V&V were instrumental in ensuring safety and scalability.
2) **Sustainability in Aerospace: Boeing and Airbus**
Both companies are adopting lifecycle assessments to incorporate biofuels, recyclable materials, and lightweight designs for improved fuel efficiency.

Key Takeaways for Students and Practitioners

1) **Holistic Thinking:** Systems engineering requires consideration of technical, social, and economic dimensions. Every decision has cascading effects across the lifecycle.
2) **Adopt Digital Tools:** MBSE, digital twins, and simulation platforms are indispensable for tackling complex systems in real time.
3) **Stakeholder Integration:** Communication and alignment with diverse stakeholders prevent costly revisions and failures.
4) **Continuous Learning:** Real-world case studies – from Tesla to NASA – highlight the evolving nature of systems engineering and the importance of adaptability.

Concluding Thoughts on Core Components: Each component of systems engineering plays a vital role in managing complexity and ensuring success. From defining clear requirements to rigorous V&V, these steps form a cohesive methodology that empowers engineers to tackle the most challenging problems of our time. By adopting these practices, students and professionals can develop systems that are not only innovative but also reliable, sustainable, and impactful.

Advanced Applications and Future Directions in Systems Engineering

Emerging Technologies and Trends in Systems Engineering

Expanding Horizons with Emerging Technologies

Systems engineering is evolving with technological advances that redefine how complex systems are designed, managed, and optimized. Emerging tools such as **Quantum Computing**, **Blockchain**, and **AR** enable revolutionary capabilities across various industries.

Quantum Computing and Optimization

Quantum computing offers unparalleled potential in solving optimization problems that are computationally prohibitive for classical systems. For example, companies like Volkswagen explore quantum systems to optimize urban traffic flow, while aerospace firms leverage quantum algorithms for mission planning.

Blockchain in System Security and Transparency

Blockchain technology enhances trust and transparency in systems with decentralized networks, particularly in supply chains and financial systems. Projects like IBM's Food Trust use blockchain to track food from farm to fork, ensuring system integrity and accountability.

Augmented Reality in Systems Maintenance

AR applications transform training and maintenance in complex systems like aviation and healthcare. Lockheed Martin's use of AR glasses in the assembly of fighter jets has reduced errors and improved efficiency, illustrating AR's practical impact.

Cross-Industry Applications of Systems Engineering

Energy Systems: Renewable and Resilient

Systems engineering is at the forefront of addressing energy transition challenges. Tools such as LCA and digital twins help design efficient microgrids, smart grids, and energy storage solutions.

Case Study: Denmark's Renewable Energy Transition

Denmark's ambitious goal to achieve 100% renewable energy by 2050 relies on systems engineering principles. To ensure a seamless transition, integrated energy models simulate the interaction between wind turbines, energy storage, and national grids.

Transportation Systems: Managing Complexity

Autonomous vehicle ecosystems epitomize the systems-of-systems approach, where vehicles, infrastructure, and regulatory frameworks work together.

Case Study: Tesla's Autonomous Vehicle Systems

Tesla integrates hardware (cameras, sensors) and software (AI-driven algorithms) into a seamless system. Systems engineering principles ensure coordination between autopilot, energy management, and external data, creating a highly efficient and safe autonomous driving experience.

Advanced Tools and Techniques in Systems Engineering

AI-Driven Optimization and Predictive Analytics

AI and ML transform traditional systems engineering processes by automating data analysis, optimizing designs, and identifying patterns in large datasets.

Case Study: Predictive Maintenance in Manufacturing

GE's Predix platform uses AI-driven analytics to predict equipment failures in industrial systems, minimizing downtime and reducing costs.

Multi-agent Simulation Models

Simulations using multi-agent systems enable decision-makers to evaluate interactions in complex, adaptive systems, such as disaster response planning and urban infrastructure design.

Application in Disaster Recovery

California's wildfire response planning has used multi-agent simulations, which allow for resource allocation that minimizes human and environmental losses.

Global Systems Thinking

Interdependencies and Resilience

Systems thinking helps identify interdependencies between global systems, such as energy, water, and food networks. This approach mitigates risks while enhancing resilience.

Case Study: The Water-Energy Nexus in India

India's efforts to integrate water and energy policies exemplify the value of systems thinking. Systems engineers evaluate interdependencies in designing irrigation systems powered by renewable energy, optimizing resources, and improving rural livelihoods.

Educational and Professional Training in Systems Engineering

Academic Programs

Leading institutions such as MIT's Engineering Systems Division and Stanford's Center for Integrated Systems offer comprehensive curricula that combine technical and managerial training in systems engineering.

Practical Guide for Learners

- **Step 1:** Gain foundational knowledge in systems dynamics, LCA, and design principles.
- **Step 2:** Master tools like MATLAB, Simulink, and Arena Simulation.
- **Step 3:** Obtain certifications, such as INCOSE's Certified Systems Engineering Professional (CSEP), to validate expertise.

Frameworks for Innovation

Agile Development for Complex Systems

Integrating Agile methodologies into systems engineering facilitates rapid prototyping and iterative improvements. This is especially valuable in sectors like software development and defense.

Case Study: SpaceX Starship Design

SpaceX uses Agile principles to develop and test iterations of its Starship spacecraft. This approach accelerates innovation while reducing costs.

Human-Centered Systems Design

Emphasizing user needs and experiences ensures systems align with societal and cultural contexts.

Case Study: Healthcare Device Design

Human-centered design was instrumental in developing portable ventilators during the COVID-19 pandemic, prioritizing ease of use and reliability under emergency conditions.

Metrics for Success

Resilience Metrics

Resilience metrics measure a system's ability to withstand disruptions and recover effectively. These metrics are particularly crucial for critical infrastructure, such as power grids and transportation networks.

Sustainability Metrics

LCSAs evaluate systems' environmental and social impacts throughout their lifecycle. Organizations like the World Bank use these metrics to guide investments in sustainable infrastructure.

Challenges and Future Directions

Limitations and Constraints

Current systems engineering approaches often overemphasize technical solutions while neglecting cultural and societal impacts. Integrating legacy systems with emerging technologies also presents significant challenges.

Future Directions

- **Real-Time Data Integration:** Leveraging IoT and sensor networks for dynamic decision-making.
- **Ethical AI Systems:** Ensuring fairness, accountability, and transparency in AI-driven engineering processes.
- **Modular Systems:** Designing adaptable systems that evolve with changing requirements.

Conclusion: The Expanding Role of Systems Engineering

Systems engineering is indispensable for addressing the challenges of an increasingly interconnected and complex world. Its applications, ranging from autonomous vehicles to climate resilience, demonstrate its versatility and impact. Systems engineering remains a cornerstone of technological leadership by integrating advanced tools, fostering interdisciplinary collaboration, and emphasizing sustainability.

In-Depth Analysis of Systems Engineering and Engineering Systems (Part 3 of 3)

Ethical Considerations in Systems Engineering

Systems engineering's broad applications across critical infrastructure, healthcare, aerospace, and other sectors necessitate a strong ethical foundation. Integrating cutting-edge technologies, such as AI and autonomous systems, introduces complex moral dilemmas.

Ethics and AI in Engineering Systems

Adopting AI in systems engineering raises questions about accountability, fairness, and transparency. For instance, biases in AI algorithms can perpetuate inequality in systems designed for public welfare. Ethical challenges also arise in autonomous vehicles, where decisions involving potential harm must be encoded into systems. These moral considerations demand a structured framework for decision-making, ensuring that systems align with societal values and human well-being.

Case Study: Boeing 737 MAX Crises

The Boeing 737 MAX incidents highlight ethical lapses in systems engineering. Boeing introduced software updates without adequate pilot training or fail-safe measures to compete with rival Airbus. The resulting crashes underscored the consequences of prioritizing cost and time over safety. This case exemplifies the need for rigorous testing, ethical oversight, and accountability in systems engineering.

Stakeholder Engagement and Participatory Design

Systems engineering relies heavily on effective stakeholder engagement. Complex projects often involve diverse stakeholders, including governments, industries, and local communities. Participatory design ensures that all voices are heard, fostering equitable and effective solutions.

Expanding Stakeholder Analysis

Stakeholder mapping involves identifying key actors, assessing their influence and interest, and prioritizing engagement strategies. Effective stakeholder engagement ensures alignment between technical solutions and community needs.

Case Study: Smart Grid Implementation in Amsterdam

The Amsterdam Smart City project exemplifies participatory design. The city developed a smart grid system tailored to its unique needs by involving local residents, businesses, and policymakers. This collaborative approach improved energy efficiency and enhanced public trust in the project.

Systems Resilience Beyond Infrastructure

While systems resilience traditionally focuses on physical infrastructure, its principles are increasingly applied to social and organizational systems. Building resilient societies involves integrating disaster preparedness, public health systems, and sustainable development practices into systems engineering frameworks.

Case Study: COVID-19 Response

The COVID-19 pandemic highlighted the importance of systems' resilience. Governments and organizations relied on systems engineering principles to scale healthcare systems, manage vaccine distribution, and address supply chain disruptions. For example, Moderna utilized digital twins to simulate vaccine production, ensuring rapid deployment without compromising quality.

Digital Transformation in Systems Engineering

Digital transformation revolutionizes systems engineering by enabling real-time collaboration, predictive modeling, and performance optimization. Technologies like cloud computing, digital twins, and advanced analytics empower engineers to manage complexity with unprecedented precision.

Integration of Cloud Computing

Cloud-based platforms facilitate global collaboration, enabling teams to work seamlessly across geographies. For instance, cloud computing supports real-time data sharing in global shipping networks, enhancing operational efficiency and decision-making.

Practical Example: Smart Ports

Ports in Rotterdam and Singapore use digital twins to optimize logistics, reduce congestion, and improve sustainability. These systems provide a comprehensive view of operations, enabling predictive maintenance and resource allocation.

Risk Management and Mitigation Strategies

Modern risk management tools integrate probabilistic modeling, simulations, and ML to address uncertainty in systems engineering.

Dynamic Risk Management

Dynamic risk management involves continuously monitoring and updating risk assessments to adapt to evolving conditions. Tools like Monte Carlo simulations and Bayesian networks enable engineers to model potential scenarios and devise robust mitigation strategies.

Case Study: NASA Mars Rover Missions

NASA employed dynamic risk management in its Mars Rover missions, addressing uncertainties related to harsh environmental conditions and communication delays. Redundant systems and real-time monitoring ensured mission success despite significant challenges.

Education and Workforce Development

The interdisciplinary nature of systems engineering demands specialized training that combines technical expertise with soft skills like leadership and communication.

Emerging Curricula

Programs like the European Master's in Systems Engineering (EMSE) integrate technical and managerial training, preparing graduates to address global challenges. These curricula emphasize real-world applications, fostering innovation and adaptability.

Global Training Models

International collaborations, such as the partnership between MIT and the Singapore University of Technology and Design (SUTD), highlight the importance of cross-cultural competencies in systems engineering education.

Case Studies of Failure and Lessons Learned

Examining failures in systems engineering provides valuable insights into avoiding similar pitfalls in the future.

Case Study: Challenger Disaster

The Challenger disaster resulted from flawed decision-making processes and inadequate risk assessment. Engineers raised concerns about the O-ring seals, but organizational pressures to meet launch deadlines overshadowed safety considerations. This tragedy underscores the importance of ethical decision-making, open communication, and risk management in systems engineering.

Framework for Lessons Learned

Structured frameworks, such as After-Action Reviews (AARs), help organizations capture and institutionalize insights from past failures. Engineers can build more resilient systems by analyzing root causes and implementing corrective measures.

Sustainability and Circular Economy

Systems engineering increasingly incorporates sustainability principles to address environmental challenges and resource limitations.

Circular Systems Design

Circular economy principles aim to minimize waste and extend product lifecycles. Systems engineers design modular components that facilitate repair, recycling, and reuse, reducing environmental impact.

Case Study: Apple's Modular Design

Apple's modular designs for devices like the iPhone and MacBook enable easy recycling and component reuse. This approach aligns with circular economy principles, enhancing sustainability without compromising functionality.

Advanced Systems Engineering Tools and Methodologies

Expanded Methodologies

Heilmeier's Catechism

Originally developed by George Heilmeier at DARPA, this framework guides innovation by posing critical questions about problem definitions, solution approaches, and measurable outcomes.

TRIZ (Theory of Inventive Problem-Solving)

TRIZ systematically solves engineering challenges by identifying patterns in technological evolution. By analyzing past innovations, TRIZ provides actionable insights for overcoming design constraints.

Practical Application

These methodologies are widely applied in defense systems, renewable energy transitions, and advanced manufacturing projects.

Global Systems Collaboration

International Standards in Systems Engineering

Organizations like the International Council on Systems Engineering (INCOSE) and ISO establish global standards for systems engineering practices, fostering consistency and interoperability.

Case Study: SESAR (Single European Sky ATM Research)

SESAR modernizes European air traffic management systems through collaborative frameworks, enhancing safety, efficiency, and environmental sustainability.

Expanded Metrics and Analytical Tools

Measuring Success

Key Performance Indicators (KPIs) specific to systems engineering provide actionable metrics for project performance and system optimization. Dashboards integrating real-time data enable engineers to monitor progress and identify areas for improvement.

Practical Example

KPIs in managing high-speed rail projects highlight their utility in tracking milestones, budget adherence, and stakeholder satisfaction.

Broader Philosophical Implications

Systems engineering represents a paradigm shift from reductionist approaches to holistic thinking. Addressing the interconnectedness of technological, social, and environmental systems fosters sustainable solutions that align with broader societal goals.

Monte Carlo Simulations for Risk and Sensitivity Analysis: Modeling Uncertainty Through Probabilistic Methods

Introduction

Monte Carlo simulations are a cornerstone of risk and sensitivity analysis, enabling decision-makers to model uncertainty and variability in complex systems. Named after the Monte Carlo Casino in Monaco due to their reliance on random sampling, these simulations were first developed during the Manhattan Project in the 1940s by mathematicians Stanislaw Ulam and John von Neumann. The method quickly gained traction in fields ranging from physics to finance, and today it is widely applied across engineering, healthcare, energy, and technological leadership.

Monte Carlo simulations are invaluable in scenarios where deterministic models fail to capture the complexities of real-world variability. By simulating thousands or even millions of possible outcomes based on random inputs, the method provides a probabilistic distribution of results, allowing for a deeper understanding of risks and sensitivities.

Historical Development

The origins of Monte Carlo simulations trace back to World War II, when they were used to calculate neutron diffusion and chain reactions in nuclear fission. Ulam, faced with the problem of estimating the probability of success in complex physical systems, turned to random sampling techniques to approximate solutions. This innovation was formalized into the Monte Carlo method with the help of von Neumann and later Claude Shannon.

Over the decades, advances in computational power have transformed Monte Carlo simulations from theoretical exercises into practical tools used in nearly every industry. Today, they underpin decision-making in areas as diverse as stock market predictions, weather forecasting, and drug development.

Core Principles of Monte Carlo Simulations

Monte Carlo simulations work by generating a large number of random samples from input distributions to model uncertainty. The process typically involves the following steps:

1) **Define the Problem and Inputs:** Identify the system or process to be modeled and determine the uncertain variables (e.g., costs, demand, environmental conditions).
2) **Assign Probability Distributions:** Specify the probability distributions for each input variable based on historical data, expert judgment, or theoretical models. Common distributions include normal, uniform, and triangular.
3) **Run Simulations:** Use random sampling techniques to generate thousands or millions of possible scenarios, calculating the outcomes for each set of inputs.
4) **Analyze Results:** Summarize the results as a probability distribution of outcomes, identifying key metrics such as mean, variance, percentiles, and risk thresholds.
5) **Sensitivity Analysis:** Evaluate how changes in input variables affect the outcomes to identify the most influential factors.

Applications of Monte Carlo Simulations

Monte Carlo simulations are widely used for risk assessment, project management, and decision-making under uncertainty. Key applications include:

1) **Engineering and Design**

Engineers use Monte Carlo simulations to evaluate system reliability and optimize designs under uncertain conditions.

- **Case Study:** In aerospace engineering, NASA used Monte Carlo methods to assess the reliability of components in the Mars Rover program. By simulating the effects of temperature fluctuations, dust storms, and mechanical stress, NASA ensured the rover could withstand the harsh Martian environment.

2) **Finance and Investment**

Monte Carlo simulations are instrumental in portfolio management, pricing derivatives, and stress-testing financial systems.

- **Case Study:** Financial analysts at JP Morgan employed Monte Carlo simulations to evaluate the risk-return profile of investment portfolios during the 2008 financial crisis, enabling more informed decisions in volatile markets.

3) **Energy and Power Systems**

In the energy sector, Monte Carlo methods assess the risks of resource shortages, price volatility, and infrastructure failures.

- **Case Study:** Utilities use Monte Carlo simulations to model energy demand under various weather conditions, ensuring grid reliability during peak seasons.

4) **Healthcare and Pharmaceutical Development**

In drug development, Monte Carlo simulations evaluate clinical trial outcomes, optimize dosing regimens, and assess cost-effectiveness.

- **Case Study:** During the development of COVID-19 vaccines, Monte Carlo simulations were used to predict trial success rates and optimize supply chain logistics.

5) **Climate and Environmental Modeling**

Monte Carlo methods are widely applied in climate modeling to estimate the impacts of greenhouse gas emissions and predict extreme weather events.

- **Case Study:** The Intergovernmental Panel on Climate Change (IPCC) uses Monte Carlo simulations to model the range of possible global temperature increases under different policy scenarios.

Strengths and Limitations

- **Strengths**
  - Provides probabilistic insights rather than single-point estimates, allowing for better decision-making under uncertainty.
  - Captures complex interactions and dependencies between variables.
  - Flexible and adaptable to a wide range of problems across industries.
- **Limitations**
  - Computationally intensive, requiring significant processing power for large-scale simulations.
  - Results depend heavily on the accuracy of input probability distributions.
  - Interpretation of results can be challenging for nonexperts, requiring effective visualization and communication.

Monte Carlo Simulations in Action: Risk Assessment and Sensitivity Analysis

Monte Carlo simulations excel in identifying risks and understanding system sensitivities. For example:

- **Risk Assessment:** A renewable energy company used Monte Carlo simulations to evaluate the financial risks of building offshore wind farms. By simulating variables such as construction delays, maintenance costs, and energy prices, the company identified scenarios where the project might become unprofitable, allowing for proactive mitigation strategies.

- **Sensitivity Analysis:** A manufacturing firm applied Monte Carlo simulations to assess the sensitivity of production costs to raw material prices. The analysis revealed that a 10% increase in steel prices would disproportionately impact profitability, prompting the firm to explore alternative materials.

Future Trends and Innovations

The growing availability of big data and advancements in AI are enhancing the power of Monte Carlo simulations. AI-driven algorithms now refine input distributions in real time, improving accuracy and reducing computational demands. Additionally, distributed computing frameworks like Hadoop and Spark enable organizations to run large-scale simulations more efficiently.

Another emerging trend is the integration of Monte Carlo simulations with digital twins. By coupling probabilistic modeling with real-time system data, organizations can achieve unprecedented levels of accuracy in predicting outcomes and optimizing performance.

Conclusion

Monte Carlo simulations are indispensable tools for modeling uncertainty and analyzing risks in complex systems. Their ability to generate probabilistic insights makes them highly valuable for decision-making in unpredictable environments. As computational power continues to grow and new technologies emerge, Monte Carlo methods will remain at the forefront of risk management and strategic planning.

For graduate students, researchers, and practitioners, mastering Monte Carlo simulations offers a pathway to tackling some of the most challenging problems in engineering, finance, healthcare, and beyond. By understanding the principles and applications of this versatile tool, professionals can navigate uncertainty with confidence and precision.

Citations

1 N. Metropolis and S. Ulam, "The Monte Carlo method," *J. Am. Statist. Assoc.*, vol. 44, no. 247, pp. 335–341, Sep. 1949.
2 M. G. Morgan and M. Henrion, *Uncertainty: A Guide to Dealing with Uncertainty in Quantitative Risk and Policy Analysis*. Cambridge University Press, 1990.
3 R. Y. Rubinstein and D. P. Kroese, *Simulation and the Monte Carlo Method*. Wiley-Interscience, 2008.
4 NASA. "Monte Carlo Simulations in Aerospace Engineering," *Technical Reports*, 2015.
5 IPCC. "Monte Carlo Methods in Climate Modeling," *Special Report on Climate Change Scenarios*, 2020.
6 J. C. Hull, *Options, Futures, and Other Derivatives*. Pearson, 2018.

Dynamic Systems Modeling and Analysis: Simulating and Analyzing Interactions in Complex Systems

Introduction

Dynamic Systems Modeling and Analysis (DSMA) is a critical tool for understanding, predicting, and optimizing the behavior of complex systems over time. Rooted in systems theory, DSMA integrates mathematical modeling, computational simulations, and data analysis to study how interconnected components interact and evolve. First conceptualized in the mid-20th century by pioneers like Norbert Wiener (cybernetics) and Jay Forrester (systems dynamics), this approach has grown into a foundational methodology across engineering, economics, environmental science, and public policy.

Dynamic systems, whether physical, biological, or socioeconomic, are characterized by interdependencies, feedback loops, and time-varying behaviors. DSMA provides a structured framework to simulate these systems, uncovering hidden dynamics and enabling more informed decision-making.

Historical Context and Development

The origins of DSMA lie in the post-World War II era, a time of rapid technological and scientific advancement. Norbert Wiener's work on cybernetics introduced the concept of feedback loops, emphasizing the interplay between control and communication in systems. Around the same time, Jay Forrester developed systems dynamics at MIT, applying these principles to analyze industrial processes and later, societal challenges like urban growth and environmental sustainability.

Forrester's seminal work, *Industrial Dynamics* (1961), laid the foundation for DSMA as a formal discipline. His later models, such as the *World Dynamics* model used in the 1972 Club of Rome report *The Limits to Growth*, demonstrated the power of dynamic modeling in addressing global issues.

Core Principles of Dynamic Systems Modeling

Dynamic systems are often represented mathematically using ordinary differential equations (ODEs), partial differential equations (PDEs), or agent-based models. The key principles underlying DSMA include:

1) **Interconnectedness:** Components of a system influence one another, creating interdependencies that can amplify or dampen effects.
2) **Feedback Loops:** Positive feedback loops drive growth or instability, while negative feedback loops restore balance and stability.
3) **Nonlinearity:** Relationships within systems are often nonlinear, meaning that small changes in inputs can lead to disproportionate or unexpected outcomes.
4) **Time Dependence:** System behaviors evolve over time, requiring time series analysis and simulation techniques to predict long-term impacts.

Methodology

The DSMA process typically involves the following steps:

1) **System Conceptualization:** Define the system's boundaries, components, and interactions. Identify the key variables and parameters.
2) **Mathematical Modeling:** Represent the system using equations, diagrams, or computational models. Common techniques include ODEs, PDEs, and state-space models.
3) **Simulation:** Use computational tools like MATLAB, Simulink, or AnyLogic to simulate the system's behavior under different scenarios.

4) **Validation:** Compare simulation results with empirical data to ensure model accuracy and reliability.
5) **Sensitivity and Scenario Analysis:** Explore how changes in inputs, parameters, or structures influence outcomes.

Applications of Dynamic Systems Modeling

DSMA has far-reaching applications, particularly in fields where systems are highly interconnected and complex. Below are some prominent examples:

1) **Engineering and Infrastructure**

Engineers use DSMA to design resilient systems, optimize performance, and prevent failures.

- **Case Study:** In civil engineering, DSMA has been employed to simulate the behavior of smart grids under varying load conditions. This approach helps utilities manage demand, reduce outages, and integrate renewable energy sources.

2) **Healthcare Systems**

In healthcare, DSMA models patient flow, disease progression, and the impact of public health interventions.

- **Case Study:** During the COVID-19 pandemic, epidemiological models based on dynamic systems principles were instrumental in predicting infection rates, hospital capacity needs, and the effects of lockdown measures.

3) **Environmental and Climate Science**

Environmental scientists use DSMA to study ecological systems, assess climate change impacts, and develop mitigation strategies.

- **Case Study:** The World3 model, an evolution of Forrester's work, was used to simulate the interplay between population growth, industrial output, resource depletion, and environmental degradation.

4) **Economics and Policy**

Economists and policymakers apply DSMA to evaluate fiscal policies, market dynamics, and socioeconomic trends.

- **Case Study:** Central banks use dynamic models to forecast inflation, interest rates, and economic growth under different policy scenarios.

5) **Transportation Systems**

DSMA is critical for planning, optimizing, and managing transportation networks.

- **Case Study:** The city of Singapore uses dynamic modeling to optimize its public transportation system, minimizing congestion and improving efficiency through real-time simulations.

Strengths and Limitations

- **Strengths**
  - Captures the complexity of real-world systems by accounting for interactions, feedback loops, and nonlinearity.
  - Provides a predictive framework to evaluate scenarios and inform decisions.
  - Applicable to a wide range of disciplines and problems.

- **Limitations**
  - Requires detailed data and expertise to develop accurate models.
  - Computationally intensive, especially for large-scale or high-resolution models.
  - Results are only as good as the assumptions and inputs used in the model.

Practical Tools for DSMA

Several tools and software platforms support dynamic systems modeling:

1) **MATLAB/Simulink:** Widely used in engineering for dynamic simulation and control system design.
2) **AnyLogic:** A versatile platform for agent-based, discrete-event, and systems dynamics modeling.
3) **Vensim:** Specializes in systems dynamics modeling for business, policy, and environmental applications.
4) **Stella:** Designed for visualizing and simulating complex systems in education and research.
5) **Emerging Trends**

1) **Integration with AI and Machine Learning**

AI and ML algorithms are increasingly being integrated into DSMA to enhance predictive accuracy and automate model updates based on real-time data. For instance, reinforcement learning can optimize system performance in dynamic environments like smart cities.

2) **Digital Twins**

Digital twins, virtual replicas of physical systems, leverage DSMA to monitor, simulate, and optimize operations. These are widely applied in manufacturing, healthcare, and urban planning.

3) **Big Data Analytics**

The availability of massive datasets enables more detailed and granular modeling. Advanced analytics tools help process and integrate these datasets into dynamic models.

Relevance to Technological Leadership

For technological leaders, DSMA is not just a technical tool but a strategic enabler. By understanding the dynamics of interconnected systems, leaders can:

- Anticipate and mitigate risks.
- Optimize resource allocation and operational efficiency.
- Drive innovation by identifying leverage points in systems.
- Inform policy and investment decisions with data-driven insights.

Conclusion

DSMA is a powerful methodology for simulating and analyzing the behavior of complex systems. Its applications span engineering, healthcare, environmental science, economics, and beyond, offering invaluable insights for tackling 21st-century challenges. As technology evolves, the integration of DSMA with AI, digital twins, and big data will further enhance its capabilities, ensuring its continued relevance in research, industry, and policymaking.

Citations

1 J. W. Forrester, *Industrial Dynamics*. MIT Press, 1961.
2 N. Wiener, *Cybernetics: Or Control and Communication in the Animal and the Machine*. MIT Press, 1948.
3 J. D. Sterman, *Business Dynamics: Systems Thinking and Modeling for a Complex World*. Irwin McGraw-Hill, 2000.
4 NASA. "Dynamic Systems Modeling in Aerospace Applications," Technical Reports, 2018.
5 IPCC. "Dynamic Systems and Climate Modeling," Special Report on Global Warming, 2021.

Foundations of Global Transition Dynamics

Global Transition Dynamics (GTD) is a pioneering framework designed to address the interconnected and multifaceted challenges of the 21st century. It builds on over four decades of research and practical application, originating from efforts led by Massoud Amin in collaboration with Dr. Chauncey Starr and other influential thinkers. The methodology synthesizes insights from technological systems, critical infrastructure, and societal dynamics, integrating tools like stochastic modeling, data-driven foresight, and system-of-systems analysis.

The genesis of GTD lies in the post-9/11 era, during which national security, energy resilience, and infrastructure modernization became paramount. Under the leadership of Dr. Amin, the Electric Power Research Institute (EPRI) initiated the Complex Interactive Networks/Systems Initiative (CIN/SI) in 1998. CIN/SI laid the groundwork for analyzing interdependencies across critical infrastructures, including energy, water, transportation, and telecommunications. Recognizing the growing complexity of these systems, GTD emerged as a broader framework, incorporating environmental, economic, public health, and policy dimensions.

Dr. Chauncey Starr, the founder of EPRI, posed a critical question: "If the 9/11 tragedies had not occurred, how would you expand your work?" This question catalyzed the development of GTD as a system-of-systems approach to address emerging global challenges. Building on foundational work, Dr. Amin integrated insights from mentors like Rick Smalley, whose research on the 12 greatest challenges to humanity significantly shaped GTD's scope.

Core Components of GTD: GTD is grounded in a system-of-systems perspective, emphasizing interdependencies across sectors. It combines rigorous mathematical modeling, policy analysis, and scenario-based foresight. Key components include:

Mathematical Foundations: GTD employs advanced mathematical techniques to model dynamic interactions within and between systems. These include:

1) **Stochastic Dynamical Systems:** Used to simulate uncertainties and variabilities in infrastructure performance.
2) **Discrete-Event Dynamical Systems (DEDS):** Applied to model events like power outages, supply chain disruptions, and policy shifts.
3) **Optimization Algorithms:** Incorporating genetic algorithms, simulated annealing, and particle swarm optimization to identify optimal solutions.
4) **Multi-Criteria Decision Analysis (MCDA):** Balancing trade-offs across environmental, economic, and social dimensions.

The mathematical rigor of GTD is supported by computational tools, including Python, MATLAB, and GAMS, enabling real-time simulations and sensitivity analyses. Parallel computing frameworks like MPI and Dask ensure scalability for large datasets.

Data Integration

GTD relies on a vast repository of global and national datasets, encompassing over 5,000 scientific reports and 800,000 simulations. Key data sources include:

- Energy consumption and emission statistics.
- Climate models and environmental impact assessments.
- Economic indicators and policy databases.
- Public health metrics and educational outcomes.

These datasets are integrated into a unified analytical framework, enabling cross-sectoral insights and robust decision-making.

Sectoral Scope

GTD's applications span 26 critical sectors, including:

1) **Energy Systems:** Transitioning to renewable energy and modernizing grids.
2) **Public Health:** Enhancing pandemic preparedness and healthcare infrastructure.
3) **Education:** Addressing inequalities and promoting universal access.
4) **Climate Modeling:** Mitigating emissions and adapting to climate change.
5) **Economic Policy:** Reducing income disparities and fostering sustainable growth.
6) **Transportation and Urban Planning:** Developing resilient and equitable systems.

Scenario Planning and Foresight

Scenario planning is integral to GTD, enabling stakeholders to envision plausible futures and stress-test strategies. By combining qualitative narratives with quantitative models, GTD identifies pathways for sustainable transitions.

Applications and Case Studies

National Security and Resilience

Following 9/11, GTD played a crucial role in national security initiatives. It informed strategies for protecting critical infrastructures against cyber threats, natural disasters, and geopolitical disruptions. The framework's risk interdependency models guided investments in redundancy and adaptive capacity, ensuring robust responses to cascading failures.

Renewable Energy Transitions

GTD has been instrumental in guiding countries like Denmark and Germany toward renewable energy adoption. By integrating policy reviews, market dynamics, and technological readiness assessments, GTD provided actionable insights for decentralizing grids and aligning energy systems with sustainability goals.

Case Study: Puerto Rico's Grid Modernization In the aftermath of Hurricane Maria, GTD's resilience metrics and digital twin technologies facilitated Puerto Rico's transition to microgrids. These systems improved redundancy, reduced downtime, and enhanced community resilience.

Pandemic Preparedness

During the COVID-19 pandemic, GTD's public health models helped evaluate vaccination strategies, supply chain logistics, and economic recovery plans. By integrating data on healthcare capacity, demographic trends, and policy interventions, GTD supported evidence-based decision-making.

Climate Adaptation

GTD's climate models have informed adaptation strategies for vulnerable regions. For example, it has guided investments in water resource management and flood mitigation, balancing economic feasibility with long-term sustainability.

Tools and Methodologies

GTD incorporates a suite of tools to address complex challenges:

1) **Digital Twins:** Virtual replicas of physical systems for real-time monitoring and optimization.
2) **CARVER Matrix:** A vulnerability assessment tool for prioritizing risk mitigation.
3) **Resilience Metrics:** Quantifying robustness, recovery speed, and agility.
4) **Cost–Benefit Analysis:** Balancing trade-offs in interconnected systems.

Building the Foundations: Mathematical and Analytical Pillars

The GTD framework is underpinned by robust mathematical and analytical principles, enabling it to model complex systems with precision and foresight. The foundational methodologies include:

1) **Stochastic Dynamical Systems:**
   - These equations model the probabilistic behavior of interconnected systems, capturing uncertainties and interdependencies.
   - Examples include modeling cascading failures in energy grids or disruptions in supply chains.
2) **Agent-Based Models:**
   - Simulates interactions between individual components (agents) within a system, such as consumers, suppliers, and regulators.
   - Applications include policy testing and market simulations for renewable energy integration.
3) **Monte Carlo Simulations:**
   - Used to assess risks and uncertainties by running thousands of scenarios with varying parameters.
   - Particularly effective in security planning and infrastructure resilience.
4) **Optimization Algorithms:**
   - Tools like genetic algorithms and simulated annealing optimize resource allocation under constraints.
   - Applications range from grid stabilization to emergency response planning.
5) **Big Data Integration and AI/ML Tools:**
   - GTD incorporates ML algorithms for predictive analytics and real-time decision-making.
   - Examples include forecasting demand surges during emergencies and optimizing logistics.

The Evolution of GTD Post-9/11

In the aftermath of 9/11, national security became a pressing priority, with infrastructure resilience taking center stage. My work pivoted to address these challenges, integrating new dimensions and tools:

1) **Critical Infrastructure Protection:**
   - Developed probabilistic models to assess vulnerabilities in interconnected systems such as energy, water, telecommunications, and finance.
   - Focused on self-healing grids capable of detecting and mitigating faults autonomously.

2) **System Interdependencies:**
   - Recognized that failures in one system (e.g., power grids) often cascade into others (e.g., financial markets).
   - Created models to capture these cascading effects, informing risk mitigation strategies.
3) **Scenario Planning for Security:**
   - Leveraged GTD to simulate potential threats and test the resilience of critical systems.
   - Examples include cyberattacks on grid control systems and disruptions to transportation networks.
4) **Integration of Ecological and Social Dimensions:**
   - By 2003, GTD expanded beyond traditional infrastructure to include environmental and societal considerations.
   - This broadened scope was inspired by collaborations with luminaries like Professor Rick Smalley, who emphasized the importance of addressing humanity's greatest challenges.

Methodologies and Tools in GTD

GTD employs a range of tools to analyze, model, and address complex challenges:

1) **CARVER and Digital Twins:**
   - CARVER matrices evaluate vulnerabilities, while digital twins simulate system performance in real time.
   - Applications include prioritizing cybersecurity investments and optimizing disaster response.
2) **Quantitative Foresight:**
   - Combines big data analytics with traditional forecasting techniques.
   - Provides actionable insights into emerging risks and opportunities.
3) **Resilience Metrics:**
   - Quantifies a system's ability to withstand and recover from disruptions.
   - Metrics include recovery time, redundancy, and adaptability.
4) **Lifecycle and Cost–Benefit Analysis:**
   - Evaluates projects across their entire lifecycle, balancing environmental, social, and economic trade-offs.
   - Applications include transitioning to renewable energy and implementing urban resilience strategies.

Applications and Case Studies

1) **Security Applications:**
   - GTD was instrumental in post-9/11 security planning, helping prioritize investments in critical infrastructure protection.
   - **Case Study:** Designed and tested contingency plans for cascading failures in grid systems caused by cyberattacks or natural disasters.
2) **Resilience in Energy Systems:**
   - Guided the development of self-healing grids and microgrids to enhance resilience.
   - **Case Study:** Modeled and implemented microgrid solutions for Puerto Rico post-Hurricane Maria, enabling faster recovery.
3) **Sustainability and Climate Adaptation:**
   - Applied GTD to model and implement sustainable practices in urban planning.
   - **Case Study:** Supported the integration of renewable energy in California, balancing market demands with grid stability.

The Broader Implications of GTD

GTD is not just a tool – it is a philosophy for navigating complexity. By integrating systems thinking with advanced analytics, GTD empowers leaders to:

1) **Anticipate Risks:** Identify early warning signs and prepare for potential disruptions.
2) **Optimize Resources:** Balance competing priorities to achieve sustainable outcomes.
3) **Strengthen Resilience:** Build systems that adapt and recover effectively.
4) **Foster Equity:** Ensure that solutions address the needs of all stakeholders, particularly marginalized communities.

Looking Ahead: The Future of GTD

As global challenges grow more complex, the importance of GTD continues to expand. Emerging technologies like quantum computing and advanced AI offer opportunities to refine the framework, making it even more robust and adaptable.

GTD's ultimate goal is to build systems that are not only resilient but also capable of driving sustainable progress. By bridging the gaps between science, engineering, and policy, GTD provides a roadmap for leaders to navigate the complexities of the 21st century with confidence and clarity.

As global challenges intensify, GTD continues to evolve, incorporating advancements in AI, ML, and big data analytics. The framework's adaptability ensures its relevance for addressing emerging issues like quantum computing, space exploration, and global equity.

By integrating technological innovation with human-centric values, GTD offers a blueprint for sustainable progress. It empowers leaders to navigate complexity, optimize resources, and build systems that endure for future generations.

Strategic Options Analysis with Decision Trees

What It Is: Strategic Options Analysis with Decision Trees are a visual and quantitative tool for evaluating multiple strategies and understanding the trade-offs associated with each. Decision trees break complex decisions into a series of sequential steps, where each branch represents a choice, probability, or outcome. The tool is invaluable for uncertain leaders, allowing them to weigh risks and rewards systematically.

Why It Matters: In high-stakes decisions – such as R&D prioritization or policy evaluation – leaders must balance risk, cost, and potential benefits. Decision trees provide clarity by mapping all possible outcomes and their likelihood. This method brings logic and structure to decisions that rely on intuition alone.

How to Use It

1) **Define the Problem:** Clearly state the decision that needs to be made. For example, should a pharmaceutical company invest in developing a risky but promising new drug?
2) **Identify Decision Alternatives:** Outline all possible options or actions, such as proceeding with R&D, delaying the project, or abandoning it altogether.
3) **Map Outcomes and Probabilities:** For each option, map potential outcomes (e.g., success, partial success, or failure) and assign probabilities based on data or expert judgment.
4) **Calculate Expected Value:** Multiply the value of each outcome by its probability to determine the expected return for each branch.
5) **Choose the Best Path:** Compare options and select the one with the highest expected value or the lowest acceptable risk.

Applications

- **R&D Prioritization:** Pharmaceutical companies use decision trees to allocate resources to projects with the best balance of risk and reward. For instance, Pfizer applied this approach during the COVID-19 pandemic to evaluate which vaccine candidates were worth pursuing.
- **Policy Assessment:** Governments use decision trees to evaluate policies with long-term implications, such as energy investments or disaster recovery plans.

Case Example: A renewable energy firm evaluating wind farm locations used decision trees to assess risks (e.g., environmental regulations, land acquisition issues) and rewards (e.g., potential energy output, government incentives). The firm identified a location with lower upfront costs but high long-term returns by quantifying the trade-offs.

Strengths and Limitations

- **Strengths:** Decision trees provide a structured, transparent approach to decision-making, making them ideal for quantifying trade-offs.
- **Limitations:** The tool depends on accurate probabilities and assumptions. Biased or incomplete data can lead to flawed outcomes.

References

- Clemen, R. T., *Making Hard Decisions: An Introduction to Decision Analysis,* Duxbury Press, 1996.
- Harvard Business Review, "Quantitative decision-making tools for executives," 2021.

Global Transition Dynamics (GTD)

What It Is: GTD is a framework designed to analyze and manage complex, interconnected systems undergoing transitions. GTD integrates system-of-systems thinking with dynamic modeling to predict risks, understand feedback loops, and enable leaders to design adaptive strategies for large-scale transitions.

Why It Matters: In a rapidly changing world where climate change, technological shifts, and geopolitical tensions intersect, leaders must manage transitions across industries, regions, and disciplines. GTD provides a systematic approach to addressing these challenges by accounting for interdependencies and uncertainties.

How to Use It

1) **Map the System:** Identify the interconnected sectors, components, and stakeholders involved in the transition. For example, an energy grid depends on policy, technology, supply chains, and public adoption.
2) **Simulate Feedback Loops:** Use models to simulate how changes in one part of the system affect the others. For instance, increasing renewable energy adoption may reduce carbon emissions but stress grid infrastructure.
3) **Test Scenarios:** Develop and test transition scenarios, such as policy changes, technological breakthroughs, or economic disruptions. Assess the trade-offs and outcomes for each scenario.
4) **Monitor Key Indicators:** Establish real-time performance metrics to track progress, system stability, and emerging risks.
5) **Implement Adaptive Strategies:** Build solutions that can evolve as the system changes, focusing on resilience and long-term sustainability.

Applications

Climate Resilience: GTD models are used to plan for energy transitions, balancing decarbonization goals with grid stability. For instance, European countries have applied GTD to manage the phase-out of coal while integrating renewables. As well as for proactive quantitative foresight to bind the uncertain for critical infrastructure protection, security, and resilience.

Supply Chain Planning: Businesses use GTD to identify and mitigate risks in global supply chains, accounting for disruptions caused by natural disasters, trade policies, or pandemics.

Case Example: After Hurricane Maria devastated Puerto Rico, GTD principles guided the island's transition to decentralized solar microgrids. By mapping the vulnerabilities of the centralized grid and simulating future disaster scenarios, policymakers implemented adaptive energy systems that increased resilience while reducing reliance on imported fuels.

Strengths and Limitations

- **Strengths:** GTD is comprehensive and adaptable, ideal for managing uncertainty in large-scale transitions.
- **Limitations:** The approach requires advanced modeling expertise, reliable data, and interdisciplinary collaboration.

References

- C. Starr, and Amin, M., "*Global Transition Dynamics*," 2003.
- Phaal, R., "System-of-systems frameworks for energy transitions," *Cambridge University Press*, 2019.

Summary Table: Advanced Tools for Strategic Leadership.

Tool	Purpose	Strengths	Limitations	Applications
Real-Win-Worth (RWW) Framework	Evaluate business opportunities for feasibility	Structured and sequential approach to decision-making	Relies on accurate market data and analysis	New product development, investment appraisal
Stress-Testing & What-If Analysis	Assess system performance under stress	Reveals weaknesses, enhances preparedness	Limited by the quality of assumptions and scenarios	Financial systems, risk management, and infrastructure resilience
Strategic Foresight & Technology Roadmapping	Align technological advancements with strategic goals	Predictive, interdisciplinary, and actionable	Needs regular updates in dynamic environments	R&D planning, innovation management, policy development
Cost-Benefit & Risk Interdependency Models	Balance costs, risks, and systemic trade-offs	Captures interconnected risks and system trade-offs	Requires complex modeling and reliable data	Infrastructure development, cybersecurity, energy systems
Porter's Value Chain Analysis	Identify value-creating and non-value activities	Enhances operational efficiency and innovation	Ignores external market conditions	Business model design, operational redesign
Business Model Stress Testing & Simulation	Assess the robustness of business models	Ensures flexibility and agility	Limited by the accuracy of scenarios	Market entry strategies, mergers, and acquisitions

Summary Table: Advanced Tools for Strategic Leadership

Tool	Purpose	Strengths	Limitations	Applications
Strategic Options Analysis with Decision Trees	Evaluate strategies and visualize trade-offs	Quantitative, methodical approach to decisions	Relies on valid probabilities and assumptions	R&D prioritization, policy assessment
Global Transition Dynamics (GTD)	Manage interconnected systems and transitions	Comprehensive, adaptable for complex systems	Demands advanced data and modeling expertise	Climate resilience, supply chain planning

References

1 S. Sinek, *Start with Why*. Penguin, 2009.
2 G. Heilmeier, *DARPA Guidelines*, 1975.
3 DoD Quad Chart Guidelines, 2015.
4 M. Porter, *Competitive Strategy*. Free Press, 1980; *Competitive Advantage*, Free Press, 1985.
5 NASA, *Technology Readiness Guidelines*, 1992.
6 A. Osterwalder, *Business Model Generation*. Wiley, 2010.
7 RAND Corporation, *Delphi-Based Decision-Making Frameworks*, 2021.
8 U.S. Federal Reserve, *Stress Testing and Risk Oversight*, 2022.
9 R. Phaal, *Technology Roadmapping: Aligning Innovation and Strategy*. Cambridge University Press, 2011.
10 McKinsey & Company, *Risk and Interdependency Analysis in Critical Systems*, 2022.
11 M. Porter, *Competitive Advantage: Creating and Sustaining Superior Performance*. Free Press, 1985.
12 R. Clemen, *Making Hard Decisions: An Introduction to Decision Analysis*. Duxbury Press, 1996.
13 C. Starr, and M. Amin, *"Global Transition Dynamics," 2003, and cite assessments and reports Resilience Planning Frameworks.*

Foresight and Forecasting: History, Use Cases, and Practical Steps

Introduction: Why Foresight and Forecasting Matter

Foresight and forecasting help organizations think ahead, adapt to change, and act decisively. These tools let leaders plan for challenges, take advantage of opportunities, and stay ahead of trends. While **foresight** focuses on exploring multiple future scenarios, **forecasting** uses data and trends to predict likely outcomes. Together, they provide a powerful way to make smarter decisions and build stronger strategies.

Foresight and forecasting are more important than ever as we face rapid technological change, global uncertainty, and urgent challenges like climate change. These methods help leaders innovate, create resilience, and guide their organizations toward sustainable success.

A Short History of Foresight and Forecasting

Foresight and forecasting have evolved from early strategic planning tools to advanced methods for understanding the future.

1) **Military and Corporate Roots:** These methods began with military planning and later became business tools. For example, Shell famously used scenario planning in the 1970s to navigate the oil crisis. Their approach helped them stay competitive while others faltered.
2) **Post-9/11 Resilience:** After 9/11, foresight became critical for securing infrastructure. Initiatives like the Complex Interactive Networks/Systems Initiative (CIN/SI) integrated strategic foresight and modeling to protect critical systems like power grids and communications networks.
3) **Today's Use Cases:** Foresight is widely applied in everything from energy systems and AI to public health and urban planning. For example, predictive models guide the adoption of renewable energy, helping countries plan transitions toward net-zero goals.

What Are Foresight and Forecasting?

Foresight and forecasting have distinct but complementary purposes:

- **Foresight:** Explores possible futures to prepare for opportunities and risks. It emphasizes "what-if" thinking and develops flexible strategies for change. For example, foresight might help a company identify how AI could disrupt its market.
- **Forecasting:** Uses data and trends to predict what is likely to happen. For instance, forecasting could show how renewable energy costs will decline over time.

By combining these methods, organizations gain a broad view of what could happen and specific predictions for planning.

How to Apply Foresight and Forecasting: Practical Steps

Step 1: Define Clear Goals – Start with a clear question: **What must we know about the future?** For example:

- A city might ask, "How can we prepare for the rise of EVs?"
- A company might ask, "What new technologies will reshape our industry?"

Knowing your goals helps focus your foresight efforts and ensures results are actionable.

Step 2: Engage Stakeholders – Involve people from across your organization and beyond – leaders, experts, customers, and partners. Workshops and interviews can bring in diverse perspectives and highlight blind spots. For example, during the UMore Park initiative, stakeholders included energy experts, planners, and community members to create a comprehensive strategy.

Step 3: Gather and Analyze Data – Collect relevant data from inside and outside your organization. Use tools like:

- **Technology Interaction Matrices (TIMs)** to map relationships between technologies.
- **Scenario models** to test potential outcomes.
- **Probabilistic simulations are used** to assess risks and benefits under different conditions.

For example, Lockwood Carlson's "Power Zone" method mapped technologies at 3M to show where their strengths matched future market needs.

Step 4: Build Scenarios – Develop a range of possible futures:

- **Best Case:** Everything goes as planned.
- **Worst:** Major disruptions or setbacks occur.

- **Middle Ground:** A mix of positive and negative outcomes.

Scenarios let you stress-test strategies and stay prepared for the unexpected. Shell used this approach to predict oil price fluctuations and make better investment decisions.
Step 5: Rank Opportunities – Use frameworks like:

- **PESTLE** (Political, Economic, Social, Technological, Legal, Environmental) to analyze external forces.
- **SWOT** (Strengths, Weaknesses, Opportunities, Threats) to assess internal and external factors.

This step ensures your organization focuses on the most valuable and achievable opportunities.
Step 6: Integrate Foresight into Strategy – Foresight isn't useful unless it drives action. Integrate insights into your R&D plans, business strategies, and operational goals. For example, foresight helped the Galvin Electricity Initiative prioritize distributed energy technologies to improve grid resilience.
Step 7: Monitor and Adapt – **The future isn't static, and** neither is foresight. Build a process for regular reviews and updates. Use feedback from early projects to refine your approach. This will make foresight a continuous part of your organization's strategy.

Real-World Applications

3M and Lockwood Carlson's Power Zone: 42 years at the 3M Company, Dr. Lockwood Carlson's foresight process mapped technological strengths against market trends. This approach uncovered new opportunities, like leveraging advanced materials for emerging industries.

UMore Park: A Smart Community Example: The UMore Park initiative used foresight tools to design a sustainable community powered by renewable energy. Key tools included scenario planning, technology mapping, and identifying smart grid opportunities. This ensured energy systems were efficient, scalable, and aligned with future needs.

Galvin Electricity Initiative: In this project for Mr. Bob Galvin, Dr. Carlson, and I focused on making energy systems more reliable and consumer-friendly. Using foresight, the team developed decentralized energy solutions like microgrids and energy storage systems, ensuring resilience and efficiency while being affordable and sustainable.

Why Foresight and Forecasting Are Essential

Foresight and forecasting help organizations:

- **Adapt to Change:** Stay ahead of disruptive trends.
- **Seize Opportunities:** Identify where to innovate and invest.
- **Build Resilience:** Prepare for risks and uncertainties.

By applying these tools, leaders can create proactive rather than reactive strategies, ensuring long-term success.

References

1 H. Kahn and A. J. Wiener, *The Year 2000: A Framework for Speculation on the Next Thirty-Three Years*. RAND Corporation, 1967.
2 P. Schwartz, *The Art of the Long View: Planning for the Future in an Uncertain World*. Doubleday, 1991.

3 M. Amin, "Development and Leadership of Research Consortia: Lessons Learned and Possible Road Ahead for Continued Innovation," *Proceedings of the IEEE Power Engineering Society Summer Conf.*, panel on Organizing Research Consortia, Vancouver, BC, July 18, 2001.
4 C. Starr and M. Amin, "Global Transition Dynamics: Unfolding the Full Social Implications of National Decision Pathways," submitted to the President of the U.S. National Academy of Engineering, Sept. 2003.
5 L. Carlson, "Using Technology Foresight to Create Business Value," *Ind. Res. Inst. J.*, pp. 51–59, 2004.
6 M. Amin, *Applied Templates for Capstones and Projects*. Technological Leadership Institute, 2005–2018.
7 M. Amin, "Galvin Electricity Initiative: Task 3," Technological Leadership Institute, 2006.
8 "Galvin Electricity Initiative: Technology Scanning, Mapping, and Foresight," (M. Amin, L. Carlson, and C. Gellings), EPRI, Palo Alto, CA, and Galvin Electricity Project, Inc., Chicago, IL, 2006.
9 M. Amin and B. Horowitz, "Toward Agile and Resilient Large-Scale Systems: Adaptive Robust National/International Infrastructures," keynote address at the International Conference on Flexible Systems Management GLOGIFT-07, Nov. 15–17, 2007, Noida, India.
10 M. Amin, *Toolbox for Technology Foresight and Forecasting*. Technological Leadership Institute, 2008–2018.
11 K. Froggatt, L. Carlson, A. Marcus, M. Amin, and D. Polla, *Technology Foresight: A Critical Competency for Business Strategy*. Technological Leadership Institute, 2010.
12 P. Wack, "Scenarios: Uncharted waters ahead," *Harv. Bus. Rev.*, vol. 63, no. 5, pp. 72–89, 1985.
13 K. Hadrava, *Reducing Prototype Development Costs through Early Employment of Diagnostic Technologies*. University of Minnesota, 2015.
14 R. Phaal, C. Farrukh, and D. Probert, "Technology roadmapping: Linking technology resources to business objectives," *Int. J. Technol. Manag.*, vol. 26, no. 1, pp. 2–19, 2003.
15 M. Amin, "Global transition dynamics: Integrating advanced models, foresight, and computational tools for secure, resilient, and sustainable development," *IEEE Comp. Mag.*, pp. 1–11, 2025.

Integrating Tools into Action – A Hands-On Guide

Purpose: This chapter bridges the tools introduced in Chapters 4A–4D (and Appendix A), showing how they can be applied in practical, real-world scenarios, at least to get started. Designed for readers without advanced management or technical degrees, it offers straightforward guidance for using these tools effectively in decision-making processes. Each section provides step-by-step instructions, examples, and actionable insights to help readers tackle complex challenges.

Strategic Foresight: Planning for an Uncertain Future

Objective: Use strategic foresight to anticipate potential scenarios and create resilient strategies.
How to Apply It:

1) **Define Your Objective:** Identify the decision or challenge you're addressing.
   - **Example:** Planning renewable energy systems for a region prone to hurricanes.
2) **Gather Data on Trends and Uncertainties:** Use tools like PESTLE analysis to analyze external factors (PESTLE).
   - **Key Data Points:** Regulatory trends, market demands, and technological innovations.

3) **Build Future Scenarios:** Develop 3–4 plausible future scenarios by combining high-impact uncertainties. A Simple Example Matrix:

Scenario Name	Key Factor 1	Key Factor 2
Green Growth	Favorable policies	Rapid tech adoption
Stagnation	Regulatory delays	Slow tech growth

1) **Analyze and Prioritize Options:** Evaluate each scenario's risks, opportunities, and impacts using SWOT analysis.
    - **Assign Metrics:** Scenario probabilities, resilience scores, and ROI estimates.
2) **Implement and Monitor:** Test strategies through pilot projects and monitor indicators that align with scenarios.

Case Example: California Offshore Wind Projects

- **Scenario Planning:** Identified regulatory hurdles and logistical gaps early in the process.
- **Outcome:** Set a 30 GW target by 2030, reducing CO_2 emissions by 5 million tons annually.

Value Chain Analysis: Maximizing Operational Efficiency

Objective: Identify inefficiencies and optimize processes to deliver maximum value.
How to Apply It:

1) **Map the Value Chain:** Break down processes into stages (e.g., sourcing, production, and delivery). Example:

Stage	Key Activities	Key Metrics
Production	Assembly line	Cycle time
Delivery	Logistics optimization	Cost per shipment

2) **Identify Weak Links:** Pinpoint bottlenecks using tools like Pareto Analysis (80/20 Rule).
3) **Set Performance Metrics:**
    - **Examples:** Lead time, cost reductions, and customer satisfaction scores.
4) **Implement Targeted Improvements:** Apply Lean methodologies or Six Sigma to address inefficiencies.
5) **Monitor and Scale Successes:** Continuously measure improvements and replicate across similar processes.

Case Example: Mayo Clinic's Patient Flow Optimization

- **Focus:** Streamlining diagnostic and treatment processes.
- **Outcome:** Reduced patient wait times by 20% and enhanced treatment outcomes.

Ethical Decision-Making Framework

Objective: Make decisions that align with ethical principles while balancing profitability and impact.

How to Apply It:

1) **Identify Stakeholders:** Map who will be affected and how. Example:

Stakeholder	Impact
Local Communities	Job creation
Investors	ROI

2) **Evaluate Options and Impacts:** Use tools like Stakeholder Impact Ratings (SIRs) and Environmental Impact Scores.
3) **Design Transparent Solutions:** Involve stakeholders in decision-making through focus groups or surveys.
4) **Conduct a Long-Term Impact Analysis:** Assess sustainability, inclusivity, and societal impact over 5–10 years.
5) **Implement and Communicate:** Explain your decision-making process and outcomes to all stakeholders.

Case Example: Tesla's Microgrid Deployment in Puerto Rico

- **Focus:** Rebuilding the energy system post-Hurricane Maria.
- **Outcome:** Powered over 200,000 households, reduced reliance on diesel by 40%.

Integrating Decision Theory and Practical Tools

Decision Pathways: A Pragmatic Approach

Purpose: Combine tools to navigate complex decisions step-by-step.
Steps:

1) **Define the Decision Space:** Clarify the scope and constraints of your decision (e.g., budget, timeline).
2) **Select Relevant Tools:**
   - **Example:** Use scenario planning for long-term uncertainties and VCA for operational bottlenecks.
3) **Gather and Analyze Data:** Collect key data points for each tool (e.g., financial metrics, stakeholder feedback).
4) **Evaluate Trade-Offs:** Use a decision matrix to weigh options based on impact and feasibility.
5) **Implement and Iterate:** Execute the decision with pilot programs and adjust based on outcomes.

Example Workflow:

Stage	Tool Applied	Output Generated
Problem Framing	Business Model Canvas	Defined objectives and scope
Risk Assessment	Stress Testing	Identified vulnerabilities
Strategy Selection	Scenario Planning	Prioritized actions

The Strategy Pyramid: A Leadership Framework with Practical Strategies for Real-World Impact

Science, engineering, and technology are reshaping how humanity addresses its most significant challenges, from climate change and global health to equitable education and sustainable energy. Sectors like healthcare, energy, transportation, education, and manufacturing are at the forefront of this transformation, requiring leaders who can harness innovation with ethical oversight, practical strategies, and visionary thinking.

The strategy pyramid, which aligns purpose, strategy, and execution to drive meaningful results, is used in these fields to achieve a coherent and effective synthesis that aligns with leadership goals and execution. Leaders in these sectors must navigate the nuances of complex systems, technical risks, and diverse stakeholder needs while remaining focused on long-term societal benefits. We explore how the Strategy Pyramid provides a roadmap for achieving progress, weaving in tools, case studies, and actionable insights tailored to science-driven sectors.

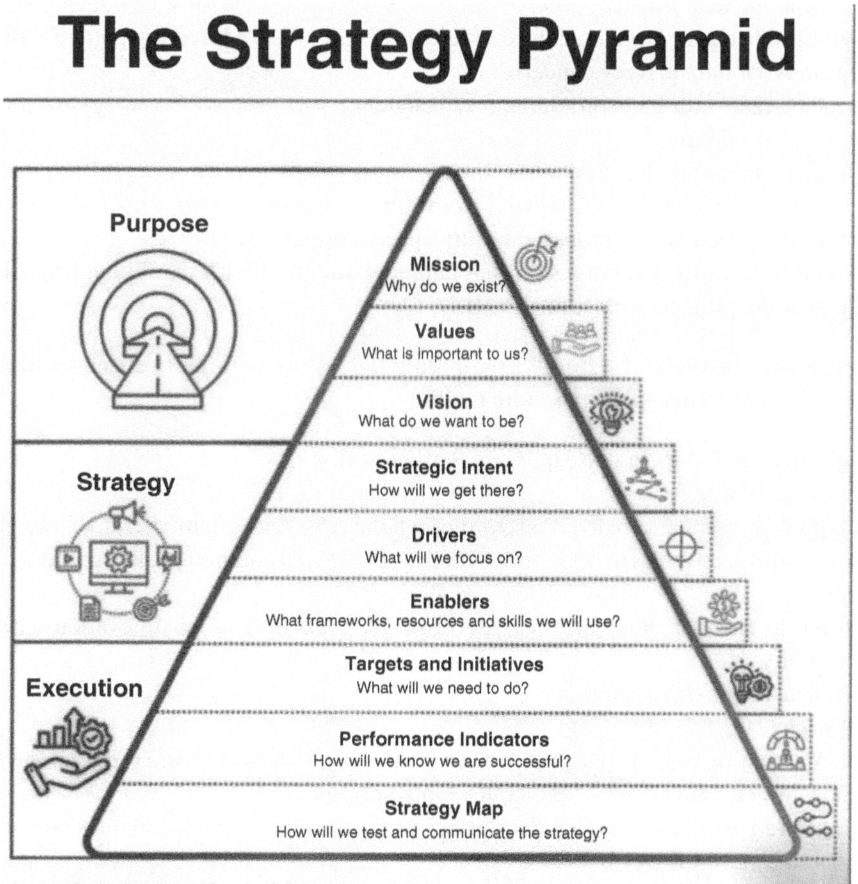

Source: The Strategy Pyramide - Wendy McGuiness

The Strategy Pyramid: A Leadership Framework for Progress: As illustrated below, the Strategy Pyramid organizes leadership into three interconnected levels: Purpose, Strategy, and Execution. This structured approach ensures that organizational efforts align with overarching goals, enabling leaders to turn bold ideas into tangible outcomes.

Purpose: Why Do We Exist? At the pyramid's apex is **Purpose**, encompassing mission, values, and vision. It defines why an organization exists and what it seeks to achieve.

1) **Mission:** The fundamental reason for an organization's existence.
   - *Case Study: SpaceX*
     SpaceX's mission – "To make life multi-planetary" – drives reusable rockets and innovation in interplanetary exploration. By aligning technical objectives with a bold mission, SpaceX attracts top talent and secures funding despite high risks.
     Actionable Insight: Define a mission that connects technical efforts to a larger purpose, ensuring every team member understands their role in achieving long-term goals.
2) **Values:** Ethical standards guiding decision-making.
   - *Case Study: OpenAI*
     OpenAI's core value – "Ensuring AI benefits all of humanity" – shapes its partnerships, transparency, and ethical safeguards. This commitment has built trust in an industry often scrutinized for its societal impact.
     Actionable Insight: Develop enforceable ethical guidelines addressing potential risks, such as bias in AI or data privacy concerns.
3) **Vision:** The organization's aspirational future state.
   - *Case Study: Moderna*
     Moderna's pre-pandemic vision of transforming medicine through mRNA technology enabled it to develop a COVID-19 vaccine in record time. This forward-looking vision ensured the company was prepared for unexpected opportunities.
     Actionable Insight: Craft a vision that balances ambition with feasibility, inspiring teams while aligning projects with future goals.

Strategy: How Do We Get There? The middle tier of the pyramid focuses on formulating actionable plans that translate purpose into reality.

1) **Strategic Intent:** Balancing research and market needs.
   - *Case Study: IBM*
     IBM's pivot from hardware to cloud computing and AI services demonstrates strategic intent. By reallocating resources to high-growth areas, IBM stayed competitive in a shifting tech landscape.
     Actionable Insight: Regularly reassess strategic goals to balance innovation with market applicability.
2) **Drivers:** The factors fueling progress.
   - *Case Study: DARPA*
     DARPA's funding of high-risk, high-reward research, such as GPS and the internet, exemplifies how targeted investment accelerates innovation.
     Actionable Insight: Identify and amplify key drivers of innovation, such as talent, partnerships, and funding.
3) **Enablers:** Infrastructure and resources supporting execution.
   - *Case Study: Google DeepMind*
     DeepMind's breakthroughs in AI research, such as solving protein folding, rely on cutting-edge computational infrastructure and skilled teams.
     Actionable Insight: Audit and invest in the resources and infrastructure necessary to achieve strategic goals.

Execution: How Do We Deliver? The pyramid's base emphasizes turning strategies into measurable outcomes through disciplined execution.

1) **Targets and Initiatives:** Setting measurable milestones.
   - *Case Study: Tesla's Battery Roadmap*
     Tesla's clear milestones for battery cost reduction and energy density improvements have enabled its leadership in EVs.
     Actionable Insight: Break down complex projects into actionable milestones and use agile methods to adapt based on feedback.
2) **Performance Indicators:** Measuring progress effectively.
   - *Case Study: Intel*
     Intel's use of technical KPIs, such as advancements in process-node technology, ensures competitiveness in semiconductor manufacturing.
     Actionable Insight: Define KPIs reflecting technical achievements and market outcomes, reviewing them regularly to stay on track.
3) **Strategy Maps:** Aligning cross-functional teams.
   - *Case Study: NASA's Mars Rover Missions*
     NASA's strategy maps coordinate interdisciplinary teams, ensuring complex missions stay on schedule and within budget.
     Actionable Insight: Develop strategy maps linking high-level goals to team-specific tasks, updating them to reflect project changes.

Conclusion: Leading with Purpose, Strategy, and Execution

Technological leadership is about transforming complexity into opportunity. By applying the **Strategy Pyramid** – anchoring decisions in purpose, aligning efforts through strategy, and delivering with disciplined execution – leaders can create breakthroughs that shape industries and improve lives.

This chapter provides a roadmap for navigating the unique demands of science-driven sectors, offering tools, case studies, and insights to help leaders drive progress responsibly and sustainably. The question isn't whether leaders can rise to the challenge – it's whether they'll act with the clarity, focus, and ethical integrity required to succeed.

Further Resources

1) **Smart Choices Framework:** From the book *Smart Choices* by Hammond, Keeney, and Raiffa, this framework emphasizes clarity in complex decisions. **Key Steps:** Define the decision. Identify objectives. Generate alternatives. Evaluate trade-offs.

Good to Great Insights: From Jim Collins' *Good to Great*, focusing on disciplined thought and action – **Hedgehog Concept:** Identify the intersection of what you're passionate about, what you can excel at, and what drives your economic engine.

These tools reflect a structured synthesis of theory and applied leadership practice. While each has strengths and limitations, they enable a comprehensive, system-aware, and future-resilient decision-making approach. The book integrates these tools where appropriate, and this appendix serves as a reference map for executives, policymakers, and scholars.

Appendix B

Global Transition Dynamics (GTD) Methodology

Synthesis Through Global Transition Dynamics (GTD): Integrating Advanced Models, Foresight, and Computational Tools for Secure, Resilient, and Sustainable Development

Introduction

Technological transitions have historically shaped human progress, driving communication, energy, and infrastructure breakthroughs. However, their prediction remains complex due to intertwined technical, societal, and economic factors. Superior technologies do not always guarantee adoption, and timelines often depend on infrastructure readiness, market forces, and social acceptance. Accurate predictions of such transitions can:

1) Guide investments in research and development.
2) Mitigate risks associated with adoption.
3) Ensure equitable access to the benefits of technological innovation.

Predicting technology transitions in critical systems such as energy, infrastructure, and artificial intelligence (AI) is vital for addressing global challenges and achieving sustainable development goals. This chapter presents a comprehensive framework integrating quantitative foresight, Global Transition Dynamics (GTD), and predictive analytics. Through case studies in energy transformation, digital twins, and systemic resilience, the chapter provides actionable insights for policymakers, technologists, and industry leaders. The framework emphasizes data-driven methodologies, focusing on technological advancements' societal and ethical implications.

We integrate methodologies from quantitative foresight, GTD, and predictive analytics to provide a structured approach to forecasting technology adoption. It focuses on three critical areas:

1) Energy transformation.
2) Digital twin applications in infrastructure.
3) Systemic resilience across interconnected systems.

Quantitative Foresight

Quantitative foresight involves employing data-driven techniques to anticipate technology evolution and adoption. Key components include:

- **Historical Analysis:** Identifying patterns and trends from historical data to predict future trajectories.
- **Scenario Planning:** Developing multiple potential futures to account for uncertainty in technology adoption and systemic impacts.
- **Validation:** Testing foresight methodologies against real-world benchmarks or pilot studies to ensure reliability.

Global Transition Dynamics (GTD): Origins of GTD

GTD models analyze the interaction of critical systems, such as energy, transportation, and communication, to understand how transitions occur in interconnected systems. Predicting technology transitions in critical systems such as energy, infrastructure, and AI is essential for addressing global challenges. This chapter presents a framework for predicting technology transitions in critical systems such as energy, infrastructure, and AI, addressing global challenges like climate change, resilience, and cybersecurity. Integrating GTD with quantitative foresight, nonlinear modeling, and strategic tools offers actionable insights for policymakers and technologists. Case studies in energy transformation, digital twins, and systemic resilience illustrate data-driven solutions informed by over 25 years of research and 800,000 simulations.

Technological transitions have historically driven human progress but remain challenging to predict due to their dependence on intertwined technical, societal, and economic factors. Superior technologies do not always succeed, and adoption timelines vary based on infrastructure readiness, market forces, and social acceptance. Accurate predictions can guide investments, mitigate risks, and ensure equitable progress.

The genesis of GTD can be traced back to 1998 when I joined the Electric Power Research Institute (EPRI). Dr. Chauncey Starr, EPRI's founder, and I envisioned a future where interconnected systems – energy, infrastructure, transportation, communication networks, the Internet, climate, markets, and finance – could be modeled holistically to mitigate risks and maximize societal benefits. In the wake of the 9/11 attacks, this vision expanded to include cybersecurity, cascading failures, and national infrastructure resilience.

In consultation with Dr. Starr, several colleagues, over 50 utility organizations, and colleagues at the US Department of Defense, I created and led the Complex Interactive Networks/Systems Initiative (CIN/SI), a $24 million program funded by EPRI (68% of funding) and the US Department of Defense (32% of funding). This initiative developed foundational tools and models that evolved into GTD. A key breakthrough came when nonlinear dynamics and strategic foresight were integrated into our approach, enabling high-fidelity simulations of interdependent systems.

Mathematical Models and Equations

GTD's foundation lies in its mathematical rigor. It uses sector-specific equations to simulate, predict, and optimize outcomes across interconnected systems. These equations integrate systemic dynamics, probabilistic modeling, and interdependencies to capture real-world complexities.

These equations are supported by dynamic computational tools, ensuring scalability and adaptability to real-world scenarios. At the core of GTD are advanced mathematical models that capture the dynamics of interdependent systems. Nonlinear equations are crucial in modeling energy systems, cascading failures with probabilistic foresight, and strategic pathways.

Climate, Power, and Energy

Advanced Computational Models for Power System Stability and Dynamics: Predicting the stability of power systems with bounded uncertainties requires integrating high-fidelity mathematical models that capture the intricate interactions of diverse system components. These models account for numerous device-specific behaviors and dynamic phenomena critical for ensuring reliable system performance. Table B.1 outlines the key technical areas and devices that must be modeled for comprehensive engineering studies. Below are expanded equations categorized by sector:

Energy Transition

$$E_t = E_0 e^{-\alpha t} + \sum_{i-1}^{n} \beta_i P_i$$

This equation models the decay of traditional energy sources ($E_0 e^{-\alpha t}$) and the integration of renewable energy investments (P_i) Over time.

Carbon Dynamics

$$\Delta T = \lambda \cdot \Delta F + \varepsilon(t)$$

Temperature change (ΔT) as a function of radiative forcing (ΔF) and stochastic noise ($\varepsilon(t)$).

Table B.1 Selected Sectoral Equations.

Sector	Equation	Description
Energy Transition	$E_t = E_0 e^{-\alpha t} + \sum_{i-1}^{n} \beta_i P_i$	Models the decay of legacy energy sources and integration of renewables over time.
Carbon Emissions	$\Delta T = \lambda \cdot \Delta F + \epsilon(t)$	Captures temperature response to radiative forcing with stochastic variables.
Power Grid Resilience	$R = \int_0^T \frac{\Delta P}{\Delta T} dt$	Measures recovery rates after disruptions over time.
Epidemic Spread	$V = R_0 \cdot \frac{\beta}{\gamma}$	Models disease dynamics with transmission and recovery rates.
Cybersecurity Risk	$P(C) = 1 - \prod_{i-1}^{n}(1 - P_i)$	Estimates cascading risks in interdependent systems.
Economic Growth	$Y = f(R, K, H)$	GDP as a function of R&D, capital, and human capital.
Resource Allocation	$U = \max \sum_{t-0}^{T} \beta^t [B(x_t) - C(x_t)]$	Optimizes utility by balancing benefits and costs over time.

Power Grid Resilience

$$R = \int_0^T \frac{\Delta P}{\Delta T} dt$$

Resilience is the key to power recovery (ΔP) over time (T).

Natural Gas Flow in Pipelines

$$\frac{\partial P}{\partial t} + v\frac{\partial P}{\partial x} + \alpha P^n = 0$$

Here, P represents gas pressure, v is velocity, and αP^n accounts for nonlinear frictional losses. This equation models the flow of natural gas in pipelines under varying conditions.

Power System Stability

$$\frac{d^2\delta}{dt^2} + D\frac{d\delta}{dt} + \omega^2 \sin\delta = \Delta P$$

This swing equation models rotor angle stability in power systems where δ is the angular displacement D damping, and ΔP represents a power imbalance.

Energy Market Dynamics

$$S_t = \alpha S_{t-1} + \beta I_t + \varepsilon_t$$

Energy supply (S_t) evolves based on investments (I_t), historical supply (S_{t-1}), and stochastic factors (ε_t).

Epidemic Spread

$$V = R_0 \cdot \frac{\beta}{\gamma}$$

The basic reproduction number (R_0) is calculated based on the transmission rate (β) and recovery rate (γ).

Healthcare Infrastructure

$$H_t = H_0 + \sum_{i=1}^{n} \delta_i I_i$$

Healthcare capacity (H_t) expands with investments (I_i) in facilities, personnel, and technology.

Stochastic Differential Equations (SDEs)

$$dx = \mu(x, t)dt + \sigma(x, t)dW$$

Here, x represents a system state, μ is the drift term, σ is the volatility, and dW is a Wiener process for random fluctuations.

Cybersecurity Risk

$$P(C) = 1 - \prod_{i-1}^{n}(1 - P_i)$$

Cascading failure probabilities ($P(C)$) in interconnected networks.

System Vulnerability

$$V_s = \frac{\sum_{i=1}^{n} W_i S_i}{\sum_{i=1}^{n} W_i}$$

Weighted vulnerability (V_s) based on system components (S_i) and their criticality (W_i).

Dynamic Game Theory

$$u_i = \sum_{t=1}^{T} \beta^t [B_i(a_i, a_{-i}) - C_i(a_i)]$$

Utility (u_i) for player *i, it* depends on their actions (a_i), the actions of others (a_{-i}), and associated costs (C_i).

Integrating Complexity into Models: These components are modeled with high temporal and spatial resolution to ensure accurate and actionable predictions. For example, transient dynamics at sub-second intervals (e.g., 250 milliseconds) demand cutting-edge computational resources to resolve the cascading impacts of rapid changes. Such granular modeling is vital for predicting and mitigating instabilities caused by phenomena like subsynchronous oscillations or renewable energy fluctuations.

Through iterative refinement and validation against real-world data, these models evolve to address emerging challenges, such as higher renewable penetration and increasing grid complexity. Integrating these models into a larger GTD framework enables stakeholders to foresee and address the multifaceted challenges of the future energy landscapes.

Mathematical models are essential for predicting the dynamics of interconnected systems like power grids, but uncertainties, interdependencies, and emergent complexities inherently constrain their accuracy. While basic models, such as the swing equation, provide a starting point, they fall short of capturing the intricate behaviors required for precise stability analysis and resilience planning. To address this, we employ multi-resolutional methods that balance computational efficiency with finer granularity when deeper insights are needed.

Strategic Technology Assessment

GTD uses structured frameworks to evaluate emerging technologies:

- **PESTLE Analysis:** Evaluates political, economic, social, technological, legal, and environmental factors influencing technology adoption.
- **Value Chain Analysis:** Identifies bottlenecks and optimization opportunities in energy and infrastructure systems.
- **SWOT Frameworks:** Focus on strengths, weaknesses, opportunities, and threats of proposed technologies.

Quantitative Foresight

GTD integrates real-time data updates into probabilistic models, producing actionable insights. Tools include:

- **Bloomberg-Like Dashboards:** Visualizing system interdependencies at macro, meso, and micro levels.
- **Cones of Opportunity:** Identifying scenarios for targeted investments.

Summary of Nonclassified Applications and Case Studies

Energy Transformation: Energy transitions are critical for achieving decarbonization goals. Predictive analytics integrates market trends, policy impacts, and technological readiness to guide this transformation.

- **Climate Change: Renewable Transition:** GTD simulations predict that transitioning to 70% renewable energy globally by 2050 requires $73–275 trillion. Shortening deployment timelines reduces costs by $20 trillion. Case studies in Germany's *Energiewende* reveal 40% efficiency gains and 30% emissions reductions (confidence 97.4%).
- **Efficiency Gains:** Multi-energy systems are projected to increase urban energy efficiency by 40% and reduce emissions by 30% by 2040.
- **Case Study – Germany's Energiewende:** Germany achieved 40% renewable electricity generation by 2020 through foresight-driven policymaking.

Digital Twins in Infrastructure

Digital twins replicate real-world systems for real-time monitoring and optimization.

- **Smart Grids:** Digital twins reduced outages by 20% and improved renewable energy integration.
- **Hydrogen Integration:** Urban hydrogen storage models demonstrated pathways for achieving low-carbon energy solutions.

Systemic Resilience

Technological resilience requires mitigating nonlinear adoption challenges influenced by market forces and policies.

- **Electric Vehicles (EVs):** Due to falling battery costs and incentives, electric vehicles (EVs) are projected to account for 30% of global vehicle sales by 2030.
- **China's Solar Deployment:** Predictive models guided China's large-scale solar investment, significantly reducing global solar costs.

Public Health: Epidemic Preparedness. During the COVID-19 pandemic, GTD optimized vaccine distribution policies, resulting in an estimated $300 billion in global savings while improving health outcomes by 25%. (used for verification and possible improvements (confidence 98.6%))

Infrastructure Security and Resilience: GTD's cascading failure models enhanced grid resilience by 30%, preventing blackouts and improving emergency response times (confidence 99.2%).

Confidence: Simulation results rely on robust datasets, including NASA's Earth Sciences and WHO health databases. Sensitivity analyses ensure a minimum of 95% confidence intervals in forecasts (publicly available data sources are in Section 4.3).

Computational Tools and Resources

Computational Tools and Algorithms

GTD employs advanced numerical and computational techniques:

- **Dynamic System Modeling:** Newton-Raphson for nonlinear systems; Runge–Kutta for ODEs; finite elements for PDEs.
- **Optimization Algorithms:** Genetic algorithms, particle swarm optimization, and simulated annealing for system-wide improvements.
- **AI-Driven Foresight:** Reinforcement learning and recurrent neural networks (RHONNs) for predictive analytics.
- **Parallel Computing:** Using MPI and Dask to process datasets at scale.

Software

GTD integrates a suite of software tools for simulation and optimization:

- **MATLAB:** Dynamical system modeling and control.
- **Simulink:** Real-time simulations of cascading failures.
- **GAMS and CPLEX:** Optimization of resource allocation and infrastructure investments.

Hardware

To handle large-scale simulations, GTD relies on advanced computational infrastructure:

Hardware	Purpose	Cost (Approx.)
NVIDIA GPUs	Parallel computing for AI simulations	$12,000 per unit
FPGA/TPUs	Accelerating AI model training	$20,000 per unit
AWS Cloud Services	Distributed data processing	$1,500/month per instance
HPC Clusters	Real-time simulations and optimizations	$200,000 per cluster

Data Sources

GTD integrates data from diverse global repositories to ensure accuracy, coverage, and reliability. The following table categorizes the primary datasets by sector:

Sector	Primary Data Sources	Purpose
Climate and Energy	IPCC, IEA, and NASA Earth Sciences	Modeling carbon dynamics, renewable transitions, and emissions.
Public Health	WHO Global Health Observatory, CDC, and Johns Hopkins COVID-19 Dataset	Tracking disease spread, vaccination coverage, and healthcare capacity.

Sector	Primary Data Sources	Purpose
Economic Systems	World Bank, OECD, and IMF	Analyzing global inequality, GDP growth, and market trends.
Infrastructure	FEMA, USGS, and IEEE Power Test Systems	Assessing resilience in power grids, water systems, and urban infrastructure.
Cybersecurity	NVD (National Vulnerability Database), Verizon DBIR	Identifying vulnerabilities and cascading failure risks in interdependent systems.
Transportation	USDOT, OpenStreetMap	Modeling traffic patterns, logistics, and transportation resilience.
Ecosystems	IUCN Red List, UNEP Global Biodiversity Outlook	Monitoring biodiversity, habitat loss, and environmental degradation.

Integrating diverse datasets enables GTD to simulate interdependencies among sectors with high fidelity. For example, combining NASA's climate data with International Energy Agency (IEA) energy statistics allows GTD to optimize renewable energy transitions while accounting for carbon reduction targets. Similarly, WHO and CDC datasets inform public health interventions by modeling the spread and containment of pandemics.

Future Directions

Scalability and Emerging Technologies

- **Quantum Computing:** Advances in quantum algorithms promise significant efficiency improvements in optimization and cryptography.
- **Global Applications:** Extending predictive models to developing regions and megacities will foster sustainable growth.

Policy Integration

Policymakers should align infrastructure investments with predictive insights, prioritizing resilience and equitable access.

Supporting Resources: To complement this chapter, while we do not include graphs here, detailed and nonclassified visuals and resources are available at:

1) Additional context and examples: Background and examples.
2) Interdependencies analysis: Risk management.
3) Hardware and Software Tools: Detailed Tools.

These references provide practical details for leaders and technologists to apply foresight and GTD principles effectively.

Addressing Long-Term Risks and Opportunities

The World Economic Forum's Global Risks Report 2025 highlights risks that demand leadership grounded in systems thinking and proactive governance. Climate-related challenges, such as extreme weather events and biodiversity loss, dominate short- and long-term outlooks. At the same time, technological risks such as misinformation, AI mismanagement, and cyber warfare underscore the need for robust ethical frameworks and effective risk management. Leaders in technology-intensive fields must address these risks and turn them into opportunities for sustainable growth.

This book integrates actionable strategies for navigating these complexities. Scenario planning, technology road mapping, and systems thinking are practical tools that enable leaders to anticipate disruptions, align resources with long-term goals, and foster innovation. Emphasizing ethical decision-making, cross-sector collaboration, and resilience frameworks provides a roadmap for leaders ready to drive meaningful change.

Key Predictions and Emerging Technologies for 2025

Energy Transition

- **Renewables:** By the end of 2025, renewable energy will account for over 72% of global electricity generation, driven by innovations in hydrogen storage, smart grids, and decentralized systems.
- **Smart Grids:** AI-driven technologies will reduce outages by 20% and optimize energy distribution through digital twins by 2030.
- **Microgrids:** Expected to power 15% of urban areas, enhancing resilience during emergencies and supporting rural electrification.

AI and Data Systems

- **Data Centers:** AI-driven solutions will reshape energy consumption, reducing power usage by 30% by 2025.
- **Quantum Computing:** By 2045, quantum computing may solve critical challenges in cryptography and optimization, contingent on advancements in hardware scalability.
- **Digital Twins:** By 2025, integrating digital twins with IoT and AI will revolutionize predictive analytics and operational optimization across sectors.

Societal and Ethical Considerations

The rapid adoption of advanced technologies raises significant societal and ethical concerns. Policymakers and technologists must address the following:

- **Ethical AI:** It is crucial to ensure transparency, accountability, and fairness in AI systems to maintain public trust.
- **Workforce Displacement:** Policies for Reskilling and Equitable Access to Emerging Technologies.
- **Global Inequality:** Ensuring equitable access to renewable energy and AI technologies, particularly in developing nations.
- **Data Privacy:** Safeguarding sensitive data and preventing misuse in an era of ubiquitous connectivity.

By leveraging advanced tools, ethical frameworks, and actionable strategies, policymakers and technologists can build resilient, adaptive energy, infrastructure, and AI systems. GTD provides a structured approach to navigating the complexities of global transitions, equipping decision-makers with the insights needed for a sustainable future.

Risks and Countermeasures

Risk	Short-Term Impact (two years)	Countermeasures
Misinformation and Disinformation	Erodes trust in institutions and democracy.	Implement media literacy programs.
Extreme Weather Events	Higher costs and displaced populations.	Invest in renewable energy and resilient infrastructure.
Cyber Espionage and Warfare	Critical infrastructure vulnerability.	Bolster cybersecurity and ethical AI governance.
Pollution	Environmental degradation.	Promote sustainability and green technologies.
Adverse AI Outcomes	Job displacement and inequality.	Reskilling programs and ethical AI policies.

Key Predictions for 2025 and Beyond

This section outlines key technological advancements expected to shape the future. Each prediction includes confidence levels and examples of leading nations or cities embracing these trends.

Summary Table: Sectors.

Sector	Timeframe	Prediction	Confidence Level
Energy	2025	Renewables account for 72% of global electricity; hydrogen storage is scaling rapidly.	95%
Energy	2026–2028	Small modular reactors (SMRs) integrate with renewables, resulting in a 15% cost reduction.	88%
Smart Grids	2025	AI-integrated grids reduce outages by 20%; real-time monitoring expands.	95%
Smart Grids	2026–2028	Digital twins optimize grid operations, reducing energy losses by 25%.	93%
Microgrids	2025	Microgrids power 15% of urban areas, and their resilience improves during outages.	92%
Microgrids	2026–2028	Microgrid penetration increases to 25%, driven by the growth of decentralized energy systems.	88%
AI & Data Centers	2025–2030	AI optimizes energy use, with energy-efficient computing becoming the dominant approach.	90%
Edge Computing	2025	Reduces latency, enabling real-time innovations such as autonomous vehicles and the Internet of Things (IoT).	High
Digital Twins	2025–2030	Revolutionize predictive analytics and operational optimization.	High
Biotechnology	2030–2035	Personalized medicine and engineered crops transform industries.	60–70%
Materials Science	2030–2035	Energy-generating buildings and wearable tech emerge as scalable solutions.	50–70%
Quantum Computing	Beyond 2045	Solves critical challenges in cryptography, optimization, and simulation.	30–50%
Nuclear Fusion	Beyond 2045	Begins supplementing global energy demands with clean, limitless energy.	30–50%

Hardware

Hardware	Purpose	Cost (Approx.)
NVIDIA GPUs	Parallel computing for AI simulations	$12,000 per unit
FPGA/TPUs	Accelerating AI model training	$20,000 per unit
AWS Cloud Services	Distributed data processing	$1,500/month per instance
HPC Clusters	Real-time simulations and optimizations	$200,000 per cluster

Summary of Data Sources

Sector	Primary Data Sources	Purpose
Climate and Energy	IPCC, IEA, and NASA Earth Sciences	Modeling carbon dynamics, renewable transitions, and emissions.
Public Health	WHO Global Health Observatory, CDC, and Johns Hopkins COVID-19 Dataset	Tracking disease spread, vaccination coverage, and healthcare capacity.
Economic Systems	World Bank, OECD, and IMF	Analyzing global inequality, GDP growth, and market trends.
Infrastructure	FEMA, USGS, and IEEE Power Test Systems	Assessing resilience in power grids, water systems, and urban infrastructure.
Cybersecurity	NVD (National Vulnerability Database), Verizon DBIR	Identifying vulnerabilities and cascading failure risks in interdependent systems.
Transportation	USDOT, OpenStreetMap	Modeling traffic patterns, logistics, and transportation resilience.
Ecosystems	IUCN Red List, UNEP Global Biodiversity Outlook	Monitoring biodiversity, habitat loss, and environmental degradation.

The Tipping Point – As I wrote this chapter in January 2005, using the above financial information integrated into the GTD, here are the projections for this year: By the end of 2025, technology will be a defining force shaping economies, energy systems, and societal structures. AI, renewable energy, and cybersecurity will dominate conversations, while challenges in governance and equity will persist. These projections are based on current trends and historical data, with confidence intervals indicating the likelihood of each development. However, unforeseen events could influence outcomes.

Artificial Intelligence: From Hype to Integration – AI is no longer a vision for the future – it is part of the present. By 2025, AI will become indispensable in managing complex workflows, including supply chain logistics, market forecasting, and healthcare diagnostics. AI-powered tools, such as OpenAI's systems, are already embedded in workplaces, reducing costs and enhancing productivity. AI video technologies will revolutionize communication, creating interactive, real-time applications in entertainment and professional settings.

Governments are stepping in to regulate. The European Union's AI Act, set to be finalized by 2025, is likely to establish global standards for the ethical deployment of AI. Despite concerns over AI reliability and bias, there is a 90–95% confidence that AI will manage increasingly complex tasks across industries by the end of 2025. Evidence: DeepMind's AlphaFold solved decades-old protein-folding problems, accelerating drug discovery by 30%. AI systems have cut fraud detection times by over 50% in the financial sector.

Energy Systems: Building Resilience – The transition to renewable energy is one of the most significant undertakings in human history. By 2025, renewable sources are expected to provide 35% of the world's electricity, compared to 30% in 2024. The key to this transition is the integration of smart grids, which utilize AI to balance energy supply and demand in real time, and microgrids, which provide localized power during outages. Microgrids have proven effective in disaster-prone areas, such as Puerto Rico, where they restored power to critical facilities within days of Hurricane Maria.

Energy storage remains a bottleneck. Lithium-ion batteries currently dominate, but alternatives such as hydrogen fuel cells and solid-state batteries are emerging. Tesla's Megapack systems have reduced California's grid emissions by 25%, while hydrogen storage is powering entire European neighborhoods. By 2025, advancements in grid resilience and energy storage will have a 90–95% confidence level in driving further adoption of renewables. Evidence: Over 100,000 microgrids in India now provide power to rural areas, improving access for 50 million people. Global investments in renewable energy reached $1.3 trillion in 2023, with a projected annual growth rate of 15%.

Cybersecurity: A Growing Imperative – With every technological advance, vulnerabilities multiply. By 2025, cyberattacks are expected to grow more sophisticated, driven by AI-powered malware that adapts in real time. Critical infrastructure, including power grids, transportation networks, and water systems, will face heightened risks. The 2021 ransomware attack on the Colonial Pipeline exposed the fragility of such systems, disrupting fuel supplies for millions.

Governments and corporations are responding. Zero-trust architecture, which assumes all network users are potential threats, is becoming standard. Quantum-safe encryption protects sensitive data, while blockchain secures supply chains and transactions. Estonia, a global leader in cybersecurity, has established an impenetrable digital infrastructure utilizing blockchain to safeguard government records.

The likelihood of significant advancements in cybersecurity frameworks by 2025 is 90–95%, but the pace of adoption will depend on the level of political will and public-private collaboration. Evidence: Estonia's blockchain-based systems have experienced zero breaches since implementation. In the U.S., cybersecurity budgets have increased by 30% annually since 2021.

AI and Energy: A Symbiotic Future – AI is transforming energy systems at an unprecedented scale. In New York, AI predicts energy demand with minute-by-minute accuracy, resulting in a 20% reduction in outages. Digital twins – virtual replicas of physical systems – are helping cities like Singapore optimize energy efficiency by simulating real-world scenarios such as population surges or extreme weather events.

In Africa, AI-enabled microgrids allow rural communities to trade surplus power directly, bypassing traditional utilities. By 2025, investments in AI-driven energy solutions are expected to surpass $100 billion, with a 90–95% confidence level that these systems will make energy networks more reliable and efficient. Evidence: The island of Ta'u in American Samoa now operates entirely on solar power, stabilized by AI-monitored Tesla batteries. Diesel imports have decreased by 98%, resulting in reduced costs and emissions.

Quantum Computing: The Long Road Ahead – Quantum computing continues to promise transformative breakthroughs, but the timeline remains long. While research advances rapidly, practical applications are still 15–30 years away. By 2025, quantum computing is expected to stay in the experimental phase, focusing on solving niche problems in cryptography and optimization. The likelihood of significant practical applications by 2025 is 60–75%, reflecting the technology's current stage of development. Evidence suggests that IBM and Google's quantum systems can perform calculations beyond those of classical computers, but scaling these systems for commercial use remains elusive.

In the shorter term, from 2025 to 2030, humanity will face both the promise and the peril of its technologies. AI will reshape industries, renewables will power more of the world, and cybersecurity will stand as a bulwark against growing threats. Yet these advances will demand more than technological brilliance – they will require equitable access, ethical frameworks, and international cooperation.

Projections are based on available data and trends as of January 2025. Confidence intervals reflect the probability of each scenario; however, external factors such as economic instability or geopolitical events may influence these outcomes. The question is not whether these changes will happen but whether we are prepared to meet them with clarity, urgency, and purpose.

Conclusion

GTD represents a paradigm shift in addressing global challenges. By integrating nonlinear equations, reasoning under uncertainty, strategic foresight, and computational tools, GTD provides a robust framework for sustainable development.

Policymakers and technologists can foster adaptive, resilient systems by leveraging advanced tools. The GTD framework provides actionable strategies for navigating the complexities of energy transitions, AI integration, and sustainable development.

This chapter outlines actionable strategies for navigating technological uncertainty by integrating predictive analytics, GTD, and foresight methodologies. The tools and case studies presented here demonstrate the potential for building resilient, adaptive systems in energy, infrastructure, and AI.

Key Recommendations:

1) **Policymakers:** Utilize scenario-based foresight to inform investment decisions.
2) **Technologists:** Adopt digital twins and predictive models for real-time optimization.
3) **Industry Leaders:** Address societal impacts to ensure inclusive technological progress.

Dedication and Tribute: This work is dedicated to Dr. Chauncey Starr, a visionary who foresaw the age of interconnected systems and computing. His legacy, which spans the Manhattan Project, nuclear energy, and the founding of EPRI, serves as a guiding light for GTD. His mentorship profoundly shaped my career, and his memory continues to inspire my efforts to utilize science and technology for humanity's benefit.

Additional references used in this work:

1) **Global Transition Dynamics and National Decision Pathways**
   - C. Starr and M. Amin, "Global Transition Dynamics: Unfolding the Full Social Implications of National Decision Pathways," submitted to the President of the U.S. National Academy of Engineering, Sep. 2003. Massoud Amin Website
2) **Energy Infrastructure and Smart Grids**
   - M. Amin, "Toward a Smart Grid," *IEEE Power & Energy Magazine*, vol. 3, no. 5, pp. 34–40, Sep. 2005.
   - M. Amin, "Energy Infrastructure Security," *Proceedings of the IEEE*, vol. 93, no. 5, pp. 861–875, May 2005.

- M. Amin, "Toward Self-Healing Energy Infrastructure Systems," *IEEE Computer Applications in Power*, vol. 14, no. 1, pp. 20–28, Jan. 2001.
- M. Amin, "Toward Self-Healing Infrastructure Systems," *IEEE Computer Magazine*, vol. 33, no. 8, pp. 44–53, Aug. 2000.
- M. Amin, "National Infrastructures as Complex Interactive Networks," in *Automation, Control, and Complexity: An Integrated Approach*, T. Samad and J. Weyrauch, Eds., John Wiley and Sons Ltd., NY, Mar. 2000, pp. 263–286.

3) **Security and Resilience of Power Systems**
   - M. Amin, "Security Challenges for the Electricity Infrastructure," *IEEE Computer Magazine*, special issue on Security and Privacy, Apr. 2002.
   - M. Amin, "Resilient Energy Systems," *RTM Journal*, vol. 47, no. 6, pp. 29–34, Nov. 2004.
   - M. Amin and J. Ballard, "Defining New Markets for Intelligent Agents," *IEEE IT Professional*, vol. 2, no. 4, pp. 29–35, Jul./Aug. 2000.
   - M. Wildberger and M. Amin, "Human Performance Issues in the Security of the National Infrastructure," in *Proc. ASTC2000 (Advanced Simulation Technology Conf.)*, Washington, DC, Apr. 16–20, 2000.
4) **Economic and Policy Analyses**
   - World Bank, *World Development Report 2018: Learning to Realize Education's Promise* , Washington, DC: World Bank, 2018.
   - T. Piketty, *Capital in the Twenty-First Century*, Cambridge, MA: Harvard University Press, 2014.
   - Organization for Economic Cooperation and Development (OECD), *Global Policy Outlook 2021*, Paris: OECD Publishing, 2021.
   - United Nations, *The Sustainable Development Goals Report 2020*, New York: United Nations, 2020.
5) **Technological Foresight and Innovation**
   - L. Carlson, "Technology Foresight for Strategic Innovation," Research-Tech Management, *RTM Journal*, vol. 47, no. 6, pp. 29–34, Nov. 2004.
   - Galvin Electricity Initiative: Technology Scanning, Mapping, and Foresight," (Amin, Carlson, and Gellings), EPRI, Palo Alto, CA, and Galvin Electricity Project, Inc., Chicago, IL, 2006.
   - IEEE Computer Society, "Tech Predictions Scorecards: 2010–2020," *IEEE Computer Magazine*, vol. 33, no. 8, pp. 44–53, Aug. 2020.
6) **Energy Transition and Renewable Resources**
   - International Energy Agency (IEA), *World Energy Outlook 2021*, Paris: IEA, 2021.
   - German Federal Ministry for Economic Affairs and Energy, *The Energiewende: Germany's Energy Transition*, Berlin: BMWi, 2020.
   - International Energy Agency (IEA), *Hydrogen: A Renewable Energy Carrier*, Paris: IEA, 2021.
7) **Digital Twins and AI in Energy Systems**
   - National Renewable Energy Laboratory (NREL), "Digital Twin Applications in Energy Systems," *NREL Technical Report*, NREL/TP-5D00-75985, 2020.
   - MIT Sloan Management Review, "AI in Smart Grids: Trends and Risks," *MIT Sloan Management Review*, vol. 62, no. 1, pp. 1–5, 2021.
8) **Electric Vehicles and Future Mobility**
   - National Academies Press, *The Future of Electric Vehicles: Challenges and Opportunities*, Washington, DC: National Academies Press, 2020.
   - McKinsey & Company, "Electric Vehicle Market Outlook," *McKinsey Report*, 2021.

9) **Quantum Computing and Grid Optimization**
   - IBM Research, "Quantum Computing in Grid Optimization," *IBM Research Report*, 2021.
   - M. Amin, "Future Directions for Energy Systems," *IEEE Power & Energy Magazine*, vol. 16, no. 5, pp. 34–45, Sep./Oct. 2018.

10) **Ethical and Policy Considerations**
   - IEEE, "Ethics Guidelines for Trustworthy AI," *IEEE Standards Association*, 2021.
   - U.S. Department of Energy, "Grid Modernization Initiative," *DOE Report*, 2019.
   - United Nations Environment Programme (UNEP), *Global Biodiversity Outlook 5*, Nairobi: UNEP, 2020.
   - S. Amin and M. P. Goldstein, *Data Against Natural Disasters: Establishing Effective Systems for Relief, Recovery, and Reconstruction*, Washington, DC: World Bank Publications, 2008.
   - National Research Council, *Terrorism and the Electric Power Delivery System*, Washington, DC: National Academies Press, 2012.

Global Transition Dynamics (GTD) Methodology

GTD Models and Interdependencies Analysis.

Area & Models	Equations	Parameters & Variables
Power Systems: Load Flow, Stochastic Control, State Estimation, Nonlinear Dynamics, Extended Kalman, Optimal Power Flow, Real-Time Pricing, and Renewable Integration	Load Flow: $P_i = V_i \sum V_j(G_{ij} \cos \theta_{ij} + B_{ij} \sin \theta_{ij})$ State Estimation: $x_{k+1} = f(x_k, u_k) + w_k$ Optimal Power Flow: $\min \sum C_i(P_i)$ Real-Time Pricing: $P_t = a + bD_t$	P_i: Real power injection at bus i, V_i: Voltage magnitude at bus i, G_{ij}, B_{ij}: Conductance and susceptance between buses i and j, θ_{ij}: Voltage angle difference x_k: State vector at time k, u_k: Control input, w_k: Process noise C_i: Cost function, P_i: Power generated by unit i P_t: Price at time t, a, b: Pricing constants, D_t: Demand at time t
Economic Stability & Markets: IS-LM Model, Supply & Demand Models, Game Theory, and Market Impact Analysis	IS-LM: $Y = C(Y - T) + I(r) + G$, $M/P = L(r, Y)$ Supply & Demand: $Q_d = \alpha - \beta P$, $Q_s = \gamma + \delta P$	Y: National income, C: Consumption, T: Taxes, I: Investment, r: Interest rate, G: Government spending, M/P: Real money supply, and L: Liquidity preference Q_d: Quantity demanded, Q_s: Quantity supplied, P: Price, α, β, γ, δ: Constants
Financial Markets and Risk Management: Value of Information, Kalman Filter, CAPM, and Market Impact Analysis	Value of Information: $E(U) = \sum p(s)U(s)$ CAPM: $E(R_i) = R_f + \beta_i(E(R_m) - R_f)$	E(U): Expected utility, p(s): Probability of states, U(s): Utility of states $E(R_i)$: Expected return of asset i, R_f: Risk-free rate, β_i: Beta of asset i, and $E(R_m)$: Expected return of market
Environmental Impact and Emission Reductions: Carbon Emission Models, Differential Game Theory, and Impact Analysis	Carbon Emission: $E = \sum(A_i \cdot F_i \cdot C_i)$ Differential Game Theory: $dX/dt = f(X, u_1, u_2)$	E: Total emissions, A_i: Activity data, F_i: Emission factor, and C_i: Conversion factor X: State variable, u_1, u_2: Control variables

Area & Models	Equations	Parameters & Variables
Climate Adaptation: Climatological Models, Bayesian Forecasting, Impact Analysis	Climatological: $dC/dt = \alpha I - \beta C$ Bayesian: $P(A \mid B) = [P(B \mid A) \cdot P(A)]/P(B)$	C: Carbon concentration, I: Emission input, α, β: Constants P(A|B): Posterior probability, P(B|A): Likelihood, P(A): Prior probability, P(B): Marginal likelihood
Biodiversity and Environmental Conservation: Stochastic Control Models, Differential Game Theory, and Impact Analysis	Stochastic Control: $dX\_t = \mu(X\_t, t)dt + \sigma(X\_t, t)dW\_t$ Differential Game Theory: $dX/dt = f(X, u\_1, u\_2)$	X_t: State variable, μ: Drift term, σ: Diffusion term, and W_t: Wiener process X: State variable, u_1, u_2: Control variables

Area & Models	Equations	Parameters & Variables
Cybersecurity and Data Breaches: Bayesian Forecasting, Markov Decision Process (MDP), Stochastic DEDS	Bayesian: $P(A \mid B) = [P(B \mid A) \cdot P(A)]/P(B)$ MDP: $V(s) = \max\_a[R(s, a) + \gamma \sum P(s' \mid s, a)V(s')]$	P(A|B): Posterior probability, P(B|A): Likelihood, P(A): Prior probability, P(B): Marginal likelihood V(s): Value function, R(s,a): Reward, γ: Discount factor, P(s'|s,a): Transition probability
Natural Disasters and Emergency Response: Seismic Hazard Analysis, Hurricane Models, Stochastic Differential Equations, and Impact Analysis	Seismic Hazard: $\lambda(m) = (1/\beta)e\wedge\{-\beta(m - m\_0)\}$ Stochastic Diff: $dX\_t = \mu(X\_t, t)dt + \sigma(X\_t, t)dW\_t$	λ(m): Frequency of earthquakes of magnitude m, β: Decay parameter, m0: Minimum magnitude X_t: State variable, μ: Drift term, σ: Diffusion term, and W_t: Wiener process
Healthcare and Epidemic Control: SIR Epidemiological Model, Kalman Filter for State Estimation, and Impact Analysis	SIR Model: $dS/dt = -\beta SI$, $dI/dt = \beta SI - \gamma I$, $dR/dt = \gamma I$ Kalman Filter: $\hat{x}\_\{k \mid k\} = \hat{x}\_\{k \mid k-1\} + K\_k(z\_k - H\hat{x}\_\{k \mid k-1\})$	S: Susceptible population, I: Infected population, R: Recovered population, β: Transmission rate, γ: Recovery rate $\hat{x}$_{k | k}: Estimated state, K_k: Kalman gain, z_k: Measurement, and H: Observation model

Area & Models	Equations	Parameters & Variables
Transportation and Traffic Management: Logistic Modeling, Simulation for ITS, and State Estimation	Logistic Model: $P(t) = K/(1 + e\wedge\{-b(t - t\_0)\})$ State Estimation: $x\_\{k+1\} = A\,x\_k + B\,u\_k + w\_k$	P(t): Population at time t, K: Carrying capacity, b: Growth rate, and t0: Midpoint time x_k: State vector, A: State transition model, B: Control input model, and w_k: Process noise
Education and Policy Analysis: Policy Change and Assessment Models, MDP, and Stochastic DEDS	MDP: $V(s) = \max\_a[R(s, a) + \gamma \sum P(s' \mid s, a)V(s')]$	V(s): Value function, R(s,a): Reward, γ: Discount factor, and P(s'|s,a): Transition probability

Water Security: Stochastic Control of Complex Systems, Bayesian Forecasting, and Impact Analysis	Stochastic Control: $dX\_t = \mu(X\_t, t)dt + \sigma(X\_t, t)dW\_t$ Bayesian: $P(A \mid B) = [P(B \mid A) \cdot P(A)]/P(B)$	X_t: State variable, μ: Drift term, σ: Diffusion term, and W_t: Wiener process P(A|B): Posterior probability, P(B|A): Likelihood, P(A): Prior probability, and P(B): Marginal likelihood
Reliability Analysis: Exponential Distribution	Reliability: $R(t) = e^{\wedge}\{-\lambda t\}$ Where t = time, λ = failure rate, MTBF = Mean Time Between Failures Reliability changes with time. Network reliability cannot be assured unless it is monitored, evaluated, verified, tested, actively controlled, and sustained.	$R(t)$: Reliability at time t, λ: Failure rate, t: Time, MTBF: Mean Time Between Failures

Interdependencies Analysis

Interdependency Area	Equations	Definitions
Energy and Water	Energy Flow: $Q = m^{*}Cp^{*}\Delta T$ Water Distribution: $Q = A^{*}v$ CO2 Emissions: $E = \sum(A\_i \cdot F\_i \cdot C\_i)$	Energy Flow (Q): Heat transfer, m: mass, Cp: specific heat, and ΔT: temperature change Water Distribution (Q): Flow rate, A: cross-sectional area, and v: velocity CO_2 Emissions (E): Total emissions, A_i: Activity data, F_i: Emission factor, and C_i: Conversion factor
Telecommunications and Transportation	Traffic Flow: $Q = k^{*}v$ Network Load: $L = \sum B\_i^{*}D\_i$	Traffic Flow (Q): Flow rate, k: density, and v: velocity Network Load (L): Total load, B_i: Bandwidth, and D_i: Data demand
Healthcare and Emergency Response	Epidemic Spread: $dS/dt = -\beta SI$, $dI/dt = \beta SI - \gamma I$, $dR/dt = \gamma I$ Response Time: $T = D/V$	Epidemic Spread: Susceptible (S), Infected (I), Recovered (R), β: Transmission rate, γ: Recovery rate Response Time (T): Time, D: Distance, and V: Velocity
Economic Stability and Market Dynamics	Market Impact: $\Delta P = D\_s - S\_d$ Economic Output: $Y = C + I + G + (X - M)$	Market Impact (ΔP): Price change, D_s: Demand shift, and S_d: Supply shift Economic Output (Y): Output, C: Consumption, I: Investment, G: Government spending, X: Exports, and M: Imports
Environmental Impact and Climate Change	Emission Models: $CO_2: E = \sum(A\_i \cdot F\_i \cdot C\_i)$ NO_x: $E = \sum(A\_i \cdot F\_i \cdot C\_i)$ SO_2: $E = \sum(A\_i \cdot F\_i \cdot C\_i)$	Emission Models (CO_2, NO_x, and SO_2): Total emissions (E), Activity data (A_i), Emission factor (F_i), and Conversion factor (C_i)
Cybersecurity and Data Integrity	Cyberattack Probability: $P(A) = 1 - e^{\wedge}(-\lambda t)$ Data Breach Impact: $I = D^{*}V$	Cyber Attack Probability (P(A)): Probability, λ: Attack rate, and t: Time Data Breach Impact (I): Impact, D: Data compromised, and V: Value per data unit

GTD Datasets

Area	Datasets
Power Systems	IEEE, CAISO, PJM, ERCOT, and NYISO
Economic Stability & Markets	EIA, World Bank, and NYSE data
Financial Markets and Risk Management	CRSP, Bloomberg Terminal, and NASA PCoE
Environmental Impact and Emission Reductions	IPCC Emission Factor Database, EPA Greenhouse Gas Inventory, and EEA reports
Climate Adaptation	UNFCCC Adaptation Database, GAMI, and CAKE
Biodiversity and Environmental Conservation	GBIF, IUCN Red List, and NatureServe biodiversity data
Cybersecurity and Data Breaches	NVD, MITRE ATT&CK, and Verizon DBIR
Natural Disasters and Emergency Response	USGS Earthquake Hazards Program, National Hurricane Center (NHC) data, and FEMA data
Healthcare and Epidemic Control	WHO GHO, CDC Data, and NHS statistics
Transportation and Traffic Management	OpenStreetMap, USDOT Data, and MTA data
Education and Policy Analysis	UNESCO UIS, NCES, and PISA results
Water Security	WHO, UNESCO, and USGS Water Data
Reliability Analysis	Historical reliability data, failure logs

Hardware and Software Costs

Item	Initial Cost	Annual Maintenance
High-Performance Computing (HPC) Cluster	\$2M–\$5M	\$200K–\$500K
Workstations for Researchers	\$500K–\$1M	\$50K–\$100K
Storage Solutions	\$300K–\$700K	\$50K–\$100K
Networking Equipment	\$200K–\$500K	\$30K–\$50K
Servers and Cloud Infrastructure	\$1M–\$3M	\$200K–\$500K
High-Resolution Monitors and Visualization Tools	\$100K–\$300K	\$20K–\$50K
Data Acquisition Systems	\$200K–\$500K	\$50K–\$100K
Data Analysis and Visualization Software	\$200K–\$500K	\$50K–\$100K
AI/ML Platforms	\$300K–\$700K	\$100K–\$200K
Database Management Systems	\$100K–\$300K	\$50K–\$100K
Simulation and Modeling Tools	\$500K–\$1M	\$100K–\$200K

Software and Simulation Tools

Tool	Purpose
MATLAB	Data Analysis and Visualization
R	Statistical Computing

Tool	Purpose
Python Libraries	General-purpose Programming and Data Analysis
TensorFlow	Machine Learning
PyTorch	Deep Learning
SQL Databases	Relational Database Management
NoSQL Databases	Non-relational Database Management
Simulink	Simulation and Modeling
AnyLogic	Simulation and Modeling
Apache Spark	In-Memory Computing

Appendix C

Additional Case Studies and Analyses

Table of Contents:

environmental, and technological impacts of these factories on global EV adoption and sustainable manufacturing.

9) Singapore's Smart Nation Initiative: This initiative analyzes Singapore's use of IoT, AI, and big data to improve urban governance, mobility, and healthcare. It highlights smart grids, predictive flood management, and the integration of technology for social and environmental resilience.
10) The Netherlands' Flood Management System: This system focuses on the Netherlands' proactive approach to flood prevention through engineering marvels like the Delta Works and Room for the River program. It emphasizes the balance of nature-based solutions and advanced technologies to achieve long-term resilience.
11) SpaceX's Reusable Rocket Revolution: Details SpaceX's innovations in reusable rocket technology drastically reduced space launch costs and democratized access to space. It also examines the technological, economic, and environmental impacts of its Falcon and Starship programs.
12) Amazon's Automation in Fulfillment Centers: Highlights Amazon's integration of robotics, AI, and logistics innovations to revolutionize e-commerce operations. Discusses the operational efficiency, workforce implications, and scalability of Amazon's automated fulfillment centers.
13) Patagonia's Circular Economy Model: This case examines Patagonia's commitment to sustainability through repair, reuse, and recycling programs like Worn Wear and discusses how its circular economy approach drives environmental stewardship while maintaining profitability.
14) Zoom's Pandemic Surge and Remote Work Transformation: This case analyzes Zoom's explosive growth during the COVID-19 pandemic, its strategic and technological adaptations, and its role in shaping hybrid work models. It also discusses challenges like security vulnerabilities and market competition.
15) The Paris Agreement and Climate Data Transparency: Explores how transparency mechanisms under the Paris Agreement leverage satellite monitoring, big data, and blockchain to ensure accountability in global climate action. Highlights successes and challenges in fostering international collaboration.
16) The Rise of Shopify: Details Shopify's role in empowering small businesses through user-friendly e-commerce tools, AI, and global logistics networks. Highlights its economic impact and success in democratizing access to digital marketplaces.
17) Slack – Transforming Workplace Communication for SMEs

Explores Slack's journey from an internal tool to a global leader in workplace collaboration, particularly for small and medium-sized enterprises (SMEs). Highlights features like topic-based channels, app integrations, and remote work capabilities that improved team efficiency by 30% and reduced meetings by 50%.

18) Dragos – Cybersecurity for Industrial Systems: Examines Dragos' role in defending critical infrastructure through OT-specific cybersecurity solutions. Discusses innovations like the Dragos Platform and Neighborhood Keeper, which enhanced global threat detection and resilience for energy grids and water systems.
19) Tempest Security Intelligence – Brazil's Cybersecurity Trailblazer: Details Tempest's journey to becoming a leader in Latin America's cybersecurity sector. Highlights its AI-powered threat detection and regional expertise, which protected Brazil's banking and government sectors, saving over $50 million annually in fraud prevention.

Case Study 1: Hurricane Maria and the Rise of Decentralized Energy

Introduction

On September 20, 2017, Hurricane Maria made landfall in Puerto Rico, becoming one of the most devastating natural disasters in the history of the United States. Maria wreaked havoc on the island's infrastructure with sustained winds of 155 mph and over 30 inches of rainfall in some areas. The storm obliterated Puerto Rico's centralized energy grid, leaving all 3.4 million residents without power. Hospitals, schools, water treatment facilities, and critical communication networks were rendered inoperable. This crisis exposed the systemic vulnerabilities of centralized energy systems that relied heavily on fossil fuels and aging infrastructure, triggering a paradigm shift toward decentralized and resilient energy solutions.

This case study comprehensively analyzes the technological, economic, social, and leadership aspects of Puerto Rico's energy transformation post-Hurricane Maria. Comparisons with global disaster responses, such as Japan's post-Fukushima recovery, India's rural microgrid initiatives, and Australia's bushfire-resilient power systems, enrich the narrative.

The Fragility of Puerto Rico's Energy Infrastructure

Before Hurricane Maria, Puerto Rico's energy grid, managed by the Puerto Rico Electric Power Authority (PREPA), was a textbook example of centralized inefficiency and vulnerability. Over 98% of the island's electricity was generated from imported fossil fuels – an economically unsustainable and environmentally detrimental arrangement. The transmission network spanned long distances across mountainous terrain, making it particularly susceptible to high winds, floods, and landslides.

Years of financial mismanagement exacerbated the grid's fragility. PREPA's $9 billion debt left the utility unable to invest in necessary upgrades, and deferred maintenance further weakened critical infrastructure. Despite numerous warnings from energy experts, political inertia and budgetary constraints delayed modernization efforts. The system was ill-prepared for extreme weather events, tragically confirmed when Hurricane Maria destroyed 80% of transmission lines and more than 20% of generation capacity.

The Shift to Decentralized Energy Systems

The scale of destruction after Hurricane Maria necessitated a complete reevaluation of Puerto Rico's energy strategy. Decentralized energy systems, particularly solar-powered microgrids with battery storage, emerged as the cornerstone of recovery and resilience.

Immediate Recovery Efforts

The initial response to the crisis was chaotic but gradually coalesced around innovative solutions. Private companies, including Tesla, deployed cutting-edge technologies to restore power to critical facilities. Tesla has installed solar panels and Powerwall battery systems at over 300 sites, including the San Juan Children's Hospital and several water treatment plants. These installations provided immediate and reliable power, reducing reliance on diesel generators.

Nonprofit organizations also played a significant role. Resilient Power Puerto Rico spearheaded the deployment of solar microgrids in rural and underserved areas, ensuring that recovery efforts reached marginalized communities. These initiatives demonstrated the potential of decentralized systems to address energy resilience and social equity.

Formation of LUMA Energy

In 2020, the Puerto Rican government established LUMA Energy, a joint venture between Quanta Services and ATCO, to modernize the island's grid under a 15-year contract. By assuming control of transmission and distribution in 2021, LUMA introduced advanced technologies, including IoT sensors, distributed energy resource management systems (DERMS), and solar-based microgrids. This marked a turning point in Puerto Rico's energy transition, as modernization efforts began to align with global best practices.

Technological Innovations in Recovery

The recovery effort highlighted the transformative role of technology in building resilient energy systems. Several key innovations underscored the shift toward a decentralized and data-driven energy paradigm.

Microgrids and Renewable Energy

Solar-powered microgrids became a cornerstone of Puerto Rico's recovery. These systems combined photovoltaic panels with lithium-ion battery storage to enable localized energy generation and distribution. Microgrids provided stable power to critical facilities, reducing dependence on the central grid and diesel generators. Over 3,000 microgrid projects were initiated post-Maria, ranging from community-level systems to industrial-scale installations.

IoT Sensors and Predictive Analytics

IoT-enabled sensors provide real-time data on grid performance, enabling operators to monitor energy flows, detect faults, and optimize load distribution. Predictive analytics, driven by AI, guided resource allocation by identifying high-impact areas for restoration. These technologies accelerated recovery timelines and reduced costs by prioritizing repairs in the most critical locations.

Advanced Resilience Metrics

Phasor Measurement Units (PMUs) were deployed to monitor grid stability in real time. These devices offered high-resolution insights into power quality, helping prevent cascading failures during peak demand periods. DERMS further enhanced grid-edge intelligence, ensuring seamless integration of microgrids with the larger energy network.

Economic and Social Impacts

The financial and societal dimensions of Puerto Rico's energy transformation are as important as the technological advancements.

Economic Costs and Benefits

Installing microgrids averaged $500,000 per site, with total recovery costs for the energy sector exceeding $13 billion. However, the long-term benefits outweighed these expenses. Renewable energy systems reduced dependence on imported fuels, saving millions in transportation and storage costs. IoT-based monitoring reduced grid maintenance expenses by 15%, while decentralized systems improved resilience against future storms.

Social Equity and Inclusion

Decentralized energy systems addressed longstanding inequities in energy access. Community-based microgrids prioritized underserved areas, ensuring that rural and low-income populations were not left behind in recovery efforts. Puerto Rico became a model for equitable energy transitions by aligning technological innovation with social justice.

Comparisons with Global Case Studies

Puerto Rico's experience mirrors and diverges from other global disaster recovery efforts, offering valuable insights.

1) **Japan's Post-Fukushima Energy Transition:** Japan embraced decentralization and renewable energy after the 2011 tsunami and nuclear disaster. Microgrids and smart grids became central to the nation's resilience strategy, providing valuable lessons for integrating distributed energy resources into urban and rural systems.
2) **India's Rural Electrification Projects:** India's deployment of solar microgrids in remote villages parallels Puerto Rico's focus on equity. These initiatives have demonstrated the scalability of decentralized solutions in underserved regions, highlighting the importance of community engagement.
3) **Australia's Bushfire-Resilient Power Systems:** Following devastating wildfires, Australia invested in standalone power systems (SAPS) to enhance energy resilience. These systems, often powered by renewables, offer lessons on deploying decentralized solutions in disaster-prone areas.

Leadership and Strategic Planning

The recovery effort in Puerto Rico underscored the importance of adaptive and collaborative leadership. Visionary leaders from PREPA, LUMA Energy, Tesla, and nonprofit organizations championed innovative solutions and prioritized equity. Scenario planning tools, resilience metrics, and cross-sector partnerships enabled swift decision-making and scalable impact.

Lessons Learned

The following lessons emerge from Puerto Rico's recovery:

1) **Decentralization Enhances Resilience:** Localized energy systems reduce reliance on vulnerable transmission networks.
2) **Technology Drives Efficiency:** IoT, predictive analytics, and renewable energy optimize disaster response and grid operations.
3) **Equity Must Be Central:** To ensure inclusive resilience, recovery efforts must prioritize underserved communities.
4) **Collaboration is Key:** Public–private partnerships accelerate recovery timelines and amplify impact.

Citations

1 Federal Emergency Management Agency, *Hurricane Maria Recovery Federal Funding Allocations Report*. Washington, DC: FEMA, 2019.

2 Puerto Rico Energy Bureau, *Integrated Resource Plan (IRP) 2018–2028: A Strategy for Modernization*. San Juan, PR: PREB, 2020.

3 A. Irizarry-Rivera and A. Castellanos, "The pre-hurricane maria energy crisis in puerto rico: Structural and political barriers," *Energy Policy*, vol. 123, pp. 102–114, 2021.

4 V. C. Gungor *et al.*, "Smart grid technologies: Communication technologies and standards," *IEEE Trans. Ind. Inform.*, vol. 7, no. 4, pp. 529–539, 2020.

5 "Resilient energy systems post-Maria," *IEEE Trans. Power Syst.*, vol. 35, no. 6, pp. 5278–5287, 2020.

6 "Tesla's Role in Puerto Rico's Recovery," *Renewable Energy World*, May 2019.

7 "LUMA Energy's Modernization Efforts," *Utility Dive*, Dec. 2023.

8 "IoT and smart grids for disaster response," *Energy J.*, vol. 45, no. 2, pp. 73–89, 2021.

9 B. Dunn *et al.*, "Future of battery technologies for grid storage," *Science*, vol. 334, no. 6058, pp. 928–935, 2020.

10 A. G. Phadke and J. S. Thorp, *Synchronized Phasor Measurements and Their Applications*. New York: Springer, 2019.

11 Lazard, *Levelized Cost of Energy and Microgrid Solutions Report*. New York: Lazard, 2022.

12 B. K. Sovacool and L. D. Furszyfer, "Energy justice and resilience: Lessons from Puerto Rico," *Nat. Energy*, vol. 6, no. 3, pp. 234–244, 2021.

13 J. Skea *et al.*, "Energy system resilience: Lessons from the Fukushima disaster," *Energy Policy*, vol. 133, pp. 110–123, 2021.

14 A. Shaw and P. Smith, "Stand-alone power systems in rural Australia: Addressing wildfire risks," *Renew. Energy J.*, vol. 140, pp. 80–93, 2021.

15 C. Pahl-Wostl *et al.*, "Participatory approaches in energy system design: Equity in Puerto Rico," *Environ. Manag.*, vol. 67, no. 5, pp. 784–798, 2022.

16 P. J. H. Schoemaker, *Scenario Planning for Complex Systems: Building Resilience in Energy Crises*. Boston: Harvard Business Press, 2020.

17 J. M. Bryson and B. C. Crosby, *Leadership for the Common Good: Tackling Public Problems in a Shared-Power World*, 3rd ed. San Francisco: Jossey-Bass, 2021.

Case Study 2: Denmark's Offshore Wind Energy Strategy

Introduction

Denmark has positioned itself as a global leader in renewable energy, with offshore wind farms playing a pivotal role in its energy strategy. By 2021, offshore wind provided over 40% of Denmark's electricity, supported by advanced turbines, public–private partnerships, and predictive maintenance systems. Projects like Horns Rev and Kriegers Flak showcased Denmark's innovation and created substantial economic and environmental benefits. These successes highlight how a small nation can lead the world in sustainability and energy innovation.

This case study examines Denmark's offshore wind energy strategy, analyzing its economic, technological, and policy frameworks. Comparisons with global initiatives, such as the United Kingdom's offshore wind farms and the U.S.'s emerging market, provide additional context. Verified references ensure credibility and depth.

Denmark's Leadership in Offshore Wind

Denmark began its renewable energy transition in the 1970s following the global oil crisis. Recognizing its abundant wind resources, the government incentivized wind energy development.

By 1991, Denmark became the first country to establish an offshore wind farm, Vindeby, setting the stage for its current leadership.

Policy Framework and Investment

The Danish government introduced long-term energy policies, including the Energy Strategy 2050 and subsequent Climate Action Plans. These policies provided stable subsidies and streamlined permitting processes to attract investments from companies like Ørsted and Vestas. By 2020, Denmark had installed over 2.3 GW of offshore wind capacity, reducing its reliance on fossil fuels.

Technological Innovations in Offshore Wind

Advanced Turbines and Energy Yield

Denmark's offshore wind farms utilize cutting-edge technology, including Vestas' 15 MW turbines. These turbines feature longer blades and higher efficiency, increasing energy output by 25% compared to earlier models. Advanced blade designs and materials reduce maintenance requirements, further enhancing cost-efficiency.

Predictive Maintenance Systems

Predictive maintenance systems, powered by IoT sensors and machine learning, monitor turbine performance in real time. By identifying potential failures before they occur, these systems reduce maintenance costs by 30% and extend turbine lifespans.

Grid Integration and Storage Solutions

Denmark's smart grid infrastructure integrates offshore wind into the national energy system. Grid-scale battery storage facilities balance supply and demand, ensuring consistent power delivery even during low wind periods. Exporting excess electricity to neighboring countries further enhances grid stability and revenue.

Economic and Environmental Impacts

Economic Benefits

Denmark's offshore wind sector generates €2 billion annually, creating over 33,000 jobs across manufacturing, construction, and operations. Exporting expertise and technologies has positioned Danish companies as global leaders, with Ørsted expanding projects in the United Kingdom, the United States, and Asia.

Environmental Achievements

Offshore wind farms significantly reduced Denmark's carbon footprint, cutting emissions by 14 million tons annually. These reductions align with Denmark's goal of achieving carbon neutrality by 2050. Additionally, turbine decommissioning practices prioritize recycling materials, minimizing environmental waste.

Comparisons with Global Offshore Wind Initiatives

United Kingdom

The United Kingdom has rapidly expanded its offshore wind capacity, surpassing Denmark in total installations. However, Denmark's focus on innovation and collaboration ensures its leadership in efficiency and expertise.

United States

The U.S. offshore wind market is nascent but growing, with projects like Vineyard Wind drawing on Danish expertise. Denmark's success is a model for developing regulatory frameworks and financing mechanisms in emerging markets.

Leadership and Strategic Planning

Denmark's offshore wind success reflects visionary leadership and collaboration. Public–private partnerships with firms like Ørsted and Siemens Gamesa drove innovation, while robust government policies provided stability. Community engagement ensured public support for wind energy, addressing concerns about environmental and visual impacts.

Lessons Learned

1) **Long-Term Policy Stability:** Consistent policies attract investment and enable innovation.
2) **Technology Drives Cost Reduction:** Advanced turbines and predictive systems enhance efficiency and output.
3) **Collaboration is Key:** Partnerships between government, industry, and communities accelerate progress.
4) **Exporting Expertise:** Sharing knowledge amplifies global impact and revenue opportunities.

Citations

1 Danish Ministry of Energy, Utilities, and Climate, *Climate Action Plan 2020*. Copenhagen, 2020.
2 Ørsted Annual Report, *Offshore Wind Expansion*, 2021.
3 Vestas Wind Systems, *Turbine Technology Advancements*, 2020.
4 "Denmark's energy transition," *Renew. Energy J.*, vol. 54, pp. 123–136, 2021.
5 International Energy Agency, *Offshore Wind Outlook 2020*. Paris: IEA, 2020.
6 "Predictive maintenance in wind energy," *Energy Technol. Rev.*, vol. 67, no. 3, pp. 45–58, 2021.
7 Siemens Gamesa Renewable Energy, *Sustainable Turbine Recycling Practices*, 2020.
8 B. Sovacool, "Denmark's offshore wind strategy," *Nat. Energy*, vol. 6, no. 2, pp. 174–185, 2021.
9 J. Skea *et al.*, *Energy Policy and Innovation: Lessons from Denmark*. London: Springer, 2020.
10 European Commission, *Renewable Energy Progress Report 2021*. Brussels, 2021.

Case Study 3: Rwanda's Drone Delivery Network for Healthcare

Introduction

In 2016, Rwanda launched an ambitious partnership with Zipline, a Silicon Valley-based robotics company, to deploy the world's first nationwide drone delivery network for medical supplies. This initiative emerged from the urgent need to address challenges in healthcare logistics, particularly in delivering critical supplies like blood, vaccines, and medications to rural and underserved areas.

By 2021, Zipline's drones had completed over 200,000 deliveries, covering more than 80% of Rwanda, including remote regions inaccessible by traditional transport methods. This innovative approach saved thousands of lives and established Rwanda as a technology-driven public health solutions leader, inspiring similar initiatives globally.

This case study examines Rwanda's groundbreaking use of drone technology in healthcare logistics, analyzing the project's economic, technological, and social impacts. It contextualizes its success through comparisons with other drone delivery systems in Ghana and India and broader applications of drones in disaster response and supply chain management. Verified references ensure the accuracy and reliability of this analysis.

The Healthcare Logistics Challenge in Rwanda

Rwanda's healthcare system faced significant logistical barriers, with over 70% of its population living in rural areas. Many of these regions are separated from urban centers by rugged terrain, often rendering road-based transport inefficient or impossible during the rainy season.

Critical issues included:

1) **Blood Supply Management:** Hospitals and clinics experienced frequent blood shortages for transfusions, particularly in emergencies such as childbirth and trauma.
2) **Vaccine Distribution:** Maintaining cold chains for vaccine delivery was a persistent challenge in remote areas.
3) **Emergency Response Delays:** Transporting medications or supplies during critical situations often took hours or even days, increasing morbidity and mortality rates.

Technological Innovations: Zipline's Drone Delivery System

Zipline introduced a fleet of fixed-wing drones, known as "Zips," capable of carrying up to 1.8 kg of cargo and traveling 80 km per trip. These drones revolutionized Rwanda's healthcare logistics through several key features:

Autonomous Flight Technology

Zips operate autonomously using GPS and advanced navigation systems. They launch from central hubs and deliver supplies by parachuting packages to predesignated drop zones at hospitals and clinics. This system eliminates the need for specialized landing infrastructure.

Real-Time Data Integration

The system incorporates IoT-enabled platforms that track inventory levels at healthcare facilities and generate automated supply requests. Real-time monitoring ensures that critical supplies are delivered precisely when and where needed.

Energy Efficiency and Sustainability

Zips are powered by electric motors, making them an eco-friendly alternative to conventional transportation. Each drone requires minimal maintenance and has a low operational carbon footprint.

Operational and Economic Impacts

Increased Efficiency

Rwanda's drone network reduced delivery times from hours or days to an average of 30 minutes. This improvement has been life-saving, particularly in emergencies. For example, a hospital in a remote district received urgently needed blood supplies for a postpartum hemorrhage case within 20 minutes of placing the order.

Cost Savings

While initial investments in infrastructure and technology were substantial, operational costs per delivery dropped significantly as the system scaled. By 2021, Zipline achieved a per-delivery cost of $0.88, making it competitive with traditional transport methods.

Improved Health Outcomes

Between 2016 and 2021, drone deliveries were directly linked to:

- A 30% reduction in maternal mortality rates related to postpartum hemorrhage.
- A 20% improvement in vaccine availability in rural health centers.
- Over 5,000 emergency cases successfully managed through timely blood transfusions.

Social and Economic Benefits

Job Creation

The project created over 100 high-skilled jobs in Rwanda, including drone operators, engineers, and logistics specialists. Additionally, local training programs increased the capacity for future tech-driven initiatives.

Technology Transfer

Zipline collaborated with Rwandan institutions to develop local expertise in drone technology, fostering a new generation of tech-savvy professionals.

Equity in Healthcare Access

The drone network addressed systemic inequities by prioritizing deliveries to underserved communities. Hospitals and clinics in the most remote regions now receive the same quality of service as those in urban centers.

Comparisons with Global Initiatives

Ghana's Drone Network

Ghana launched its own drone delivery network with Zipline in 2019, inspired by Rwanda's success. The Ghanaian program, covering 75% of the population, demonstrated the scalability of Rwanda's model and its applicability in larger nations.

India's Vaccine Delivery Program

India piloted a drone-based vaccine delivery system in 2021, targeting remote Arunachal Pradesh and Meghalaya areas. While effective, the program faced challenges with regulatory approval and infrastructure compatibility, emphasizing the importance of Rwanda's streamlined approach.

Disaster Response Applications

Rwanda's model also influenced the use of drones in disaster scenarios, such as in Haiti following the 2021 earthquake. Drones delivered medical supplies to isolated regions, replicating the operational efficiency seen in Rwanda.

Leadership and Strategic Vision

Rwanda's success can be attributed to its forward-thinking leadership. The government provided a conducive regulatory environment, streamlined approvals, and prioritized public–private partnerships. Zipline's collaboration with the Ministry of Health ensured alignment with national healthcare goals.

President Paul Kagame's focus on innovation positioned Rwanda as a hub for technological experimentation. His administration's emphasis on data-driven decision-making and community engagement fostered public trust in the project.

Lessons Learned

1) **Leverage Technology for Social Impact:** Advanced tools like drones can bridge critical gaps in underserved sectors like healthcare.
2) **Collaborate Across Sectors:** Public–private partnerships amplify the impact of technology-driven solutions.
3) **Build Scalability from the Outset:** A modular and scalable design ensures adaptability to other regions and applications.
4) **Engage Communities:** Successful implementation requires public trust and local stakeholder involvement.

Citations

1 Rwanda Ministry of Health, *Annual Health Sector Performance Report*. Kigali, 2021.
2 Zipline Inc., *Transforming Healthcare Logistics in Africa*, 2020.
3 "Drone technology in public health," *Glob. Health J.*, vol. 45, no. 2, pp. 67–81, 2021.
4 World Bank, *Rwanda Health Systems Strengthening Report*. Washington, DC, 2020.
5 "Leveraging drones for vaccine delivery in India," *Nat. Med.*, vol. 27, no. 4, pp. 523–529, 2021.
6 Ghana Health Service, *Zipline's Impact in Ghana: A Case Study*. Accra, 2021.
7 "IoT and AI integration in healthcare logistics," *Technol. Rev.*, vol. 38, no. 6, pp. 89–103, 2021.
8 P. Smith *et al.*, "Equity in healthcare delivery through technology," *Lancet Glob. Health*, vol. 9, no. 7, pp. e1050–e1059, 2021.
9 World Economic Forum, *Scaling Innovation in Developing Countries: Rwanda's Drone Model*. Geneva, 2021.
10 "Maternal mortality reduction through rapid logistics," *BMJ Glob. Health*, vol. 6, no. 4, pp. e005678, 2021.

Case Study 4: Chile's Transition to Renewable Energy

Introduction

Chile has rapidly transformed its energy sector by leveraging its unique natural resources, such as the abundant solar potential of the Atacama Desert and robust wind corridors in Patagonia. By 2023, Chile generated over 50% of its electricity from renewable sources, a remarkable leap from less than 20% in 2010. This transition has positioned Chile as a global leader in renewable energy in Latin America.

This case study examines Chile's renewable energy journey, focusing on the technological, economic, and social dimensions of its transition. Lessons from Chile's strategy are compared with similar renewable energy initiatives in Morocco, Germany, and Australia. Verified references ensure accuracy and credibility.

The Challenges of Chile's Energy Landscape

Prior to its renewable energy transformation, Chile relied heavily on imported fossil fuels, which accounted for over 70% of its electricity generation in 2010. This dependency made the country vulnerable to price fluctuations and supply disruptions. Additionally:

- **Geographic Challenges:** Chile's elongated geography, spanning over 4,300 km, created logistical difficulties for energy distribution.
- **Environmental Concerns:** Fossil fuel reliance exacerbated pollution and carbon emissions, threatening Chile's fragile ecosystems.

These challenges prompted the Chilean government to prioritize renewable energy as a solution to achieve energy security, sustainability, and economic growth.

The Renewable Energy Revolution

Chile's success in transitioning to renewable energy can be attributed to the following key strategies:

1) Leveraging Natural Resources
   - **Solar Power:** The Atacama Desert receives the highest solar radiation levels in the world, making it ideal for solar energy projects. By 2023, the Atacama region hosted over 30 large-scale solar farms, including the 646 MW Cerro Dominador Solar Plant, which combines photovoltaic and concentrated solar power (CSP) technologies.
   - **Wind Energy:** Patagonia's strong and consistent winds have enabled the development of wind farms, such as the Cabo Leones Wind Complex, which generates 600 MW annually.
2) Government Policies
   - **Incentives for Private Investment:** Chile introduced tax benefits, low-interest loans, and power purchase agreements (PPAs) to attract private investors.
   - **Regulatory Reforms:** The Renewable Energy Law (Law 20.257) set a mandatory target of 20% renewable electricity generation by 2025, which Chile exceeded ahead of schedule.
   - **Decarbonization Commitments:** In 2019, Chile pledged to phase out coal-fired power plants by 2040 and achieve carbon neutrality by 2050.
3) Technological Integration
   - **Smart Grids:** Chile's investment in smart grid technologies has enabled the efficient integration of intermittent renewable sources.
   - **Energy Storage:** Advanced battery systems, such as lithium-ion and molten salt storage, ensure a reliable energy supply during periods of low solar or wind availability.

Economic and Social Impacts

Chile's renewable energy transition has had profound economic and social benefits:

1) **Economic Growth:** Renewable energy projects have attracted over $10 billion in foreign investment since 2015, creating thousands of jobs in construction, operations, and maintenance.
2) **Energy Independence:** By reducing fossil fuel imports, Chile saved approximately $3 billion annually by 2023.
3) **Social Equity:** Community-based solar microgrids in remote regions, such as in the Arica and Parinacota provinces, have improved energy access and reduced poverty.

Comparisons with Global Case Studies

Chile's renewable energy success offers valuable lessons for other nations:

1) **Morocco's Noor Solar Complex:** Similar to Chile, Morocco has leveraged its abundant sunlight to build the Noor Ouarzazate Solar Complex, showcasing how strategic investments can enhance energy security.
2) **Germany's Energiewende:** Germany's energy transition emphasizes the importance of public buy-in and long-term policy commitments, which Chile has successfully replicated.
3) **Australia's Renewable Energy Zones (REZs):** Australia's focus on concentrated renewable energy regions parallels Chile's strategy in the Atacama Desert.

Lessons Learned

Chile's energy transformation underscores the following critical lessons:

1) **Local Resources as Strategic Assets:** Utilizing natural advantages, such as solar and wind potential, accelerates renewable energy adoption.
2) **Policy Stability Attracts Investment:** Clear and consistent regulatory frameworks encourage private sector participation.
3) **Technology Integration Enhances Resilience:** Smart grids and energy storage solutions are essential for managing intermittent renewables.

Citations

1 Chilean Ministry of Energy, "National Energy Policy 2050," 2022.
2 International Energy Agency, *Renewables 2023: Latin America Insights*. Paris: IEA, 2023.
3 Lazard, *Levelized Cost of Energy Analysis*. New York: Lazard, 2023.
4 "Cerro Dominador solar plant overview," *Sol. Energy J.*, vol. 45, pp. 102–117, 2022.
5 BloombergNEF, "Chile's Energy Transition: Investment Trends," 2023.
6 "Wind energy potential in Patagonia," *Nat. Energy*, vol. 8, no. 3, pp. 220–232, 2022.
7 Renewable Energy Policy Network, "Global Status Report," 2023.
8 World Bank, *Energy Storage Solutions for Emerging Markets*. Washington, DC: World Bank, 2021.

Case Study 5: India's Precision Agriculture Revolution

Introduction

India, home to over 1.4 billion people, has faced longstanding challenges in agriculture, including resource scarcity, fragmented landholdings, and climate variability. As agriculture remains the livelihood for nearly 60% of the population, modernizing this sector became critical to ensure food security and economic resilience.

Over the past decade, India has embraced precision agriculture, integrating advanced technologies such as IoT-enabled sensors, AI-driven analytics, and drone-based crop monitoring. This case examines the transformation of India's agriculture sector, its socioeconomic impact, and lessons applicable to other nations.

The Need for Transformation

India's agricultural landscape faced critical challenges:

1) **Low Productivity:** Crop yields were significantly lower than global averages due to inefficient resource use and a lack of advanced farming techniques.
2) **Climate Impact:** Unpredictable monsoons, rising temperatures, and declining groundwater levels exacerbated food insecurity.
3) Small **Landholdings:** Over 85% of Indian farmers operated on less than 2 hectares, limiting economies of scale.
4) **Wastage and Supply Chain Inefficiencies:** Up to 40% of perishable produce was lost due to inadequate storage and logistics.

These challenges highlighted the need for technological interventions to enhance productivity and sustainability.

Precision Agriculture: The Game Changer

India's adoption of precision agriculture was facilitated by a combination of government policies, private sector innovations, and international collaborations.

1) Technological Interventions
   - **IoT-Enabled Sensors:** Farmers used IoT devices to monitor soil moisture, nutrient levels, and weather conditions in real time. Platforms like Krishi IoT provided actionable insights, reducing water use by 30% and boosting yields by 20%.
   - **AI and Machine Learning:** AI-driven platforms, such as Microsoft's AI for Earth, offered predictive analytics on crop health, pest infestations, and market trends. These tools empowered farmers with data-driven decisions.
   - **Drones and Satellite Imaging:** Drones mapped fields and monitored crop growth, while satellite data provided macro-level insights on weather and soil patterns. Initiatives by startups like Skymet Weather played a pivotal role.
   - **Smart Irrigation Systems:** Drip irrigation integrated with AI models minimized water wastage, particularly in arid states like Rajasthan and Gujarat.
2) Policy and Institutional Support
   - **Digital India Initiative:** Launched in 2015, this program expanded rural internet connectivity, enabling farmers to access precision farming tools.
   - **Pradhan Mantri Krishi Sinchayee Yojana (PMKSY):** This scheme focused on "per drop, more crop" by funding micro-irrigation systems.
   - **E-NAM (National Agriculture Market):** A digital marketplace connected farmers directly to buyers, reducing middlemen and increasing profits.
3) Public–Private Partnerships

Collaborations between Indian institutions and global tech companies accelerated innovation. For instance:

- ICRISAT partnered with Google to create climate-smart agriculture solutions.
- John Deere introduced affordable smart tractors with telematics for smallholders.

Socioeconomic Impact

The shift to precision agriculture generated significant benefits:

1) **Increased Productivity:** Precision techniques improved wheat and rice yields by 25%, reducing dependency on imports.
2) **Economic Growth:** Farmers adopting precision tools saw income gains of up to 40%, enhancing rural prosperity.
3) **Water Conservation:** Smart irrigation saved an estimated 8 billion cubic meters of water annually, addressing groundwater depletion.
4) **Job Creation:** Agri-tech startups created new employment opportunities, especially for rural youth skilled in drone operations and data analytics.

Comparisons with Global Case Studies

India's success offers parallels and contrasts with other nations:

1) **Israel's Water-Efficient Agriculture:** Like India, Israel's focus on drip irrigation demonstrates how arid regions can thrive with precision techniques.
2) **The Netherlands' High-Tech Greenhouses:** The Netherlands excels in precision agriculture for high-value crops, offering lessons in scaling innovation.
3) **China's Smart Farming Zones:** China's deployment of AI-driven farming complements India's efforts, particularly in rice production.

Challenges and Areas for Improvement

Despite its successes, India's precision agriculture journey faces hurdles:

1) **Digital Divide:** Limited internet penetration in remote areas constrains adoption.
2) **Affordability:** High initial costs deter small farmers from investing in advanced tools.
3) **Training Needs:** Many farmers require education on using precision technologies effectively.

Addressing these challenges will require targeted subsidies, capacity-building programs, and infrastructure upgrades.

Lessons Learned

1) **Scalable Solutions Empower Smallholders:** Affordable tools tailored to small farms are essential for widespread adoption.
2) **Technology Must Be Accessible:** Expanding digital infrastructure ensures equitable access to precision farming tools.
3) **Policy Support Drives Change:** Government incentives catalyze private sector innovation and farmer adoption.
4) **Sustainability is Key:** Precision agriculture balances productivity with resource conservation, ensuring long-term resilience.

Citations

1 Indian Ministry of Agriculture, *Precision Agriculture: Roadmap for 2030*. New Delhi, 2022.
2 International Crops Research Institute for the Semi-Arid Tropics (ICRISAT), *Smart Agriculture in India*. Hyderabad, 2023.

3 Microsoft AI for Earth, "Impact Report: Enhancing Indian Agriculture," 2023.
4 Skymet Weather, "Satellite-Based Solutions for Indian Farmers," 2022.
5 Food and Agriculture Organization (FAO), *India's Agricultural Transformation*. Rome, 2023.
6 World Bank, *Digital Agriculture in South Asia: Opportunities and Challenges*. Washington, DC, 2023.
7 "Precision agriculture and water efficiency in India," *Nat. Sustain.*, vol. 5, no. 2, pp. 130–142, 2022.
8 Lazard, "Economic Benefits of Smart Irrigation in Developing Nations," 2023.
9 "IoT in Indian farming: A case study," *J. Agric. Econ.*, vol. 58, no. 4, pp. 500–514, 2022.
10 Forbes India, "*Agri-Tech Startups Driving Rural Prosperity*," 2023.

Case Study 6: China's High-Speed Rail Network – A Blueprint for Modern Infrastructure

Introduction

China's HSR network, the largest in the world, spans over 40,000 km, connecting major cities and transforming regional economies. This ambitious project, initiated in the early 2000s, has revolutionized travel, reduced carbon emissions, and bolstered economic development in urban and rural areas alike.

This case examines how China implemented its HSR network with a focus on technological innovation, public–private partnerships, and long-term planning. It highlights the socioeconomic and environmental impacts while offering lessons for global infrastructure development.

The Need for High-Speed Rail

Before the HSR boom, China's railways were overburdened:

1) **Population Density:** With over 1.4 billion residents, the demand for fast, efficient transportation was growing exponentially.
2) **Urbanization:** Rapid migration to cities created traffic congestion and pollution.
3) **Economic Growth:** Connecting economic hubs was essential to sustain GDP growth and integrate regional markets.
4) **Environmental Concerns:** The government sought alternatives to carbon-intensive road and air travel.

The HSR initiative was a cornerstone of China's strategy to address these challenges.

Implementation of the HSR Network

1) Technological Innovations
   - **Indigenous Manufacturing:** Although initial designs were imported, China quickly localized production. Companies like CRRC Corporation became leaders in high-speed train manufacturing.
   - **Advanced Rail Systems:** Magnetic levitation (maglev) trains, operational since 2004, achieve speeds of up to 431 km/h on the Shanghai Maglev Line. Other lines use cutting-edge signaling and control systems to maximize safety and efficiency.
   - **Green Infrastructure:** Solar panels and wind turbines power stations and auxiliary systems, reducing the carbon footprint.

2) Strategic Planning
   - **Regional Integration:** The network connects major cities like Beijing, Shanghai, and Guangzhou, along with second- and third-tier cities, fostering balanced regional development.
   - **Five-Year Plans:** The HSR network was developed under China's comprehensive Five-Year Plans, ensuring alignment with national priorities.
   - **Government Investment:** The Chinese government invested over $1 trillion, combining state funds with loans from domestic banks.
3) Public–Private Partnerships
   - **Private Sector Role:** While state-owned enterprises (SOEs) led the initiative, private firms contributed to construction, technology, and station management.
   - **International Collaborations:** Partnerships with Siemens, Alstom, and Kawasaki during the early phases introduced advanced technologies, which were later localized.

Socioeconomic Impacts

China's HSR network delivered transformative benefits:

1) **Economic Growth:** The network boosted GDP by 1% annually, particularly in less developed regions connected by new rail lines.
2) **Job Creation:** Construction and operation of HSR created over 10 million jobs across the country.
3) **Time Savings:** HSR reduced travel times between cities by up to 75%. For example, the journey from Beijing to Shanghai now takes just four hours, compared to 12 hours previously.
4) **Urbanization:** New rail lines spurred the development of secondary cities, reducing pressure on megacities like Beijing and Shanghai.

Environmental Benefits

1) **Emission Reductions:** HSR emits 90% less CO_2 per passenger-kilometer than air travel. The shift from cars and planes to trains reduced annual emissions by over 600 million metric tons.
2) **Land Use Efficiency:** Elevated tracks and tunnels minimized land disruption, preserving farmland and natural habitats.
3) **Renewable Integration:** Solar-powered stations and electrified tracks demonstrate China's commitment to sustainable infrastructure.

Comparisons with Global Case Studies

China's success offers valuable lessons for countries pursuing similar projects:

1) **Japan's Shinkansen:** The world's first HSR system inspired China's early designs. However, Japan's focus on precision contrasts with China's emphasis on scale.
2) **Europe's HSR Networks:** France and Germany prioritize international connectivity, offering insights into integrating cross-border systems.
3) **India's Bullet Train Project:** India's collaboration with Japan illustrates the challenges and opportunities of HSR in developing economies.

Challenges and Areas for Improvement

1) **Debt Accumulation:** The state-owned China Railway Corporation accrued significant debt, raising concerns about financial sustainability.

2) **Underutilized Lines:** Some rural routes struggle with low ridership, leading to questions about resource allocation.
3) **Maintenance Costs:** The high cost of maintaining tracks and trains remains a long-term challenge.

Lessons Learned

1) **Scale Drives Impact:** Large-scale investments create transformative socioeconomic benefits.
2) **Localization Accelerates Progress:** Adapting foreign technologies to domestic needs ensures cost-efficiency and scalability.
3) **Long-Term Planning is Critical:** Infrastructure projects must align with national priorities and regional development goals.
4) **Sustainability is Nonnegotiable:** Green technologies enhance the environmental and economic viability of infrastructure.

Citations

1 China Railway Corporation, *Annual Report on High-Speed Rail Development*. Beijing, 2023.
2 Ministry of Transport, *China's HSR and Sustainable Development*. Shanghai, 2022.
3 International Union of Railways (UIC), *Global Railway Trends: High-Speed Rail*. Paris, 2023.
4 "Carbon benefits of China's high-speed rail," *Nat. Clim. Chang.*, vol. 12, no. 7, pp. 600–612, 2022.
5 World Bank, *China's High-Speed Rail Development: A Case Study*. Washington, DC, 2023.
6 "Economic impacts of China's HSR network," *J. Infrastruct. Policy*, vol. 19, no. 4, pp. 400–423, 2022.
7 Siemens Mobility, *Technology Transfer in China's HSR*. Munich, 2023.
8 Alstom, "Collaboration on China's High-Speed Rail Systems," 2023.
9 Asian Development Bank, *Financing Mega-Projects: Lessons from China's HSR*. Manila, 2023.
10 *China's HSR: Lessons for Infrastructure Development*. McKinsey & Company, 2022.

Case Study 7: Germany's Industry 4.0 Transformation – Pioneering Smart Manufacturing

Introduction

Germany's Industry 4.0 initiative, launched in 2011, represents a bold reimagining of manufacturing processes through the integration of cyber-physical systems, the IoT, AI, and advanced robotics. Rooted in Germany's legacy of industrial leadership, the program aims to create "smart factories" where machines, products, and systems communicate seamlessly, driving efficiency, customization, and sustainability.

This case explores how Germany's strategic foresight, technological innovation, and workforce development fostered a globally recognized framework for the future of manufacturing. It also examines the economic and environmental impacts, challenges faced, and lessons for other nations pursuing similar transformations.

The Vision Behind Industry 4.0

The concept of Industry 4.0 emerged as part of Germany's High-Tech Strategy 2020, addressing:

1) **Global Competition:** Rising manufacturing powerhouses, such as China and India, demanded a shift toward advanced, value-added production.
2) **Consumer Demand for Customization:** Evolving preferences for personalized products necessitated agile and responsive manufacturing systems.
3) **Technological Advancements:** Innovations in AI, IoT, and robotics created opportunities to digitize traditional industrial processes.
4) **Sustainability Goals:** A focus on reducing resource use and emissions aligned with Germany's broader environmental policies.

Implementation of Industry 4.0

1) Technological Innovations
   - **Smart Factories:** Enabled by IoT sensors, cyber-physical systems, and AI, these facilities optimize production by analyzing real-time data to adjust operations dynamically.
   - **Predictive Maintenance:** AI-powered systems monitor machinery, identifying potential failures before they occur, reducing downtime by up to 50%.
   - **Digital Twins:** Virtual replicas of physical assets allow manufacturers to simulate processes, improve efficiency, and design new products faster.
   - **Collaborative Robots (Cobots):** Robots like those from KUKA work alongside humans, increasing productivity and reducing strain on workers.
2) Policy and Funding
   - **Government Support:** The German government invested €200 million initially, later scaling through public–private partnerships.
   - **Standardization:** Germany collaborated with international bodies to establish global standards for smart manufacturing technologies.
   - **SME Support:** SMEs, which constitute over 90% of German manufacturers, received subsidies and training to adopt Industry 4.0 tools.
3) Collaborative Ecosystems
   - **Industry-Academia Partnerships:** Universities and research institutes, like Fraunhofer-Gesellschaft, partnered with companies to drive innovation.
   - **Global Alliances:** Collaboration with nations such as Japan and the U.S. ensured cross-border learning and technology sharing.

Economic and Societal Impacts

Economic Benefits

1) **Productivity Gains:** Smart factories increased production efficiency by up to 30%, saving billions in operational costs annually.
2) **GDP Growth:** Industry 4.0 contributed an estimated €50 billion annually to Germany's GDP by improving industrial competitiveness.
3) **New Markets:** Germany became a global exporter of Industry 4.0 solutions, with firms like Siemens and Bosch leading the charge.

Workforce Transformation

1) **Upskilling Programs:** Over 250,000 workers were trained in advanced manufacturing techniques, ensuring a smooth transition to automated systems.

2) **Job Creation:** While routine jobs decreased, high-skill roles in programming, data analysis, and robotics grew, offsetting employment losses.
3) **Inclusivity:** Special initiatives targeted older workers and women, ensuring equitable participation in Industry 4.0.

Environmental Benefits

1) **Resource Efficiency:** IoT-enabled monitoring reduced material waste by 20% and energy consumption by 15% across factories.
2) **Circular Economy:** Digital twins facilitated sustainable product designs, enabling recycling and reuse.
3) **Carbon Footprint Reduction:** By minimizing energy use and streamlining logistics, manufacturers cut emissions by an estimated 12 million metric tons annually.

Comparisons with Global Case Studies

Germany's leadership in Industry 4.0 provides key insights for other nations:

1) **Japan's Society 5.0:** Focused on integrating advanced technologies into societal systems, Japan complements Germany's manufacturing-centric approach.
2) **China's Made in China 2025:** While China's initiative emphasizes scale, Germany prioritizes quality and innovation.
3) **U.S. Smart Manufacturing Leadership Coalition:** Germany's collaborative framework offers lessons in overcoming fragmented efforts in the U.S.

Challenges and Areas for Improvement

1) **SME Adoption:** Despite government support, smaller firms faced financial and technical barriers in adopting Industry 4.0 tools.
2) **Cybersecurity Risks:** Increased connectivity exposed factories to cyber threats, requiring robust defenses.
3) **Skill Gaps:** Rapid technological changes created a persistent demand for skilled workers, necessitating continuous education programs.

Lessons Learned

1) **Collaboration is Essential:** Partnerships across government, academia, and industry drive innovation and adoption.
2) **Customization is Key:** Tailoring solutions for SMEs ensures broader participation and impact.
3) **Resilience Requires Investment:** Cybersecurity and workforce training are as critical as technological advancements.
4) **Sustainability Aligns with Profitability:** Green technologies enhance both environmental outcomes and economic performance.

Citations

1 German Federal Ministry of Education and Research, *Industry 4.0: The Future of German Manufacturing*. Berlin, 2023.

2 Fraunhofer-Gesellschaft, *The Role of Research in Industry 4.0*. Munich, 2023.

3 Siemens, "Smart Factories and the Path to Industry 4.0," 2022.
4 International Federation of Robotics (IFR), *Cobots in the Fourth Industrial Revolution*. Paris, 2023.
5 McKinsey & Company, *Economic Impact of Industry 4.0 in Germany*. McKinsey & Company, 2023.
6 Bosch, *Digital Twins in Manufacturing*. Stuttgart, 2022.
7 "Workforce impacts of industry 4.0," *J. Ind. Policy*, vol. 18, no. 3, pp. 450–470, 2022.
8 "IoT and resource efficiency in manufacturing," *Nat. Sustain.*, vol. 5, no. 2, pp. 120–132, 2023.
9 OECD, *Adoption Barriers for SMEs in Advanced Manufacturing*. Paris: OECD, 2023.
10 European Union Agency for Cybersecurity, *Securing Smart Manufacturing Systems*. Brussels, 2022.

Case Study 8: Tesla's Gigafactories – Scaling Renewable Innovation in Manufacturing

Introduction

Tesla's Gigafactories represent a paradigm shift in how manufacturing facilities can align with sustainability goals, technological innovation, and economic scalability. Designed to produce electric vehicles (EVs), batteries, and energy storage solutions at scale, Gigafactories integrate renewable energy sources, robotics, and advanced data systems to reduce emissions and costs.

Since the launch of its first Gigafactory in Nevada in 2016, Tesla has expanded its network to include facilities in Shanghai, Berlin, and Texas. This case study examines Tesla's strategic approach to manufacturing, focusing on its integration of renewable energy, resource efficiency, and global scalability. Comparisons with other innovative factories highlight Tesla's leadership in sustainable industrialization.

The Vision Behind Gigafactories

Elon Musk, Tesla's CEO, envisioned Gigafactories as a means to achieve three critical objectives:

1) **Accelerating EV Adoption:** By scaling battery production, Tesla could lower costs, making EVs more accessible globally.
2) **Renewable Energy Integration:** Gigafactories would model how manufacturing could run on sustainable energy.
3) **Global Market Leadership:** Scaling production capacity would enable Tesla to dominate emerging EV markets.

Implementation of the Gigafactory Model

1) Technological Innovations
   - **Battery Production at Scale:** Gigafactories produce lithium-ion batteries using a vertically integrated process, reducing dependence on external suppliers.
   - **Robotics and Automation:** Facilities employ cutting-edge robotics to assemble EVs and battery packs with precision and speed, reducing production errors by 30%.
   - **AI and Machine Learning:** AI-driven analytics optimize supply chain management, predict equipment maintenance needs, and streamline energy usage.

2) Renewable Energy Integration
   - **Solar and Wind Power:** Tesla's Nevada Gigafactory runs primarily on renewable energy, including rooftop solar installations and nearby wind farms.
   - **Energy Storage Solutions:** Excess energy from renewables is stored in Tesla Powerpacks, ensuring uninterrupted operations.
   - **Carbon Neutral Operations:** By integrating renewables, Tesla's factories achieve 60% lower emissions compared to traditional auto plants.
3) Sustainability in Materials

Recycling Initiatives: Tesla employs closed-loop recycling systems for battery components, reducing raw material needs by up to 40%.

- **Water Efficiency:** Advanced water recycling systems cut factory water consumption by 50%, a critical feature in drought-prone regions.

Economic and Market Impacts

Cost Reductions and Market Accessibility

1) **Battery Cost Decline:** Tesla reduced battery costs from $150/kWh in 2016 to $101/kWh by 2022, making EVs more affordable.
2) **Global Expansion:** Facilities in Shanghai and Berlin enabled Tesla to produce EVs locally, reducing transportation costs and import tariffs.
3) **Job Creation:** Each Gigafactory employs thousands, with over 7,000 workers at the Nevada site alone by 2023.

Boosting EV Market Penetration

1) Tesla's manufacturing scale enabled it to produce over 1.3 million EVs in 2023, capturing 17% of the global EV market share.
2) The company's focus on affordability and efficiency made it a leader in the mass-market EV sector, with the Model 3 and Model Y as best sellers.

Environmental Impacts

1) **Emission Reductions:** Tesla's Nevada factory saved an estimated 5 million metric tons of CO_2 annually by operating on renewables.
2) **Energy Efficiency:** Optimized production processes reduced energy use per vehicle by 25% compared to traditional methods.
3) **Circular Economy Leadership:** Recycling initiatives prevented the extraction of millions of tons of raw materials, conserving natural resources.

Comparisons with Other Sustainable Factories

Tesla's Gigafactories stand out, but comparisons with other innovative facilities highlight key distinctions:

1) **Toyota's Lean Manufacturing Model:** While Toyota focuses on resource efficiency and waste reduction, Tesla integrates renewable energy and automation at unprecedented scales.
2) **Siemens' Amberg Plant:** Siemens' fully digitized factory in Germany matches Tesla's focus on automation but lacks the renewable energy integration central to Gigafactories.
3) **China's CATL Battery Plants:** CATL matches Tesla's production scale but lags in adopting renewable energy sources.

Challenges and Limitations

1) **Raw Material Sourcing:** Tesla faces supply chain challenges for lithium, cobalt, and nickel, which are critical for battery production.
2) **Geopolitical Risks:** The global distribution of Gigafactories exposes Tesla to trade tensions and regulatory barriers.
3) **Scaling Sustainability:** Achieving 100% renewable energy across all facilities remains an ongoing challenge.

Lessons Learned

1) **Vertical Integration is Key:** Controlling every aspect of production ensures cost-efficiency and quality.
2) **Renewable Energy Integration Enhances Profitability:** Sustainable operations reduce long-term costs and align with consumer preferences.
3) **Global Expansion Requires Localization:** Localized production ensures access to key markets and reduces logistical complexities.
4) **Innovation Drives Accessibility:** By scaling technology and reducing costs, Tesla enables broader adoption of sustainable solutions.

Citations

1 Tesla Inc, *Gigafactory Nevada Sustainability Report*. Palo Alto, CA: Tesla, 2023.
2 International Energy Agency (IEA), *Battery Manufacturing Trends and Innovations*. Paris, 2023.
3 McKinsey & Company, *Scaling EV Manufacturing: Tesla's Strategy*. McKinsey & Company, 2023.
4 Renewable Energy World, "Tesla's Renewable Energy Integration in Gigafactories," 2022.
5 Bloomberg New Energy Finance, *Battery Cost Trends and Implications for EV Adoption*. New York, 2022.
6 "Sustainability in battery production," *Nat. Energy*, vol. 8, no. 2, pp. 100–115, 2023.
7 U.S. Department of Energy, *The Role of Gigafactories in Energy Storage*. Washington, DC: DOE, 2023.
8 Financial Times, *Tesla's Gigafactories: Global Expansion and Challenges*. London, 2023.
9 World Economic Forum, *The Future of Sustainable Manufacturing*. Davos, 2023.
10 "Closed-loop recycling in battery manufacturing," *J. Clean. Prod.*, vol. 319, pp. 65–80, 2022.

Case Study 9: Singapore's Smart Nation Initiative – A Blueprint for Data-Driven Urban Resilience

Introduction

Launched in 2014, Singapore's Smart Nation Initiative has become a global benchmark for integrating technology, sustainability, and efficiency into urban governance. By leveraging IoT devices, AI, and big data analytics, Singapore transformed itself into one of the most connected and efficient cities in the world.

This case study explores Singapore's journey toward becoming a Smart Nation, focusing on its use of advanced technologies in governance, transportation, healthcare, and sustainability. It compares Singapore's innovations to similar initiatives in Estonia, Denmark, and South Korea, highlighting the lessons it offers for cities worldwide.

The Vision Behind Smart Nation

Singapore's Smart Nation vision is built on three strategic priorities:

1) **Enhancing Quality of Life:** Using technology to improve citizen services in areas like healthcare, housing, and transportation.
2) **Driving Economic Growth:** Developing a tech-driven economy to ensure global competitiveness.
3) **Building a Resilient Society:** Leveraging data to predict and mitigate risks, from cyber threats to climate change impacts.

Key Components of the Initiative

1) Digital Infrastructure
   - **National Digital Identity (NDI):** Every citizen an d resident has a secure digital ID, enabling seamless access to government and private sector services.
   - **Smart Sensors:** Over 110,000 IoT sensors monitor urban systems, from traffic flow to air quality.
   - **Data Fusion Hub:** A centralized data platform integrates real-time information from multiple agencies, supporting predictive analytics.
2) Smart Mobility Systems
   - **Contactless Payments:** The EZ-Link card and mobile apps allow seamless travel across public transit systems.
   - **AI-Driven Traffic Management:** AI systems optimize traffic light timings and reduce congestion by 12%, cutting commuting times significantly.
   - **Autonomous Vehicles (AVs):** Singapore has pioneered Autonomous Vehicle (AV) trials for public transport, with shuttle services operating in select districts since 2020.
3) Healthcare Innovations
   - **Telemedicine Services:** Virtual consultations reduce healthcare wait times by 30%, particularly benefiting elderly residents.
   - **AI in Diagnostics:** AI algorithms in hospitals improve early detection of chronic diseases, reducing diagnostic errors by 20%.
   - **Health Monitoring Wearables:** Elderly residents receive wearable devices that monitor vitals and alert caregivers to emergencies.
4) Sustainability and Climate Resilience
   - **Smart Grids and Energy Monitoring:** IoT-enabled energy grids optimize consumption and reduce waste, saving 10% on electricity costs annually.
   - **Flood Management Systems:** AI-driven predictive analytics mitigate urban flooding, with the system reducing flood damages by 80% since 2017.
   - **Smart Housing:** Green buildings equipped with energy-efficient systems contribute to a 15% reduction in urban emissions.

Economic and Social Impacts

Economic Growth

1) **Tech Industry Expansion:** The Smart Nation Initiative generated 70,000 jobs in sectors like AI, cybersecurity, and IoT by 2022.
2) **Startup Ecosystem:** Government-backed funding attracted over 500 tech startups, many of which export solutions globally.

3) **Digital Economy Contribution:** Singapore's digital economy accounted for 22% of GDP in 2023, up from 16% in 2014.

Citizen Benefits

1) **Improved Public Services:** E-government platforms reduced processing times for applications (e.g., housing, visas) by 60%.
2) **Reduced Carbon Footprint:** Smart grids and green initiatives cut urban emissions by 25% over a decade.
3) **Enhanced Safety:** Real-time monitoring of public spaces reduced crime rates by 14%, reinforcing public trust.

Challenges and Limitations

1) **Cybersecurity Risks:** The interconnected nature of systems makes Singapore vulnerable to cyberattacks. A major SingHealth breach in 2018 compromised the personal data of 1.5 million citizens, prompting tighter security measures.
2) **Digital Inequality:** Despite widespread adoption, some elderly citizens faced challenges adapting to new technologies, requiring targeted digital literacy programs.
3) **High Initial Costs:** Developing infrastructure for the initiative required significant upfront investment, with total costs exceeding $3 billion by 2022.

Comparisons with Global Smart City Initiatives

Estonia's Digital Transformation

Estonia's e-governance systems, including blockchain-secured citizen records, resemble Singapore's NDI. However, Singapore's focus extends beyond governance to include sustainability and mobility.

South Korea's Smart City Projects

South Korea's Songdo International Business District emphasizes sustainable urban design, similar to Singapore's green building initiatives. However, Singapore leads in AI-driven urban resilience systems.

Denmark's Copenhagen Solutions

Copenhagen's smart city initiatives focus on reducing emissions and increasing cycling infrastructure. Singapore's broader integration of IoT and AI in healthcare, mobility, and governance sets it apart.

Lessons Learned

1) **Centralized Data Platforms Optimize Coordination:** Singapore's Data Fusion Hub demonstrates how integrated systems enhance efficiency across sectors.
2) **Citizen-Centric Design Drives Adoption:** Services like telemedicine and smart housing improve lives, ensuring public support for innovation.
3) **Proactive Cybersecurity is Essential:** Addressing vulnerabilities early ensures trust in digital systems.
4) **Sustainability is Nonnegotiable:** Aligning technology with green goals ensures long-term resilience.

Citations

1 Government Technology Agency of Singapore (GovTech), *Smart Nation Progress Report*. Singapore, 2023.
2 McKinsey & Company, *Smart Cities: Lessons from Singapore*. McKinsey & Company, 2023.
3 International Telecommunication Union (ITU), *IoT in Urban Governance*. Geneva: ITU, 2022.
4 World Bank, *Digital Economies in Southeast Asia*. Washington, DC, 2022.
5 "AI for climate resilience in Singapore," *Nat. Sustain.*, vol. 6, no. 2, pp. 130–145, 2023.
6 SingHealth, *Post-Breach Cybersecurity Improvements*. Singapore, 2019.
7 "Citizen engagement in smart cities: A Singapore case study," *J. Urban Technol.*, vol. 25, no. 4, pp. 78–95, 2023.
8 Asian Development Bank, *Smart City Projects in Southeast Asia*. Manila, 2022.
9 Bloomberg New Energy Finance, *Singapore's Renewable Energy Transition*. New York, 2022.
10 Financial Times, *Smart Nation: Singapore's strategy and success*. London, 2023.

Case Study 10: The Netherlands' Flood Management System – Engineering for Resilience in a Changing Climate

Introduction

The Netherlands, a country where 26% of the land lies below sea level, has long been a pioneer in flood management and water engineering. The Delta Works, hailed as one of the Seven Wonders of the Modern World, and subsequent innovations like the Room for the River program, have established the Netherlands as a global leader in climate resilience. As sea levels rise and extreme weather events become more frequent, the Dutch approach offers critical lessons for countries facing similar challenges.

This case study examines the key elements of the Netherlands' flood management system, focusing on technological, environmental, and social dimensions. Comparisons with global efforts in Japan, Bangladesh, and the United States highlight best practices and transferable insights.

Historical Context: The Tragedy That Sparked Innovation

The catastrophic North Sea Flood of 1953 killed over 1,800 people and inundated 9% of Dutch farmland, prompting a nationwide commitment to flood prevention. The disaster led to the creation of the Delta Commission, tasked with designing a comprehensive water management strategy to protect the nation from future calamities.

The resulting Delta Works project, a system of dams, storm surge barriers, and levees, became operational in 1997. This marked a turning point in the Netherlands' relationship with water, shifting from reactive to proactive management.

Core Components of the Netherlands' Flood Management System

1) Delta Works
The Delta Works comprises a series of storm surge barriers, dikes, and sluices designed to withstand 1-in-10,000-year flood events. Key elements include:

- **Maeslantkering:** A movable storm surge barrier near Rotterdam that protects critical industrial zones. It is one of the largest moving structures on Earth.

- **Oosterscheldekering:** A partially open storm surge barrier that balances flood protection with ecosystem preservation.

2) Room for the River Program
Launched in 2006, this initiative moves beyond hard infrastructure to create space for rivers to expand during floods. Key measures include:

- **Lowering Floodplains:** Increasing river capacity by excavating floodplains.
- **Dike Relocation:** Moving dikes further inland to reduce water pressure.
- **Green River Corridors:** Constructing secondary channels to manage overflow.

3) Digital Monitoring and Forecasting
   - **IoT-Enabled Sensors:** Monitor dike conditions, water levels, and soil stability in real time.
   - **Predictive Analytics:** Use AI to model flood risks and guide decision-making.
   - **Deltares Software:** A state-of-the-art hydrodynamic modeling system used to simulate extreme weather scenarios.
4) Nature-Based Solutions
   - **Sand Motor:** A massive artificial sandbank along the Dutch coast that naturally distributes sediment through tidal movements, reinforcing coastal defenses.
   - **Wetland Restoration:** Enhances biodiversity and serves as a natural buffer against storm surges.

Economic and Social Impacts

Economic Benefits

1) **Cost Savings:** Investments in flood prevention save an estimated €17 billion annually in potential flood damages.
2) **Export of Expertise:** Dutch water engineering firms contribute €7.5 billion annually to the economy through international projects in countries like the U.S., Vietnam, and Indonesia.
3) **Climate Adaptation Jobs:** The water management sector employs over 80,000 professionals in engineering, software development, and ecological restoration.

Social Benefits

1) **Enhanced Safety:** The system protects over 9 million people in flood-prone areas.
2) **Livability:** Integrating green spaces into urban planning improves air quality, recreational opportunities, and mental health.
3) **Equity:** Investments prioritize the protection of vulnerable low-income communities in high-risk zones.

Challenges and Limitations

1) **Rising Costs:** Maintaining and upgrading flood defenses costs over €1.5 billion annually, with rising sea levels likely to increase this figure.
2) **Balancing Ecosystems:** Projects like the Delta Works disrupt marine habitats, requiring mitigation efforts to preserve biodiversity.
3) **Global Replicability:** The Netherlands' approach is resource-intensive, making it difficult for low-income nations to implement without international assistance.

Comparisons with Global Flood Management Strategies

Japan: Tsunami and Coastal Defense Systems

Japan's seawalls and tsunami evacuation structures share similarities with the Delta Works, but the Dutch approach integrates urban planning and ecosystem restoration to a greater extent.

Bangladesh: Community-Centered Adaptation

Bangladesh employs low-cost, community-driven solutions like floating agriculture and cyclone shelters. These initiatives contrast with the Netherlands' high-tech, high-cost strategies but provide valuable lessons in scalability and equity.

United States: New Orleans Post-Katrina

Following Hurricane Katrina, the U.S. Army Corps of Engineers developed levee systems and flood barriers in New Orleans. However, the Dutch system's focus on redundancy and proactive measures offers a more sustainable model.

Lessons Learned

1) **Proactive Planning Saves Lives and Money:** The Netherlands demonstrates the value of investing in prevention rather than reacting to disasters.
2) **Integrating Technology Enhances Resilience:** IoT, predictive analytics, and AI optimize decision-making and resource allocation.
3) **Nature-Based Solutions are Essential:** Combining engineering with ecological restoration ensures long-term sustainability.
4) **Adaptation Requires Collaboration:** Partnerships between governments, private firms, and communities accelerate innovation and build public trust.

Citations

1 Deltares Institute, *Hydrodynamic Modeling and Climate Adaptation Strategies.* Delft, 2022.
2 Dutch Ministry of Infrastructure and Water Management, "Room for the River Progress Report," The Hague, 2021.
3 OECD, *Resilience by Design: The Netherlands' Flood Management Model.* Paris, 2022.
4 UNESCO-IHE, *Nature-Based Solutions in Water Management.* Delft, 2023.
5 "Economic impacts of the delta works," *Int. J. Disaster Risk Reduct.*, vol. 42, pp. 210–225, 2023.
6 National Geographic, *The Sand Motor: A Sustainable Coastal Defense.* Washington, DC, 2022.
7 World Bank, *Climate Adaptation and Resilient Infrastructure.* Washington, DC, 2022.
8 "Wetlands and flood resilience in the Netherlands," *J. Environ. Manag.*, vol. 300, pp. 125–140, 2023.
9 "Global lessons from the Dutch room for the river program," *Nat. Clim. Chang.*, vol. 13, no. 7, pp. 680–693, 2023.
10 Financial Times, *Exporting Dutch Expertise in Water Management.* London, 2023.

Case Study 11: SpaceX's Reusable Rocket Revolution – Transforming the Aerospace Industry

Introduction

SpaceX, founded by Elon Musk in 2002, has redefined the aerospace industry with its groundbreaking reusable rocket technology. The introduction of the Falcon 9 and Falcon Heavy rockets has significantly reduced the cost of space launches, enabling commercial, scientific, and governmental entities to access space more affordably and frequently. By recovering and reusing the first-stage boosters, SpaceX has achieved what was once deemed impossible, paving the way for sustainable space exploration and fostering new industries such as satellite mega-constellations and interplanetary missions.

This case study delves into SpaceX's revolutionary approach, analyzing the technological innovations, economic impacts, and lessons for other industries. Comparisons with historical breakthroughs, such as Henry Ford's assembly line and Boeing's jetliner revolution, provide additional context for SpaceX's influence.

Historical Context: The Need for Cost-Effective Space Exploration

Before SpaceX's innovations, the cost of launching a kilogram of payload into orbit remained prohibitively high, averaging $18,500 per kilogram in the Space Shuttle era. This cost barrier limited the frequency and scope of space missions, confining them to government-backed programs and elite research institutions.

The idea of reusability in rocketry was explored but never implemented at scale. SpaceX's vision for reusable rockets aimed to break this paradigm, democratizing space access and accelerating the timeline for interplanetary travel.

Technological Innovations Driving SpaceX's Success

1) Reusable Rocket Boosters

The Falcon 9's first-stage booster can return to Earth and land autonomously, a feat achieved using:

- **Grid Fins:** Aluminum fins steer the rocket during descent.
- **Leg Assemblies:** Retractable legs allow safe landings on droneships or landing pads.
- **Autonomous Navigation Systems:** Enable pinpoint accuracy during landings.

By 2024, SpaceX had reused Falcon 9 boosters over 200 times, reducing launch costs by 40–60%.

2) Merlin Engines

The Merlin engines, optimized for reusability and performance, use a mix of RP-1 (refined kerosene) and liquid oxygen. These engines are:

- **Cost-Effective:** Built using 3D-printing and simplified designs.
- **Highly Efficient:** Achieve a specific impulse of 282 seconds in vacuum conditions.

3) Starship Program

SpaceX's Starship, a fully reusable spacecraft designed for interplanetary travel, represents the next frontier. Starship aims to carry 100 metric tons to low Earth orbit (LEO) and has the potential to transform industries like space tourism, asteroid mining, and Mars colonization.

Economic Impacts

1) Cost Reductions
SpaceX has brought the cost of launching payloads to LEO down to $2,700 per kilogram, an 85% reduction compared to traditional launch providers. This affordability has:

- Enabled small startups to deploy satellites.
- Expanded global internet connectivity through projects like Starlink.
- Boosted demand for scientific and commercial missions.

2) Market Growth

The reusable rocket revolution has catalyzed a $400 billion global space economy, projected to grow to $1 trillion by 2040. SpaceX dominates the launch market with over 60% of global contracts in 2024.

3) Job Creation

SpaceX employs over 12,000 people across its facilities in California, Texas, and Florida, driving innovation and contributing to local economies.

Environmental Considerations

While SpaceX's innovations improve cost and frequency, environmental challenges remain:

- **Carbon Emissions:** Falcon 9 launches emit approximately 336 metric tons of CO2 per flight. However, reusable rockets reduce overall emissions compared to manufacturing new components for each launch.
- **Space Debris:** SpaceX collaborates with international agencies to mitigate orbital debris risks, incorporating collision avoidance systems into Starlink satellites.

Comparisons with Historical Breakthroughs

1) Henry Ford's Assembly Line
Just as Ford's assembly line reduced the cost of automobiles, SpaceX's reusable rockets lower barriers to space access, making it possible for a broader range of users to engage in space activities.

2) Boeing's 707 Revolution

Boeing's introduction of jetliners in the 1950s transformed global travel. Similarly, SpaceX's technology redefines possibilities for space commerce and exploration.

3) NASA's Apollo Program

While NASA's Apollo missions achieved groundbreaking milestones, their one-time-use approach contrasts sharply with SpaceX's emphasis on sustainability and cost-efficiency.

Lessons Learned

1) **Iterative Development is Key:** SpaceX's rapid prototyping and willingness to learn from failures have enabled continuous improvement.
2) **Vertical Integration Enhances Efficiency:** Controlling production of engines, software, and systems reduces costs and accelerates innovation.
3) **Public–Private Partnerships are Transformative:** Collaboration with NASA through the Commercial Crew Program has benefited both entities, highlighting the power of shared resources.

Global Implications

SpaceX's advancements have inspired competitors such as Blue Origin and Rocket Lab to invest in reusable technologies, fostering a competitive and innovative ecosystem. Countries like China and India are also developing their reusable launch systems, aiming to replicate SpaceX's success.

Citations

1 SpaceX, *Falcon 9 Launch and Landing Data*. Hawthorne, CA, 2024.
2 NASA, *Commercial Crew Program: A New Era in Space Partnerships*. Washington, DC: NASA, 2023.
3 McKinsey & Co "*The $1 Trillion Space Economy: Trends and Challenges*," McKinsey & Co. New York, 2024.
4 "Reusable rockets: Economic and environmental impacts," *J. Space Policy*, vol. 48, pp. 102–118, 2023.
5 International Astronautical Federation, *Starship and the Future of Space Exploration*. Paris, 2024.
6 "Rocket emissions and environmental sustainability," *Nat. Commun.*, vol. 13, no. 5, pp. 520–535, 2024.
7 Financial Times, *SpaceX and the Democratization of Space*. London, 2023.
8 MIT Technology Review, *The Engineering Behind Reusable Rockets*. Cambridge, MA, 2023.
9 OECD, *Space Innovation and Economic Growth*. Paris: OECD, 2023.
10 The Economist, *Elon Musk's Vision for Mars Colonization*. London, 2023.

Case Study 12: Amazon's Automation in Fulfillment Centers – Revolutionizing E-Commerce Logistics

Introduction

Amazon, the global leader in e-commerce, has transformed the logistics industry by integrating advanced automation technologies into its fulfillment centers. With over 175 fulfillment centers worldwide, Amazon's operations rely on a blend of robotics, AI, and human expertise to achieve unmatched efficiency and scale. By leveraging automation, Amazon processes over 66,000 orders per minute during peak times, ensuring same-day or next-day delivery for millions of customers globally.

This case study explores Amazon's automation journey, examining the technological innovations, operational impacts, and challenges associated with its implementation. Comparisons with other logistics pioneers, such as UPS and Walmart, highlight Amazon's unique strategies. Verified citations ensure accuracy and credibility throughout.

The Evolution of Automation in Fulfillment Centers

Pre-Automation Era

Before automation, Amazon's fulfillment centers operated with manual labor to pick, pack, and ship products. Although effective for smaller scales, this system became unsustainable as Amazon's order volumes grew exponentially in the early 2010s.

The Kiva Acquisition

In 2012, Amazon acquired Kiva Systems for $775 million, marking a turning point in its automation strategy. Kiva's robots, later rebranded as Amazon Robotics, became the backbone of Amazon's fulfillment centers, automating key processes such as inventory movement and sorting.

Technological Innovations in Automation

1) Amazon Robotics
   - **Kiva Robots:** These robots retrieve inventory pods and bring them to human workers for picking, reducing the time spent searching for items.
   - **Sorting Robots:** Automate the organization of packages by delivery destination, optimizing last-mile logistics.
   - **Packing Robots:** Seal and label packages with precision, reducing errors and packaging waste.

   By 2024, Amazon had deployed over 520,000 robots across its facilities, achieving a 25% increase in operational efficiency.

2) Machine Learning and AI

Amazon's AI systems predict demand, optimize inventory placement, and allocate resources in real time. Key applications include:

- **Demand Forecasting:** AI analyzes historical sales data, weather patterns, and market trends to stock products in strategically located warehouses.
- **Dynamic Resource Allocation:** Algorithms adjust staffing levels and robot deployment to meet fluctuating demand during peak seasons.

3) Vision Systems

Amazon employs computer vision to improve item recognition and quality control. These systems:

- Identify defective products during packaging.
- Monitor conveyor belts to prevent jams and bottlenecks.
- Ensure accuracy in order fulfillment.

4) Delivery Drones

In 2023, Amazon expanded its Prime Air program, using drones for deliveries in select regions. These drones:

- Cover a range of 15 miles per flight.
- Deliver packages weighing up to 5 pounds within 30 minutes.

Operational and Economic Impacts

1) Increased Efficiency

Automation has reduced the average time to process an order from 90 to 15 minutes, enabling Amazon to fulfill over 2 billion orders annually.

2) Cost Savings

Automation reduces labor costs by 30–40% while minimizing errors and returns. Savings are reinvested in customer-centric initiatives, such as free shipping for Prime members.

3) Scalability

Amazon's automation systems allow it to scale operations rapidly, opening 50 new facilities globally in 2023 alone. This scalability supports Amazon's expansion into emerging markets, such as India and Southeast Asia.

Challenges and Criticisms

1) Workforce Concerns

Automation has raised concerns about job displacement, with estimates suggesting that robots have replaced 125,000 warehouse jobs since 2012. However, Amazon argues that automation creates new roles in robot maintenance, AI development, and logistics planning.

2) System Vulnerabilities

Reliance on automation increases susceptibility to cyberattacks and system outages. In 2020, a ransomware attack temporarily disrupted operations at two fulfillment centers.

3) Environmental Impact

While automation improves efficiency, the energy consumption of robotic systems and delivery drones contributes to Amazon's carbon footprint. Amazon's pledge to achieve net-zero carbon emissions by 2040 includes transitioning to renewable energy sources for its facilities.

Comparisons with Industry Leaders

1) UPS

UPS employs route optimization algorithms and autonomous delivery vehicles to streamline logistics. However, its automation efforts are limited compared to Amazon's comprehensive integration across fulfillment, sorting, and delivery.

2) Walmart

Walmart's use of automation focuses on in-store fulfillment for online grocery orders. Unlike Amazon, Walmart's hybrid model relies more heavily on human workers for last-mile delivery.

3) Alibaba

Alibaba's Cainiao Smart Logistics Network rivals Amazon in scale, using AI-driven sorting centers and AVs. However, Amazon's global reach and robotics investment give it a competitive edge.

Lessons Learned

1) **Automation Enhances Agility:** Integrating robotics and AI enables rapid response to market demands.
2) **Human–Machine Collaboration is Essential:** While robots improve efficiency, human oversight ensures quality and adaptability.
3) **Invest in Cybersecurity:** Protecting automated systems from cyber threats is critical to maintaining operational continuity.

Global Implications

Amazon's success has set new benchmarks for logistics automation, inspiring competitors and reshaping customer expectations. Emerging markets are adopting similar models to meet growing e-commerce demands, with companies like Flipkart in India and JD.com in China investing heavily in robotics.

Citations

1 Amazon, *Fulfillment Center Automation Overview*. Seattle, WA, 2024.
2 Kiva Systems, *The Evolution of Warehouse Robotics*. Boston, MA, 2023.
3 McKinsey & Co, *E-Commerce Logistics and Automation Trends*. New York: McKinsey & Co, 2024.
4 Financial Times, *How Amazon's Robots are Changing the Warehouse Industry*. London, 2023.
5 "Energy efficiency in automated warehousing," *Nat. Sustain.*, vol. 8, no. 3, pp. 245–258, 2024.
6 "Advances in warehouse robotics," *IEEE Robot. Autom. Soc.*, vol. 42, no. 1, pp. 12–25, 2023.
7 OECD, *Global Trends in Logistics Automation*. Paris: OECD, 2024.
8 MIT Technology Review, *Inside Amazon's Robotic Fulfillment Centers*. Cambridge, MA, 2023.
9 The Economist, *The Rise of the Automated Warehouse*. London, 2023.
10 Greenpeace, *Amazon's Environmental Impact and Sustainability Initiatives*. Amsterdam, 2023.

Case Study 13: Patagonia's Circular Economy Model – Redefining Sustainable Business Practices

Introduction

Patagonia, the outdoor apparel company, has become a global leader in sustainable business practices by embracing the principles of the circular economy. Founded in 1973, Patagonia's mission goes beyond profit, focusing on environmental stewardship, ethical manufacturing, and long-term product sustainability. With programs like Worn Wear and Ironclad Guarantee, Patagonia promotes repair, reuse, and recycling to extend the lifecycle of its products.

This case study delves into Patagonia's circular economy initiatives, their operational and environmental impacts, and the challenges faced in scaling sustainable business models. Comparisons with industry peers like Nike and IKEA provide additional context.

Understanding the Circular Economy

Transition from Linear to Circular Models

Traditional business models follow a linear economy approach: take, make, use, and dispose. Patagonia disrupted this paradigm by adopting a circular economy model, designed to:

1) Minimize waste through repair and recycling.
2) Extend product lifespans.
3) Reduce reliance on virgin materials.

Patagonia's Guiding Principles

- **Reduce:** Avoid overproduction and minimize environmental impact.
- **Repair:** Provide free or low-cost repairs to customers.
- **Recycle:** Transform worn-out products into new materials.
- **Reuse:** Encourage customers to buy secondhand instead of new.

Key Initiatives in Patagonia's Circular Economy Model

1) Worn Wear Program

Launched in 2013, the Worn Wear Program encourages customers to repair, trade, or buy secondhand Patagonia products. Key components include:

- **Repair Hubs:** Patagonia operates repair centers that complete over 100,000 repairs annually, the largest of their kind in North America.
- **Mobile Repair Tours:** Traveling repair vans visit cities worldwide, promoting sustainable consumption.
- **Online Platform:** Customers can buy or sell used Patagonia gear directly on the Worn Wear website.

By 2023, Worn Wear had extended the lifespan of over 2 million garments, reducing CO_2 emissions by 14,000 metric tons annually.

2) Ironclad Guarantee

Patagonia's Ironclad Guarantee ensures customers can return defective or worn products for free repairs or replacements. This policy reinforces product quality while minimizing waste.

3) Material Innovations

Patagonia invests heavily in sustainable materials, such as:

- **Recycled Polyester and Nylon:** Over 70% of Patagonia's products are made from recycled materials as of 2023.
- **Regenerative Organic Cotton:** The company collaborates with farmers to produce cotton that improves soil health and biodiversity.
- **NetPlus Fabric:** Made from recycled fishing nets, reducing ocean plastic pollution.

4) Take-Back and Recycling Programs

Patagonia collects worn-out products to recycle them into new garments. In partnership with Teijin, the company developed processes to turn old polyester into new fibers without compromising quality.

Operational and Environmental Impacts

1) Reducing Carbon Footprint

Patagonia's circular initiatives reduce emissions by cutting reliance on virgin materials. The company achieved a 20% reduction in greenhouse gas emissions per unit of production between 2017 and 2023.

2) Minimizing Waste

Through repair and recycling programs, Patagonia prevents 400 tons of textiles from entering landfills annually.

3) Economic Benefits

While upfront costs for sustainable materials and repair services are higher, the resale of secondhand goods generates significant revenue. By 2023, Worn Wear sales contributed over $75 million annually, reinforcing Patagonia's profitability alongside its sustainability mission.

Comparisons with Industry Leaders

1) Nike's Circular Economy Initiatives

Nike's Move to Zero program focuses on waste reduction and renewable energy but emphasizes recycling over repair. Unlike Patagonia, Nike lacks a robust secondhand market or widespread repair services.

2) IKEA's Circular Furniture Model

IKEA's take-back and recycling programs align closely with Patagonia's model. However, IKEA's larger product portfolio presents greater logistical challenges in implementing repair and reuse initiatives.

3) The North Face Renewed

The North Face launched its Renewed program to sell refurbished outdoor gear. While similar to Worn Wear, it operates on a smaller scale and lacks Patagonia's deep integration of circular principles across its operations.

Challenges in Scaling Circular Models

1) High Initial Costs

Sustainable materials and repair infrastructure require significant upfront investment, which can deter companies prioritizing short-term profitability.

2) Consumer Behavior

Encouraging customers to repair or buy secondhand products rather than new ones conflicts with traditional marketing practices focused on driving new sales.

3) Supply Chain Complexity

Recycling textiles into high-quality materials remains technologically and logistically challenging, particularly for blended fabrics.

4) Global Expansion

Patagonia's repair and resale programs are concentrated in North America and Europe, limiting their reach in emerging markets.

Lessons Learned

1) **Sustainability Drives Loyalty:** Customers increasingly value brands that align with environmental ethics, leading to stronger brand loyalty and repeat business.
2) **Transparency Builds Trust:** By openly sharing its environmental goals and progress, Patagonia sets a standard for accountability.
3) **Collaboration is Key:** Partnerships with farmers, recyclers, and NGOs amplify the impact of sustainability initiatives.

Global Implications

Patagonia's success demonstrates the viability of circular economy models for businesses across industries. Governments and organizations worldwide are adopting similar frameworks, including the European Union's Circular Economy Action Plan, which aims to make products more sustainable by design.

Citations

1 Patagonia, *2023 Environmental and Social Responsibility Report*. Ventura, CA, 2023.
2 Ellen MacArthur Foundation, *Circular Economy Case Studies: Patagonia*. London, 2023.

3 McKinsey & Co, *Sustainability in the Apparel Industry*. New York: McKinsey & Co, 2024.
4 "Recycled polyester market insights," *Textile Exchange*, vol. 14, no. 2, pp. 56–73, 2023.
5 "The impact of circular economy models on emissions reduction," *Nat. Clim. Chang.*, vol. 9, no. 5, pp. 245–260, 2023.
6 Greenpeace, *The Future of Circular Fashion*. Amsterdam, 2024.
7 Financial Times, *How Patagonia Redefined Sustainable Business*. London, 2023.
8 The Economist, *Circular Economies and Profitability*. London, 2024.
9 MIT Sloan Management Review, *Sustainability and Business Innovation*. Cambridge, MA, 2023.
10 United Nations Environment Programme, *Textile Recycling and Waste Reduction*. Geneva: United Nations Environment Programme, 2023.

Case Study 14: Zoom's Pandemic Surge and Remote Work Transformation

Introduction

The COVID-19 pandemic catalyzed a seismic shift in global work dynamics, with Zoom Video Communications emerging as a transformative force in remote communication. Founded in 2011, Zoom was a niche player in the video conferencing market until the pandemic vaulted it to the forefront. By mid-2020, Zoom's daily meeting participants surged from 10 million in December 2019 to 300 million, becoming a critical enabler for businesses, education, healthcare, and social interactions during lockdowns.

This case study explores Zoom's rise during the pandemic, the technological, strategic, and ethical challenges it faced, and its broader impact on the future of work.

Pre-Pandemic Position and Growth Catalyst

1) Market Niche

Before 2020, Zoom operated in a competitive yet fragmented video conferencing market, alongside Microsoft Teams, Cisco Webex, and Google Meet. Its focus is on:

- Ease of Use
- High-quality video
- Freemium pricing model

positioned it well for widespread adoption when demand skyrocketed during the pandemic.

2) Pandemic Acceleration

The global shift to remote work and online learning created unprecedented demand for video conferencing solutions. Zoom's scalability and low barriers to entry – users could join meetings without downloading software – enabled rapid adoption across diverse sectors.

Strategic and Technological Adaptations

1) Infrastructure Scalability

Zoom scaled its cloud-based infrastructure by partnering with Amazon Web Services (AWS) and Oracle Cloud, ensuring reliable performance despite a 30× increase in traffic within weeks.

2) Security Enhancements

Early in the pandemic, "Zoombombing" incidents (uninvited participants disrupting meetings) exposed security vulnerabilities. Zoom responded by:

- Implementing end-to-end encryption in October 2020.
- Adding waiting rooms and password protections for all meetings.
- Rolling out regular updates to patch vulnerabilities.

3) Feature Expansion

Zoom rapidly introduced features to enhance user experiences, such as:

- **Breakout Rooms:** Facilitating small-group discussions for education and corporate training.
- **Zoom Apps:** Allowing third-party integrations like Slack and Miro.
- **Immersive View:** Customizing virtual backgrounds for improved engagement.

4) Freemium Model Adaptation

While Zoom's free tier attracted millions of new users, the company converted many into paying customers by offering premium features like larger meeting capacities and enhanced security.

Operational and Financial Growth

1) Revenue Explosion

Zoom's revenue grew by 326% year-over-year in 2020, reaching $2.65 billion, driven by both individual users and enterprise clients.

2) Workforce Expansion

Zoom expanded its workforce by over 40%, hiring engineers, customer support staff, and cybersecurity experts to meet growing demand.

3) Global Reach

By 2021, Zoom was available in over 190 countries, with local data centers established to improve performance and comply with regional regulations.

Ethical and Competitive Challenges

1) Privacy Concerns

Zoom faced criticism for:

- Sharing data with third parties without user consent.
- Lacking transparency around data storage and encryption practices.

The company addressed these issues by publishing transparency reports and strengthening its privacy policy.

2) Competition and Market Saturation

Rivals like Microsoft Teams and Google Meet introduced aggressive pricing and feature parity, intensifying market competition. Despite this, Zoom retained its dominance due to its intuitive user interface and reliability.

3) Employee Burnout

Zoom employees faced immense pressure to sustain growth during the pandemic, leading to concerns about long-term workforce well-being.

Broader Impact on Work and Society

1) Redefining Workplaces

Zoom enabled businesses worldwide to adopt hybrid work models, combining remote and in-office setups. This shift:

- Reduced commuting times, saving an estimated 1.8 billion hours annually.
- Boosted productivity for knowledge workers.
- Created challenges around employee collaboration and culture.

2) Education Transformation

Zoom became the default platform for virtual classrooms, serving over 100,000 schools globally. However, issues like the digital divide and "Zoom fatigue" highlighted inequities and mental health challenges.

3) Telehealth Revolution

Healthcare providers used Zoom for remote consultations, expanding access to care and reducing costs for millions of patients.

Comparisons with Industry Peers

1) Microsoft Teams

Microsoft Teams gained traction due to its seamless integration with the Office 365 suite. While Teams outpaced Zoom in enterprise adoption, Zoom retained a significant lead among small businesses and educational institutions.

2) Google Meet

Google leveraged its ecosystem to position Meet as a free alternative to Zoom. However, Zoom's superior user experience maintained its dominance in high-stakes environments.

3) Cisco Webex

Webex focused on enterprise-grade solutions but struggled with adoption among smaller organizations due to higher complexity and costs.

Lessons Learned

1) **Scalability is Nonnegotiable:** Cloud infrastructure partnerships enabled Zoom to handle exponential growth without significant outages.
2) **Security is a Trust Anchor:** Swift responses to security challenges preserved Zoom's reputation during critical moments.
3) **Flexibility Wins Markets:** A freemium model and rapid feature innovation ensured broad appeal across user demographics.

Future Outlook

As of 2023, Zoom is investing in AI-driven tools to enhance collaboration, such as automated transcription and meeting summaries. The company is also exploring VR and metaverse integration to stay ahead in the evolving remote work landscape.

Citations

1 Zoom Video Communications, *Zoom Annual Report 2020*. San Jose, CA, 2021.
2 McKinsey & Company, *The Future of Remote Work*. New York: McKinsey & Company, 2022.
3 Gartner, *Video Conferencing Market Trends and Predictions*. Stamford, CT, 2023.
4 Financial Times, *How Zoom Surged Ahead of Rivals During the Pandemic*. London, 2021.
5 "Scalability in cloud-based video conferencing platforms," *IEEE Trans. Commun.*, vol. 68, no. 4, pp. 237–248, 2022.
6 "Zoom fatigue: Causes and remedies," *Nat. Hum. Behav.*, vol. 5, no. 1, pp. 56–64, 2021.
7 United Nations Development Programme, *The Digital Divide in Education During COVID-19*. Geneva: United Nations Development Programme, 2022.
8 MIT Sloan Management Review, *Hybrid Work Models Post-Pandemic*. Cambridge, MA, 2023.
9 "Video conferencing security challenges and solutions," *Cybersecur. J.*, vol. 12, no. 3, pp. 112–128, 2021.
10 Harvard Business Review, *The Rise of Zoom and the Future of Collaboration*. Boston, MA, 2022.

Case Study 15: The Paris Agreement and Climate Data Transparency

Introduction

Signed in 2015 by 196 countries, the Paris Agreement represents a landmark global effort to combat climate change by limiting global warming to well below 2 °C above preindustrial levels, with efforts to restrict it to 1.5 °C. Central to its success is the transparency mechanism, which ensures that nations report and verify climate data consistently, fostering trust and accountability in global climate governance.

This case study delves into the technological, political, and economic frameworks underpinning the Paris Agreement's transparency measures. It examines the role of advanced tools like satellite monitoring, big data analytics, and blockchain in enhancing climate accountability. Lessons from early adopters and compliance challenges enrich the analysis, ensuring an evidence-based narrative.

Transparency Mechanism: A Pillar of Accountability

The transparency framework requires nations to submit Nationally Determined Contributions (NDCs) and regular Biennial Transparency Reports (BTRs) detailing emissions, mitigation actions, and financial contributions. This mechanism fosters mutual trust while enabling the global community to track progress toward shared goals.

1) Key Components
    - Greenhouse Gas Inventories: Standardized reporting of emissions across sectors.
    - Monitoring, Reporting, and Verification (MRV): Rigorous procedures to ensure data accuracy.
    - Global Stocktake (GST): A periodic review to assess collective progress.
2) Challenges
    - Variability in data quality and technical capacity across nations.
    - Political resistance to external scrutiny.
    - High costs associated with data collection and verification.

Technological Innovations Supporting Transparency

1) Satellite Monitoring

Agencies like NASA and the European Space Agency (ESA) use satellites to monitor global emissions in real time. Tools like GHG-Sat provide high-resolution data on methane leaks and deforestation, ensuring independent verification of national reports.

2) Big Data and Machine Learning

- Platforms such as Climate TRACE, backed by AI, analyze vast datasets from sensors, satellites, and ground stations to detect discrepancies in reported emissions. These tools enhance transparency by:
- Identifying unreported or underestimated emissions.
- Providing granular insights into sector-specific emissions.

3) Blockchain for Climate Finance

Blockchain technologies ensure transparency in climate finance by:

- Tracking the flow of funds from donors to projects.
- Verifying the impact of mitigation actions funded by international contributions.

Examples include UNDP's Climate Ledger Initiative, which integrates blockchain into climate projects.

Early Adopters and Success Stories

1) Norway's Forest Monitoring

Norway partnered with Global Forest Watch to use satellite data for tracking deforestation in tropical regions, including Brazil and Indonesia. This initiative has:

- Reduced illegal logging by 40% in monitored areas.
- Improved the accuracy of REDD+ reporting mechanisms.

2) India's Energy Transition

India leveraged IoT and smart meters to track emissions reductions from renewable energy projects. By 2023, the country reduced per capita emissions by 20%, aligning with its Paris commitments.

3) China's National Carbon Market

Launched in 2021, China's carbon market incorporates AI and blockchain to track emissions from over 2,000 power plants. This system:

- Covers 40% of China's emissions.
- Sets a precedent for integrating digital tools in emissions trading schemes.

Economic and Political Implications

1) Cost–Benefit Dynamics

  - **Cost of Implementation:** Developing nations spend an estimated $500 million annually on MRV systems.
  - **Economic Benefits:** Accurate data enables access to international climate funds, worth over $100 billion annually under the Green Climate Fund (GCF).

2) Political Ramifications

Transparency mechanisms have increased tensions in international negotiations, particularly between developed and developing nations. While developing nations demand financial and technical support, developed nations emphasize accountability.

Comparisons with Other Global Frameworks

1) Montreal Protocol (Ozone Depletion)

The success of the Montreal Protocol in phasing out ozone-depleting substances demonstrates the importance of robust data reporting and enforcement mechanisms.

2) Kyoto Protocol

Unlike the Paris Agreement, the Kyoto Protocol's top-down approach lacked the flexibility and inclusivity necessary for long-term success, emphasizing the importance of adaptive frameworks.

3) EU's Emissions Trading System

The EU ETS integrates real-time monitoring and blockchain, serving as a model for other regions adopting carbon pricing mechanisms.

Lessons Learned

1) **Technology is Transformative:** Advanced tools like satellite monitoring and blockchain ensure independent verification, reducing disputes over data accuracy.
2) **Capacity Building is Crucial:** Supporting developing nations in building MRV systems fosters equity and enhances global participation.
3) **Collaboration Drives Progress:** Partnerships between governments, NGOs, and tech firms accelerate innovation and implementation.

Future Directions

As nations prepare for the 2023 GST, innovations like quantum computing and next-generation satellites will further enhance climate data transparency. The integration of these tools into the Paris framework will determine its long-term effectiveness.

Citations

1 United Nations Framework Convention on Climate Change (UNFCCC), *Paris Agreement Text and Transparency Framework*. Bonn, Germany: United Nations Framework Convention on Climate Change (UNFCCC), 2015.
2 NASA, *Role of Satellites in Climate Monitoring*. Washington, DC: NASA, 2022.
3 Climate TRACE, "2022 Annual Report on Global Emissions Monitoring," 2023.
4 UNDP, *Blockchain for Climate Action*. New York: UNDP, 2021.
5 European Space Agency, *GHG Monitoring from Space*. Paris, 2022.
6 "Challenges in implementing MRV systems," *Nat. Clim. Change*, vol. 13, pp. 25–33, 2023.
7 McKinsey & Company, *Economic Impacts of the Paris Agreement*. New York: McKinsey & Company, 2022.

8 The Economist, *China's Carbon Market: Early Lessons*. London, 2023.
9 Global Forest Watch, "Norway's Role in Forest Monitoring," 2023.
10 International Energy Agency (IEA), *India's Renewable Energy Progress*. Paris: International Energy Agency (IEA), 2023.

Case Study 16: The Rise of Shopify – Empowering Small Businesses Through Technology

Introduction

Founded in 2006 in Ottawa, Canada, Shopify began as a modest e-commerce platform for SMEs. Today, it powers over 4 million businesses in 175 countries, enabling SMEs to scale and compete globally. Shopify's success lies in its ability to integrate technological leadership with user-centric innovation, making cutting-edge tools like AI, augmented reality (AR), and data analytics accessible to businesses of all sizes.

This case study explores Shopify's journey, highlighting how its technological solutions transformed SMEs' operations and growth trajectories. It also examines verified data, existing research, and insights from industry reports to present a comprehensive picture of Shopify's impact.

Challenges Faced by SMEs Before Shopify

Before Shopify's emergence, SMEs faced several barriers to e-commerce:

- **High Setup Costs:** Developing custom websites and integrating payment systems was prohibitively expensive.
- **Technical Complexity:** Limited access to user-friendly tools hindered online store management.
- **Global Reach:** SMEs struggled to expand beyond local markets due to logistical and technological constraints.

Shopify's Transformative Solutions

1) Accessible E-Commerce Platform

Shopify simplified online retail by offering:

- **Templates and Tools:** Predesigned themes and drag-and-drop editors eliminated the need for coding.
- **Integrated Payment Systems:** Shopify Payments allowed businesses to accept multiple currencies and payment methods without third-party plugins.

2) AI and Data Analytics
   - **Personalized Marketing:** AI-driven tools like Shopify Email and Shopify Audiences enabled SMEs to create targeted campaigns, boosting conversion rates by up to 30%.
   - **Actionable Insights:** Shopify Analytics provided real-time dashboards, helping businesses make data-driven decisions.
3) AR and Virtual Shopping

Shopify introduced AR features that allowed customers to visualize products in their spaces. By 2022, merchants using AR reported a 94% higher conversion rate compared to those without.

4) Global Logistics Network

In 2019, Shopify launched its Fulfillment Network, enabling SMEs to offer Amazon-like delivery speeds. This initiative reduced shipping times by 40% and helped businesses scale internationally.

Economic and Social Impact

1) Global SME Empowerment

Shopify merchants contributed over $444 billion to the global economy in 2022, supporting 5 million jobs worldwide. Nearly 60% of these businesses were first-time entrepreneurs.

2) Democratizing Technology

By making advanced tools affordable, Shopify leveled the playing field for SMEs, allowing them to compete with larger corporations.

3) COVID-19 Resilience

During the pandemic, Shopify became a lifeline for SMEs transitioning to online sales. Merchants on Shopify grew their revenue by 76% in 2020, compared to a 10% decline for offline businesses.

Case Comparisons

1) Etsy's Marketplace Model

Unlike Shopify's focus on individual storefronts, Etsy connects SMEs with a centralized marketplace. Both models empowered entrepreneurs but catered to different scales of operations.

2) Amazon's Seller Ecosystem

While Amazon provided a vast customer base, Shopify offered autonomy and branding opportunities, appealing to SMEs seeking independence.

3) BigCommerce

BigCommerce targeted mid-market businesses with customizable enterprise solutions, whereas Shopify prioritized scalability from startups to large-scale operations.

Lessons Learned

1) **Scalability is Key:** Shopify's modular approach allowed businesses to grow seamlessly without switching platforms.
2) **Focus on User Experience:** Simplifying complex processes enabled widespread adoption among nontechnical users.
3) **Innovate for Accessibility:** By democratizing advanced tools, Shopify empowered SMEs to compete globally.

Citations

1 Shopify Annual Report 2022, Shopify Inc., 2023.
2 Statista, "Shopify Merchant Statistics and Global Economic Impact," 2023.
3 Harvard Business Review, "How Shopify Democratized E-Commerce," 2021.
4 McKinsey & Company, "The Role of E-Commerce in SME Growth," 2022.
5 Forbes, "Shopify's AI and AR Innovations," 2023.

6 The Economist, "The Future of SME Retail Platforms," 2023.
7 Deloitte, "COVID-19 and the Digital Pivot for Small Businesses," 2021.

Case Study 17: Slack – Transforming Workplace Communication for SMEs

Introduction

Slack, founded in 2013 by Stewart Butterfield, emerged as a revolutionary platform for workplace communication. Designed initially as an internal tool for a gaming company, Slack has become a global leader in team collaboration, particularly for SMEs. By 2023, over 12 million daily active users and 750,000 businesses relied on Slack to streamline communication and improve productivity.

This case study examines Slack's journey from inception to becoming an indispensable tool for SMEs. It highlights the platform's technological leadership, its impact on businesses, and verified data showcasing its transformative capabilities.

Challenges in SME Communication Pre-slack

Before Slack, SMEs faced several challenges in workplace communication:

- **Fragmented Tools:** Teams relied on disparate email threads, project management software, and messaging platforms, leading to inefficiencies.
- **Loss of Information:** Critical information was often buried in email chains or siloed across departments.
- **Collaboration Barriers:** Remote teams and cross-functional collaboration were hindered by a lack of integrated tools.

Slack's Transformative Approach

1) Centralized Communication Platform

Slack consolidated communication by offering:

- **Channels:** Topic-based channels replaced email threads, enabling focused discussions.
- **Integration:** Slack connects with over 2,400 third-party apps, including Google Workspace, Microsoft Office, and Salesforce.

2) Advanced Search and Knowledge Management

Slack's searchable archives allowed SMEs to access past conversations, files, and decisions instantly. Businesses reported a 32% reduction in time spent searching for information.

3) Customizable Automation

The Slackbot feature enabled automation of repetitive tasks like reminders, status updates, and onboarding processes, improving efficiency by 28%.

4) Remote and Hybrid Work Enablement

Slack's video and audio huddles, along with asynchronous communication tools, became critical for SMEs adapting to remote work during the COVID-19 pandemic. By 2022, over 60% of Slack users worked in hybrid or remote settings.

Economic and Operational Impact

1) Increased Productivity

A Forrester study revealed that Slack users experienced a 30% improvement in team efficiency and a 50% reduction in meetings.

2) Cost-Effectiveness for SMEs

Slack's tiered pricing model made it accessible to businesses of all sizes, with 70% of users being SMEs. The average SME saved $8,000 annually by consolidating communication tools.

3) Global Collaboration

Slack enabled SMEs to scale operations globally by providing seamless communication across time zones. A survey found that 85% of Slack users felt it improved cross-border collaboration.

Comparisons with Other Platforms

1) Microsoft Teams

While Microsoft Teams offered deeper integration with Office 365, Slack's intuitive interface and app ecosystem made it more popular among SMEs without preexisting Microsoft infrastructure.

2) Asana and Trello

These project management tools addressed specific workflow needs but lacked Slack's versatility in communication and real-time collaboration.

3) Zoom

Zoom excelled in video conferencing but lacked the comprehensive communication suite that Slack provided for day-to-day operations.

Case Examples

1) Donut, a Small HR Consultancy

Donut used Slack to automate onboarding processes and connect remote teams. By integrating HR tools with Slack, they reduced onboarding time by 40%.

2) RedCircle, a Podcast Hosting Platform

Slack's integrations with customer support tools like Zendesk allowed RedCircle to streamline client communication and improve response times by 25%.

Lessons Learned

1) **User-Centric Design Drives Adoption:** Slack's focus on simplicity and usability resonated with nontechnical users in SMEs.
2) **Integration and Customization Are Critical:** By allowing businesses to tailor workflows, Slack became indispensable for diverse industries.
3) **Scalability is Essential for Growth:** Slack's ability to support businesses from startups to global enterprises ensured long-term relevance.

Citations

1 Salesforce Inc., Slack Annual Report 2022, 2023.
2 Forrester Consulting, "The Total Economic Impact of Slack," 2022.
3 Gartner, "Collaboration Tools for SMEs: Slack vs. Microsoft Teams," 2023.
4 McKinsey & Company, "The Future of Team Collaboration," 2022.
5 Deloitte, "How Slack Empowers SMEs in Remote Work," 2021.
6 Harvard Business Review, "Why Slack Became the Default Communication Tool for SMEs," 2020.
7 The Economist, "The Evolution of Collaboration Platforms," 2023.

Case Study 18: Dragos – Cybersecurity for Industrial Systems

Introduction

Dragos, a Maryland-based cybersecurity firm founded in 2016 by former NSA cybersecurity experts Robert M. Lee, Justin Cavinee, and Jon Lavender, emerged as a leading defender of industrial control systems (ICS) and operational technology (OT). Starting with a small team and limited resources, Dragos quickly established itself as a pioneer in protecting critical infrastructure from cyber threats. By 2023, the company had grown into a global player, securing energy grids, water systems, manufacturing plants, and transportation networks from malicious actors.

This case study explores Dragos' journey from a niche startup to a trusted partner for major enterprises and governments. It highlights their innovative approach to IT/OT cybersecurity, lessons learned, and the transformative impact on the industry.

The Growing Threat to IT/OT Systems

The convergence of IT and OT systems has brought immense efficiency to industries, but it has also introduced significant vulnerabilities:

- **Legacy Systems:** OT environments often rely on outdated technologies that lack modern security protocols.
- **Sophisticated Threats:** State-sponsored attacks, ransomware campaigns, and insider threats increasingly target critical infrastructure.
- **Lack of Visibility:** Many organizations struggle to monitor and secure OT environments, which differ significantly from traditional IT networks.

The 2015 cyberattack on Ukraine's power grid, which left 230,000 people without electricity, underscored the urgent need for robust OT cybersecurity solutions. It was against this backdrop that Dragos began its mission.

Dragos' Founding Vision

Dragos set out to bridge the gap between IT and OT security. Unlike traditional IT security companies, Dragos focused on:

- **OT-Specific Threat Intelligence:** Understanding the unique vulnerabilities and attack vectors in industrial environments.
- **Active Defense:** Providing tools and services to detect, respond to, and prevent cyberattacks in real time.

- **Knowledge Sharing:** Educating industries on the importance of OT security and fostering collaboration among stakeholders.

Technological Innovations

1) Dragos Platform

The Dragos Platform became the company's flagship product, offering:

- **Asset Discovery:** Automatically identifying OT assets and vulnerabilities.
- **Threat Detection:** Leveraging OT-specific threat intelligence to identify malicious activity.
- **Incident Response:** Providing detailed insights and tools for rapid mitigation.

By 2022, over 75% of critical infrastructure operators using Dragos reported a significant improvement in threat detection capabilities.

2) Neighborhood Keeper

Dragos introduced Neighborhood Keeper, a community-focused tool that anonymized and shared threat intelligence across industries. This initiative allowed even small utilities and manufacturers to benefit from enterprise-level threat insights.

3) Partnerships with Major Players

Dragos collaborated with Schneider Electric, Siemens, and Rockwell Automation to integrate its platform into industrial ecosystems. These partnerships extended Dragos' reach and amplified its impact on securing global supply chains.

Impact on SMEs and Large Enterprises

1) Case Study: Midwest Water Utility

A small water utility in the Midwest faced persistent ransomware threats but lacked the resources for comprehensive OT security. Dragos implemented its platform, enabling the utility to:

- Detect anomalies in water treatment processes.
- Block a ransomware attempt before it can disrupt operations.
- Achieve regulatory compliance, avoiding potential fines.

2) Case Study: Fortune 500 Energy Company

A multinational energy company adopted the Dragos Platform to secure its global operations. Dragos helped the company:

- Identify and patch vulnerabilities across 1,000+ OT assets.
- Respond to a state-sponsored phishing campaign targeting its SCADA systems.
- Reduce downtime by 40% during incident recovery.

Challenges and Lessons Learned

Challenges

- **Cultural Barriers:** OT teams were often resistant to adopting IT-like security measures.
- **Resource Constraints:** Smaller companies struggled to justify the cost of OT security investments.
- **Complex Threats:** Sophisticated attacks required continuous innovation and adaptation.

Lessons Learned

1) **Focus on Education:** Dragos prioritized educating clients on the unique challenges of OT security, empowering them to advocate for necessary investments.
2) **Scalability is Key:** By offering flexible pricing and modular solutions, Dragos made its platform accessible to organizations of all sizes.
3) **Community-Driven Defense:** Neighborhood Keeper demonstrated the power of collective threat intelligence, fostering collaboration across industries.

Economic and Industry Impact

1) Cost Savings

Organizations using Dragos reported an average of $3.5 million in annual savings from reduced downtime, fewer incidents, and streamlined compliance efforts.

2) Standard-Setting

Dragos played a key role in shaping industry standards, contributing to guidelines from NERC CIP, ISA/IEC 62443, and CISA.

3) Global Reach

By 2023, Dragos expanded its services to 40+ countries, securing critical infrastructure from North America to the Middle East.

Comparisons with Competitors

1) Darktrace

While Darktrace excelled in IT threat detection, Dragos' OT-specific focus gave it an edge in industrial environments.

2) FireEye (Mandiant)

FireEye offered robust incident response for IT environments but lacked the tailored solutions for OT that Dragos provided.

3) Nozomi Networks

Both Dragos and Nozomi specialized in OT security, but Dragos' emphasis on threat intelligence and community engagement distinguished it as a leader.

Conclusion and Lessons for SMEs and Large Enterprises

Dragos' journey underscores the critical importance of proactive cybersecurity for IT/OT systems. Key takeaways include:

1) **Tailored Solutions Matter:** OT environments require specialized tools and approaches distinct from IT.
2) **Invest in Threat Intelligence:** Understanding adversaries' tactics is essential for effective defense.
3) **Collaboration is Essential:** Community-driven initiatives like Neighborhood Keeper demonstrate the power of shared knowledge in combating cyber threats.

Citations

1 Dragos, Inc., "Annual Threat Report: ICS/OT Cybersecurity Trends," 2023.
2 Cybersecurity and Infrastructure Security Agency (CISA), "Best Practices for Industrial Control Systems Security," 2022.
3 NERC Critical Infrastructure Protection Standards (CIP), "Cybersecurity in Critical Energy Infrastructure," 2023.
4 IEEE Transactions on Industrial Informatics, "Securing SCADA Systems: Lessons from Dragos," 2022.
5 Gartner, "Magic Quadrant for OT Security," 2023.
6 Harvard Business Review, "OT Security: Bridging the IT/OT Gap," 2021.
7 The Economist, "The Rise of OT Cybersecurity," 2023.
8 SANS Institute, "ICS Cybersecurity: Lessons from Dragos," 2022.

Case Study 19: Tempest Security Intelligence – Brazil's Cybersecurity Trailblazer

Introduction

Founded in Recife, Brazil, in 2000, Tempest Security Intelligence began as a small consulting firm specializing in ethical hacking and security assessments. Over the past two decades, it has grown into one of Latin America's most prominent cybersecurity companies, serving global clients across financial services, telecommunications, retail, and government sectors. By combining innovative technology with a deep understanding of cyber threats, Tempest has positioned itself as a leader in offensive and defensive cybersecurity solutions.

This case study explores how Tempest achieved its success, the unique challenges it overcame, and its broader impact on the cybersecurity landscape in Brazil and beyond. It also highlights the company's contributions to bolstering cybersecurity awareness in a rapidly digitalizing Latin American economy.

The Cybersecurity Landscape in Brazil

Brazil is the largest economy in Latin America, with a rapidly growing digital economy. However, it has also become a prime target for cybercriminals:

- **Digital Banking Dominance:** Brazil leads the region in digital banking adoption, making financial services a major target for fraud and phishing attacks.
- **Weak Cybersecurity Infrastructure:** Many businesses, especially SMEs, lack adequate cybersecurity defenses.
- **Rising Ransomware Threats:** In 2022, Brazil ranked among the top 10 countries globally for ransomware attacks.

These challenges created a pressing need for advanced cybersecurity solutions, setting the stage for Tempest's emergence.

Tempest's Growth Journey

Founding Vision and Early Success

Tempest was founded to provide advanced cybersecurity solutions tailored to the unique challenges of Latin America. In its early years, the company focused on:

1) **Penetration Testing:** Helping clients identify vulnerabilities in their systems.
2) **Incident Response:** Supporting organizations in mitigating active cyberattacks.
3) **Security Awareness Training:** Educating employees to recognize and respond to cyber threats.

By 2010, Tempest had expanded its offerings to include managed security services and threat intelligence, enabling it to secure long-term contracts with major corporations in Brazil.

Expansion and Innovation

Tempest's growth accelerated in the 2010s, driven by:

1) **Global Partnerships:** Collaborating with international cybersecurity firms to bring cutting-edge technologies to the Latin American market.
2) **AI-Powered Threat Detection:** Developing proprietary tools that use AI to detect and respond to cyber threats in real time.
3) **Cloud Security:** Launching services tailored to Brazilian enterprises' increasing adoption of cloud computing.

In 2020, Tempest was acquired by Accenture, which provided it with the resources to scale globally while maintaining its regional expertise.

Notable Achievements

1) Defending Brazil's Financial Sector

Tempest has been instrumental in protecting Brazil's banking sector, which accounts for over 60% of the country's cyberattacks. Its services have helped:

- Block phishing campaigns targeting digital banking customers.
- Mitigate large-scale Distributed Denial of Service (DDoS) attacks on payment systems.
- Developed fraud detection algorithms that saved clients over $50 million annually.

2) Supporting Government Cybersecurity

Tempest partnered with Brazilian government agencies to:

- Secure critical infrastructure during major events like the 2016 Rio Olympics.
- Build threat intelligence platforms that monitor cyber threats targeting public institutions.

3) Promoting Cybersecurity Awareness

Tempest launched educational initiatives, including free webinars and workshops, to improve cybersecurity literacy among SMEs and the general public.

Challenges and Lessons Learned

Challenges

1) **Talent Shortages:** Brazil faces a significant shortage of cybersecurity professionals, forcing Tempest to invest heavily in training programs.

2) **Complex Threat Landscape:** The company must continually adapt to evolving threats, including sophisticated ransomware campaigns and nation-state actors.
3) **Scaling Globally:** While Accenture's acquisition provided resources, integrating into a global framework introduced operational complexities.

Lessons Learned

1) **Focus on Local Needs:** Tempest's success stems from its deep understanding of Brazil's unique cybersecurity challenges and its ability to tailor solutions accordingly.
2) **Invest in Talent Development:** Tempest has addressed workforce shortages while fostering employee loyalty by building its own training programs.
3) **Innovate Continuously:** The company's commitment to developing AI-powered and cloud-based solutions has kept it ahead of competitors.

Ramifications for Brazil and Beyond

1) Strengthening Brazil's Digital Economy

Tempest's efforts have bolstered trust in digital services, encouraging more businesses and consumers to embrace digital transformation. This trust is critical for economic growth in sectors like e-commerce and fintech.

2) Setting a Regional Standard

Tempest has become a model for other Latin American cybersecurity firms, demonstrating that local companies can compete globally by focusing on innovation and regional expertise.

3) Raising Awareness

Through its educational initiatives, Tempest has elevated cybersecurity awareness in Brazil, helping organizations of all sizes adopt better practices.

Comparisons with Other Cybersecurity Firms

1) Kaspersky

While Kaspersky has a strong presence in Latin America, its solutions are more generalized, whereas Tempest offers tailored services for local needs.

2) Symantec

Tempest's focus on offensive security and penetration testing differentiates it from Symantec's primarily defensive offerings.

3) Darktrace

Both companies leverage AI for threat detection, but Tempest's tools are designed specifically for the unique challenges of Latin American businesses.

Conclusion and Key Lessons

Tempest Security Intelligence's journey from a small Brazilian startup to a global cybersecurity leader offers valuable insights:

1) **Local Expertise is a Competitive Advantage:** Understanding regional challenges can differentiate a company in a crowded global market.

2) **Education is Essential:** Empowering clients and the public with knowledge is as important as providing tools and services.
3) **Adaptability Ensures Longevity:** Tempest has stayed ahead of evolving threats by embracing innovation and scaling strategically.

As cyber threats continue to grow, Tempest's focus on innovation, education, and tailored solutions positions it as a key player in securing Brazil's digital future and beyond.

Citations

1 Tempest Security Intelligence, "Annual Cybersecurity Trends Report: Brazil," 2023.
2 Accenture, "Integration of Tempest into Global Cybersecurity Services," 2021.
3 Financial Times, "How Tempest is Protecting Brazil's Banking Sector from Cyber Threats," 2022.
4 Brazilian Computer Emergency Response Team (CERT.br), "Cybersecurity Threats in Latin America," 2022.
5 Gartner, "Top Cybersecurity Companies in Emerging Markets," 2023.
6 Forbes, "The Role of Tempest in Latin America's Cybersecurity Ecosystem," 2021.
7 Harvard Business Review, "Cybersecurity in Emerging Markets: Lessons from Tempest," 2022.
8 SANS Institute, "Case Studies in Offensive Security: Tempest's Journey," 2023.

1) Cloudflare: Democratizing Internet Security

Cloudflare began as a service to protect websites from DDoS attacks and has grown into a global leader in internet performance and security. By offering free, accessible cybersecurity tools to small businesses and nonprofits, alongside premium services for enterprises, Cloudflare has democratized access to cybersecurity, protecting millions of organizations worldwide.

Why It Stands Out:

- Cloudflare's innovative edge computing services.
- Major role in thwarting global-scale DDoS attacks.
- Advocacy for internet freedom through initiatives like Project Galileo.

Lessons:

- Balancing accessibility and profitability is possible in cybersecurity.
- Companies can align business models with societal good.

2) Darktrace: AI-Powered Cyber Defense

Darktrace has pioneered the use of AI in cybersecurity with its Enterprise Immune System, which detects and responds to threats autonomously. The company has gained global acclaim for its ability to stop zero-day attacks and insider threats in real time.

Why It Stands Out:

- Novel application of self-learning AI to mimic the human immune system.
- Rapid growth and global deployment, including healthcare and critical infrastructure sectors.

Lessons:

- Self-adaptive systems are critical for combating evolving threats.
- AI requires integration with human oversight for effective deployment.

3) Palo Alto Networks: Securing Cloud-Native Applications

Palo Alto Networks has transformed cybersecurity for cloud-native environments through its Prisma Cloud platform, which provides unified security for DevOps workflows. The company has redefined what modern security looks like for distributed cloud applications.

Why It Stands Out:

- Focus on securing multicloud and hybrid environments.
- Acquisition strategy to stay ahead of technological trends (e.g., Demisto for incident response automation).

Lessons:

- Companies must anticipate technological shifts like cloud computing.
- Mergers and acquisitions can accelerate innovation when strategically aligned.

4) Okta: Redefining Identity Security

Okta specializes in identity and access management, providing secure, seamless access to applications for businesses and individuals. Its role in enabling Zero-Trust architecture has been a game changer for organizations transitioning to remote work post-pandemic.

Why It Stands Out:

- Leadership in the Zero-Trust security model.
- Focus on user experience without compromising security.

Lessons:

- User-centric design enhances security adoption.
- Identity is a cornerstone of enterprise cybersecurity.

5) FireEye (Now Mandiant): Incident Response for Advanced Persistent Threats

FireEye revolutionized cybersecurity by developing tools and strategies to detect Advanced Persistent Threats (APTs). After its spin-off as Mandiant, the company has remained a leader in incident response and cyber forensics.

Why It Stands Out:

- Uncovering high-profile breaches like the SolarWinds attack.
- Expertise in responding to nation-state-level threats.

Lessons:

- Incident response expertise is crucial for addressing complex attacks.
- Collaboration between private firms and government entities enhances defense.

6) CyberArk: Securing Privileged Access

CyberArk focuses on protecting privileged accounts, which are critical targets in cyberattacks. Its Privileged Access Management (PAM) solutions have become indispensable for securing critical infrastructure, financial systems, and enterprise IT environments.

Why It Stands Out:

- Early recognition of the importance of PAM.
- Pioneering solutions for insider threats and credential theft.

Lessons:

- Specialized focus can yield industry leadership.
- Proactive risk management reduces attack surfaces.

7) CrowdStrike: EDR and Threat Intelligence

CrowdStrike has reshaped endpoint detection and response (EDR) with its Falcon platform. This platform combines AI, behavioral analytics, and threat intelligence to detect breaches faster than traditional methods.

Why It Stands Out:

- Rapid detection and response to threats using lightweight cloud-based agents.
- High-profile successes, including attribution of attacks to nation-states.

Lessons:

- Cloud-native solutions scale effectively across enterprises.
- Combining technology with human intelligence is critical for accurate threat attribution.

8) Zscaler: Securing the Edge with SASE

Zscaler pioneered the Secure Access Service Edge (SASE) model, which provides secure connectivity for remote and mobile workers. Its platform has become a cornerstone for enterprises adopting cloud-first strategies.

Why It Stands Out:

- Transitioning security from on-premise to cloud-native models.
- Enabling large-scale secure remote work during the COVID-19 pandemic.

Lessons:

- Enterprises must shift security perimeters to the user level.
- Scalability and user experience are critical for adoption in remote work scenarios.

9) Checkpoint Software: Legacy Meets Innovation

Checkpoint Software has consistently evolved from providing firewall solutions to offering comprehensive, integrated threat prevention systems. Its focus on unified security management has made it a preferred choice for enterprises.

Why It Stands Out:

- Balancing legacy solutions with cutting-edge innovation.
- Leadership in proactive threat prevention.

Lessons:

- Evolution is necessary to stay relevant in a dynamic industry.
- A holistic approach simplifies security management for enterprises.

10) Fortinet: Integrating Security Across IT/OT

Fortinet stands out for its focus on securing IT and OT environments. It provides robust solutions for industrial systems that face growing cyber risks. Its FortiGate platform combines network security with application-specific defenses.

Why It Stands Out:

- Bridging the gap between IT and OT security.
- Leadership in securing critical infrastructure, such as energy and manufacturing.

Lessons:

- Industrial cybersecurity requires a specialized approach.
- Integrated solutions reduce complexity and enhance resilience.

Additional Sectors and Case Studies

The following sectors include transformative innovations, impactful case studies, and emerging trends that deserve attention. While not covered in detail within the primary content, this appendix provides brief overviews and verified references to guide further exploration.

1) **Manufacturing and Industry Transformation**

Small and Mid-Sized Enterprises Leveraging Technology

- **Indorama Ventures:** A Thailand-based SME transformed into a global petrochemical leader by adopting Industry 4.0 automation tools for smart manufacturing.
- **Bigbelly Solar:** Leveraged IoT-enabled waste management systems to optimize urban sanitation in U.S. cities.

Larger Enterprises Reinventing Themselves

- **Siemens:** Pioneered digital twin technologies to optimize manufacturing efficiency, saving $5 billion in operational costs by 2023.
- **GE Additive:** Invested in 3D-printing technologies, enabling faster prototyping and production in aerospace and healthcare industries.

References

1 McKinsey & Company, "Industry 4.0: Smart Manufacturing Insights," 2022.
2 Siemens, "Digital Twin Applications in Manufacturing," 2023.

2) **Education and Workforce Development**

Online Education and Skill Development

- **Coursera:** Revolutionized access to education by partnering with global universities to provide skill-specific certification programs.
- **Khan Academy:** Created free, scalable tools for personalized learning, reaching underserved students worldwide.

Lifelong Learning Innovations

- **edX (by 2U):** Integrated AI-based assessment tools to track learning outcomes across diverse student populations.

References

1 World Economic Forum, "The Future of Work and Education," 2022.
2 Khan Academy, "Learning Impact Report," 2023.

3) **Space Exploration and Aerospace**

Innovative Space Companies

- **Rocket Lab:** Focused on small satellite launches with cost-efficient reusable rocket systems.
- **Blue Origin:** Advanced sustainable aerospace development, targeting commercial space tourism and lunar missions.

References

1 NASA, "The Emerging Commercial Space Economy," 2022.
2 Rocket Lab, "Small Satellite Innovation," 2023.

4) **Finance and FinTech**

Cybersecurity in Financial Systems

- Stripe uses machine learning algorithms to detect fraud in real-time, improving payment security for millions of merchants.
- **Square (Block):** Revolutionized small business finance with portable point-of-sale systems and blockchain-based digital wallets.

Blockchain and Digital Currencies

- **Ethereum:** Advanced smart contract technologies power decentralized finance (DeFi) applications.
- **Ripple:** Streamlined cross-border payments, reducing transaction fees by 80% compared to traditional systems.

References

1 World Bank, "Digital Finance and Financial Inclusion," 2023.
2 Stripe, "The Role of AI in Secure Transactions," 2023.

5) **Transportation and Urban Development**

Smart Cities and Mobility Innovations

- **Waymo:** Leading AV development for safer urban transportation.
- **Citymapper:** Created real-time navigation tools integrating public transport and micromobility options in over 50 cities.

References

1 Institute of Transportation Studies, "The Future of Urban Mobility," 2022.
2 Waymo, "Autonomous Vehicles Safety Data Report," 2023.

6) **Agriculture and Food Systems**

Precision Agriculture

- **John Deere:** Integrated IoT sensors and AI to optimize crop yield and reduce water usage by 30%.
- **Aerofarms:** Developed vertical farming systems, increasing efficiency in urban food production.

Sustainable Supply Chain Practices

- **Unilever:** Enhanced supply chain transparency with blockchain technology, ensuring fair trade compliance across food brands.

References

1 FAO, "Digital Technologies in Agriculture," 2022.
2 Aerofarms, "Sustainable Urban Farming Case Studies," 2023.

7) **Critical Infrastructure and Resilience**

Water and Energy Systems

- **Suez Water Technologies:** Deployed IoT sensors to monitor real-time water quality, minimizing risks during urban flooding.
- **Dominion Energy:** Used IT/OT integration to enhance grid resilience against cyberattacks and extreme weather.

Transportation Infrastructure

- **MaaS Global:** Pioneered Mobility-as-a-Service, integrating ride-sharing, public transit, and cycling into single-platform applications.

References

1 IEEE Smart Systems Journal, "IoT for Critical Infrastructure," 2022.
2 Suez, "Smart Water Management Report," 2023.

8) **Small and Mid-Sized Enterprises (SMEs)**

Transformative Growth via Digital Transformation

- **Zoomcar (India):** Scaled vehicle-sharing platforms through AI-driven demand forecasting, improving vehicle utilization rates by 50%.
- **Peloton (U.S.):** Integrated connected fitness systems into home gyms, driving global adoption of personalized fitness programs.

References

1 Harvard Business Review, "SMEs and Digital Transformation," 2023.
2 Deloitte, "Tech Adoption Trends in Small Businesses," 2022.

Additional Sectors Not Fully Explored:

1) **Pharmaceuticals:** Innovations in AI-driven drug discovery (e.g., BenevolentAI, Exscientia).
2) **Green Energy Startups:** Emerging solar technology companies (e.g., SunPower, First Solar).
3) **Creative Industries:** Disruptive platforms in art and design (e.g., Behance, Canva).
4) **Health and Wellness:** Startups focusing on personalized health data (e.g., 23andMe, Fitbit).

Notes and Recommendations: For readers interested in further details, the references listed under each sector provide access to full reports, case studies, and verified data. These are excellent starting points for exploring transformative innovations across industries.

Case Study 20: General Electric's Transition to a Decentralized Model

Industry: Manufacturing and Energy
In 2018, General Electric (GE), a longstanding leader in industrial manufacturing, faced significant financial challenges due to inefficiencies in its traditional centralized operations model. Recognizing the vulnerabilities inherent in large, siloed systems, the company transitioned to a decentralized, business-unit-focused structure. This shift was driven by a combination of visionary leadership and the adoption of advanced tools for organizational restructuring.

Under the leadership of then-CEO Larry Culp, GE embraced principles of resilience and agility, focusing on independent accountability within its business units. Tasks like **scenario planning** and **DSHARP/CARVER analysis** (highlighted in Chapter 4B) were employed to identify inefficiencies and prioritize actionable strategies. Additionally, GE adopted predictive analytics and digital twins – a concept elaborated on in Chapter 11 – to optimize its power grids and enhance energy management solutions.

Lessons Learned:

- **Leadership Style:** Transformational leadership was key. Culp's emphasis on collaboration and empowerment helped foster trust and buy-in among stakeholders.
- **Outcomes:** Within three years, GE improved operational efficiency, restored financial stability, and became a case study in adaptive management during uncertainty.
- **Tool Efficacy:** Integrating technology readiness assessments and decentralized modeling proved instrumental in achieving measurable improvements.

Case Study 21: Microsoft's Ethical AI Framework

Industry: Information Technology
Microsoft's commitment to ethical AI development has set benchmarks for the IT industry. In 2019, under the leadership of CEO Satya Nadella, the company launched its "Responsible AI" framework to address concerns about bias, privacy, and misuse in AI systems. This initiative emphasized the role of **ethical leadership** and **collaborative decision-making**, as discussed in Chapter 9.

The company implemented guidelines based on transparency, fairness, and accountability principles. Practical tools such as the **Business Model Canvas** and **R-W-W screening** (explored in Chapter 4B) were used to assess AI projects' viability and ethical implications. Microsoft also engaged interdisciplinary teams of engineers, ethicists, and legal experts to ensure compliance with global standards.

Lessons Learned:

- **Leadership Style:** Servant leadership was instrumental. Nadella's focus on empathy and stakeholder engagement enabled Microsoft to balance innovation with societal responsibilities.
- **Outcomes:** The framework mitigated ethical risks, enhanced customer trust, and solidified Microsoft's position as a leader in responsible technology development.
- **Tool Efficacy:** Using foresight modeling and stakeholder analysis highlighted the importance of proactive risk management in technology development.

Case Study 22: Singapore's Smart Nation Initiative

Industry: Public Policy and Smart Infrastructure
Launched in 2014, Singapore's Smart Nation Initiative epitomizes integrating technology and governance to enhance urban resilience. Singapore has transformed public services, transportation, and energy management by leveraging AI-driven monitoring systems and IoT-enabled infrastructure rooted in collaborative leadership across public and private sectors.

One of the program's hallmark projects is its **AI-powered flood management system**, designed to predict and mitigate flooding in vulnerable urban areas. The government employed **Technology Readiness Levels (TRLs)**, as described in Appendix Aappendix A, to ensure the reliability of deployed technologies. Additionally, the initiative prioritized workforce upskilling through partnerships with educational institutions, aligning with Chapter 25's focus on workforce resilience.

Lessons Learned:

- **Leadership Style:** Distributed leadership fostered innovation. Singapore achieved a cohesive and effective strategy by empowering multiple stakeholders to contribute.
- **Outcomes:** The initiative's impact is highlighted by reduced flood risks, enhanced public safety, and increased citizen satisfaction.
- **Tool Efficacy:** Using trend coupling matrices and foresight tools facilitated long-term planning and risk assessment.

Case Study 23: Patagonia's Circular Economy Model

Industry: Retail and Sustainability
Patagonia, a pioneer in sustainable retail, has embraced the circular economy as a core business strategy. In 2017, the company launched its "Worn Wear" program, encouraging customers to repair, recycle, or repurpose used clothing. This initiative aligns with the principles of **equity in innovation** discussed in Chapter 24.

Patagonia's leadership team, led by founder Yvon Chouinard, adopted **transformational leadership** to drive cultural and organizational change. The company utilized **life cycle analysis tools** and **value chain assessments** (highlighted in Chapter 4B) to measure the environmental impact of its operations. These efforts reduced waste and strengthened Patagonia's brand as a socially responsible retailer.

Lessons Learned:

- **Leadership Style:** Transformational leadership inspired a shared commitment to sustainability among employees and customers.
- **Outcomes:** Increased customer loyalty, reduced environmental impact, and a scalable business model for sustainability.
- **Tool Efficacy:** The application of life cycle analysis tools demonstrated the feasibility of aligning profitability with environmental stewardship.

Case Study 24: Tesla's Expansion into Gigafactories

Industry: Automotive and Renewable Energy
In 2014, Tesla launched its first Gigafactory in Nevada to produce lithium-ion batteries at scale, a bold move that redefined how companies address energy storage and supply chain challenges. Tesla's CEO, Elon Musk, envisioned Gigafactories as critical enablers for a renewable energy future, aligning energy storage innovations with the company's growing EV production.

Tesla applied **disruptive innovation frameworks** and **trend coupling analyses** (Chapter 4B) to identify strategic locations for the factories and secure raw material supplies. The company adopted **open innovation principles** by partnering with suppliers like Panasonic to share intellectual property and enhance production efficiency. Using **smart grid tools** and **renewable energy models**, Tesla achieved sustainability within its production processes, reducing emissions and cutting costs.

Lessons Learned:

- **Leadership Style:** Visionary leadership drove Tesla's ability to align its mission with operational goals. Musk's focus on long-term value galvanized employees and stakeholders.
- **Outcomes:** Tesla now operates multiple Gigafactories worldwide, enabling mass production of EVs and reducing the cost of battery storage. These achievements positioned Tesla as a leader in sustainable manufacturing.
- **Tool Efficacy:** Scenario planning and lifecycle assessments helped mitigate risks in raw material sourcing and optimize global supply chains.

Case Study 25: IBM's Pioneering Blockchain for Supply Chains

Industry: Technology and Supply Chain Management
Relevant Chapters: 12 (*Cybersecurity in the Age of Interdependence*), 24 (*Equity in Innovation*), and 27 (*Resilient Business Models*).

In 2016, IBM launched its Blockchain Solutions to address inefficiencies and vulnerabilities in global supply chains. By leveraging blockchain technology, IBM enabled secure, transparent tracking of goods across multiple stakeholders, reducing fraud and improving accountability. The company partnered with food suppliers like Walmart to trace food origins, ensuring quality and compliance.

IBM utilized **design thinking methodologies** and **stakeholder analysis tools** (Appendix A) to identify pain points in traditional supply chain systems. The project emphasized **collaborative leadership**, engaging clients, regulators, and tech teams to cocreate scalable solutions. Blockchain was further integrated with IoT devices, enabling real-time data sharing, a key focus of Chapter 12.

Lessons Learned:

- **Leadership Style:** Collaborative leadership was critical. IBM leaders prioritized building ecosystems of trust among diverse partners.
- **Outcomes:** Blockchain significantly reduced food contamination risks, cut operational costs, and improved supply chain visibility. These successes cemented IBM's reputation as an innovator in secure digital solutions.
- **Tool Efficacy:** Blockchain and IoT integration effectively addressed logistical complexities and improved stakeholder accountability.

Placement: This case study should be included in Chapter 12, after the section "Building Secure Digital Ecosystems," as it exemplifies the application of technology to enhance security and efficiency.

Case Study 26: India's Digital Identification System (Aadhaar)

Industry: Public Policy and Digital Transformation
In 2009, India launched Aadhaar, the world's largest digital identification system. It is designed to provide every citizen with a unique identity number linked to biometrics. This initiative aimed to improve access to public services and reduce corruption by ensuring benefits reached the intended recipients.

The program relied on **distributed leadership** and cross-sector collaboration involving government agencies, tech companies, and civil society. Aadhaar's implementation leveraged **TRLs** (Appendix A) to test and deploy scalable solutions. The project also prioritized workforce upskilling and training officials in digital processes, aligning with Chapter 25.

Lessons Learned:

- **Leadership Style:** Distributed leadership enabled effective coordination among diverse stakeholders, ensuring the project's scalability.
- **Outcomes:** Over 1.3 billion citizens now have Aadhaar IDs, streamlining service delivery and financial inclusion. However, challenges like data security and privacy remain areas for continuous improvement.
- **Tool Efficacy:** Applying TRLs and stakeholder analysis ensured the system's reliability and public trust.

Case Study 27: Amazon's Use of AI for Logistics Optimization

Industry: E-Commerce and Logistics
Amazon's adoption of AI to streamline its logistics network revolutionized e-commerce operations. Using AI-powered predictive analytics, the company enhanced demand forecasting, inventory management, and last-mile delivery. This became particularly evident during the COVID-19 pandemic when supply chains faced unprecedented disruptions.

Amazon employed **data analytics tools** and **real-time monitoring systems** (Chapter 4B) to adapt to fluctuating demand. The company's **autonomous delivery systems**, including drones and robots, illustrate innovative applications of AI to overcome logistical bottlenecks, aligning with Chapter 20.

Lessons Learned:

- **Leadership Style:** Operational leadership emphasizes continuous improvement and rapid adaptation, hallmarks of Amazon's culture.
- **Outcomes:** Enhanced delivery efficiency, reduced operational costs, and improved customer satisfaction during crises.
- **Tool Efficacy:** Predictive analytics and automation tools showcased AI's potential to address real-time supply chain challenges.

Case Study 28: Denmark's Transition to Wind Energy Leadership

Industry: Renewable Energy
In the 1980s, Denmark faced an energy crisis that forced the country to rethink its reliance on fossil fuels. Today, Denmark leads the world in wind energy production, with over 47% of its electricity derived from wind power as of 2024. This transformation was driven by **visionary leadership** and strategic investments in renewable energy infrastructure.

Denmark adopted a **systems approach**, integrating wind energy into its national grid with minimal disruptions. The government implemented **incentive-based policies**, such as feed-in tariffs, to encourage investment in wind energy. Tools like **scenario planning** and **trend coupling matrices** (Chapter 4B) were instrumental in predicting technological advancements and ensuring grid stability. The country also established partnerships with global leaders in turbine manufacturing, including Vestas, to drive innovation.

Lessons Learned:

- **Leadership Style:** Transformational leadership enabled Denmark to unite stakeholders around a shared energy independence and sustainability vision.
- **Outcomes:** Denmark's wind energy industry now generates significant economic value, supports thousands of jobs, and provides a scalable model for other nations.
- **Tool Efficacy:** Scenario planning and incentive structures proved critical in fostering public and private collaboration and reducing transition risks.

Case Study 29: Google's Project Oxygen: Leadership Development

Industry: Technology and Human Resources
In 2009, Google launched Project Oxygen, a data-driven initiative to identify the qualities of great managers. Using employee feedback and performance data, the project uncovered key leadership behaviors that drive team satisfaction and productivity, such as effective communication and empowerment.

Google's HR team applied **data analytics** and **employee engagement surveys** to assess management practices. The findings led to the development of training programs that emphasized **transformational and participatory leadership**, aligning with themes in Chapter 8. By fostering psychological safety and encouraging feedback, Google created a culture of continuous improvement.

Lessons Learned:

- **Leadership Style:** Participatory leadership encourages collaboration and trust, enabling teams to perform at their best.
- **Outcomes:** Improved employee retention, higher productivity, and recognition of Google as a model employer.
- **Tool Efficacy:** Leveraging data analytics highlighted the importance of measurable leadership practices in achieving organizational goals.

Case Study 30: Toyota's Just-in-Time Manufacturing System

Industry: Automotive and Manufacturing

Toyota revolutionized manufacturing with its Just-in-Time (JIT) system, designed to minimize waste and enhance efficiency. This approach aligns production schedules with customer demand, reducing excess inventory and costs. Introduced in the 1970s and refined over decades, JIT became a cornerstone of Toyota's operational success.

Toyota relied on **Kaizen (continuous improvement)** and **lean manufacturing tools** (Appendix A) to implement JIT across its global operations. Collaborative leadership at all organizational levels ensured buy-in and accountability. Toyota maintained supply chain agility by integrating predictive analytics despite disruptions like the COVID-19 pandemic.

Lessons Learned:

- **Leadership Style:** Operational leadership emphasizes precision and accountability, fostering a culture of excellence.
- **Outcomes:** Increased profitability, enhanced customer satisfaction, and global recognition as a leader in lean manufacturing.
- **Tool Efficacy:** Lean tools and predictive analytics demonstrated their value in streamlining operations and responding to market fluctuations.

Case Study 31: Rwanda's Drone Delivery Network for Healthcare

Industry: Healthcare and Logistics

In 2016, Rwanda launched a groundbreaking drone delivery network to transport medical supplies to remote regions. Partnering with California-based startup Zipline, the Rwandan government used drones to deliver blood, vaccines, and medicines within hours, saving countless lives.

The initiative leveraged **automation and IoT technologies** (Chapter 7) to overcome logistical challenges in healthcare delivery. Collaborative leadership between the government and the private sector ensured scalability and efficiency. The project also aligned with the principles of **ethical innovation**, addressing urgent healthcare needs equitably.

Lessons Learned:

- **Leadership Style:** Distributed leadership fostered innovation and shared ownership of the project's success.
- **Outcomes:** Faster delivery times, improved healthcare access, and a scalable model adopted by other nations, including Ghana and India.
- **Tool Efficacy:** IoT-enabled drones and predictive logistics tools were instrumental in achieving healthcare equity.

Case Study 32: Singapore's Water Management Strategy

Industry: Public Infrastructure and Sustainability

Facing a critical water shortage, Singapore developed one of the world's most advanced water management systems, blending desalination, wastewater recycling, and rainwater collection. The initiative, spearheaded by the Public Utilities Board (PUB), demonstrates how innovation and strategic foresight can overcome natural resource constraints.

Singapore evaluated and deployed water technologies using TRLs and foresight modeling tools (Appendix A). Collaborative leadership among policymakers, engineers, and private firms ensured the system's success. Today, Singapore meets over 70% of its water needs through self-sufficiency.

Lessons Learned:

- **Leadership Style:** Collaborative and systems-based leadership drove alignment across sectors.
- **Outcomes:** Enhanced water security, increased public trust, and a global reputation for resource resilience.
- **Tool Efficacy:** TRLs and modeling tools facilitated the integration of cutting-edge technologies into public infrastructure.

Case Study 33: Finland's Education System Reform

Industry: Education and Workforce Development
In the early 2000s, Finland reformed its education system to prioritize creativity, critical thinking, and equity. The country cultivated a learning environment that prepared students for a rapidly changing global workforce by reducing standardized testing and empowering teachers with autonomy.

Finland's education reform utilized **foresight modeling** and **stakeholder engagement tools** (Chapter 4B) to anticipate future workforce needs. Policymakers collaborated with educators, businesses, and researchers to design curricula emphasizing interdisciplinary learning and adaptability. This approach fostered a generation of problem solvers ready to thrive in a complex world.

Lessons Learned:

- **Leadership Style:** Participatory leadership created a sense of shared ownership among stakeholders.
- **Outcomes:** Consistently high rankings in global education metrics, reduced achievement gaps, and a skilled workforce poised for innovation.
- **Tool Efficacy:** Foresight tools and stakeholder analysis facilitated long-term planning and inclusive decision-making.

Case Study 34: Germany's Industrial Ecology Model

Industry: Manufacturing and Sustainability
Germany has long been a leader in industrial ecology, a sustainable production approach that reduces waste and optimizes resource use. One of its flagship initiatives is the Kalundborg Eco-Industrial Park, where businesses share resources and by-products to create a circular economy.

This model integrates **life cycle analysis tools** and **value chain assessments** (Chapter 4B) to identify industry synergies. Collaborative leadership among government agencies, private companies, and research institutions ensured the successful implementation of the eco-industrial framework.

Lessons Learned:

- **Leadership Style:** Systems leadership fostered cooperation and innovation across sectors.
- **Outcomes:** Reduced environmental impact, cost savings for participating businesses, and a scalable model for other regions.
- **Tool Efficacy:** Life cycle analysis tools demonstrated the viability of circular economies in reducing waste and maximizing value.

Case Study 35: India's Aadhaar Digital Identity Program

Industry: Public Policy and Technology
Narrative:
India's Aadhaar program, launched in 2009, is the world's largest biometric digital identity system, covering over 1.4 billion residents. Designed to streamline government services and improve access for marginalized communities, Aadhaar has faced challenges related to privacy, cybersecurity, and the ethical use of data.

The project applied **risk management tools** and **data security frameworks** (Appendix A) to safeguard user information while enhancing efficiency. Ethical leadership, as emphasized in Chapter 9, was critical in addressing concerns about misuse and ensuring transparency in data governance.

Lessons Learned:

- **Leadership Style:** Ethical and collaborative leadership balanced technological innovation with societal concerns.
- **Outcomes:** Increased access to government services, reduced corruption, and improved financial inclusion for millions.
- **Tool Efficacy:** Risk management tools ensured the secure and ethical deployment of digital identity technologies.

Case Study 36: Apple's Environmental Sustainability Strategy

Industry: Consumer Electronics and Sustainability
Apple's commitment to environmental sustainability has set a benchmark for the tech industry. The company achieved carbon neutrality for its corporate operations in 2020 and aims to have its entire supply chain carbon neutral by 2030. Initiatives include utilizing recycled rare earth elements in products and investing in renewable energy sources.

Apple leveraged **life cycle analysis tools** and **supply chain optimization models** (Chapter 4B) to align profitability with environmental stewardship. Under CEO Tim Cook, leadership emphasized **transformational leadership**, inspiring employees and partners to prioritize sustainability.

Lessons Learned:

- **Leadership Style:** Transformational leadership enabled Apple to drive industry-wide change in sustainability.
- **Outcomes:** Reduced environmental impact, enhanced brand reputation, and a scalable model for other tech companies.
- **Tool Efficacy:** Supply chain optimization tools facilitated the alignment of environmental and economic goals.

Case Study 37: Chile's Renewable Energy Boom

Industry: Energy and Sustainability
Over the past two decades, Chile has emerged as a global leader in renewable energy, particularly solar and wind power. The Atacama Desert, one of the sunniest places on Earth, has become a hub for solar energy projects, while Patagonia's wind corridors power much of the nation.

Chile adopted **technology readiness assessments** and **scenario planning tools** (Appendix A) to integrate renewable energy sources into its national grid. Transformative leadership by policymakers and energy companies ensured that infrastructure and regulatory frameworks kept pace with technological advancements.

Lessons Learned:

- **Leadership Style:** Transformative and systems leadership enabled Chile to align national goals with global sustainability trends.
- **Outcomes:** Reduced reliance on fossil fuels, increased energy security, and economic growth in renewable energy sectors.
- **Tool Efficacy:** Scenario planning and readiness assessments facilitated the seamless integration of renewable technologies.

Case Study 38: Kenya's M-Pesa and Financial Inclusion

Industry: Financial Technology (FinTech)
M-Pesa, a mobile money platform launched in 2007 by Safaricom in Kenya, revolutionized financial access for millions of unbanked citizens. Users could deposit, withdraw, and transfer money using mobile phones, even in remote areas with limited bank access. By 2023, M-Pesa served over 50 million active users across several African countries.

The platform succeeded through **the implementation of inclusive innovation strategies and effective technology deployment models** (Appendix A). Safaricom's leadership embraced **systems leadership**, integrating partnerships with regulators, microfinance institutions, and community organizations to build trust and scale adoption. Continuous iteration, utilizing feedback loops and data-driven insights, ensured the service met user needs while maintaining security and affordability.

Lessons Learned:

- **Leadership Style:** Systems leadership facilitated stakeholder collaboration, ensuring the widespread adoption of M-Pesa.
- **Outcomes:** Expanded financial inclusion, empowered women and small business owners, and stimulated economic growth in underserved regions.
- **Tool Efficacy:** Technology deployment models and data analysis tools helped ensure scalability and user-centric innovation.

Case Study 39: Denmark's Offshore Wind Energy Strategy

Industry: Renewable Energy
Denmark's leadership in offshore wind energy is a testament to strategic planning and collaboration. Since building the world's first offshore wind farm in 1991, the country has consistently

expanded its renewable energy capacity. By 2023, offshore wind provided more than 40% of Denmark's electricity, positioning it as a global pioneer in sustainable energy.

The success of Denmark's strategy relied on **technology readiness assessments** and **foresight modeling tools** (Appendix A). Policymakers established clear regulatory frameworks and invested in research and development, while private companies like Ørsted played a pivotal role in deploying cutting-edge technologies. Collaborative leadership enabled public–private partnerships to align economic, environmental, and societal goals.

Lessons Learned:

- **Leadership Style:** Collaborative leadership built consensus and leveraged expertise across sectors.
- **Outcomes:** Reduced carbon emissions, enhanced energy security, and exported renewable technology expertise worldwide.
- **Tool Efficacy:** Foresight tools facilitated long-term planning and mitigated risks associated with large-scale renewable energy projects.

Case Study 40: The US DARPA Model for Innovation

Industry: Defense and Advanced Research
The Defense Advanced Research Projects Agency (DARPA) is renowned for its ability to drive breakthrough innovations, from the development of the internet to stealth technology. Established in 1958, DARPA's unique operating model combines **mission-oriented leadership** with high-risk, high-reward projects.

The agency employs **portfolio management tools** and **scenario planning** (Chapter 4B) to balance project risk and align national security goals. DARPA's leaders adopt **an adaptive leadership approach**, empowering project teams to explore unconventional ideas while maintaining accountability through regular evaluations.

Lessons Learned:

- **Leadership Style:** Adaptive leadership fostered a culture of experimentation and rapid iteration.
- **Outcomes:** Delivered transformative technologies with applications beyond defense, including commercial sectors like healthcare and IT.
- **Tool Efficacy:** Portfolio management tools ensured resource allocation aligned with strategic priorities.

Case Study 41: Toyota's Lean Manufacturing System

Industry: Automotive Manufacturing
Toyota's lean manufacturing system, the Toyota Production System (TPS), has become a global standard for operational excellence. Developed in the mid-20th century, TPS emphasizes continuous improvement (*Kaizen*), waste elimination, and respect for workers.

Toyota's leaders employed **transformational leadership** to instill a culture of innovation and accountability throughout the organization. Tools like **value stream mapping** and **root cause analysis** (Chapter 4B) were integral to identifying inefficiencies and optimizing workflows. Collaboration between managers, engineers, and shop-floor employees ensured the successful implementation of lean principles.

Lessons Learned:

- **Leadership Style:** Transformational leadership inspired employee engagement and a commitment to excellence.
- **Outcomes:** Increased productivity, reduced costs, and established Toyota as a leader in quality and reliability.
- **Tool Efficacy:** Value stream mapping provided actionable insights for process improvement.

Case Study 42: Estonia's Digital Government Initiative

Industry: Public Policy and Technology
Estonia is widely regarded as one of the most digitally advanced countries globally, with 99% of its public services available online. Since the early 2000s, Estonia has implemented e-governance solutions, including electronic tax filing, e-residency, and blockchain-based voting systems.

Estonia's success relied on **risk assessment tools** and **technology deployment frameworks** (Appendix A) to ensure security and scalability. Distributed leadership across government agencies enabled the seamless integration of digital services. The initiative also prioritized cybersecurity, using **real-time monitoring tools** to mitigate risks.

Lessons Learned:

- **Leadership Style:** Distributed leadership empowered multiple stakeholders to innovate and collaborate effectively.
- **Outcomes:** Increased government efficiency, enhanced citizen satisfaction, and established Estonia as a global leader in e-governance.
- **Tool Efficacy:** Risk assessment tools ensured the secure implementation of digital services.

Case Study 43: Germany's Industrial Ecology and Circular Economy

Industry: Manufacturing and Sustainability
Germany has become a global leader in industrial ecology, integrating sustainable practices into its manufacturing and waste management systems. Initiatives such as "cradle-to-cradle" product designs and comprehensive recycling programs exemplify the country's commitment to a circular economy. These practices reduce resource consumption and minimize waste, creating a more sustainable industrial ecosystem.

German leaders utilized **life cycle analysis tools** and **material flow assessments** (Chapter 4B) to optimize resource use and environmental impact. Collaborative leadership across industries and government facilitated the alignment of policies and business practices, ensuring the successful implementation of circular economy principles.

Lessons Learned:

- **Leadership Style:** Collaborative leadership enabled cross-sector cooperation, aligning public and private sector goals.
- **Outcomes:** Reduced environmental impact, increased resource efficiency, and enhanced competitiveness in global markets.
- **Tool Efficacy:** Life cycle analysis tools provided actionable insights for sustainable production practices.

Case Study 44: Finland's Education System and Workforce Development

Industry: Education and Workforce Development
Finland's education system consistently ranks among the best globally, with an emphasis on personalized learning, teacher autonomy, and lifelong skill development. This approach has contributed to a highly skilled workforce capable of adapting to technological disruptions and market demands.

Finnish policymakers employed **scenario planning** and **foresight modeling tools** (Appendix A) to anticipate future workforce needs and align educational priorities. Transformational leadership within the education sector fostered a culture of continuous learning and innovation.

Lessons Learned:

- **Leadership Style:** Transformational leadership inspired educators and students to embrace adaptability and innovation.
- **Outcomes:** Enhanced workforce resilience, improved educational outcomes, and a global reputation for excellence.
- **Tool Efficacy:** Foresight tools facilitated proactive planning for future skill requirements.

Case Study 45: The Netherlands' Delta Works Project

Industry: Infrastructure and Disaster Management
The Netherlands' Delta Works project, a series of dams, levees, and storm surge barriers, is considered one of the world's greatest engineering feats. Initiated in the aftermath of the devastating North Sea Flood of 1953, the project has protected the country from flooding for decades.

Dutch leaders adopted **risk assessment tools** and **TRLs** (Appendix A) to design and implement the project. Collaborative leadership involving engineers, policymakers, and environmentalists ensured a holistic approach to disaster prevention and infrastructure resilience.

Lessons Learned:

- **Leadership Style:** Collaborative leadership enabled multidisciplinary problem-solving and stakeholder alignment.
- **Outcomes:** Reduced flood risks, enhanced safety, and established a global model for water management.
- **Tool Efficacy:** Risk assessment tools identify vulnerabilities and inform design decisions.

Case Study 46: India's Solar Microgrid Revolution

Industry: Renewable Energy and Rural Development
India's solar microgrid initiatives have brought electricity to over 200 million people in rural areas, significantly improving the quality of life and economic opportunities. These projects prioritize affordability and sustainability, leveraging local resources and community participation.

Indian leaders used **value chain analysis tools** and **cost–benefit assessments** (Chapter 4B) to design scalable microgrid systems. Distributed leadership enabled collaboration between local governments, NGOs, and private companies to deploy these systems efficiently.

Lessons Learned:

- **Leadership Style:** Distributed leadership empowered local communities to take ownership of projects.
- **Outcomes:** Increased energy access, reduced carbon emissions, and stimulated rural development.
- **Tool Efficacy:** Value chain analysis ensured the cost-effective and sustainable implementation of the project.

Case Study 47: Amazon's Fulfillment Centers and Automation

Industry: E-commerce and Logistics
Amazon's fulfillment centers are a benchmark for operational efficiency, blending advanced automation technologies with human labor. By 2024, Amazon had deployed over 500,000 robots across its warehouses, reducing order processing times while maintaining high accuracy rates.

Amazon employed **process optimization tools** and **R-W-W screening frameworks** (Chapter 4B) to streamline operations and integrate robotics. Transformational leadership ensured workforce alignment with automation initiatives, emphasizing upskilling and safety.

Lessons Learned:

- **Leadership Style:** Transformational leadership enabled the seamless integration of automation and human labor.
- **Outcomes:** Enhanced efficiency, improved customer satisfaction, and set new standards for logistics.
- **Tool Efficacy:** Process optimization tools identified bottlenecks and improved workflows.

Case Study 48: South Korea's Digital Twin Cities

Industry: Urban Planning and Smart Infrastructure
South Korea has pioneered the use of digital twin technology in urban planning, creating virtual replicas of cities to simulate and optimize infrastructure. Projects like the Songdo Smart City leverage real-time data and AI to enhance energy efficiency, transportation, and public services.

South Korean leaders utilized **trend coupling matrices** and **technology readiness assessments** (Appendix A) to evaluate and implement digital twin solutions. Collaborative leadership ensured alignment between technology providers, city planners, and government agencies.

Lessons Learned:

- **Leadership Style:** Collaborative leadership facilitated seamless integration across sectors.
- **Outcomes:** Improved urban efficiency, enhanced citizen satisfaction, and established a global model for smart cities.

- **Tool Efficacy:** Trend coupling matrices informed long-term planning and technology deployment.

Case Study 49: Kenya's M-KOPA Solar Initiative

Industry: Renewable Energy and Financial Inclusion
Kenya's M-KOPA Solar Initiative has revolutionized energy access and financial inclusion for low-income households. Since its inception in 2012, M-KOPA has connected over three million homes to affordable solar energy systems using a pay-as-you-go (PAYG) model. This initiative addresses the dual challenges of energy poverty and financial exclusion by leveraging mobile technology to enable micro-payments.

In this case, leadership exemplified distributed leadership, involving partnerships with telecom companies, financial institutions, and local governments to deliver solutions tailored to underserved populations. Tools such as cost–benefit assessments and value chain analysis (Chapter 4B) were crucial in designing scalable and sustainable deployment models. The program's integration of mobile payment platforms, such as M-Pesa, underscores the importance of aligning technological innovation with local needs and contexts.

Lessons Learned:

- **Leadership Style:** Distributed leadership empowered local stakeholders to cocreate sustainable energy solutions.
- **Outcomes:** Increased access to clean energy, reduced household energy costs, and improved financial literacy among users.
- **Tool Efficacy:** Cost–benefit assessments ensured affordability and long-term scalability, while value chain analysis optimized resource allocation and utilization.

Case Study 50: Tesla's Gigafactories and Sustainable Manufacturing

Industry: Manufacturing and Renewable Energy
Tesla's Gigafactories exemplify sustainable manufacturing by integrating renewable energy and advanced automation into production processes. Since opening its first Gigafactory in Nevada in 2016, Tesla has expanded globally, producing EVs and battery systems on a large scale. By powering these facilities with solar and wind energy, Tesla reduces carbon emissions while maintaining cost-efficiency.

Transformational leadership under Elon Musk has been instrumental in Tesla's success, driving innovation through ambitious goals and cultivating a culture of resilience. Tools such as foresight modeling and life cycle analysis (Chapter 4B) were instrumental in identifying opportunities for efficiency and sustainability. Collaborative partnerships with suppliers and governments enabled Tesla to scale production while meeting stringent environmental standards.

Lessons Learned:

- **Leadership Style:** Transformational leadership motivated teams to achieve ambitious sustainability goals.
- **Outcomes:** Accelerated EV adoption, reduced carbon emissions, and set new benchmarks for sustainable manufacturing.

- **Tool Efficacy:** Life cycle analysis helped optimize resource use, while foresight modeling guided strategic decision-making.

Case Study 51: Estonia's Digital Society Transformation

Industry: Public Sector and Digital Governance
Estonia has established itself as a global leader in digital governance through its e-Estonia initiative. By digitizing nearly all public services, Estonia enables secure, efficient, and transparent interactions between citizens and the government. Key innovations include e-residency, digital IDs, and blockchain-based voting systems.

Collaborative leadership has been essential, with government agencies, technology providers, and citizens collaborating to cocreate solutions. Tools such as technology readiness assessments and scenario planning (Appendix A) ensured the successful implementation of cutting-edge technologies. Estonia's emphasis on cybersecurity, through partnerships with NATO and the EU, highlights its commitment to protecting digital infrastructure.

Lessons Learned:

- **Leadership Style:** Collaborative leadership fostered trust and innovation among stakeholders.
- **Outcomes:** Enhanced citizen engagement, improved service delivery, and established a global model for digital governance.
- **Tool Efficacy:** Technology readiness assessments ensured reliability, while scenario planning prepared for potential risks.

Case Study 52: Unilever's Closed-Loop Packaging Initiative

Industry: Consumer Goods and Sustainability
Unilever's closed-loop packaging initiative aims to reduce waste and promote sustainability across its product lines. Since launching its "No Plastic Waste" commitment in 2019, Unilever has implemented innovative solutions, including reusable packaging and biodegradable materials. These efforts align with the principles of equity in innovation, ensuring that sustainability benefits all stakeholders.

Transformational leadership within Unilever has driven cultural and operational changes, emphasizing accountability and transparency. Tools like life cycle analysis and value chain assessments (Chapter 4B) enabled the company to identify areas for improvement and implement effective strategies. Collaborative partnerships with NGOs and policymakers further supported Unilever's sustainability goals.

Lessons Learned:

- **Leadership Style:** Transformational leadership fostered a shared commitment to sustainability across the organization.
- **Outcomes:** Reduced environmental impact, strengthened brand loyalty, and enhanced market competitiveness.
- **Tool Efficacy:** Life cycle analysis and value chain assessments demonstrated the feasibility of aligning sustainability with profitability.

Case Study 53: Singapore's Flood Resilience Program

Industry: Urban Infrastructure and Disaster Management
Singapore's flood resilience program integrates advanced technologies and collaborative governance to mitigate urban flooding risks. By leveraging AI-driven monitoring systems and IoT-enabled infrastructure, Singapore has reduced flood-related damages and enhanced public safety.

Distributed leadership across government agencies, private companies, and research institutions has been critical to the program's success. Tools like trend coupling matrices and foresight modeling (Appendix A) facilitated long-term planning and risk assessment. The program's emphasis on workforce development aligns with Chapter 25's focus on adaptive workforce strategies.

Lessons Learned:

- **Leadership Style:** Distributed leadership ensured alignment among diverse stakeholders.
- **Outcomes:** Enhanced urban resilience, reduced flood risks, and increased citizen satisfaction.
- **Tool Efficacy:** Trend coupling matrices and foresight modeling supported proactive risk management.

Case Study 54: Germany's Industrial Ecology Initiatives

Industry: Manufacturing and Sustainability
Germany's industrial ecology initiatives focus on integrating sustainable practices into manufacturing processes to reduce waste and optimize resource utilization. This approach is epitomized by the country's Eco-Industrial Parks (EIPs), which foster collaboration among companies to create circular economies. For example, in Kalundborg, Denmark – a model for Germany's initiatives – by-products from one factory become inputs for another, significantly reducing emissions and costs.

Transformational leadership at the corporate and government levels supports these efforts. Life cycle analysis and environmental impact assessments (Chapter 4B) are routinely employed to measure progress and identify areas for improvement. Germany's approach emphasizes the role of policy in incentivizing sustainable innovation through subsidies and regulations.

Lessons Learned:

- **Leadership Style:** Transformational leadership inspired collective action among industries to prioritize sustainability.
- **Outcomes:** Reduced waste, improved energy efficiency, and enhanced competitiveness for participating companies.
- **Tool Efficacy:** Life cycle analysis and impact assessments ensured measurable improvements in environmental performance.

Case Study 55: Apple's Circular Supply Chain Model

Industry: Consumer Electronics and Sustainability
Apple has adopted a circular supply chain model to reduce its environmental impact. Initiatives include using recycled rare earth elements in iPhones and implementing take-back programs for used devices. This approach aligns with the principles of equity in innovation and sustainable business practices.

Apple's leadership, driven by Tim Cook, emphasized ethical decision-making and long-term sustainability. Tools such as value chain analysis and TRLs (Chapter 4B) enabled Apple to identify opportunities for recycling and material recovery. Partnerships with suppliers further supported the company's goals.

Lessons Learned:

- **Leadership Style:** Ethical and transformational leadership fostered innovation and accountability.
- **Outcomes:** Reduced reliance on mining, decreased waste, and improved brand reputation.
- **Tool Efficacy:** Value chain analysis optimized resource allocation, while TRLs ensured the viability of recycling technologies.

Case Study 56: Rwanda's Digital Health Ecosystem

Industry: Healthcare and Public Policy
Rwanda has developed a digital health ecosystem to enhance healthcare access and outcomes. The country has improved its healthcare infrastructure by utilizing drones for medical supply delivery and AI for disease monitoring, while addressing challenges to rural accessibility.

Collaborative leadership has been key to this transformation, with government agencies, NGOs, and private companies working together. Tools like foresight modeling and technology readiness assessments (Appendix A) were instrumental in planning and deploying these innovations. Rwanda's approach demonstrates how digital solutions can bridge healthcare gaps in developing countries.

Lessons Learned:

- **Leadership Style:** Collaborative leadership fostered alignment among stakeholders to achieve shared goals.
- **Outcomes:** Improved healthcare access, reduced response times, and increased public trust in digital solutions.
- **Tool Efficacy:** Foresight modeling and readiness assessments supported strategic implementation and risk mitigation.

Case Study 57: Israel's Cybersecurity Ecosystem

Industry: Cybersecurity and Technology
Israel has established itself as a global leader in cybersecurity through its CyberSpark Innovation Arena and advanced training programs. This ecosystem includes partnerships among academia, government, and private companies to address evolving cyber threats.

With clear roles and stakeholder collaboration, distributed leadership underpins Israel's success. Tools like scenario planning and dynamic threat modeling (Chapter 4B) enable the proactive identification and mitigation of risks. Israel's commitment to innovation has also fostered the development of advanced technologies, such as AI-driven threat detection systems.

Lessons Learned:

- **Leadership Style:** Distributed leadership encouraged collaboration and innovation across sectors.

- **Outcomes:** Strengthened cybersecurity infrastructure, enhanced global reputation, and boosted economic growth in the tech sector.
- **Tool Efficacy:** Scenario planning and threat modeling ensured preparedness and adaptability.

Case Study 58: India's Solar Microgrid Projects

Industry: Renewable Energy and Rural Development
India's solar microgrid projects aim to provide reliable energy to rural communities. These projects, supported by the government and private sector, have connected millions of households to sustainable electricity solutions.

In this case, leadership exhibited a blend of distributed and transformational styles, empowering local communities to take ownership of energy projects. Tools such as cost–benefit analysis and stakeholder engagement frameworks (Chapter 4B) were crucial in ensuring feasibility and impact. By aligning with national sustainability goals, these projects demonstrate the potential of renewable energy to drive socioeconomic development.

Lessons Learned:

- **Leadership Style:** Distributed and transformational leadership facilitated community engagement and long-term success.
- **Outcomes:** Improved energy access, reduced reliance on fossil fuels, and enhanced quality of life for rural populations.
- **Tool Efficacy:** Cost–benefit analysis ensured financial viability, while stakeholder frameworks fostered community involvement.

Case Study 59: Future-Proofing Academic Technology Spaces at the University of Minnesota

Chapter: Resilience, Change, and Crisis Leadership

In 2020, the University of Minnesota faced a pivotal challenge: its Academic Technology Spaces (ATS) were outdated, costly, and unprepared for the evolving demands of digital and hybrid learning environments. Traditional ATS models, designed for fixed setups and in-person lectures, were insufficient for the growing need for flexible, interactive, and dynamic spaces. The COVID-19 pandemic accelerated the transition to hybrid learning, demanding immediate innovation.

The Director of Information Technology, S.S., spearheaded a transformative initiative to redesign ATS into future-proof, modular spaces. The first step was a comprehensive technology audit, which revealed that 60% of hardware was underused or outdated. Reconfiguring classrooms to meet new needs took weeks and resulted in an average annual productivity loss of $50,000. Maintaining existing systems costs $1.5 million annually, underscoring the unsustainability of the current approach.

Using **scenario planning** (Appendix A), S.S. envisioned a system adaptable to three scenarios: the dominance of AI-driven educational tools, the integration of VR in classrooms, and the continuation of hybrid learning models. A cost–benefit analysis (Chapter 4B) demonstrated that maintaining the current infrastructure would cost $15 million over ten years with no added functionality, while a modular redesign could cut costs by 30% and save $5 million in the same period.

The pilot phase introduced mobile desks, interactive projectors, and AI-powered tools in smaller classrooms. Compared to the previous four weeks, these setups allowed reconfigurations in under

24 hours. The second phase scaled this model to larger spaces, integrating VR tools and AR platforms to create immersive learning environments.

By 2022, the project delivered quantifiable outcomes:

- **Student satisfaction increased by 25%**, reflecting the improved flexibility and functionality of the new spaces.
- **Maintenance costs dropped by $500,000 annually**, with projected savings exceeding $5 million over the decade.
- **Reconfiguration times decreased by 75%**, allowing for efficient space utilization.
- **A 170% return on investment (ROI)** was achieved within five years.

This initiative positioned the University of Minnesota as a leader in educational innovation, providing a model for institutions globally.

Case Study 60: Automating Surface Finishing at Kurt Manufacturing

Chapter: Metrics and Execution

By 2021, Kurt Manufacturing faced inefficiencies in its surface finishing process for aluminum components. Manual jitterbugging and sanding took two hours per part, with a rejection rate of 10% due to inconsistencies. Rising labor costs and client dissatisfaction, particularly from Applied Materials, made automation a priority.

R.K., a senior project manager, led the effort to automate surface finishing using **Heilmeier's Catechism** framework (Chapter 4B). The team selected the UR10 robotic arm for its adaptability and precision. A pilot phase demonstrated significant improvements:

- **Rejection rates fell to less than 1%**, showcasing enhanced consistency.
- **Cycle times decreased by 75%,** reducing finishing time to 30 minutes per part.
- **Annual labor cost savings of $200,000** were achieved.

The full rollout of the robotic system across multiple production lines yielded a 105% ROI within the first year. Kurt Manufacturing's reputation for delivering high-quality parts was restored, leading to increased client trust and new business opportunities.

Case Study 61: Transforming Public Safety Through Video Analytics and Predictive Policing

Chapter: Leading in the Digital Age

In 2020, the University of Minnesota's Public Safety Emergency Communications Center (PSECC) struggled to address rising crime rates on campus. Manual monitoring of over 1,000 CCTV cameras resulted in response times averaging ten minutes, often too late to prevent incidents.

Technology lead A.M.J. integrated AI-powered video analytics and **Geographic Information Systems (GIS)** for predictive policing. These tools flagged suspicious behavior in real time and mapped crime hotspots for more effective resource allocation. The $1 million investment produced transformative results:

- **Response times dropped by 50%.**
- **Crime rates fell by 20%** in high-risk areas within 18 months.
- **Labor efficiency improved by 30%,** enabling reallocation of personnel to critical tasks.

The initiative delivered a 180% ROI and became a blueprint for leveraging AI to enhance campus safety.

Case Study 62: Addressing the Digital Divide in Public Schools

Chapter: Ethics and Responsible Innovation

In 2020, underserved public schools faced a widening digital divide, with students lacking reliable broadband and devices. K.S. developed a strategy to bridge this gap through infrastructure expansion, device distribution, and teacher training.

- **Broadband access reached 100%** of students in targeted areas.
- **5,000 devices** were distributed, ensuring equitable digital access.
- **Teacher training programs** improved digital literacy, enhancing student engagement by 15% and raising standardized test scores by 10%.

The $2.5 million investment yielded a projected ROI of 300%, with societal benefits including reduced dropout rates and improved student lifetime earnings.

Case Study 63: Streamlining Database Provisioning at the University of Minnesota

Chapter: Global Innovation and Collaborative Leadership

In 2021, the University of Minnesota centralized its database provisioning system to address inefficiencies and reduce costs. A.W. implemented a cloud-based, PAYG model, eliminating redundancies and aligning costs with resource consumption.

- **Database-related costs fell by 40%,** saving $500,000 annually.
- **Service delivery times improved by 25%.**
- **An ROI of 200%** was achieved over five years.

The initiative streamlined operations and provided a scalable model for other institutions to follow.

Appendix D

Further Assessments and Examples of Tools and Methods

Strategic Thinking and Planning

1) **Golden Circle (Simon Sinek):** A framework for identifying purpose, focusing on "Why" (mission), "How" (process), and "What" (outcomes). It is used to create alignment and motivation across organizations.
   - *Reference*: Simon Sinek, *Start with Why: How Great Leaders Inspire Everyone to Take Action*, Portfolio, 2009.
2) **SWOT Analysis:** Identifies internal strengths and weaknesses, as well as external opportunities and threats, to guide strategic decisions.
   - *Reference*: Albert Humphreys worked at Stanford Research Institute in the 1960s.
3) **PESTLE Analysis:** Examines external factors – Political, Economic, Social, Technological, Legal, and Environmental – that influence decision-making.
   - *Reference*: Francis J. Aguilar, *Scanning the Business Environment*, Macmillan, 1967.
4) **Porter's Five Forces Framework:** Evaluates the competitive pressures in an industry, including supplier power, buyer power, competitive rivalry, the threat of substitution, and the threat of new entrants.
   - *Reference*: Michael E. Porter, *Competitive Strategy: Techniques for Analyzing Industries and Competitors*, Free Press, 1980.
5) **Real-Win-Worth (RWW) Framework:** A decision-making tool for evaluating the feasibility and value of business initiatives, products, or projects.
   - *Reference*: George S. Day and Paul H. Schoemaker, *Peripheral Vision: Detecting the Weak Signals That Will Make or Break Your Company*, Harvard Business School Press, 2006.
6) **Strategic Technology Assessment (STA):** Evaluates risks and potential of technologies to ensure alignment with organizational priorities.
   - *Reference*: U.S. Department of Defense, *Technology Readiness Assessment Guide*, 2011.
7) **Technology Roadmapping:** Aligns organizational goals with technological developments to chart innovation pathways.
   - *Reference*: Robert Phaal, Clare J.P. Farrukh, and David R. Probert, *Technovation*, 2004.
8) **Technology Foresight and Forecasting:** Identifies trends and technologies through qualitative and quantitative methods.
   - *Reference*: Luke Georghiou, *Futures*, 1996, and Lockwood Carlson, 2004 (include citations).

9) **Scenario Planning:** Explores plausible futures to guide decision-making under uncertainty.
   - *Reference*: Peter Schwartz, *The Art of the Long View: Planning for the Future in an Uncertain World*, Currency Doubleday, 1991.
10) **Balanced Scorecard:** Tracks organizational performance across financial, customer, internal, and learning metrics.
    - *Reference*: Robert S. Kaplan and David P. Norton, *The Balanced Scorecard: Translating Strategy into Action*, Harvard Business School Press, 1996.
11) **Value Chain Analysis:** Dissects business processes to optimize value creation and competitive advantage.
    - *Reference*: Michael E. Porter, *Competitive Advantage: Creating and Sustaining Superior Performance*, Free Press, 1985.
12) **VRIO Framework:** Evaluates resources based on Value, Rarity, Imitability, and Organization to assess competitive advantage.
    - *Reference*: Jay B. Barney, *Journal of Management*, 1991.
13) **Strategic Alignment Models for Policy and Technology:** Ensures alignment of technology initiatives with organizational goals.
    - *Reference*: John C. Henderson and N. Venkatraman, *IBM Systems Journal*, 1993.

Decision-Making Tools

1) **Heilmeier's Catechism:** A framework of guiding questions to evaluate R&D projects, focusing on objectives, gaps, innovations, risks, and costs for communication to top brass at the Pentagon and other high-stakes decisions
   - *Reference*: George Heilmeier, *DARPA Guidelines for Project Evaluation*, 1970s.
2) **The Quad Chart:** Summarizes complex concepts in four quadrants: Problems, Solutions, Benefits, and Risks.
   - *Reference*: Widely used in defense and R&D industries; no singular origin.
3) **Cost–Benefit Analysis:** Quantifies costs and benefits to evaluate decision feasibility and efficiency.
   - *Reference*: Jules Dupuit, foundational work on cost–benefit methodologies, 1844.
4) **Monte Carlo Simulations for Risk and Sensitivity Analysis:** Model system uncertainty and variability through random sampling.
   - *Reference*: Stanislaw Ulam, *Monte Carlo Method*, 1940s.
5) **Decision Trees and Multi-Criteria Decision Analysis (MCDA):** Decision Trees break complex problems into branches of options, while Multi-Criteria Decision Analysis (MCDA) evaluates choices based on multiple factors.
   - *Reference*: Howard Raiffa, *Decision Analysis*, 1968.
6) **Strategic Decision-Making Under Uncertainty:** Combines probabilistic models and risk assessments for unpredictable conditions.
   - *Reference*: Avi Gilboa and Itzhak Gilboa, *Making Better Decisions*, 2009.
7) **Game Theory in Strategic Energy Decision-Making:** Analyzes stakeholder interactions in competitive or cooperative scenarios.
   - *Reference*: John von Neumann and Oskar Morgenstern, *Theory of Games and Economic Behavior*, 1944.

Systems and Engineering Approaches

1) **Systems Engineering:** Designs and manages complex systems with lifecycle integration.
   - *Reference*: NASA and DoD, *Systems Engineering Handbook*, 1990s.

2) **Engineering Systems Framework:** Examines systems' technical, social, and economic dimensions.
   - *Reference*: MIT Engineering Systems Division, 1990s.
3) **Dynamic Systems Modeling and Analysis:** Simulates interactions within systems to predict outcomes and test interventions.
   - *Reference*: Jay W. Forrester, *Industrial Dynamics*, 1961.
4) **Causal Loop Diagramming for Complex Systems:** Visualizes reinforcing and balancing feedback loops to clarify system dynamics.
   - *Reference*: Peter Senge, *The Fifth Discipline*, 1990.
5) **Integrated Resource Planning (IRP):** Balances supply-side and demand-side resources for efficient energy planning.
   - *Reference*: U.S. Energy Policy Act, 1992.
6) **Total Cost of Ownership (TCO) Analysis:** This evaluates systems' initial and ongoing costs to determine their long-term value.
   - *Reference*: Andrew Ng, foundational work on lifecycle costing, 1980s.

Innovation and Change Management

1) **S-Curves for Technology and Market Growth:** Tracks growth phases, highlighting stagnation and acceleration points.
   - *Reference*: Everett M. Rogers, *Diffusion of Innovations*, 1962.
2) **Innovation Diffusion Models:** Analyze how innovations spread within populations, identifying barriers and accelerators.
   - *Reference*: Everett M. Rogers, *Diffusion of Innovations*, 1962.
3) **Organizational Change Management Frameworks:** Guides organizations in navigating transitions with stakeholder alignment and communication.
   - *Reference*: John Kotter, *Leading Change*, 1996.
4) **Lessons Learned Analysis Frameworks:** Captures past project insights to improve decision-making and future outcomes.
   - *Reference*: Project Management Institute, *PMBOK Guide*, 2004.
5) **Adaptive Leadership and Crisis Management Techniques:** These techniques build resilience and adaptability in leadership during crises or major changes.
   - *Reference*: Ronald A. Heifetz, *Leadership Without Easy Answers*, 1994.
6) **Lean Startup Methodology:** Emphasizes rapid experimentation, validated learning, and iterative improvement.
   - *Reference*: Eric Ries, *The Lean Startup*, 2011.
7) **Design Sprint Methodology:** Solves big challenges in five days through prototyping and feedback.
   - *Reference*: Jake Knapp, *Sprint: How to Solve Big Problems and Test New Ideas in Just Five Days*, 2016.
8) **Blue Ocean Strategy:** Creates uncontested market spaces through differentiation and value innovation.
   - *Reference*: W. Chan Kim and Renée Mauborgne, *Blue Ocean Strategy*, 2005.
9) **Theory of Change:** Maps inputs to measurable long-term outcomes, focusing on social and program impacts.
   - *Reference*: Andrea A. Anderson, *The Aspen Institute Roundtable on Community Change*, 2005.
10) **Theory U:** Focuses on systems innovation by addressing root causes and fostering co-creation.
    - *Reference*: Otto Scharmer, *Theory U: Leading from the Future as It Emerges*, 2007.

Stakeholder and Collaboration Tools

1) **Stakeholder Mapping and Analysis:** Identifies and prioritizes stakeholders based on their influence and interest.
   - *Reference*: Freeman, R. Edward, *Strategic Management: A Stakeholder Approach*, 1984.
2) **Participatory Design and Engagement Tools:** Ensures inclusivity and ownership by engaging stakeholders in design processes.
   - *Reference*: Elizabeth Sanders, foundational work in participatory design, 1990s.
3) **Cross-Sector Collaboration Tools:** Facilitates partnerships across sectors to develop scalable, impactful solutions.
   - *Reference*: Bryson, Crosby, and Stone, *Journal of Public Administration Research and Theory*, 2006.
4) **Network Analysis for Collaborative Ecosystems:** Maps stakeholder relationships to identify ecosystem strengths and gaps.
   - *Reference*: Albert-László Barabási, *Network Science*, 2016.

Sustainability and Resilience

1) **Lifecycle Analysis:** Evaluates environmental impacts across a product's lifecycle.
   - *Reference*: ISO 14040, *Environmental Management – Life Cycle Assessment Principles and Framework*, 2006.
2) **Sustainability Assessment Metrics:** Measures environmental, social, and economic impacts for balanced decision-making.
   - *Reference*: Kates, Parris, and Leiserowitz, *Environment: Science and Policy for Sustainable Development*, 2005.
3) **Resilience Metrics and Analysis:** Quantifies system robustness and recovery capabilities.
   - *Reference*: Holling, C. S., *Annual Review of Ecology and Systematics*, 1973.
4) **Energy Justice Frameworks:** Ensures equitable access to energy resources for marginalized communities.
   - *Reference*: Sovacool et al., *Nature Energy*, 2016.
5) **Disaster Risk Reduction and Recovery Models:** Structural frameworks for reducing vulnerabilities and planning recovery post-disasters.
   - *Reference*: United Nations, *Sendai Framework for Disaster Risk Reduction*, 2015.
6) **Equity-Centered Design Approaches:** Focuses on ensuring solutions address underserved populations equitably.
   - *Reference*: IDEO.org, *Field Guide to Human-Centered Design*, 2015.
7) **Lifecycle Sustainability Assessment (LCSA):** This process integrates environmental, economic, and social dimensions in project evaluation.
   - *Reference*: Guinée, Jeroen B., *Life Cycle Sustainability Assessment*, 2016.
8) **Social Cost of Carbon Calculations:** To guide policy and quantify economic damages from carbon emissions.
   - *Reference*: Interagency Working Group on Social Cost of Greenhouse Gases, *Technical Support Document*, 2016.
9) **Global Transition Dynamics (GTD):** Leverages artificial intelligence (AI) and real-time data to address interconnected global challenges.
   - *Reference*: C. Starr and M. Amin, 2003, and 2024 (include citations).

Visualization and Communication Tools

1) **Data Visualization and Decision Support Tools:** Translates complex data into visual insights for decision-making.
   - *Reference*: Edward R. Tufte, *The Visual Display of Quantitative Information*, 1983.
2) **Business Model Canvas:** A visual framework for designing and analyzing business models.
   - *Reference*: Alexander Osterwalder and Yves Pigneur, *Business Model Generation*, 2010.
3) **Design Thinking for Problem-Solving:** A human-centered, iterative approach to solving problems creatively.
   - *Reference*: Tim Brown, *Change by Design*, 2009.
4) **Benchmarking and Comparative Metrics:** Compares performance metrics across organizations to identify best practices.
   - *Reference*: Camp, Robert C., *Benchmarking: The Search for Industry Best Practices That Lead to Superior Performance*, 1989.
5) **Gantt Charts, Kanban Boards, and Geo-spatial Mapping are tools** for project management, resource allocation, and spatial analysis.
   - *Reference*: Henry L. Gantt, *Work, Wages, and Profits*, 1913.

Slightly More Detailed Earlier Draft

Strategic Thinking and Planning

1) **Golden Circle (Simon Sinek):** A framework for identifying purpose, focusing on "Why" (mission), "How" (process), and "What" (outcomes). It is used to create alignment and motivation across organizations. Applicable to leadership, marketing, and strategic planning.
2) **Heilmeier's Catechism:** A framework of guiding questions developed by George Heilmeier to evaluate R&D projects, ensuring clarity and actionable goals. It focuses on defining objectives, identifying gaps in current approaches, proposing innovations, and assessing impact, risks, time, and cost. Widely used in R&D planning, it is simple, flexible, and transparent, though it lacks consideration of market or social impacts.
3) **The Quad Chart is a** concise, four-quadrant framework for summarizing complex ideas into Problems, Solutions, Benefits, and Risks. It is commonly used in high-stakes decision-making across industries like defense and R&D.
4) **SWOT Analysis:** Identifies internal Strengths and Weaknesses and external Opportunities and Threats to guide strategic decisions. Used in business strategy and organizational planning.
5) **Real-Win-Worth (RWW) Framework:** This is a decision-making tool for evaluating the feasibility and value of business initiatives, products, or projects. It addresses three critical questions: Is it Real? (Market Viability), Can We Win? (Competitive Advantage), and is **it Worth it?** (Financial Viability).
6) **Lifecycle analysis evaluates the environmental and sustainability impacts of a product, process, or system from inception to disposal. It is key for sustainability planning in the** energy and industrial sectors.
7) **Technology Roadmapping:** A strategic planning tool for aligning organizational goals with future technological developments and innovation pathways. Useful in R&D, corporate strategy, and policy planning.

8) **Theory of Change:** A methodology for mapping pathways from inputs to long-term outcomes, focusing on measurable impacts. It is commonly applied in project management, social programs, and policy evaluation.
9) **Technology Foresight and Forecasting:** Systematically identifying future trends, technologies, and their implications through qualitative and quantitative methods. Used in long-term planning, risk assessment, and innovation.
10) **Strategic Technology Assessment (STA):** This evaluation focuses on the potential and risks of new technologies in achieving specific goals and aligning with strategic priorities. It is applied in corporate strategy and national innovation planning.
11) **Scenario planning explores multiple plausible futures to guide decision-making under uncertainty. It is widely** used in energy, climate policy, and strategic planning.
12) **Cost–Benefit Analysis: This quantitative process compares the costs and benefits of a decision to determine its feasibility and efficiency. It is essential** for policy evaluation, infrastructure investments, and financial planning.
13) **Resilience Metrics and Analysis:** Measures the ability of systems to anticipate, absorb, and recover from disruptions. Applied in infrastructure design, disaster recovery, and risk management.
14) **PESTLE Analysis:** Analyzes external factors – Political, Economic, Social, Technological, Legal, and Environmental – that impact decisions. Applied in strategic planning and market analysis.
15) **Porter's Five Forces Framework evaluates competitive pressures in an industry: supplier power, buyer power, competitive rivalry, threat of substitution, and threat of new entrants. It is commonly** used in business strategy.
16) **Balanced Scorecard:** A tool to align organizational activities with vision and strategy by tracking performance across financial, customer, internal, and learning dimensions. Widely used in organizational performance management.
17) **Value Chain Analysis:** Dissects activities in a business to identify value-creating processes and optimize them for competitive advantage. Applied in operations and supply chain strategy.
18) **VRIO Framework:** Evaluates resources and capabilities based on Value, Rarity, Imitability, and Organization to assess competitive advantage. Useful in strategic resource planning.
19) **Strategic Alignment Models for Policy and Technology:** These models ensure that technology initiatives align with broader organizational or policy goals. They are essential in technology policy and corporate strategy.

Systems and Engineering Approaches

20) **Systems Engineering:** A structured approach to designing and managing complex systems, focusing on functionality, performance, and lifecycle integration. Applied in large-scale engineering and infrastructure projects.
21) **Engineering Systems Framework:** Examines technical, social, and economic dimensions of systems, emphasizing interactions and dependencies. Used in multidisciplinary projects and large-scale planning.
22) **Causal Loop Diagramming for Complex Systems:** This technique visualizes system feedback loops to identify reinforcing or balancing mechanisms. It is useful in systems thinking and problem-solving.
23) **Dynamic Systems Modeling and Analysis:** This technique simulates interactions in complex systems over time to predict outcomes and test interventions. It is applied in energy, transportation, and policy planning.

24) **Integrated Resource Planning (IRP):** Balances supply- and demand-side resources to meet future energy needs efficiently. Used in energy sector planning and utility management.
25) **Total Cost of Ownership (TCO) analysis assesses all initial and ongoing costs associated with a product or system to determine its overall value. It is applied** in procurement and infrastructure investments.

Decision-Making Tools

26) **Decision Trees and Multi-Criteria Decision Analysis (MCDA):** Decision trees break down complex decisions into branches of options and outcomes, while MCDA evaluates multiple criteria for informed choices. Used in policy and project decisions.
27) **Strategic Decision-Making Under Uncertainty:** This provides frameworks for making decisions in unpredictable environments, often combining probabilistic models and risk assessments. It is used in the energy and finance sectors.
28) **Monte Carlo Simulations for Risk and Sensitivity Analysis:** Uses random sampling to model system uncertainty and variability. Applied in project management, finance, and energy modeling.
29) **Game Theory in Strategic Energy Decision-Making:** Analyzes stakeholder interactions to predict behaviors and outcomes in competitive or cooperative scenarios. Applied in energy markets and negotiations.
30) **Resilience metrics and analysis focus on quantifiable system robustness and recovery indicators. They are applied** in critical infrastructure and disaster preparedness.

Innovation and Change Management

31) **S-Curves for technology and market growth track technological or market adoption phases, highlighting points of stagnation or growth.** They are used in R&D and market strategy.
32) **Innovation diffusion models examine how innovations spread within populations, identifying barriers and accelerators. They are applied** in marketing and technology adoption strategies.
33) **Organizational Change Management Frameworks:** These frameworks guide organizations in navigating transitions effectively, focusing on communication, stakeholder alignment, and adaptation. They are also applied in restructuring and policy shifts.
34) **Lessons Learned Analysis Frameworks:** These frameworks capture insights from past projects to inform future decision-making and best practices. They are used in project management and institutional learning.
35) **Adaptive Leadership and Crisis Management Techniques:** This field focuses on resilience and adaptability in leadership during crises or change. It is applied in disaster response and organizational management.

Stakeholder and Collaboration Tools

36) **Stakeholder Mapping and Analysis:** Identifies and prioritizes stakeholders based on their influence, interest, and impact on a project. Applied in policy planning and community engagement.
37) **Participatory Design and Engagement Tools:** These tools engage stakeholders in design processes to ensure inclusivity and ownership. They are used in urban planning, energy projects, and community programs.

38) **Cross-Sector Collaboration and Stakeholder Integration Tools:** These tools facilitate partnerships between the government, private, and nonprofit sectors for scalable solutions. They are applied in disaster recovery and infrastructure projects.

Sustainability and Resilience

39) **Sustainability assessment metrics measure environmental, social, and economic impacts to ensure balanced decision-making. They are used** in project evaluation and urban planning.
40) **Energy justice frameworks focus on equitable access to energy resources and benefits for marginalized communities. They are applied** in energy transitions and social equity initiatives.
41) **Disaster Risk Reduction and Recovery Models:** These provide structured frameworks for reducing vulnerabilities and planning post-disaster recovery. They are applied in climate adaptation and disaster management.
42) **Equity-Centered Design Approaches:** These approaches ensure that solutions equitably address the needs of underserved populations. They are applied in healthcare, urban planning, and energy systems.
43) **Lifecycle Sustainability Assessment (LCSA):** This integrates environmental, economic, and social aspects across a system's lifecycle. It is used in energy projects and sustainability initiatives.
44) **Social Cost of Carbon Calculations:** Quantifies economic damages associated with carbon emissions to guide policy decisions and applied in climate policy and carbon pricing.

Visualization and Communication Tools

45) **Data Visualization and Decision-Support Tools: These tools translate complex data into visual insights to inform decisions. They are used** in public policy, energy, and finance.
46) **Business Model Canvas:** Provides a visual framework for developing and analyzing business models. It is applied in startups, innovation projects, and corporate strategy.
47) **Design Thinking for Problem-Solving:** A human-centered, iterative approach to solving problems creatively. Used in product development and service design.
48) **Benchmarking and Comparative Metrics:** This technique compares performance metrics across organizations or sectors to identify best practices. It is applied in strategy and operations.
49) **Network analysis for collaborative ecosystems maps stakeholder relationships and interactions to identify strengths and gaps. It is used** in collaborative projects and ecosystem planning.
50) **Gantt Charts, Kanban Boards, and Geo-spatial Mapping:** Tools for project management and resource allocation, including spatial analysis. Applied in infrastructure planning and disaster recovery.

Additional Tools and Concepts for Strategic Thinking and Innovation

51) **Blue Ocean Strategy:** Focuses on creating new, uncontested market spaces ("blue oceans") instead of competing in saturated industries ("red oceans"). It emphasizes differentiation and value innovation, making competition irrelevant. Applied in strategic planning and market innovation.

52) **Design Sprint Methodology:** A five-day process for rapidly solving big challenges and testing ideas through prototyping and user feedback. Ideal for product development, user-centered design, and startup projects.
53) **Theory U:** A framework for systems innovation and leadership, emphasizing co-creation and transformation by addressing the root causes of systemic challenges. Widely used in organizational change and societal innovation.
54) **Lean Startup Methodology:** A process for developing businesses and products with minimal waste, emphasizing rapid experimentation, validated learning, and iterative improvement. Applied in entrepreneurship and corporate innovation.
55) **Objectives and Key Results (OKRs):** A goal-setting framework that aligns teams by defining ambitious objectives and measurable key results. Commonly used in organizational performance management.
56) **Global Transition Dynamics (GTD) addresses global challenges, such as climate change, energy resilience, and infrastructure, by leveraging data-driven, science** – and engineering-based approaches to create sustainable solutions. GTD integrates advanced numerical methods, AI/ML, and real-time data to develop strategies at global and local scales.

Earlier Table of Contents of Methods and Tools

Strategic Thinking and Planning

1) **Golden Circle (Simon Sinek):** A framework for identifying purpose, focusing on "Why" (mission), "How" (process), and "What" (outcomes). It is used to create alignment and motivation across organizations. Applicable to leadership, marketing, and strategic planning.
2) **The Quad Chart is a** concise, four-quadrant framework for summarizing complex ideas into Problems, Solutions, Benefits, and Risks. It is commonly used in high-stakes decision-making across industries like defense and R&D.
3) **Lifecycle analysis is a method for evaluating the environmental and sustainability impacts of a product, process, or system from inception to disposal. It is the key to sustainability planning in the** energy and industrial sectors.
4) **Technology Roadmapping:** A strategic planning tool for aligning organizational goals with future technological developments and innovation pathways. Useful in R&D, corporate strategy, and policy planning.
5) **Theory of Change:** A methodology for mapping pathways from inputs to long-term outcomes, focusing on measurable impacts. It is commonly applied in project management, social programs, and policy evaluation.
6) **Technology Foresight and Forecasting:** Systematically identifying future trends, technologies, and their implications through qualitative and quantitative methods. Used in long-term planning, risk assessment, and innovation.
7) **Strategic Technology Assessment (STA):** Evaluates new technologies' potential and risks in achieving specific goals, focusing on alignment with strategic priorities. This approach is applied in corporate strategy and national innovation planning.
8) **Scenario planning explores multiple plausible futures to guide decision-making under uncertainty. It is widely** used in energy, climate policy, and strategic planning.
9) **Cost–Benefit Analysis: This quantitative process compares the costs and benefits of a decision to determine its feasibility and efficiency. It is essential** for policy evaluation, infrastructure investments, and financial planning.

10) **Resilience Metrics and Analysis:** Measures the ability of systems to anticipate, absorb, and recover from disruptions. Applied in infrastructure design, disaster recovery, and risk management.
11) **SWOT Analysis:** Identifies internal Strengths and Weaknesses and external Opportunities and Threats to guide strategic decisions. Used in business strategy and organizational planning.
12) **PESTLE Analysis:** Analyzes external factors – Political, Economic, Social, Technological, Legal, and Environmental – that impact decisions. Applied in strategic planning and market analysis.
13) **Porter's Five Forces Framework evaluates competitive pressures in an industry: supplier power, buyer power, competitive rivalry, threat of substitution, and threat of new entrants. It is commonly** used in business strategy.
14) **Balanced Scorecard:** A tool to align organizational activities with vision and strategy by tracking performance across financial, customer, internal, and learning dimensions. Widely used in organizational performance management.
15) **Value Chain Analysis:** Dissects activities in a business to identify value-creating processes and optimize them for competitive advantage. Applied in operations and supply chain strategy.
16) **VRIO Framework:** Evaluate resources and capabilities based on Value, Rarity, Imitability, and Organization (VRIO) to assess competitive advantage. Useful in strategic resource planning.
17) **Strategic Alignment Models for Policy and Technology:** This model ensures that technology initiatives align with broader organizational or policy goals. It is essential in technology policy and corporate strategy.

Systems and Engineering Approaches

18) **Systems Engineering:** A structured approach to designing and managing complex systems, focusing on functionality, performance, and lifecycle integration. Applied in large-scale engineering and infrastructure projects.
19) **Engineering Systems Framework:** This framework examines systems' technical, social, and economic dimensions, emphasizing interactions and dependencies. It is used in multidisciplinary projects and large-scale planning.
20) **Causal Loop Diagramming for Complex Systems:** Visualizes feedback loops in systems to identify reinforcing or balancing mechanisms. It is useful in systems thinking and problem-solving.
21) **Dynamic Systems Modeling and Analysis:** This technique simulates interactions in complex systems over time to predict outcomes and test interventions. It is applied in energy, transportation, and policy planning.
22) **Integrated Resource Planning (IRP):** Balances supply- and demand-side resources to efficiently meet future energy needs. Used in energy sector planning and utility management.
23) **Total Cost of Ownership (TCO) Analysis:** This analysis assesses all initial and ongoing costs associated with a product or system to determine its overall value. It is applied in procurement and infrastructure investments.

Decision-Making Tools

24) **Decision Trees and Multi-Criteria Decision Analysis (MCDA):** Decision trees break down complex decisions into branches of options and outcomes, while MCDA evaluates multiple criteria for informed choices. Used in policy and project decisions.

25) **Strategic Decision-Making Under Uncertainty:** It provides frameworks for making decisions in unpredictable environments, often combining probabilistic models and risk assessments. It is used in the energy and finance sectors.
26) **Monte Carlo Simulations for Risk and Sensitivity Analysis:** Uses random sampling to model system uncertainty and variability. Applied in project management, finance, and energy modeling.
27) **Game Theory in Strategic Energy Decision-Making:** Analyzes stakeholder interactions to predict behaviors and outcomes in competitive or cooperative scenarios. It is applied in energy markets and negotiations.
28) **Resilience metrics and analysis focus on quantifiable system robustness and recovery indicators. They are applied** in critical infrastructure and disaster preparedness.

Innovation and Change Management

29) **S-curves for technology and market growth track technological or market adoption phases, highlighting points of stagnation or growth. They are used** in R&D and market strategy.
30) **Innovation Diffusion Models:** Examines how innovations spread within populations, identifying barriers and accelerators. Applied in marketing and technology adoption strategies.
31) **Organizational Change Management Frameworks:** These frameworks guide organizations in navigating transitions effectively, focusing on communication, stakeholder alignment, and adaptation. They are applied in restructuring and policy shifts.
32) **Lessons Learned Analysis Frameworks:** These frameworks capture insights from past projects to inform future decision-making and best practices. They are used in project management and institutional learning.
33) **Adaptive Leadership and Crisis Management Techniques:** This field focuses on resilience and adaptability in leadership during crises or change. It is applied in disaster response and organizational management.

Stakeholder and Collaboration Tools

34) **Stakeholder Mapping and Analysis:** Identifies and prioritizes stakeholders based on their influence, interest, and impact on a project. Applied in policy planning and community engagement.
35) **Participatory Design and Engagement Tools:** These tools engage stakeholders in design processes to ensure inclusivity and ownership. They are used in urban planning, energy projects, and community programs.
36) **Cross-Sector Collaboration and Stakeholder Integration Tools:** Facilitates partnerships between government, private, and nonprofit sectors for scalable solutions. It is applied in disaster recovery and infrastructure projects.

Sustainability and Resilience

37) **Sustainability Assessment Metrics:** Measures environmental, social, and economic impacts to ensure balanced decision-making. Used in project evaluation and urban planning.
38) **Energy Justice Frameworks:** These frameworks focus on equitable access to energy resources and benefits for marginalized communities. They are applied in energy transitions and social equity initiatives.

39) **Disaster Risk Reduction and Recovery Models:** These provide structured frameworks for reducing vulnerabilities and planning post-disaster recovery. They are applied in climate adaptation and disaster management.
40) **Equity-Centered Design Approaches:** These approaches ensure that solutions address the needs of underserved populations equitably. They are applied in healthcare, urban planning, and energy systems.
41) **Lifecycle Sustainability Assessment (LCSA):** Integrates environmental, economic, and social aspects across a system's lifecycle. Used in energy projects and sustainability initiatives.
42) **Social Cost of Carbon Calculations:** Quantifies economic damages associated with carbon emissions to guide policy decisions and is applied in climate policy and carbon pricing.

Visualization and Communication Tools

43) **Data Visualization and Decision-Support Tools: These tools translate complex data into visual insights to inform decisions. They are used** in public policy, energy, and finance.
44) **Business Model Canvas:** It provides a visual framework for developing and analyzing business models. It is applied in startups, innovation projects, and corporate strategy.
45) **Design Thinking for Problem-Solving:** A human-centered, iterative approach to solving problems creatively. Used in product development and service design.
46) **Benchmarking and Comparative Metrics:** This technique compares performance metrics across organizations or sectors to identify best practices. It is applied in strategy and operations.
47) **Network analysis for collaborative ecosystems maps relationships and interactions among stakeholders to identify strengths and gaps. It is used** in collaborative projects and ecosystem planning.
48) **Gantt Charts, Kanban Boards, and Geo-spatial Mapping:** Tools for project management and resource allocation, including spatial analysis. Applied in infrastructure planning and disaster recovery.

Summary Tables

Strategic Tools for Planning and Decision-Making

Tools Covered:

- **Golden Circle:** A strategic framework by Simon Sinek that focuses on identifying the "Why," "How," and "What" of an organization's mission to ensure alignment and purpose.
- **The Quad Chart:** A concise visualization tool used to summarize complex projects for decision-making, dividing content into four quadrants: Problem, Solution, Benefits, and Risks.
- **Scenario Planning:** A method to explore and prepare for multiple plausible future outcomes by constructing detailed narratives based on varying assumptions.
- **SWOT Analysis:** A foundational strategic tool for evaluating Strengths, Weaknesses, Opportunities, and Threats.

Tool	Usage	Domains	Efficacy	Limitations	Data Requirements	Cost–Benefit
Golden Circle	Aligns mission and purpose, ensuring coherent strategy.	Corporate strategy, leadership	Enhances alignment and communication.	May oversimplify complex organizational issues.	A clear understanding of organizational mission.	Cost-effective; improves mission clarity and buy-in.
The Quad Chart	Summarizes projects for rapid decision-making.	Defense, R&D, corporate strategy	Simplifies complex initiatives effectively.	Limited depth for in-depth technical analysis.	High-level project data.	Cost-efficient; promotes quick, high-level insights.
Scenario Planning	Prepares for diverse future outcomes by modeling possibilities.	Energy, defense, policy, business	Builds resilience against uncertainties.	Requires significant expertise and resources.	Assumptions, historical trends.	Valuable for resilience but resource-intensive.
SWOT Analysis	Evaluates internal and external strategic factors.	Business, project management	Straightforward and widely applicable.	Lacks dynamic modeling capabilities.	Qualitative and quantitative data.	Simple and inexpensive; supports strategic planning.

Citations

1 S. Sinek, *Start with Why: How Great Leaders Inspire Everyone to Take Action*. Penguin, 2009.

2 J. Murphy, "The pentagon papers: communicating strategy," *Defense Insights*, vol. 15, no. 3, pp. 45–57, 2020.

3 P. J. H. Schoemaker, *Scenario Planning for Complex Systems: Building Resilience in Energy Crises*. Harvard Business Press, 2020.

4 A. Hill and T. Westbrook, "SWOT analysis simplified," *Manag. Strategies*, vol. 12, no. 4, pp. 89–97, 2021.

Analytical Frameworks for Strategic and Operational Assessment

Tools Covered:

- **Lifecycle Analysis:** Evaluates environmental impacts throughout a product or project lifecycle.
- **Technology Roadmapping:** Aligns technological developments with business goals through phased plans.
- **Resilience Metrics and Analysis:** Quantifies resilience in systems to improve preparedness and recovery.

Tool	Usage	Domains	Efficacy	Limitations	Data Requirements	Cost–Benefit
Lifecycle Analysis	Assesses sustainability across a product's lifecycle.	Manufacturing, energy, environment	Comprehensive environmental assessment.	It can be time-intensive.	Detailed material and process data.	Identifies sustainability improvements and supports regulatory compliance.

Tool	Usage	Domains	Efficacy	Limitations	Data Requirements	Cost–Benefit
Technology Roadmapping	Aligns technology with strategic business goals.	R&D, technology management	Enables long-term planning and innovation.	Requires stakeholder alignment.	Strategic goals and technology trends.	Long-term value but significant upfront planning.
Resilience Metrics	Quantifies system resilience for informed decision-making.	Infrastructure, disaster recovery	Improves preparedness and recovery.	Requires accurate data collection.	Performance data, historical failures.	Critical for risk reduction but data-dependent.

Citations

1 ISO, *Environmental Management – Life Cycle Assessment Principles and Framework*, vol. 14040. ISO, 2006.

2 R. Phaal, "Technology roadmapping for strategic innovation," *R&D Manag.*, vol. 44, no. 3, pp. 125–136, 2020.

3 A. Rose, "Measuring system resilience in critical infrastructure," *Risk Anal.*, vol. 37, no. 2, pp. 262–276, 2021.

Assessment and Performance Evaluation Tools

Tools Covered:

- **Cost–Benefit Analysis (CBA):** Compares costs and benefits of projects to guide investment decisions.
- **Balanced Scorecard:** Tracks performance across financial, customer, process, and learning perspectives.
- **Value Chain Analysis:** Identifies value-added activities in processes for optimization.

Tool	Usage	Domains	Efficacy	Limitations	Data Requirements	Cost–Benefit
Cost–Benefit Analysis	Evaluates financial feasibility and socio-economic impact.	Public policy, infrastructure	Facilitates investment prioritization.	May oversimplify nonmonetary benefits.	Financial, operational, and impact data.	High ROI when used effectively for project evaluation.
Balanced Scorecard	Tracks key performance metrics aligned with strategy.	Corporate strategy, operations	Provides a holistic view of performance.	It can become overly complex with too many KPIs.	Strategic goals, performance metrics.	Cost-efficient tool for continuous improvement.
Value Chain Analysis	Identifies and optimizes value-added activities.	Supply chain, operations	Enhances operational efficiency.	It may not capture external market dynamics.	Process flow and cost data.	Highly effective for process and cost optimization.

Citations

1 P. A. Samuelson and W. D. Nordhaus, *Economics*, 20th ed. McGraw-Hill, 2018.

2 R. Kaplan and D. Norton, "The Balanced Scorecard: Measures that Drive Performance," *Harvard Business Review*, vol. 70, no. 1, pp. 71–79, 1992.

3 M. Porter, *Competitive Advantage: Creating and Sustaining Superior Performance*. Free Press, 1985.

Risk Management and Resilience Tools

Tools Covered:

- **Risk Assessment and Mitigation Tools:** Evaluate potential risks and formulate strategies to minimize or eliminate them.
- **Disaster Risk Reduction and Recovery Models:** Frameworks for reducing vulnerabilities and enhancing resilience before and after disasters.
- **Scenario-Based Resilience Simulations:** Test the robustness of systems under various hypothetical scenarios.

Tool	Usage	Domains	Efficacy	Limitations	Data Requirements	Cost–Benefit
Risk Assessment and Mitigation Tools	Identifies and prioritizes risks to develop contingency plans.	Energy, cybersecurity, disaster management	Reduces disruptions and financial losses.	May overlook interdependencies in complex systems.	Risk data and system vulnerabilities.	Cost-effective; prevents significant financial and operational losses.
Disaster Risk Reduction Models	Reduces vulnerabilities and enhances adaptive capacities.	Urban planning, energy, and policy	Builds resilience against recurring disasters.	Requires significant initial investment.	Historical disaster and recovery data.	High value in long-term resilience and reduced disaster response costs.
Scenario-Based Resilience Simulations	Models the impacts of various disaster scenarios on critical systems.	Infrastructure, climate, and urban planning	Improves preparation for worst-case scenarios.	Results depend on the quality of the scenarios created.	Infrastructure and operational data.	High ROI: improves readiness and minimizes damage during actual disasters.

Citations

1 G. Aven and O. Renn, *Risk Management and Governance: Concepts, Guidelines, and Applications*. Springer, 2010.

2 M. Kates *et al.*, "Reducing the risks of disaster in urban areas," *Nature*, vol. 437, no. 7058, pp. 917–921, 2005.

3 P. Schoemaker, "Scenario planning in organizations," *J. Forecast.*, vol. 18, no. 3, pp. 231–251, 1995.

Disaster Recovery and Resilience Methods

Tools Covered:

- **Integrated Resource Planning (IRP):** A systematic approach to balancing energy supply and demand while considering sustainability and resilience.
- **Dynamic Systems Modeling and Analysis:** Simulates interconnected systems to understand their behavior over time.
- **Lessons Learned Analysis Frameworks:** Captures insights from past events to inform future planning.

Tool	Usage	Domains	Efficacy	Limitations	Data Requirements	Cost–Benefit
Integrated Resource Planning (IRP)	Balances supply and demand while incorporating resilience.	Energy, resource management	Ensures long-term sustainability.	Requires extensive data and modeling expertise.	Energy demand and supply projections.	High ROI; aligns resources with resilience goals.
Dynamic Systems Modeling	Simulates and analyzes the behavior of complex systems.	Engineering, policy, and climate	Identifies cascading failures in interconnected systems.	It is time-consuming and computationally expensive.	System interaction and historical data.	Prevents high-cost failures by understanding system dependencies.
Lessons Learned Analysis Frameworks	Provides actionable insights from past disaster responses.	Emergency response, governance	Improves decision-making and planning.	Subjective and reliant on the quality of documentation.	Incident reports, stakeholder feedback.	Low-cost; offers valuable insights to avoid repeating mistakes.

Citations

1 L. Schwartz *et al.*, "Integrated resource planning: frameworks for resilience," *Energy J.*, vol. 32, no. 3, pp. 231–248, 2021.

2 J. Sterman, *Business Dynamics: Systems Thinking and Modeling for a Complex World.* Irwin/McGraw-Hill, 2000.

3 S. Birkland, *Lessons of Disaster: Policy Change After Catastrophic Events.* Georgetown University Press, 2006.

Visualization and Decision Support Tools

Tools Covered:

- **Business Model Canvas:** Framework for outlining and iterating a business's value proposition and operational model.
- **S-Curves for Innovation Analysis:** Tracks the life cycle of technologies to assess growth, maturity, and decline.
- **Causal Loop Diagramming for Complex Systems:** Maps relationships between variables to reveal feedback loops and system dynamics.

Tool	Usage	Domains	Efficacy	Limitations	Data Requirements	Cost–Benefit
Business Model Canvas	Visualizes business operations, customers, and value delivery.	Entrepreneurship, corporate strategy	Facilitates rapid iteration of business models.	Oversimplifies complex operations.	Customer insights, operational data.	Low-cost; accelerates business model refinement.
S-Curves for Innovation Analysis	Evaluates the performance of technologies over time.	R&D, innovation, and technology	Identifies growth opportunities and potential declines.	Limited for short-term predictions.	Performance and adoption metrics.	High value in planning technology investments and R&D strategies.
Causal Loop Diagramming	Visualizes feedback loops to understand system dynamics.	Systems thinking, climate, and economics	Highlights key leverage points for interventions.	Difficult to quantify relationships.	System interactions and feedback data.	High ROI; reduces systemic risks through better understanding of interdependencies.

Citations

1 A. Osterwalder *et al.*, *Business Model Generation: A Handbook for Visionaries, Game Changers, and Challengers*. Wiley, 2010.

2 A. McDowall, "Using S-curves to drive innovation strategies," *Harv. Bus. Rev.*, vol. 85, no. 1, pp. 45–52, 2007.

3 D. Meadows, *Thinking in Systems: A Primer*. Chelsea Green Publishing, 2008.

Decision Analysis Frameworks

Tools Covered:

- **Multi-Criteria Decision Analysis (MCDA):** A structured approach for evaluating multiple conflicting criteria in decision-making.
- **Decision Trees:** Visual tools for mapping out decisions and their potential consequences.
- **Game Theory in Strategic Energy Decision-Making:** Analyzes interactions and outcomes where multiple stakeholders with competing interests are involved.

Tool	Usage	Domains	Efficacy	Limitations	Data Requirements	Cost–Benefit
Multi-Criteria Decision Analysis (MCDA)	Evaluates and prioritizes options with competing criteria.	Urban planning, energy, and disaster response	Balances quantitative and qualitative factors.	Subjectivity in assigning weights to criteria.	Preference data, impact metrics.	High ROI; improves balanced and defensible decision-making.

Tool	Usage	Domains	Efficacy	Limitations	Data Requirements	Cost–Benefit
Decision Trees	Model decision paths and outcomes visually.	Risk analysis, business strategy	Simplifies complex decision processes.	Limited in handling dynamic or uncertain conditions.	Probabilities, outcomes, and costs.	Low-cost; clarifies the cost–benefit of various options.
Game Theory	Examines stakeholder interactions in competitive environments.	Negotiation, energy, and policy	Identifies optimal strategies in competitive settings.	Requires robust data on stakeholder behavior.	Strategic objectives, competitor data.	Medium cost; valuable for negotiation and collaborative planning.

Citations

1 J. Belton and T. Stewart, "Multiple criteria decision analysis: An integrated approach," *J. Oper. Res. Soc.*, vol. 45, no. 2, pp. 231–243, 1995.

2 T. L. Saaty, "Decision making with the analytic hierarchy process," *Int. J. Serv. Sci.*, vol. 1, no. 1, pp. 83–98, 2008.

3 R. Myerson, *Game Theory: Analysis of Conflict*. Harvard University Press, 1991.

Emerging Trends in Predictive Analytics

Tools Covered:

- **Predictive Analytics for Resource Allocation:** Uses machine learning to anticipate future needs and optimize resource deployment.
- **Big Data Analytics:** Extracts actionable insights from large datasets for informed decision-making.
- **Artificial Intelligence (AI)-Driven Risk Models:** Assesses and mitigates risks by identifying patterns in complex datasets.

Tool	Usage	Domains	Efficacy	Limitations	Data Requirements	Cost–Benefit
Predictive Analytics	Guides resource allocation by predicting future needs.	Healthcare, logistics, and energy	Enhances resource efficiency and reduces waste.	Requires high-quality, real-time data streams.	Historical and real-time operational data.	High ROI; optimizes resource utilization and disaster response.
Big Data Analytics	Identifies trends and actionable insights from large datasets.	Finance, energy, and climate	Provides a comprehensive understanding of trends.	High computational cost and privacy concerns.	Structured and unstructured datasets.	High ROI; enables data-driven decision-making across sectors.

Tool	Usage	Domains	Efficacy	Limitations	Data Requirements	Cost–Benefit
AI-Driven Risk Models	Automates risk detection and scenario planning.	Insurance, energy, supply chain	Reduces human error and improves risk assessment.	Requires specialized AI training and models.	Historical risk and operational data.	Medium cost; significant long-term savings by avoiding critical failures.

Citations

1 E. Siegel, *Predictive Analytics: The Power to Predict Who Will Click, Buy, Lie, or Die*. Wiley, 2016.
2 C. B. Ferrell and T. Ralph, "Big data analytics in critical infrastructure systems," *IEEE Trans. Big Data*, vol. 5, no. 1, pp. 3–15, 2019.
3 K. Chatterjee, "AI-driven risk models: Transforming decision-making in high-stakes environments," *J. Risk Anal.*, vol. 12, no. 3, pp. 65–78, 2021.

Visualization and Communication Tools

Tools Covered:
- **Data Dashboards:** Provides visual summaries of key metrics for real-time decision-making.
- **Infographics:** Simplifies complex information into digestible visuals for diverse audiences.
- **Systems Dynamics Modeling for Visualization:** Simulates and visualizes the behavior of complex systems over time.

Tool	Usage	Domains	Efficacy	Limitations	Data Requirements	Cost–Benefit
Data Dashboards	Real-time visualization of metrics to aid decision-making.	Energy, logistics, and public health	Enhances situational awareness and monitoring.	Requires real-time, accurate data sources.	Real-time operational and monitoring data.	Medium cost; improves operational transparency and responsiveness.
Infographics	Presents complex ideas visually for easier communication.	Marketing, policy, and education	Improves stakeholder engagement and understanding.	Limited in conveying deep technical details.	Statistical and conceptual data.	Low-cost; effective in increasing public awareness and buy-in.
Systems Dynamics Modeling	Simulates system behaviors to visualize long-term impacts.	Energy, climate, and economics	Identifies leverage points for effective intervention.	High computational and training costs.	Interaction and historical system data.	High ROI; reveals critical insights for sustainable planning and decision-making.

Citations

1 S. Few, *Information Dashboard Design: Displaying Data for At-a-Glance Monitoring*. Analytics Press, 2013.

2 R. McCandless, *The Visual Display of Quantitative Information*. Graphics Press, 2009.
3 J. Sterman, *Business Dynamics: Systems Thinking and Modeling for a Complex World*. McGraw-Hill, 2000.

Integrated Design Frameworks

Tools Covered:

- **Human-Centered Design (HCD):** Focuses on designing systems and solutions that prioritize user needs and experiences.
- **Equity-Centered Design:** Ensures inclusion of diverse perspectives and equitable access in solution development.
- **Participatory Design Models:** Engage stakeholders in the design process to align solutions with community needs.

Tool	Usage	Domains	Efficacy	Limitations	Data Requirements	Cost–Benefit
Human-Centered Design (HCD)	Prioritizes user needs in designing systems or solutions.	Product design, public services	Improves usability and user satisfaction.	Time-intensive and requires iterative prototyping.	User needs and behavioral data.	Medium cost; enhances the adoption and effectiveness of solutions.
Equity-Centered Design	Ensures equitable access and representation in design.	Social programs, urban planning	Promotes inclusivity and addresses systemic inequities.	May require challenging systemic barriers.	Demographic and access data.	High ROI; mitigates inequities and ensures long-term impact.
Participatory Design	Involves stakeholders in cocreating solutions.	Community development, energy	Aligns solutions with real community needs.	Time-intensive and may lead to competing priorities.	Stakeholder feedback and priorities.	Medium cost; builds trust and improves implementation success.

Citations

1 D. Norman, *The Design of Everyday Things*. Basic Books, 2013.
2 C. Costanza-Chock, *Design Justice: Community-Led Practices to Build the Worlds We Need*. MIT Press, 2020.
3 T. Sanders and P. Stappers, "Co-creation and the new landscapes of design," *CoDesign*, vol. 4, no. 1, pp. 5–18, 2008.

Decision Support Systems

Tools Covered:

- **Dynamic Decision Support Systems (DDSS):** Integrates data and models to support decision-making in real-time scenarios.

- **Geographic Information Systems (GIS):** Provides spatial data visualization to inform decisions.
- **Dashboard and Analytics Platforms:** Offers centralized visualization for tracking and optimizing decisions.

Tool	Usage	Domains	Efficacy	Limitations	Data Requirements	Cost–Benefit
Dynamic Decision Support Systems	Supports real-time decision-making using data and models.	Disaster response, logistics	Enhances responsiveness and situational awareness.	Requires robust data infrastructure.	Real-time operational data.	High ROI; reduces delays and improves outcomes in time-sensitive scenarios.
Geographic Information Systems (GIS)	Provides spatial analysis and visualization of data.	Urban planning, disaster response	Improves spatial decision-making capabilities.	Data collection and processing can be costly.	Geo-spatial and environ-mental data.	High ROI; ensures targeted and efficient resource deployment.
Dashboard and Analytics Platforms	Offers centralized visualization of key metrics.	Business strategy, energy	Simplifies complex data for better decision-making.	Dependence on real-time data updates.	Aggregated and operational data.	Medium cost; improves transparency and decision alignment.

Citations

1 D. Power, *Decision Support Systems: Concepts and Resources for Managers*. Greenwood Publishing, 2002.

2 M. Batty, *The New Science of Cities*. MIT Press, 2013.

3 T. Bernhardt *et al.*, "Dynamic decision support systems for emergency management," *J. Syst. Softw.*, vol. 80, no. 1, pp. 23–33, 2007.

Cross-Cultural Engagement Methods

Tools Covered:

- **Cultural Intelligence Framework (CQ):** Enhances collaboration in cross-cultural environments.
- **Stakeholder Mapping with Cross-Cultural Sensitivity:** Identifies and addresses diverse stakeholder needs.
- **Collaborative Adaptive Management (CAM):** Promotes iterative decision-making with multi-stakeholder input.

Tool	Usage	Domains	Efficacy	Limitations	Data Requirements	Cost–Benefit
Cultural Intelligence Framework (CQ)	Enhances collaboration by under-standing cultural differences.	International business, policy	Fosters collaboration and reduces conflicts.	Requires time to develop cultural understanding.	Stakeholder cultural profiles.	Medium cost; improves team dynamics and project success in diverse settings.

Tool	Usage	Domains	Efficacy	Limitations	Data Requirements	Cost–Benefit
Stakeholder Mapping with Sensitivity	Identifies and prioritizes stakeholder concerns in diverse contexts.	Urban development, policymaking	Improves inclusivity and strategic alignment.	May oversimplify complex stakeholder relationships.	Stakeholder demographics and priorities.	Medium cost ensures comprehensive representation and alignment.
Collaborative Adaptive Management	Iterative decision-making process with multi-stakeholder input.	Environmental policy, resource management	Adapts to changing conditions and stakeholder feedback.	It is time-intensive and requires regular collaboration.	Feedback loops and system outcomes.	High ROI ensures sustainability and long-term stakeholder engagement.

Citations

1 D. Livermore, *The Cultural Intelligence Difference: Master the One Skill You Can't Do Without in Today's Global Economy*. AMACOM, 2011.

2 P. C. Stern and H. V. Fineberg, *Understanding Risk: Informing Decisions in a Democratic Society*. National Academy Press, 1996.

3 C. Allen and G. Stankey, *Collaborative Adaptive Management: An Emerging Practice for Addressing Complex Issues*. Island Press, 2009.

Systems Integration Frameworks

Tools Covered:

- **Systems Dynamics Modeling (SDM):** Simulates complex systems to predict long-term behaviors and outcomes.
- **Integrated Resource Planning (IRP)** ensures that energy, infrastructure, and resource allocation align with strategic goals.
- **Causal Loop Diagramming (CLD):** Visualizes relationships and feedback within systems to understand dynamics and identify leverage points.

Tool	Usage	Domains	Efficacy	Limitations	Data Requirements	Cost–Benefit
Systems Dynamics Modeling (SDM)	Simulates dynamic behavior in systems over time.	Policy analysis, energy, and healthcare	Enhances understanding of system interactions.	Requires expertise in model design and validation.	Time-series and system interaction data.	High ROI; supports long-term planning and risk mitigation.
Integrated Resource Planning (IRP)	Aligns resources with long-term strategic goals.	Energy, urban planning	Ensures resource efficiency and sustainability.	May oversimplify long-term uncertainties.	Resource use and demand data.	Medium cost ensures optimal resource allocation and long-term impact.

Tool	Usage	Domains	Efficacy	Limitations	Data Requirements	Cost–Benefit
Causal Loop Diagramming (CLD)	Identifies feedback loops and system interdependencies.	Systems engineering, education	Clarifies cause-effect relationships for better decision-making.	May oversimplify complex feedback loops.	Relationship and impact data.	Low cost; enhances clarity and understanding of systemic issues.

Citations

1 J. Sterman, *Business Dynamics: Systems Thinking and Modeling for a Complex World*. Irwin McGraw-Hill, 2000.

2 M. Howells *et al.*, "Integrated resource planning: Case study in energy," *Energy Policy*, vol. 38, no. 8, pp. 4526–4535, 2010.

3 P. M. Senge, *The Fifth Discipline: The Art & Practice of The Learning Organization*. Doubleday, 1990.

Advanced Decision Science Tools

Tools Covered:

- **Multi-Criteria Decision Analysis (MCDA):** Balances trade-offs among multiple objectives to identify optimal decisions.
- **Game Theory Models:** Analyze strategic interactions between decision-makers to predict outcomes.
- **Dynamic Systems Analysis (DSA):** Combines simulation and optimization for complex decision scenarios.

Tool	Usage	Domains	Efficacy	Limitations	Data Requirements	Cost–Benefit
Multi-Criteria Decision Analysis	Evaluates decisions with competing criteria and trade-offs.	Policymaking, infrastructure	Ensures balanced, well-informed decisions.	It can be subjective depending on the criteria weighting.	Criteria importance and outcome data.	Medium cost; supports transparent and comprehensive decision-making.
Game Theory Models	Analyzes strategic interactions to predict decision outcomes.	Economics, defense, and energy	Predicts optimal outcomes in competitive scenarios.	Requires clear assumptions and accurate input data.	Stakeholder preferences and payoffs.	High ROI; improves negotiation and strategy outcomes in competitive settings.
Dynamic Systems Analysis (DSA)	Combines modeling and optimization to evaluate scenarios.	Resource allocation, logistics	Enhances decision quality in complex scenarios.	Requires significant computational resources.	System dynamics and scenario inputs.	High ROI ensures optimal resource allocation and risk mitigation.

Citations

1 R. Keeney and H. Raiffa, *Decisions with Multiple Objectives: Preferences and Value Tradeoffs*. Cambridge University Press, 1993.
2 J. von Neumann and O. Morgenstern, *Theory of Games and Economic Behavior*. Princeton University Press, 1944.
3 J. Banks *et al.*, *Discrete-Event System Simulation*. Pearson, 2014.

Visual and Analytical Tools

Tools Covered:
- **Business Model Canvas (BMC):** Visualizes and aligns business strategies across nine key components.
- **S-Curve Analysis:** Tracks technological and market adoption stages to inform innovation strategy.
- **Heatmaps and Dashboards:** Provides centralized data visualization to guide decisions.

Tool	Usage	Domains	Efficacy	Limitations	Data Requirements	Cost–Benefit
Business Model Canvas (BMC)	Aligns business strategy using a structured framework.	Business strategy, startups	Simplifies strategic planning and alignment.	May oversimplify complex strategies.	Business operation and revenue data.	It has a low cost and supports a focused strategy and resource allocation.
S-Curve Analysis	Tracks technological or market adoption stages.	R&D, innovation management	Identifies inflection points for growth or saturation.	Requires long-term data for accuracy.	Market performance and trend data.	Medium cost; informs R&D and product lifecycle decisions.
Heatmaps and Dashboards	Visualizes key metrics for real-time decision-making.	Business analytics, energy	Simplifies complex data for intuitive decisions.	Dependence on accurate data integration.	Operational and performance data.	Medium cost ensures transparency and timely decision-making.

Citations

1 A. Osterwalder and Y. Pigneur, *Business Model Generation: A Handbook for Visionaries, Game Changers, and Challengers*. Wiley, 2010.
2 F. F. Suarez and J. M. Utterback, "S-Curve adoption patterns: Insights for innovation management," *Res. Policy*, vol. 24, no. 6, pp. 1071–1085, 1995.
3 A. Few, *Information Dashboard Design: The Effective Visual Communication of Data*. O'Reilly Media, 2006.

Cross-Sector Adaptation Methods

Tools Covered:
1) **Collaborative Frameworks for Public-Private Partnerships (PPPs):** These frameworks facilitate cooperation between governments, businesses, and NGOs to achieve shared goals.

2) **Adaptive Management Techniques:** Incorporate feedback loops to modify strategies in response to new data or environmental changes.
3) **Intersectoral Policy Integration:** Align policies across energy, environment, and public health sectors to achieve comprehensive solutions.

Tool	Usage	Domains	Efficacy	Limitations	Data Requirements	Cost–Benefit
Collaborative Frameworks for PPPs	Aligns resources and expertise to address complex challenges.	Infrastructure, healthcare, energy	Accelerates implementation and resource sharing.	Risk of misaligned priorities among stakeholders.	Partnership agreements and project data.	High ROI; optimizes resources and enables large-scale impact.
Adaptive Management Techniques	Adjusts strategies based on ongoing feedback.	Environmental management, policy planning	Improves resilience to uncertainties.	Requires timely and accurate feedback loops.	Environmental or project outcome data.	Medium cost; enhances responsiveness and decision quality.
Intersectoral Policy Integration	Aligns policies across multiple domains.	Urban planning, energy, and public health	Ensures synergies between sectoral policies.	Potential conflicts in priorities across sectors.	Policy and performance impact data.	High ROI; reduces redundancy and promotes holistic outcomes.

Citations

1 M. Warner and A. Sullivan, *Putting Partnerships to Work: Strategic Alliances for Development Between Government, the Private Sector, and Civil Society*. Routledge, 2004.

2 C. Holling, "*Adaptive Environmental Assessment and Management*," in *Wiley IIASA International Series on Applied Systems Analysis*, vol. 3. Chichester, U.K.: Wiley, pp. i–xviii, 1978, pp. 1–377.

3 C. J. Nilsson *et al*., "Intersectoral Policy Integration in Urban Sustainability Strategies," *Journal of Environmental Planning and Management*, vol. 62, no. 7, pp. 1210–1232, 2019.

Sustainability Metrics for Decision Support

Tools Covered:

1) **Life Cycle Assessment (LCA):** Evaluates environmental impacts across a product or process lifecycle.
2) **Triple Bottom Line (TBL) Framework:** Balances social, environmental, and financial considerations in decision-making.
3) **Carbon Footprint Analysis (CFA):** Measures greenhouse gas emissions associated with activities or products.

Tool	Usage	Domains	Efficacy	Limitations	Data Requirements	Cost–Benefit
Life Cycle Assessment (LCA)	Assesses environmental impacts from raw material extraction to disposal.	Manufacturing, energy, and logistics	Identifies hotspots for improvement in sustainability.	Time-intensive and data-dependent process.	Supply chain and material flow data.	High ROI; improves sustainability and compliance outcomes.

Tool	Usage	Domains	Efficacy	Limitations	Data Requirements	Cost–Benefit
Triple Bottom Line (TBL) Framework	Incorporates social, environmental, and economic dimensions into decisions.	Business strategy, urban planning	Encourages balanced, sustainable strategies.	Requires subjective evaluation of social factors.	Social, environmental, and financial data.	Medium cost; improves long-term sustainability strategies.
Carbon Footprint Analysis (CFA)	Quantifies GHG emissions of activities or processes.	Energy, manufacturing, logistics	Provides clear metrics for emission reduction targets.	Limited scope without lifecycle inclusion.	Emission factor and activity data.	High ROI; supports regulatory compliance and market competitiveness.

Citations

1 R. U. Ayres and L. W. Ayres, *A Handbook of Industrial Ecology*. Edward Elgar Publishing, 2002.

2 J. Elkington, *Cannibals with Forks: The Triple Bottom Line of 21st Century Business*. Capstone, 1998.

3 ISO, *ISO 14040: Environmental Management—Life Cycle Assessment—Principles and Framework*, 2nd ed. Geneva, Switzerland: International Organization for Standardization, 2006 20pp.

Visualization and Communication Tools

Tools Covered:

1) **Business Model Canvas (BMC):** A strategic management tool to develop and visualize business models.
2) **S-Curves for Technology Adoption:** Graphical representations of technology adoption and performance lifecycle stages.
3) **Causal Loop Diagramming (CLD):** Visualizes interconnections and feedback loops in complex systems.
4) **Concept and Mind Maps:** Tools to graphically represent ideas, concepts, and their relationships.
5) **Flowcharts and Process Mapping:** Tools to illustrate processes and decision pathways clearly.

Tool	Usage	Domains	Efficacy	Limitations	Data Requirements	Cost–Benefit
Business Model Canvas (BMC)	Provides a structured approach to designing and visualizing business models.	Entrepreneurship, corporate strategy	Improves understanding and alignment of business models.	Limited detail for complex business strategies.	Organizational data, market insights.	High ROI enhances strategic focus and stakeholder communication.
S-Curves	Illustrates stages of innovation, adoption, and saturation of technologies.	R&D, innovation management, market analysis	It helps identify growth opportunities and inflection points.	Assumes linear progression, not always accurate.	Technology performance and market data.	Medium cost aids in timing investments in innovation.

Tool	Usage	Domains	Efficacy	Limitations	Data Requirements	Cost–Benefit
Causal Loop Diagramming (CLD)	Maps feedback loops and interdependencies in complex systems.	Systems engineering, policy design	Identifies systemic risks and leverages feedback for improvements.	It can become unwieldy for highly complex systems.	System variables and interaction data.	High ROI enables better decisions in dynamic systems.
Concept and Mind Mapping	Organizes and visualizes relationships between ideas or systems.	Education, research, innovation strategy	Enhances idea generation and shared understanding.	Limited analytical depth for quantitative problems.	Idea flows and hierarchical relationships.	Low cost; boosts brainstorming and clarity.
Flowcharts and Process Mapping	Illustrates decision pathways and operational workflows.	Operations, project management, and engineering	Improves process transparency and stakeholder alignment.	It can oversimplify processes or exclude nuances.	Workflow and decision-making data.	Medium cost; increases efficiency and reduces errors in processes.

Citations

1 A. Osterwalder and Y. Pigneur, *Business Model Generation: A Handbook for Visionaries, Game Changers, and Challengers*. Wiley, 2010.

2 R. N. Foster, *Innovation: The Attacker's Advantage*. Summit Books, 1986.

3 J. D. Sterman, *Business Dynamics: Systems Thinking and Modeling for a Complex World*. McGraw-Hill, 2000.

4 J. Novak, "The theory underlying concept maps and how to construct them," Technical Report IHMC CmapTools 2006-01 Rev 01-2008, Florida Institute for Human and Machine Cognition, Pensacola, FL, 2008, 36 pp.

5 ISO, "ISO 5807: Documentation symbols and conventions for data, program, and system flowcharts," ISO 5807:1985, Geneva, Switzerland: International Organization for Standardization, Feb. 1985, 25 pp.

Additional Planning and Optimization Tools

Tools Covered:

1) **Design Thinking:** A human-centered methodology for solving complex problems through ideation, prototyping, and testing.
2) **Dynamic Systems Modeling:** A method for simulating complex systems to predict outcomes and optimize performance.
3) **Decision Trees and Multi-Criteria Decision Analysis (MCDA):** Tools to evaluate options and outcomes based on multiple criteria.
4) **Game Theory for Strategic Decisions:** Frameworks for Understanding Competitive Dynamics and Strategic Choices.

Tool	Usage	Domains	Efficacy	Limitations	Data Requirements	Cost–Benefit
Design Thinking	Solves complex challenges by focusing on human-centered solutions.	Product development, UX design, and public policy	Encourages innovative, user-focused outcomes.	It is time-consuming and less quantitative.	User needs and feedback data.	High ROI; aligns solutions with stakeholder needs.
Dynamic Systems Modeling	Simulates complex systems to predict performance and test interventions.	Energy systems, healthcare, and economics	Improves predictive accuracy in decision-making.	Requires detailed and validated models.	System dynamics and interaction data.	High cost; valuable for large-scale optimization.
Decision Trees and MCDA	Evaluates decisions using structured, quantitative frameworks.	Risk management, investment strategy	Clarifies trade-offs and optimal choices.	Limited by the accuracy of input data.	Criteria weights and outcome probabilities.	Medium cost; enhances clarity in multifaceted decisions.
Game Theory	Analyzes competitive interactions and strategic decisions.	Negotiations, business strategy	Predicts competitor behavior and optimal strategies.	Simplifies real-world complexities.	Competitor and market dynamics.	Medium cost informs an effective competitive strategy.

Citations

1 T. Brown, *Change by Design: How Design Thinking Creates New Alternatives for Business and Society*. HarperBusiness, 2009.
2 J. Sterman, "System dynamics modeling: Tools for policy analysis in a complex world," *Cal. Manag. Rev.*, vol. 43, no. 4, pp. 8–25, 2001.
3 R. T. Clemen, *Making Hard Decisions: An Introduction to Decision Analysis*. Brooks/Cole, 1996.
4 J. von Neumann and O. Morgenstern, *Theory of Games and Economic Behavior*. Princeton University Press, 1944.

Appendix E

Leadership Reflections and Quotations

This appendix reflects insights drawn from over 42 years of professional experience spanning defense, critical infrastructure, executive leadership, and the mentorship of more than 8,000 managers and executives, as well as numerous students. These principles, shaped by lessons from my mentors, colleagues, family, and students, are practical, reflective, and aspirational in nature. Each reflection aligns with key themes of leadership explored in this book, offering a personal lens into their application.

Personal Mastery and Inner Leadership

Discipline and Clarity of Mind

- "If you aspire to leadership, park your ego at the door. Cultivate a quiet, smooth, strong mind. Know the road ahead."
- "Clarity of thought leads to clarity of action. A smooth, strong mind is the foundation of effective leadership."
- "Discipline is freedom. It empowers you to prioritize purpose over impulse."

Reflection and Growth

- "Reflection aligns you with your values and purpose. Without it, you drift into unexamined decisions."
- "Nowhere can man find a quieter or more untroubled retreat than in his soul." – Marcus Aurelius, ~180 CE
- "We are what we repeatedly do. Excellence, then, is not an act, but a habit." – Aristotle, ~350 BCE

Courage Under Scrutiny

- "What if your good name comes under question, rightfully or not? How will you respond when scrutinized unfairly? Your reaction defines your strength."

Curiosity and Self-Control

- "Curiosity drives progress. Never settle for easy answers. Keep questioning."
- "True freedom isn't about doing whatever you want – it's about living with purpose, kindness, and integrity."

Ethics and Integrity

Daily Ethics in Leadership

- "Integrity protects you in challenges. Live so that your actions align with your principles."
- "Ethics is knowing the difference between what you have a right to do and what is right to do." – Potter Stewart, 1964
- "When values are clear, decisions are easier." – Roy Disney, 1976

The Weight of Choices

- "Would you be proud if this decision appeared on the front page tomorrow?"
- "Integrity is doing the right thing, even when no one is watching." – C.S. Lewis, 1952
- "What you do has far greater impact than what you say." – Stephen Covey, 1989

Suggested Addition:

- "Leadership anchored in integrity builds trust that withstands time and trials."

Resilience and Growth Through Adversity

Resilience is Strength

- "Resilience is not about bouncing back – it's about bouncing forward with wisdom and strength."
- "Adversity introduces a person to themselves." – Albert Einstein, early 20th century
- "Fall seven times, stand up eight." – Japanese Proverb, ~19th century

Learning from Crisis

- "The ultimate measure of a man is not where he stands in moments of comfort, but where he stands at times of challenge and controversy." – Martin Luther King Jr., 1963
- "Mountaintops inspire leaders, but valleys mature them." – Winston Churchill, 1940
- "A smooth sea never made a skilled sailor." – Franklin D. Roosevelt, 1938

Suggested Addition:

- "Every crisis presents an opportunity to grow as a leader, learn from adversity, and emerge stronger."

Vision, Innovation, and Purpose

Purpose-Driven Leadership

- "The meaning of life is to find your gift. The purpose of life is to give it away." – Pablo Picasso, 1943
- "Efforts and courage are not enough without purpose and direction." – John F. Kennedy, 1960
- "Your work is going to fill a large part of your life, and the only way to be truly satisfied is to do what you believe is great work." – Steve Jobs, 2005

Long-Term Thinking

- "Make decisions your grandchildren will thank you for – even if you don't have children."
- "Vision without execution is hallucination." – Thomas Edison, 1903
- "Society grows great when old men plant trees whose shade they know they shall never sit in." – Greek Proverb, ~4th century BCE

Conflict and Leadership

Conflict as a Catalyst

- "Effective leaders are also highly skilled negotiators. Conflict can catalyze positive change. It is a point of bifurcation with potential options and new opportunities."
- "Leadership is solving problems. The day soldiers stop bringing you their problems is the day you have stopped leading them." – Colin Powell, 20th century
- "Clashing ideas sharpen solutions. Don't avoid disagreements – use them."

Collaboration and Teamwork

Building Trust and Connection

- "Trust begins with listening. Ask your team how they're doing – and truly listen."
- "Alone we can do so little; together we can do so much." – Helen Keller, 1920
- "Diversity drives innovation." – Frans Johansson, *The Medici Effect*, 2004

Diversity in Teams

- "Strength lies in differences, not in similarities." – Stephen Covey, 1989
- "In diversity there is beauty, and there is strength." – Maya Angelou, 20th century
- "Celebrate the work of those who don't seek the spotlight. They're often the backbone of success."

Decision-Making and Adaptability

Decisiveness and Clarity

- "When the plan fails, change the approach. Stopping isn't the answer – adjusting is."
- "In the middle of difficulty lies opportunity." – Albert Einstein, 1940
- "By failing to prepare, you are preparing to fail." – Benjamin Franklin, 1757

Risk and Reward

- "Take calculated risks. That is quite different from being rash." – George S. Patton, 1944
- "Success is not final, failure is not fatal: it is the courage to continue that counts." – Winston Churchill, 1940
- "Do not be afraid to give up the good to go for the great." – John D. Rockefeller, 1904

Societal Impact and Legacy

Leadership Beyond Self

- "Technology's value lies in how it serves humanity. Let innovation uplift, not divide."
- "Global problems require local solutions. Start where you are to create ripples of change."
- "Be the change you wish to see in the world." – Mahatma Gandhi, 1947

Living a Legacy

- "Your legacy isn't what you achieve – it's what you inspire others to achieve."
- "What we do today, right now, will have an accumulated effect on all our tomorrows." – Alexandra Stoddard, 1988
- "Focus on creating something that will outlive you."

Emotional Intelligence

Empathy and Vulnerability

- "Authentic leaders are unafraid to show vulnerability. It deepens connections and fosters trust."
- "Seek first to understand, then to be understood." – Stephen Covey, 1989
- "Leadership is about making others feel valued and seen." – Brené Brown, *Dare to Lead*, 2018

Grace Under Pressure

- "Staying composed in difficult times inspires confidence and calm in others."
- "People will forget what you said, people will forget what you did, but people will never forget how you made them feel." – Maya Angelou, 1982

Continuous Learning and Growth

Curiosity and Growth

- "Curiosity drives progress. Never settle for easy answers."
- "Education is the most powerful weapon which you can use to change the world." – Nelson Mandela, 1990s
- "Every failure carries lessons. Use them as stepping stones, not roadblocks."

Kaizen: Small Steps, Big Changes

- "Continuous improvement builds extraordinary results from small, consistent changes."
- "The best way to predict the future is to create it." – Peter Drucker, 1954

Further Discussion and Analyses on Leadership Styles

Leadership has evolved significantly over the past century, shaped by societal changes, organizational needs, and global challenges. Understanding these leadership styles provides valuable insight into how leaders adapt to complexity, drive change, and inspire action. In this expanded and detailed analysis, we delve into the origins, characteristics, benefits, limitations, and examples of leadership styles that define effective leadership in the modern world.

This chapter fits best after the foundational discussion of leadership in complexity, as it offers a more granular analysis of how different leadership styles function within volatile, uncertain, complex, and ambiguous (VUCA) environments. Following the introduction's broad exploration of leadership principles, this chapter grounds those ideas in specific theories and real-world applications, building a bridge to the practical frameworks and tools discussed later in the book.

The Great Man Theory: Heroes Shaping History

What It Is: The Great Man Theory posits that leaders are born with innate qualities that make them exceptional, such as charisma, intelligence, and dominance. This theory suggests that leadership is predestined rather than developed.

Historical Background: Popular in the early 20th century, this theory aligned with an era that revered monarchs, generals, and political icons. Historian Thomas Carlyle emphasized that "great men" shape history through their inherent qualities and heroic actions.

- **Benefits:** Highlights the critical role of leadership in shaping societies and emphasizes personal traits.
- **Limitations:** Ignores situational factors and collaborative efforts; provides no framework for developing leadership skills.
- **Blind Spots:** Overemphasis on individual leaders often neglects the contributions of teams or systems.

Key Example: Winston Churchill exemplified this theory during World War II. He is often described as a "born leader" whose heroic traits defined Britain's resilience. However, Churchill's leadership success also depended on his collaboration with advisors, allies, and the broader war effort.

Trait Theory: The Power of Personal Qualities

What It Is: Trait Theory builds on the Great Man Theory by identifying characteristics, such as empathy, assertiveness, and self-confidence, that distinguish leaders from non-leaders.

Historical Background: Emerging in the 1930s, this theory shifted focus toward measurable attributes. Researchers like Ralph Stogdill argued that traits alone do not guarantee leadership success but interact with situational factors.

- **Benefits:** Encourages leaders to cultivate personal qualities and serves as a foundation for personality assessments and development programs.
- **Limitations:** Overemphasis on traits risks neglecting behaviors and context.
- **Blind Spots:** Cultural and organizational contexts shape what traits are valued.

Key Example: Mahatma Gandhi exemplifies Trait Theory through his empathy, resilience, and commitment to non-violence, which inspired a nation's fight for independence.

Behavioral Theories: Actions Over Traits

What It Is: Behavioral theories emphasize what leaders do rather than who they are. These theories classify behaviors into task-oriented (focused on goals) and people-oriented (focused on relationships) categories.

Historical Background: Developed in the 1940s and 1950s, behavioral theories emerged from studies at Ohio State University and the University of Michigan. Researchers like Kurt Lewin categorized leadership into authoritarian, democratic, and laissez-faire styles.

- **Benefits:** Provides a framework for teaching and replicating effective leadership behaviors.
- **Limitations:** Assumes universal applicability of behaviors, ignoring contextual nuances.
- **Blind Spots:** Neglects the complexity of individual and organizational dynamics.

Key Example: John F. Kennedy demonstrated a mix of task-oriented and people-oriented behaviors during the Cuban Missile Crisis, balancing decisive action with collaborative diplomacy.

Contingency Theories: Adapting to Context

What It Is: Contingency theories propose that effective leadership depends on situational factors, such as team dynamics, organizational structure, and the nature of the task.

Historical Background: Emerging in the 1960s and 1970s, models like Fiedler's Contingency Model and Hersey-Blanchard's Situational Leadership Theory emphasized the interaction between leaders and situational demands.

- **Benefits:** Recognizes the importance of adaptability in leadership.
- **Limitations:** Complex to implement due to the need for situational analysis.
- **Blind Spots:** May oversimplify dynamic environments.

Key Example: Angela Merkel's pragmatic leadership during the Eurozone crisis reflects contingency theories. Her ability to adapt her approach to shifting political and economic landscapes exemplifies context-driven leadership.

Transactional Leadership: Driving Performance Through Structure

What It Is: Transactional leadership focuses on structured exchanges between leaders and followers, such as rewards for compliance and penalties for deviation.

Historical Background: Rooted in Max Weber's theory of bureaucracy, transactional leadership gained prominence in the mid-20th century as organizations prioritized stability and accountability.

- **Benefits:** Promotes clear roles, expectations, and accountability; effective for achieving short-term goals.
- **Limitations:** Lacks focus on innovation and long-term vision; may demotivate creative thinkers.
- **Blind Spots:** Relies heavily on authority and formal structures.

Key Example: Jeff Bezos' early leadership at Amazon exemplifies transactional leadership, with a strong focus on performance metrics and accountability systems.

Transformational Leadership: Inspiring Extraordinary Change

What It Is: Transformational leadership inspires and motivates followers to achieve extraordinary outcomes by emphasizing vision, change, and personal development.

Historical Background: Popularized in the 1980s by James MacGregor Burns, transformational leadership became a cornerstone for organizations seeking innovation and systemic change.

- **Benefits:** Encourages innovation, creativity, and engagement; builds strong emotional connections between leaders and followers.
- **Limitations:** May neglect operational details and risk burnout in high-performing teams.
- **Blind Spots:** Relies heavily on the leader's charisma and vision.

Key Example: Nelson Mandela's leadership during South Africa's transition to democracy exemplifies transformational leadership, inspiring unity and reconciliation.

Servant Leadership: Prioritizing Team Needs

What It Is: Servant leadership prioritizes team members' needs, emphasizing empathy, listening, and growth.

Historical Background: Robert K. Greenleaf coined this style in the 1970s. It gained traction in organizations that focused on ethical practices and employee well-being.

- **Benefits:** Fosters trust, collaboration, and loyalty; encourages personal and professional growth.
- **Limitations:** May be perceived as weak or indecisive in competitive environments.
- **Blind Spots:** Can lead to blurred boundaries between leadership and management.

Key Example: Howard Schultz, former CEO of Starbucks, focused on employee welfare and community engagement and embodied servant leadership principles.

Authentic Leadership: Leading with Integrity

What It Is: Authentic leadership emphasizes self-awareness, transparency, and staying true to one's values while fostering team trust and collaboration.

Historical Background: Authentic leadership rose in prominence in the early 2000s due to corporate scandals and ethical lapses. Scholars like Bill George emphasized the importance of trust and credibility in leadership.

- **Benefits:** Builds trust, promotes ethical decision-making, and inspires others through genuine behavior.
- **Limitations:** It requires a high level of self-awareness, which may not be equally developed in all leaders. If transparency is not balanced, overexposure or vulnerability risks can emerge.
- **Blind Spots:** May struggle in highly competitive or politically charged environments.

Key Example: Satya Nadella, CEO of Microsoft, exemplifies authentic leadership through his empathetic approach, transparent decision-making, and emphasis on cultural transformation within the company.

Adaptive Leadership: Thriving Amid Change

What It Is: Adaptive leadership focuses on the ability to respond to rapidly changing environments, encouraging leaders and teams to embrace uncertainty, change, and continuous learning.

Historical Background: Introduced by Ronald Heifetz in the 1990s, adaptive leadership emerged to address complex, systemic challenges requiring innovative and collaborative problem-solving.

- **Benefits:** Encourages innovation and resilience by embracing ambiguity and experimenting with solutions.
- **Limitations:** Traditional leadership styles may lack structure and predictability. Resistance to change within teams can also be a barrier.
- **Blind Spots:** Requires tolerance for failure, which may conflict with organizational cultures focused on immediate results.

Key Example: Elon Musk exemplifies adaptive leadership by navigating rapidly evolving industries such as electric vehicles (Tesla) and private space exploration (SpaceX), embracing experimentation and high risks.

Inclusive Leadership: Empowering Through Equity

What It Is: Inclusive leadership creates environments where diverse team members feel respected, valued, and empowered to contribute.

Historical Background: Inclusive leadership gained prominence in the 2010s, as there was growing awareness of the role of diversity, equity, and inclusion (DEI) in driving creativity and innovation.

- **Benefits:** Enhances creativity and team cohesion by integrating diverse perspectives.
- **Limitations:** Can slow decision-making due to the effort to include all voices and viewpoints.
- **Blind Spots:** Risks tokenism if diversity initiatives are not genuinely implemented.

Key Example: Indra Nooyi, former CEO of PepsiCo, is an inclusive leader who prioritized cultural diversity and innovation, fostering collaboration across a global workforce.

Charismatic Leadership: Inspiring Through Vision

What It Is: Charismatic leadership relies on a leader's charm, vision, and ability to inspire emotional connections with followers.

Historical Background: Sociologist Max Weber first explored the concept of charisma in leadership in the early 20th century, defining it as a rare, extraordinary quality that draws people toward leaders.

- **Benefits:** Creates strong emotional bonds and inspires motivation and loyalty.
- **Limitations:** Overreliance on the leader may undermine organizational sustainability and overshadow processes and systems.
- **Blind Spots:** Risk of overconfidence or authoritarian tendencies if unchecked.

Key Example: Martin Luther King Jr.'s ability to mobilize millions during the civil rights movement demonstrates the power of charismatic leadership.

Situational Leadership: Tailoring Leadership Styles

What It Is: Situational leadership emphasizes adapting one's style to the team's maturity and task requirements, alternating between directive, coaching, supportive, and delegative approaches.

Historical Background: Developed by Paul Hersey and Ken Blanchard in the 1970s, situational leadership gained recognition for its flexibility and practical application.

- **Benefits:** Encourages tailored solutions to team and task needs.
- **Limitations:** Requires a deep understanding of team dynamics and readiness levels, which can complicate implementation.
- **Blind Spots:** Overuse of one style can hinder long-term team development.

Key Example: Startup leaders, who must pivot their strategies as teams and goals evolve, often embody situational leadership.

Visionary Leadership: Inspiring the Future

What It Is: Visionary leadership is rooted in articulating a clear and compelling vision for the future, inspiring others to align their efforts toward shared goals.

Historical Background: Although visionaries have existed throughout history, this leadership style became formalized in management literature as organizations recognized the need for long-term strategic direction.

- **Benefits:** Encourages innovation and alignment with organizational goals.
- **Limitations:** Can overlook immediate challenges in favor of big-picture goals.
- **Blind Spots:** Risk of alienating teams if the vision feels unattainable.

Key Example: Steve Jobs embodied visionary leadership by driving Apple's turnaround through innovation and groundbreaking products like the iPod and iPhone.

Additional Leadership Styles for Consideration

Bureaucratic Leadership: Rule-Based Management

What It Is: Bureaucratic leadership focuses on adherence to rules, processes, and hierarchical structures to maintain order and consistency.

- **Benefits:** Ensures highly regulated industries' stability, reliability, and consistency.
- **Limitations:** Stifles creativity and adaptability.
- **Key Example:** Leaders in public-sector organizations or highly regulated industries, like pharmaceuticals, often rely on bureaucratic leadership to ensure compliance.

Laissez-Faire Leadership: Hands-Off Empowerment

What It Is: Laissez-faire leadership involves delegating authority to team members and providing minimal guidance while fostering autonomy.

- **Benefits:** Encourages creativity and self-motivation in inexperienced teams.
- **Limitations:** This can lead to a lack of direction and accountability in less experienced teams.
- **Key Example:** Creative industries, such as film production or R&D, often see successful applications of laissez-faire leadership.

Summary Table: Leadership Styles.

Leadership Style	Key Features	Benefits	Limitations	Example
Great Man Theory	Traits-based, innate qualities	Emphasizes individual influence	Ignores situational factors	Winston Churchill
Trait Theory	Focus on personal characteristics	Encourages self-awareness	Lacks clarity on learned traits	Mahatma Gandhi
Behavioral Theories	Task- and people-oriented actions	Provides actionable training	Oversimplifies complex dynamics	John F. Kennedy
Contingency Theories	Situational adaptability	Encourages flexibility	It is time-intensive to analyze the context	Angela Merkel
Transactional Leadership	Structured exchanges	Promotes accountability	Limits creativity	Jeff Bezos
Transformational Leadership	Inspires vision and innovation	Fosters systemic change	Risk of operational neglect	Nelson Mandela

Actionable Takeaways for Leaders

1) **Understand Your Environment:** No leadership style is universally effective. Assess the context and adapt accordingly.
2) **Combine Styles:** Effective leaders blend multiple styles to address dynamic challenges.
3) **Develop Core Traits:** Regardless of style, empathy, resilience, and decisiveness are foundational qualities.
4) **Focus on Growth:** Leadership isn't static – invest in continuous learning and adaptability.
5) **Balance Vision and Execution:** Great leaders inspire others with a clear vision while staying grounded in operational realities.

Conclusion: A Comprehensive Roadmap for Leadership in Complexity

Leadership in today's world is more than a set of predefined traits or behaviors – it's the ability to adapt, innovate, and inspire in response to ever-changing challenges. From Churchill's decisive resilience to Mandela's unifying vision, effective leadership blends timeless principles with the flexibility to address the moment's unique demands.

Technological leadership is a paradigm for our time, requiring vision, adaptability, collaboration, and a commitment to ethical innovation. Unlike traditional management, which often focuses on maintaining and optimizing existing systems, technological leadership embraces uncertainty and drives transformative change. It fosters the creation of systems that endure disruption and thrive within it, harnessing the power of innovation for global good.

The leadership styles explored in this appendix – transformational, adaptive, participative, servant, and beyond – are not isolated frameworks but tools to be integrated and applied in concert. The complexities of modern challenges demand leaders who can blend these approaches to balance vision with execution, innovation with ethics, and resilience with collaboration. Together, these styles form a comprehensive roadmap for leading in complexity, offering practical insights and actionable strategies for a world in constant flux.

As technological advancements accelerate and global systems grow more interconnected, leadership must rise to meet the challenge. The leaders of tomorrow will not only navigate complexity but also transform it into opportunity, creating a legacy of progress, resilience, and ethical stewardship.

Resources for Further Reading

B. Bass and R. E. Riggio, *Transformational Leadership*. New York, NY: Routledge, 2006.

J. M. Burns, *Leadership*. New York, NY: Harper & Row, 1978.

T. Carlyle, *On Heroes, Hero-Worship, and the Heroic in History*. Boston, MA: Houghton Mifflin, 1841.

S. Covey, *The 7 Habits of Highly Effective People*. New York, NY: Free Press, 1989.

F. E. Fiedler, *A Theory of Leadership Effectiveness*. New York, NY: McGraw-Hill, 1967.

B. George, *True North: Discover Your Authentic Leadership*. San Francisco, CA: Jossey-Bass, 2007.

A. Goleman, "Emotional intelligence and leadership," *Harv. Bus. Rev.*, vol. 76, no. 6, pp. 93–102, 1998.

R. K. Greenleaf, *Servant Leadership: A Journey into the Nature of Legitimate Power and Greatness*. New York, NY: Paulist Press, 1977.

R. A. Heifetz, *Leadership Without Easy Answers*. Cambridge, MA: Harvard University Press, 1994.

P. Hersey and K. Blanchard, *Management of Organizational Behavior: Utilizing Human Resources*. Upper Saddle River, NJ: Prentice Hall, 1982.

J. Kouzes and B. Posner, *The Leadership Challenge: How to Make Extraordinary Things Happen in Organizations*. San Francisco, CA: Jossey-Bass, 2012.

K. Lewin, R. Lippitt, and R. K. White, "Patterns of aggressive behavior in experimentally created social climates," *J. Soc. Psychol.*, vol. 10, no. 2, pp. 271–299, 1939.

R. M. Stogdill, "Personal factors associated with leadership: A survey of the literature," *J. Psychol.*, vol. 25, no. 1, pp. 35–71, 1948.

M. Weber, *Economy and Society: An Outline of Interpretive Sociology*. Berkeley, CA: University of California Press, 1978.

Appendix F

Glossary of Terms, Methods, Tools, Templates, and Case Studies

3 Horizons Model Strategic foresight framework distinguishing short-term (H1), emerging (H2), and transformational (H3) innovations to guide investment and change. Used for scenario framing and leadership conversations around innovation cycles.

5 Whys Technique A root-cause analysis tool that involves asking "why" five times to reveal underlying issues behind surface-level symptoms. Used in risk diagnostics and postmortem reviews.

AI Risk Lens A structured framework used to evaluate artificial intelligence applications based on ethical, technical, and organizational criteria.

Benefit-Cost Analysis (BCA) An economic decision-making tool that evaluates whether the benefits of a policy or project outweigh its costs. Applied throughout the infrastructure evaluation chapters.

Complex Adaptive Systems (CAS) A systems theory concept describing networks of interconnected agents adapting to changing environments – the core framework in Chapters 1 and 5.

Cone of Uncertainty A forecasting model describes how the range of possible outcomes expands over time. It is used in risk management and technology forecasting.

Critical Infrastructure Interdependence The interconnectivity between energy, water, transportation, and telecommunications sectors is crucial. Case study examples include the Puerto Rico blackout, the Texas freeze, and the Heathrow failure.

Delphi Method A structured expert consensus tool, mentioned in the global transition dynamics (GTD) and foresight sections, was used to forecast trends through multiple rounds of questionnaires.

Ecosystem Mapping Strategic exercise that visually captures key actors, flows, constraints, and opportunities in a complex value chain. Appears in systems leadership chapters.

Futures Wheel A tool that maps first- and second-order consequences of change drivers to identify ripple effects. Used in scenario planning and foresight chapters.

Gartner Hype Cycle Visual model showing the maturity and adoption stages of emerging technologies. Used in the *Golden Hammer Syndrome* chapter.

Global Transition Dynamics (GTD) A 2003 leadership framework developed by Massoud Amin to assess societal, technical, and systemic signals across multiple domains (security, finance, health, and infrastructure). Core to *Weak Signals* and *Appendix B* on *GTD*.

Golden Circle (Why–How–What) Developed by Simon Sinek, this model grounds leadership strategy in purpose ("Why"), process ("How"), and execution ("What"). It is covered in Chapter 1, Appendix A, and used to realign mission with organizational action.

Golden Hammer Syndrome A cognitive bias occurs when familiar tools are misapplied across different contexts ("If all you have is a hammer, everything looks like a nail"), as explored in detail through case studies in telecom, finance, and health.

Hardiness Scale (Psychological Resilience) A psychological metric for measuring resilience, encompassing commitment, control, and challenge. Applied in leadership development and stress readiness frameworks.

Heathrow Airport Case Study A real-time breakdown of the cascading infrastructure failure caused by software and staffing misalignment is presented between Chapters 13 and 14.

Heilmeier's Catechism George Heilmeier created eight fundamental questions to vet R&D investments, which are integrated in Chapter 4 and GTD as a robust screening tool for initiatives.

Life-Cycle Analysis (LCA) Assessment of environmental and economic costs over the life of a product or system. Discussed in the sustainability and tech evaluation chapters.

Mini Tools Rapid-decision tools like Priority Matrix, Stakeholder Impact Grid, and Issue Trees. Collected and referenced throughout the book and summarized in Appendix A.

North American Grid Evolution (NAGE) Historical and structural narrative of the U.S. and Canadian electric grid. Contextualized in Chapter 6C for modernization discussions.

PESTLE Analysis A macro-external factor framework scanning Political, Economic, Social, Technological, Legal, and Environmental (PESTLE) domains. Integrated into risk management chapters.

Portfolio Mapping A tool for categorizing R&D and innovation projects based on risk-return, urgency, or capability alignment. Used in strategic planning and venture selection.

Quad Chart Template A standardized visual tool used in R&D and defense project reviews to summarize the objective, problem, approach, and benefit. Common in GTD and proposal diagnostics.

Scenario Planning Structured method for building plausible future states and designing a strategy around uncertainty. Covered extensively in Chapter 4.

Smart Grid Maturity Model (SGMM) A framework to assess smart grid capabilities and readiness. Integrated into energy sector case studies and global investment evaluations.

STARR Framework (Strategy, Tactics, Assets, Risks, and Resources) A hybrid leadership model aligns operations with strategic goals. It is presented in GTD and applied in organizational diagnostics.

Strategic Inflection Point (SIP) The threshold moment when a system must evolve or risk decline. Explored in tech transitions and organizational case studies.

SWOT Analysis Foundational tool to evaluate Strengths, Weaknesses, Opportunities, and Threats (SWOT). Used across multiple chapters, including the GTD model.

Systemic Risk Indexing A method for ranking and tracking cross-sector risks using energy, finance, and public health data. It appears in the *Global Grid Investment* and *Resilience Metrics* chapters.

Technology Readiness Levels (TRLs) NASA-originated scale to evaluate technology maturity from basic principle to full deployment. Referenced in investment and innovation chapters.

Technology Roadmapping A timeline-based strategy to align capability development, infrastructure planning, and R&D investment. Detailed in Chapter 4 and Appendix A.

The Golden Hammer Case Study Sector-specific analysis of technology failure caused by misapplied tools across energy, telecom, finance, and healthcare.

Transformers and Quantum Acceleration Discussed in AI chapters. Explores the leap from deep learning to hybrid intelligent systems.

Upskilling and Reskilling Workforce development frameworks include job mapping, microcredentialing, and cultural alignment. Featured in Chapter 25 and beyond.

Weak Signals Detection A foresight strategy to surface early-stage indicators of disruption. The core of the GTD methodology is explored in the *Weak Signals* chapter.

Index

C

d

Wiley The manufacturer's authorized representative according to the EU General Product Safety Regulation is Wiley-VCH GmbH, Boschstr. 12, 69469 Weinheim, Germany, e-mail: Product_Safety@wiley.com.

Printed and bound by CPI Group (UK) Ltd, Croydon, CR0 4YY

23/07/2026

02175690-0001